COASTAL OCEANOGRAPHY OF WASHINGTON AND OREGON

FURTHER TITLES IN THIS SERIES

1 J.L. MERO
THE MINERAL RESOURCES OF THE SEA
2 L.M. FOMIN
THE DYNAMIC METHOD IN OCEANOGRAPHY
3 E.J.F. WOOD
MICROBIOLOGY OF OCEANS AND ESTUARIES
4 G. NEUMANN
OCEAN CURRENTS
5 N.G. JERLOV
OPTICAL OCEANOGRAPHY
6 V. VACOUIER
GEOMAGNETISM IN MARINE GEOLOGY
7 W.J. WALLACE
THE DEVELOPMENTS OF THE CHLORINITY/
SALINITY CONCEPT IN OCEANOGRAPHY
8 E. LISITZIN
SEA-LEVEL CHANGES
9 R.H. PARKER
THE STUDY OF BENTHIC COMMUNITIES
10 J.C.J. NIHOUL (Editor)
MODELLING OF MARINE SYSTEMS
11 O.I. MAMAYEV
TEMPERATURE SALINITY ANALYSIS OF WORLD OCEAN WATERS
12 E.J. FERGUSON WOOD and R.E. JOHANNES
TROPICAL MARINE POLLUTION
13 E. STEEMANN NIELSEN
MARINE PHOTOSYNTHESIS
14 N.G. JERLOV
MARINE OPTICS
15 G.P. GLASBY
MARINE MANGANESE DEPOSITS
16 V.M. KAMENKOVICH
FUNDAMENTALS OF OCEAN DYNAMICS
17 R.A. GEYER
SUBMERSIBLES AND THEIR USE IN OCEANOGRAPHY AND OCEAN ENGINEERING
18 J.W. CARUTHERS
FUNDAMENTALS OF MARINE ACOUSTICS
19 J.C.J. NIHOUL (Editor)
BOTTOM TURBULENCE
20 P.H. LEBLOND and L.A. MYSAK
WAVES IN THE OCEAN
21 C.C. VON DER BORCH (Editor)
SYNTHESIS OF DEEP-SEA DRILLING RESULTS IN THE INDIAN OCEAN
22 P. DEHLINGER
MARINE GRAVITY
23 J.C.J. NIHOUL (Editor)
HYDRODYNAMICS OF ESTUARIES AND FJORDS
24 F.T. BANNER, M.B. COLLINS and K.S. MASSIE (Editors)
THE NORTH-WEST EUROPEAN SHELF SEAS: THE SEA BED AND THE SEA IN MOTION
25 J.C.J. NIHOUL (Editor)
MARINE FORECASTING
26 H.G. RAMMING and Z. KOWALIK
NUMERICAL MODELLING MARINE HYDRODYNAMICS
27 R.A. GEYER (Editor)
MARINE ENVIRONMENTAL POLLUTION
28 J.C.J. NIHOUL (Editor)
MARINE TURBULENCE
29 M. M. WALDICHUK, G.B. KULLENBERG and M.J. ORREN (Editors)
MARINE POLLUTANT TRANSFER PROCESSES
30 A. VOIPIO (Editor)
THE BALTIC SEA
31 E.K. DUURSMA and R. DAWSON (Editors)
MARINE ORGANIC CHEMISTRY
32 J.C.J. NIHOUL (Editor)
ECOHYDRODYNAMICS
33 R. HEKINIAN
PETROLOGY OF THE OCEAN FLOOR
34 J.C.J. NIHOUL (Editor)
HYDRODYNAMICS OF SEMI-ENCLOSED SEAS
35 B. JOHNS (Editor)
PHYSICAL OCEANOGRAPHY OF COASTAL AND SHELF SEAS
36 J.C.J. NIHOUL (Editor)
HYDRODYNAMICS OF THE EQUATORIAL OCEAN
37 W. LANGERAAR
SURVEYING AND CHARTING OF THE SEAS
38 J.C.J. NIHOUL (Editor)
REMOTE SENSING OF SHELF SEA HYDRODYNAMICS
39 T. ICHIYE (Editor)
OCEAN HYDRODYNAMICS OF THE JAPAN AND EAST CHINA SEAS
40 J.C.J. NIHOUL (Editor)
COUPLED OCEAN–ATMOSPHERE MODELS
41 H. KUNZEDORF (Editor)
MARINE MINERAL EXPLORATION
42 J.C.J. NIHOUL (Editor)
MARINE INTERFACES ECOHYDRODYNAMICS
43 P. LASSERRE and J.M. MARTIN (Editors)
BIOGEOCHEMICAL PROCESSES AT THE LAND-SEA BOUNDARY
44 I.P. MARTINI (Editor)
CANADIAN INLAND SEAS
45 J.C.J. NIHOUL and B.M. JAMART (Editors)
THREE-DIMENSIONAL MODELS OF MARINE AND ESTUARIN DYNAMICS
46 J.C.J. NIHOUL and B.M. JAMART (Editors)
SMALL-SCALE TURBULENCE AND MIXING IN THE OCEAN

Elsevier Oceanography Series, 47

COASTAL OCEANOGRAPHY OF WASHINGTON AND OREGON

Edited by

MICHAEL R. LANDRY
Department of Oceanography and Hawaii Institute of Geophysics, University of Hawaii, Honolulu, HI 96822, U.S.A.

and

BARBARA M. HICKEY
School of Oceanography, University of Washington, Seattle, WA 98195, U.S.A.

ELSEVIER

Amsterdam — Oxford — New York — Tokyo 1989

ELSEVIER SCIENCE PUBLISHERS B.V.
Sara Burgerhartstraat 25
P.O. Box 211, 1000 AE Amsterdam, The Netherlands

Distributors for the United States and Canada:

ELSEVIER SCIENCE PUBLISHING COMPANY INC.
655, Avenue of the Americas
New York, NY 10010, U.S.A.

ISBN 0-444-87308-2 (Vol. 47)
ISBN 0-444-41623-4 (Series)

Printed in The Netherlands

ACKNOWLEDGEMENTS

We express our deep gratitude to the Department of Energy for their unflagging support during the collection and analysis of data presented in this book. The comprehensive nature of the data is directly attributable to the continuity of funding provided by this agency, which continues to be a leader in providing long-term support for basic research on the marine ecosystems of United States' coastal regions. Even with this funding base, however, a significant portion of the results presented would not have been possible without the support of individual projects by the National Science Foundation, the National Aeronautics and Space Administration, the U.S. Army Corps of Engineers and the Sea Grant Program.

The Department of Energy initiated the idea to synthesize in a book the oceanographic research performed over the last decade and a half off the Washington-Oregon coasts. DOE provided support for the authors to complete their analyses and to prepare their chapters for inclusion in the book. The material contained in this book has, due to the possibility of offshore drilling in this region, recently become extremely important to Washington and Oregon States. Accordingly, to accelerate the publication date of this material, additional support for final editing of the material into book format was provided by the Washington Sea Grant Program as part of a study commissioned by the Washington State Legislature.

We are indebted to numerous technicians, students and volunteers who collected the raw data for this book, often under difficult conditions at sea. We also thank John Osborne and Dan Clapp, who were responsible for the layout and construction of the book; Don Doyle, who performed the majority of the cartography; Dick Cromoga, who was responsible for the photography; and our secretaries, Su Fagerberg and Marian Peacock, who patiently labored through all our drafts. Sue Geier provided general support and assistance. We gratefully acknowledge the contribution of Dr. George C. Anderson, who coordinated DOE studies in the Pacific Northwest from 1979 until his retirement from the University of Washington in 1982. Finally, we extend our warm appreciation to those colleagues who helped review the scientific content of the chapters and encouraged us throughout the duration of this project.

CONTRIBUTORS

David A. Armstrong School of Fisheries, University of Washington, Seattle, Washington 98195
Karl Banse School of Oceanography, University of Washington, Seattle, Washington 98195
James P. Bolger School of Oceanography, University of Washington, Seattle, Washington 98195
Louis Botsford Department of Wildlife and Fisheries Biology, University of California, Davis, California 95616
Roy Carpenter School of Oceanography, University of Washington, Seattle, Washington 98195
Quay Dortch Louisiana Universities Marine Consortium, Chauvin, Louisiana 70344
Albert J. Hermann School of Oceanography, University of Washington, Seattle, Washington 98195
Barbara M. Hickey School of Oceanography, University of Washington, Seattle, Washington 98195
Peter A. Jumars School of Oceanography, University of Washington, Seattle, Washington 98195
Nancy B. Kachel School of Oceanography, University of Washington, Seattle, Washington 98195
Michael R. Landry Department of Oceanography and Hawaii Institute of Geophysics, University of Hawaii, Manoa, Honolulu, Hawaii 96822
Joyce R. Lewin School of Oceanography, University of Washington, Seattle, Washington 98195
Carl J. Lorenzen School of Oceanography, University of Washington, Seattle, Washington 98195
David M. Nelson College of Oceanography, Oregon State University, Corvallis, Oregon 97331
Jean Newman Jet Propulsion Laboratory, Pasadena, California 91109
Hasong Pak College of Oceanography, Oregon State University, Corvallis, Oregon 97331
Mary Jane Perry School of Oceanography, University of Washington, Seattle, Washington 98195
Michael L. Peterson School of Oceanography, University of Washington, Seattle, Washington 98195
Willis K. Peterson School of Oceanography, University of Washington, Seattle, Washington 98195
James R. Postel School of Oceanography, University of Washington, Seattle, Washington 98195
Charles T. Schaefer School of Oceanography, University of Washington, Seattle, Washington 98195
Jonathan M. Shenker University of California, Bodega Marine Laboratory, P.O. Box 247, Bodega Bay, California 94923
Lawrence F. Small College of Oceanography, Oregon State University, Corvallis, Oregon 97331
J. Dungan Smith Geophysics Program, University of Washington, Seattle, Washington 98195
Chris S. Weimer College of Oceanography, Oregon State University, Corvallis, Oregon 97331
Donald F. Winter Department of Engineering, University of Redlands, Redlands, California 92373

PREFACE

The United States Atomic Energy Commission, precursor to the present Department of Energy, began its support of oceanographic research off Washington and Oregon in 1960. Early studies were linked to the presence of nuclear reactors on the Columbia River and limited by the comparatively primitive oceanographic instrumentation and techniques of the day. Consequently, they mainly described seasonal variations in the coastal environment directly influenced by the Columbia River effluent. The results of this work are summarized in a book entitled "The Columbia River Estuary and Adjacent Ocean Waters" (Pruter and Alverson, 1972). Even before this book was published, however, programmatic considerations had shifted the focus of the research to oceanic, offshore waters and to mechanisms relating to the penetration of radionuclides to the deep ocean. It was not until 1974 that understanding the interaction of physical, chemical and biological processes in the dynamic shelf ecosystem became an important goal in its own right as a basis for future energy-related policy decisions in the U.S. Pacific Northwest. The present book synthesizes the information which grew out of this later focus.

The Washington/Oregon coastal zone is a classical Eastern Boundary Current region. The area is extremely productive, the productivity dependent on near-shore infusions of nutrients into surface layers during wind-driven coastal upwelling. In this respect the area contrasts dramatically with other U.S. coastal regions, such as the California Bight and the southeast continental shelf, where offshore boundary currents rather than coastal wind forcing are primarily responsible for regional characteristics in productivity. The signature of coastal upwelling is clearly apparent in the event captured in Figure i.1. The zone of cold, upwelled water roughly parallels the coastline from northern Washington to mid-Oregon, yet mesoscale features are apparent, in particular, in the lee of coastal promontories and at the offshore edge of the coastal water. Near Haceta Banks (~ 44° N) off southern Oregon, where the shelf doubles in width over a short distance, the recently-upwelled water is deflected offshore in what appears to be a dramatic coastal squirt.

The Washington-Oregon coastline is much more regular than areas off California or off the East Coast, where large capes lend complexity to both the physical environment and the ecosystem response. The relatively straight coastline and broad, deep shelf greatly simplify the physical environment, so that processes responsible for much of the variance are more easily identified. The system response from mid-Oregon northward, although not strictly two-dimensional, is more so than many other coastal areas. Consequently, the system is amenable to the testing of relatively simple models integrating wind forcing with physical, chemical and biological responses in the upper water column.

The Columbia River, a major distinguishing feature of the Pacific Northwest coast, is evident in Figure i.1 as a warm surface plume that emanates from the shore near 46° N. The Columbia River provides the majority of freshwater and also lithogenic sediment to the coastal region. Understanding sediment movement on the shelf is simplified by the fact that large quantities of lithogenic sediment are deposited annually almost entirely by one source.

This book emphasizes the interdisciplinary nature of oceanographic research. Physical processes along the coasts of Washington and Oregon not only determine the direction and speed of water movement, they strongly constrain nutrient and chemical inputs, magnitudes and variability of biological stocks and production rates, as well as deposition and transport of sediments on the shelf. Biological and physical interactions are particularly important in understanding sediment transport mechanisms. Physical, chemical and biological processes interact to produce unusual sedimentary inventories of particle-reactive chemicals and productive beach ecosystems.

Chapter 1 describes regional, seasonal and interannual patterns in the distributions of water-column properties and their relationship to time-variable physical and biological processes. Chapter 2 presents a detailed discussion of physical circulation and its forcing mechanisms based on the results obtained from modern measurement techniques and analyses. The following three chapters describe the mean magnitude and control of primary production on the Washington shelf (Chapter 3), the interaction between phytoplankton production and nutrient inputs and cycling (Chapter 4) and the utilization of phytoplankton stocks by zooplankton (Chapter 5). Chapter 6 deals with the dynamics of physical, chemical and biological processes during coastal upwelling events. The next three chapters involve various aspects of particle movement on the shelf and slope. Seasonal distribution of suspended particle fields off Oregon are described in Chapter 7. Chapter 8 presents a physically-driven model for sediment transport on the Washington shelf. Chapter 9 discusses the implications of benthic populations on sediment transport mechanisms. Chapter 10 presents an extensive review of chemical inputs to the water column and sediments and their relationships to physical and biological processes. The needs and challenges of developing an understanding of the influence of oceanographic processes on commercially-fished species are discussed in Chapter 11. Chapter 12 elucidates the physical and biological mechanisms that result in uniquely productive communities of surf-zone diatoms and razor clams along the beaches of Washington and Oregon. Lastly, Chapter 13 summarizes and integrates the information from preceding chapters and suggests areas for future research.

Seattle, Washington
September 30, 1988

M.R. Landry
B.M. Hickey

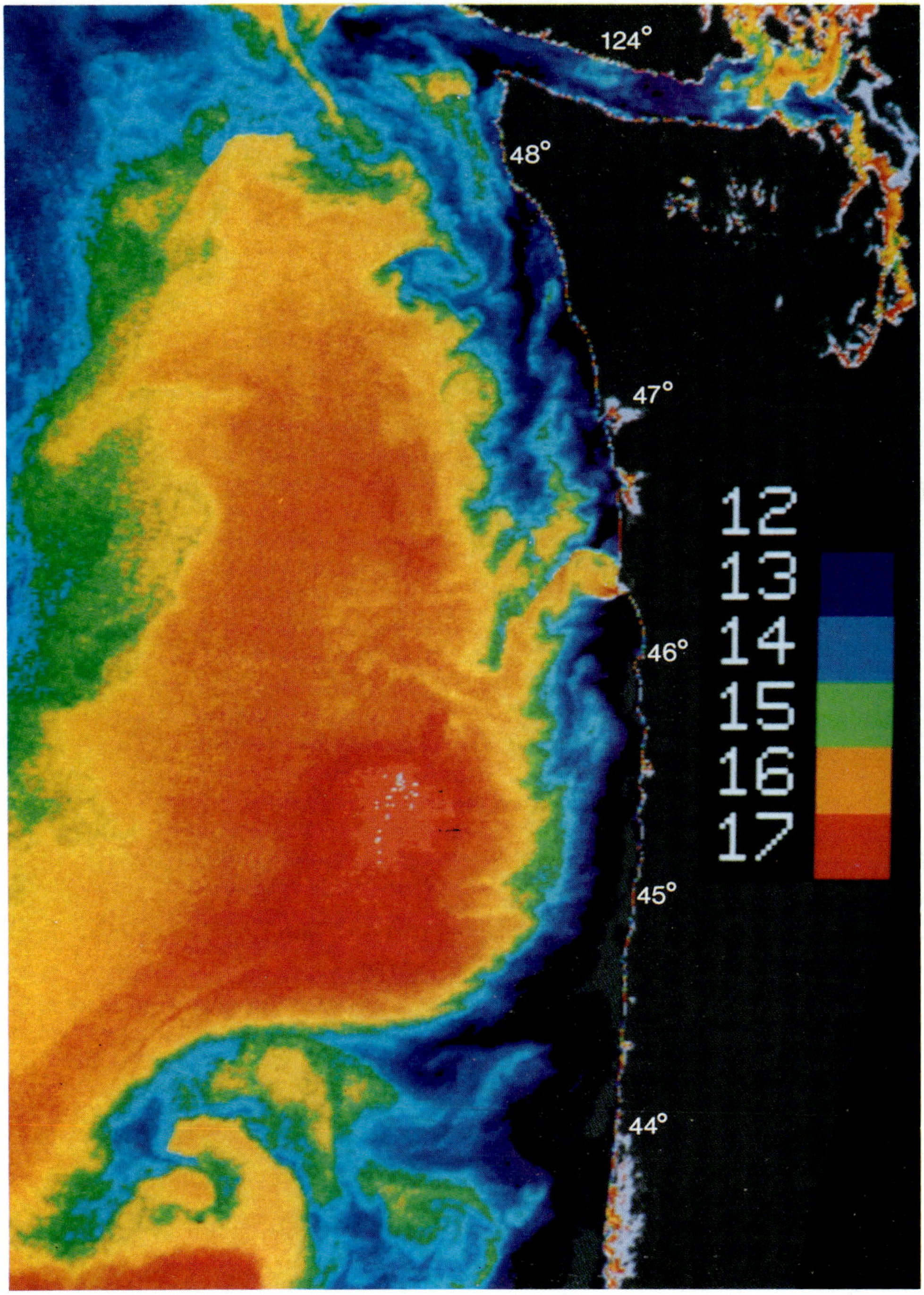

Fig. i.1 Satellite-derived SST on 10 August, 1981 for the coasts of Washington and Oregon. The color bar on the right defines the temperature in °C.

TABLE OF CONTENTS

Chapter 4: Phytoplankton-Nitrogen Interactions

Chapter 5: Abundance, Distribution, and Grazing Impact of Zooplankton on the Washington Shelf

Chapter 6: Coastal Upwelling Dynamics

Chapter 7: Seasonal Dynamics of Suspended Particulate Matter

Chapter 7: Seasonal Dynamics of Suspended Particulate Matter (continued)

Chapter 8: Sediment Transport and Deposition on the Washington Continental Shelf

Chapter 9: Benthos and its Interaction with Bottom Boundary Layer Processes

Chapter 10: Chemical Cycling in Washington's Coastal Zone

Chapter 11: Oceanographic Influences on the Dynamics of Commercially-Fished Populations

Chapter 11: Oceanographic Influences on the Dynamics of Commercially-Fished Populations (continued)

CHAPTER 1

BROAD-SCALE DISTRIBUTIONAL PATTERNS OF HYDROGRAPHIC VARIABLES ON THE WASHINGTON/OREGON SHELF

M.R. Landry[1], *J.R.Postel*[2], *W.K. Peterson*[2] *and J. Newman*[3]
[1] *Department of Oceanography and Hawaii Institute of Geophysics, University of Hawaii, Honolulu, HI 96822*
[2] *School of Oceanography, University of Washington, Seattle, WA 98195*
[3] *Jet Propulsion Laboratory, Pasadena, CA 91109*

1.1 INTRODUCTION

The continental shelf region of Washington and Oregon is characterized by an intensely dynamic pelagic ecosystem. Water column properties are continuously changing due to *in situ* biological, chemical and physically-mediated processes, regional cycles in solar radiation, river run-off, prevailing winds, periodic storm events and global-scale effects on ocean circulation. This chapter deals with the net effects of these influences – the broad-scale temporal and spatial patterns in distributions of physical, chemical and biological properties of the environment. Our goal, within the context of this book, is to provide a general, descriptive background for the more dynamical and process-orientated chapters which follow. Beyond this immediate goal, we have also endeavored to meet a regional need for the compilation and synthesis of hydrographic information in a format which can be useful in interpretating the results of past research programs and in designing new initiatives.

Despite more than three decades of oceanographic research on the Washington/Oregon coast and considerable progress in understanding regional dynamics, broad-scale distributional features are still poorly known. This reflects, in part, the lack of a dedicated survey program for the systematic collection and dissemination of regional hydrographic data such as the CalCOFI (California Cooperative Oceanic Fisheries Investigation) Program off California. The little published information which does exist is limited in temporal and spatial scope. For instance, the suite of papers published in the Columbia River volume (Pruter and Alverson, 1972) provides regional and seasonal information on temperature, salinity, density, nutrient and chlorophyll distributions but the data are largely from hydrographic surveys in 1961 and 1962. Similarly, Small and Menzies' (1981) detailed analyses of local upwelling effects on distributions of phytoplankton biomass and primary production relate to cruises off Oregon in 1972-74. On the other hand, Huyer's (1977) more general description of seasonal cycles in temperature, salinity and density is based on a 10-y (1961-1970), but spatially-limited (three stations off central Oregon), data set.

This chapter represents an effort to compile and synthesize hydrographic data from the Washington/Oregon shelf region from all of its various sources. If we have been even moderately successful in this effort, it has only been through the generosity and cooperation of numerous investigators at Oregon State University, the University of Washington, the National Oceanic and Atmospheric Association and the National Marine Fisheries Service as well as

National Oceanographic Data Collection tapes. Most of the data has not been published previously; however, all known published material is included.

1.2 DATA MANAGEMENT AND REDUCTION

We consider 11 parameters which have a long history of routine use in characterizing the pelagic environment: temperature, salinity, density (sigma-t), dissolved oxygen content and percent saturation, dissolved nutrients – nitrate, nitrite, ammonium, phosphate and silicate and phytoplankton chlorophyll *a*. Since collections of these measurements were made following the varied interests of individual research programs rather than a systematic plan, the resulting data montage has its expected flaws and limitations. First, spatial and temporal coverage is not uniform. Predictably, summer and spring months are better represented than winter months. Seasonal and regional coverage is most complete during the interval from 1958 through 1972 which spans something of a "Golden Age" of oceanographic survey research during which the Washington/Oregon shelf was the site of the Atomic Energy Commission's Columbia River Program, the National Science Foundation's Coastal Upwelling Experiment, as well as a long time-series of hydrographic observations off of Newport, Oregon supported by the Office of Naval Research. As a result of changing resource levels and research interests, the more recent studies have involved detailed, process-oriented work at given points in space and time in lieu of broad-scale efforts. Second, not all variables were measured at all times. More information was available for temperature and salinity, parameters routinely measured in all oceanographic programs; dissolved oxygen and nutrients and plant pigment data, on the other hand, were available only from cruises with a chemical or biological orientation. Lastly, the technologies and methodologies of making standard oceanographic measurements have advanced considerably over the past 30 to 40 y and the complied data set incorporates the new developments and procedural changes to varying extents. In effect, the data set lacks some of the desirable features of a designed regional survey program such as CalCOFI (e.g., uniformity in temporal and spatial sampling and methodological consistency). However, the suite of measured parameters is greater than that the CalCOFI program and over the time interval of interest (1950-1984) the temporal coverage (i.e., months in which measurements were made) is comparable to that of the CalCOFI cruises (Table 1.1).

The composite data set for the Washington-Oregon coast consisted of over 13,000 hydrographic stations occupied between 1950 and 1984. These data were edited to eliminate duplication and to minimize errors. Questionable data were identified from notations in original data reports and from obvious inconsistencies among data from adjacent depths or stations. Most data errors seemed to be keypunching mistakes. They were corrected from the original data when possible or omitted from subsequent analyses.

In presenting the data, we attempted to strike a balance between a level of detail adequate for describing general distributional features and a useful level for designing future research programs. Even so, the appearance of some figure excess and redundancy is unavoidable since each of the 11 parameters varies in four dimensions – time, depth, latitude and distance from shore – and only two dimensions can be represented in any given figure. We chose the

TABLE 1.1 Distribution of hydrographic sampling effort off Washington and Oregon by month of year (1950-1984). Tabulated numbers are the number of years in which the parameters were sampled in the composite data set.

			Winter						Summer			
Parameter	J	F	M	A	M	J	J	A	S	O	N	D
Washington												
Temperature	10	11	10	16	19	15	16	16	16	14	8	7
Salinity	10	11	10	16	20	15	16	16	16	14	8	7
Density	10	11	10	16	19	15	16	16	16	14	8	7
Oxygen	8	7	7	11	11	12	13	12	11	10	3	4
Nitrate	6	5	6	7	4	9	9	10	9	8	2	2
Phosphate	6	5	6	8	4	9	10	10	9	8	2	2
Silicate	6	5	6	8	3	9	9	10	9	8	2	2
Ammonium	2	1	2	3	0	2	5	4	4	4	0	0
Nitrite	2	1	2	3	0	2	5	4	4	4	0	0
Chlorophyll	5	4	5	7	3	8	7	9	8	8	2	2
Oregon												
Temperature	13	13	12	14	14	17	17	16	14	14	11	11
Salinity	12	12	11	13	14	17	17	15	12	12	10	10
Density	12	12	11	13	14	17	17	15	12	12	10	10
Oxygen	10	9	9	10	12	14	13	13	9	9	9	8
Nitrate	10	7	8	9	9	15	13	12	10	10	5	5
Phosphate	10	7	9	9	10	15	14	13	10	11	6	6
Silicate	9	6	7	7	8	15	13	12	10	10	5	5
Ammonium	0	0	0	0	0	1	2	0	0	0	0	0
Nitrite	0	0	0	0	0	1	2	0	0	0	0	0
Chlorophyll	5	3	5	7	9	12	11	14	8	9	4	4

following analytical procedures and graphical formats as those which best illustrate the distributional features:

Regional contour maps are used to represent the general features of summertime *vs.* wintertime distributions of surface parameters in the area bounded by 124° to 128° W longitude and 43° to 49° N latitude. The maps were constructed from all available surface (about 1-m depth) measurements made during each of the two seasons – here defined as July-September for summer and January-March for winter based on parameter similarities in between-month analyses of variance. Data were organized by coordinate position into a matrix of 15'x15' grid squares. The averaged grid values were contour mapped using the SURFACE II graphics system of the Kansas Geological Survey (Sampson, 1984).

Time-distance contour plots are used to describe the temporal dynamics of surface parameter distributions in cross-shelf transects off Washington and Oregon. These plots are based on repeated historical sampling of transects off of Newport, Oregon and Copalis Beach, Washington; included are all data within the narrow latitudinal boundaries of 44°35' N to 44°45' N and 47°02' N to 47°12' N, respectively. Data were organized and averaged by month in 5-km incremental distances from shore before contouring.

Seasonal, cross-shelf depth profiles are used to illustrate the connections between the temporal dynamics of regional surface features and vertical distributions. As above, the plots incorporate all measurements from 44°35' N to 44°45' N off the Oregon coast (the Yaquina Head line) and 47°02' N to 47°12' N off Washington (the Copalis Beach line). Data from summer and winter seasons were averaged in 5-km distances from shore before contouring.

Time-depth contour plots are used to show the temporal dynamics of water column properties on the Washington and Oregon mid-shelf regions – here defined as those areas with a total water column depth of 70 to 130 m bounded by 44°20' N to 45°20' N off Oregon and 46°40' N to 47°40' N off Washington. Mean parameter values were computed at 10-m depth increments from all depth profiles made in a given month and year for each mid-shelf region. Contour plotting was based on the averaged monthly profiles for all years in which the parameters were measured. In addition to time-depth contour plots of this data, monthly-averaged temperature and salinity values computed as above were graphed for two depths (surface and 70 m) overlying the Oregon and Washington mid-shelves. These graphs depict, in a more quantitative way, the differences in seasonal environmental conditions on the two shelf regions.

Interannual anomaly plots are used to summarize low frequency changes in regional environmental conditions. Following standard procedures in analyses of anomaly patterns from CalCOFI data (e.g., Chelton *et al.*, 1982), we based our calculations on regionally-averaged parameter values from the 10-m depth. Data from the central coasts of Oregon and Washington were organized into 10 areas of similar hydrography (narrow, longshore sections following bathymetric contours in the nearshore and broader regions offshore; Fig. 1.1). Within each area, we averaged all parameter measurements at the 10-m depth for each month of each year. Interannual monthly means for each area were computed as the unweighted average of monthly values for all years in which data were collected. Regional parameter anomalies for a given month and year were computed as the mean of all individual anomalies relative to the monthly means for the areas in which the measurements were made. We present these regional anomalies as quarterly averages (i.e., the means of three consecutive winter, spring, summer, or autumn months). For comparative purposes, quarterly anomalies for the CalCOFI region and the Subarctic Pacific were similarly computed from data provided by D. Chelton and from S. Tabata and the Canadian Marine Environmental Data Service, respectively. Nonparametric, rank-difference analysis (Tate and Clelland, 1957) was used to test correlations among anomaly patterns. This test is less powerful (i.e., more conservative) than the conventional least-squares approach but avoids the distributional and relational assumptions implicit in parametric analyses.

1.3 ANNUAL PATTERNS

1.3.1 Solar Radiation, Wind and Precipitation

Solar radiation, wind and precipitation substantially influence regional hydrography of the U.S. Pacific Northwest. The seasonal pattern in solar irradiation affects the warming of surface waters and the timing and magnitude of primary production. Local wind stress contributes to the mixing of water-column properties and controls the direction and speed of surface water movement, hence the occurrence of upwelling events (Hickey, 1989, Chapter 2). Regional precipitation, which occurs in excess of evaporation, influences the salt content of all surface waters but more dramatically at the points of river discharge.

The annual patterns in incident solar radiation presented in Figure 1.2 (upper panel) are adapted from long-term means reported by Nelson and Husby (1983). The values represent net incoming short-wave radiation from sun and sky, measured in units of heat flux (W m $^{-2}$), corrected for cloud cover and sea surface albedo and averaged for a 24-h day. Monthly mean values are plotted for the 1° squares centered at 47° N, 125° W and 45° N, 125° W for the

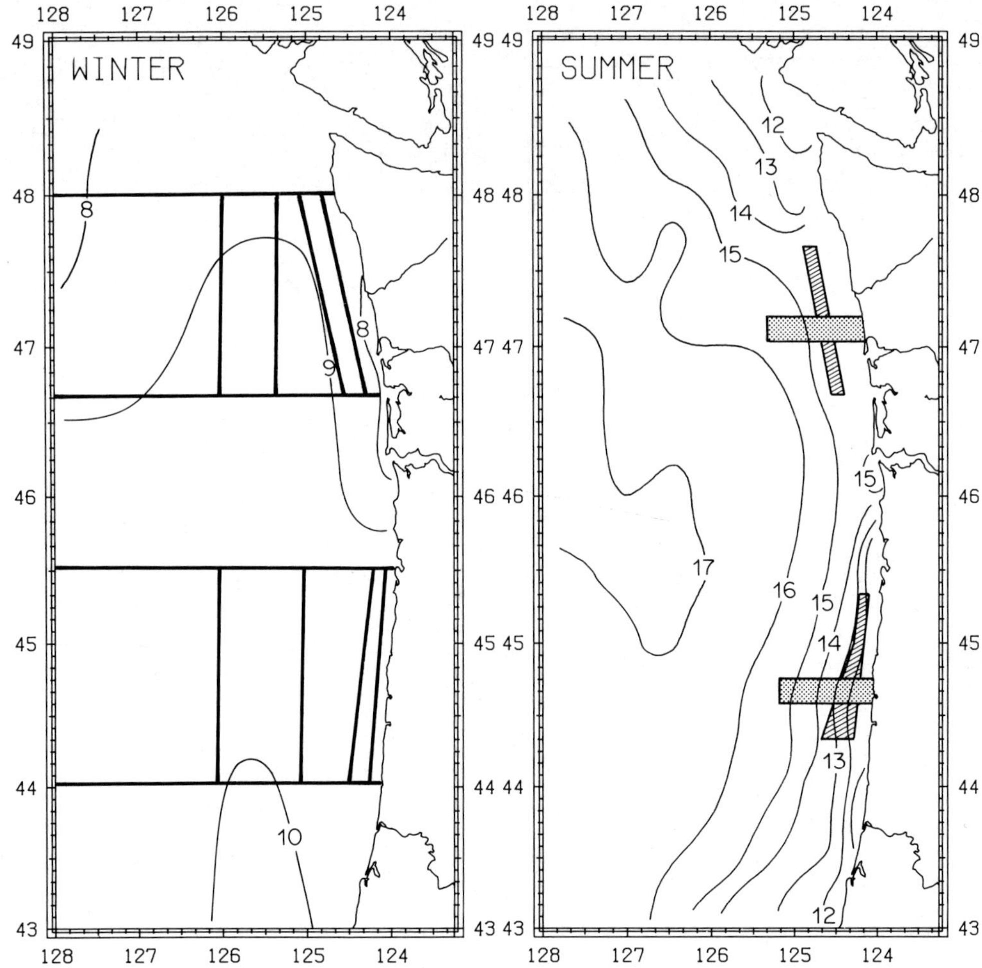

Fig. 1.1 Mean summer and winter distributions of temperature (°C) in surface waters off Washington and Oregon. Contours are fit to composite measurements made between 1950-1984 averaged in 15' x 15' grid squares. Stippled, cross-shelf transects depict areas of intensive sampling detailed in Figures 1.5, 1.6, 1.7, 1.11, 1.12, 1.17, 1.18, 1.19, 1.20, 1.21, 1.24, and 1.25. Hatched, longshore regions depict areas of intensive sampling detailed in Figures 1.8, 1.9, 1.13, 1.22, and 1.26. In the winter distribution, the ten areas between 44° N and 45° 30 'N off Oregon and between 46° 40' N and 48° N off Washington define regions from which data for computations of interannual anomalies were drawn.

Washington and Oregon shelf regions, respectively. Clearly, there is little distinguishing these areas in terms of the timing and magnitude of solar irradiation. We can roughly convert incident solar radiation measured in units of heat flux (W m^{-2}) to units of light intensity relevant to primary production (μE m^{-2} s^{-1}) by the relationship:

$$1 \text{ W m}^{-2} = 0.239 \text{ cal m}^{-2} \text{ s}^{-1} = (\text{PAR})\,(4.60\ \mu\text{E m}^{-2} \text{ s}^{-1})$$

where PAR is fraction of photosynthetically active radiation in the wavelength range of 400-700 nm (Morel and Smith, 1974). PAR varies with solar altitude, cloud cover and other atmospheric conditions. It can be as low as 35% or as high as 80%, but a value of 50% is generally

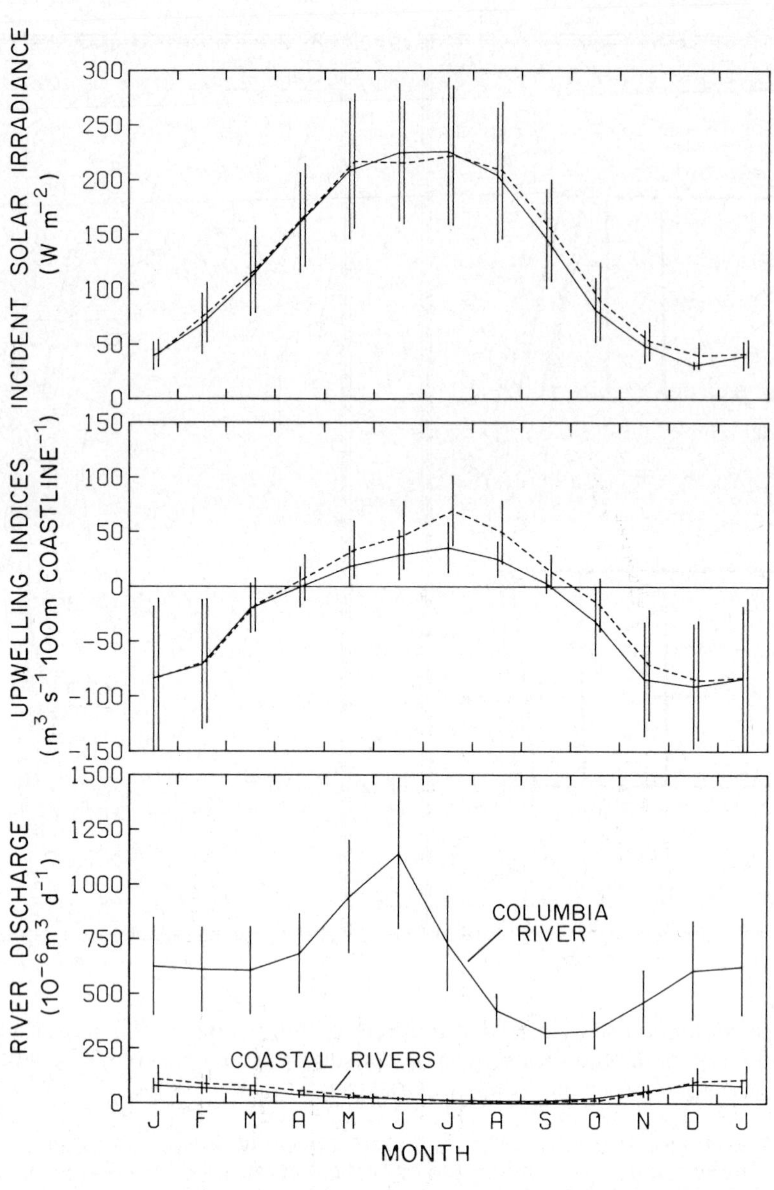

Fig. 1.2 Annual patterns of solar radiation, wind-forced upwelling, and river runoff influences on the coastal environment of Washington and Oregon. Dashed and solid lines indicate Oregon and Washington data, respectively. Vertical lines denote one standard deviation.

considered reasonable (e.g., Strickland, 1965; Parsons *et al.*, 1977). If we make this assumption and account for seasonal differences in daylength, photosynthetically active radiation at the sea surface ranges seasonally during daylight hours from a low of about 200 μE m $^{-2}$ s $^{-1}$ in December to a high of about 780 μE m $^{-2}$ s $^{-1}$ in June and July.

The mean annual pattern of wind-forced coastal upwelling is presented in Figure 1.1 (middle panel) as mean-monthly Bakun Indices at 45° N, 125° W off Oregon and 48° N, 125° W off Washington during the period from 1946 to 1984 (Bakun 1973, 1975, personal communication). Index values were computed from wind stress estimates based on observed atmospheric pressure fields. Upwelling, the offshore movement of surface waters and onshore movement of deep water, is driven by the southward component of the wind stress and denoted by positive values of the index. Downwelling, the onshore movement of surface waters, is driven by the northward component of the wind stress and denoted by negative values. While both upwelling and downwelling occur as discrete events any time of the year, upwelling predominates on average during late spring and summer months (May through August and generally into September) and downwelling dominates during autumn and winter months (October through March). The severity of winter storm events (generally southerly winds) accounts for the greater magnitude of winter downwelling as opposed to summer upwelling. The influence of winter winds are comparable on the Oregon and Washington coasts. In contrast, differences in spring and summer winds are substantial, leading to approximately double the upwelling effect off Oregon.

River discharge data over the period from 1961 to 1981 (Fig. 1.1 lower panel) were obtained from annual reports prepared by the U. S. Geological Survey, Water Resource Division (1961-1982, incl.) and from Orem (1968). We divided mean monthly river runoff into three components, the contributions of Washington and Oregon coastal rivers and the Columbia River discharge. Runoff from the small coastal rivers of Washington and Oregon is comparable in both magnitude and timing. Washington coastal discharge (18.7×10^9 m^3 y^{-1}) represents the composite runoff from the Quinault, Hoh and Queets Rivers and the various rivers emptying through Willapa Bay and Grays Harbor. Approximately 43% of this outflow occurs through the Grays Harbor estuary. Oregon coastal discharge (17.4×10^9 m^3 y^{-1}) is the composite of the Rogue, Umpqua, Siletz, Wilson and Nehalem Rivers. The Umpqua (40%) and Rogue (32%) Rivers are of greatest relative importance. The seasonal pattern in small river discharge, which peaks during the winter and is minimal during the summer, largely reflects the pattern in regional rainfall. In contrast, Columbia River discharge, which peaks on average in June, is dominated by the timing of the annual snow melt. Columbia River discharge (228×10^9 m^3 y^{-1}) is the dominant source of freshwater runoff to the Washington/Oregon coast during all months, but particularly in late spring and early summer.

1.3.2 Temperature, Salinity and Density

Seasonal surface maps of temperature, salinity and sigma-t (Figs. 1.1, 1.3 and 1.4) emphasize the impact of the Columbia River on regional hydrography. In summer months, isotherms approach the coastline in the river mouth area and bow away from the coastline both to the north and south of the river. Concave isohalines extend out from the river mouth and intersect the coastline to the north and south. River water is evident as a tongue of low salinity, low density surface water which lies offshore and to the south in the summer. During winter, however, the river is a low temperature, low salinity feature compressed against the coastline to the north. Surface temperature throughout the winter falls generally within the range of 8 to 10° C with colder waters characteristic of the subarctic Pacific north of 47° N and west of 128° W.

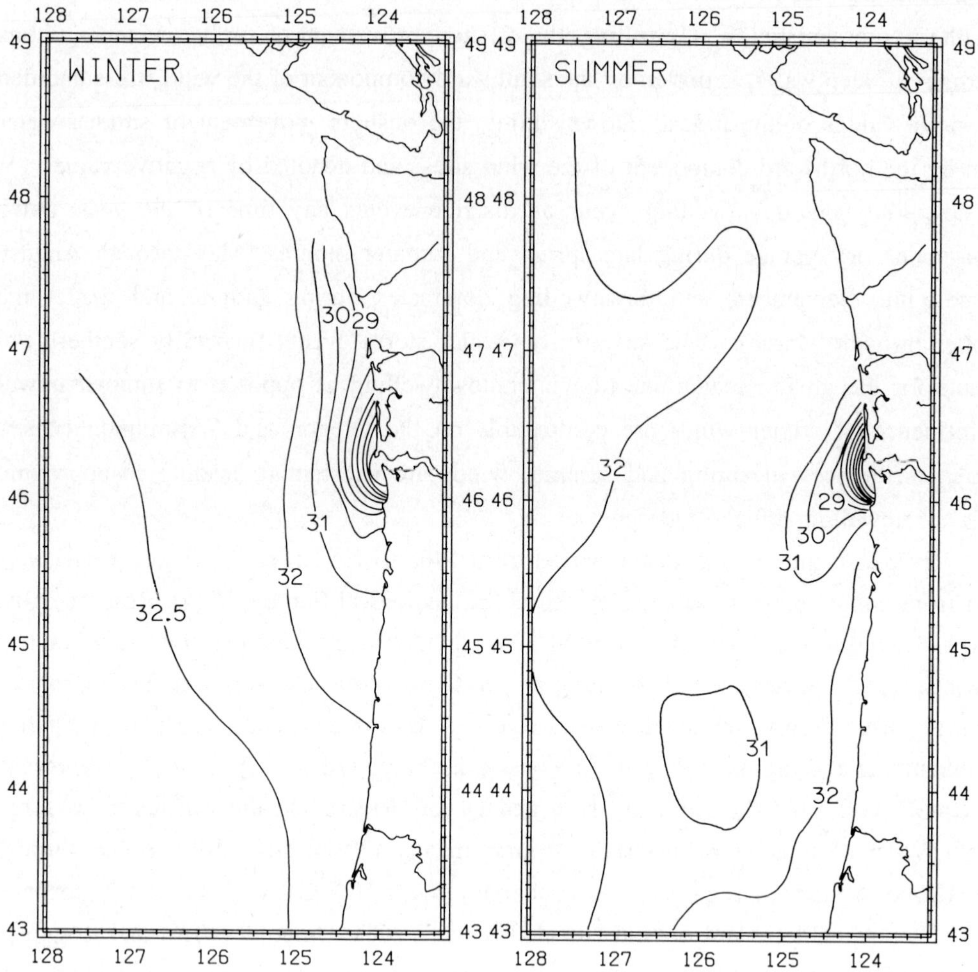

Fig. 1.3 Mean summer and winter distributions of salinity (ppt) in surface waters off Washington and Oregon.

Surface waters adjacent to the coast of Oregon during summer are colder, saltier and denser than those off Washington (Figs. 1.1, 1.3 and 1.4). Washington-Oregon differences in the timing and development of surface features and in the subsurface distributions of variables are detailed in Figures 1.5 to 1.7. These figures synthesize all parameter measurements made within the two cross-shelf transect areas depicted in Figure 1.1.

Patterns in the distributions of temperature, salinity and density are similar, for the most part, off Washington and Oregon at distances greater than about 50-km offshore, close to the shelf break. Surface temperatures and salinities are in excess of 9° C and 32 ppt in both offshore regions during winter and depth distributions of isotherms and isohalines are comparable (Figs. 1.5 and 1.6). Surface warming occurs rapidly through April and May, while cooling begins in September. August is the month of closest shoreward penetration of the 15° and 16° C isotherms. The major distinction between the offshore distributional features off the two states is the occurrence of the June salinity minimum off Oregon (Fig. 1.6, upper right panel). This

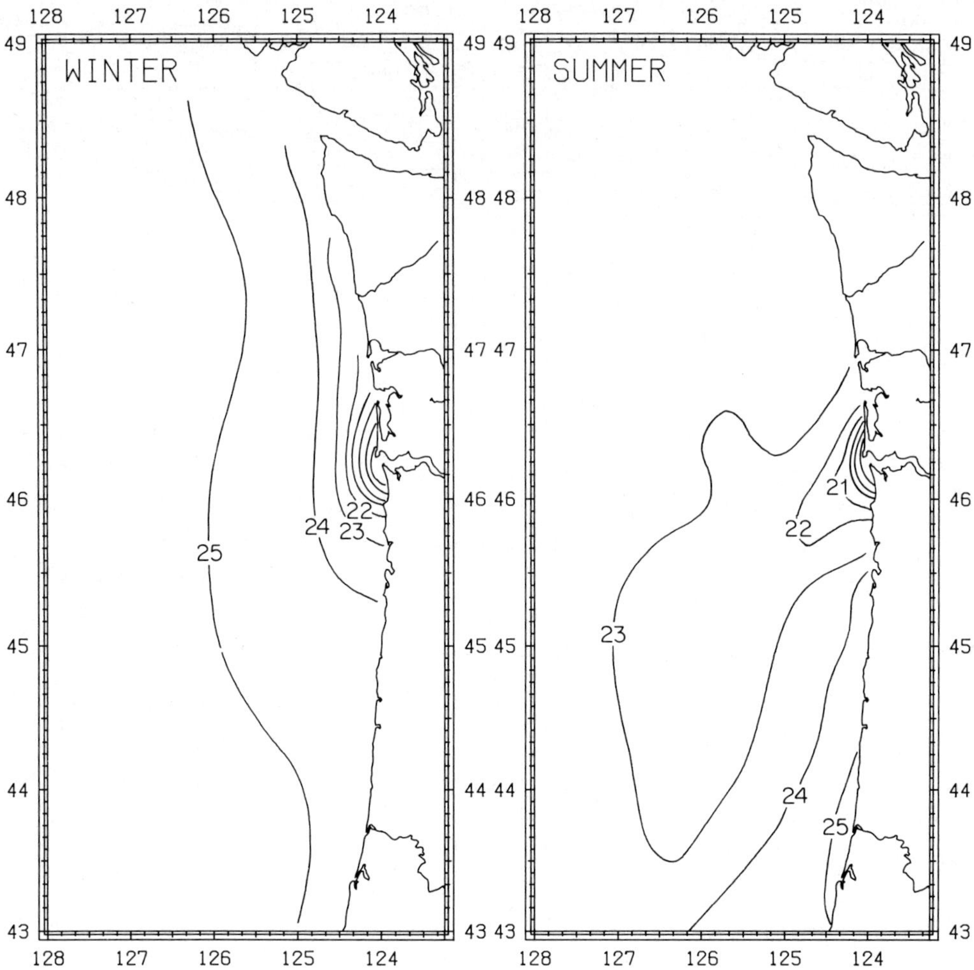

Fig. 1.4 Mean summer and winter distributions of density (sigma-t) in surface waters off Washington and Oregon.

salinity minimum coincides with the maximum Columbia River discharge and the seasonal shift to northerly winds (Fig. 1.2) which drives the river discharge offshore and to the south. The combination of river and wind impacts are clearly evident in the temporal development of surface salinity and density distributions (Figs. 1.6 and 1.7, upper panels). The low salinity, low density water of the Columbia River plume lies inshore off Washington during autumn and winter. The plume is driven offshore during the beginning of northerly dominant winds in May and reappears as the offshore salinity/density minimum off Oregon during summer.

Although similar physical processes impact the Washington and Oregon coasts, they result in substantially different patterns in nearshore distributional features. For example, prevailing southerly winds during winter (Fig. 1.2) result in downwelling, evident in Figures 1.5 and 1.6 (lower panels) as downward-sloping isotherms and isohalines. Downwelling compresses river water toward the coastline where it is carried northward by the prevailing currents. The result is a winter, nearshore salinity minimum off both coasts; however, the effect is more

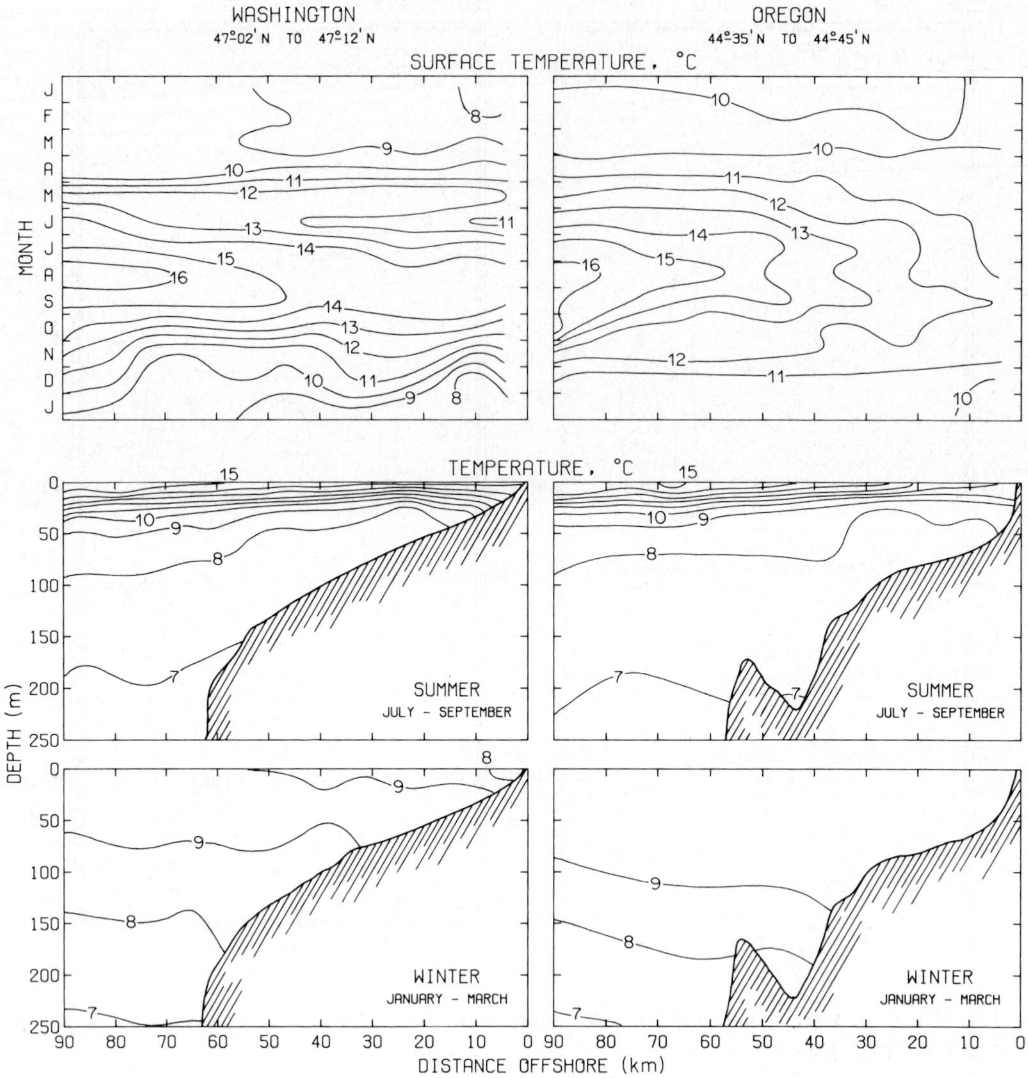

Fig. 1.5 Development of surface temperature features and seasonal depth structure of temperature in waters overlying the shelf regions of Washington and Oregon. Contours are based on composite measurements made during 1950-1984 from the cross-shelf transect areas (Copalis Head, Wa. and Yaquina Head, Ore.) depicted in Figure 1.1. Data were averaged for months and 5-km distances from shore prior to contouring.

dramatic off Washington since the Columbia River outflow, which lies along the Washington coast, vastly exceeds runoff from the smaller rivers along the Oregon coast. Relatively cold Columbia River water overlies warmer, oceanic water off Washington during winter, resulting in a substantial temperature inversion in the nearshore zone (Fig. 1.5). Upwelling occurs with the seasonal shift to northerly winds. General upwelling conditions during summer are apparent in the upward sloping isotherms, isohalines and isopycnals over much of the water column off both Washington and Oregon (Figs. 1.5 to 1.7, middle panels). Maximum upwelling, as indicated by the timing of summertime minima in temperature and maxima in salinity and density in

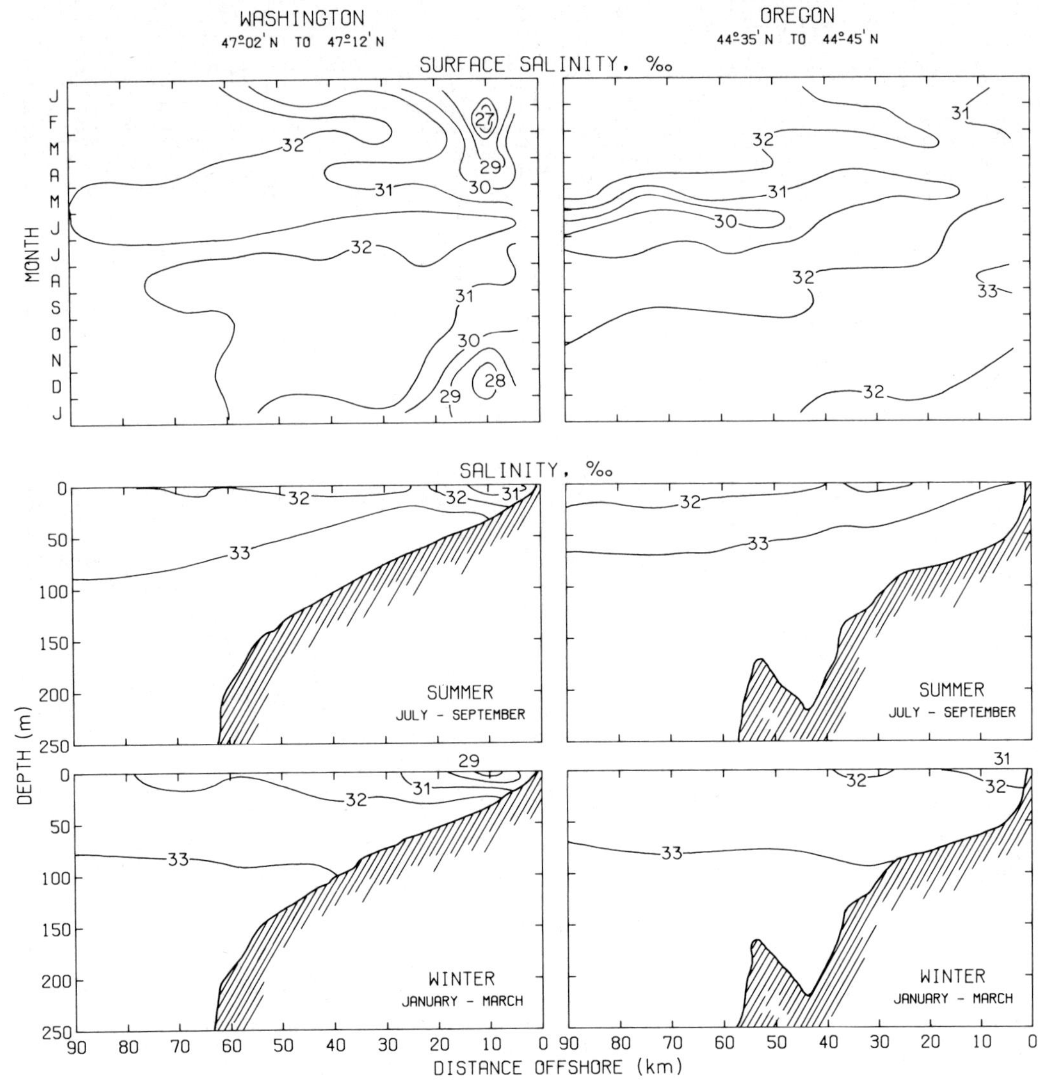

Fig. 1.6 Development of surface salinity features and seasonal depth structure of salinity in waters overlying the shelf regions of Washington and Oregon.

nearshore surface features (Figs. 1.5 to 1.7, upper panels), occurs in June off Washington and in July and August off Oregon. Differences in the magnitudes of coastal upwelling in the two areas are obvious in both surface features and depth profiles. For example, the 11° C-isotherm, the 33-ppt isohaline and the 25 sigma-t isopycnal intersect nearshore surface waters off Oregon but not off Washington (Figs. 1.5 to 1.7, middle panels). In fact, downwelling is indicated for the nearshore zone off Washington. These differences may be related to observed differences in the magnitudes of upwelling-related winds in the two areas (Fig. 1.2 middle panel).

The development and magnitude of subsurface temperature and salinity features in the mid-shelf regions of Washington and Oregon are compared in Figures 1.8 and 1.9. The figure synthesizes all parameter measurements from the two longshore sampling areas depicted in

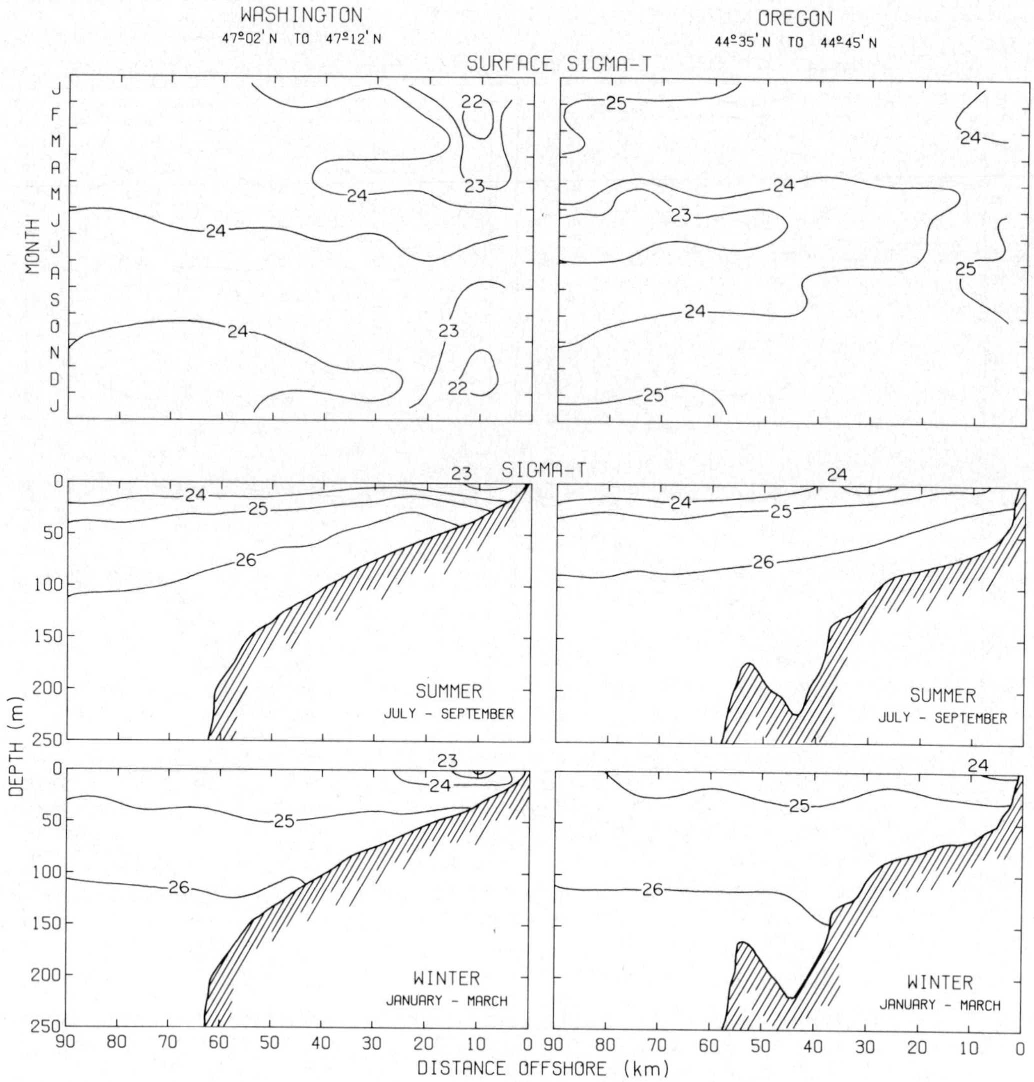

Fig. 1.7 Development of surface density (sigma-t) features and seasonal depth structure of sigma-t in waters overlying the shelf regions of Washington and Oregon.

Figure 1.1. The time-depth profile of monthly-averaged temperature for the Oregon shelf (Fig. 1.8) shows the dual maxima in near-surface values observed previously by Huyer (1977). In contrast, a single maximum occurs on the Washington shelf in mid-summer at about the time of summer minimum temperatures off Oregon. Both areas show the doming of cold, high salinity, high density water in mid-summer. Off Oregon, upwelled water breaks through to the surface in temperature, salinity and sigma-t isograms (Figs. 1.5 to 1.7). Off Washington, the effect is subsurface and is overridden by the seasonal temperature maximum in the surface layer.

Monthly means and standard deviations of temperatures and salinities at the surface and 70-m depth are compared in Figure 1.9. Standard deviations (i.e., interannual variability) for mean surface temperatures are in the range of 0.6 to 1° C in the winter and 1 to 2° C in late

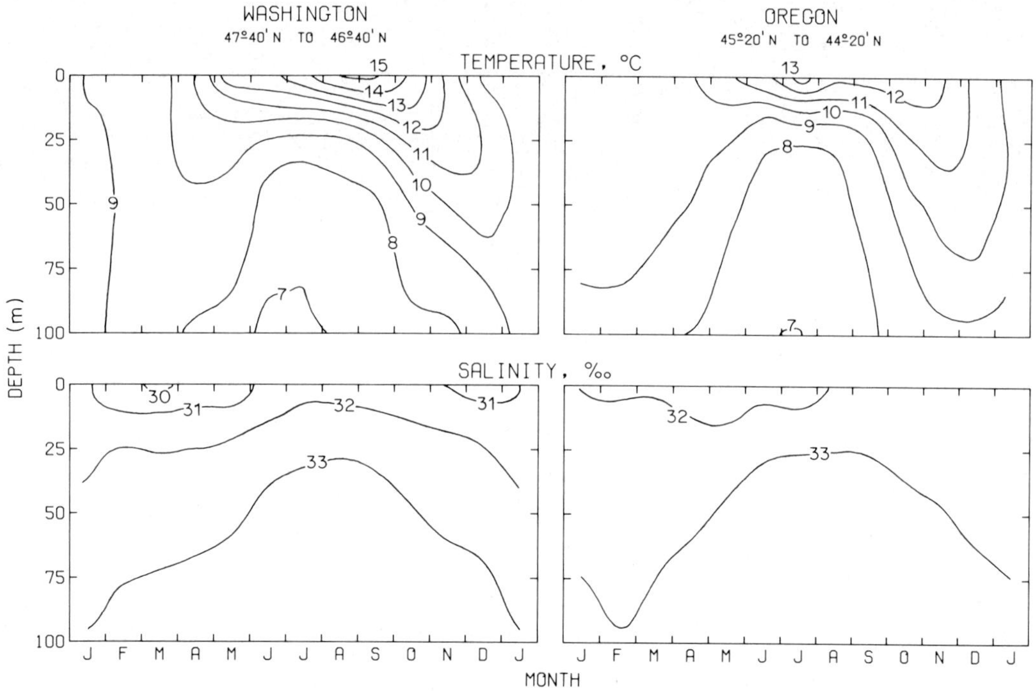

Fig. 1.8 Mean-monthly development of depth structure in mid-shelf temperature and salinity off Washington and Oregon. Contours are based on composite measurements made during 1950-1984 from the longshore transect areas depicted in Figure 1.1. These areas are defined by total water column depths in the range of 70 to 130 m.

spring through early autumn. Variability in surface salinity off Washington is greatest during the winter and early spring when Columbia River water is advected to the north (std. dev. = 1 to 1.3 ppt) relative to the summer and autumn (std. dev. = 0.5 to 0.9 ppt). In contrast, maximum interannual variability in surface salinity off Oregon occurs during peak river discharge in June and July (std. dev. = 1.1 to 1.5 ppt), when Columbia River water is advected to the south. Relatively low variability occurs during the remainder of the year (monthly std. dev. = 0.2 to 1.0 ppt). As might be expected, variability in salinity directly offshore of the Columbia River is extreme, particularly during June and July (std. dev. = 5 to 7 ppt) when mid-shelf, surface salinities as low as 15.5 ppt have been measured. Interannual variability in monthly mean temperature at 70 m is generally low (std. dev. = 0.2-0.8° C) except during November and December (1.0 to 1.4° C) which are similar off Washington and Oregon. Variability in monthly mean salinity values at 70 m is low (std. dev. = 0.1 to 0.4 ppt).

Despite the magnitude of interannual variabilities, certain aspects of the seasonal patterns differ significantly (i.e., at a probability level of $p \leq .05$) between areas. Surface temperatures, for instance, are significantly warmer off Oregon than off Washington from November to April due to the presence of the cold Columbia River plume off Washington. Surface temperatures are also significantly colder off Oregon than off Washington in the late summer (August and September) due to greater Oregon upwelling. As a result, the mean annual variation in surface

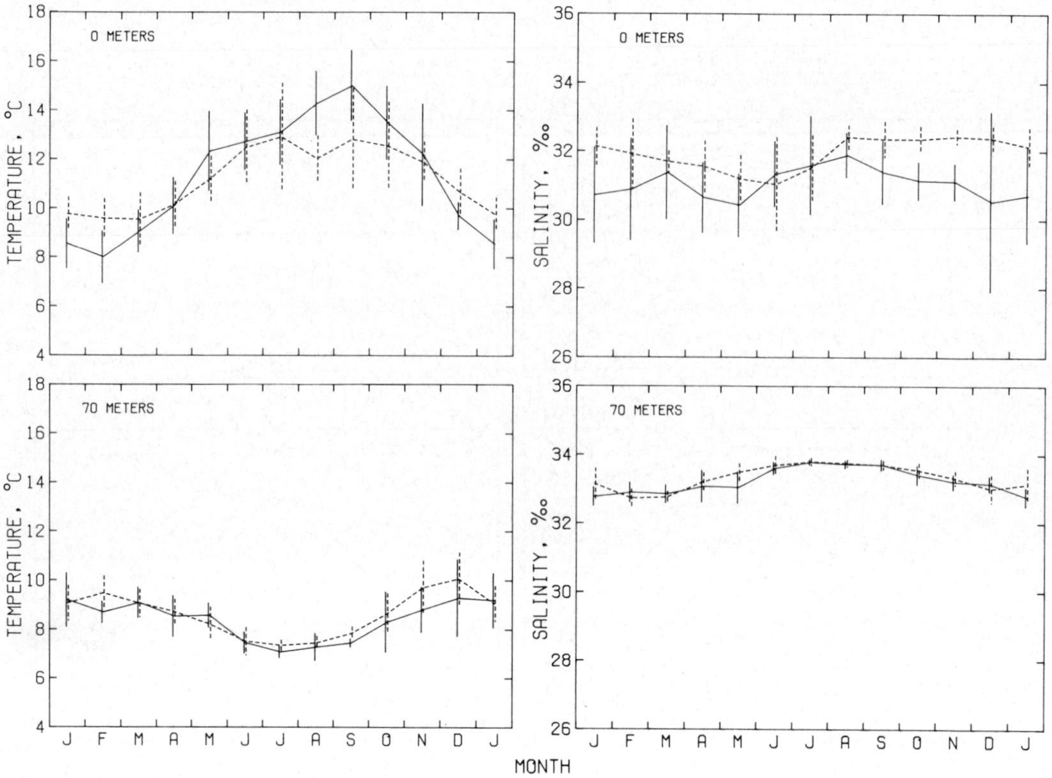

Fig. 1.9 Mean-monthly temperature and salinity values at the surface and 70 m for the mid-shelf regions off Washington (solid line) and Oregon (dashed line) defined in Figure 1.8. Vertical lines denote standard deviations of monthly means for all measurements made between 1950 and 1984.

temperature on the Oregon mid-shelf is 3.2° C as opposed to 5.7° C for the Washington mid-shelf. The reduced annual temperature range and the local temperature minimum in August off Oregon are the two features distinguishing seasonal variation in surface water temperature on the Washington and Oregon shelves. Surface salinities are consistently 0.6 to 1.0 ppt higher off Oregon than off Washington except during the early summer when Columbia River discharge is maximum and prevailing currents direct the plume southward to Oregon.

The annual amplitudes of mid-shelf temperature and salinity variations at 70 m are about 2.5 to 3.0° C and 1.0 ppt for both Washington and Oregon. As previously described by Huyer (1977), minimum temperature and maximum salinity values occur below the surface layer during summer months due to the upwelling of deep, offshore water onto the shelf.

1.3.3 Dissolved Oxygen

Although patterns in the distribution of dissolved oxygen must be interpreted with some caution since they are influenced by both biological and physical processes, they generally conform to the patterns observed for the more conservative variables – temperature, salinity and density. On a regional scale, for instance, summertime distributions of percent oxygen saturation are consistent with stronger upwelling of deep (low oxygen saturation) water off

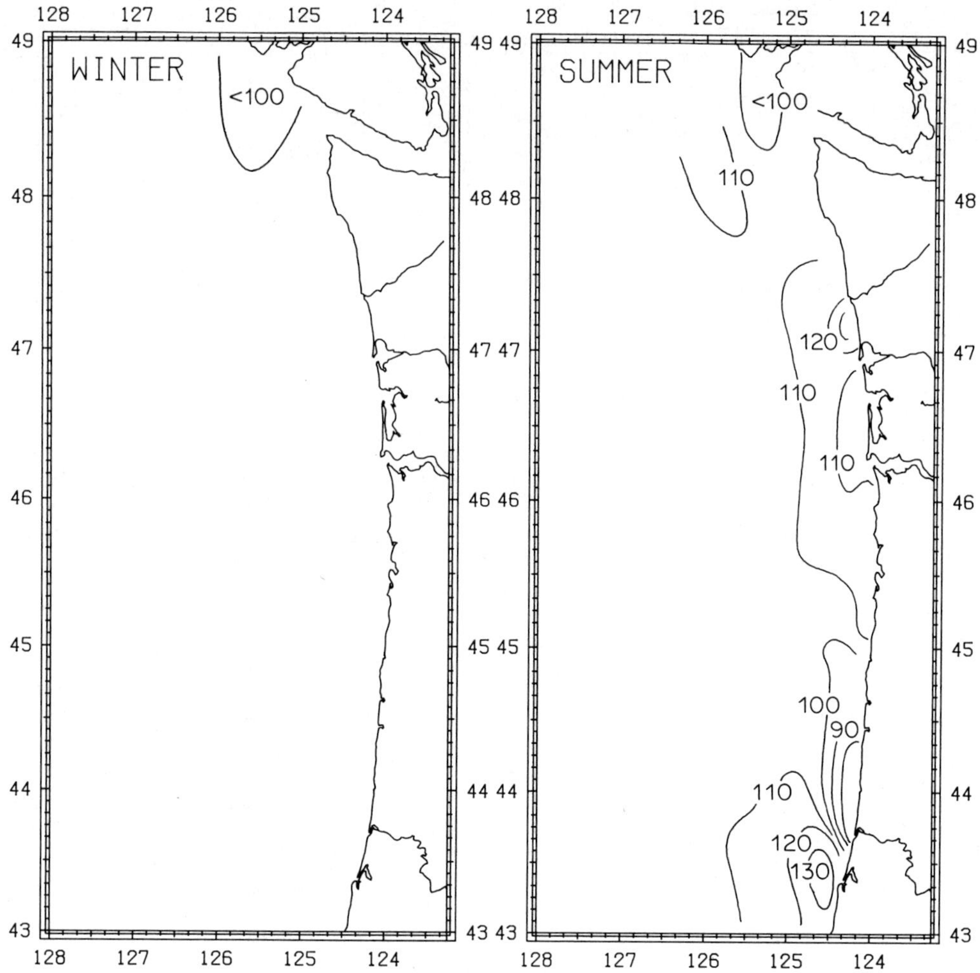

Fig. 1.10 Mean summer distribution of percent oxygen saturation in surface waters off Washington and Oregon. Contours are fit to composite measurements made between 1950-1984 averaged in 15' x 15' grid squares.

Oregon than off Washington (Fig. 1.10). Percent saturation in surface waters of the nearshore, mid-coast region off Washington is greater than 110% compared to less than 100% for the comparable region off Oregon. However, the southern-most section of the Oregon coast, where the shelf expands into the Hecata Banks, seems to be an area of oxygen supersaturation in surface waters similar to that off Washington. This section of the Oregon coast could therefore be an area of reduced upwelling, enhanced surface warming and/or higher biological production. The Heceta Banks region is not in the mid-shelf and cross-shelf areas described in Figures 1.11 to 1.13.

The seasonal transition from winter to summer distributions of surface oxygen off the Washington coast is clearly represented in Figure 1.11 (upper left). Percent oxygen saturation over the width of the shelf surpasses 110% during late March and early April suggesting the occurrence of the spring phytoplankton bloom and/or surface stratification and warming. The

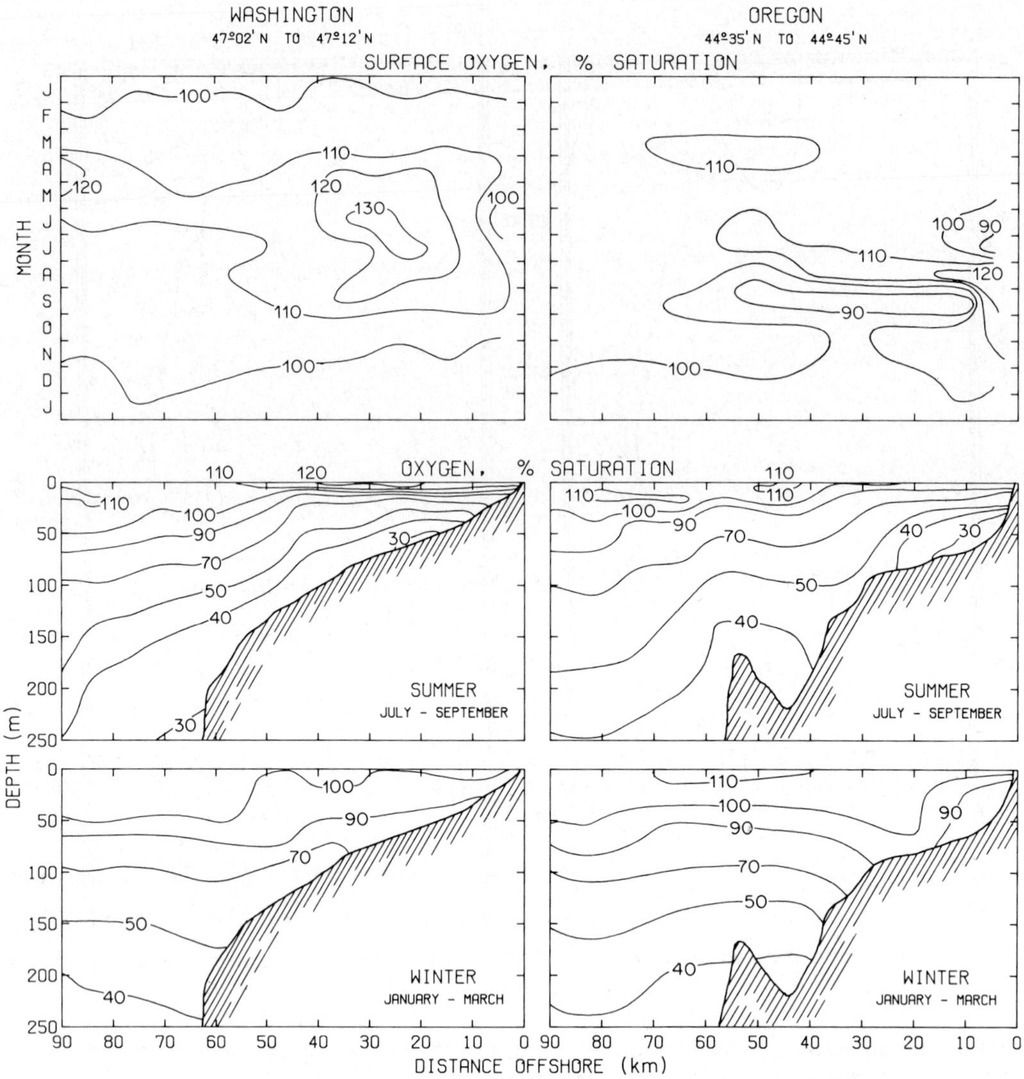

Fig. 1.11 Development of surface oxygen saturation features and seasonal depth structure of percent oxygen saturation in waters overlying the shelf regions of Washington and Oregon. Contours are based on composite measurements made during 1950-1984 from the cross-shelf transect areas (Copalis Head, Washington and Yaquina Head, Oregon) depicted in Figure 1.1. Data were averaged for months and 5-km distances from shore prior to contouring.

bloom explanation, i.e., biogenic production of oxygen, is favored by the observation that surface oxygen concentration also increases at this time (Fig. 1.13). Surface oxygen remains supersaturated at 120 to 130% over the mid-shelf region throughout summer even as the nearshore presence of oxygen-unsaturated water in June marks the peak of seasonal upwelling. The transition from typical summer conditions to typical winter conditions begins in September. In contrast to the rather gradual and persistent features off Washington, the surface distribution of percent oxygen saturation off Oregon (Fig. 1.11, top right) suggests dramatic, month-to-month

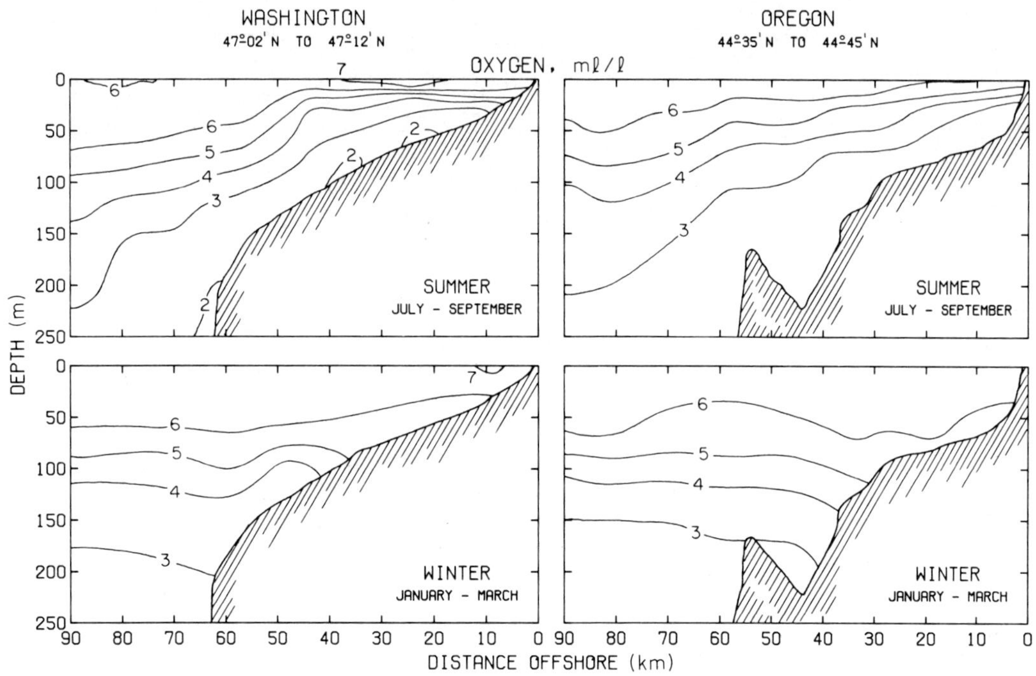

Fig. 1.12 Mean summer and winter distributions of oxygen content in waters overlying the shelf regions of Washington and Oregon defined in Figure 1.1.

transitions. The magnitudes of these swings – from nearshore subsaturation in June and July to shelf-wide supersaturation in August back to subsaturation in September – are undoubtedly influenced by chance sampling of different upwelling states in the different months (e.g., Small and Menzies, 1981). This may also reflect a more dynamic coastal upwelling system off Oregon than off Washington.

Profiles of cross-shelf oxygen content (Fig. 1.12) and percent oxygen saturation (Fig. 1.11) are indicative of general upwelling during summer months except in the nearshore region off Washington (upward-sloping isograms) and downwelling during winter months (downward-sloping isolines at 100 m and below). Water of low oxygen saturation (30-40%) overrides the bottoms of both shelves during summer, the result of upwelling and benthic metabolism. These water column profiles underrepresent the metabolic contribution to oxygen depletion near the water-sediment interface. The greater magnitude of summer upwelling off the mid-coast of Oregon is demonstrated by the surfacing of 6 ml O_2 l^{-1} and 100% O_2 saturation isograms off Oregon but not off Washington. Cross-shelf profiles from Washington and Oregon show similar subsurface maxima in percent oxygen saturation (> 110%) seaward of 70-80 km offshore (Fig. 1.11, middle panels). The major difference in the cross-shelf profiles is the persistent cap of highly-oxygenated water which occurs during summer over the mid-shelf off Washington but only weakly over Oregon (Figs. 1.11, middle; 1.13). This difference can be attributed to three interactive factors. First, the Oregon shelf experiences more summer

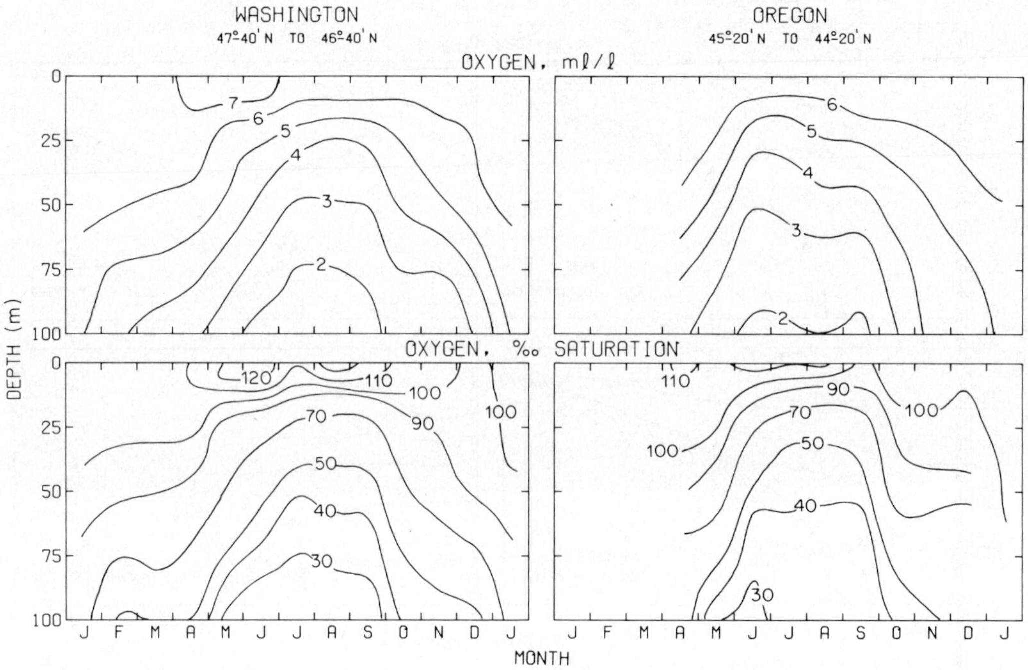

Fig. 1.13 Mean-monthly development of depth profiles in mid-shelf oxygen content and percent saturation off Washington and Oregon. Contours are based on composite measurements made during 1950-1984 from the longshore transect areas depicted in Figure 1.1. These areas are defined by total water column depths in the range of 70 to 130 m.

upwelling than the Washington shelf. Therefore, a greater volume of deep water, with low oxygen concentration and low percentage saturation, is advected into Oregon surface waters. Second, the surface waters off Washington are strongly stratified and experience greater summer warming than Oregon surface waters. Warming reduces the solubility of oxygen, hence elevates percent saturation for a given oxygen concentration. Third, the production of oxygen by phytoplankton in near-surface waters may be somewhat higher, on average, off Washington. This is suggested from differences in surface chlorophyll concentrations in the two shelf regions (see below; Figs. 1.24 to 1.26), particularly during late spring and early summer when the dissolved oxygen difference between Oregon and Washington surface waters is greatest.

The temporal development of surface and subsurface oxygen structures off Oregon and Washington are compared in Figure 1.13. On the Washington mid-shelf, an advanced spring bloom in phytoplankton production is evident as high surface oxygen content and percent saturation in April, following the start of seasonal temperature stratification in March (Fig. 1.8). A weak maximum in surface oxygen is also apparent off Oregon during April; however, insufficient data from Oregon through winter and early spring make it impossible to compare the timing and magnitude of the spring bloom to that off Washington. Upwelling-induced doming of oxygen isograms occurs at both locations throughout the summer following the patterns observed for temperature and salinity (Fig. 1.8).

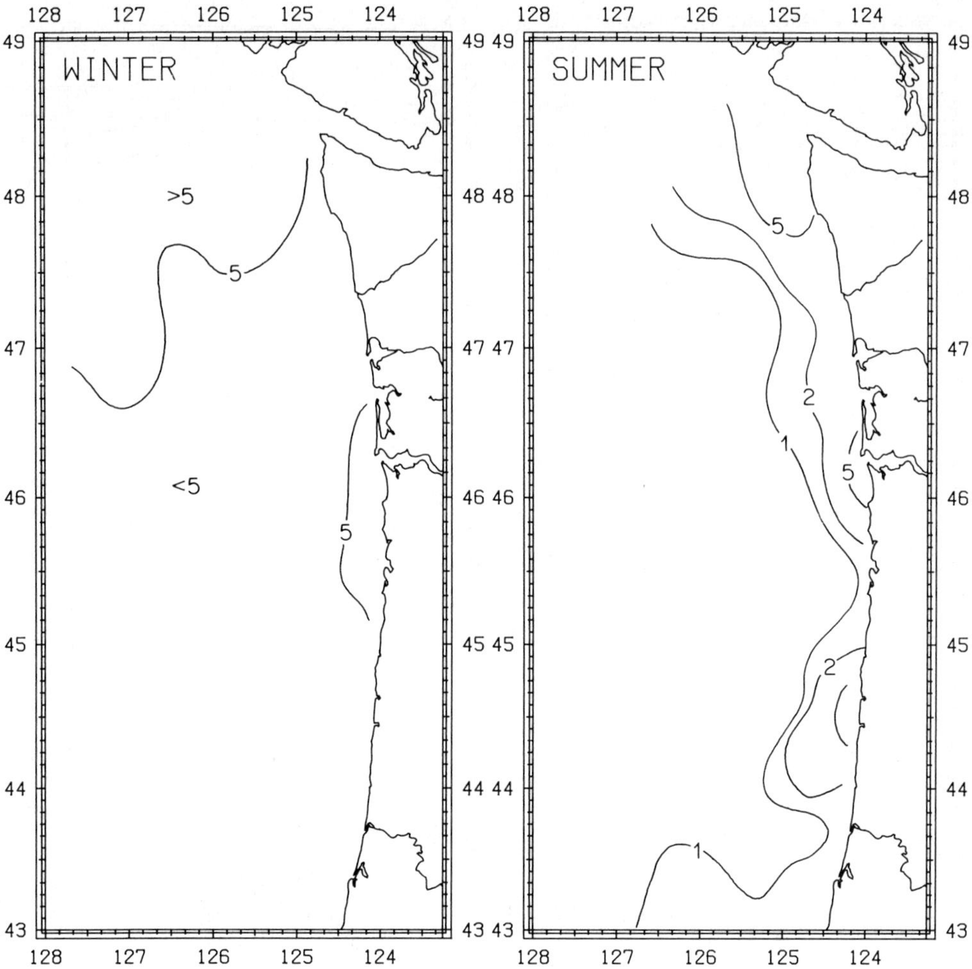

Fig. 1.14 Mean summer and winter distributions of nitrate concentration in surface waters off Washington and Oregon. Contours are fit to composite measurements made between 1950-1984 averaged in 15' x 15' grid squares.

1.3.4 Dissolved Nutrients

Distributional features of dissolved nutrients (nitrate, nitrite, ammonium, phosphate and silicate) are described in Figures 1.14 through 1.22. Monthly means are not supported by as many years of data as temperature and salinity (Table 1.1); therefore, more attention should be focused on recurrent patterns in nutrient concentrations rather on the apparent differences in mean concentrations between months.

On a regional scale, surface nutrient concentrations are generally high everywhere during winter but are higher near the coast during summer (Figs. 1.14 to 1.16). High nutrient water from the subarctic Pacific is evident in the northwest corner of the region during winter. The Columbia and Fraser (via the Strait of Juan de Fuca; upper right) Rivers are sources of high nitrate (> 5 μM), phosphate (> 1 μM) and silicate (> 20 μM) input to the nearshore pelagic environment in both winter and summer. Silicate distributions appear to be heavily influenced

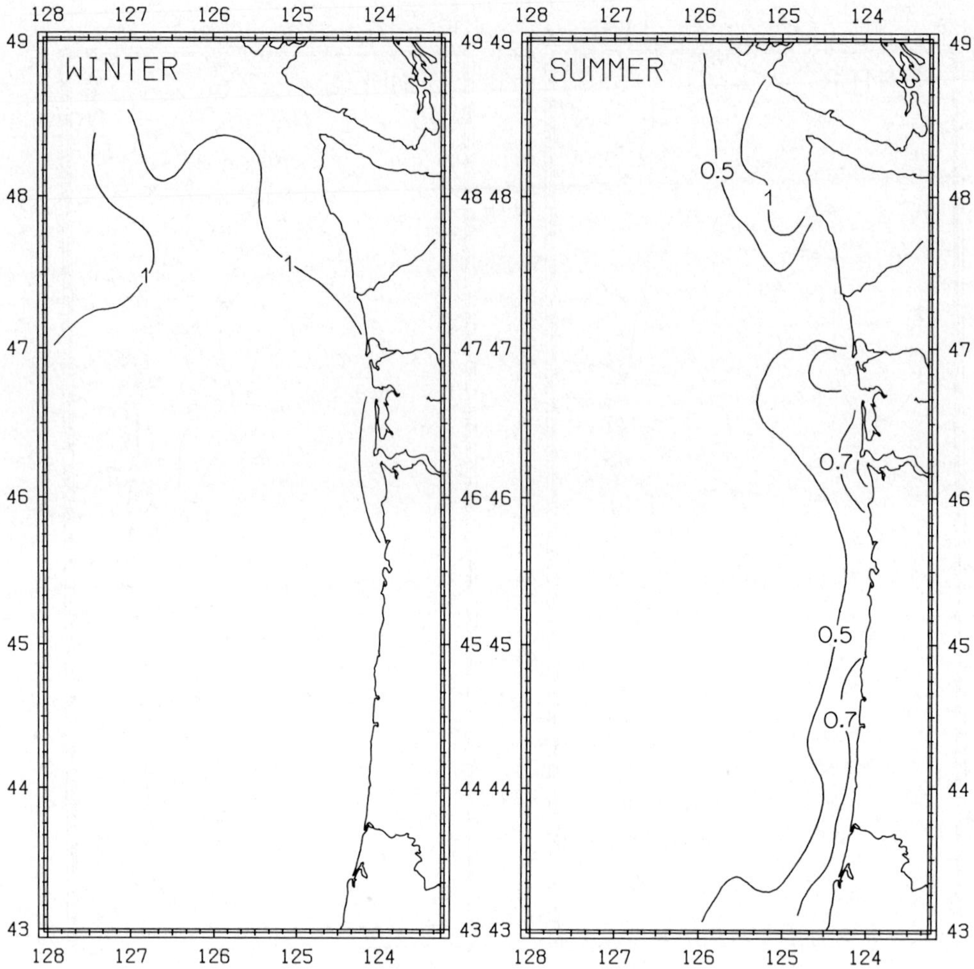

Fig.1.15 Mean summer and winter distributions of phosphate concentration in surface waters off Washington and Oregon.

by the plumes of the Columbia and Fraser Rivers. The effects of summer upwelling are more apparent for surface nitrate and phosphate distributions, particularly off central Oregon. However, nearshore, upwelling-related features for all nutrients are not clearly resolved in these regional maps due to the coarse scale of averaging and smoothing. These features are better represented in the cross-shelf sections described below.

Monthly surface distributions of nutrients in the cross-shelf sections off Washington and Oregon are affected by three seasonal phenomena, wintertime outflow of the Columbia River, the timing of the spring bloom and summer upwelling (Figs. 1.17 to 1.19, upper panels). The Columbia River influence is particularly apparent as a winter silicate maximum close to the Washington coast (Fig. 1.19). The river effect is less evident in the phosphate and nitrate distributions (Figs. 1.17 and 1.18). The effect of the spring bloom in phytoplankton is most dramatic for nitrate concentrations which fall precipitously during April off both Washington

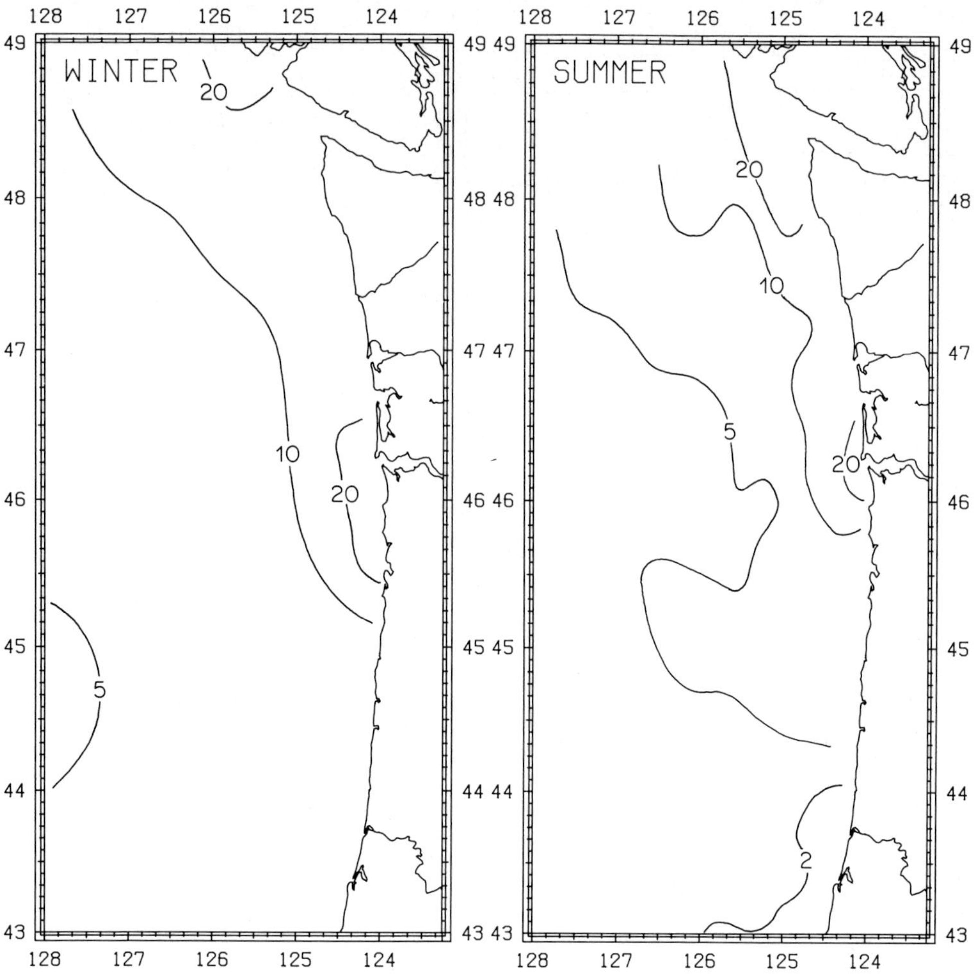

Fig. 1.16 Mean summer and winter distributions of silicate concentration in surface waters off Washington and Oregon.

and Oregon (Fig. 1.17). Phosphate and silicate also decline during the bloom but not to the same extent as nitrate. Consequently, the reintroduction of nutrients to the surface waters via upwelling is more pronounced for the nearshore gradient in summer nitrate than for the phosphate or silicate gradients although upwelling clearly influences the latter two nutrients as well. The peak in nearshore nutrient concentrations due to upwelling occurs during June off Washington while upwelling leads to enhanced nutrient levels (e.g., 5-10 μM nitrate) in the nearshore off Oregon throughout the summer. This is consistent with stronger and more persistent summer upwelling off Oregon.

The depth distributions of nutrients during summer and winter months emphasize different sources of input. For example, the high inshore concentrations of phosphate and particularly silicate during winter are indicative of river runoff (Figs. 1.18 and 1.19). Similarly, the upward sloping isograms for nitrate, phosphate and silicate during summer follow the pattern

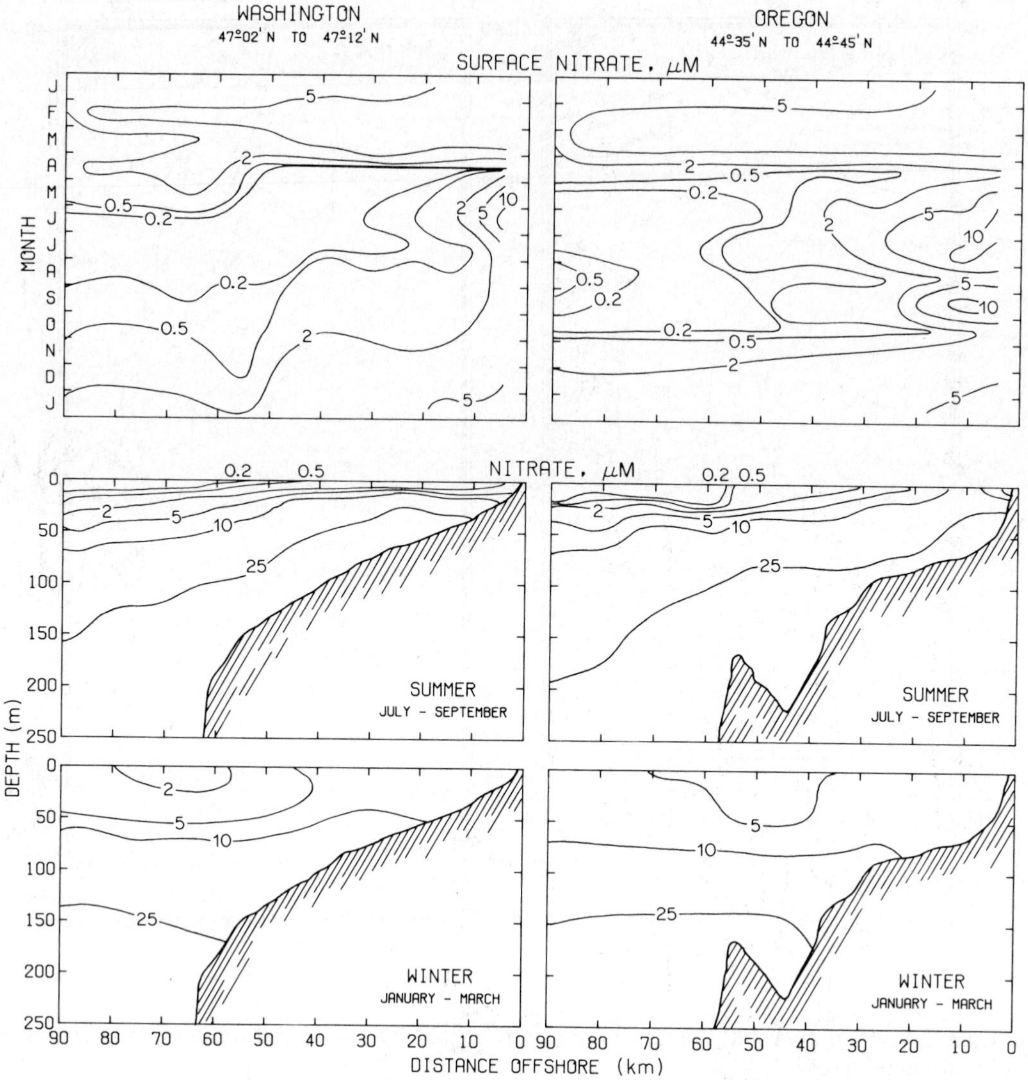

Fig. 1.17 Development of surface nitrate features and seasonal depth structure of nitrate concentration in waters overlying the shelf regions of Washington and Oregon. Contours are based on composite measurements made during 1950-1984 from the cross-shelf transect areas (Copalis Head, Wa. and Yaquina Head, Ore.) depicted in Figure 1.1. Data were averaged for months and 5-km distances from shore prior to contouring.

expected from seasonal upwelling. The greater magnitude of upwelling off Oregon is evident from the observations that higher nutrient isograms (10 μM nitrate, 1 μM phosphate and 20 μM silicate) intersect the surface off Oregon but not off Washington.

Whereas nitrate, phosphate and silicate are injected into surface waters by upwelling and river runoff, distributions of nitrite and ammonium point to nearshore and subsurface areas of nitrogen regeneration (Figs. 1.20 and 1.21). A well-developed maximum in ammonium concentration occurs below the thermocline over the Washington shelf during summer. Highest

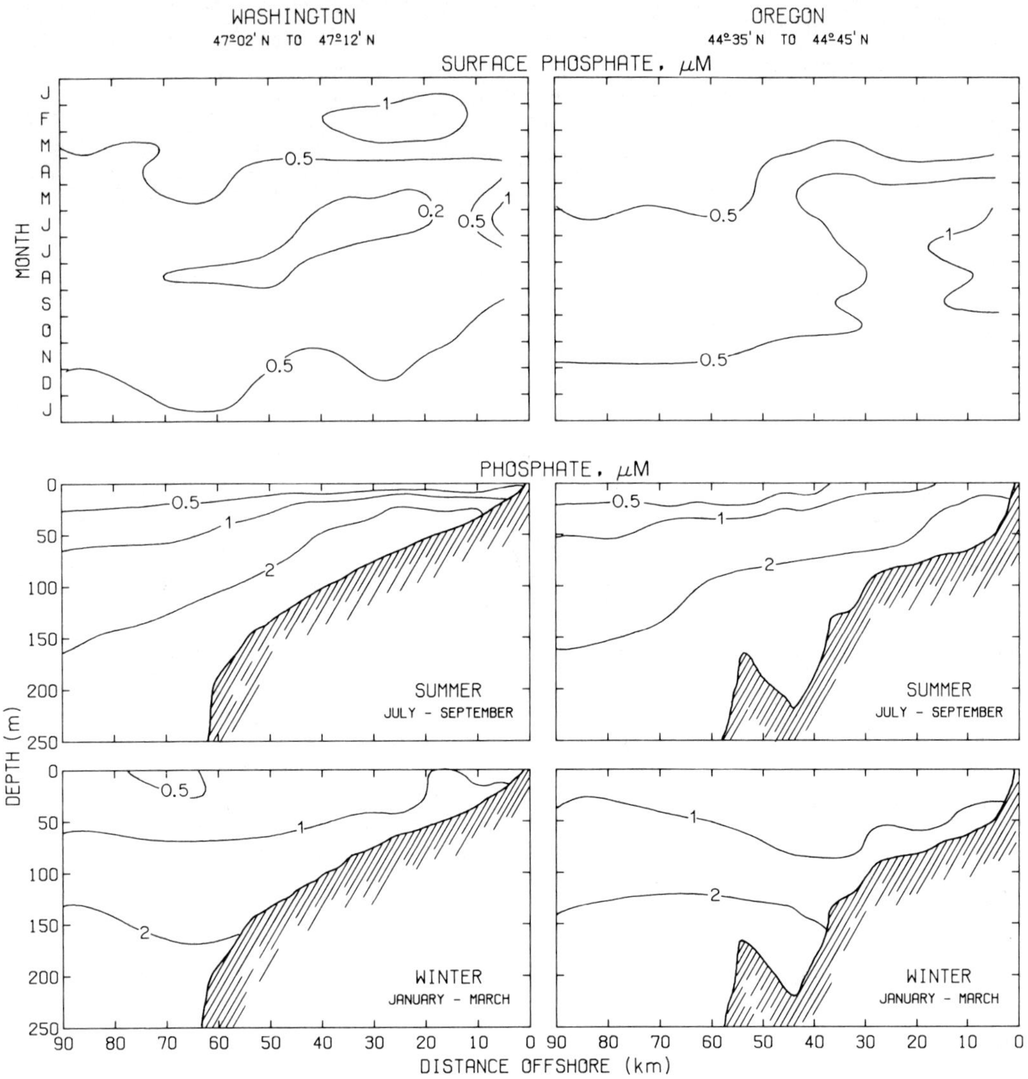

Fig. 1.18 Development of surface phosphate features and seasonal depth structure of phosphate concentration in waters overlying the shelf regions of Washington and Oregon.

concentrations occur over the nearshore sediments in both summer and winter (Fig. 1.20). The position of the maximum reflects a balance between ammonium utilization, which is highest in the euphotic zone due to phytoplankton uptake and ammonium production, which is highest below the euphotic zone and in shallow sediments where concentrations of organic substrate are highest. Oxidation of ammonium to nitrite by nitrifying bacteria is photoinhibited in near-surface waters and proportional to substrate (i.e., ammonium) concentration in deeper water (Ward *et al.*, 1984). Consequently, the prominent subsurface (30-60 m; Fig. 1.21) maximum in nitrite on the Washington and Oregon shelves is closely associated with the ammonium maximum.

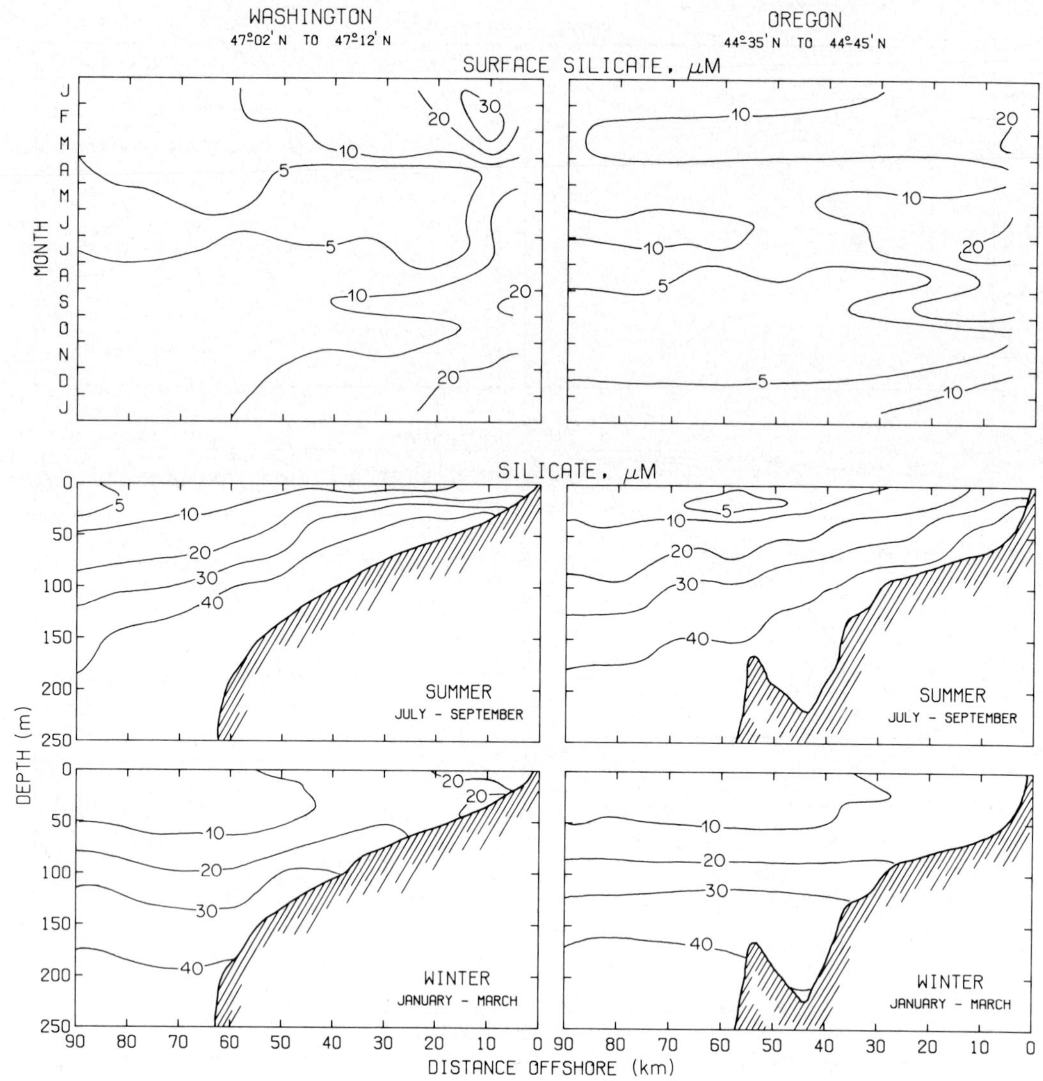

Fig. 1.19 Development of surface silicate features and seasonal depth structure of silicate concentration in waters overlying the shelf regions of Washington and Oregon.

Monthly variations in the depth structure of nitrate, phosphate and silicate in waters overlying the mid-shelf areas of the Washington and Oregon coasts are presented in Figure 1.22. The major features in the distributions are the doming of isograms due to summer upwelling and the relative depletion of surface nutrients during spring and summer months. The mid-shelf areas are well offshore of the regions most strongly influenced by upwelling; consequently, isograms of high nutrient concentrations (10 μM nitrate, 1 μM phosphate and 20 μM silicate) do not break the surface. Nutrient depletion is clearly greatest for nitrate on the Washington mid-shelf. Surface nutrients decline rapidly during the month of April, presumably due to uptake by the spring phytoplankton bloom. A secondary decline in surface nitrate occurs off Oregon in September. This coincides with the seasonal decline in upwelling-favorable winds (Fig. 1.2).

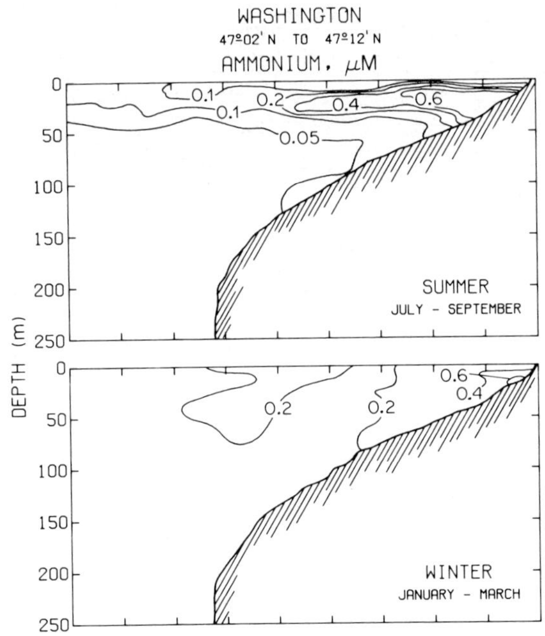

Fig. 1.20 Mean summer and winter distributions of ammonium concentration in waters overlying the Washington shelf. A corresponding figure could not be constructed for the Oregon coast due to insufficient data.

1.3.5 Chlorophyll *a*

One feature in the regional maps of surface chlorophyll *a* is the similarity between offshore winter and summer values (Fig. 1.23). Mean nearshore concentrations of chlorophyll vary by about an order of magnitude between seasons and a sharp chlorophyll gradient exists between inshore and offshore regions during the summer. Seasonal surface maps, however, do not reflect the true differences between winter and summer concentrations of phytoplankton for a number of reasons. Winter levels are biased by including data from the month of March. March generally conforms to winter conditions for other variables; however, March cruises sometimes coincided with the beginning of spring phytoplankton blooms. Also, summer surface maps average short-lived variations in chlorophyll ranging from relatively low, winter values to blooms in excess of 20 μg Chl *a* l^{-1} (June, 1982; Washington mid-shelf; Hermann *et al.*, 1989, Chapter 6, Fig. 6.7). Furthermore, maximum concentrations of phytoplankton pigments frequently occur as subsurface structures during summer months and are, therefore, not included in surface maps.

The seasonal development and structure of chlorophyll features off Washington and Oregon are compared in Figures 1.24 and 1.25. Winter chlorophyll levels in the nearshore environment range from 0.3 to 1.0 mg Chl *a* m^{-3} (< 20 mg Chl *a* m^{-2}). In the offshore region, surface concentrations are lowest during mid- to late summer. The increase of chlorophyll values over the entire Washington shelf region during March and April corresponds to the spring phytoplankton bloom implied previously by the increase in oxygen content and decreases in surface nutrient concentrations observed during the same period. The reduced offshore concentration of chlorophyll in May, about the same level as observed during wintertime, coincides with the period of low surface nutrient levels (e.g., Fig. 1.17). A strong spring bloom is not apparent from the available data off Oregon, but it may simply have been missed. As noted previously, a marked decline in surface nutrient concentrations during April and May occurs off Oregon as well as Washington providing indirect evidence for similar bloom timing in the two regions.

The mean cross-shelf distributions of chlorophyll are similar off Washington and Oregon during summer months (Fig. 1.24, middle panels). The main differences appear during late spring. Seasonally high concentrations of chlorophyll, on the order of 6 to 10 mg Chl *a* m^{-3}

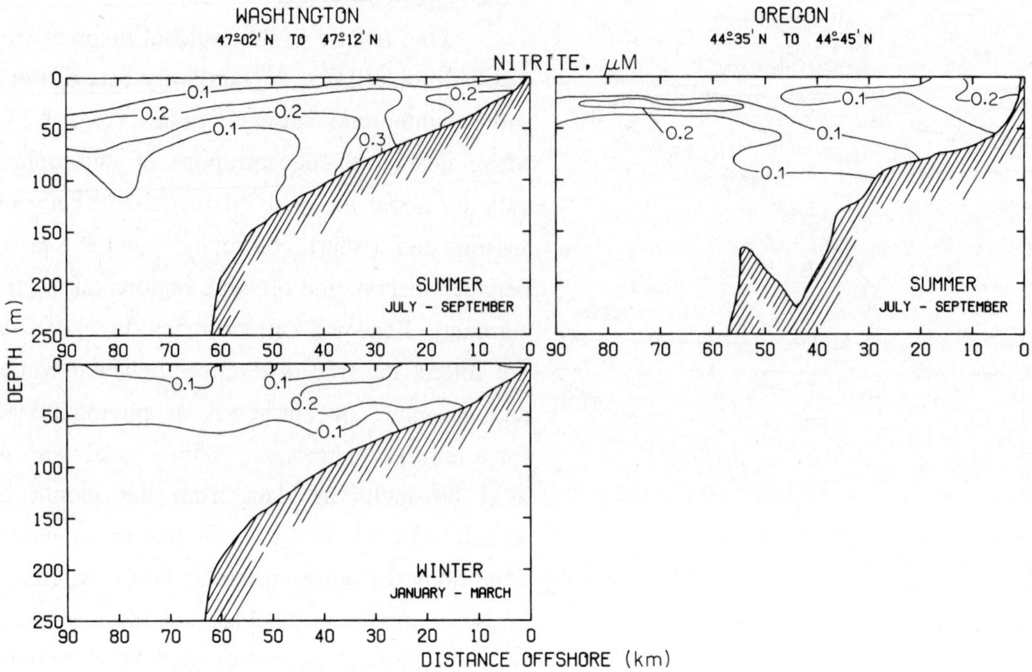

Fig. 1.21 Mean summer and winter distributions of nitrite concentration in waters overlying the shelf regions of Washington and Oregon.

(100-200 mg Chl *a* m $^{-2}$), occur over the inner shelf of Washington during the June peak in upwelling. Mean chlorophyll concentrations shelf decline gradually through the summer and autumn although variability is high with standard deviations comparable to mean values. Off Oregon, episodes of high upwelling intensity occur throughout the summer. Consequently mean summer chlorophyll concentrations, in the range of 3 to 6 mg Chl m $^{-3}$, are more or less evenly distributed across the 40-km shelf from mid-May through August.

Maximum concentrations of chlorophyll *a* generally occur inshore of the mid-shelf regions off Washington and Oregon. Small and Menzies (1981) and Hermann *et al.* (1989, Chapter 6) have illustrated that the highest chlorophyll levels develop over the inner shelf within 20 km from shore during the downwelling phase of the intermittent upwelling cycles. This pattern is also apparent from the monthly means in surface and water-column integrated chlorophyll in Figures 1.24 and 1.25. Thus the mid-shelf distributions of chlorophyll concentration depicted in Figure 1.26 only capture the outer fringes of maximum shelf structures. Nonetheless, they demonstrate the March-April spring bloom, the May depression off Washington, the asymmetrical summer maximum off Washington and the more evenly distributed summer distributions off Oregon as described previously (Figs. 1.24 and 1.25).

1.4 INTERANNUAL VARIABILITY

Interannual variations in the physical environment of coastal Washington and Oregon are substantial with monthly, near-surface temperatures and salinities varying over ranges

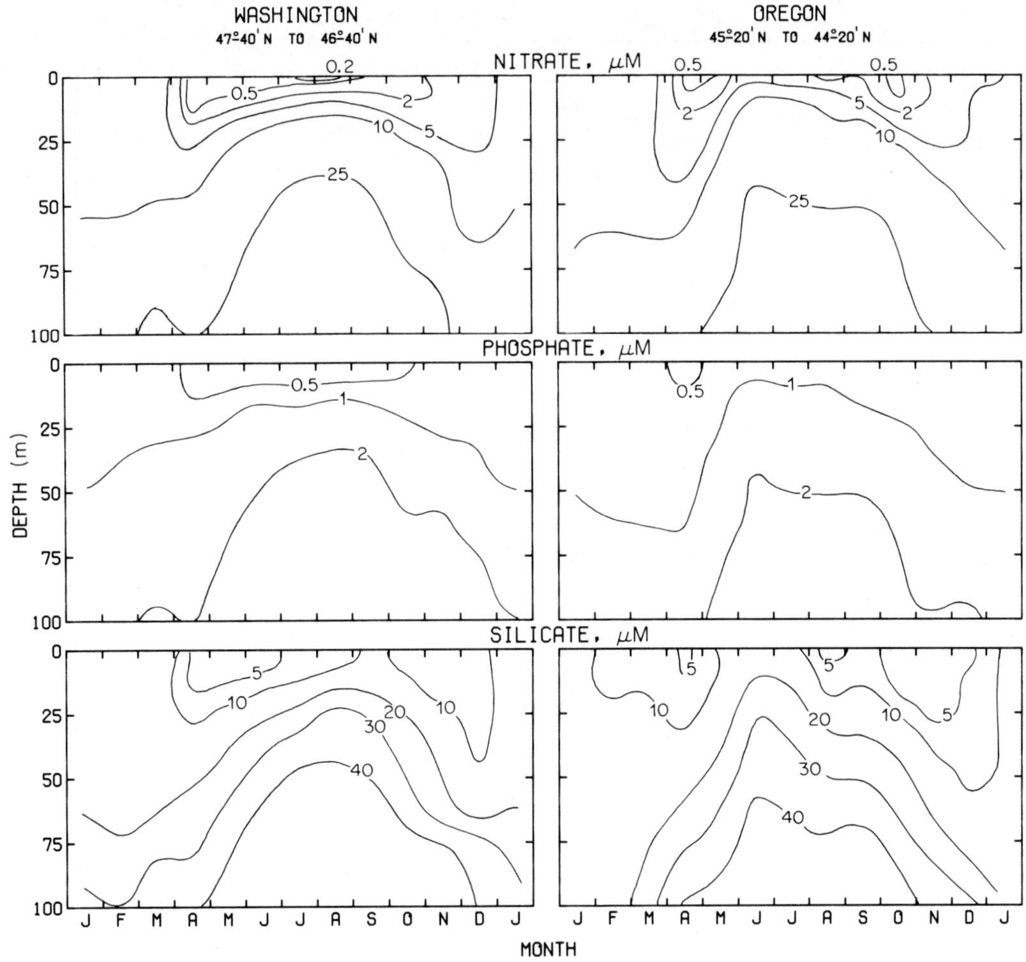

Fig. 1.22 Mean-monthly development of depth structure in mid-shelf nitrate, phosphate, and silicate concentrations off of Washington and Oregon. Contours are based on composite measurements made during 1950-1984 from the long-shore transect areas depicted in Figure 1.1. These areas are defined by total water column depths in the range of 70 to 130 m.

comparable to mean seasonal signals (i.e., 5 to 6° C and 3 to 4 ppt – exclusive of the Columbia River plume). Interannual differences in winds and currents vary over similar ranges (Hickey, 1989, Chapter 2). One consequence of this high level of variability is that "average" conditions are known only within large limits of uncertainty. Therefore, in a given year, we would expect little chance of observing patterns exactly as depicted in long-term averages. The variability also appears to be an important ecological parameter in its own right. Although linking mechanisms are still poorly understood, year-to-year variations in the abiotic environment play a significant role in the recruitment success of many commercial stocks in the Northeast Pacific (e.g., Wooster, 1983; Botsford *et al.*, 1989, Chapter 11).

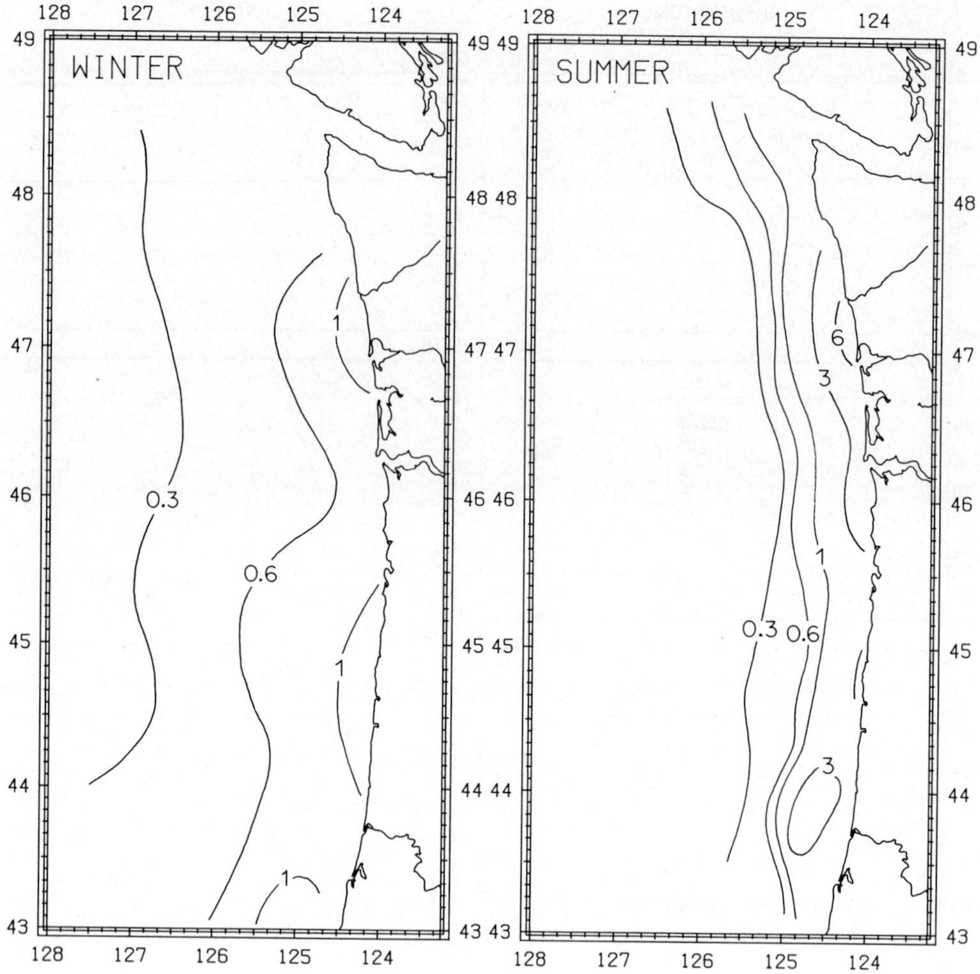

Fig. 1.23 Mean summer and winter distributions of chlorophyll *a* concentration in surface waters off Washington and Oregon. Contours are fit to composite measurements made between 1950-1984 averaged in 15' x 15' grid squares.

Much of the interannual variability in ocean temperature and climate throughout the Pacific Ocean has been ascribed to periodic ocean-atmosphere perturbations along the Equator. These perturbations, known as "El Niño-Southern Oscillation" (ENSO) events, arise when trade winds relax and waters in the eastern tropical Pacific and along coastal North and South America become anomalously warm. The position of the tropical temperature anomaly also affects global patterns in atmospheric circulation and precipitation. In the Northeast Pacific, for instance, ENSO events are marked by a winter intensification of the Aleutian Low (i.e., strong cyclonic circulation over the Gulf of Alaska) which further facilitates the northeastward advection of warm water toward the coasts of Oregon, Washington and British Columbia (Hamilton and Emery, 1985).

The influences of global-scale phenomena on the physical environment of the Washington/Oregon continental margin should also be expressed in anomaly patterns in adjacent

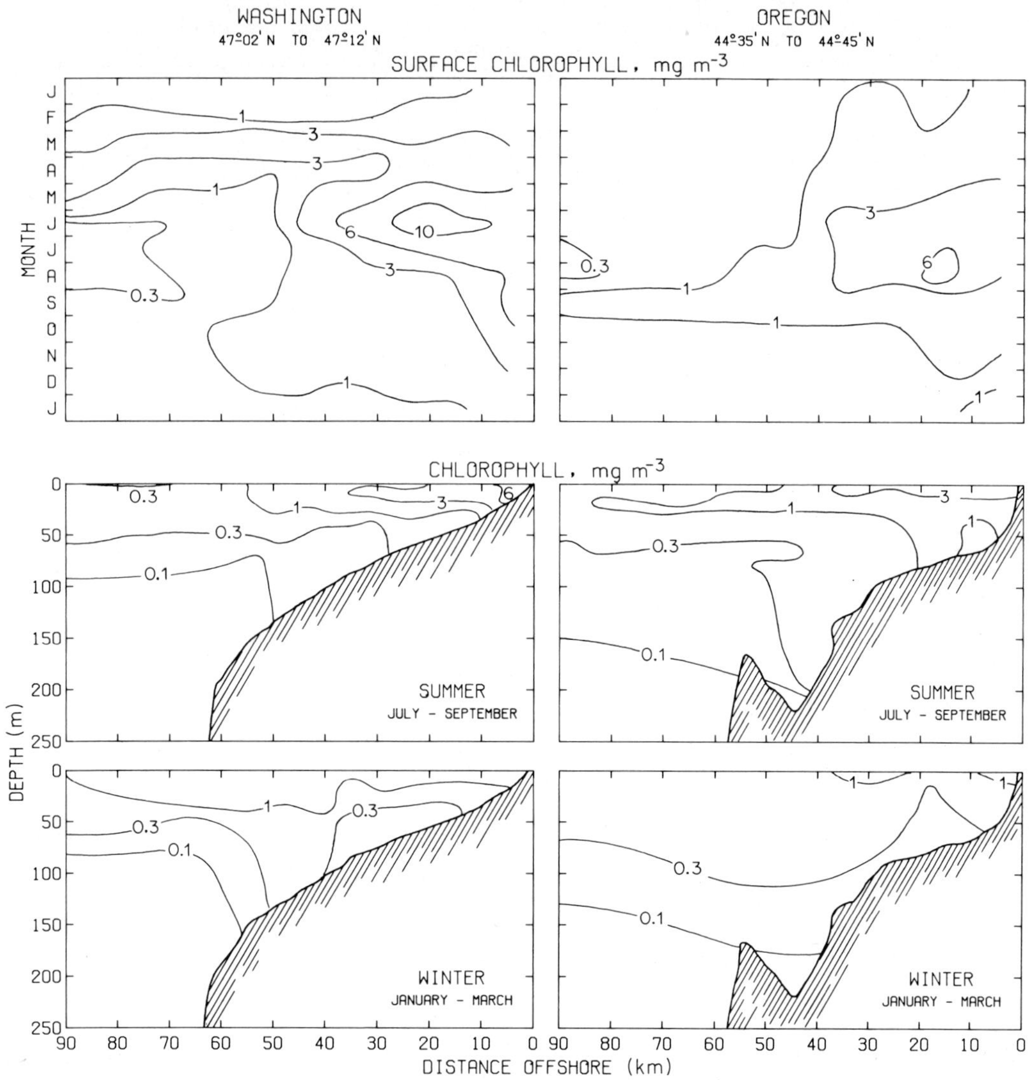

Fig. 1.24 Development of surface chlorophyll *a* features and seasonal depth structure of chlorophyll in waters overlying the shelf regions of Washington and Oregon. Contours are based on composite measurements made during 1950-1984 from the cross shelf transect areas (Copalis Head, Wa. and Yaquina Head, Ore.) depicted in Figure 1.1. Data were averaged for months and 5-km distances from shore prior to contouring.

regions. The longest time-series of temperature measurements from the oceanic North Pacific is the more than 30 y (1950-1981) of data from the Canadian weathership program at station P (50° N, 145° W) in the Gulf of Alaska (Tabata, 1983). The CalCOFI program off Southern California and Northern Mexico covers a similar time period but has large sampling gaps during the 1970s. Both of these data sets emphasize sampling stations with bottom depths greater than 500 m whereas the Washington-Oregon data largely consists of measurements at shelf stations. This difference, we will suggest, significantly affects the signature of global-scale processes in the Washington-Oregon data.

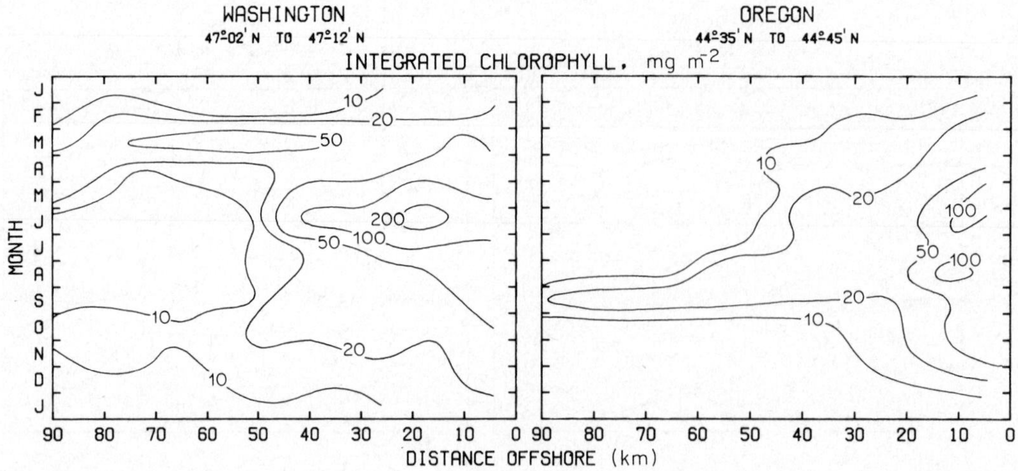

Fig. 1.25 Mean-monthly distribution of integrated chlorophyll *a* in waters overlying the shelf regions of Washington and Oregon. Pigment profiles were integrated to the depth of the 1% light level or the water-sediment interface.

Major temperature anomalies from the Washington/Oregon coast are coherent with data from both the subarctic Pacific and CalCOFI regions (Fig. 1.27). Sea surface temperatures were generally cold during the period from 1950 through 1956 except for brief warming during weak ENSO episodes in late 1951 and 1953. The warm-water period in 1957 and 1958 was a well-documented, major Niño event along the west coast of North America (Sette and Isaacs, 1960). Periods of warmer or cooler water occurred during the 1960s, but the frequency of temperature reversals was greater and the amplitudes of the anomalies generally lower (i.e., $< 1°$ C) than in the previous decade. In addition, changes in the oceanic subarctic during the 1960s appeared to precede changes on the continental margins of Washington and California by one or two quarters. The 1965-66 El Niño, which had a catastrophic impact on the Peruvian anchovetta fishery, appears as an insignificant perturbation in the three data sets from the Northeast Pacific.

The nine-year interval from 1969 to 1977 was an anomalously cold-water period throughout the northeastern Pacific interrupted only by the 1972-73 Niño event. This trend was reversed with a large negative swing of the Southern Oscillation Index during 1976 and 1977 (Niebauer, 1985) resulting in a recent succession of anomalously warm years except for a brief reversal in 1982. A major Niño event occurred during 1983 in the midst of this general warm trend (Wooster and Fluharty, 1985).

The coastal coherence of the 1983 Niño event as well as general trends of the past 13 y are apparent in Figure 1.28 which is based on 3° latitude-longitude averaged anomalies from Baja California to Vancouver Island (Norton *et al.*, 1985). Figure 1.28 also illustrates some interregional differences in the time-series data sets. For instance, while the 1971-72 ENSO was evident as a 1972 warm-water anomaly off California (and the Subarctic Pacific – Fig. 1.27), the

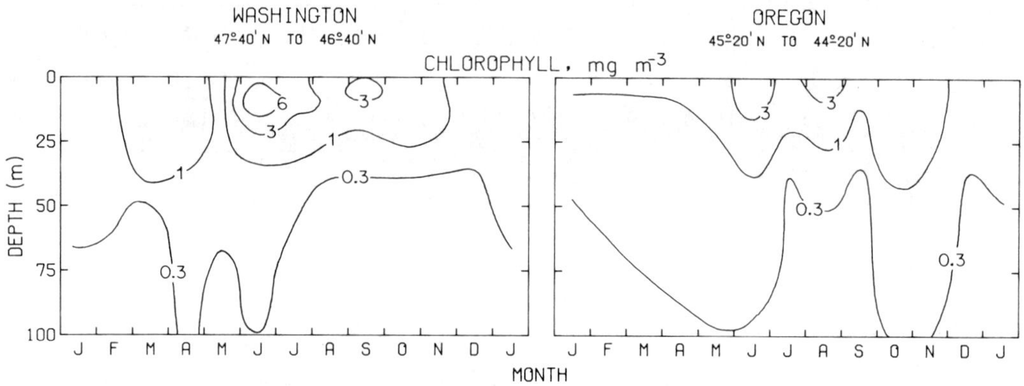

Fig. 1.26 Mean-monthly development of depth structure in mid-shelf chlorophyll *a* off of Washington and Oregon. Contours are based on composite measurements made during 1950-1984 from the longshore transect areas depicted in Figure 1.1. These areas are defined by total water column depths in the range of 70 to 130 m.

warming apparently did not extend into Oregon and Washington waters until early 1973. Additionally, 1982 was a cold-water year for the Northwest coast and Subarctic Pacific data sets but not for southern waters.

Despite some apparent differences in the timing and magnitude of temperature-related events in the three regions, quarterly anomalies from the Washington/Oregon coast are significantly correlated with the anomalies observed off California ($n = 67$; $r = 0.48$; $p < 0.01$ by nonparametric rank-difference analysis) and in the Subarctic Pacific ($n = 95$; $r = 0.32$; $p < 0.05$). This result is consistent with the view that global-scale processes control an important component of the interannual variability in the ocean's physical environment throughout the northeast Pacific. While the percentage of variability explained by these correlations is not impressive (less than 23%), perhaps no more can be expected from data sets collected on such different temporal and spatial scales (e.g., different times of sampling within a quarter, different base years for computing anomalies and variable sampling locations for the Washington/Oregon data). Step-wise, multiple regression analysis of temperature anomalies, lagged 0, 1, 2, or 3 months, against regional upwelling, sea surface height (a proxy for advective mass transport; Chelton *et al.*, 1982) and Columbia River discharge accounts for a maximum of 52% of the variability in the time-series.

Time series of salinity anomalies from the Subarctic Pacific, the Washington/Oregon coast and the CalCOFI Program are compared in Figure 1.29. Variability is markedly greater in the Pacific Northwest than in the CalCOFI and Subarctic data, probably due to interannual variability in the location of the Columbia River plume and the magnitude and timing of river runoff. The CalCOFI and subarctic Pacific data exhibit only gradual, low magnitude changes in salinity with time. The three data sets share at least one feature – a period of anomalously high salinity from late 1957 through 1961 which begins with a major ENSO event. Overall, quarterly

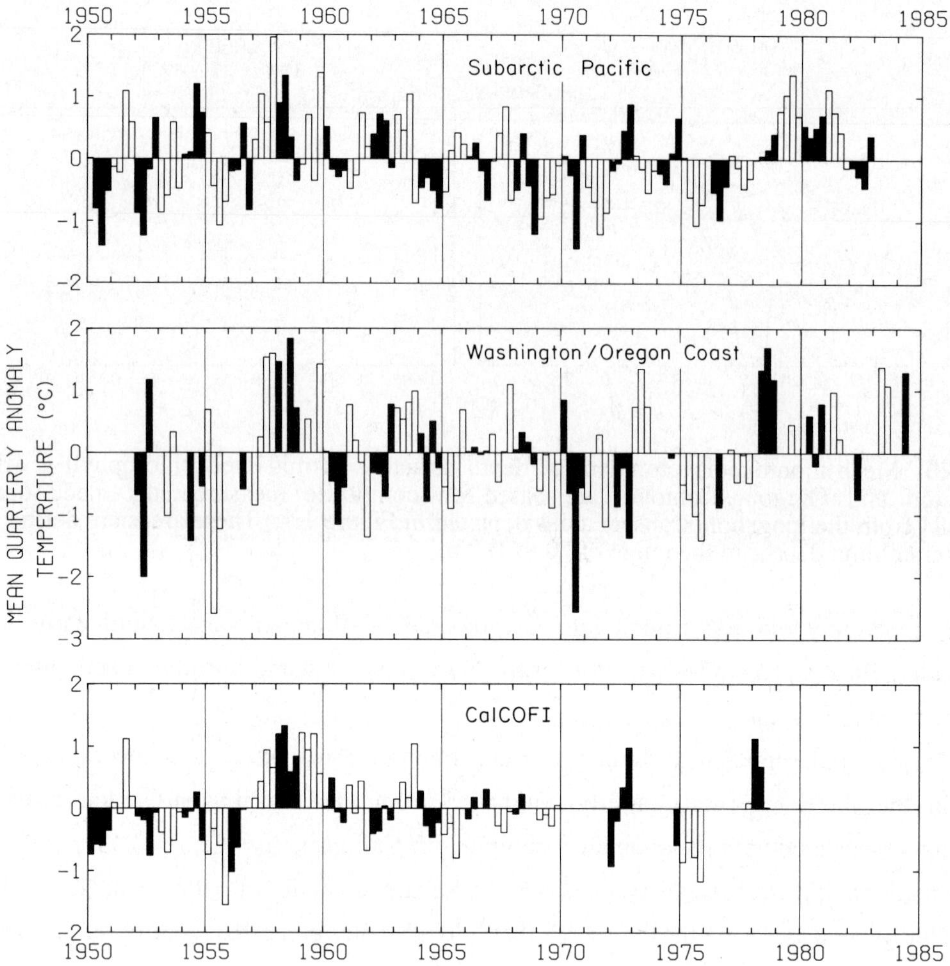

Fig. 1.27 Anomalies of mean quarterly, near-surface temperature in the oceanic subarctic Pacific and along the coastal margins of Washington/Oregon and Southern California/ Northern Mexico. Even-numbered years are shaded. Data from the subarctic Pacific were provided by S. Tabata; California data from the CalCOFI Program were provided by D. Chelton and J. McGowan.

salinity anomalies for the Washington/Oregon coast are positively correlated with those from the CalCOFI Program ($r = 0.34$; $p < 0.05$) but are not significantly correlated with anomalies in the subarctic Pacific ($r < 0.1$; $p > 0.90$).

Chelton *et al.* (1982) used the relationship between temperature and salinity anomalies to distinguish the effects of large-scale, horizontal advection from local upwelling in the CalCOFI data. They argued that temperature was not useful by itself since increased southward flow of the California Current bringing colder water from the North would also result in an upward tilting of isopycnals and thus near-shore cooling similar to wind-driven upwelling. However, salinity anomalies should vary positively with temperature if longshore advection was the dominant process because temperature and salinity both increase north to south along the west coast of

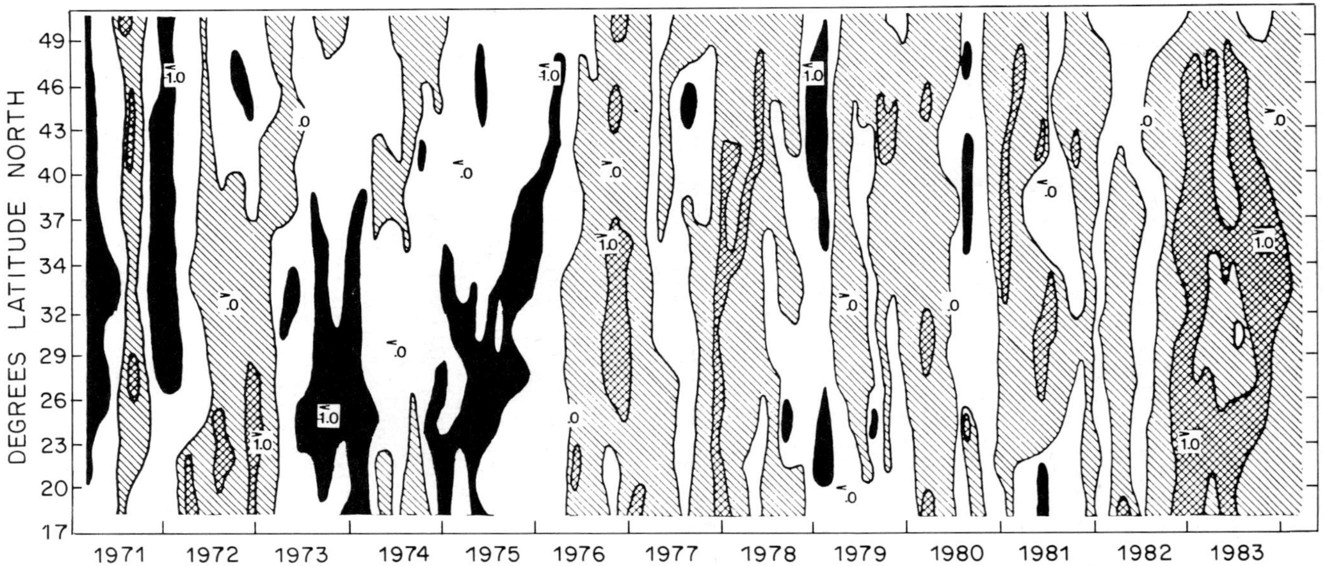

Fig. 1.28 Anomalies of monthly mean sea-surface temperature along the west coast of North America from Baja California, Mexico to Vancouver Island, Canada. Hatched areas represent anomalies between 0 and +1°C; cross-hatched areas indicate anomalies > +1°C. Unshaded areas denote anomalies between 0 and −1°C; black areas indicate anomalies < −1 C. Data from mean sea-surface temperatures in 3° x 3° blocks of latitude and longitude computed from sub-means for 1° blocks. Figure reprinted with permission from Norton *et al.* (1985).

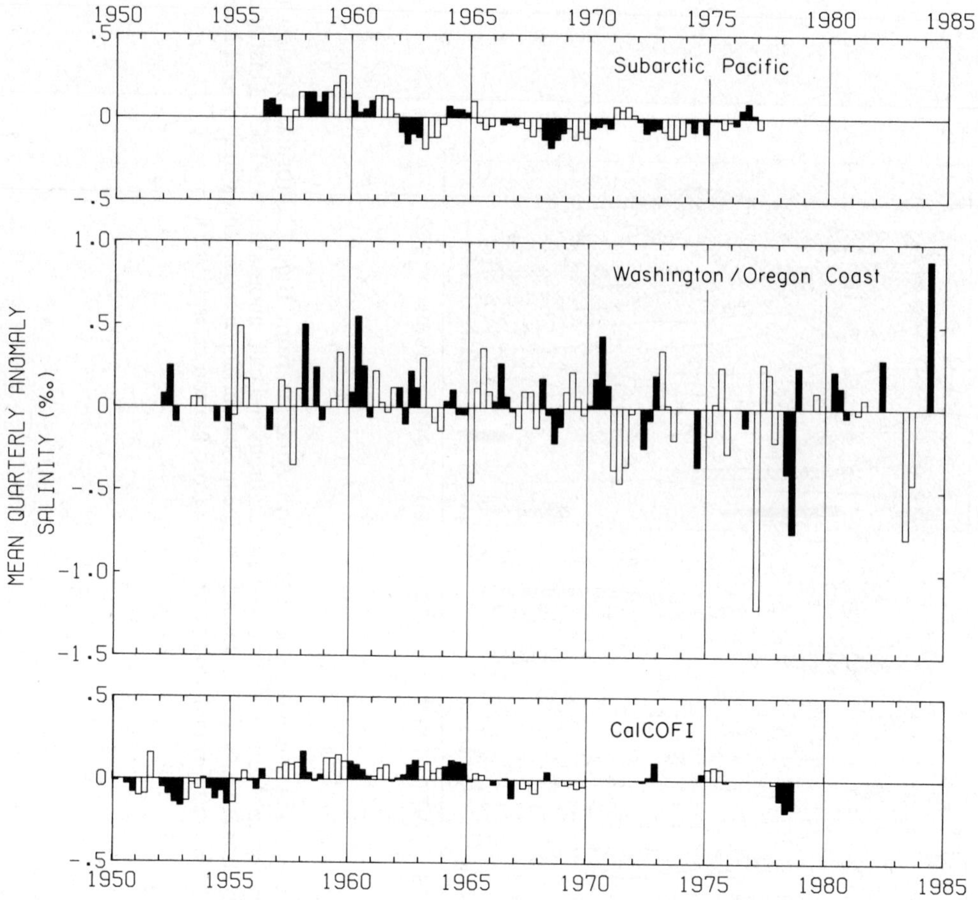

Fig. 1.29 Anomalies of mean quarterly, near-surface salinity in the oceanic subarctic Pacific and along the coastal margins of Washington/Oregon and Southern California/Northern Mexico. Even-numbered years are shaded. Data from the subarctic Pacific were provided by the Canadian Marine Environmental Data Service; California data from the CalCOFI Program were provided by D. Chelton and J. McGowan.

North America. In contrast, the vertical advection of high-salinity, low-temperature water from depth during upwelling should result in a negative correlation between temperature and salinity anomalies. As previously noted, the Columbia River influence adds to the variability in salinity in the Pacific Northwest. It also accentuates the upwelling signal. Under downwelling conditions the low salinity river water lies along the coast, particularly over the Washington shelf. During periods of upwelling, the river water moves offshore and is replaced at the surface by deeper water with a much higher salt content.

Chelton *et al.* (1982) observed that salinity anomalies were positively, though not significantly, correlated with temperature anomalies in the CalCOFI data. This data set is dominated by stations with water depths greater than 500 m and, therefore, does not specifically address shelf dynamics. For the Washington/Oregon shelf, we find a negative relationship between salinity and temperature anomalies. The relationship is not significant (n = 97;

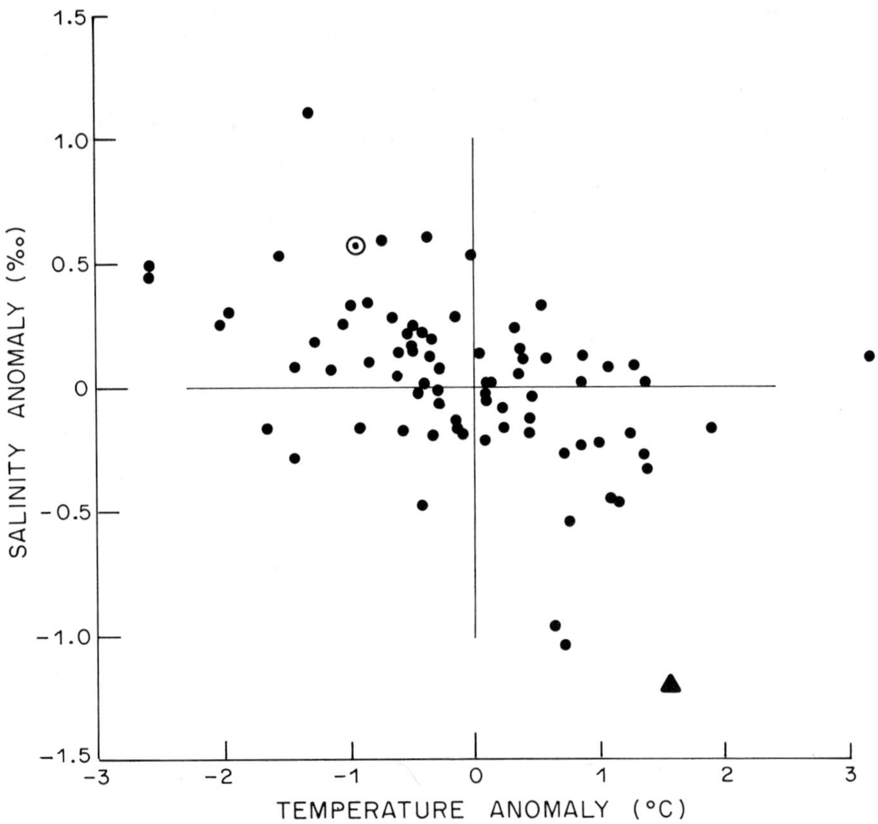

Fig. 1.30 Scatter-plot relationship between summer (May-August) anomalies in near-surface salinity and temperature in the mid-shelf region of Washington and Oregon. Data were computed from monthly mean measurements in the regions defined in Figure 1.1. The circled point defines data from June 1982, a period of high upwelling described in Hermann *et al.* (1989, Chapter 6). The triangle denotes data from May 1983 during a major ENSO event.

$r = -0.20$) for the quarterly-averaged anomalies presented in Figures 1.28 and 1.29. However, a relatively strong, negative correlation exists between monthly-averaged temperature and salinity anomalies from cruises during the May through August, upwelling season ($n = 80$; $r = -0.52$; $p < 0.01$: Fig. 1.30). This result implies that wind-driven upwelling may contribute substantially to the observed interannual variability of temperature and salinity on the Washington/Oregon shelf. In fact, while Chelton *et al.* (1982) concluded that interannual temperature variations in the large-scale California Current are uncorrelated with wind forcing off California, we find that mid-shelf temperature anomalies from May through August are negatively correlated ($n = 61$; $r = -0.47$; $p < 0.01$) and salinity anomalies are positively correlated ($r = 0.32$; $p < 0.05$) with anomalies in the monthly Upwelling Index in the Pacific Northwest (Fig. 1.31). These results are consistent with the view that much of the interannual variability in coastal sea level is driven by variations in coastal, alongshore wind stress (Hickey, 1989, Chapter 2). As might be expected from the influence of winds on the orientation of the Columbia River plume, the

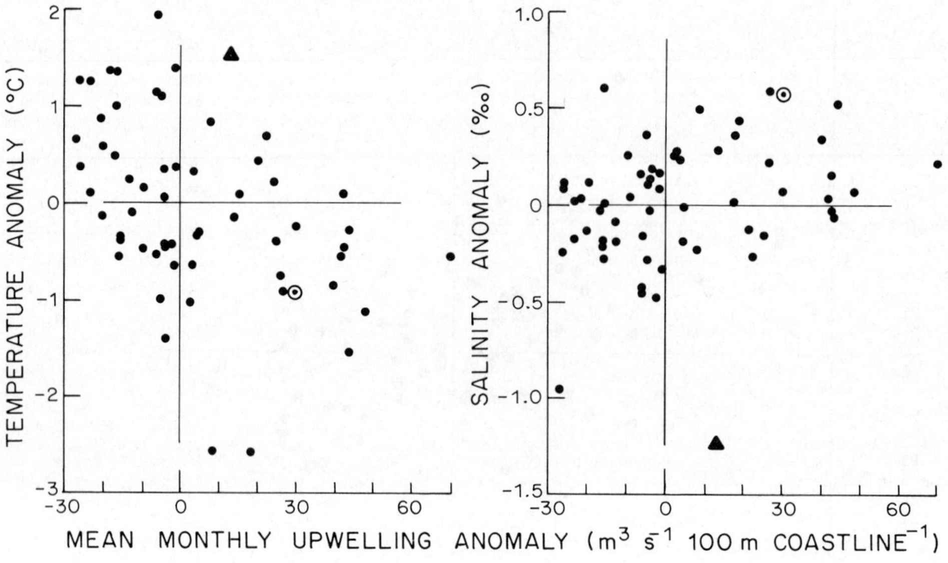

Fig. 1.31 Scatter-plot relationships between anomalies in the Bakun Upwelling Index and near-surface temperature and salinity in the mid-shelf regions of Washington and Oregon. Temperature and salinity data are from monthly means during the typical upwelling season (May through August). June 1982 and May 1983 data are identified as in Figure 1.30.

relationship between salinity and upwelling anomalies improves when the analysis considers only the Washington mid-shelf region ($n = 48$; $r = 0.42$; $p < 0.02$).

With physical forcing by either large-scale advection or local upwelling, high concentrations of nutrients (e.g., nitrate) in surface waters of the Pacific Northwest should be associated with anomalously low temperatures. However, since nitrate is a nonconservative property of seawater due to its rapid utilization by phytoplankton, one might expect difficulty in observing a clearcut relationship. Nonetheless, monthly nitrate anomalies in the Washington mid-shelf region are strongly correlated with temperature anomalies for the months of April to October ($n = 50$; $r = -0.56$; $p < 0.01$). Similarly, during peak upwelling months in summer (May through August), surface nitrate concentrations are positively related to the Upwelling Index ($n = 28$; $r = 0.47$; $p < 0.02$: Fig. 1.32). Nitrate concentrations are always very low when upwelling is weak while higher, but substantially more variable, concentrations are observed during months of moderate to high upwelling (Fig. 1.32). Low nitrate concentrations are sometimes encountered during periods of high upwelling, an inevitable consequence of the fact that the physical process of upwelling and the biological responses to upwelling (nutrient uptake and phytoplankton growth) occur on different time scales. For example, the relatively low nitrate concentrations observed during strong upwelling in June, 1982 (Hermann *et al.*, 1989, Chapter 6) co-occurred with moderate to high chlorophyll levels (Fig. 1.32) indicating that the observations were made after upwelled nutrients were largely depleted by phytoplankton (Small and Menzies, 1981).

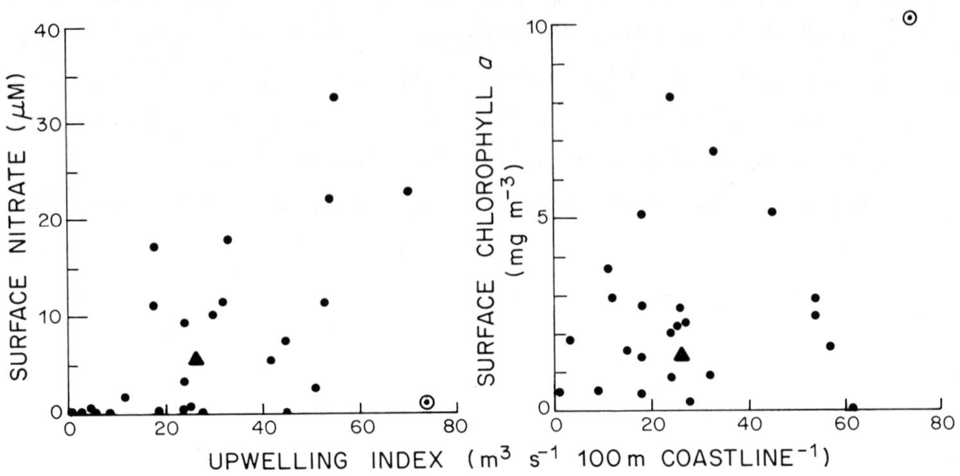

Fig. 1.32 Scatter-plot relationships between the mean-monthly Bakun Upwelling Index and surface nitrate and chlorophyll concentrations in the mid-shelf region of the Washington coast. Chlorophyll and nitrate data are cruise means for all measurements made during May through August within the region defined in Figure 1.1. June 1982 and May 1983 data are identified as in Figure 1.30.

Surface concentrations of chlorophyll *a* during summer are positively, but not significantly, correlated with the Upwelling Index by the nonparametric rank-difference test ($n = 26$; $r = 0.16$). However, the correlation is marginally significant ($r = 0.35$; $p = 0.084$) when computed by the more powerful, least-squares regression approach (Fig. 1.32).

1.5 SUMMARY

Hydrographic data collected over the past three decades has provided a relatively coherent view of mean seasonal dynamics in the pelagic environment on the Washington/Oregon shelf. Surface waters are cold and nutrient-rich during winter months. Prevailing winds are from the south, resulting in general downwelling and causing the Columbia River effluent to lie to the north and along the Washington coast. Chlorophyll concentrations off Washington illustrate that the spring bloom develops during March and continues into April when it becomes evident in increasing oxygen concentration of surface waters and declining nutrient levels. Low nutrient concentrations persist in the euphotic zone from the later half of April through early May when wind conditions become favorable for coastal upwelling. The seasonal shift to upwelling (northerly) winds causes the Columbia River (maximum discharge during June) influence to migrate from the Washington nearshore to the Oregon offshore. Although periodic upwelling events affect the dynamics of both nearshore environments during summer, wind effects are strongest off Oregon. Maximum upwelling influences on nearshore temperature, salinity, oxygen, nutrient levels and chlorophyll concentration occur during June off Washington. A strong thermocline develops over the Washington shelf region during mid- to late summer and northern excursions of low-salinity, river water during periodic wind reversals further contribute to water-column stability in the nearshore environment. Isograms of high nutrient content (e.g., > 5 μM nitrate) rarely break the surface during upwelling favorable conditions past mid-summer off Washington. In contrast, the Oregon shelf appears to be more dynamic, with strong upwelling more evenly distributed throughout the summer. Phytoplankton standing stock, measured as chlorophyll, declines to moderate winter levels (about 1 μg Chl *a* l^{-1}) as the upwelling season ends in August or September. In the very nearshore, however, diatom communities thrive in the surf zone during winter (Lewin *et al.*, 1989, Chapter 12).

Variability on scales of days to decades is incorporated in the mean annual and spatial distributions of environmental parameters. The upwelling examples in Chapter 6 (Hermann *et al.*, 1989, Chapter 6) illustrate the level of variability observed over the course of several days. Much of such variability is attributable to changes in local winds and *in situ* biological processes. Low-frequency, interannual variability on the Pacific Northwest shelf has both global- and regional-scale origins. Global-scale processes are implicated by the coherence of anomaly patterns in waters off California, Oregon and Washington and the oceanic subarctic Pacific and their relations to documented ENSO events in the Equatorial Pacific. The effects of wind-forced upwelling on regional interannual variability are indicated by the negative correlation between surface temperature and the Upwelling Index and by the positive correlations between both surface salinity and nitrate concentration and the Upwelling Index. The relative contributions of global and regional forcing mechanisms to observed physical variability in the Pacific Northwest cannot be reliably ascertained by statistical methods because the processes are linked through ocean-atmosphere coupling and because about half of the variability in the data is unexplained by either mechanism.

That so much of the observed physical variability is unaccounted for by obvious forcing mechanisms may partly be explained by the random nature of the data collection (different sampling stations and times) and possibly by differences in data quality through the years. To some extent, however, the large fraction of unresolved variability is just symptomatic of the problems of sampling intensely dynamic environments from ships. It has long been understood that such sampling cannot provide synoptic views of parameter distributions over broad regions, but the extent to which the temporal and spatial scales of variability are confounded by shipboard observations was only fully appreciated after satellite remote-sensing became accessible to the oceanographic community. Satellite sensors are capable of discriminating fine-scale structure in sea-surface height, temperature, phytoplankton pigments and near-surface winds over broad areas. Therefore, they measure parameters that distinguish regional and global effects on physical and biological variability. In future work on the Washington-Oregon shelf, satellites will be indispensible tools for long-term monitoring of the coastal system, for assessing the temporal and spatial scales of regional variability and their links to global-scale processes and in providing a broader context for more detailed experimental studies conducted from ships or moored instrument arrays.

ACKNOWLEDGEMENTS

The authors benefitted from the input of B.M. Hickey, L.F. Small and M.J. Perry. The compilation of this data was supported by the Department of Energy under contract DE-AT06-

EV-10047. Contribution No. 1784 from the School of Oceanography, University of Washington.

REFERENCES

Bakun, A. 1973. Coastal upwelling indices, west coast of North America, 1946-1971. U. S. Dept. of Commerce, NOAA Tech. Rep. NMFS SSRF-671, 103 pp.

Bakun, A. 1975. Daily and weekly upwelling indices, west coast of North America, 1967-73. U. S. Dept. of Commerce, NOAA Tech. Rep. NMFS SSRF-693, 114 pp.

Botsford, L.W., D. A. Armstrong and J. M. Shenker. 1989. Oceanographic influences on the dynamics of commercial fish stocks. Pages 511-564 in: M.R. Landry and B.M. Hickey (eds.), *Coastal Oceanography of Washington and Oregon.* Elsevier,Amsterdam, Neth.

Chelton, D.B., P.A. Bernal and J.A. McGowan. 1982. Large-scale interannual physical and biological interaction in the California Current. *J. Mar. Res.*, 40: 1095-1125.

Hamilton, K. and W.J. Emery. 1985. Regional atmospheric forcing of interannual surface temperature and sea level variability in the Northeast Pacific. Pages 22-30 in: W. S. Wooster and D. L. Fluharty (eds.), *El Niño North. Niño Effects in the Eastern Subarctic Pacific Ocean.* Washington Sea Grant Prog., Univ. of Washington, Seattle, Wa.

Hermann, A.J., B.M.Hickey, M.R. Landry and D.F. Winter. 1989. Coastal upwelling dynamics. Pages 211-254 in: M.R. Landry and B.M. Hickey (eds.), *Coastal Oceanography of Washington and Oregon.* Elsevier, Amsterdam, Neth.

Hickey, B.M. 1989. Patterns and processes of circulation over the shelf and slope. Pages 41-116 in: M.R. Landry and B.M. Hickey (eds.), *Coastal Oceanography of Washington and Oregon.* Elsevier, Amsterdam, Neth.

Huyer, A. 1977. Seasonal variation in temperature, salinity and density over the continental shelf off Oregon. *Limnol. Oceanogr.*, 22: 442-453.

Lewin, J., C.T. Schaefer and D.F. Winter. 1989. Surf-zone ecology and dynamics. Pages 567-594 in: M.R. Landry and B.M. Hickey (eds.), *Coastal Oceanography of Washington and Oregon.* Elsevier, Amsterdam, Neth.

Morel, A. and R.C. Smith. 1974. Relation between total quanta and total energy for aquatic photosynthesis. *Limnol. Oceanogr.*, 19: 591-600.

Nelson, C.S. and D.M. Husby. 1983. Climatology of surface heat fluxes over the California Current region. U. S. Dept. of Commerce, NOAA Tech. Rep. NMFS SSRF-763, 155 pp.

Niebauer, H.J. 1985. Southern Oscillation/El Niño effects in the Eastern Bering Sea. Pages 116-118 in: W. S. Wooster and D. L. Fluharty (eds.), *El Niño North. Niño Effects in the Eastern Subarctic Pacific Ocean.* Washington Sea Grant Prog., Univ. of Washington, Seattle, Wa.

Norton, J., D. McLain, R. Brainard and D. Husby. 1985. The 1982-83 El Niño event off Baja and Alta California and its ocean climate context. Pages 44-72 in: W. S. Wooster and D. L. Fluharty (eds.), *El Niño North. Niño Effects in the Eastern Subarctic Pacific Ocean.* Washington Sea Grant Prog., Univ. of Washington, Seattle, Wa.

Orem, H.M. 1968. Discharge in the lower Columbia River Basin 1928-1965. U. S. Geological Survey, Circular 550, 24 pp.

Parsons, T.R., M. Takahashi and B. Hargrave. 1977. Biological oceanographic processes. Pergamon Press, New York, N.Y., 33 pp.

Pruter, A.T. and D.L. Alverson (eds.). 1972. *The Columbia River estuary and adjacent ocean waters. Bioenvironmental studies.* Univ. of Washington Press, Seattle, Wa., 86 pp.

Sampson, R.L. 1984. *Surface II graphics system.* Kansas Geological Survey, Lawrence, 24 pp.

Sette, O.E. and J.D. Isaacs (eds.). 1960. *The changing Pacific Ocean in 1957 and 1958. California Cooperative Fisheries Investigations Rep.*, Vol. 7. Calif. Dept. Fish Game, Mar. Res. Comm., 217 pp.

Small L.F. and D.W. Menzies. 1981. Patterns of primary productivity and biomass in a coastal upwelling region. *Deep-Sea Res.*, 28: 123-149.

Strickland, J.D.H. 1965. Production of organic matter in the primary stages of the marine food chain. Pages 477-610 in: Riley, J. P. and G. Skirrow (eds.), *Chemical Oceanography.* Vol. 1. Academic Press, New York., N.Y.

Tabata, S. 1983. Interannual variability in the abiotic environment of the Bering Sea and the Gulf of Alaska. Commentary. Pages 139-145 in: W. S. Wooster (ed.), *From Year to Year: Interannual Variability of the Environment and Fisheries of the Gulf of Alaska and the Eastern Bering Sea.* Univ. of Washington Press, Seattle, Wa.

Tate, M.W. and R.C. Clelland. 1957. *Nonparametric and shortcut statistics in the social, biological and medical sciences.* Interstate Printers and Publ., Inc. Danville, Ill., 171 pp.

U. S. Geological Survey, Water Resources Division. Surface water records for Oregon 1961-1982. Issued annually.

U. S. Geological Survey, Water Resources Division. Surface water records for Washington 1961-1982. Issued annually.

Ward, B.B., M.C. Talbot and M.J. Perry. 1984. Contributions of phytoplankton and nitrifying bacteria to ammonium and nitrite dynamics in coastal waters. *Cont. Shelf. Res.*, 3: 383-398.

Wooster, W.S. (ed.). 1983. From year to year. *Interannual variability of the environment and fisheries of the Gulf of Alaska and the Eastern Bering Sea.* Univ. of Washington Press, Seattle, Wa., 208 pp.

Wooster, W.S. and D.L. Fluharty (eds.). 1985. *El Niño north. Niño effects in the eastern Subarctic Pacific Ocean.* Washington Sea Grant Prog., Univ. of Washington, Seattle, Wa.

CHAPTER 2

PATTERNS AND PROCESSES OF CIRCULATION OVER THE WASHINGTON CONTINENTAL SHELF AND SLOPE

B. M. Hickey
School of Oceanography, University of Washington, Seattle, WA 98195

2.1 INTRODUCTION

A physical oceanography research program in the Washington coastal zone has been conducted by the University of Washington for the past two decades. The research has included direct current measurements, hydrographic measurements, and *in situ* as well as coastal wind measurements. The historical progression of these field experiments was dictated by three time-dependent factors: agency responsibility, state of physical oceanographic knowledge, and types and numbers of equipment available. Table 2.1 illustrates the extensive data set that has been generated. This data set has been sufficient to establish coherence scales of the fluctuating flow field over the shelf and to determine the principal forcing mechanisms of the coastal circulation.

The initial goal of the direct current measurement program, which began when internally recording current meters first became available, was to obtain information on the seasonal characteristics of the shelf circulation, with particular emphasis on sediment movement. The scope of the experiments increased as the number of instruments increased. In particular, a group of experiments was executed to determine alongshelf and cross-shelf coherence scales of the fluctuating flow field. A third group of experiments focused on the effect of submarine canyons on the circulation, on upwelling of deep nutrient-rich water, and on the removal of sedimentary material from the shelf to the deep ocean. The two most recent experiments, in which the detailed structure of the cross-shelf flow field was examined, followed the development of the vector measuring current meter, an instrument that allows accurate measurements in the wave dominated near-shore and upper ocean boundary layers.

The research program also included the development of both analytical and numerical models of shelf circulation. Since the original focus of the research was on sediment movement, which occurs primarily during winter when much of the fluctuating flow is driven by the local wind, the initial modeling efforts were two-dimensional, although time-dependent. Recently, modeling efforts have expanded to include the three-dimensional effects of remotely generated coastal-trapped waves.

2.2 LARGE-SCALE SETTING

2.2.1 Deep-Ocean Currents

The U.S. Pacific Northwest coastal zone is located at the eastern edge of the region wherein the West Wind Drift divides into the poleward flowing Alaskan Current and the equatorward flowing California Current (Fig. 2.1). Hickey (1979) described the spatial scales and

TABLE 2.1 Record lengths of selected current meter records and timing of CTD or hydrographic cross-shelf sections collected on the Pacific Northwest shelf by the University of Washington during the period 1971-1978. Symbols are used to indicate cross-shelf location, type of current meter, and the focus of the experiment for which the measurements were made (see key). Most data were obtained off mid- to southern Washington. However, the Poleward Undercurrent experiment, performed jointly with A. Huyer and R.L. Smith at Oregon State University, took place off mid-Oregon, and the Juan de Fuca experiment, performed jointly with R. Thomson at the Institute of Ocean Sciences, Patricia Bay, B.C., and P. LeBlond at the University of British Columbia, covered the region from northern Washington to mid-Vancouver Island.

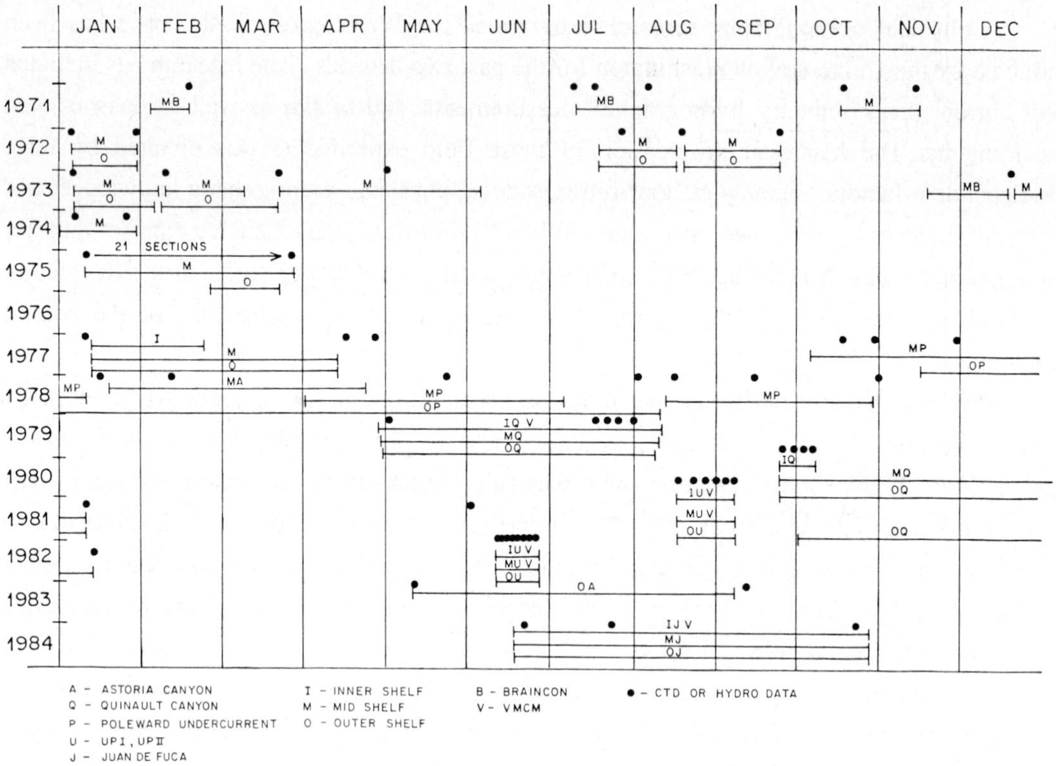

seasonal variation of the deep-ocean boundary currents that comprise the California Current system; namely, the California Current, the Davidson Current, and the California Undercurrent. That information is summarized in a schematic (Fig. 2.2). The California Current, with a typical velocity of 10 cm s $^{-1}$ and an east-west scale ~1000 km, is strongest during summer (July-August). The location of the velocity maximum appears to be well seaward of the continental shelf at most locations. The California Undercurrent, a jet-like poleward flow with a subsurface maximum whose core appears to be confined to the continental slope, also reaches maximum strength during summer. During winter in the Pacific Northwest, the California Current either moves offshore or is replaced by a poleward flow known as the Davidson Current. This flow has a larger east-west scale than the California Undercurrent and has no subsurface maximum.

The California Current system, at least off California, has interannual as well as seasonal fluctuations (Chelton, 1980). These fluctuations appear to be related to the Southern Oscillation

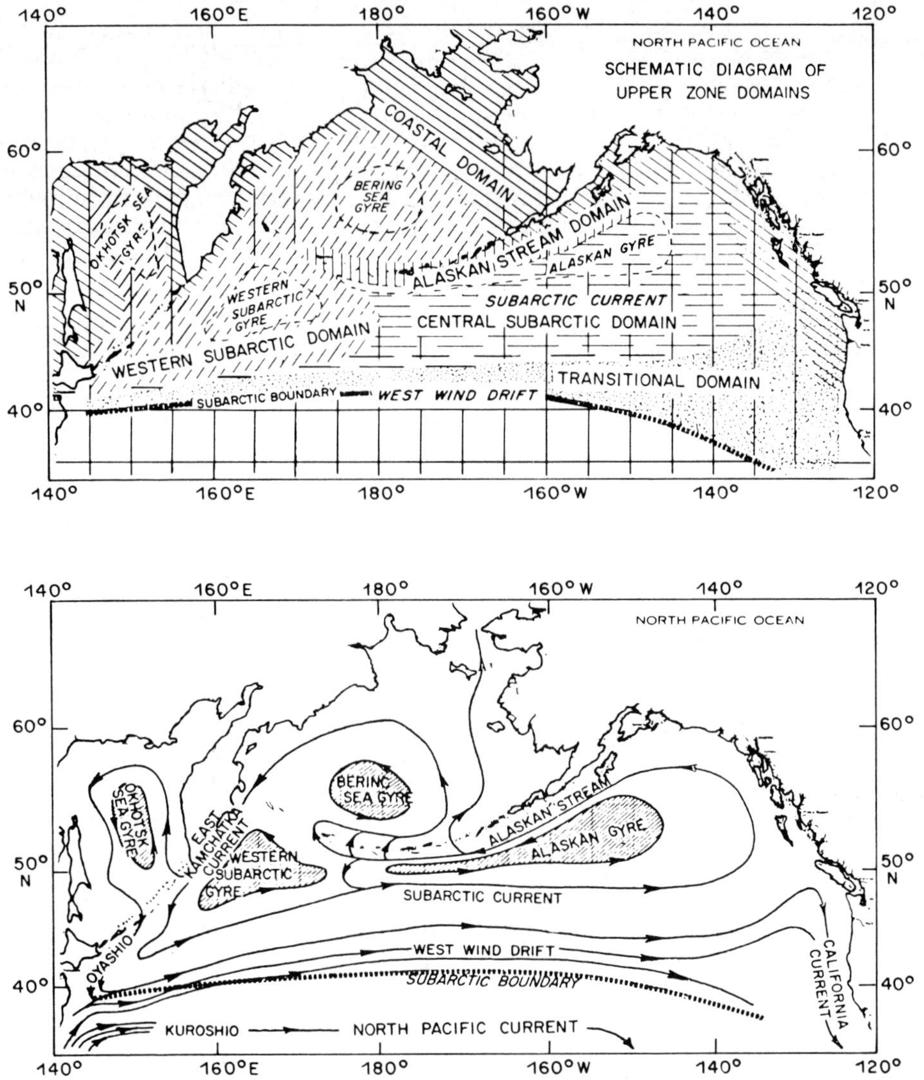

Fig. 2.1 Schematic diagram of upper zone domains in the North Pacific. Adapted from Dodimead *et al.* (1963). The upper zone includes water above the permanent halocline in the subarctic Pacific (upper panel). Schematic diagram of surface circulation of the North Pacific relative to 1,000 db (lower panel). Adapted from Dodimead *et al.* (1963).

and the El Niño phenomenon. During El Niño events, equatorward flow in the California Current is anomalously weak; during anti-El Niños, equatorward flow in the California Current is anomalously strong. Chelton *et al.* (1982) have demonstrated that interannual fluctuations in zooplankton biomass off California are related to such anomalous transport events: when transport in the California Current is anomalously weak (El Niño), zooplankton biomass is anomalously low; conversely, strong equatorward transport results in high zooplankton biomass.

A complete discussion of the mechanisms driving the eastern boundary current system off the United States Pacific Northwest is beyond the scope of this paper. Readers are referred to O'Brien *et al.* (1977), McCreary (1981), Chelton (1984) and McCreary and Chao (1985) for reviews of steady eastern boundary current models, and to the time-variable models of Philander

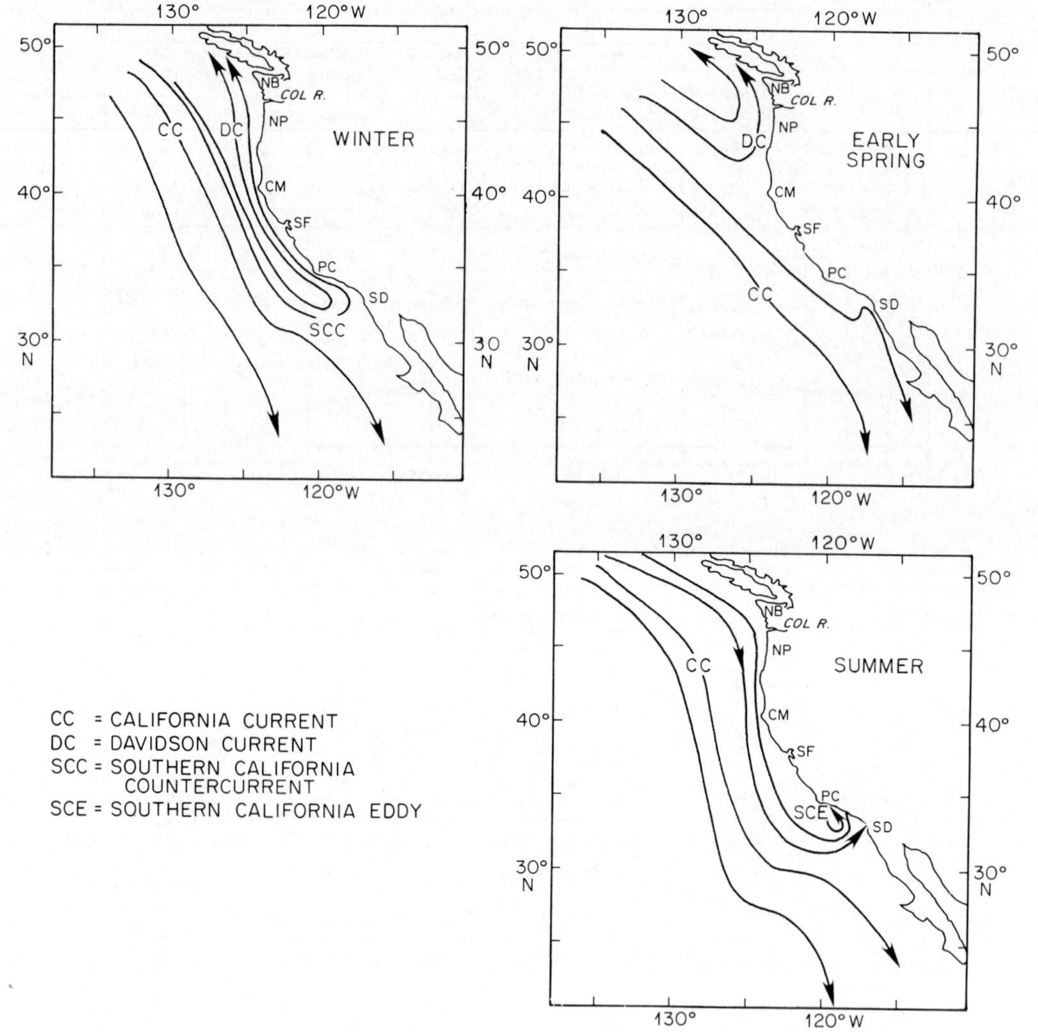

Fig. 2.2 Schematic illustrating seasonal variation of deep-ocean boundary currents off the U.S. west coast. Abbreviations along the coast signify (from north to south) Neah Bay, Newport, Cape Mendocino, San Francisco, Point Conception, and San Diego.

and Yoon (1982) and McCreary *et al.* (1987). Although some of the elements of the circulation have been reproduced by the various models, each of the models has inherent limitations, such as a flat bottom or absence of wind stress curl, that precludes evaluation of the relative efficiency of various effects.

In particular, it is unclear whether the deep-ocean boundary currents, that is, those whose velocity maxima occur seaward of the continental slope, owe their generation to the shelf currents or are generated quite independently of the shelf currents, for example, by wind stress curl or by basin-scale, zonal wind stress. Both Philander and Yoon (1982) and McCreary *et al.* (1987) address this problem explicitly. McCreary *et al.* (1987) conclude that the equatorward flowing California Current offshore of the shelf is forced by positive wind stress curl, whereas shelf currents are forced primarily by alongshelf wind stress. Philander and Yoon (1982) conclude that equatorward flow offshore of the shelf is a result of Rossby wave dispersion of the equatorward coastal jet that is forced by curl free alongshelf wind stress. Actually, several bands of poleward and equatorward flow occur in their model and the authors point out that such currents should be superimposed on results that include wind stress curl, such as those of McCreary *et al.* (1987). We note, as did Philander and Yoon (1982), that the observations provide some support for the idea that the California Current has a banded flow structure (Hickey, 1979). Thus, in one model, the offshore California Current is considered to be forced independently of the shelf currents; in the other, it is intimately tied to the shelf currents, sharing a driving force. McCreary also concludes that the deep-ocean Davidson Current is driven primarily by the curl of the wind stress and thus is largely independent of shelf flows.

Equally important questions are whether and how the Eastern Boundary deep-ocean currents, however generated, modify or even generate shelf circulation. Although evidence for meanders and eddy formation in the California Current exists (e.g., Hickey, 1979), it has never been demonstrated that the deep-ocean California Current meanders onto the shelf. On the other hand, some evidence has been presented suggesting that the pressure field associated with deep-ocean currents has an expression at the coast. Specifically, Hickey and Pola (1983) demonstrated that although much of the variance in seasonal coastal sea level in the Pacific Northwest can be accounted for by alongshelf wind stress driving of shelf currents, a significant residual is observed. The amplitude and phase of the residual is well matched with those of the offshore steric height signal. Off southern California, where the shelf is much narrower, the seasonal coastal sea level signal is almost identical to that of the offshore steric height. Middleton (1987) shows that these results are consistent with results of a barotropic model in which shelf currents and pressure gradients are driven by an alongshore pressure gradient field imposed at the shelf break. As also shown by Csanady (1978), the pressure gradient signal is damped and phase shifted as it is transmitted across the shelf, the extent and nature of the modification being a function of the width and structure of the coastal margin. For shelf widths typical of the Pacific Northwest, Middleton (1987) suggests that ~20% of the signal may be transmitted to the coast. For baroclinic currents, as are observed, the percent transmitted is likely to be less than 20%.

2.2.2 Water Masses

The water masses associated with the deep-ocean boundary currents are quite distinctive. The California Current brings low temperature, low salinity, high oxygen, and high phosphate subarctic type water from high to low latitudes. The California Undercurrent and the Davidson Current both bring warmer, saltier, low oxygen, high phosphate equatorial type water from low to high latitudes (Hickey, 1979).

Water masses are modified in the U.S. Pacific Northwest by precipitation and runoff. The principal modification is a result of Columbia River discharge, which is responsible for 77% of the total drainage into the Pacific between San Francisco Bay and the Strait of Juan de Fuca (Budinger *et al.*, 1964). Columbia River discharge peaks annually in June (see Fig. 1.1; Landry *et al.*, 1989, Chapter 1). Flow regulation in recent years has strongly decreased the natural seasonal flow variability over the last two decades, so that winter freshets are now often more intense, although shorter in duration than the spring freshet (Sherwood *et al.*, in review). The effluent, which is 5-20 m thick, generally flows poleward off Washington during the winter and equatorward off Oregon during the summer in response to the seasonal circulation (Fig. 2.3). During winter, onshore flow in the wind-driven, surface Ekman layer confines the effluent primarily to the shelf (bottom depths ≤ 200 m). During summer, on the other hand, when the surface Ekman flux is directed offshore, the effluent can extend several hundred miles off the coast. Within each season, the spatial structure of the plume exhibits considerable variability over time scales of several days in response to changes in prevailing winds and currents over the shelf (Hickey, 1983). The effluent has a significant impact on the baroclinic structure of the velocity field, causing vertical shear in both velocity components. Effluent from the Strait of Juan de Fuca is much saltier than the Columbia effluent (31.5 ‰ cf. < 20 ‰), a result of tidal mixing in the Strait. The Juan de Fuca effluent generally travels poleward along the coast of Vancouver Island with no impact on the Washington-Oregon coast (Thomson *et al.*, in press).

The density field contains a shallow (5-20 m) pycnocline that is related primarily to the presence of the Columbia River plume and a deeper permanent pycnocline (75-125 m) that is related to the permanent halocline (Fig. 2.3). During the upwelling season, isopycnals from the permanent pycnocline slope upwards to the coast and intersect the surface, forming a front (Huyer, 1984). The surface mixed layer is on the order of 10-20 m thick during summer; during winter, the mixed layer is usually relatively shallow (< 10 m) off the Washington coast in nearshore regions due to the presence of the Columbia River plume. On the outer shelf and upper slope, the mixed layer is generally 40-60 m thick during the winter season. For further details on water properties off the Washington-Oregon coasts, including their seasonal variation, the reader is referred to Landry *et al.* (1989, Chapter 1).

2.3 SHELF/SLOPE CIRCULATION

2.3.1 Introduction

Fluctuations in currents over the continental shelf and slope span time scales ranging from seconds to decades and space scales ranging from meters to thousands of kilometers. This section is organized according to temporal scale: interannual (several years), seasonal, event (several days), and shorter period (≤ 1 day). The relative amplitudes of current fluctuations in the upper 200 m of the water column over the shelf and slope at these scales are 5-10 cm s^{-1} (interannual), 5-20 cm s^{-1} (seasonal), 20-100 cm s^{-1} (event), 1-10 cm s^{-1} (tidal), and 5-40 cm s^{-1} (inertial). In deeper depths, the amplitude of event-scale fluctuations is much reduced and shorter period fluctuations contribute a much higher percentage of the variance to the flow field.

Current data were obtained at 10- to 30-min. intervals using generally Aanderaa current meters suspended from taut wire moorings. Braincon (film recording) current meters were used in several early deployments (see Table 2.1). Vector Measuring Current Meters (VMCMs)

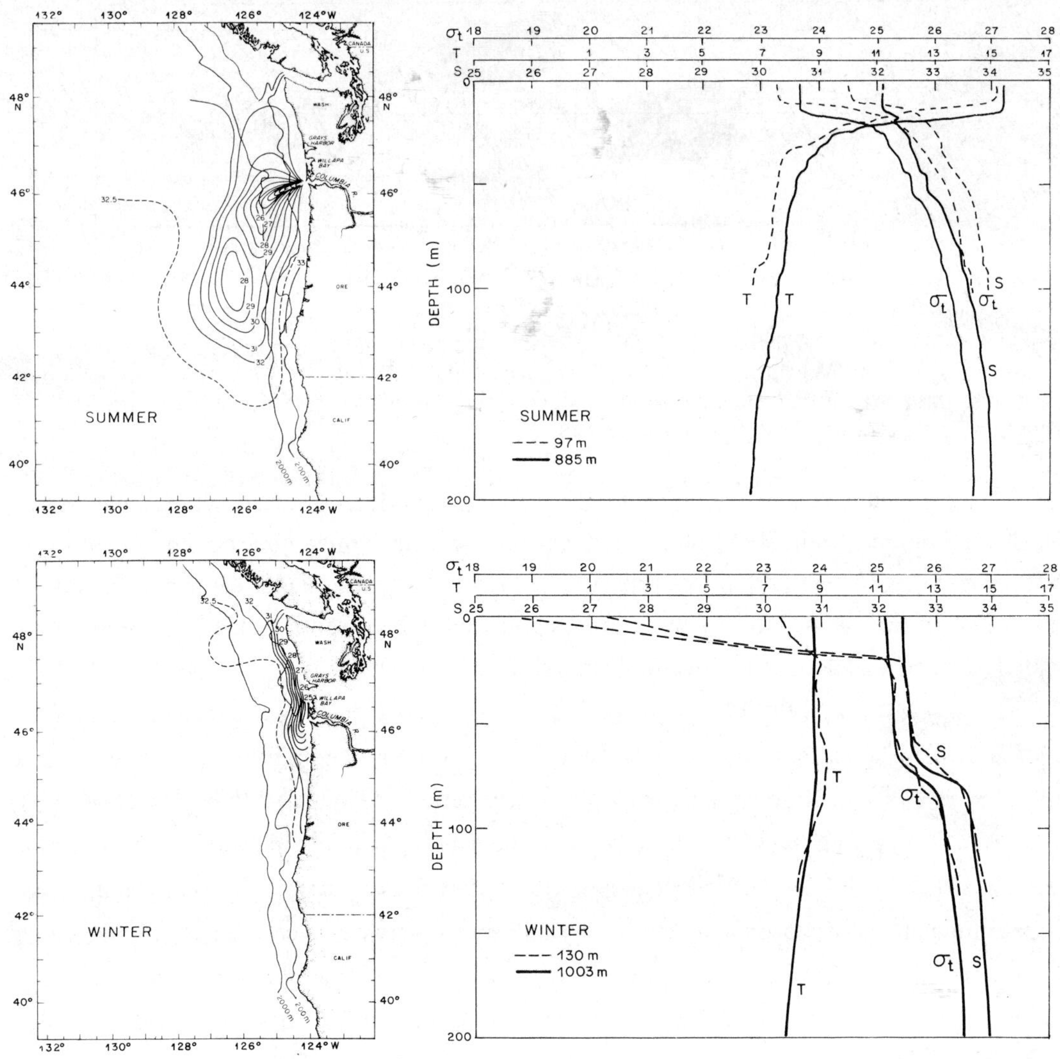

Fig. 2.3 Salinity (⁰/oo) at the sea surface during summer (upper) and winter (lower) off the Washington-Oregon coast. Adapted from Barnes *et al.* (1972). Typical summer and winter salinity (S), temperature (T), and sigma-t (σ_t) profiles near mid-shelf (bottom depths of 97 m and 130 m) and near mid-slope (bottom depths of 885 m and 1,003 m) at locations off southern Washington are also shown.

suspended from surface toroids were used in several later deployments to measure both wind and surface or very near-shore currents (see Table 2.1). The uppermost meter on the taut wire mooring was placed farther from the surface as mooring technology became more sophisticated. For the first several years, upper meters were placed as shallow as 15 or 20 m from the surface. These data are probably contaminated by wave motion and could overestimate current speed by as much as ~25%, particularly during the winter season. In later years, upper meters were placed at least 30 m from the surface during winter and 25 m during summer. Summer measurements made simultaneously with a VMCM and an Aanderaa at similar depths demonstrate that this mooring scheme is adequate for Aanderaa meters during the summer

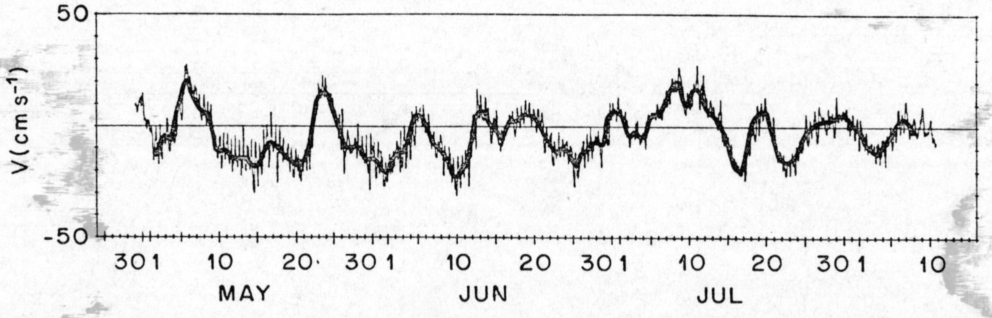

Fig. 2.4 Twenty-minute and six-hourly time series of the alongshelf component of velocity at a mid-shelf location off Washington. The six-hourly data are indicated by the heavier line. Twenty-min data are actual measurements; six-hourly data have been low-pass filtered to remove tidal and other high frequency fluctuations.

season. Unfortunately, no such comparisons are available for winter. Since the dominant waves do reach the bottom at mid-shelf, wave contamination during winter may be significant even with upper meters as deep as 30 m from the surface. However, we note that the results presented herein which utilize such data do not differ significantly in essential elements from those presented for other locations in the Pacific Northwest.

The analysis for event and longer periods is carried out on low-pass filtered (subtidal) data. Specifically, the measurements are filtered with a five-point binomial filter and then with a Cosine-Lanczos filter (half power point of 2.5 h). This series is decimated to hourly values and filtered again (half power point of 40 h) to produce the subtidal data set. Several of the data sets during the period 1971-1974 were low passed with a 25-h running mean. These data are used only for computations of seasonal statistics. The difference between measured 20-min data and the low-passed data is illustrated in Figure 2.4. It is sobering to remember that it is the 20-min rather than the smoothed velocity field that transports suspended and dissolved material in the water column.

A local Cartesian coordinate system is used for data display, that is, x is positive onshore, y is positive in the poleward direction, and z is positive downward. Since subtidal currents tend to follow the local isobath direction, current vectors have been rotated into the local isobath frame of reference. Rotations on the Washington shelf are typically on the order of 10 or 20 degrees west of north. Across-isobath and along-isobath components of velocity are denoted by u and v respectively.

Typical subtidal currents over the Washington shelf and slope during winter, spring, and summer are illustrated in Figures 2.5-2.7. Figure 2.5 emphasizes alongshelf similarities, whereas Figures 2.6 and 2.7 emphasize cross-shelf and shelf-slope (Fig. 2.7 only) differences. The most striking observation is the existence of frequent current reversals at most locations and water depths in every season. A seasonal signal is, in fact, barely discernible in these figures: it is masked by the large amplitude fluctuations that occur on scales of several days. The seasonal flow and the event-scale fluctuating flow will be treated separately in the two sections that follow.

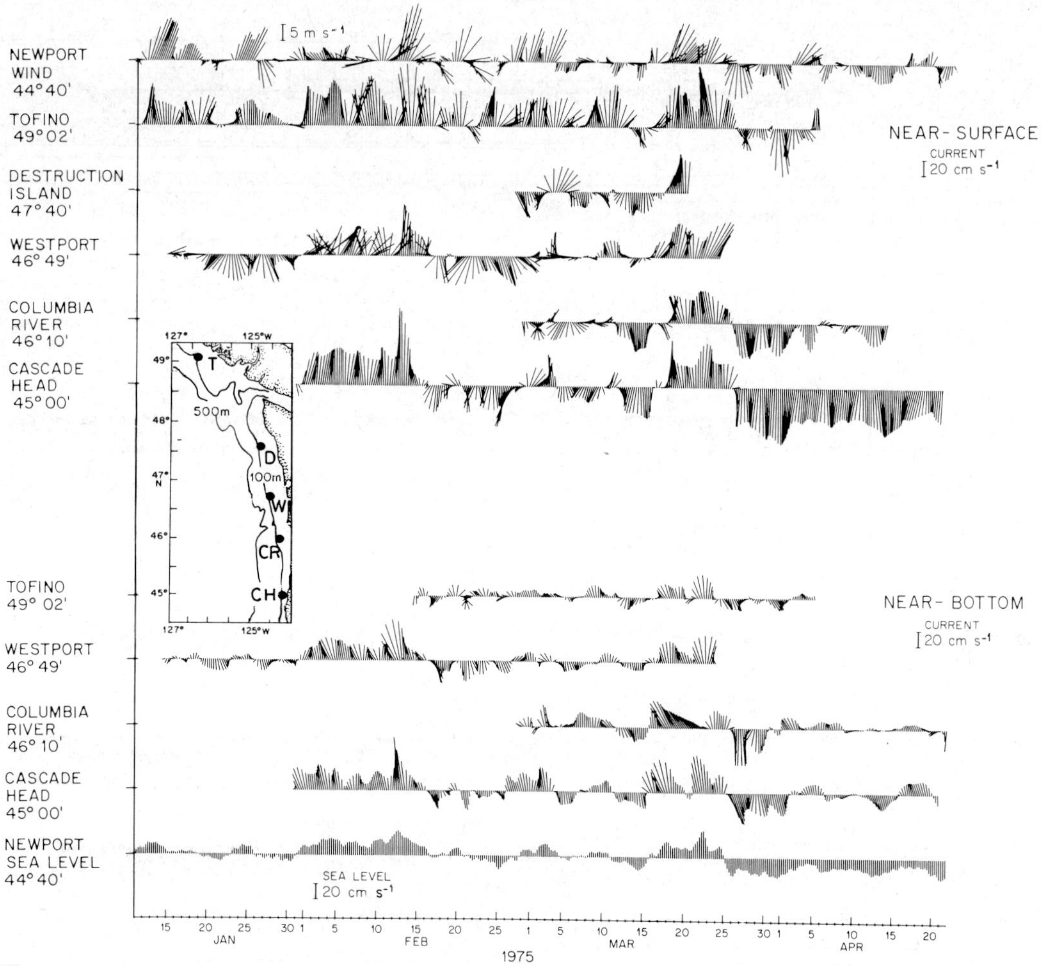

Fig. 2.5 Selected subtidal vector velocity time series at mid-shelf locations along the Pacific Northwest shelf during winter-spring 1975. Data have been rotated into a local isobath frame of reference. Mooring locations are shown on the inset map. Wind and sea level at Newport, Ore., are also shown. Near-surface data are 20-30 m from the surface; near-bottom data are 5-10 m from the bottom (in a bottom depth of about 100 m). The data illustrate the large alongshelf coherence scales that occur in the U.S. Pacific Northwest.

2.3.2 Seasonal Mean Circulation

Year-long vector time series in Figure 2.8 illustrate that there is a significant seasonal signal in currents at mid-shelf locations. The seasonal mean flow is poleward during fall and winter and equatorward during spring and early summer. The transition from poleward winter flow to equatorward spring flow occurs abruptly (see also Figs. 2.5 and 2.6), whereas the transition from equatorward summer flow to poleward flow in the fall occurs more gradually. The more gradual fall transition in currents is due in part to the seasonal development of the poleward undercurrent over the shelf and slope. This undercurrent, which begins to develop a few days after the spring transition, reaches maximum strength in late summer, when it fills the majority of the water column at mid-shelf. Note that the fall transition in water properties, on the other hand, as with the spring transition, is abrupt, dramatic, and relatively large-scale

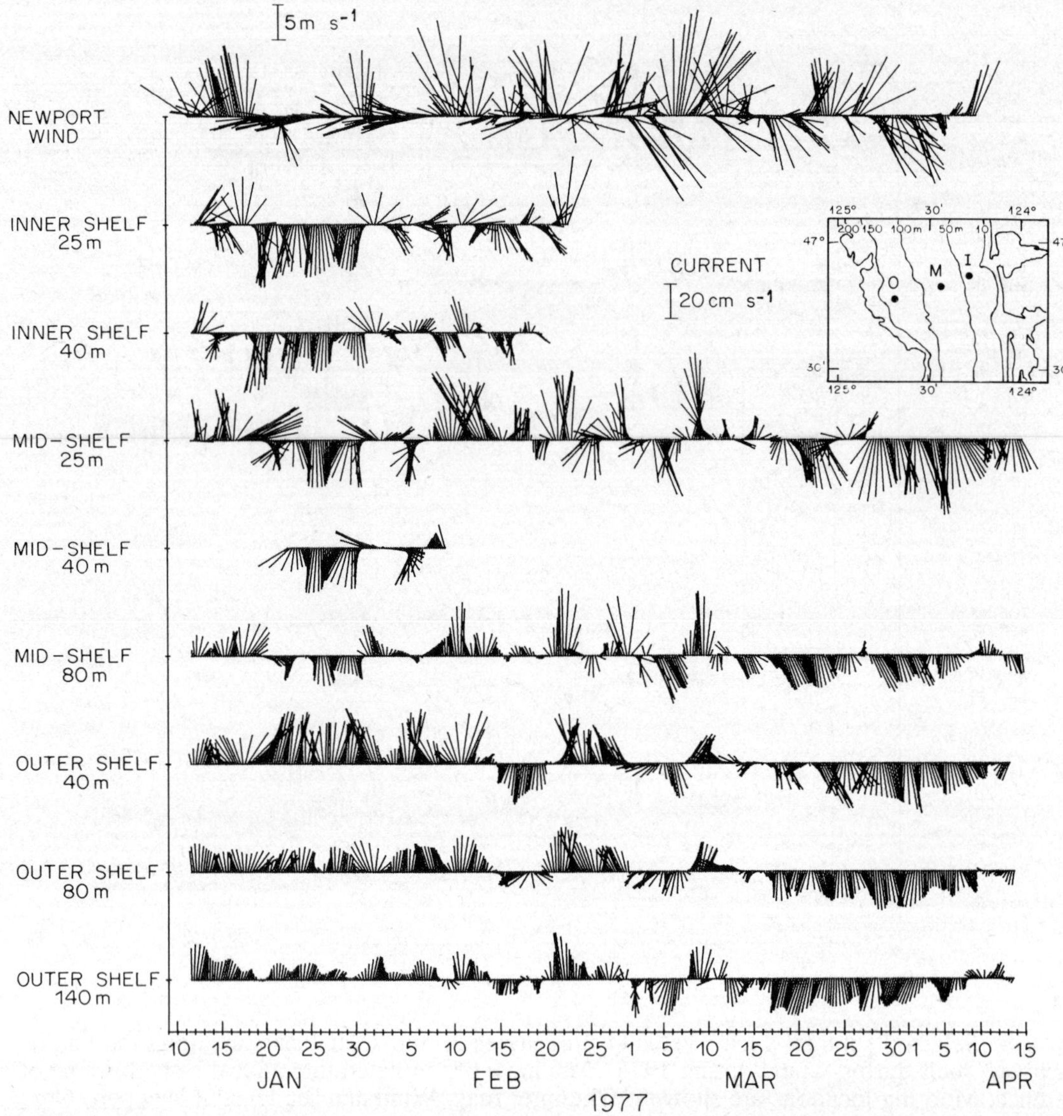

Fig. 2.6 Selected subtidal vector velocity time series on a section across the Washington shelf during winter-spring 1977. Data have been rotated into a local isobath frame of reference. Mooring locations are shown on the inset map. Wind at Newport, Ore. (~45° N) is also shown. The data demonstrate the existence of significant cross-shelf and vertical differences during winter and early spring.

(Hickey *et al.*, in review). Maximum equatorward flow on the Washington shelf occurs during spring and early summer, preceding the seasonal maximum in the offshore California Current by several months.

Seasonal statistics for the Washington shelf

To quantify the seasonal variation and seasonal variability of the flow, monthly means and standard deviations of the alongshelf component of current were calculated using all available Washington shelf data except those near the Strait of Juan de Fuca (see Table 2.1). Data was divided according to bottom depth, z_b, into inner shelf ($20 \le z_b \le 55$ m), mid-shelf ($77 \le z_b \le 100$ m), and outer shelf ($165 \text{ m} < z_b \le 190$ m) depth ranges and into several depth

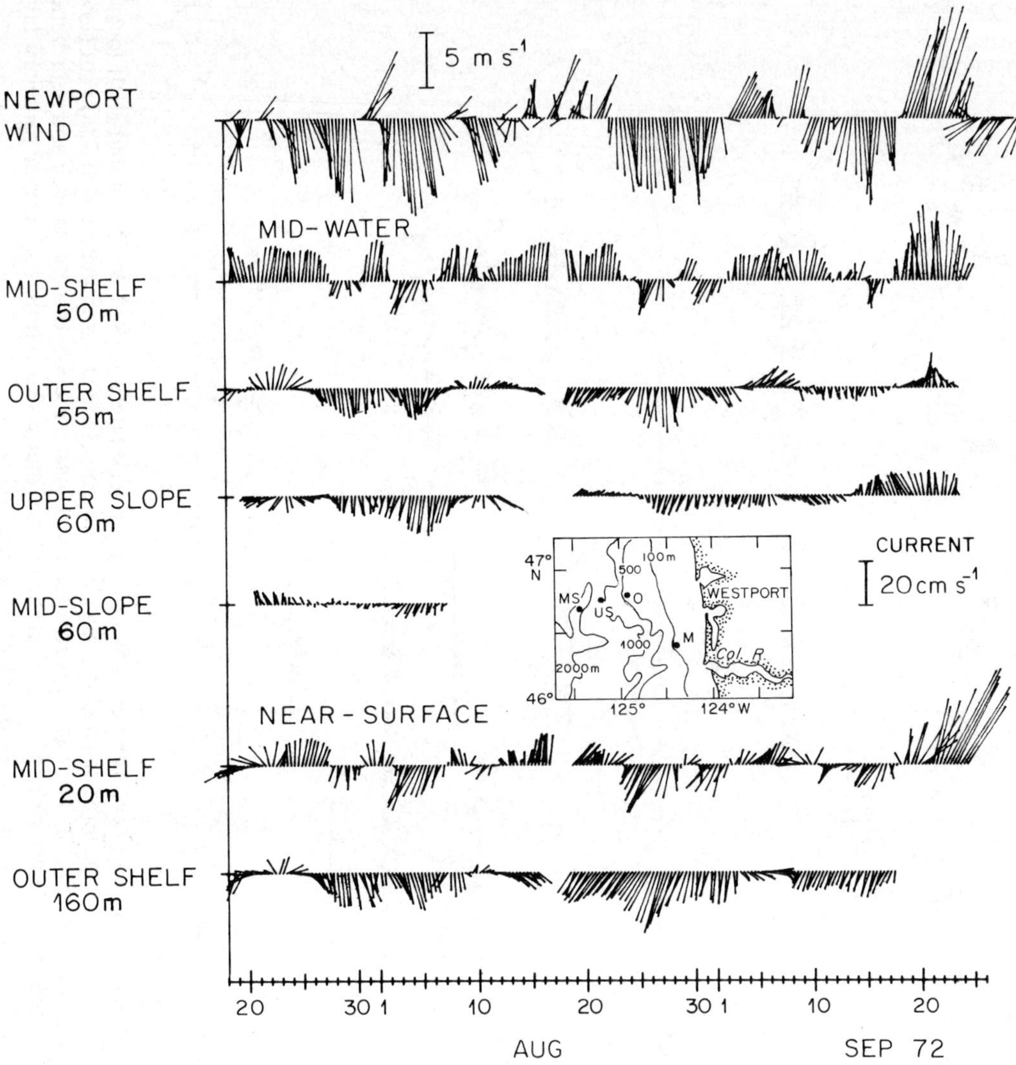

Fig. 2.7 Selected subtidal vector velocity time series on a section across the Washington shelf and slope during summer 1972. Data have been rotated into a local isobath frame of reference. Mooring locations are shown on the inset map. Wind at Newport, Ore. (~45° N) is also shown. The data demonstrate the existence of significant cross-shelf, shelf-slope, and vertical differences during summer.

strata at each shelf site. Because the scale of typical current events is on the order of three to ten days, only data whose record lengths exceeded nine days (one to three periods) were included in the calculations. Means computed using 9 to 14 data days are indicated in the figures by a special symbol. A weighted mean was also calculated for each month. The number of data days used for each weighted mean is given in Table 2.2. Standard deviation (σ_V) were computed using six-hourly subtidal data and the same criteria as for the means. The standard deviation of individual six-hourly data from the weighted monthly mean was also computed. Results are presented in Figures 2.9 (means) and 2.10 (standard deviations).

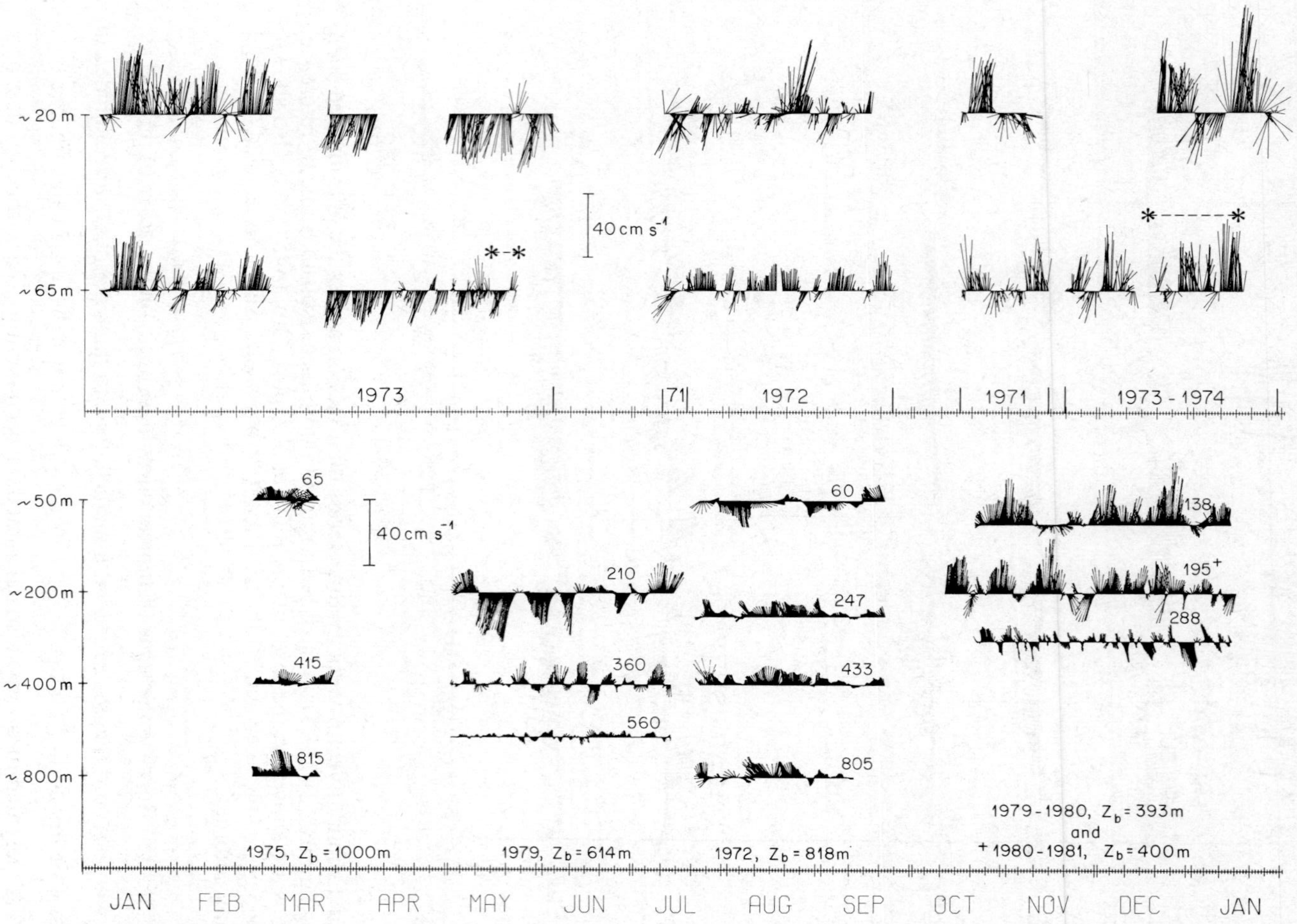

Fig. 2.8 Selected subtidal upper (~20 m) and lower (50-75 m) water column vector velocity time series during 1971-1973 at a mid-shelf location (bottom depth ~80 m) off the southern Washington coast (upper panel) and over the mid-Washington slope (bottom depths 400-1,000 m) (lower panel). Data have been rotated into a local isobath frame of reference. The data illustrate a typical seasonal cycle of near-surface and near-bottom currents over the shelf and at selected depths over the slope. An asterisk indicates that the current meter was within the bottom boundary layer. Note that winter data at 195 m are from a different year than that in time series directly above and below.

TABLE 2.2 Number of days of data used to calculate weighted monthly mean alongshelf velocities given in Figure 2.9.

	Months											
Depths	1	2	3	4	5	6	7	8	9	10	11	12
						Inner Shelf						
5-10	--	--	--	--	--	9	11	39	--	--	--	--
18-32	23	24	--	--	39	38	31	36	--	--	--	--
40-48	21	24	--	--	--	8	--	12	--	--	--	--
						Mid-shelf						
10	--	--	--	--	--	8	--	13	--	--	--	--
18-32	109	114	75	34	23	16	35	52	54	32	--	--
47-60	107	80	34	22	30	38	51	52	54	62	42	31
65-91	145	129	149	34	11	38	50	42	54	46	13	24
						Outer shelf						
18-32	44	26	51	--	--	8	13	40	24	17	--	--
47-60	78	76	98	44	39	9	14	40	24	--	--	--
85-120	71	49	46	11	--	9	13	41	24	17	13	--
132-162	64	48	46	11	29	38	44	40	24	73	73	62

The standard deviation of the monthly means about the grand mean for all data from a given month can be found by a standard statistical method (Panofsky and Brier, 1968). Examples of the 95% confidence interval for monthly means about the line in Figure 2.9 are $\overline{v} = 10.5 \pm 5.9$ cm s $^{-1}$ for the 18-32-m depth range on the mid-shelf in February, and $\overline{v} = -20.5 \pm 15.1$ cm s $^{-1}$ for the same location in June. The extent to which the weighted mean includes interannual variability differs from month to month, as can be seen by the number of points (+) for each month and location. Thus, greater confidence can be placed on the means from winter and from the mid- to outer shelf than from months and locations where data are sparser.

The middle panel in Figure 2.9 confirms the existence of a seasonal cycle at mid-shelf at all water depths. In the upper water column (18-32 m), equatorward flow occurs for almost one-half of the year, from spring through summer. In the lower water column, on the other hand, equatorward flow occurs for only a brief period during spring. The amplitude of the seasonal cycle decreases with water column depth, from ~20 cm s $^{-1}$ in the upper water column to ~10 cm s $^{-1}$ in the lower water column. There is some indication that the amplitude of the seasonal cycle may be a factor of two larger in the very near-surface layer (5-10 m depth range) where direct wind-driving is important (i.e., in the surface Ekman layer).

A significant seasonal cycle is also apparent on the outer shelf (Fig. 2.9, lower panel). As at mid-shelf, the duration of the period of equatorward flow decreases with depth. The

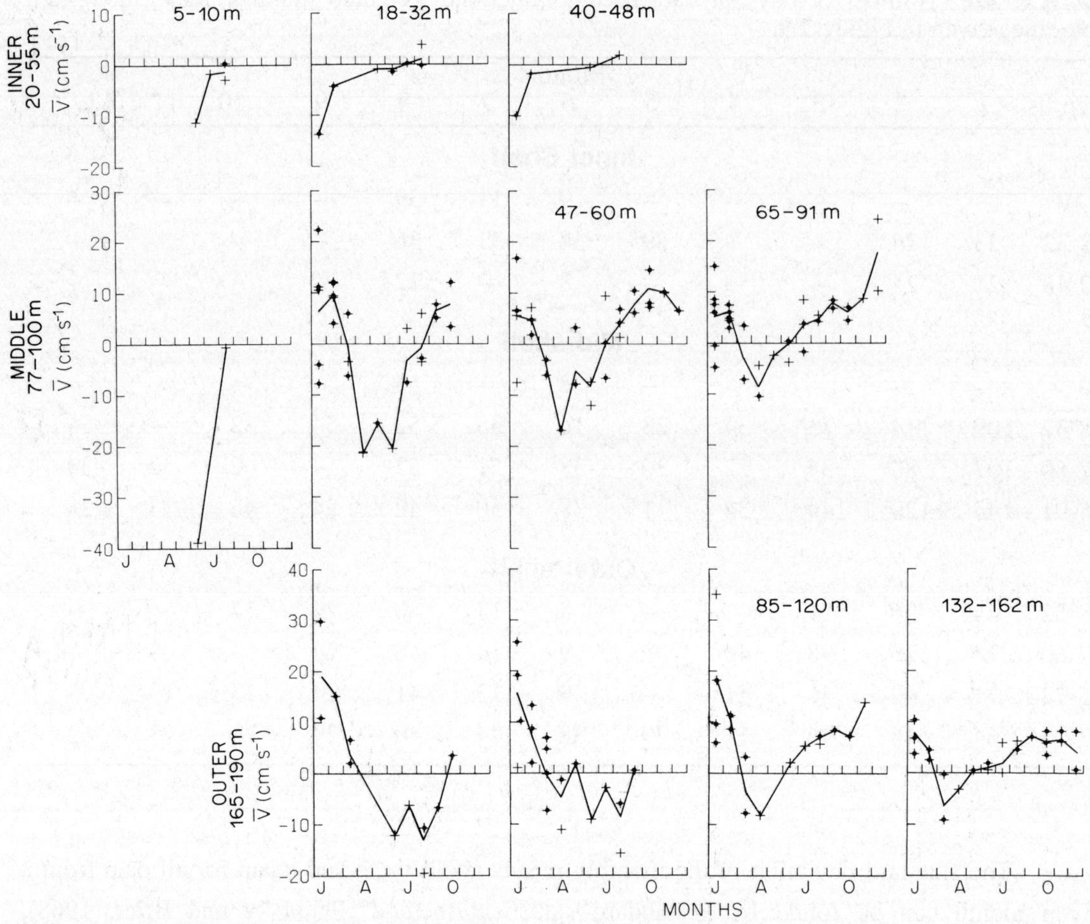

Fig. 2.9 Monthly mean alongshelf velocity in several depth intervals at inner, mid-, and outer shelf locations off the Washington coast. The weighted monthly mean is shown as a solid line. Values calculated using less than 15 d of data are shown with lighter symbols. The number of data days used for each weighted mean is given in Table 2.2.

amplitude of the seasonal cycle is comparable to that at mid-shelf. However, poleward flow during winter is strongest on the outer shelf, whereas equatorward flow during spring is strongest at mid-shelf (except near the bottom).

A seasonal cycle is not readily apparent on the inner shelf (Fig. 2.9, upper panel). The one available set of winter data indicates equatorward mean flow for the period January to March 1977 (see Fig. 2.6). Equatorward flow below a shallow surface layer is also demonstrated by geostrophic velocity sections computed from seasonally and spatially averaged σ_t data (Fig. 2.11). The equatorward flow might be related to the Columbia River plume and its associated alongshelf pressure gradient field (see discussion of alongshore differences). Seasonal mean flow over the inner shelf during summer is weakly equatorward during early summer and is weakly poleward during late summer. The records from which these results were derived are relatively short (~1 month) and may not be representative of the seasonal mean flow. However, seasonal mean geostrophic velocity sections confirm the existence of poleward flow

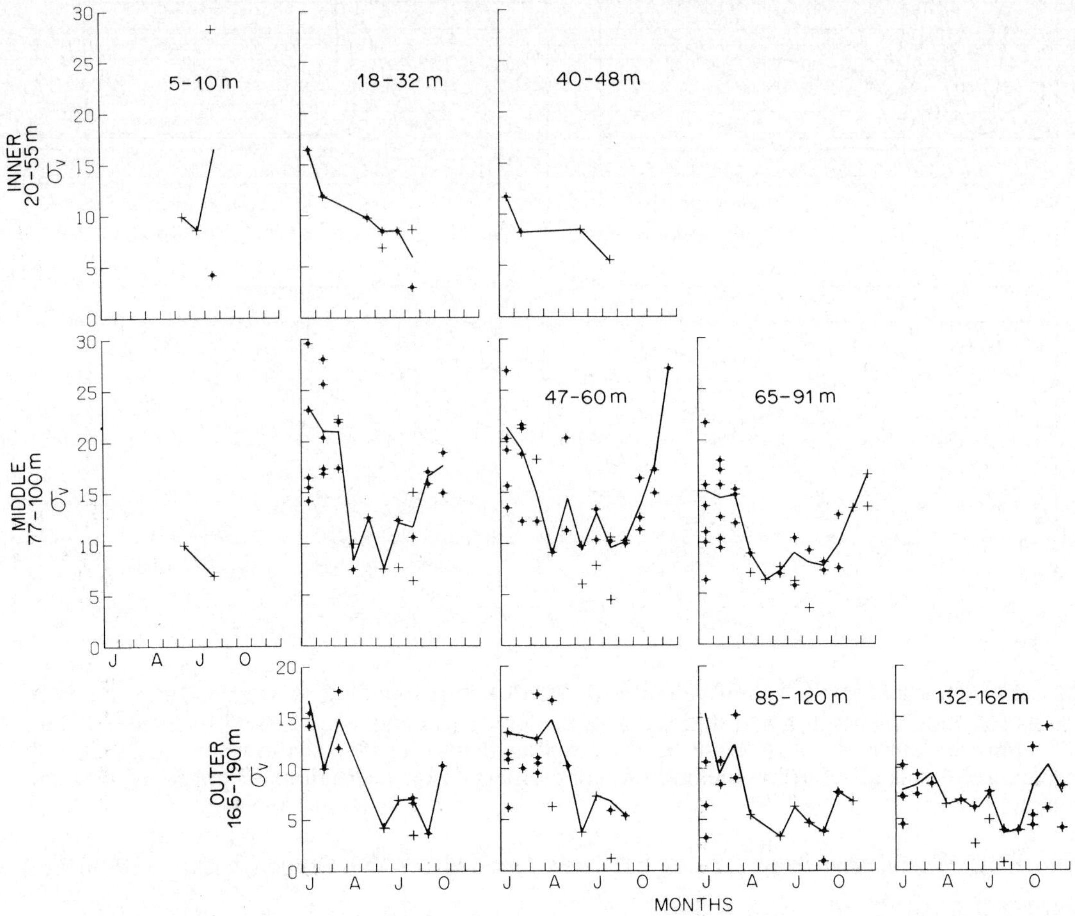

Fig. 2.10 Monthly mean standard deviation of alongshelf velocity shown in Figure 2.9. Units are cm s $^{-1}$. The solid line indicates the standard deviation computed using the weighted mean shown in Figure 2.9.

on the inner shelf during summer (Fig. 2.11). Data off the Oregon coast also suggest that mean flow on the inner shelf during summer is weakly poleward (Huyer *et al.*, 1975, Figs. 7b,c, and Kundu and Allen, 1976, Fig.3).

The phase of the seasonal cycle differs across the shelf. Maximum poleward flow occurs in late fall/early winter at most locations and depths on the mid- and outer shelf. However, maximum equatorward flow in the upper water column (0-60 m) occurs earlier at mid-shelf (spring) than over the outer shelf (summer). Thus after June, equatorward flow is strongest on the outer shelf, so that the mid-shelf coastal jet reported for the central Oregon shelf (Huyer *et al.*, 1975; Kundu and Allen, 1976) does not occur. This result is confirmed in vector plots (Fig.2.7) and in cross-shelf sections of seasonal mean alongshelf flow (Fig. 2.12). In both examples of late summer conditions, maximum equatorward flow occurs over the outer shelf and upper slope, as has also been observed off British Columbia (Freeland *et al.*, 1984). Even during June, when an upper layer coastal jet is observed, the velocity maximum occurs twice as far

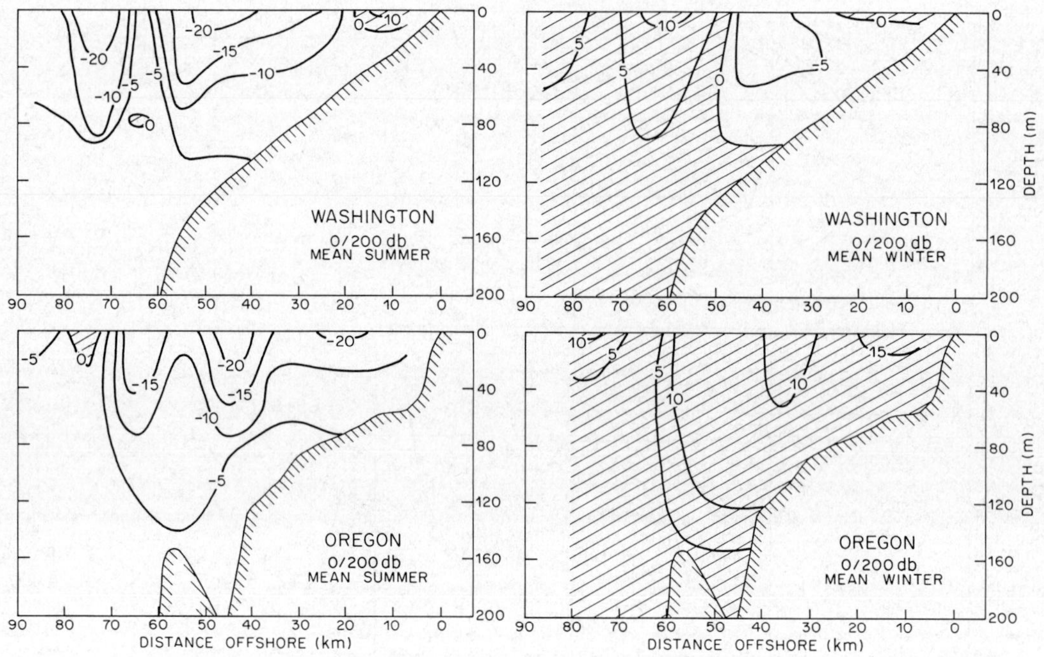

Fig. 2.11 Geostrophic alongshelf velocity (0/200 db) corresponding to seasonally-averaged σ_t sections for mid-Washington and mid-Oregon, presented in Landry *et al.* (1989, Chapter 1, Fig. 1.7). Units are cm s^{-1}. Data have been extrapolated into regions shallower than the 200 db reference level according to the method of Montgomery (1941). Regions of poleward flow are hatched.

offshore off Washington (~60 km) as has been reported for the Oregon coast (~20-30 km) (Kundu and Allen, 1976).

The event scale variability of the fluctuating flow also has significant seasonal variation at most locations (Fig. 2.10). Only during spring on the mid-shelf is the amplitude of the fluctuations less than the magnitude of the seasonal mean current. Generally, the amplitudes of the fluctuations are on the order of that of the seasonal mean. Variability is a maximum in winter and a minimum in summer. Mean variability (i.e., the standard deviation determined using the weighted mean) is generally a maximum at mid-shelf during all seasons in all depth ranges and is generally a minimum on the outer shelf. At mid-shelf during every season and on the outer shelf during winter, the amplitude of the fluctuations decreases by roughly a factor of two between the upper and lower water column.

Alongshore differences in shelf circulation

Significant differences exist between the current structure reported here for the Washington shelf and that reported for the Oregon shelf. This is not surprising, given the fact that the Columbia River plume flows generally poleward (hence over the Washington mid- and inner shelf) during winter and offshore and equatorward (hence over the Oregon mid- and outer shelf) during spring and summer. First, we note that mid- to late summer (July-Aug) equatorward flow reported at two locations on the Oregon shelf is a maximum at mid-shelf, rather than

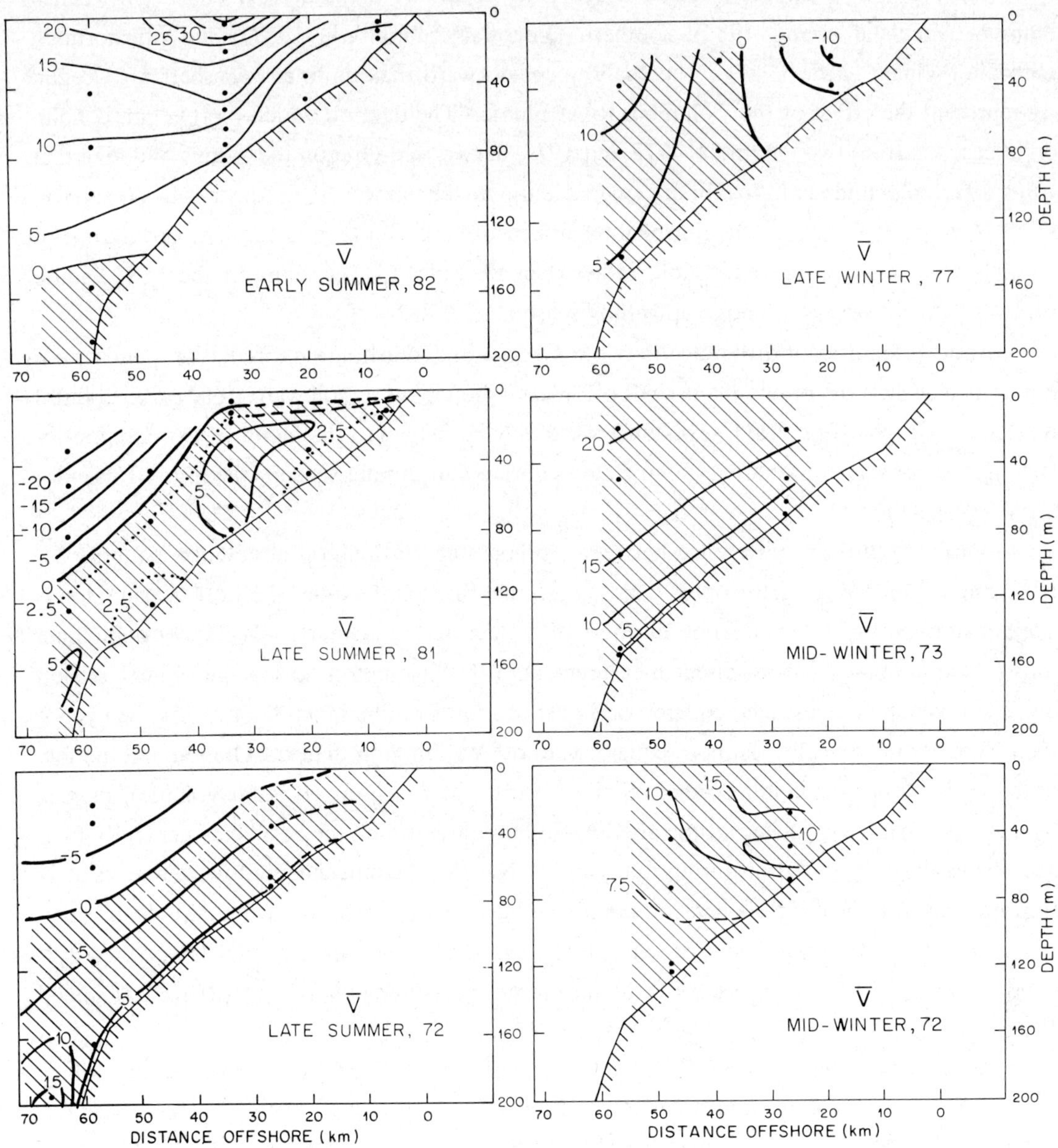

Fig. 2.12 Cross-shelf sections of mean alongshelf velocity off the Washington coast during several summers and winters. Data periods are 12-23 June 1982; 20 Aug-4 Sept 1981; 21 July-28 Aug 1972; 12 Jan-20 Feb 1977; 1-29 Jan 1972; and 7 Feb-7 Mar 1973. Locations of current meters are indicated with dots. Regions of poleward flow are hatched. Units are cm s $^{-1}$.

on the outer shelf and slope as off Washington (Huyer *et al.*, 1975; Kundu and Allen, 1976). This result may be related to the location of the Columbia plume off Oregon during that period. However, geostrophic velocity sections calculated from seasonally- and alongshelf-averaged σ_t data given in Chapter 1 of this volume, do not suggest significant Washington to Oregon differences in the location of the equatorward current maximum below the seasonal pycnocline (Fig. 2.11). In both cases, the maximum below the upper 10 m occurs over the outer shelf or upper slope, as has been observed at most other locations off the west coast, off British Columbia (Freeland *et al.*, 1984), southern Oregon (Strub *et al.*, 1987), and off northern California (Winant *et al.*, 1987). The shallow equatorward maximum at mid-shelf off Oregon may represent the effect of the Columbia River plume. The data on which the Oregon conclusion is based is from two summers, 1972 and 1973, at two mid-Oregon locations. We note that during 1973, the mid-shelf maximum occurs only in the upper 20 m (near 44° 40' N and 45° 15' N, respectively), as in the seasonal mean geostrophic data (Fig. 2.12). The occurrence of a mid-shelf jet may be the exception, rather than the rule, for locations in the U.S. Pacific Northwest, except very early in the upwelling season.

Another significant difference between Oregon and Washington circulation patterns for summer is that the flow on the inner shelf off Washington is more poleward (relative to 200 db) than that off Oregon (Fig. 2.11). This difference may be related to the factor of two Oregon-to-Washington decrease in equatorward wind stress during the summer season (see Fig. 1.1; Landry *et al.*, 1989, Chapter 1).

A third significant difference between Oregon and Washington circulation patterns occurs during winter. Mean geostrophic data suggest that flow on the inner shelf off Washington is more equatorward relative to 200 db during winter than that off Oregon. The geostrophic data confirm results obtained from direct measurements off Washington during one winter season (Fig. 2.12), which demonstrated equatorward mean currents on the inner shelf below the pycnocline. The fact that the flow is more equatorward off Washington than off Oregon and the fact that it appears to be related to a core of relatively fresh water (see Fig. 1.7, this volume), suggest that this flow feature is related to the shallow Columbia River plume, which is generally confined to the Washington coast during the winter. The flow within the shallow plume itself is poleward (Fig. 2.11). The flow beneath the plume may be equatorward during periods of relaxation of the typically strong poleward winds that occur during that season, such as occurred during winter of 1977 when the one set of direct current measurements (Fig. 2.12) were obtained (Hickey, 1984).

Last, we note that results for Washington suggest that seasonal mean flow during winter may be stronger on the outer shelf than at mid-shelf. The Washington result is demonstrated in the statistical means (Fig. 2.9) and in the means for selected winters (Fig. 2.12, 2/3 cases; see also vector plots in Figure 2.6). The alongshore difference, namely, an increase in poleward flow from the mid- to outer shelf off Washington versus a decrease off Oregon, is also consistent with the mean geostrophic velocity sections (Fig. 2.11). This contrasts with results presented for the Oregon coast for both winter 1975 and winter 1976 (Huyer *et al.*, 1978; Huyer *et al.*, 1984).

The data set used to describe the seasonal statistics off Washington includes only data over the southern half of the shelf. The statistics are not expected to differ significantly between the southern and northern portions of the shelf. However, at the northernmost tip of the shelf, within ~15 km of the mouth of the Strait of Juan de Fuca, the seasonal mean flow during the summer appears to be affected by the presence of the Strait, or possibly, by the semi-permanent Juan de Fuca eddy that occurs offshore of the Strait during the upwelling season (e.g., see Fig. 2.37). The mean flow at mid-shelf in this region during the summer is directed poleward, rather than equatorward, as occurs just 30 km to the south (Thomson *et al.*, in press; Hickey *et al.*, in review).

Seasonal patterns of flow over the slope

The data set available for the Washington slope is insufficient to allow estimation of detailed statistics. Multiple moorings were maintained over the slope only during one three-month period in summer 1972 (see Fig. 2.13 for locations). Single moorings were maintained over the slope during Feb-Mar 1975, May-July 1979, Nov 1979-Jan 1980, and Oct 1980-Jan 1981. To illustrate qualitatively the seasonal variation of flow over the Washington slope, selected data from these deployments have been combined into a single time series (Fig. 2.8). Note that data from winters 1980 and 1981 are from moorings located in shallower bottom depths than the other data (400 m vs 800 or 1,000 m).

The seasonal variation of the upper water column flow over the slope is similar to that over the outer shelf: equatorward during spring and summer and poleward during the winter. Below 200 m, the monthly mean flow over the slope is poleward generally throughout the entire year. The magnitude of the poleward flow has a seasonal cycle with minima in the spring and fall and maxima in late summer and winter. This semi-annual signal in poleward flow over the slope appears to be a relatively ubiquitous feature of the California Current System. Hickey (1979) discussed the existence of such a signal in observations at several locations from Point Conception (35° N) to Cape San Lazaro (~25° N). Off Washington, as off Point Conception, the winter maximum in poleward flow occurs at the surface, rather than at depth.

The development of the so-called 'California Undercurrent' during summer (characterized by a subsurface maximum in poleward flow) is likely a response to large-scale forcing of the Eastern Boundary Current system that includes the California Current. The Undercurrent off the Washington slope occurs as a narrow jet-like feature (~20 km) whose velocity maxima occurs just below the shelf break (Fig. 2.13). An experiment to monitor the seasonal variation of the Undercurrent off Oregon failed to locate such a well-defined high speed core with the five moorings that were maintained (Huyer *et al.*, 1984). However, near-simultaneous geostrophic velocity sections obtained along the mooring section off Oregon during the experiment and along a section off mid-Washington suggest that a narrow core does occur off Oregon. The mooring design, which was based more or less on the Washington section (left hand sections in Fig. 2.13) was simply inadequate to resolve the Oregon core, which occurs shallower and closer to shore at that location (right hand section in Fig. 2.13). Whether a narrow core, as observed off Washington, is a ubiquitous feature and how or whether the core is related to the topography has not yet been demonstrated in the observations. However, the seasonal variation of the velocity of the slope Undercurrent in the U.S. Pacific Northwest (i.e., a summertime maximum in positive vertical shear) seems reasonably well determined by the Washington data set and by the year-long Oregon data set described in Huyer *et al.* (1984).

During winter, both the data and modeling results (see next section) suggest that an equatorward undercurrent (the 'Washington Undercurrent') sometimes occurs over the Washington

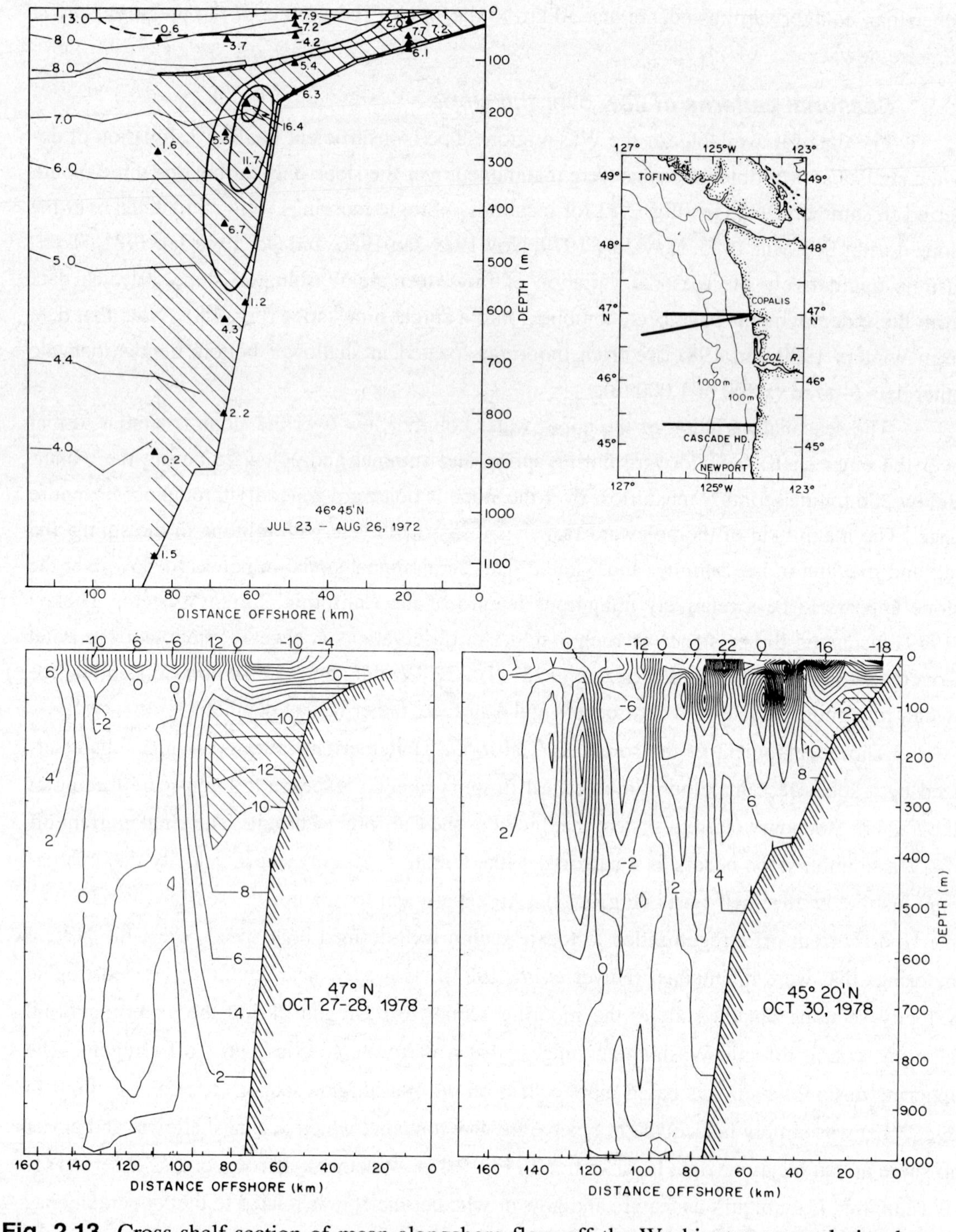

Fig. 2.13 Cross-shelf section of mean alongshore flow off the Washington coast during late summer 1972 and of temperature on 27 July 1972 (upper panel). Units of velocity and temperature are cm s^{-1} and °C, respectively. Locations of current meters are indicated by triangles. The data illustrate the cross-shelf structure of the poleward undercurrent (hatched area). Adapted from Hickey (1979). Near-simultaneous geostrophic alongshelf velocity (0/1,000 db) for mid-Washington (lower panel left) and mid-Oregon (lower panel right). Units are cm s^{-1}. Extrapolation into regions shallower than the reference level follows Montgomery (1941). Regions of strongest poleward flow are hatched to illustrate the core of the poleward undercurrent.

slope. The equatorward undercurrent maximum is deeper than the poleward undercurrent maximum (~400 m). Data in Huyer *et al.* (1984) illustrate the tendency for equatorward flow during this period at similar depths off Oregon. The horizontal structure of this undercurrent is completely unknown. Observations that support the existence of this undercurrent are discussed in more detail in Hickey (in press).

Forcing mechanisms

For a two-dimensional, cross-shelf and depth-dependent, frictional model of shelf/slope circulation, the steady state response to an imposed equatorward wind stress is a balance between offshelf flow in a surface frictional layer and net onshelf flow in a bottom frictional layer, with upwelling at the coastal wall. The response to an imposed equatorward pressure gradient force (i.e., an alongshelf sea level slope) differs only in that the offshelf flow is distributed throughout the water column rather than being confined to a surface frictional layer. In either case the net alongshelf flow is equatorward. If, however, the surface stress and pressure gradient forces are in opposite directions, the alongshelf flow reverses from top to bottom, that is, an undercurrent occurs. If the surface stress and pressure gradient forces are in the same direction, the net alongshelf flow is larger than that due to either of the individual forces because increased offshelf flow in the bottom boundary layer is required for two-dimensional mass balance. Thus, for a two-dimensional model of shelf circulation, the local values of both the alongshelf component of wind stress at the coast as well as the alongshelf sea level slope must be considered. In a three-dimensional model, the sea level slope would be predicted by the model itself.

Both the seasonal wind stress and sea level slope in the U.S. Pacific Northwest have significant seasonal variations. The large-scale, seasonal distribution of wind stress along the west coast of the United States is controlled by the relative locations of two atmospheric pressure systems, the North Pacific High and the Alaskan Low. The High is centered off northern California (~40° N) during summer, and the resulting wind stress is equatorward all along the coast with a maximum off northern California (Hickey, 1979). During winter, the High moves farther south and is replaced by the Alaskan Low off northern Washington with the result that wind stress diverges near 42° N (i.e., wind stress is poleward north of 42° N and equatorward south of 42° N). During spring, the High is somewhat farther north, resulting in weak poleward wind stress off Washington but strong equatorward stress off California with maximum stress off northern California as in summer. The seasonal changes in wind stress and atmospheric pressure systems are discussed in more detail in Hickey (1979).

Coastal sea level in the U.S. Pacific Northwest increases to the north during winter and to the south during summer (Fig. 2.14). During spring, a relative minimum occurs near San Francisco (40° N), so that sea level increases toward the north in the Pacific Northwest but to the south off California. Long-term mean sea level also has a relative minimum near 40° N. Thus, in the Pacific Northwest the alongshelf pressure gradient force is equatorward during winter and spring and poleward during summer. Only during spring are the pressure gradient force and local wind stress in the same direction.

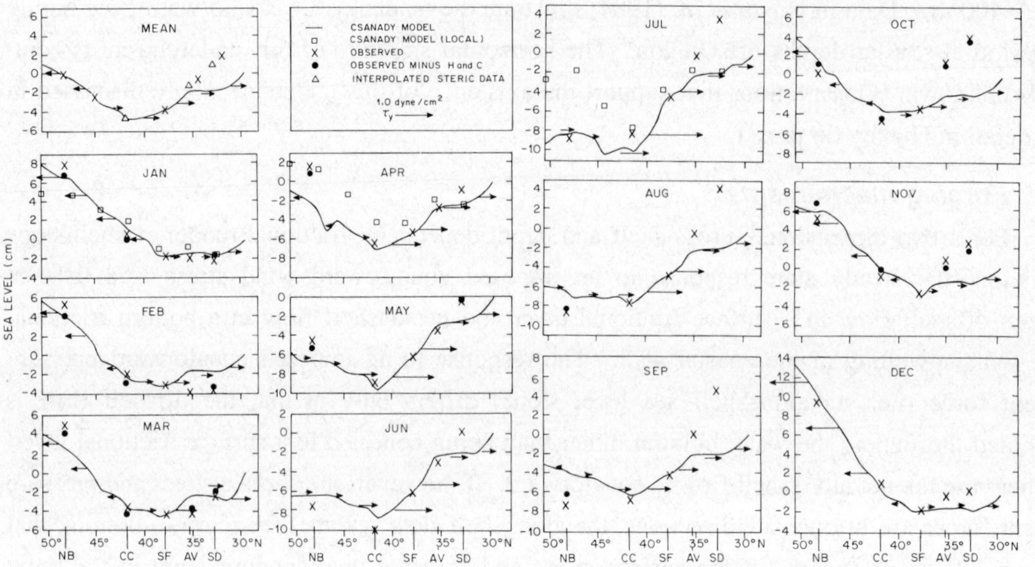

Fig. 2.14 Observed and predicted sea level elevation as a function of latitude along the U.S. west coast. Observations are shown with the annual heating and cooling effect (H and C) removed when the latter exceeds 0.4 cm. Predictions are made with the Csanady (1978) model, summing the solutions forced by individual blocks of wind stress. The local response, that is, the response to each individual wind stress block, is also shown in selected months (denoted "local"). Wind stress data used to drive the model are shown as arrows. Observations are plotted relative to elevation at San Francisco. From Hickey and Pola (1983).

The seasonal variation of coastal sea level in the Pacific Northwest is related to the seasonal variation of the coastal wind stress patterns that were previously described (Hickey and Pola, 1983). This conclusion was reached via comparison of observed sea level and predictions of the wind-forced topographic wave model of Csanady (1978). The model was forced by the alongshore component of the large-scale coastal wind stress field from the Pacific Northwest to southern California (Fig. 2.14). The model results indicate that roughly 65% of the amplitude of the seasonal sea level variation in the Pacific Northwest is due to remote, rather than local, wind stress forcing. Remote forcing is particularly important during spring and summer, when the magnitude of alongshore wind stress has a maximum near 40° N, so that sea level north of 40° N is depressed by contributions from the strong forcing near 40° N. During winter, when the gradient of wind magnitude is reversed north of 40° N, remotely forced contributions at locations north of 40° N are correspondingly smaller. Off southern California, the seasonal variation of sea level is more likely to be related to the seasonal variation of the large-scale boundary currents rather than to wind forcing (Fig. 2.15) (Hickey and Pola, 1983). Boundary current forcing may also contribute significantly to sea level in the Pacific Northwest, particularly during early spring when coastal sea level may be elevated due to the presence of the large-scale Davidson Current (Figs. 2.14 and 2.15).

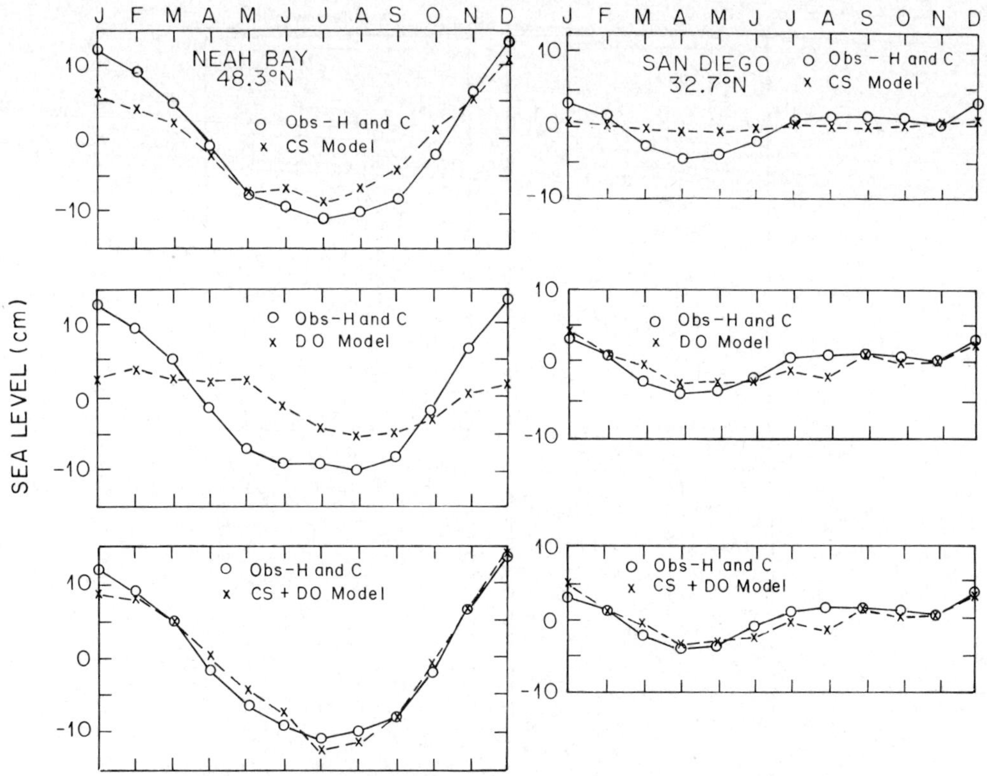

Fig. 2.15 Observed and predicted sea level elevation at two west coast tide stations. The annual heating and cooling effect (H and C) has been removed from the observations and from the deep-ocean elevation data. Predictions are made with the Csanady (1978) model (CS), with deep-ocean elevation data minus H and C (DO) and with their sums (CS + DO). From Hickey and Pola (1983).

To investigate forcing mechanisms for the seasonal mean currents in the Pacific Northwest, the observed seasonal mean pressure alongshelf gradient force was used together with observed local alongshelf wind stress to drive an alongshore-independent numerical model of the coastal circulation (Werner and Hickey, 1983). The pressure gradient force was regarded as the net result of the alongshore-dependent effects (such as alongshore-dependent wind stress), with the assumption that mass balance occurs primarily in the x-z plane. In spite of the fact that mass balance in the Pacific Northwest is not, in general, two-dimensional (see section on coastal upwelling), comparisons between model results and observations indicate that several important features of the seasonal circulation off Washington can be explained using this somewhat simplistic approach (Fig. 2.16). The model predicted the existence of both the equatorward (summer) and poleward (winter) undercurrents that appear in the observations over the slope (see next section) as well as the anomalously strong (cf. local wind stress) equatorward flow that follows the spring transition. Conversely, none of these features could be predicted without inclusion of the alongshelf pressure gradient force.

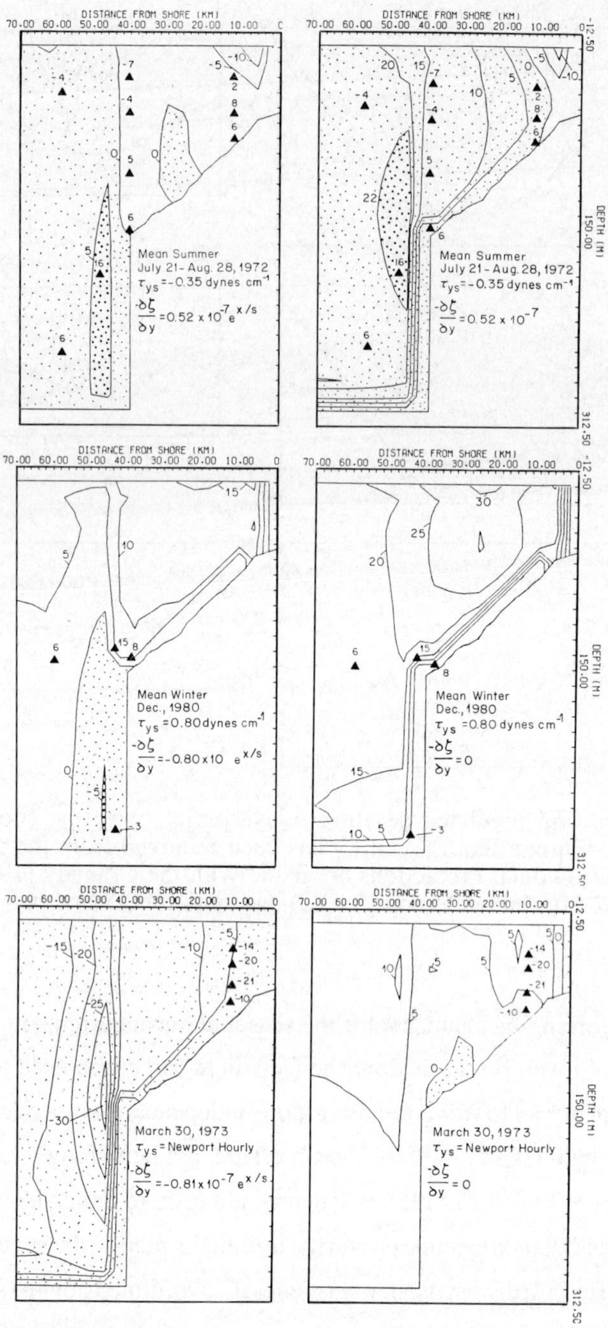

Fig. 2.16 The left hand panel gives contours of alongshelf velocity (cm s $^{-1}$) generated by observed mean wind stress and observed mean alongshelf pressure gradient force for late summer 1972, winter 1980, and spring 1973. Numbers next to triangles are observed mean velocities. The summer and winter experiments were run for seasonally-averaged data; the spring experiment was run for a specific wind event immediately following the spring transition of 1973. The right hand panel illustrates for comparison the same experiments run with an alongshelf pressure gradient force constant in the cross-shelf direction (summer) and with no alongshelf pressure gradient (winter and spring). Regions of poleward modeled flow are shaded for the summer case; regions of equatorward flow are shaded for the winter and spring cases. Adapted from Werner and Hickey (1983).

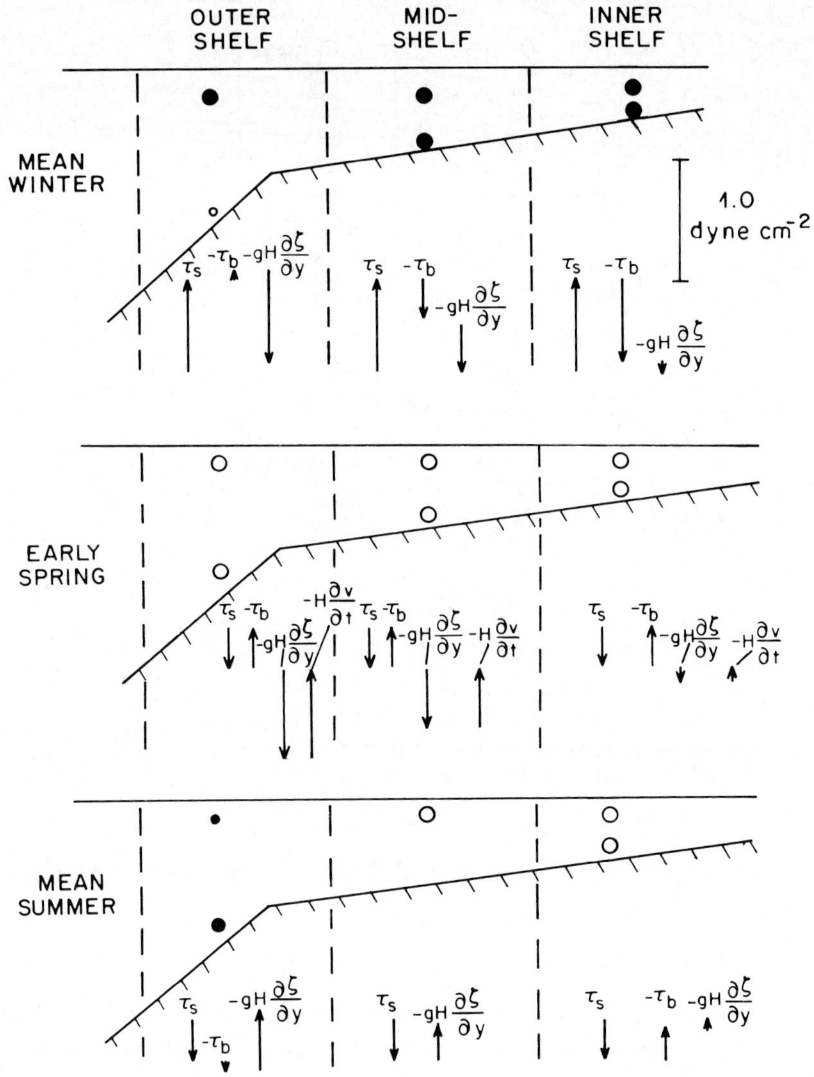

Fig. 2.17 Predicted seasonal variation of alongshelf momentum balance in the Pacific Northwest coastal zone for mean winter and mean summer conditions and for 24 March 1973, the day following the spring transition. Solid dots indicate poleward flow (into the page). Open dots indicate equatorward flow. Mean summer results are those derived with a coastal trapping factor included (Fig. 2.16, left panel). From Werner and Hickey (1983). Note that significant differences occur for the summer case between these predictions and the observations (see also Fig. 2.12). Observations suggest that the flow is poleward on the mid- and inner shelf below the surface layer, and that the flow is equatorward in the surface layer over the outer shelf and slope.

For the summer case, better results were obtained by Werner and Hickey (1983) when coastal trapping of the pressure gradient force was not assumed (Fig. 2.16). This is consistent with Hickey and Pola's (1983) conclusion that deep-ocean boundary currents make significant contributions to alongshelf sea level slope in the Pacific Northwest during summer. The most

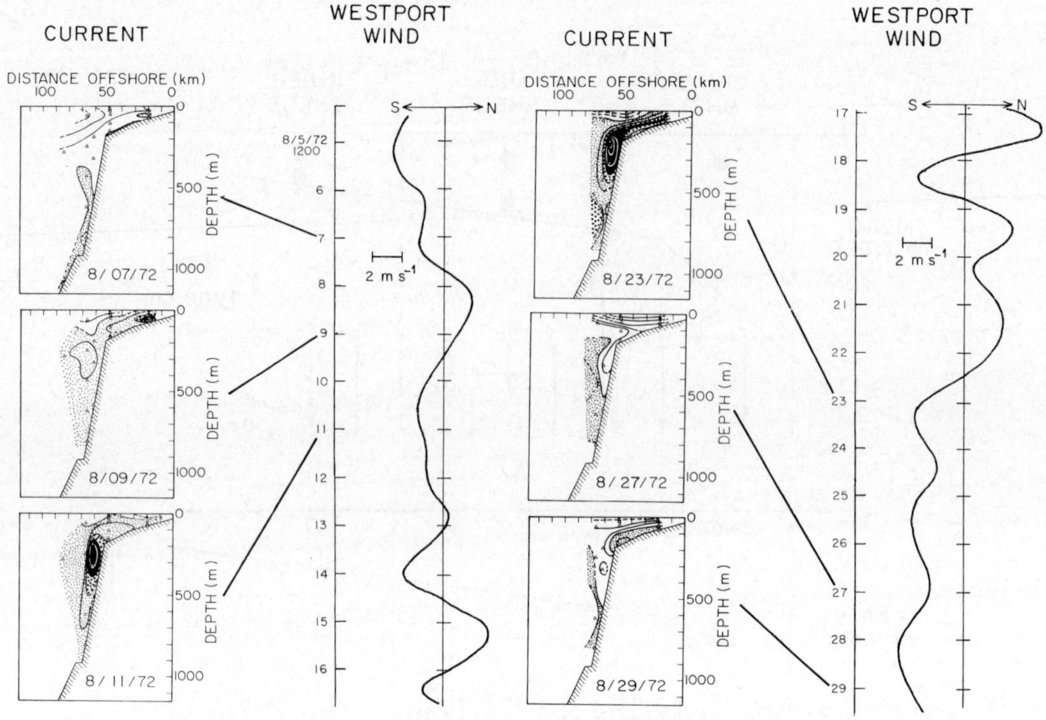

Fig. 2.18 Time series of cross-shelf structure of subtidal alongshelf velocity off the Washington coast and alongshore wind at the coast during summer 1972. Regions of poleward flow are dotted. Triangles indicate locations of current meters. The contour interval is 5 cm s^{-1}.

serious weakness of the simple, two-dimensional, *f*-plane model is its failure to predict the equatorward flow over the outer shelf and slope during summer. Isopycnals over the outer shelf and slope never tilt upward toward the coast as they must to reproduce the observed baroclinic equatorward flow (Figs. 2.11 and 2.12). However, the model was not run sufficiently long for diffusion to become important, as would be expected for seasonal scales. Horizontal and vertical diffusion results in offshore displacement of the maximum in alongshelf velocity (Hamilton and Rattray, 1978). For seasonal time scales, offshore spreading due to the β effect is also important (Philander and Yoon, 1982).

The seasonal variation of the seasonal mean alongshelf balance of momentum on the inner, mid- and outer shelf in the Pacific Northwest, based on the Werner and Hickey (1984) model analysis, is illustrated in Figure 2.17. On the shallow inner shelf, the balance is primarily frictional in every season. Over the central shelf, the alongshelf pressure gradient term is also important, opposing wind stress forcing in winter and summer and adding to it during spring. Bottom stress over the central shelf is only significant during winter and spring, when alongshelf currents are large. On the outer shelf, the alongshelf pressure gradient force makes an even greater contribution to the momentum balance.

Recently, a number of theories have been presented to account for poleward flow along eastern boundaries. These theories suggest that poleward flow can be generated by a variety of

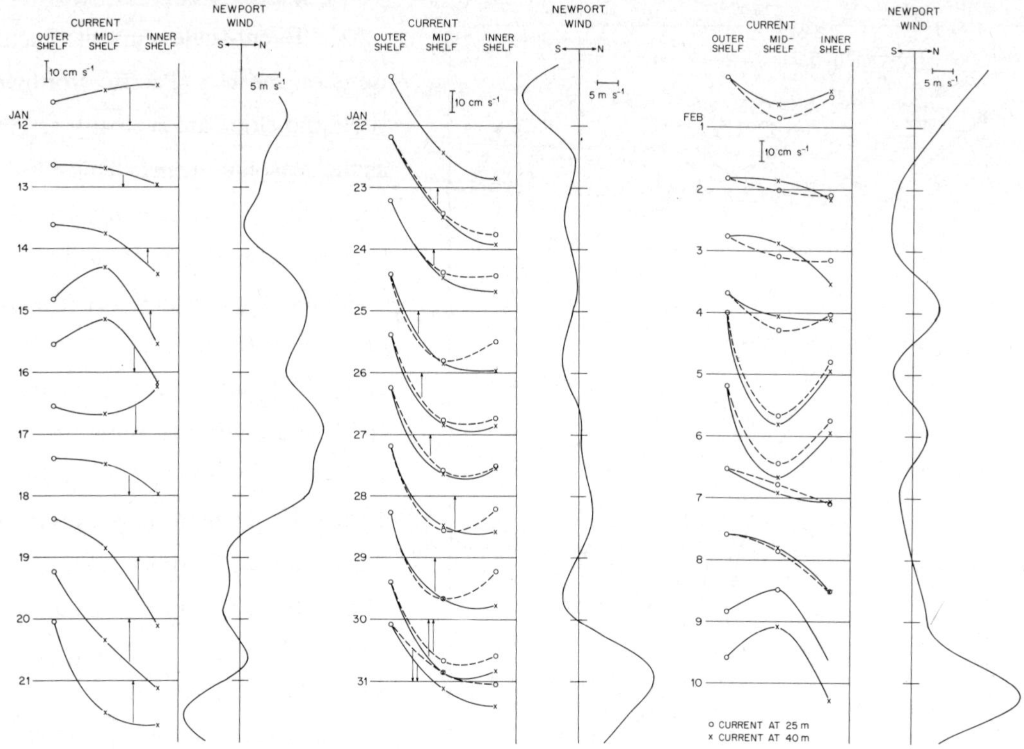

Fig. 2.19 Time series of cross-shelf structure of subtidal alongshelf velocity over the Washington shelf and alongshore wind on the Oregon coast during late winter 1977. Current profiles are given at 25 m and 40 m from the surface where data are available. The full vector data set is presented in Figure 2.6.

mechanisms such as topographic stress generated by the interaction of non-linear shelf waves or eddies with alongshore variations in topography (Holloway, 1987) or topographic form drag (Brink, 1986; Haidvogel and Brink, 1986; Samelson and Allen, 1987). These theories have not yet been developed to the state where details such as vertical structure or seasonal variation can be addressed. Thus, it is unclear what fraction of the signal in poleward flow over the shelf and slope should be attributed to each of these mechanisms and to alongshore sea level slopes (much of which ultimately derives from alongshelf wind stress structure, as previously discussed). At present, the wind stress forcing of alongshore sea level slope and poleward undercurrents is the most attractive forcing mechanism for Pacific Northwest locations since it seems to account for much of the seasonal variability in alongshore sea level slope and in the resulting shelf and slope undercurrents, both poleward and equatorward. However, south of Point Conception, sea level slope does not appear to be strongly related to alongshelf wind stress and other mechanisms that generate alongshelf slopes and/or poleward flows may make important contributions to the dynamics.

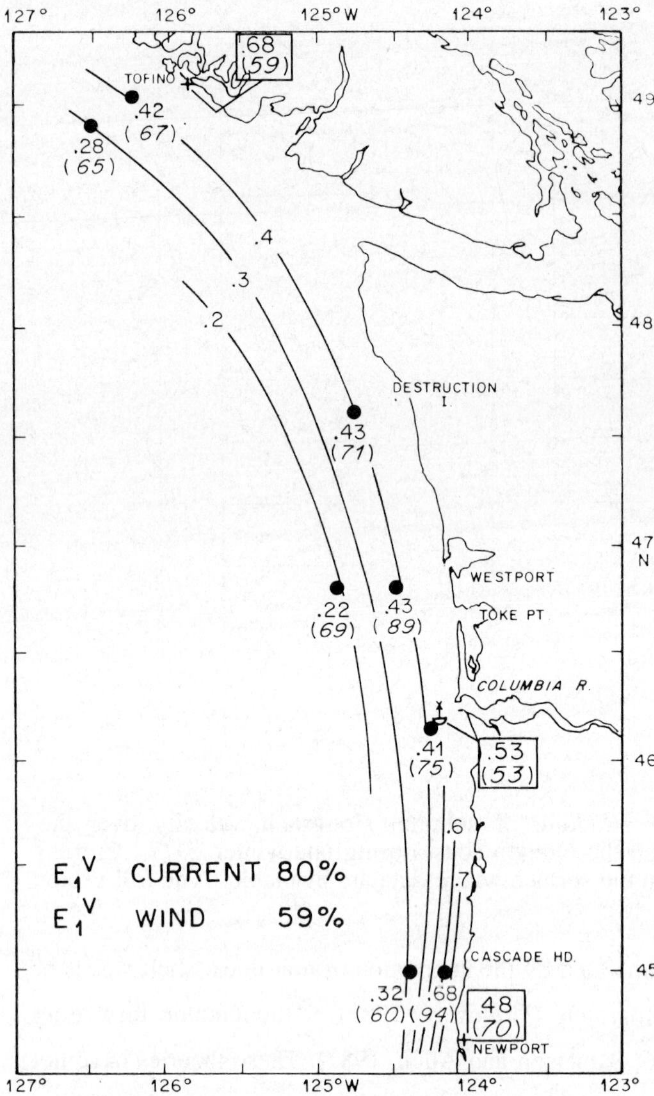

Fig. 2.20 Spatial structure of the first empirical orthogonal eigenfunctions of alongshelf current and wind and, in brackets, the percent of variance accounted for by the eigenfunction at each location off the Pacific North-west during winter-spring 1975. Wind eigenfunction amplitudes are given in boxes. Adapted from Hickey (1981).

2.3.3 Event-Scale Variability

Event-scale current fluctuations on the U.S. Pacific Northwest shelf and slope are typically as large as the seasonal means. This extreme variability is visually illustrated in Figures 2.18 and 2.19, which give contours of alongshelf velocity as a function of time across the mid- and outer shelf and slope during summer 1972 and over the inner to outer shelf during winter 1977. Although a relationship between currents and the local wind field is sometimes evident, the correlation is far from unity. Moreover, currents over the outer shelf and/or slope often oppose those on the mid- and/or inner shelves. At times, currents at mid-shelf oppose those over the inner and outer shelves. Research performed during the past decade has successfully unraveled the causes of much of this variability.

Coherence scales

Coherence scales of a flow field can provide substantial insight into the forcing mechanisms of the fluctuations in the field. During all seasons, the event-scale current fluctuations in the U.S. Pacific Northwest have large alongshelf coherence scales in the alongshelf component of velocity (Fig. 2.5). Empirical orthogonal eigenfunction (EOF) analysis over the mid-water column of the mid- and outer shelves indicate that as much as 80% of the variance can be contained in a single mode with an alongshelf scale exceeding 500 km (Fig. 2.20). This contrasts markedly with results on the east coast. The continental shelf on the U.S. east coast is generally a factor of two or three wider than the shelf on the west coast and is not as continuous in the alongshelf direction as the west coast shelf (Pietrefesa,

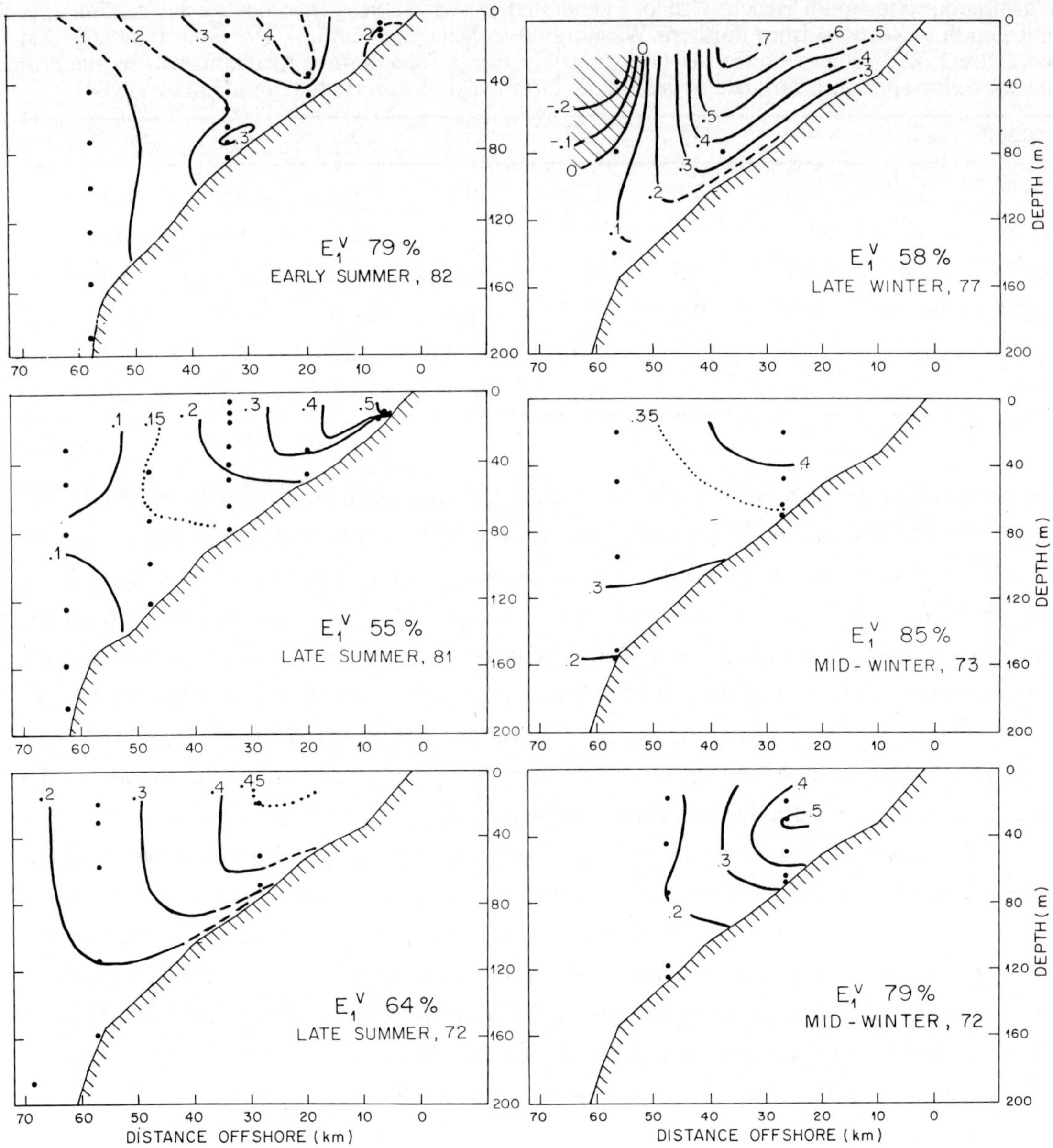

Fig. 2.21 Cross-shelf structure of the first empirical orthogonal eigenfunction (EOF) of alongshelf velocity during several winters and summers. Dots indicate location of current meters. The percent of variance accounted for by the EOF is given on each panel. Selected time series from the data sets for the late winter 1977 and late summer 1972 cases are displayed in Figures 2.6 and 2.7, respectively.

1983). Alongshelf coherence scales for the cross-shelf velocity component are much smaller, typically 200 km for near-bottom currents and < 200 km for mid- and upper water column currents (Hickey, 1981).

EOF analysis indicates that alongshelf current fluctuations are also positively coherent <u>across</u> the shelf during most periods (Fig. 2.21). The one example of negatively correlated

TABLE 2.3 The percent of variance in predicted subsurface coastal pressure at Neah Bay, Washington, and South Beach, Oregon, generated by wind stress in each coastal section (per unit length of section) from northern Washington to Baja, California. The table indicates that wind stress off northern California is responsible (via a forced-wave mechanism) for much of the subsurface pressure variance in the Pacific Northwest. From Battisti and Hickey (1984).

Section	48	46	44	42	39	37	33	31° N
NB 72	3.6	8.3	4.6	12.6	35.8	18.0	14.9	2.2
SB 72	--	--	6.3	15.2	36.4	22.1	18.9	1.1
NB 78	12.9	10.6	6.0	12.2	36.6	8.8	9.4	3.5
SB 78	--	--	9.2	16.8	46.6	12.0	13.4	1.9
NB 77	37.3	27.7	15.3	10.2	3.8	2.1	2.1	1.5
SB 77	--	--	47.5	28.0	10.5	6.2	6.0	1.8

fluctuations over the shelf break is likely a result of the occurrence of an eddy during that period (see Fig. 2.6). The first mode generally represents 60-80% of the total variance in alongshelf currents. The amplitude of fluctuations is a maximum roughly 15 km from the coast at the outer edge of the inner shelf ($z_b \sim 50$ m) during early summer, giving the appearance of a coastal jet. During the one winter for which inner shelf data are available, the amplitude maximum appears to occur farther offshore, roughly 30-40 km from the coast ($z_b \sim 100$ m). No mid-shelf coastal jet is evident during late summer; rather, the fluctuations are a maximum on the inner shelf. However, the lack of mid- and upper water column measurements on the outer portion of the inner shelf prevents a definitive conclusion for this period.

The alongshelf current fluctuations represented by the first mode are quasi-barotropic in all seasons and at all locations (with the possible exception of the outer shelf in winter, 1977); that is, fluctuations have the same sign throughout the water column. However, the fluctuations on the mid- and inner shelf have considerable vertical shear, both in the surface and bottom frictional layers and in the geostrophic interior. Vertical differences in amplitude of roughly a factor of two occur over the inner shelf in all seasons examined. Vertical shear at mid-shelf locations is strong in the interior (a factor of two amplitude difference) only during winter.

The structures of the Washington shelf EOFs are similar in most respects to those presented for the Oregon shelf (Fig. 6 in Huyer *et al.*, 1978), with the Washington early summer case being roughly equivalent to that presented for spring off Oregon. In particular, maximum vertical shear over the inner, mid-, and outer shelf occurs during winter off Oregon as well as off Washington (Huyer *et al.*, 1978).

Wind forcing

The large scales of alongshelf coherence (Fig. 2.20) as well as the offshore decay of alongshelf velocity fluctuations in the U.S. Pacific Northwest are typical of wind-forced, coastal-trapped waves. Battisti and Hickey (1984) demonstrated that the majority (up to 90%) of the coastal sea level and mid-shelf velocity fluctuations in the Pacific Northwest are a result of mode one, wind-forced coastal-trapped waves (Fig. 2.22). The mode one wave, with typical along-

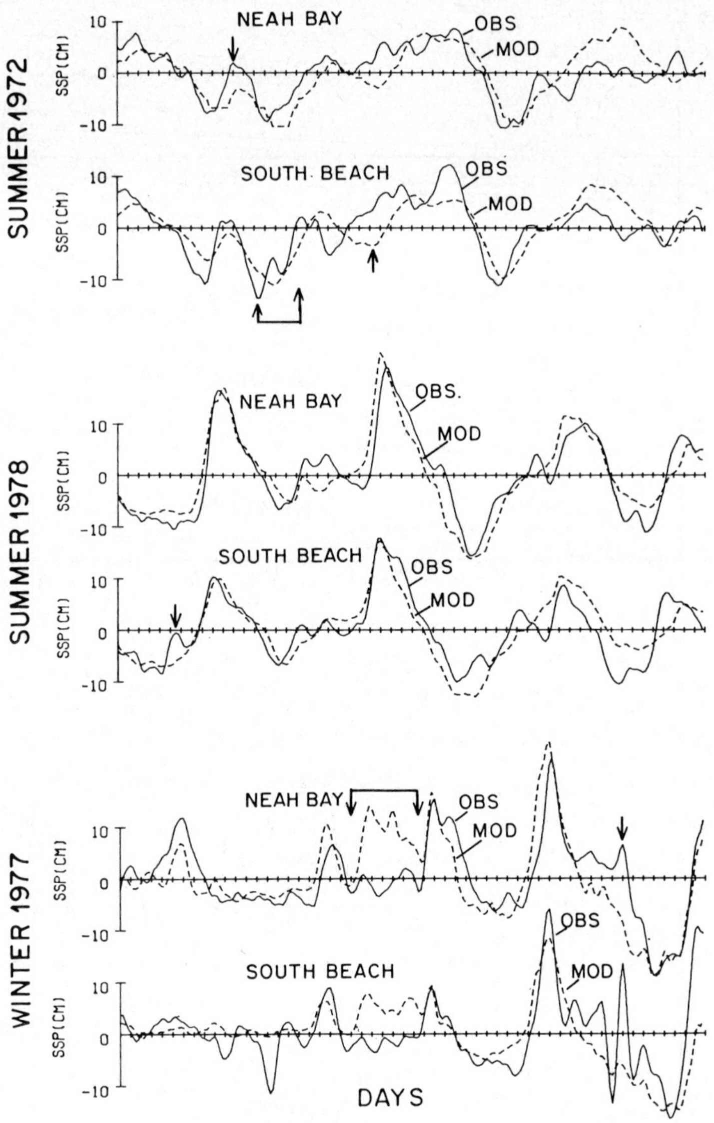

Fig. 2.22 Observed (OBS) and modeled (MOD) subsurface pressure (SSP) for Neah Bay, Washington, and South Beach, Oregon, during 21 July-15 September 1972, 15 August -15 October 1978, and 12 January-5 March 1977. Model results were generated using the dynamics of wind-forced hybrid coastal-trapped waves. Bakun wind stress at three-degree intervals along the coast from Washington to Baja California was used to drive the model. Each tick on the time axis represents one day. Vertical arrows illustrate inaccuracies in model predictions that are likely due to inadequate prescription of the local wind stress. The differences between local and large-scale (Bakun) wind stress are shown in Figure 2.23. From Battisti and Hickey (1984).

shelf scales of 2,000 km, dominates the response because the shelf is relatively narrow, the latitude is relatively high, and the stratification is relatively weak. Model results illustrate that, during summer, when wind magnitude increases equatorward from Washington to northern California, the majority of the mid-shelf current and sea level variance observed off Washington is generated off northern California (Table 2.3). Halliwell and Allen (1984) suggested that

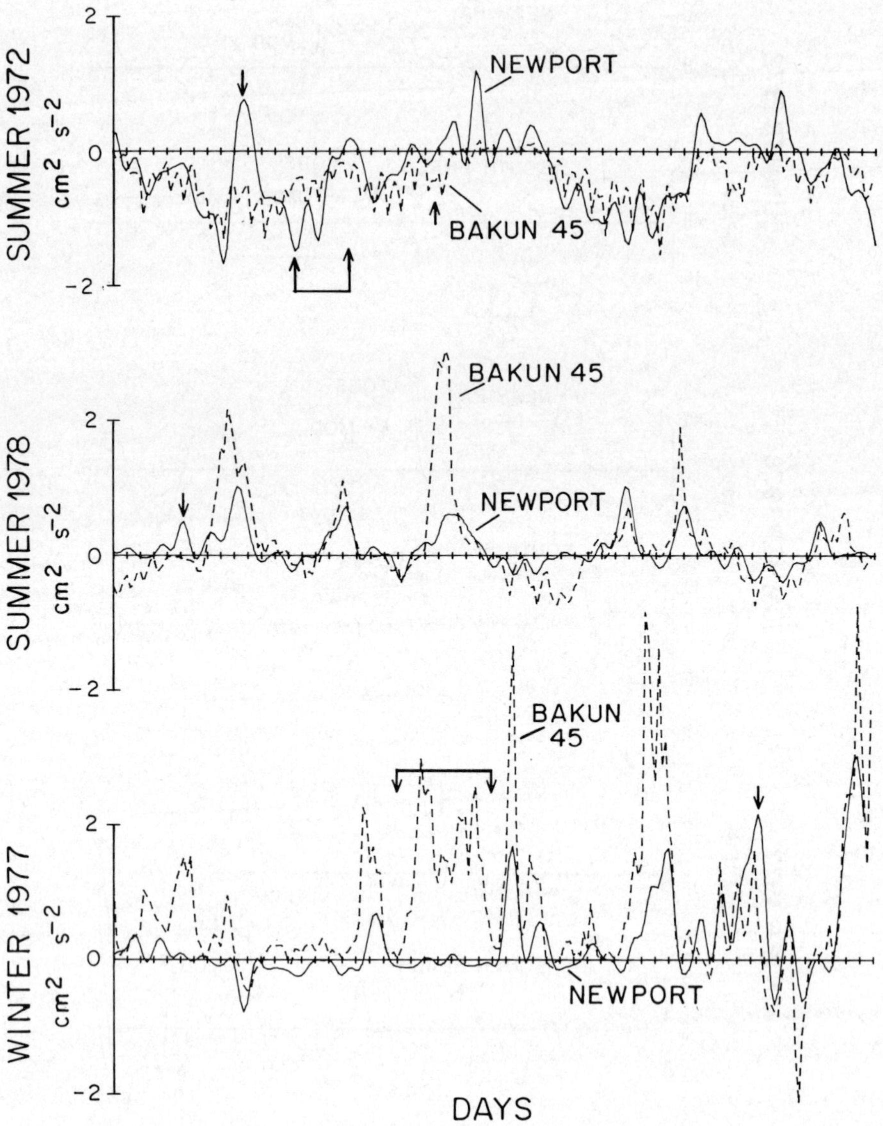

Fig. 2.23 Alongshelf wind stress at a local coastal station (Newport, Ore.) versus Bakun wind stress computed at 45° N in summers 1972 and 1978 and winter 1977. Wind stress at Newport was computed from hourly data using a constant drag coefficient of 1.3×10^{-3}. Each tick on the time axis represents one day. Vertical arrows indicate examples of relatively large differences between the two stress estimates. Corresponding arrows in Figure 2.22 illustrate that these differences appeared to degrade the accuracy of the coastal-trapped wave model. From Battisti and Hickey (1984).

fluctuations generated in a second forcing region, off the coast of northern Baja California, can also contribute to the variance off Washington. During winter, when wind stress magnitude increases poleward, the majority of the variance is generated closer to the Pacific Northwest – within ~600 km. However, the alongshelf scales of the wind field itself are large (> 500 km) so that the observed response in currents and sea level during that period also has large alongshelf scales. Battisti and Hickey (1984) also illustrated that although large-scale wind stress data such as are provided by Bakun upwelling indices are adequate for predictions when the majority of the response has been remotely generated, the accurate prescription of the smaller scale wind field is critical during periods (such as most winters) when local forcing is more important (Fig. 2.23).

Higher mode coastal-trapped waves, with shorter alongshelf wavelengths and more complicated cross-shelf modal structure (i.e., zero crossings over the shelf or slope) may contribute a significant portion of the variance of alongshelf velocity, particularly over the outer shelf and slope where the amplitude of the first mode is relatively smaller. However, unequivocable evidence for higher modes has not, to date, been presented. An energetic second mode wave has been identified off the southern British Columbia coast (Yao *et al.*, 1984). Rather than being wind-induced, it seems likely that these waves may be generated by topographic scattering from the first mode over the rugged canyon and bank region at the exit of the Strait of Juan de Fuca. Higher order coastal-trapped waves off Washington and British Columbia are being investigated in ongoing research.

The fact that mode one, coastal-trapped waves are responsible for a significant portion of the variance of alongshelf currents at mid-shelf locations suggests that the barotropic alongshelf pressure gradient at mid-shelf is responsible for much of the observed variance in the acceleration of alongshelf current. The vertical average of the alongshelf equation of motion for a coastal-trapped wave is

$$\frac{\partial v}{\partial t} = -\frac{\partial P}{\partial y} \tag{2.1}$$

where v is vertically-averaged, alongshelf velocity, t is time, P is vertically-averaged, subsurface pressure and y is distance along the shelf. The vertical integral of the Coriolis term is assumed to be small, following Hickey (1984), and the pressure field is assumed to be depth-independent (i.e., barotropic, as is the case for the mode one wave). In the more general case, surface and bottom stress also contribute significantly to the variance of acceleration, and the equation governing the acceleration of the alongshelf flow can be written as

$$\frac{\partial v}{\partial t} = \frac{\tau_s}{H} - \frac{\partial P}{\partial y} - \frac{\tau_b}{H} \tag{2.2}$$

The relative magnitudes of the individual terms in Equation 2.2 at mid-shelf locations have been computed during two winter and two summer periods, using coastal sea level and atmospheric pressure data to obtain estimates of subsurface pressure (Fig. 2.24). Both surface and bottom stress were computed from hourly data using quadratic stress laws and speed variable drag coefficients. The drag coefficient for surface stress was obtained from Large and Pond (1981)

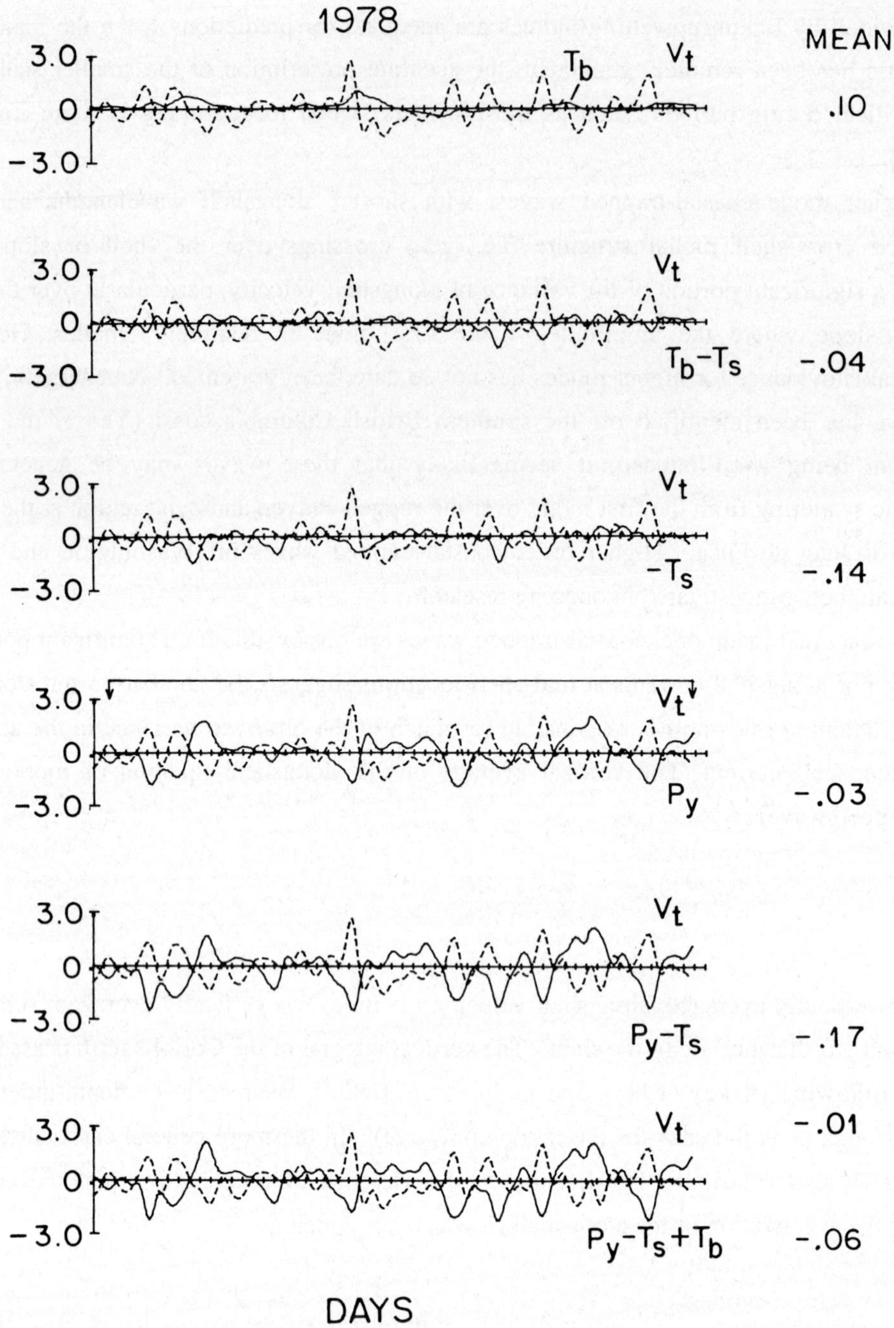

Fig. 2.24 Principal terms in the large-scale (> 300 km) vertically-averaged alongshelf momentum equation as a function of time during summer 1978 at a mid-shelf location in the U.S. Pacific Northwest. V_t, T_s, T_b, and P_y represent $\Delta v/\Delta t$, τ_s/H, τ_b/H, and $\Delta p/\Delta y$, respectively. H is water depth. Units are 10^{-4} cm s^{-2}. Each tick on the time axis represents one day. A mirror image between a pair of plots would indicate a perfect balance of terms. Vertical arrows indicate periods when P_y is particularly significant to the balance. The record mean of each term is given on the right hand side of the figure. From Hickey (1984).

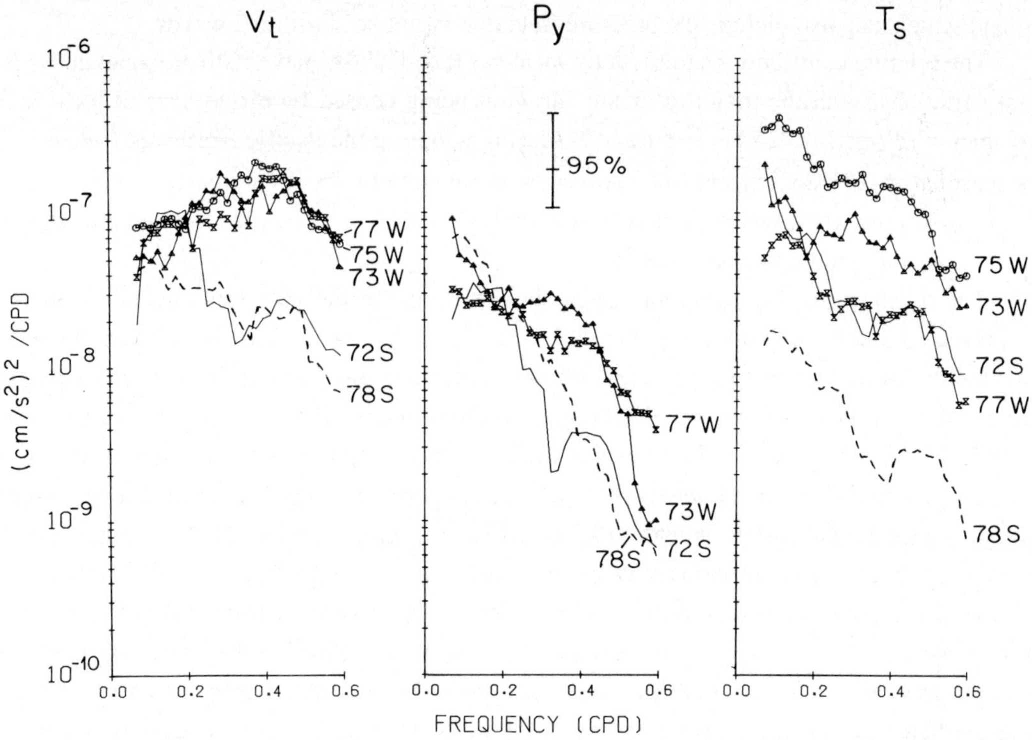

Fig. 2.25 Spectra of V_t (left), P_y (center), and T_s (right) in various years during winter (W) and summer (S). Each record length is 56 d. Spectral estimates have 14 degrees of freedom. Newport wind data were used for all stress computations, and in every case, the wind was multiplied by a factor of 1.3 to simulate the coast-to-ocean gradient in wind speed. Estimates of P_y were multiplied in each case by an offshore decay factor of 0.7. From Hickey (1984).

and modified to include a land-to-ocean gradient in stress. That for bottom stress was obtained from calculations of the sediment dynamics model described in Chapter 8 for bottom stress; this coefficient included effects of near-bottom stratification during sediment resuspension and is tailored for Washington sediment and bottom roughness.

During every two-month period, $\partial P/\partial y$ contributes significantly to the variance, consistent with forcing by coastal-trapped waves. A particular acceleration event may be caused by either τ_s/H or $\partial P/\partial y$ (Fig. 2.24). However, the most accurate balance for $\partial v/\partial t$ over any extended period is provided by the sum of τ_b/H and $\partial P/\partial y$. Bottom stress (τ_s/H) makes relatively small contributions to the variance of acceleration at mid-shelf locations. Bottom stress would be expected to be more important on the shallower inner shelf where the effective magnitude of the pressure gradient term is reduced. These results are important for prediction of currents because estimates of local τ and local $\partial P/\partial y$ are readily available from coastal sea level and atmospheric pressure stations (P is the sum of sea level height plus atmospheric pressure, converted from mb to cm). Moreover, wave model results, although important in determining the dynamics of the coastal currents, may not be as accurate as direct measurements of the pressure field, which are not affected by limitations in prescription of the wind stress field all along the coast and which can also include the pressure field due to non-wind-forced waves.

The relative contributions to $\partial v/\partial t$ by local τ_s/H and $\partial P/\partial y$ have both seasonal and interannual variations, with the majority of the variation being caused by differences in local τ_s/H rather than in $\partial P/\partial y$ (Fig. 2.25). In particular, during winter in the Pacific Northwest, when τ_s/H has a seasonal maximum, current fluctuations are dominated by local wind stress. This was the case during winter 1975, which Hickey and Hamilton (1980) and Hickey (1981) claimed to be governed by local wind-driven dynamics.

The dynamics of the mean and fluctuating currents on the inner continental shelf have not received as much attention in the literature as those at mid-shelf. This is as much due to instrumentation limitations as to a preoccupation by west coast modelers on long wave dynamics. Current and wind data gathered off Washington in the two upwelling experiments during 1981 and 1982 (Fig. 2.11) represent the only spatially comprehensive data sets available in this zone off Washington and Oregon. However, both of these experiments were of less than one month's duration. Other available data include one record each during summer off Washington (1979) and Oregon (1972) and two records during winter off Washington (1977; Fig. 2.6). In the near-shore zone, the balance of momentum is expected to be primarily frictional (Mitchum and Clarke, 1986), rather than wavelike, because the relative contribution to acceleration by the alongshelf pressure gradient term in Equation 2.2 is reduced as bottom depth decreases. The excellent predictability of coastal sea level by the large-scale model of coastal-trapped waves, as previously discussed, argues against significant and/or frequent contributions by eddies on the inner shelf. Examination of the available data on the inner shelf is the subject of ongoing research.

Eddies, meanders, squirts, and jets

Equation 2.2 is most efficient at describing the balance of momentum in mid-shelf regions in the bottom half of the water column. Even at mid-shelf, significant departures from the large-scale momentum balance occur in the near-surface layers. It is likely that these departures are indicators of small-scale eddy activity (Hickey, 1984).

On the outer shelf, the simple wind-forced wave dynamics represented by Equation 2.2 accounts for only 30-50% of the alongshelf current variance at any depth. Over the outer slope, the majority of the fluctuating flow field below the surface Ekman layer is not significantly correlated with local wind stress or with the large-scale alongshelf, barotropic pressure gradient force. The relative failure of large-scale linear dynamics over the outer shelf and slope may be indicative of the presence of meanders and/or eddies.

Data from late summer 1981 illustrate the presence of a series of parallel jets, or, alternatively, eddy-like features in the geostrophic velocity field (Fig. 2.26) both on the shelf and over the slope. Current measurements taken at the same time as the CTD sections are generally consistent with the geostrophic currents. The relatively large (~20 km scale) features over the outer shelf and slope were reproduced in five consecutive sections taken at 2-4 d intervals. The feature adjacent to the coast is consistent with the presence of a relatively freshwater eddy, which was well documented in alongshelf and cross-shelf σ_t sections: water of density < 21 σ_t

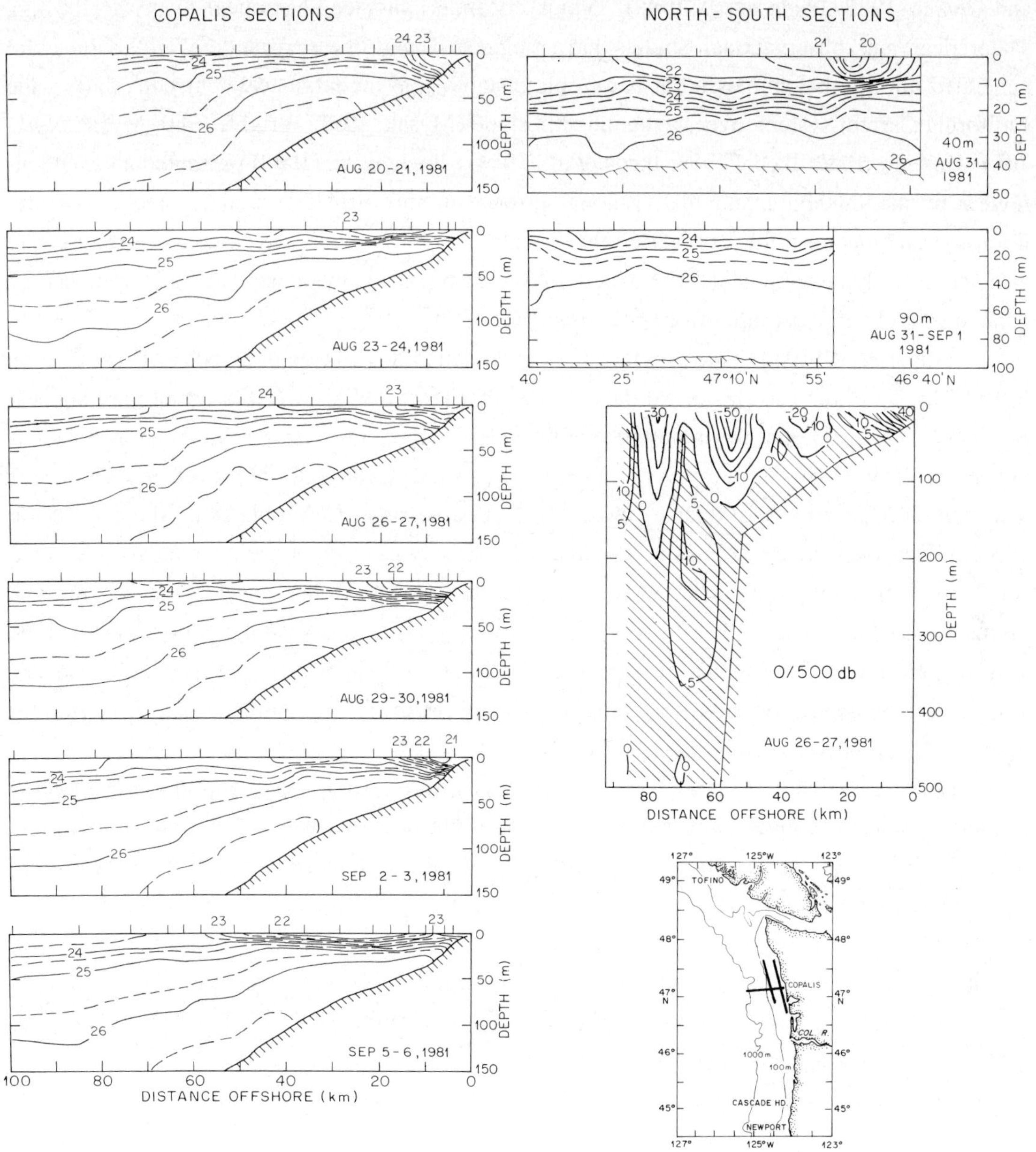

Fig. 2.26 Time series of contoured cross-shelf sections of σ_t off Copalis, Washington, during Aug-Sept 1981. Sections along the mid- and outer shelf during that period and section locations are also shown. Ticks along the top of each section indicate station locations. Geostrophic velocity (0/500 db) on Aug 26-27 is given in the middle panel. Units are cm s^{-1}. Regions of poleward flow are hatched. The data were extrapolated into regions shallower than the 500 db level of no motion according to the method of Montgomery (1941).

units is confined to within ~10 km of the coast (left panels, Fig. 2.26) and to a distance of ~15 km along the coast (upper right hand panel, Fig. 2.26).

Numerous meanders and eddies have been identified and discussed for the region off Vancouver Island (e.g., Emery and Mysak, 1980; Thomson, 1984; Ikeda *et al.*, 1984; Thomson and Gower, 1985; Ikeda *et al.*, 1985). Typical meander wavelengths appear to be 80-200 km. Theoretical and observational studies have suggested that these meanders and eddies are generated through baroclinic and/or barotropic instability of the large-scale currents, that is, the California Current system, over the continental slope (Mysak, 1977; Wright, 1980; Mysak *et al.*, 1981; Hukudu, 1982; Ikeda, 1983; Ikeda *et al.*, 1984). Ikeda *et al.* (1984) presented an excellent review of the modeling and observational efforts on this problem to date. More recently, Thomson and Gower (1985) suggested that instabilities in the wind-driven coastal currents can also lead to eddy development. The data in Figure 2.26 provide evidence that eddies can also be generated via freshwater input from river mouths.

Huyer *et al.* (1984) have provided the only detailed description of an eddy off the Oregon coast. These authors used historical data to infer that large eddies (~100-km scale) are unusual off Oregon. At this point, the importance of meanders and eddies over the Washington slope has not been fully assessed. However, since the seasonal mean currents over the slope off Washington appear to be similar to those off southern British Columbia (e.g., there is strong vertical shear over the slope in summer), and since, in any case, eddies and meanders generated off British Columbia during summer may be advected equatorward to the Washington coast, it seems reasonable to expect that the mesoscale eddy field over the Washington continental slope will be significant. The impact may be less off Oregon, as suggested by Huyer *et al.* (1984), since the vertical shear over the slope appears to be weaker off Oregon than off Washington (at least during summer and fall).

Since the advent of satellite sea surface temperature data, much attention has been given to the fact that, for a given wind stress event, upwelling of cold water is not uniform along the coast (Kelly, 1985). Rather, stronger upwelling occurs at and downstream of capes and promontories. Such upwelling 'centers' appear to generate narrow (~40 km) offshore surface (0-200 m) jets which carry filaments of coastal water several hundred kilometers offshore (Brink and Hartwig, 1985). The nature of these jets and their effect on organisms and the mixing of water masses is being examined in a series of experiments off northern California during 1987 and 1988. Since the Washington-Oregon coast has few major promontories (Fig. 2.30), such jets are expected to be of lesser importance in the U.S. Pacific Northwest than off California.

Coastal upwelling

In the simplest conceptual model of two-dimensional coastal upwelling, vertical movement of water occurs next to vertical topographic boundaries such as the shelf edge and the coast, or along a sloping bottom in response to alongshore wind stress or to an alongshore pressure gradient force (or sea level slope) (Hamilton and Rattray, 1978; Werner and Hickey, 1983; Werner, 1984). The horizontal distance over which vertical motion is significant is on the order of the internal Rossby radius of deformation, typically 10-20 km over the Washington shelf. In the case of equatorward wind stress, cold, nutrient rich water moves up the continental slope, over the sloping shelf, and up to the surface in the near-shore zone. Once upwelled water has reached the surface layer it is advected offshore in the surface Ekman layer. For two-

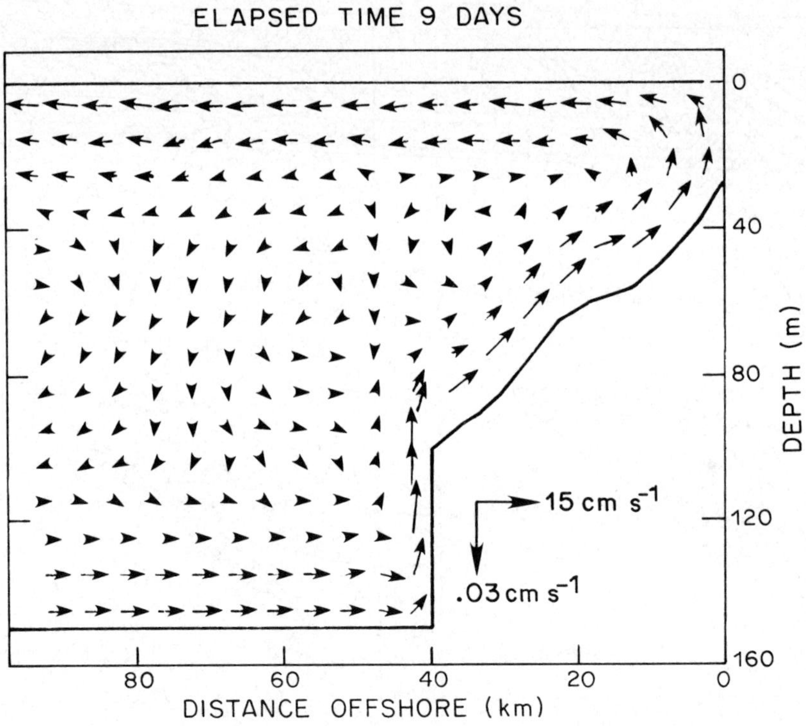

Fig. 2.27 Cross-shelf velocity vectors after nine days of alongshelf wind stress, $\tau_y = 1.5$ dynes cm^{-2}, as predicted by the time-dependent, baroclinic, two-dimensional model of Hamilton and Rattray (1978). The results illustrate a classical single-cell upwelling circulation pattern.

dimensional (cross-shelf) mass balance, the offshore mass flux in the surface layer is balanced by onshore mass flux in the remainder of the water column, either in the interior or in a frictional bottom boundary layer. If the water is stratified, as is generally the case in the Pacific Northwest, isopycnals move vertically with the upwelling water and both horizontal and vertical mixing become important. The pycnocline tilts and gradually moves offshore. Inshore of the inclined pycnocline the water column may be nearly homogeneous. Secondary circulation patterns are expected to develop in the vicinity of the inclined pycnocline (Mooers *et al.*, 1976; Hamilton and Rattray, 1978). The cross-shelf circulation pattern is further complicated by the seasonal development of the poleward undercurrent over the slope and much of the shelf: at locations where the alongshelf flow is poleward, flow in the bottom boundary layer is offshore, rather than onshore. A classical example of modeled cross-shelf upwelling circulation for a two-dimensional stratified system with bottom friction and realistic bottom topography is displayed in Figure 2.27. Note that the flow along the bottom of the figure at 140 m is an artifact of the false bottom in the model at that depth. Onshore flow at that depth has not been demonstrated with field observations. The direction of the arrows would be reversed for a downwelling situation, such as occurs in response to poleward coastal wind stress.

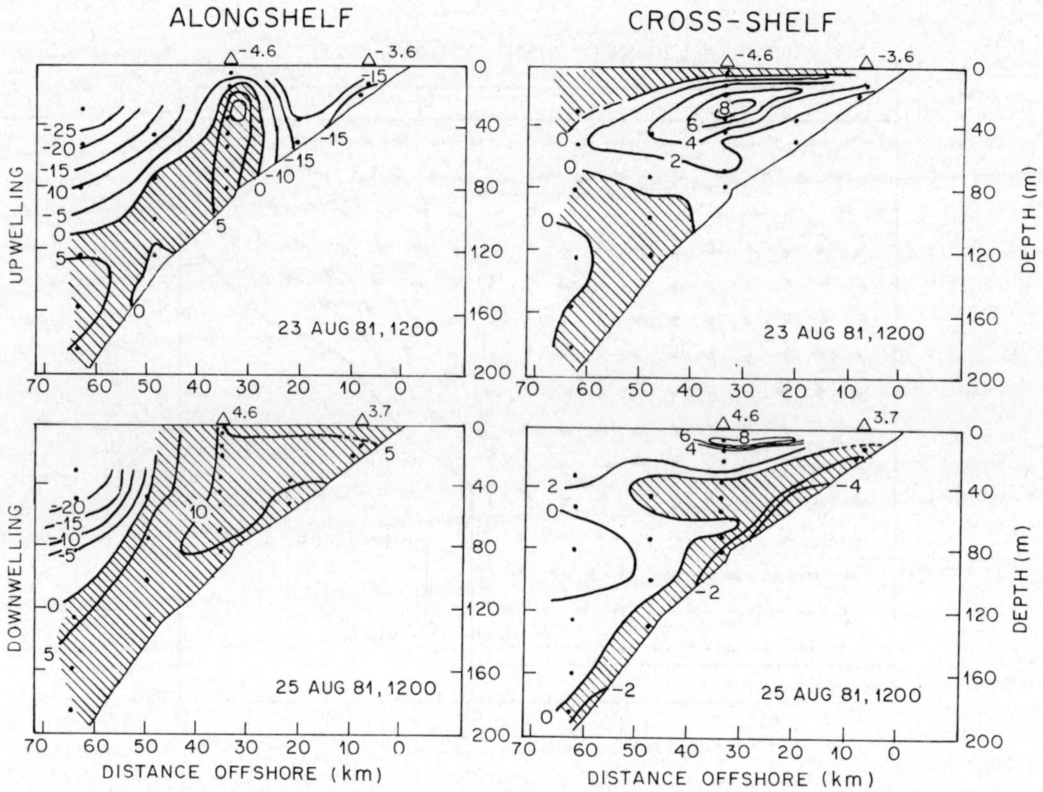

Fig. 2.28 Contoured cross-sections of alongshelf (v) and cross-shelf velocity (u) during an upwelling event (upper panel) and a downwelling event (lower panel) in late summer 1981. Hatched areas indicate regions of poleward (v) or offshore (u) flow. Dots indicate locations of current meters. Numbers at the top are the north-south component of wind measured on surface buoys. Negative values of wind denote wind from the north. Units are cm s^{-1} and m s^{-1} for current and wind, respectively.

Two coastal upwelling experiments, specifically designed to resolve the cross-shelf circulation over the Washington shelf, were conducted during 1981 and 1982. These experiments incorporated *in situ* wind measurements and also current measurements in the surface layer and on the shallow inner shelf. The experiments are described more fully by Hermann *et al.*, 1989, Chapter 6), where the data are used to calculate advective contributions to observed biological and chemical property distributions. The first experiment took place during seasonally intermittent upwelling (Aug-Sept), the second during seasonally persistent upwelling (June). Examples of observed alongshelf and cross-shelf circulation patterns during upwelling (equatorward wind stress) and downwelling (poleward wind stress) events during the intermittent upwelling period are given in Figure 2.28. In the near-shore region, equatorward flow occurs during the upwelling event and poleward flow occurs during the downwelling event just two days later. Alongshelf flow at mid-shelf reverses only in a shallow near-surface layer (≤ 15 m). Alongshelf flow over the outer shelf appears to be relatively unaffected by the wind and current reversals. The mean cross-shelf circulation patterns reverse, as expected, between

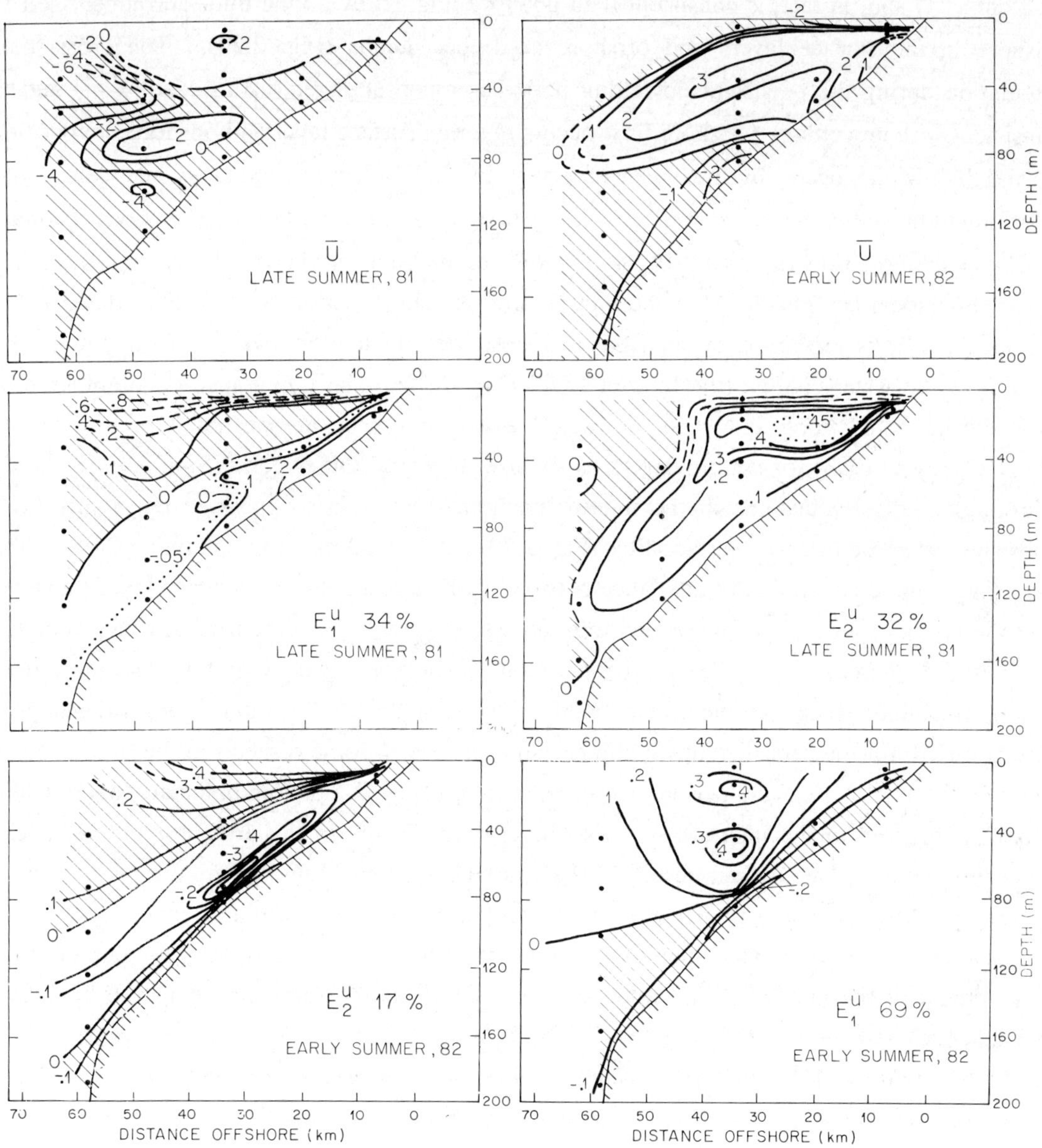

Fig. 2.29 Cross-shelf structure of the mean and the first and second EOFs of cross-shelf velocity during early summer 1982 and late summer 1981. The percent of variance accounted for by each EOF is given on each panel. Dots indicate location of current meters. For comparison, the first EOF of alongshelf velocity is given in Figure 2.21.

the upwelling and downwelling wind events. Offshore flow in the upwelling event is confined to a very shallow (< 10 m) layer. Also, a significant portion of the return onshore flow occurs not in the bottom boundary layer, but just beneath the pycnocline. During the downwelling event, onshore flow occurs in a thicker layer at mid-shelf (~40 m), but the strongest flow occurs in a thin surface layer (< 10 m). Offshore return flow is strongest in the bottom boundary layer.

The mean cross-shelf circulation during the intermittent upwelling period is relatively weak (< 2 cm s $^{-1}$) and, in fact, is consistent with downwelling: flow on the mid- and inner shelf is onshore in the surface layers and offshore at deeper depths (Fig. 2.29). The cross-shelf circulation during the persistent upwelling period is much stronger (up to 4 cm s $^{-1}$) and is consistent with upwelling. Offshore flow occurs in a thin surface layer and onshore flow occurs in and below the pycnocline, as was observed for individual upwelling events during the intermittent upwelling period (compare with upper right hand panel in Fig. 2.28). Over the mid- and outer shelf the flow reverses to offshore in the bottom boundary layer.

Statistical computations for the intermittent upwelling period confirm that much of the variance (34%) is explained by an EOF that describes fluctuations with a maximum in the interior of the water column (the Interior EOF) (Fig. 2.29). The importance of interior return flow during both strong and weak upwelling events has been demonstrated previously off Oregon (Huyer, 1983) for bottom depths as shallow as 50 m. Off Oregon (Smith, 1981) as off Washington, the fluctuations in the interior onshore flow do not appear to be significantly correlated with the offshore surface flux (Fig. 2.29). A second pattern, containing 34% of the variance, is more suggestive of a balance between surface and bottom boundary layer transport (the Boundary Layer EOF). The second EOF for the persistent upwelling period, which contains 17% of the variance, is similar in shape to the Boundary Layer EOF for the intermittent upwelling period, except that the bottom boundary layer at mid-shelf is more strongly developed. The first EOF for the persistent upwelling period is similar in some respects to the Interior EOF for the intermittent upwelling period. However, a return flow in a bottom boundary layer, which was not observed in the intermittent upwelling period, occurs in this EOF for the persistent upwelling period. The overall result is that the mean circulation pattern during persistent upwelling corresponds to a fluctuating circulation pattern during intermittent upwelling. Also, the fluctuations during both persistent and intermittent upwelling periods demonstrate involvement of the interior of the water column as well as the surface and bottom boundary layers in each case.

The cross-shelf fluctuations during both intermittent and persistent upwelling are weakly correlated with alongshelf local wind stress for the Boundary Layer EOF, consistent with the surface layer intensification of the EOF. This was not the case for the Interior EOFs, which could be the result of remotely-forced coastal-trapped waves. The fact that the Interior EOF accounts for more variance in early summer is consistent with the fact that coastal-trapped wave activity is expected to be a maximum at this time of year (June), when wind stress off California is a seasonal maximum and local wind stress is at a seasonal minimum. The time variation of the EOFs and relation to *in situ* wind is illustrated and discussed in greater detail in Hermann *et al.* (1989, Chapter 6), where advection is related to observed biological and chemical changes.

The cross-shelf circulation patterns, both in the observations and the EOF analysis, are surprisingly simple, considering the variability in the cross-shelf density field (e.g., see Fig. 2.26). In particular, the EOFs do not provide support for the existence of a two-celled circulation pattern for either the mean or the fluctuating flow field (Fig. 2.29). Evidence for a two-celled pattern, originally suggested by Mooers *et al.* (1976), has been provided by profiling current meter data (Johnson *et al.*, 1976; Johnson and Johnson, 1979) as well as by distributions of optical parameters (Pak *et al.*, 1970), phytoplankton (Peterson *et al.*, 1979; Small and Menzies, 1981), and zooplankton (Peterson *et al.*, 1979). The two-celled circulation pattern involves onshore flow in the pycnocline at mid-shelf and offshore flow beneath the pycnocline. Critiquing the profiling current meter data, Huyer (1983) concluded that the evidence for a double cell is weak. Huyer (1983) illustrated that fixed-level, moored current meter data off Oregon (as off Washington) support a single celled upwelling response during both weak and strong equatorward wind events (see her Fig. 13).

The simple upwelling picture conceptualized in Figure 2.27 is predicated on the existence of mass balance in the x-z plane. Such a balance has not, to date, been demonstrated for either the seasonal mean or the fluctuating flow field (e.g., see Figs. 2.28 and 2.29). The failure to demonstrate this balance may be due in part to errors involved in accurately calculating the cross-isobath component of velocity. In general, the fluctuating flow follows local isobaths, and since alongshelf velocities (v) are much larger than cross-shelf velocities (u), rotation into a local isobath reference frame can easily introduce significant errors in u. Alternatively, the flow may not be quasi-two-dimensional. Significant alongshore changes in the seasonal mean alongshelf velocity field have been observed (see preceding sections). For example, the fact that the alongshelf flow at mid-shelf is more poleward off Washington than off Oregon gives rise to a divergence in the alongshelf flow. Alongshore divergence was of the correct order during spring 1973 to account for the observed excess onshore mass flux that was observed (Huyer, 1983). For the fluctuating flow field, the documented existence of propogating waves, which are inherently three-dimensional, implies that the longshore dimension is important to the mass balance. In particular, coastal-trapped wave models predict no significant coastal upwelling during the passage of a wave; i.e., all of the mass balance is accomplished in the x-y plane.

2.3.4 Interannual Variability

The significant interannual variability of currents over the Washington shelf is dramatically illustrated in Figure 2.30, which depicts the subtidal mid-water column alongshelf velocity at mid-shelf locations during several winter-spring seasons. For example, currents were more continuously poleward prior to the spring transition during 1973 than during 1975. The magnitude of the equatorward currents after the transition in 1975 were roughly twice as large as those in the other years presented. The timing of the spring transition occurs within 1 to 2 wk of March 25 for three of the four years. In the fourth year (1978), the transition occurred in late April (not shown).

The discussion of subtidal current variability illustrates that the distribution of the alongshore component of wind stress at the coast all along the west coast is responsible for much of the variability in currents over the Washington/Oregon shelf. Coastal wind stress is determined primarily by the structure of atmospheric pressure patterns along the coast, and these have significant differences from year to year. For example, during some summers a high pressure system remains offshore for extended periods and coastal winds are equatorward.

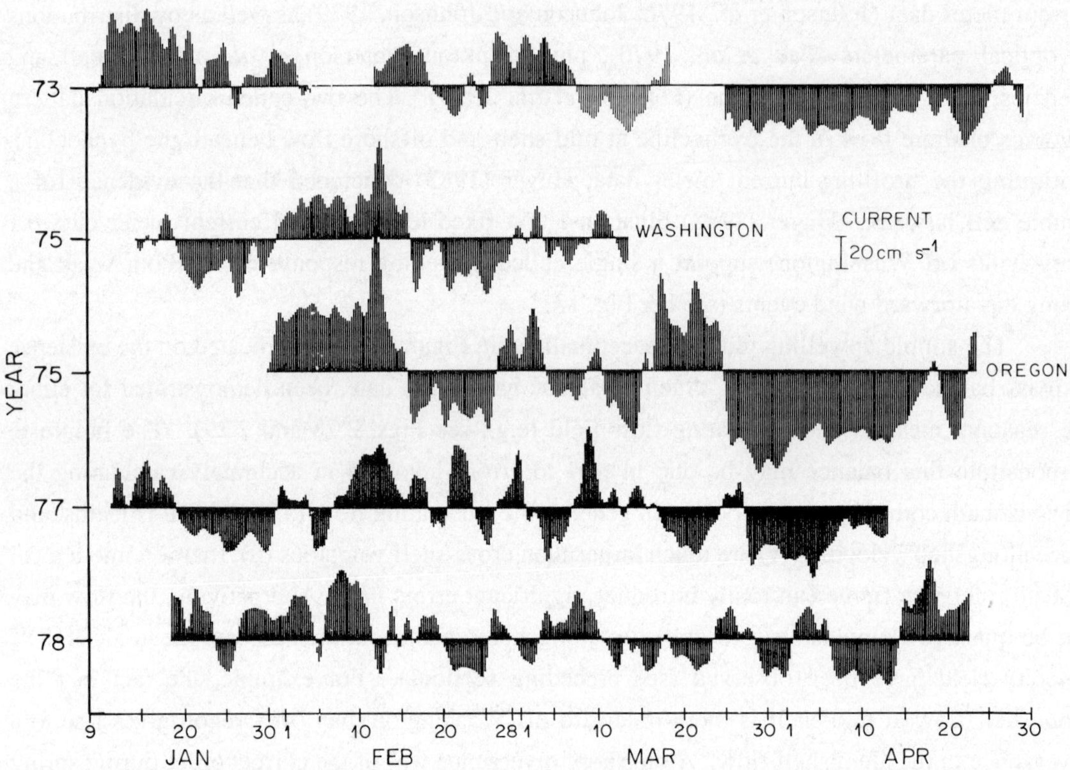

Fig. 2.30 Time series of alongshelf velocity at mid-water column depths at mid-shelf locations off Washington during several late winter-early spring seasons and off Oregon during winter-spring 1975. Meters were located at a depth of 50 m ± 3 m from the surface except during 1977, when the meter was located 25 m from the surface. The data illustrate interannual variability of the spring transition and of the structure of events preceding it.

During other summers the high is displaced by traveling low pressure systems (storms), and coastal winds alternate between being equatorward (during the high pressure period) and poleward (during the low pressure period). Interannual variability in coastal wind stress within any given season can be larger than season-to-season variability (Fig. 2.25).

The particular type of wind-driven dynamics governing the coastal system is intimately dependent upon the relative magnitudes of local wind stress and remote wind stress. Therefore, the governing dynamics vary significantly from year to year. For example, during winter 1975 when local wind stress off the Pacific Northwest coast was very strong, the dynamics were dominated by local winds; that is, alongshelf acceleration of currents would be expected to be balanced primarily by surface stress and bottom stress. During winter 1977, when local wind stress was relatively weak, remote wind forcing was more important and propagating waves dominated the dynamics. In this case, the alongshelf pressure gradient was responsible for much of the observed current acceleration. In addition to interannual variability in the magnitude of local wind stress in the Pacific Northwest, the differences in atmospheric pressure patterns cause significant differences in the along coast coherence scales of the large-scale wind field (Denbo and Allen, 1986). Thus, the principal generation regions for propagating waves (the remotely forced contribution to the variance) may vary from year to year.

In addition to interannual variability of event-scale fluctuations and their dynamics, boundary currents and coastal sea level off the west coast experience interannual variations. In particular, Chelton (1980) demonstrated that the California Current system and coastal sea level off California exhibit interannual variations related to tropical variations of sea surface temperature in the Eastern Tropical Pacific. During El Niño years, large-scale coastal currents are more poleward and coastal sea level is higher than during average years. Enfield and Allen (1980) and Chelton and Davis (1982) demonstrated that the interannual fluctuations in coastal sea level are correlated over much of the west coast. The authors suggested that equatorial fluctuations propagate poleward from the equator as coastal Kelvin waves.

Huyer and Smith (1984) examined the signature of the recent 1982 El Niño in detail for the Oregon-Washington region. The El Niño response off Oregon included anomalously high sea level, warmer coastal water temperature, and increased poleward flow. The authors used CTD data to document the presence of water of southern origin off Oregon, to distances of at least 300 km from the coast, and to water depths of 300 m. Water properties remained anomalous for at least nine months within 200 km of the coast, and even longer at greater distances from the coast. Huyer and Smith (1984) used temperature, sea level, and current data measured on the Oregon shelf, as well as basinwide sea level pressure information and sea level data off Peru, to deduce that the El Niño signal was initially transmitted to Oregon via the water as a coastal Kelvin wave. However, their data demonstrated that atmospheric perturbations in the form of an increase in magnitude and frequency of southerly winds near the coast reinforced the El Niño signal but lagged the initial changes by two to three months. In support of this latter result, data presented in Landry *et al.* (1989, Chapter 1) suggest that interannual fluctuations in water properties off Washington and Oregon are significantly related to local wind stress fluctuations.

2.3.5 High Frequency (>1 cpd) Variability

Tidal variability in the Pacific Northwest has been examined in detail only for one period, July-August 1973 (Hayes and Halpern, 1976; Torgrimson and Hickey, 1979). The data was obtained over the Oregon shelf and slope during the CUEA upwelling study. Prior to these analyses, the prevailing sentiment among observationalists, who had noted significant temporal and spatial variation in the tidal constituents, was that it was unlikely that tidal variability would ever be related to available models of tidal generation such as Mooers (1970) or Prinsenberg and Rattray (1975). The encouraging result from the summer 1973 analyses is that the diurnal and semidiurnal tidal oscillations do appear to be organized in persistent spatial patterns that are generally consistent with available models.

Torgrimson and Hickey (1979) estimated the barotropic vertical velocity from tidal elevations. Barotropic horizontal velocity was estimated using data least contaminated by the baroclinic tide (i.e., that outside the internal wave beam) which was readily identified via signal-to-noise ratios and momentum balances. The resulting baroclinic components from their

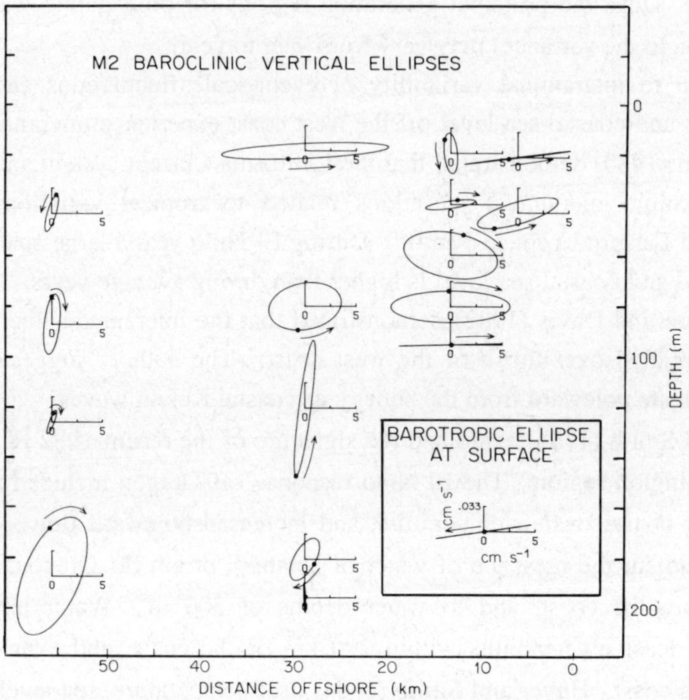

Fig. 2.31 Baroclinic and barotropic vertical current ellipses for the M_2 tide as a function of distance offshore. Locations of current meters relative to the bottom topography are shown in Figure 2.33. Solid dots indicate M_2 tidal elevation maximum at Tofino, B.C. Adapted from Torgrimson and Hickey (1979).

analysis (observed minus barotropic component) are presented in Figures 2.31 (vertical) and 2.32 (horizontal). From this analysis, it appears that the velocity fluctuations associated with the diurnal tide off Oregon are oriented in a northeast-southwest direction and are essentially barotropic (constant with depth). The amplitude of the diurnal tide (1.2 cm s^{-1}) is much smaller than that of the semidiurnal tide (~5 cm s^{-1}) in most locations. The semidiurnal tide is highly baroclinic, and baroclinic and barotropic currents are of the same order of magnitude (Figs. 2.31 and 2.32). Although the barotropic semidiurnal tidal ellipse is oriented cross-shelf, measured velocity ellipses are oriented alongshelf in regions of strong internal tide activity. The distribution of amplitudes and phases as well as momentum balances and signal-to-noise ratios are consistent with generation of the semidiurnal internal tide along the primary continental slope off Oregon (~500-1,000 m). The tidal fluctuations are focused in a wide beam (~80 m) that reflects from the sea surface and sea sea floor as it travels over the shelf, a distance of at least 50 km from the generation area (Figs. 2.31 and 2.32). The exact location of the beam varies only slightly from day to day as a function of stratification, excepth on the inner shelf (Fig. 2.33).

The internal tide is generated through the interaction of the barotropic tide with bottom topography in the presence of stratification. The barotropic tide is relatively uniform in the Pacific Northwest (Battisti and Clarke, 1982, and although bottom topography and stratification do vary along the coast, the alongshore differences are relatively minor (e.g., the shelf is

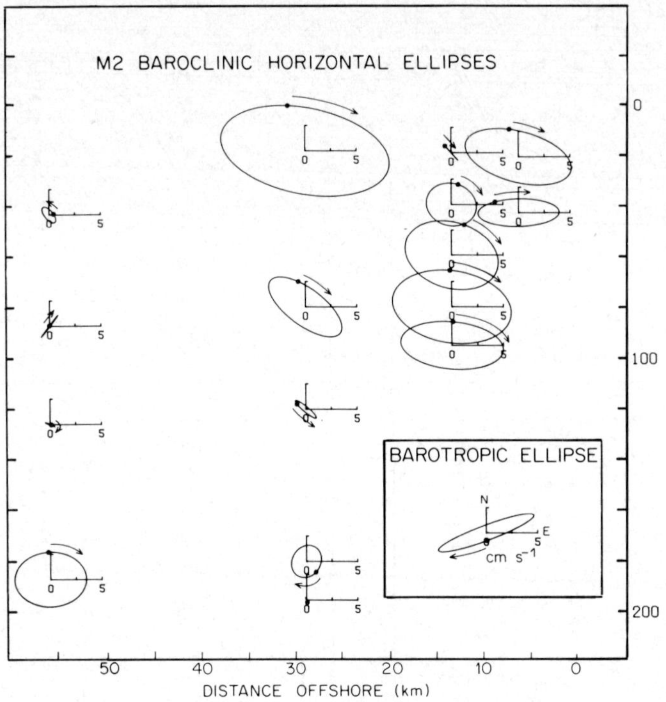

Fig. 2.32 As in Figure 2.31, for horizontal current ellipses.

one-third wider to the north). Therefore, the essence of the tidal response off the Washington coast is expected to be similar to that off Oregon. The details of the response (e.g., generation region and paths of characteristics) depend intimately upon the exact slope of the bottom and the exact stratification and will differ significantly along the coast. In addition, the Washington shelf and slope is indented with numerous submarine canyons. These topographic irregularities cause bottom intensification of the internal tide and also intensification relative to open slope locations (see Section 2.4). Thus, the structure of the internal tide over the Washington slope and shelf is expected to vary significantly in the alongshore direction.

Inertial period current fluctuations have been investigated off the Oregon coast during summer 1972 (Kindle, 1974), summer 1973 (Kundu, 1976), and winter 1977-78 (Anderson *et al.*, 1983). Results should be generally applicable to other locations in the U.S. Pacific Northwest. Oscillations at near-inertial frequencies are ubiquitous over the shelf and slope, occurring intermittently at all locations examined from the surface to 2,000 m. The disturbances tend to persist for several oscillations in time (e.g., eight days over the slope, seven days over the shelf for the winter data set; Anderson *et al.*, 1983) and several wavelengths in space (approximately three wavelengths in the winter data set; Anderson *et al.*, 1983). Horizontal coherence scales appear to be on the order of 50-100 km during winter (Anderson *et al.*, 1983) and 10-20 km during summer (Kindle, 1974). Coherence scales and wavelengths are highly time-variable.

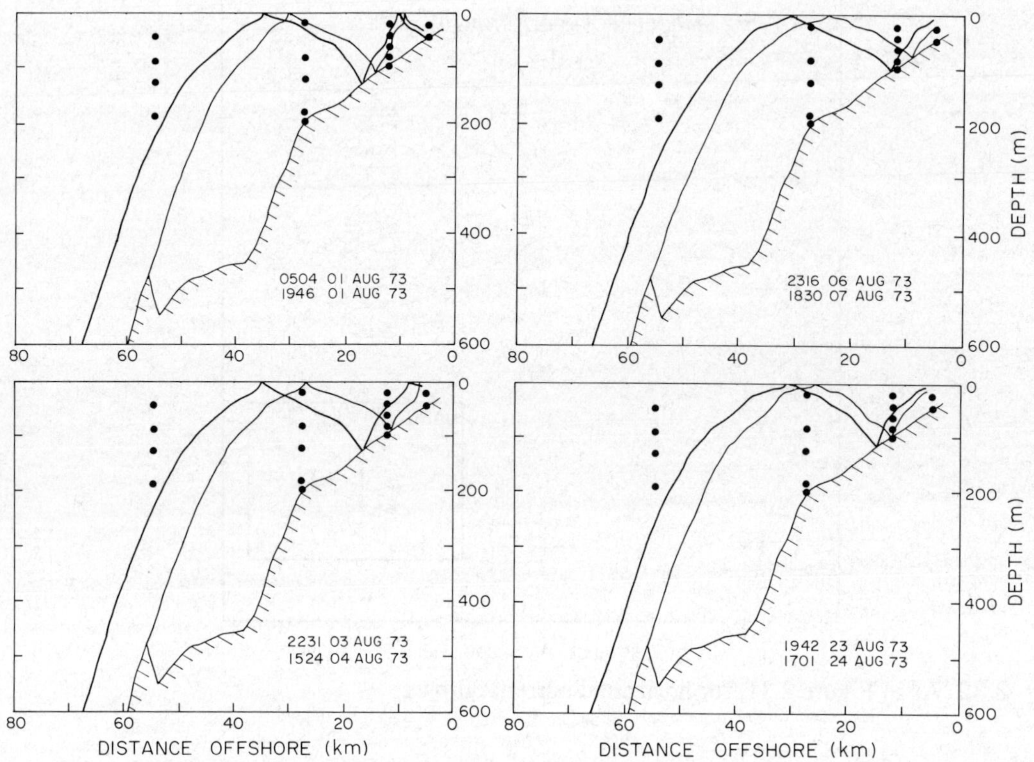

Fig. 2.33 Characteristics emanating from the principal continental slope for the M_2 frequency, August 1973. Locations of current meter data used in Figures 2.31 and 2.32 are indicated by dots. The data illustrate the relative stability of the characteristics over time over most of the shelf.

Velocity amplitudes are on the order of 10-40 cm s $^{-1}$ above 200-300 m from the surface and 5 cm s $^{-1}$ in deeper depths. Thus, in the upper water column, inertial currents can exceed currents due to the internal tide and even subtidal and seasonal mean currents. The amplitude of the oscillations is higher at a given depth from the surface over the slope than over the shelf. Maximum amplitudes observed over the shelf are on the order of 15 cm s $^{-1}$, one-half to one-third of the maximum amplitude over the slope. The slope values are similar to values observed in the open ocean (30-50 cm s $^{-1}$; Pollard, 1980). The oscillations are circular over the slope and slightly elliptical over the shelf (Anderson *et al.*, 1983).

2.4 EFFECTS OF SUBMARINE CANYONS

Submarine canyons indent the shelf edge at ~20-40 km intervals along the Washington coast (Fig. 2.34). Canyons are typically 5-20 km wide and > 1,000 m deep. One canyon (Astoria) traverses one-half the shelf; another (Juan de Fuca) traverses the entire shelf with a 200 m deep channel, allowing a direct connection between off-shelf waters and inland waters in the Juan de Fuca Strait. This contrasts with shelf topography to the south off Oregon, where

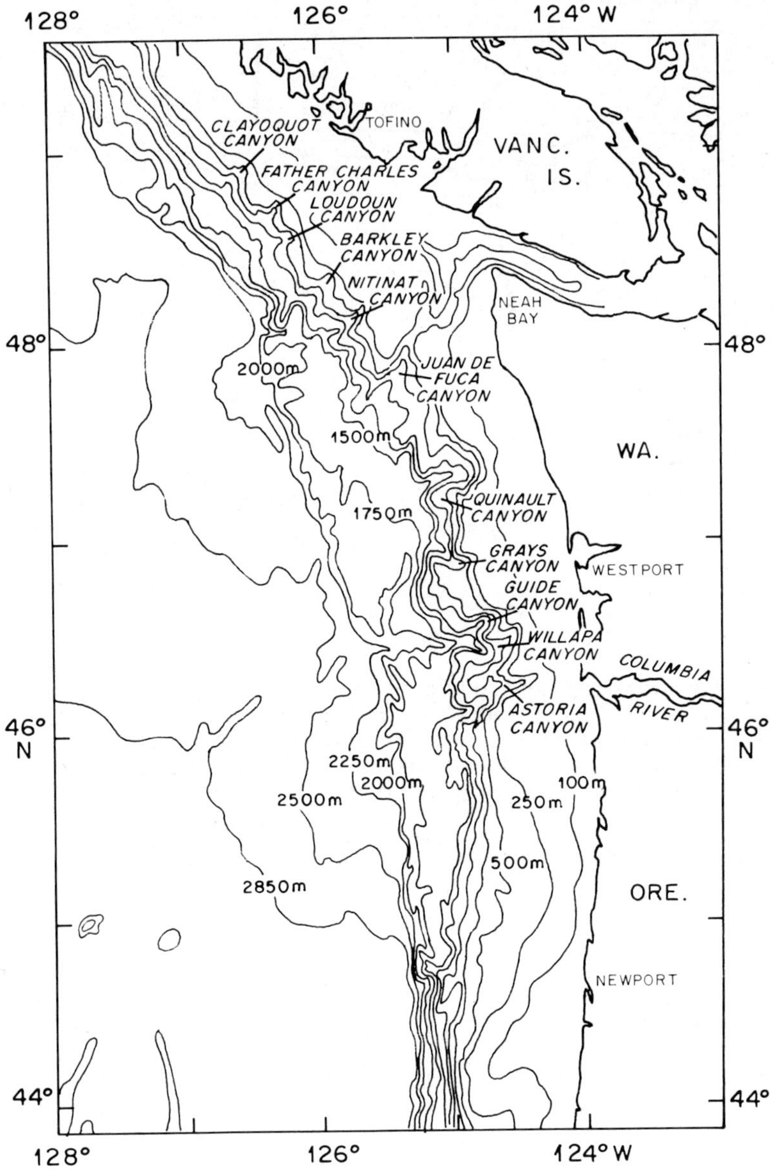

Fig. 2.34 Bathymetric map of the Washington-Oregon shelf and slope, illustrating the number and spatial scales of submarine canyons.

submarine canyons are relatively infrequent and make only small indentations into the continental shelf (Fig. 2.34).

2.4.1 Impact on Regional Circulation

The zero order effect of a submarine canyon is that the flow on the shelf, at least near the bottom, attempts to follow isobaths around the canyon. Poleward flow at shelf edge water column depths in the vicinity of the shelf edge around the ~30-km wide Quinault canyon is directed northeastward on the south side of the canyon and northwestward at the head of the canyon (Fig. 2.35). Geostrophic velocity data illustrate that these directions are reversed during periods of equatorward flow at the shelf edge (see Fig. 2.38, discussed below). Mean flow over the walls of the narrower (~5-km wide) Astoria canyon just above the lip of the canyon (as well as within the canyon) appears to circulate around the canyon in a counterclockwise eddy (Fig. 2.36). There is evidence of a similar counterclockwise, semi-permanent eddy at shelf edge depths (i.e., below the surface layer) along the walls of Juan de Fuca canyon (Fig. 2.37). Higher in the water column, the flow is less affected by the canyon topography. Indeed, flow over Astoria canyon at a depth of 60 m from the surface appears to be decoupled completely from the canyon topography (Fig. 2.36).

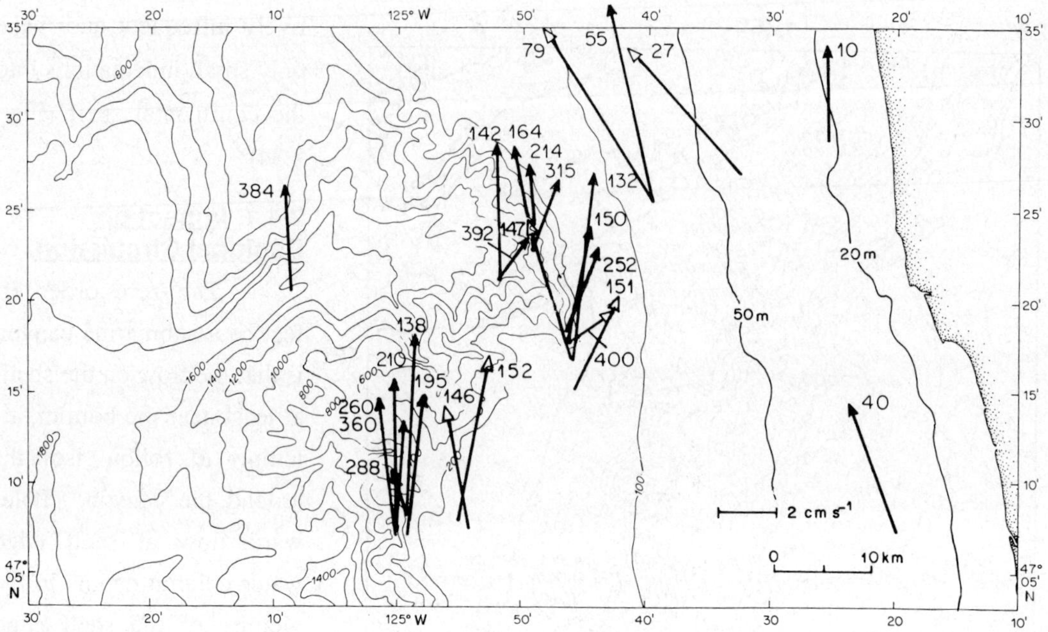

Fig. 2.35 Mean velocity vectors for northward flow conditions in the upper water column in the vicinity of Quinault submarine canyon. Depth from the surface in meters is given beside each vector. Open-ended arrows indicate that the current meter was roughly five meters above bottom (i.e., in the bottom boundary layer) and would be expected to be directed to the left of the local isobath direction. Adapted from Hickey *et al.* (1986).

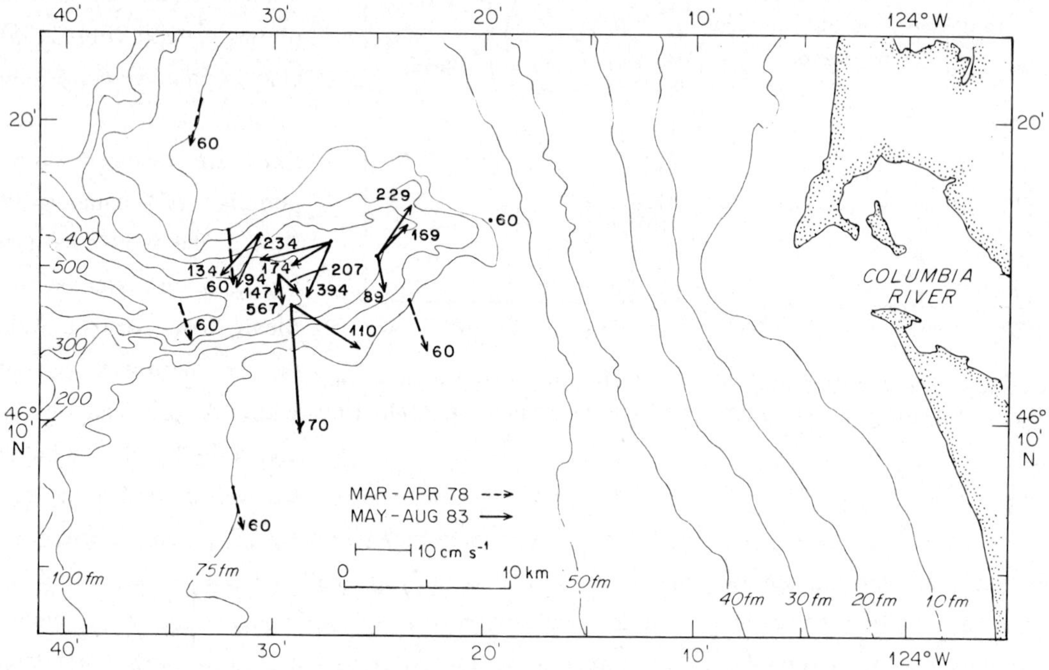

Fig. 2.36 Mean velocity vectors in the vicinity of Astoria submarine canyon during two periods of predominantly equatorward near-surface flow over the shelf. Depth from the surface in meters is given near the tip of each vector. The figure demonstrates the existence of a counterclockwise subsurface eddy in the vicinity of the canyon.

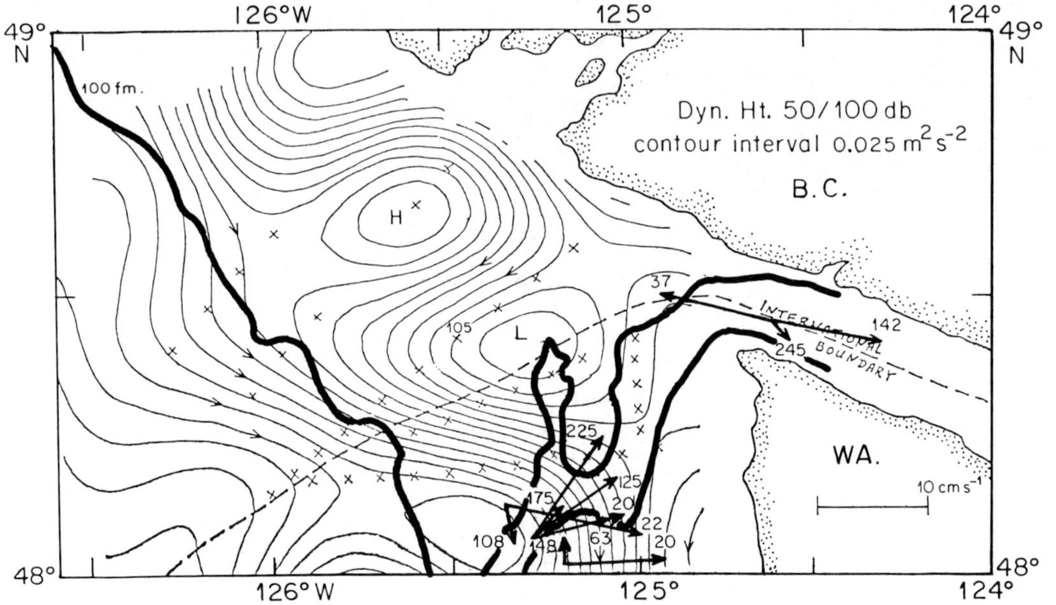

Fig. 2.37 Dynamic height (50/100 db) during September 1980, computed using objective analysis (as in Freeland and Denman, 1982). Stations are indicated by x's. The 100 f = 180 m depth contour is indicated with a heavy contour line. Mean velocity vectors observed during October-November 1971 by Cannon *et al.* (1972) are superimposed. Depth from the surface in meters is indicated at the tip of each arrow.

These results are consistent with scaling arguments, which suggest that if the flow is both geostrophic and barotropic, it is constrained by conservation of vorticity to follow the local isobaths. However, if the flow is geostrophic but baroclinic (density- dependent), stratification masks the bottom topography, so that the effects of the bottom topography are confined to a height H above the bottom. H is on the order of fL/N, where f is the Coriolis parameter, L is the scale width of the isobath perturbation, and N is Väisälä-Brunt frequency. For Astoria canyon, H is on the order of 50 m, consistent with the experimental results (Fig. 2.36). For Juan de Fuca canyon, H is about 140 m. For Quinault canyon, which narrows by an order of magnitude between the upper and lower canyon, L varies significantly as a function of isobath, but N also varies with depth. The net result is that H is about 400 m for the 200-m isobath and about 500 m for the 1,000-m isobath. Thus, flow over the shelf and along the shelf edge will tend to follow the canyon curvature throughout the interior of the water column, whereas flow at similar water depths out over the canyon (in deeper bottom depths) 'misses the turn' and crosses over canyon isobaths (e.g., in Fig. 2.35, 142 m from the surface in a water depth of 1,200 m; 384 m from the surface in a water depth of 1,600 m). This is confirmed by available geostrophic flow data (Figs. 2.38 and 2.39): the magnitude of the flow component parallel to canyon isobaths at and below the canyon lip generally decreases and/or becomes variable in direction as the axis of the canyon is approached.

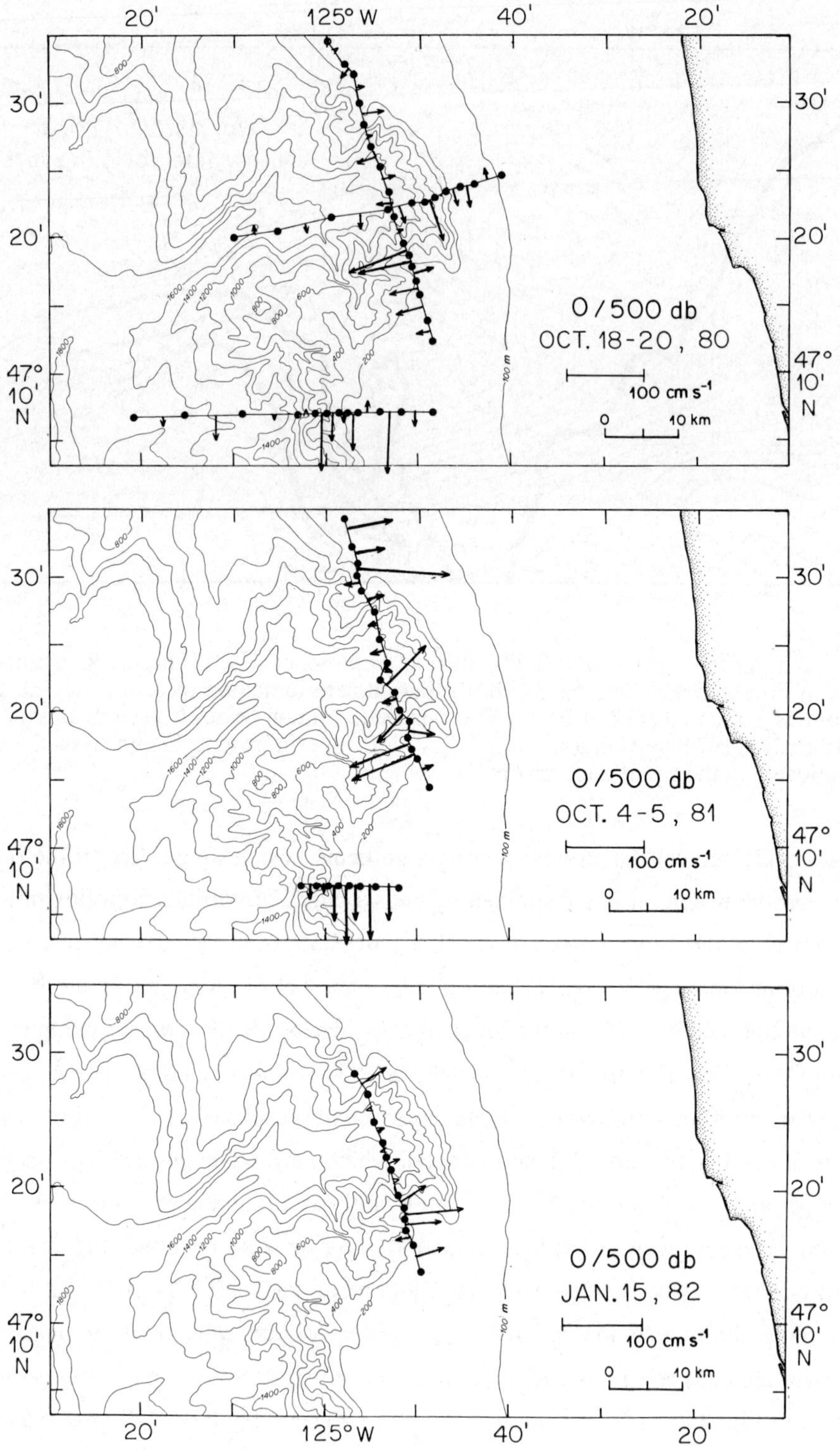

Fig. 2.38 Geostrophic velocity (0/500 db) in the vicinity of Quinault submarine canyon on three occasions during fall and winter. Data have been extrapolated into regions shallower than the reference level using the method of Montgomery (1941). Dots indicate locations of CTD stations.

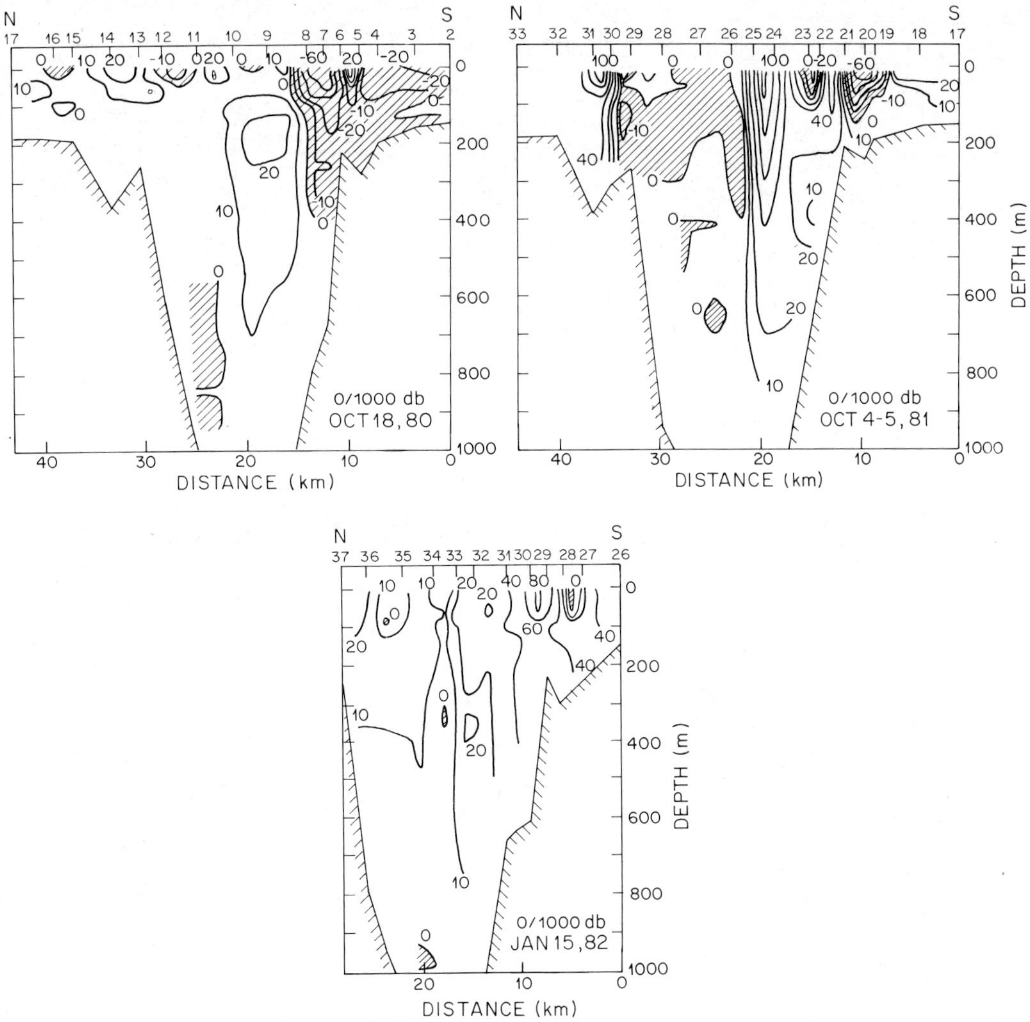

Fig. 2.39 Geostrophic velocity (0/1,000 db) along the sections across Quinault submarine canyon that are shown in Figure 2.38. Hatching indicates regions of offshore flow. Units are cm s $^{-1}$. Station numbers are given along the top of each section. Note that during January 1982 the section extends only to the canyon lip on the north side.

The masking of canyon topography by stratification is further illustrated with geostrophic velocities computed from a CTD section across the narrow lower Quinault canyon (Fig. 2.40). At the surface (0/500 db), the flow component parallel to the canyon axis has a random orientation. At 200 m (200/1,000 db), the flow begins to demonstrate a more organized pattern: into the canyon along the north wall and out along the south wall. At 500 m (500/1,000 db), this pattern is strongly developed along the two walls. Thus, the effective height of the baroclinic disturbance is on the order of 500 m as expected for the roughly 10-km wide feature for the quasi-geostrophic flow. As in the upper canyon, the flow component into or out of the canyon is strongest over the canyon walls and weak and/or variable in direction over the axis.

The cross-isobath relative or geostrophic flow field over and near the canyon has significant vertical as well as horizontal structure, particularly above canyon lip depths

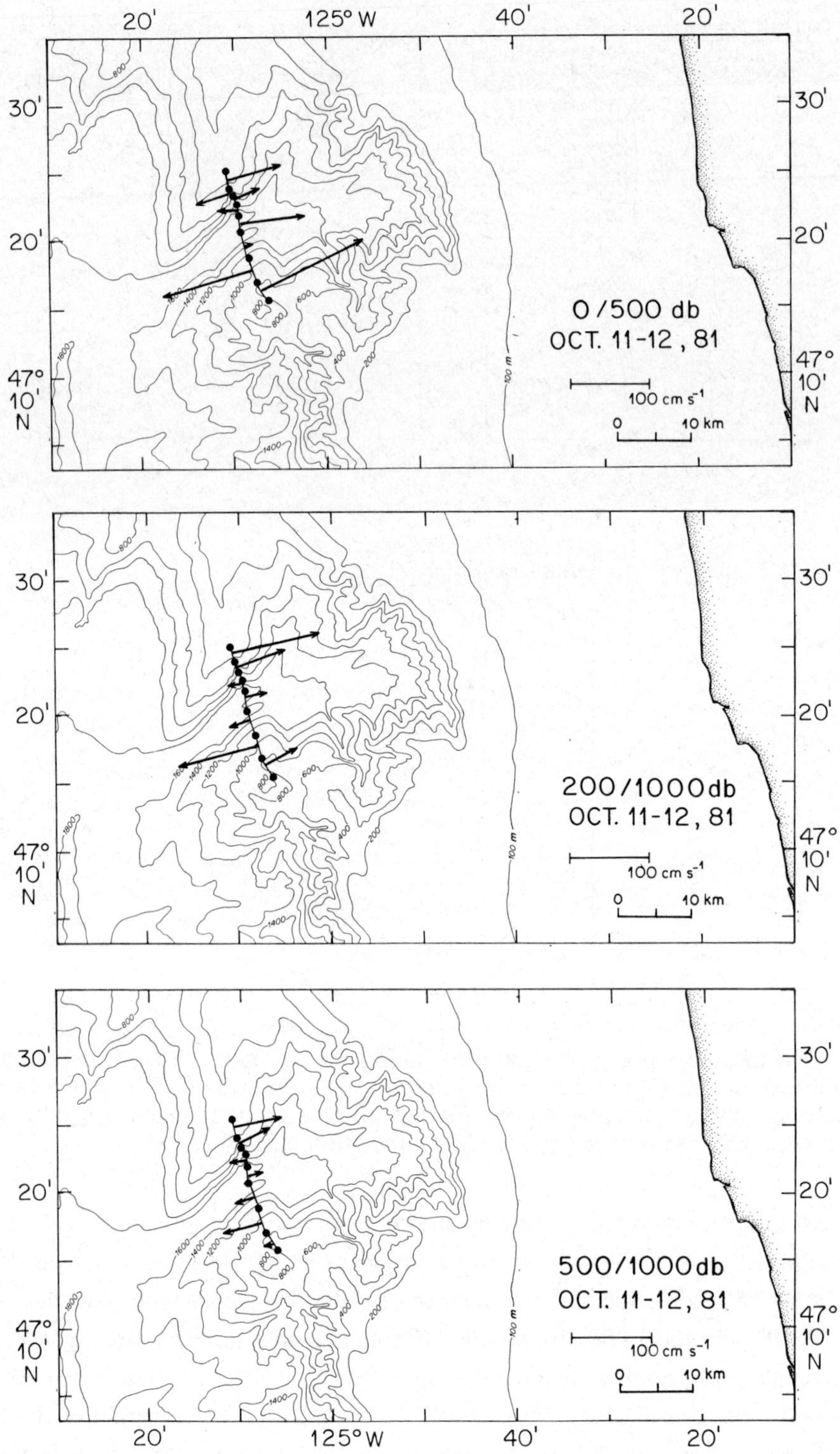

Fig. 2.40 Geostrophic velocity at various depths relative to 1,000 db along a section across the narrow lower Quinault canyon on 12 October 1982. Data have been extrapolated into regions shallower than the reference level using the method of Montgomery (1941). Dots indicate locations of CTD stations.

(Fig. 2.39). Alternating bands of onshore and offshore flow do not, at first glance, appear to be consistent with the picture of flow gently curving around and/or crossing the canyon. However, we note that most of the direct measurements have been at depths exceeding 100 m from the surface. The majority of the observed geostrophic shear occurs in the very near-surface layer where eddy-like features are known to occur. Also, some of the horizontal shear is likely due to errors introduced into the geostrophic velocity calculation by the extremely close (~1-2 km) station spacing. Filtering the hydrographic data would undoubtedly remove much of the horizontal variability, however, since no direct measurements are available to allow us to distinguish between the signal and the noise, we have presented the data here in its least massaged form.

To the extent that canyons have relatively small-scale features, specifically, features whose width is much less than the internal Rossby radius of deformation ($R = NH/f$), inertial effects can also be important. In such cases, the flow cannot adjust rapidly enough to follow the changes in the orientation of isobath direction (e.g., in Fig. 2.35, vectors with depths of 150, 252, 151, and 400 m from the surface). The moorings supporting these meters were located in and adjacent to a small submarine valley that cuts into the canyon wall. The data illustrate that the flow above shelf edge depths ignores the submarine valley. Of course, flow within the valley itself, i.e., below its lip, would be channelized by the narrow valley topography.

Model studies of the effects of alongshore perturbations in shelf topography predict complicated circulation patterns in the vicinity of canyons, principally as a result of scattering and reflection of long shelf waves and excitation of higher mode waves (Wang, 1980; Killworth, 1978). None of our experiments were designed to adequately resolve the predicted circulation patterns. However, an example of anomalous behavior was observed in the vicinity of Quinault canyon during May-August 1979. Energy at a depth of 210 m over the open slope was comparable in magnitude to that in the bottom boundary layer at mid-shelf. The energy at a similar depth at a canyon head location was an order of magnitude less than the energy at the shelf location (Fig. 2.41). Thus, energy at the open slope and canyon locations at the same depth from the surface and in the same water depth differed by an order of magnitude. Also, subtidal fluctuations over the slope were much more coherent with fluctuations over the shelf than were those within the canyon head at a similar depth (Fig. 2.41). These results are confirmed by the spatial pattern of the first empirical orthogonal eigenfunction for that period (Fig. 2.42): the amplitude of the response is reduced at all depths within the canyon head (both above and below the lip of the canyon) in comparison to the amplitude over the shelf and over the open slope. As previously discussed, the first EOF of alongshelf current generally represents the mode one large-scale, coastal-trapped wave response (Battisti and Hickey, 1984; Halliwell and Allen, 1984). Data from a winter experiment in exactly the same region, on the other hand, does not demonstrate any obvious canyon effects – the amplitude of the first EOF is relatively uniform over the slope, both inside and outside the canyon (Fig. 2.42). These conflicting results may represent real seasonal differences resulting from differences in mean flow structure and/or differences in the dynamics of the fluctuating flow (e.g., the relative proportion of remote and

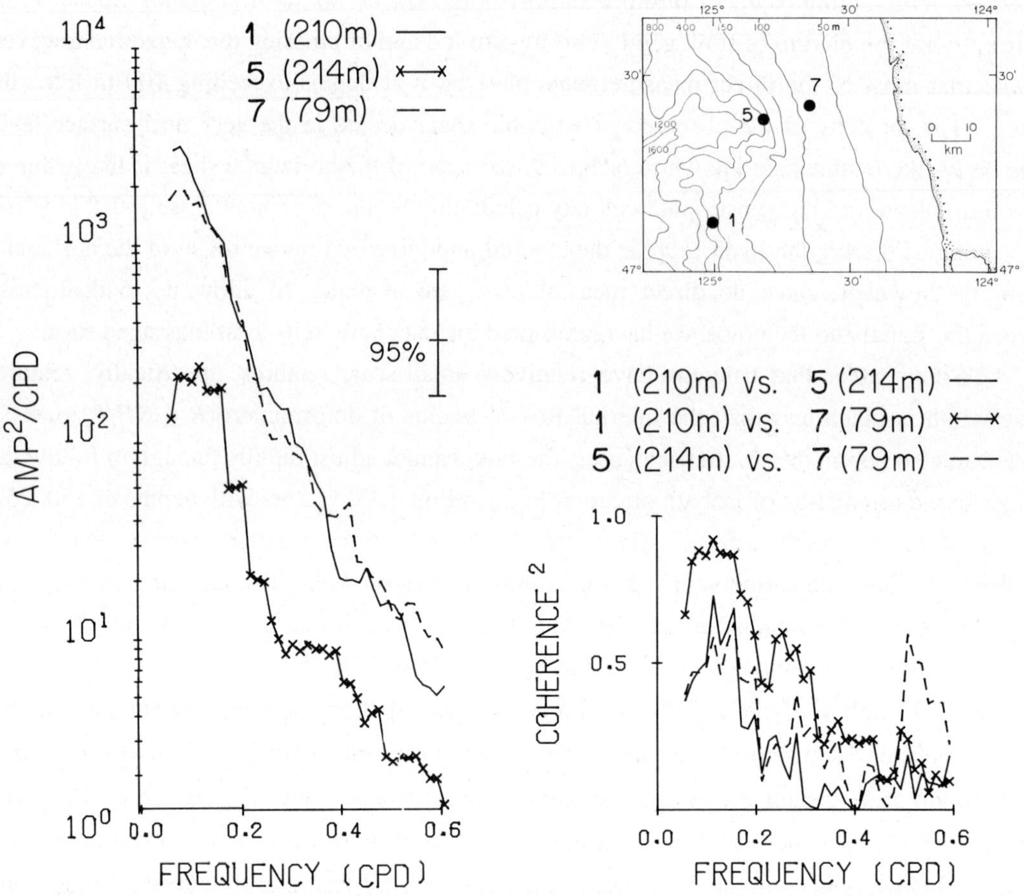

Fig. 2.41. Spectra (left) of alongshelf velocity on the shelf (location 7), and at similar depths within the head of Quinault Canyon (location 5) and over the open slope (location 1) during May-August 1979, and coherence squared estimates (right) between pairs of the time series. Station locations are shown on the inset map. Meter depths are given in brackets next to location numbers. Each estimate has 14 degrees of freedom. Confidence limits and significance levels (0.47 for the 95% level) are computed from Koopmans (1974). The spectra illustrate a marked decrease in energy over the canyon at all frequencies. The coherence estimates indicate that currents over the canyon slope are generally much less coherent with shelf currents than are currents over the open slope.

local forcing). The conflicting results could also be simply due to undersampling spatial variations in the flow field over the canyon.

2.4.2 Deep Canyon Axial Circulation

In all canyons where measurements have been made, currents on the canyon axis near the bottom are directed primarily along the canyon axis (e.g., Shepard *et al.*, 1979). Indeed, current measurements made in canyons prior to the last decade were always made along canyon axes and were usually made well below the canyon lip (Shepard *et al.*, 1979). Measurements over the axis of Quinault canyon indicate a predominantly axial seasonal mean flow within the bottom 200 m of the canyon (Fig. 2.43). The flow was directed upcanyon in the upper canyon and downcanyon in the lower canyon (except within 5 m of the bottom at one location) during two

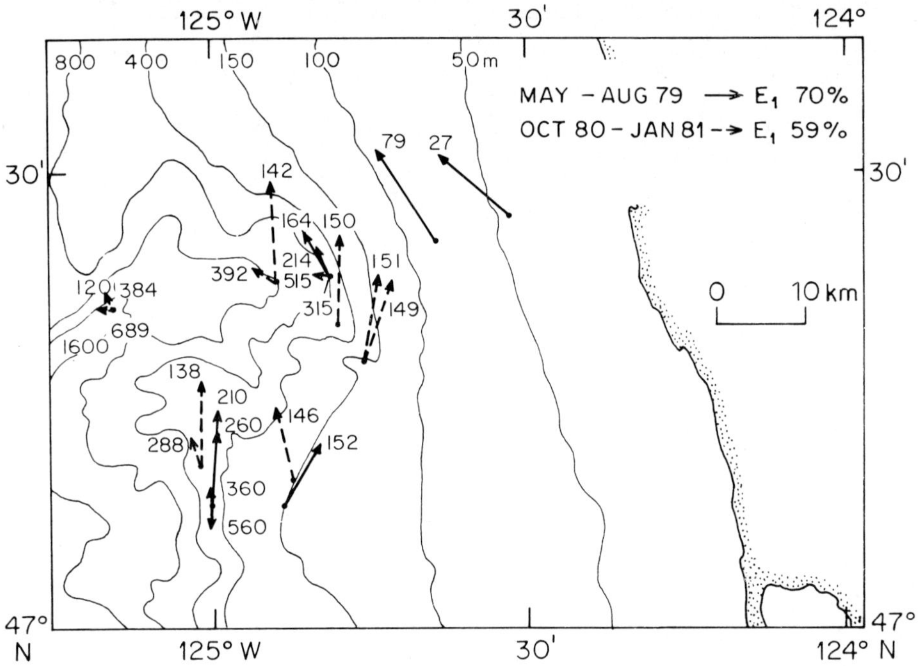

Fig. 2.42 Spatial structure of first empirical orthogonal eigenfunction for currents in the vicinity of Quinault canyon during May-August 1979 and during October 1980-January 1981. The percent of variance accounted for by the eigenfunction is also given. Depth from the surface in meters is given near the tip of each vector.

separate winter experiments (Fig. 2.43). EOF analysis illustrates that event-scale fluctuations were also axial in the bottom 200 m of Quinault canyon (Fig. 2.44). The magnitude of such fluctuations in Quinault submarine canyon is on the order of 3 to 5 cm s^{-1} and at all locations the magnitude decays away from the bottom. Tidal rectification has been suggested as the generation mechanism for low frequency axial canyon currents in some canyons. However, in Quinault canyon, measurements during two winter seasons suggest that the deep canyon axial flow is related to the circulation over the adjacent shelf and slope. Coherence between axial canyon flow, as represented by the first EOF, and the alongshelf flow at the shelf edge was significant during both measurement periods, with a dominant period of five days (Fig. 2.43). The phase lag (~180°) is consistent with the hypothesis, first suggested by Cannon (1972) for Juan de Fuca canyon, that the barotropic cross-shelf pressure gradient (in geostrophic equilibrium with the alongshelf flow) drives an axial circulation in the canyon. Thus, the pressure gradient associated with poleward flow over the shelf and slope may drive flow out of the canyon and that associated with equatorward flow may drive flow into the canyon, in the manner of surface estuarine circulation driven by a freshwater effluent.

It seems likely that this canyon mode would dominate circulation patterns (and coherence scales) in canyons at depths for which the canyon is sufficiently narrow to impede complete geostrophic adjustment to regional circulation (~5-10 km); that is, the water can no longer enter the canyon along one wall and exit on the other, rather, it is forced to move move up or down the

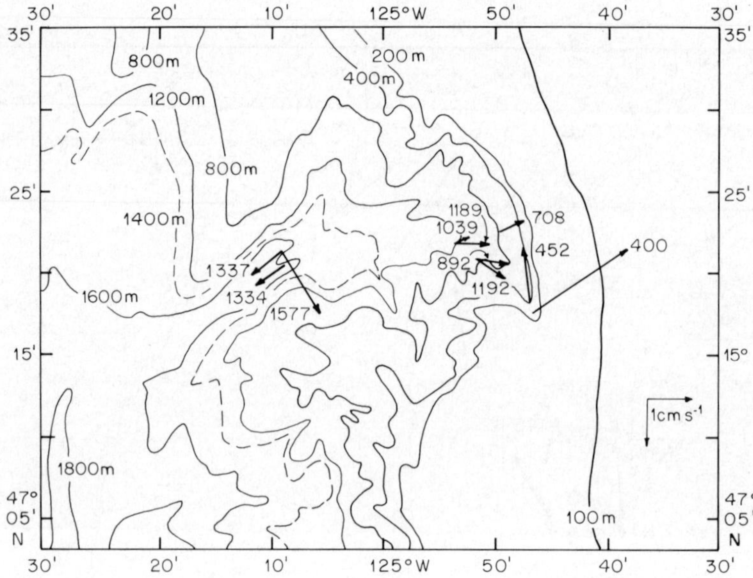

Fig. 2.43 Mean velocity at depths within 305 m of the bottom of Quinault canyon during summer 1979 (708 m), winter 1980 (892, 1,192, 400, and 1,334 m), and winter 1981 (452, 1,337, and 1,577 m). Depth from the surface in meters is given near the tip of each vector. Note that there is an apparent mismatch between chart topography and meter depth computed from pressure gauges.

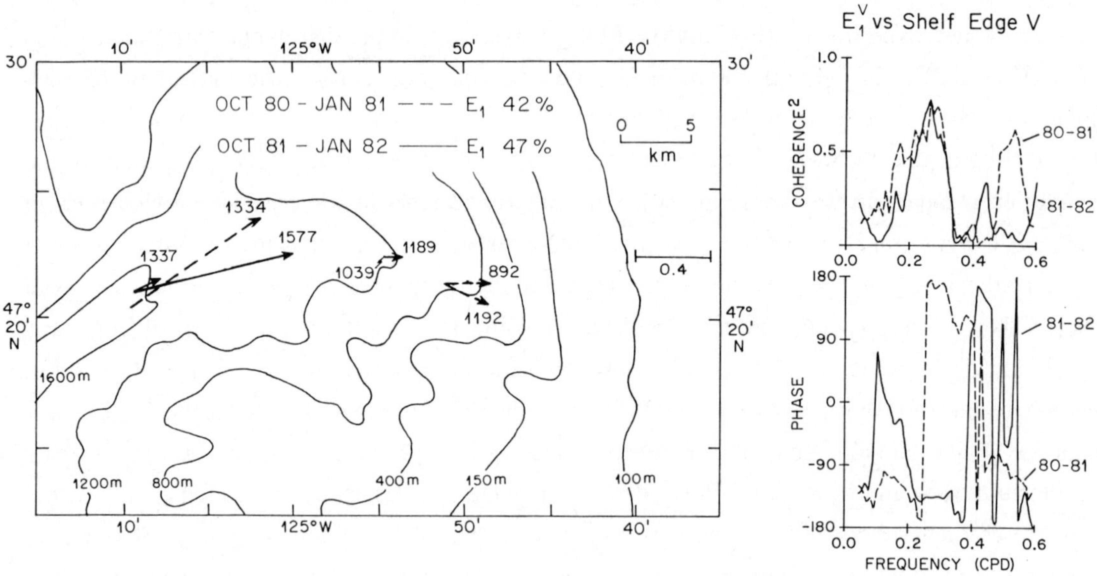

Fig. 2.44 Spatial structure of first empirical orthogonal eigenfunction (EOF) in Quinault canyon during two winters at locations within 305 m of the canyon floor (mab). Depth from the surface in meters is given near the tip of each vector. Measurements at 1,577 m and 1,192 m were made 5 mab; measurements at 1,189 m were made 55 mab.. Coherence squared and phase estimates between the time series of the first EOF and northward velocity 5 mab at the shelf edge are given on the right. A coherence squared estimate of 0.47 is significant at the 95% level (Koopmans, 1974).

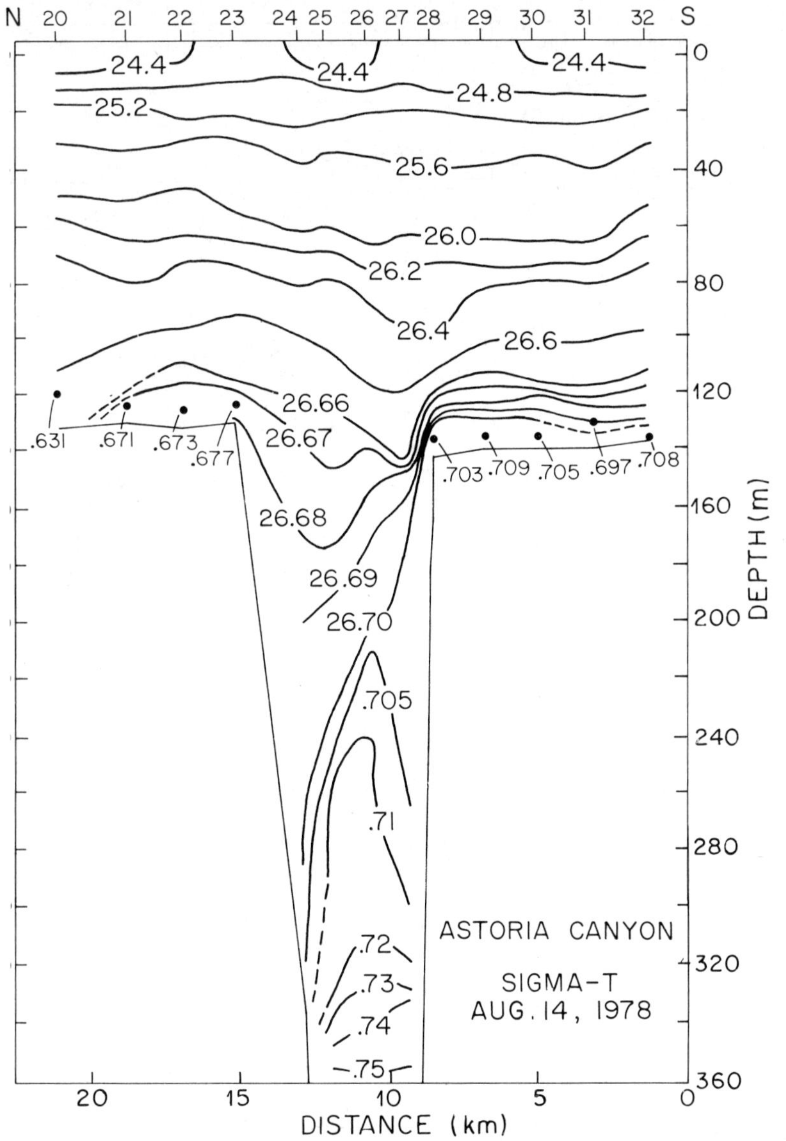

Fig. 2.48 North-south σ_t section across the head of Astoria submarine canyon during August 1978. Dots indicate location of deepest measurements on each CTD profile. Station locations are indicated along the top of the section. Data illustrate enhanced upwelling on the south side of the canyon.

Both theoretical and laboratory studies have indicated that internal tides are enhanced within submarine canyons in comparison with open slope regions (Baines, 1983). Data collected in Quinault canyon confirms these results. Both within and above the small submarine valley on the southern wall of the canyon, observed temperature fluctuations are consistent with vertical displacement of isotherms as much as ± 100 m at the semi-diurnal period (Fig. 2.50), and current speeds were on the order of 15 cm s $^{-1}$ (Hickey *et al.*, 1986). Phase relationships between temperature and the two horizontal components of velocity were consistent with the presence of the internal tide. Over the open slope, on the other hand, vertical displacements due to the

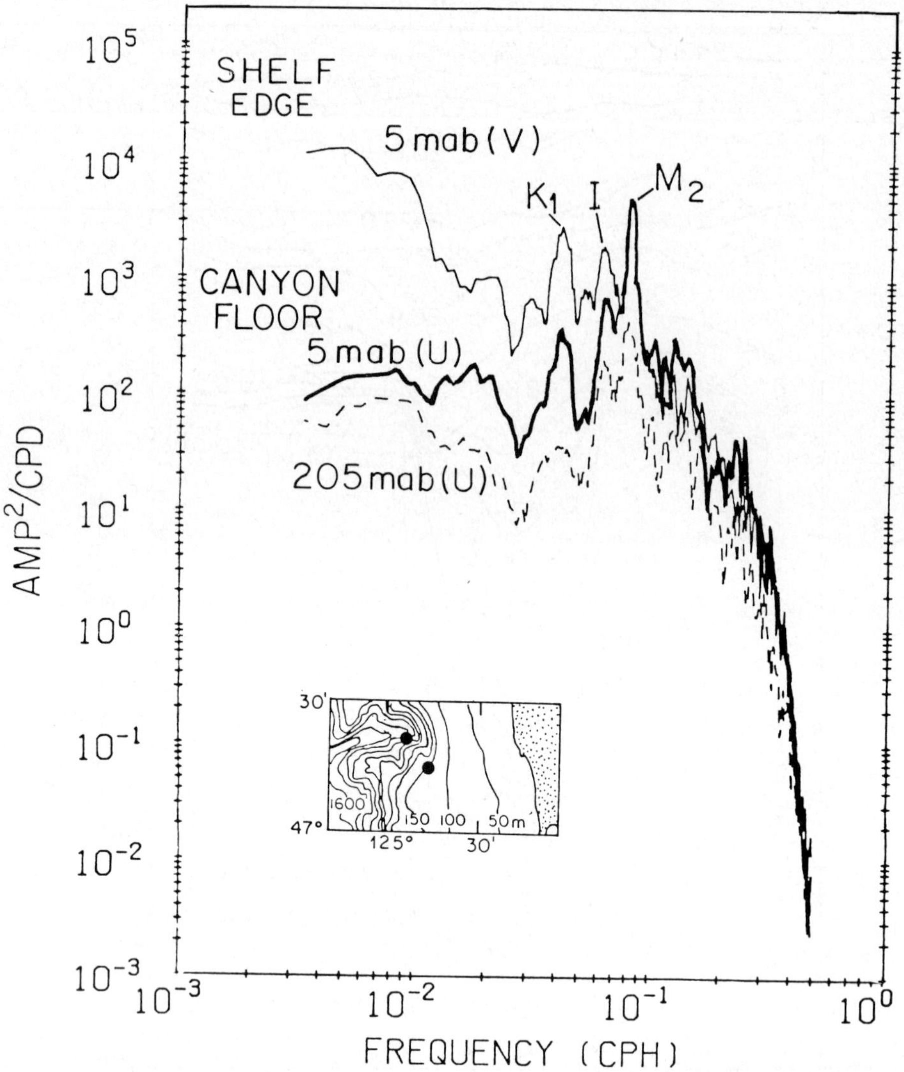

Fig. 2.49 Spectra of alongshelf (v) velocity at the shelf edge and along-axis (u) velocity over the deep canyon floor during winter 1980-81. Height above the bottom in meters (mab) is given on each plot. Mooring locations are shown on the inset map. Each estimate has 14 degrees of freedom. Confidence limits are based on Koopmans (1974). The M_2 tide (12.4 h), inertial period (I, 16.1 h) and the K_1 tide (23.8 h) spectral peaks are indicated on the figure.

internal tide are on the order of 5 m and current speeds are on the order of 5 cm s^{-1} (Torgrimson and Hickey, 1979). Such fluctuations can have a significant impact on the observed particle field at depth. When an intermediate-depth nepheloid layer is present over the canyon slope, the turbid layer moves vertically with the water in response to the internal tide (see semidiurnal fluctuations in particle concentration in Fig. 2.50). A transmissometer sensor located below the depth of local resuspension can thus record an apparent low frequency (< 1 cpd) decrease in transmission that is due not to local resuspension, but to vertical advection at tidal periods of material that has been resuspended at shallower depths.

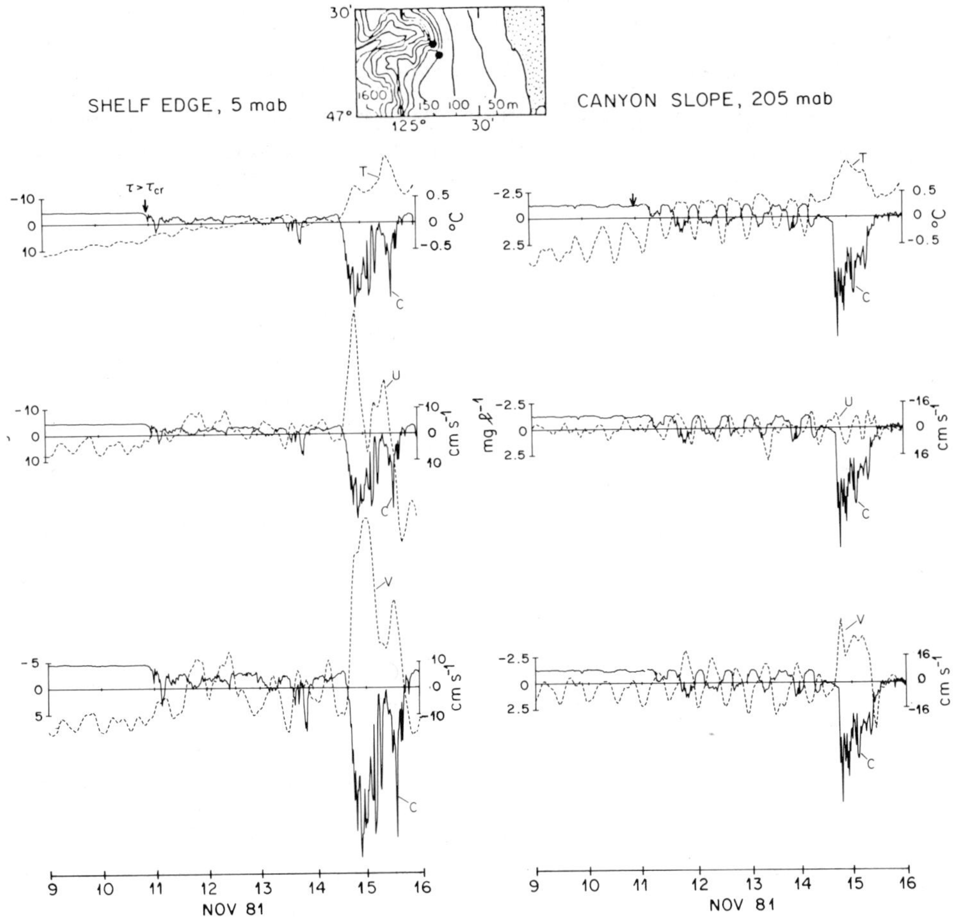

Fig. 2.50 Time series of temperature (T), concentration (C), and east-west (U) or north-south (V) velocity at the shelf edge 5 m above bottom (mab) and over the canyon slope (205 mab). Station locations are shown on the inset map. Means have been removed from the series. Concentration data are at 20-min intervals; velocity and temperature data are at hourly intervals. Data are from a major resuspension event during winter 1981. Note the difference in scales between left and right panels. The vertical arrow indicates the first initiation of local resuspension according to the threshold criterion that the sum of the individual current and wave shear stress (τ) exceeds τ_{cr} = 1.2 dynes cm $^{-1}$. Fluctuations in the data are consistent with vertical movement of water masses due to internal wave activity over the canyon. From Hickey *et al.* (1986).

2.5 SUMMARY

What have we learned during over a decade of current and water property measurements over the Washington continental shelf and slope? The most important result is the determination of the principal forcing mechanism for event-scale fluctuations in the mid- and outer shelf flow. Both models and observations illustrate that 70-90% (50%) of the variance of fluctuations in alongshelf flow at mid-shelf (outer shelf) locations and of fluctuations in coastal sea level can be explained as first mode coastal-trapped waves. The coastal-trapped waves are generated primarily off northern California during summer when coastal wind stress increases equatorward from Washington to California. The response during winter, on the other hand, is dominated by more local wind stress, local winds being much more energetic than those to the south (Battisti and Hickey, 1984).

The possible importance of freely propagating waves was acknowledged in early research on coastal sea level data in the U.S. Pacific Northwest (e.g., Mooers and Smith, 1968). By the mid-1970s, a strong relationship between alongshelf current and the alongshelf component of local coastal wind stress had been demonstrated (Smith, 1974; Huyer *et al.*, 1975) and some evidence for propagating disturbances in the velocity field had been presented (Huyer *et al.*, 1975; Kundu *et al.*, 1975). However, because wind disturbances often travel poleward in these regions at speeds comparable to that of free shelf waves, it was difficult to separate the free and forced contributions to the variance using simple statistics and known phase speeds of the free or forced waves. Additional confusion was introduced by the fact that Oregon researchers were using summer data sets obtained during CUEA, whereas Washington researchers were using primarily winter data sets obtained during DOE-sponsored work emphasizing sediment transport. Thus, Oregon investigators concluded that propagating waves were highly important in the Pacific Northwest, whereas Washington investigators insisted that local wind stress forcing was the dominant physical process. As has now been demonstrated, both groups were at least partially correct.

The percent of variance attributable to coastal-trapped waves was ultimately assessed by two independent techniques applied to the same data sets and applied as a function of season. The first technique consisted of using coastal wind stress data along the coast from southern California to Washington to model the coastal-trapped wave response over the shelf all along the coast (Battisti and Hickey, 1984). This analysis isolated the principal wave generation regions as a function of season. The second technique consisted of evaluating terms in the vertically averaged alongshelf momentum equation, $v_t = \tau_s /H - P_y - \tau_b /H$ (Hickey, 1984). This technique allows the free and forced contributions to the acceleration (hence velocity) variance to be directly estimated, since τ_s represents local wind stress forcing, and P_y, the large-scale alongshelf pressure gradient, essentially represents the sea level perturbations by coastal-trapped waves generated equatorward of the location of interest. Since coastal sea level is dominated by contributions from the first mode wave (Hsieh, 1982, 1985) and since the pressure field due to the first mode wave varies smoothly over the shelf, the alongshelf pressure gradient can, in theory, be readily estimated from coastal sea level and atmospheric pressure data. In practice, this has not been the case. This calculation was not attempted earlier because investigators believed the signal-to-noise ratio would be small. When Allen and Smith (1981) subsequently performed the calculation, the results were not encouraging: the alongshelf pressure gradient was positively rather than negatively correlated with acceleration, the opposite of what would be expected for the first mode free waves expected for this period (Wang and Mooers, 1977). The failure of the Allen and Smith (1981) attempt and the subsequent success of the Hickey (1984) attempt discussed in this chapter illustrates that sea level stations must be selected for the maximum signal-to-noise ratio (an alongshore spacing of ~400 km). Thus, a second important conclusion of the last decade of research is that local and remote wind stress forcing can be effectively separated only if local wind stress and coastal sea level data are judiciously selected.

The simplicity of the forcing mechanisms for U.S. Pacific Northwest shelves is remarkable in itself and is in stark contrast to the situation on east coast shelves at similar latitudes. The east coast shelf and shoreline is broken alongshore at intervals of 100-200 km by capes and shoals. Also the east coast shelf is wider and shallower than the west coast shelf. Thus diffraction and dissipation of coastal-trapped waves would be substantially greater on the east coast shelves. In addition, penetration onto the shelf by the strong western boundary currents, which contributes strongly to current variance on east coast shelves (Pietrafesa, 1983), is not significant on the U.S. Pacific Northwest shelf.

Another important result allowed by the decade and a half of physical measurements off the Washington shelf is the determination of the seasonal cycle of the fluctuating and mean alongshelf flow for mid- and outer shelf locations. Event-scale fluctuations in the flow are generally of the same order as the seasonal mean flow. The amplitude of these fluctuations is a maximum in fall and winter and a minimum in summer, corresponding roughly to the periods of seasonal mean poleward and equatorward flow, respectively. The timing and duration of the periods of equatorward or poleward flow are not constant either across the shelf or with depth. In particular, maximum equatorward flow occurs during spring and early summer at mid-shelf, but during late summer on the outer shelf, so that the location of the equatorward coastal jet moves progressively offshore as the upwelling season advances. The duration of the period of equatorward flow decreases with depth at most locations. The winter to spring transition occurs abruptly and simultaneously in the Pacific Northwest. The summer to fall transition, on the other hand, occurs more gradually, in part because of the seasonal development of a poleward undercurrent on the shelf. This undercurrent attains its seasonal maximum in late summer at the same time as the slope undercurrent (commonly denoted the California Undercurrent) and the large-scale California Current. The few data that are available do not suggest that comparable seasonal cycles occur on the Washington inner shelf. Available data on the Washington inner shelf during winter suggest that a significant equatorward undercurrent may be associated with the Columbia River plume. Flow over the slope during summer is equatorward to depths of ~100 m and poleward at deeper depths, with a subsurface maximum (the California Undercurrent) at ~200 m and speeds on the order of 10-20 cm s^{-1}. The equatorward flow over the slope and the California Undercurrent both attain peak speeds in late summer. Flow over the slope during winter is generally poleward, with no subsurface maximum in the poleward flow. On the contrary, poleward seasonal mean flow decreases with depth and sometimes reverses to equatorward seasonal mean flow (the Washington Undercurrent) with a velocity maximum near ~400 m.

The principal local forcing mechanisms for the seasonal mean currents over the Washington shelf and upper slope have also been determined (Werner and Hickey, 1983). During summer, equatorward alongshelf wind stress and an opposing poleward alongshelf pressure gradient force drive the equatorward near-surface flow and the poleward undercurrent over the shelf and slope. During winter, alongshelf wind stress and pressure gradient force both reverse directions, producing the poleward flow observed over the shelf and in the upper water column over the slope and the equatorward undercurrent observed over the slope. During spring, alongshelf wind stress and pressure gradient forces are both equatorward, resulting in the seasonal maximum in equatorward flow over the shelf. The strong equatorward jet that occurs over the outer shelf and upper slope in late summer in the upper 100 m of the water column was not well reproduced with the Werner and Hickey (1983) model, which does not include the effects of lateral diffusion or β on seasonal scales.

The principal forcing mechanisms for the seasonal and long-term mean alongshelf pressure gradient force along the west coast have also been determined (Hickey and Pola, 1983). In the Pacific Northwest, the sea surface slope is generated primarily by wind stress along the west coast. The forcing is largely remote (30% in winter, 100% in spring). For example, during spring in the Pacific Northwest, local wind stress and pressure gradient force are in the same direction. Elevation slope at that time is a result of the fact that equatorward wind stress is a maximum off northern California. These results were obtained by comparing observed sea level with results of the arrested topographic wave model of Csanady (1978).

Significant departures from the classical two-dimensional picture of coastal upwelling have been demonstrated by the available data sets. In particular, measurements of the cross-shelf flow structure during upwelling events are not suggestive of two-dimensional mass balance. Also, both off Oregon and off Washington, much of the cross-shelf flow beneath the pycnocline occurs not in the bottom boundary layer, but in or beneath the pycnocline. The magnitude of this interior flow can exceed that in the bottom boundary layer at mid-shelf locations.

From the relative success of the first mode, linear, long wave dynamics in accounting for observed alongshelf current variance at mid-shelf, mid-water column locations and sea level variance at the coast, we have concluded that small-scale eddies are not the principal source of variance at these locations. Eddies are expected to be of greater significance in the near-surface layers and on the outer shelf, but the percent of variance attributable to them has not been determined.

A significant amount of our research over the last decade has been focused on submarine canyons. Current measurements in both wide (Quinault) and narrow (Astoria) canyons have been obtained in five separate field programs of from two to four months' duration. This data set is unique on the west coast. Important conclusions include the fact that upwelling is enhanced in the vicinity of canyons, particularly on the downstream side of the canyon. Although vertical oscillations of isopycnals as large as 100 m are common over the canyon slope at and below shelf break depths, isopycnals do not break the surface. Upwelled water must move across the shelf to the nearshore zone to be uplifted into the near-surface layers. In the Juan de Fuca canyon system, water is upwelled directly to the base of the near-surface layer, where it can be entrained via mixing into the surface layers. The perturbation of the canyon on the flow over the shelf and slope has also been examined. The basic result is that the effect of canyons of these scales (~ the internal Rossby radius) is a function of both stratification and canyon width. For the narrow (~3-km half width) Astoria canyon, the canyon effects are confined to within 50 m of the shelf bottom. Above 50 m, stratification 'masks' the canyon from the flow field. Below 50 m, for equatorward flow, the topography induces a counterclockwise eddy over the canyon and below the canyon lip. For the wider Quinault canyon, flow at the shelf edge follows the large-scale (≥ 10 km) canyon topography whereas upper water column flow in deeper depths crosses canyon isobaths (i.e., the flow 'misses the turn').

Fluctuations in near-bottom axial flow in Quinault canyon are significantly coherent with fluctuations in the alongshelf component of current over the shelf. This 'canyon-mode' response, which occurs predominantly at a period of five days, is consistent with forcing into and out of the canyon by the large-scale, cross-shelf, barotropic pressure gradient that is in geostrophic equilibrium with the alongshelf flow.

Two of the canyon experiments were designed explicitly to determine the role of submarine canyons in the transfer of sediment from the shelf to the deep sea. Simultaneous current, transmissometer, and sediment trap data indicated that the canyons focus sediment dispersal via the preferential development of nepheloid layers over the canyon heads during shelf sediment resuspension events (Baker and Hickey, 1986). Lithogenic material is apparently scavenged from the nepheloid layers by amorphous detritus and restructured into larger parcels that sink rapidly (~200 m d^{-1}) into depths from which they cannot be resuspended.

A final result from the submarine canyon research concerns the amplification of internal tides. Semi-diurnal oscillation corresponding to vertical movements of ± 100 m were observed in the narrow valley on the south wall of upper Quinault canyon, consistent with theoretical predictions. Both horizontal velocities (~15 cm s^{-1}) and the amplitude of the vertical excursion (~100 m) were several times greater than those observed over the open slope.

2.6 CONCLUSIONS

The Pacific Northwest coastal zone provides a natural laboratory for investigation of processes affecting circulation. The region is at the inshore side of an eastern, rather than a western boundary current system, so that large-scale boundary currents are relatively weak and tend to have fewer instabilities that can generate mesoscale eddies and meanders. The coastline is relatively straight for a distance of 1,500 km, the shelf is relatively narrow and deep along the whole coast, and the shelf is continuous alongshore – it is never less than 30 km wide between Point Conception and the Strait of Juan de Fuca, a distance of about 1,500 km. River input is very well defined – the Columbia provides roughly 80% of the annual buoyant flux to the coastal ocean between San Francisco and the Strait of Juan de Fuca. Last, the coastal wind field has a relatively large alongshore scale for much of the year. This fortunate physical setting, when combined with the extensive current meter data set generated primarily under long-term DOE support as well as the CUEA data set off Oregon and an extensive monitoring array of sea level and atmospheric pressure, have allowed investigators to separate the dominant forcing mechanisms of the shelf circulation and their seasonal and even interannual variability.

Since the zero-order, mid-shelf dynamics are relatively well understood and since predictive models have been developed, current research is focused on contributions to the variance by higher order effects such as buoyancy flux driving (e.g., river inflow) and higher mode, coastal-trapped waves. In addition, the near-shore frictional zone and the offshore zone where boundary-current instabilities may be important are being presently examined.

ACKNOWLEDGEMENTS

This work could not have been accomplished without the care and patience of Ms. Sue Geier, who performed the majority of the analyses and was responsible for data display, and without the industry of Mr. Dale Ripley, who was responsible for the majority of the current meter mooring programs, without the cheerfulness of Ms. Su Fagerberg, who typed the many versions of the manuscript, and without the perseverance of D. Doyle, who patiently drafted and redrafted versions of the figures over a period of several years. The majority of the field work and much of the analysis described in this chapter were supported by the Energy Research and Development Administration under contract EV(45-1)-2225 TA 25 and the Department of Energy under contract DE-AT06-76EV71025. Additional support was provided by the National Science Foundation under grants OCE 76-21508, 77-08791, 79-19252, 80-23465 and 83-08196. This is a component of Contribution No. 1784 from the School of Oceanography, University of Washington.

REFERENCES

Allen, J. and R.L. Smith. 1981. On the dynamics of wind-driven shelf currents. *Phil. Trans. Roy. Soc. Lond.*, A302, pp: 617-634.

Anderson, I., A. Huyer and R.L. Smith. 1983. Near-inertial motions off the Oregon coast. *J. Geophys. Res.* 88(10)., pp: 5960-5972.

Baines, P.G. 1983. Tidal motion in submarine canyons – a laboratory experiment. *J. Phys. Oceanogr.*, 13: 310-328.

Baker, E.T. and B.M. Hickey. 1986. Contemporary sedimentation processes in and around an active west coast submarine canyon. *Mar. Geol.*, 71: 15-34.

Barnes, C.A., A.C. Duxbury and B.A. Morse. 1972. Circulation and selected properties of the Columbia River effluent at sea. Pages 41-80 in: D.L. Alverson and A. T. Pruter (eds.), *The Columbia River estuary and adjacent ocean regions*: *Bioenvironmental Studies*, Univ. of Washington Press, Seattle, Wa.

Battisti, D.S. and A.J. Clarke. 1982. A simple method for estimating barotropic tidal currents on continental margins with specific application to the M_2 tide off the Atlantic and Pacific coasts of the United States. *J. Phys. Oceanogr.*, 12: 8-16.

Battisti, D.S. and B.M. Hickey. 1984. Application of remote wind-forced coastal trapped wave theory to the Oregon and Washington coasts. *J. Phys. Oceanogr.*, 14: 887-903.

Brink, K.H. 1986. Topographic drag due to barotropic flow over the continental shelf and slope. *J. Phys. Oceanogr.*, 16: 2150-2158.

Brink, K.H. and E.O. Hartwig. 1985. Coastal Transition Zone Workshop Report, Naval Postgraduate School, Monterey, Calif., 67 pp.

Budinger, T.F., L.K. Coachman and B.A. Morse. 1964. Columbia River effluent in the northeast Pacific Ocean, 1961, 1962: selected aspects of physical oceanography. Univ. of Washington, Dept. of Oceanogr., Tech. Rep. No. 99, 78 pp.

Cannon, G. 1972. Wind effects on currents observed in Juan de Fuca submarine canyon. *J. Phys. Oceanogr.*, 2(3): 281-285.

Cannon, G., N.P. Laird and R.V. Ryan. 1972. Currents observed in Juan de Fuca submarine canyon and vicinity, 1971. NOAA Tech. Rep. ERL252-POL14., 57 pp.

Carpenter, R. and M.L. Peterson. 1989. Chemical cycling in Washington's coastal zone. Pages 366-510 in: M.R. Landry and B. M. Hickey (eds.), *Coastal Oceanography of Washington and Oregon*. Elsevier, Amsterdam, Neth.

Carpenter, R., M.L. Peterson and J.T. Bennett. 1982. Pb-derived sediment accumulation and mixing rates for the Washington continental slope. *Mar. Geol.*, 48: 135-164.

Chelton, D.B. 1980. Low frequency sea level variability along the west coast of North America. Ph.D. Diss., Univ. of California, San Diego, Calif., 212 pp.

Chelton, D.B. 1984. Seasonal variability of alongshore geostrophic velocity off central California. *J. Geophys. Res.*, 89: 3473-3486.

Chelton, D.B., P.A. Bernal and J.A. McGowan. 1982. Large-scale interannual physical and biological interaction in the California Current. *J. Mar. Res.*, 40(4): 1095-1125.

Chelton, D.B. and R.E. Davis. 1982. Monthly mean sea-level variability along the west coast of North America. *J. Phys. Oceanogr.*, 12: 757-784.

Csanady, G.T. 1978. The arrested topographic wave. *J. Phys. Oceanogr.*, 8: 47-62.

Denbo, D.W. and J.S. Allen. 1986. Large-scale response to atmospheric forcing of shelf currents and coastal sea level off California and Oregon: May-July, 1981 and 1982. *J. Geophys. Res.*, 92(C2): 1757-1782.

Dodimead, A.J., F. Favorite and T. Hirano. 1963. Salmon of the North Pacific Ocean, part II. Review of the oceanography of the sub-Arctic Pacific region. *Inter. N. Pac. Fish. Comm. Bul.* 13, 195 pp.

Emery, W.J. and L.A. Mysak. 1980. Dynamical interpretations of satellite-sensed thermal features off Vancouver Island. *J. Phys. Oceanogr.*, 10: 961-970.

Enfield, D.B. and J.S. Allen. 1980. On the structure and dynamics of monthly sea level anomalies along the Pacific coast of North and South America. *J. Phys. Oceanogr.*, 10(4): 557-578.

Freeland, H.J., W.R. Crawford and R.E. Thomson. 1984. Currents along the Pacific coast of Canada. *Atmosphere Ocean*, 22(2): 151-172.

Freeland, H.J. and K.L. Denman. 1982. A topographically controlled upwelling center off southern Vancouver Island. *J. Mar. Res.*, 4(4): 1069-1093.

Haidvogel, D.B. and K.H. Brink. 1986. Mean currents driven by topographic drag over the continental shelf and slope. *J. Phys. Oceanogr.*, 16: 2159-2171.

Halliwell, G.R. and J.S. Allen. 1984. Large scale sea level response to atmospheric forcing along the west coast of North America, summer, 1973. *J. Phys. Oceanogr.*, 14: 864-886.

Hamilton, P. and M. Rattray. 1978. A numerical model of the depth-dependent wind-driven upwelling circulation on a continental shelf. *J. Phys. Oceanogr.*, 8(3): 437-457.

Hayes, S.P. and D. Halpern. 1976. Observations of internal waves and coastal upwelling off the Oregon coast. *J. Mar. Res.*, 34: 247-267.

Hermann *et al.*, 1989. Coastal Upwelling Dynamics. Pages 211-254 in: M.R. Landry and B. M. Hickey (eds.), *Coastal Oceanography of Washington and Oregon.* Elsevier, Amsterdam, Neth.

Hickey, B.M. 1979. The California Current System – hypotheses and facts. *Prog. Oceanogr.*, 8: 191-279.

Hickey, B.M. 1981. Alongshore coherence on the Pacific Northwest continental shelf (January-April, 1975). *J. Phys. Oceanogr.*, 11: 822-835.

Hickey, B.M. 1983. Hydrographic observations over the Washington-Oregon continental shelf and slope during WISP (February-March, 1975). Univ. of Washington, School of Oceanogr. Special Rep. No. 86, 296 pp.

Hickey, B.M. 1984. The fluctuating longshore pressure gradient on the Pacific Northwest shelf: a dynamical analysis. *J. Phys. Oceanogr.*, 14: 276-293.

Hickey, B.M. In press. Poleward flow near the northern and southern boundaries of the U.S. west coast. In: *Lecture Note Series: Poleward Flow on Eastern Ocean Boundaries.* Springer-Verlag, New York, NY.

Hickey, B.M., E.T. Baker and N.B. Kachel. 1986. Suspended particle movement in and around Quinault submarine canyon. *Mar. Geol.*, 71: 35-85.

Hickey, B.M. and P. Hamilton. 1980. A spin-up model as a diagnostic tool for interpretation of current and density measurements on the continental shelf of the Pacific Northwest. *J. Phys. Oceanogr.*, 10: 12-24.

Hickey, B.M. and N.E. Pola. 1983. The seasonal alongshore pressure gradient on the West Coast of the United States. *J. Geophys. Res.*, 88: 7623-7633.

Hickey, B.M., R. Thomson, H. Yih and P. LeBlond. In review. A buoyancy-driven coastal current. *J. Geophys. Res.*.

Holloway, G. 1987. Systematic forcing of large-scale geophysical flows by eddy-topography interaction. *J. Fluid Mech.*, 184: 463-476.

Hotchkiss, F. and C. Wunsch. 1982. Internal waves in Hudson Canyon with possible geological implications. *Deep-Sea Res.*, 29: 415-442.

Hsieh, W.W. 1982. On the detection of continental shelf waves. *J. Phys. Oceanogr.*, 12: 414-427.

Hsieh, W.W. 1985. Modal bias in sea level and sea surface temperature, with applications to the Oregon shelf. *J. Phys. Oceanogr.*, 15(3): 351-356.

Hukuda, H. 1982. Subharmonic destabilization off Vancouver Island. *J. Phys. Oceanogr.*, 12: 285-292.

Huyer, A. 1983. Coastal upwelling in the California Current System. *Progr. Oceanogr.*, 12: 259-284.

Huyer, A., B.M. Hickey, J.D. Smith, R.L. Smith and R.D. Pillsbury. 1975. Alongshore coherence at low frequencies in currents observed over the continental shelf off Oregon and Washington. *J. Geophys. Res.*, 80: 3495-3505.

Huyer, A. and R.L. Smith. 1984. The signature of El Niño off Oregon, 1982-1983. *J. Geophys. Res.*, 90(C4): 7133-7142.

Huyer, A., R.L. Smith and B.M. Hickey. 1984. Observations of a warm-core eddy off Oregon, January to March 1978. *Deep-Sea Res.*, 31(2): 97-117.

Huyer, A., R.L. Smith and E. Sobey. 1978. Seasonal differences in low- frequency current fluctuations over the Oregon continental shelf. *J. Geophys. Res.*, 83: 5076-5089.

Ikeda, M. 1983. Linear instability of a current flowing along a bottom slope using a three-layer model. *J. Phys. Oceanogr.*, 18: 208-223.

Ikeda, M., W.J. Emery and L.A. Mysak. 1984. Seasonal variability in meanders of the California Current System off Vancouver Island. *J. Geophys. Res.*, 89(C3): 3487-3505.

Johnson, D.R. and W.R. Johnson. 1979. Vertical and cross-shelf flow in the coastal upwelling region off Oregon. *Deep-Sea Res.*, 26: 399-408.

Johnson, W.R., J.C. Van Leer and C.N.K. Mooers. 1976. A cyclesonde view of coastal upwelling. *J. Phys. Oceanogr.*, 6: 556-574.

Kelly, K.A. 1985. The influence of winds and topography on the sea surface temperature patterns over the northern California slope. *J. Geophys. Res.*, 90: 11,783-11,798.

Killworth, P.D. 1978. Coastal upwelling and Kelvin waves with small longshore topography. *J. Phys. Oceanogr.*, 8: 188-205.

Kindle, J.C. 1974. The horizontal coherence of inertial oscillations in a coastal region. *Geophys. Res. Lett.*, 1: 127-130.

Koopmans, L.H. 1974. *The Spectral Analysis of Time Series*. Academic Press, N.Y., 366 pp.

Kundu, P.K. 1976. An analysis of inertial oscillations observed near the Oregon coast. *J. Phys. Oceanogr.*, 6: 879-898.

Kundu, P.K. and J.S. Allen. 1976. Some three-dimensional characteristics of low-frequency current fluctuations near the Oregon coast. *J. Phys. Oceanogr.*, 6: 181-199.

Kundu, P.K., J.S. Allen and R.L. Smith. 1975. Modal decomposition of the velocity field near the Oregon coast. *J. Phys. Oceanogr.*, 5(4): 683-704.

Landry, M.L. *et al.* 1989. Broad-scale patterns in the distribution of hydrographic variables. Pages 1-41 in: M.R. Landry and B. M. Hickey (eds.), *Coastal Oceanography of Washington and Oregon*. Elsevier, Amsterdam, Neth.

Large, W.G. and S. Pond. 1981. Open ocean momentum flux measurements in moderate to strong winds. *J. Phys. Oceanogr.*, 11: 324-336.

McCreary, J.P., Jr. and S.Y. Chao. 1985. Three-dimensional shelf circulation along an eastern boundary. *J. Mar. Res.*, 43: 13-36.

McCreary, J.P., Jr., P.K. Kundu and S. Chao. 1987. Dynamics of the California Current system. *J. Mar. Res.*, 45: 1-32.

Middleton, J.H. 1987. Steady coastal circulation due to alongshore pressure gradients. *J. Phys. Oceanogr.*, 17(5): 604-612.

Mitchum, G.T. and A.J. Clarke. 1986. The frictional nearshore response to forcing by synoptic scale winds. *J. Phys. Oceanogr.*, 16: 934-946.

Montgomery, R.B. 1941. Transport of the Florida Current off Havana. *J. Mar. Res.*, 4: 198-220.

Mooers, C.N.K. 1970. The interaction of an internal tide with the frontal zone of a coastal upwelling region. Ph.D. Diss., Oregon St. Univ., Corvalis, Ore., 480 pp.

Mooers, C.N.K., C.A. Collins and R.L. Smith. 1976. The dynamic structure of the frontal zone in the coastal upwelling region off Oregon. *J. Phys. Oceanogr.*, 6: 3-21.

Mooers, C.N.K. and R.L. Smith. 1968. Continental shelf waves off Oregon. *J. Geophys. Res.*, 73(2): 549-557.

Mysak, L.A. 1977. On the stability of the California Undercurrent off Vancouver Island. *J. Phys. Oceanogr.*, 7: 904-917.

Mysak, L.A., E.R. Johnson and W.W. Hsieh. 1981. Baroclinic and barotropic instabilities of coastal currents. *J. Phys. Oceanogr.*, 11: 209-230.

O'Brien, J.J., R.M. Clancy, A.J. Clarke, M. Crepon, R. Ellsburg, T. Grammelsrod, M. MacVean, L. Proed and J.D. Thompson. 1977. Upwelling in the ocean: two- and three-dimensional models of upper ocean dynamics and variability. Pages 178-228 in: E. Krauss, (ed.), *Modelling and predicting the upper layers of the ocean.* Pergamon Press, N.Y.

Pak, H., G.F. Beardsley and R.L. Smith. 1970. An optical and hydrographic study of a temperature inversion off Oregon during upwelling. *J. Geophys. Res.*, 75: 629-636.

Panofsky, H.A. and G.W. Brier. 1968. Some applications of statistics to meteorology. Penn. State Univ., Univ. Park, Penn, pp 224.

Peffley, M.B. and J.J. O'Brien. 1976. A three-dimensional simulation of coastal upwelling off Oregon. *J. Phys. Oceanogr.*, 6: 164-179.

Peterson, W.F., C.B. Miller and A. Hutchinson. 1979. Zonation and maintenance of copepod populations in the Oregon upwelling zone. *Deep-Sea Res.*, 26A: 467-494.

Philander, S.G.H. and J.H. Yoon. 1982. Eastern boundary currents and coastal upwelling. *J. Phys. Oceanogr.*, 12: 862-879.

Pietrafesa, L.J. 1983. Shelfbreak circulation, fronts and physical oceanography: east and west coast perspectives. Pages 233-250 in: D.S. Stanley and G.J. Moore, (eds.), *The Shelfbreak: Critical Interface on Continental Margins.* Soc. EC. Pal. Special Publ. No. 33.

Pollard, R.T. 1980. Properties of near-surface inertial oscillations. *J. Phys. Oceanogr.*, 10: 385-398.

Prinsenberg, S. and M. Rattray, Jr. 1975. Effects of continental slope and variable Brunt-Vaisala frequency on the coastal generation of internal tides. *Deep-Sea Res.*, 22: 251-263.

Samelson, R.M. and J.S. Allen. 1987. Quasigeostrophic topographically generated mean flow over the continental margin. *J. Phys. Oceanogr.*, 17(11): 2043-2064.

Shepard, F.P., N.F. Marshall, P.A. McLoughlin and G.G. Sullivan. 1979. Currents in submarine canyons and other sea valleys. *Am. Assoc. Pet. Geol.*, Tulsa, Okla. Studies in Geology No. 8, 173 pp.

Sherwood, C.R., D.A. Jay, P. Hamilton, R.B. Harvey and C.A. Simenstad. In review. Historical changes in the Columbia River Estuary. *Prog. Oceanogr.*

Small, L.F. and D.W. Menzies. 1981. Patterns of primary productivity and biomass in a coastal upwelling region. *Deep-Sea Res.*, 28A: 123-149.

Smith, R.L. 1974. A description of current, wind and sea level variations during coastal upwelling off the Oregon coast, July-August 1972. *J. Geophys. Res.*, 79(3): 435-443.

Smith, R.L. 1981. A comparison of the structure and variability of the flow field in the three coastal upwelling regions: Oregon, Northwest Africa, and Peru. Pages 107-118 in: F.A. Richards, (ed.), *Coastal upwelling*. American Geophysical Union, Washington, D.C.

Strub, P.T., J.S. Allen, A. Huyer, R.L. Smith and R.C. Beardsley. 1987. Seasonal cycles of currents, temperatures, winds and sea level over the northeast Pacific continental shelf. *J. Geophys. Res.*, 92(C2): 1507-1526.

Thomson, R.E. 1984. A cyclonic eddy over the continental margin of Vancouver Is.: evidence for baroclinic instability. *J. Phys. Oceanogr.*, 14(8): 1326-1348.

Thomson, R.E., B.M. Hickey and P.H. LeBlond. In press. The Vancouver Island Coastal Current: fisheries barrier and conduit. *Can. Fish. Bull.*

Thomson, R. and J.F.R. Gower. 1985. A wind-induced mesoscale eddy over the Vancouver Island continental slope. *J. Geophys. Res.*, 90(C5): 8981-8993.

Torgrimson, G.M. and B.M. Hickey. 1979. Barotropic and baroclinic tides over the continental slope and shelf off Oregon. *J. Phys. Oceanogr.*, 9: 945-961.

Wang, D.P. 1980. Diffraction of continental shelf waves by irregular alongshore geometry. *J. Phys. Oceanogr.*, 10: 1187-1199.

Wang, D.P. 1982. Effects of continental slope on the mean shelf circulation. *J. Phys. Oceanogr.*, 12: 1524-1526.

Wang, D.P. and C.N.K. Mooers. 1977. Long coastal trapped waves off the west coast of the United States, summer 1973. *J. Phys. Oceanogr.*, 7: 856-864.

Werner, F.E. 1984. Finite element computations of spin-up and wind-driven coastal ocean circulation. Ph.D. Diss., School of Oceanogr., Univ. of Washington, Seattle, Wa., 130 pp.

Werner, F.E. and B.M. Hickey. 1983. The role of a longshore pressure gradient in Pacific Northwest coastal dynamics. *J. Phys. Oceanogr.*, 13: 395-410.

Winant, C.D., R.C. Beardsley and R.E. Davis. 1987. Moored wind, temperature and current observations made during the coastal ocean dynamics experiments 1 and 2 over the northern California continental shelf and upper slope. *J. Geophys. Res.*, 92(C2): 1569-1604.

Wright, D.G. 1980. On the stability of a fluid with specialized density stratification-II. Mixed baroclinic-barotropic instability with application to the Northeast Pacific. *J. Phys. Oceanogr.*, 10: 1307-1322.

Yao, T., H.J. Freeland and L.A. Mysak. 1984. A comparison of low-frequency current observations off British Columbia with coastal-trapped wave theory. *J. Phys. Oceanogr.*, 14: 22-34.

CHAPTER 3

PRIMARY PRODUCTION IN WASHINGTON COASTAL WATERS

M. J. Perry, J.P. Bolger and D.C. English,
School of Oceanography, University of Washington, Seattle, WA 98195

3.1 INTRODUCTION

The distribution of phytoplankton and their productivity in the waters overlying the Washington shelf and slope vary in space and time in response to the variability in the near-field and far-field physical forcing. The dominant physical forcings on phytoplankton processes at this mid-latitude region include the strong seasonal pattern in solar radiation, the predominantly northerly winds in the spring and summer which regulate the upwelling of nutrient-rich water and its offshore transport and the presence of a rich coastal mesoscale-eddy field. The wind-driven flux of nutrients into the euphotic zone is high, although episodic, between April and October (Landry *et al.*, 1989, Chapter 1; Hickey, 1989, Chapter 2). This almost continual replenishment of nutrients in euphotic waters during a period when flux of solar radiation to the sea surface is high and the days are long is responsible for the persistent high phytoplankton standing stocks and rates of production that are characteristic of this region.

In this chapter we describe the spatial patterns of phytoplankton and their primary production as directly determined from the shipboard program conducted between 1974 and 1982 and as inferred from Coastal Zone Color Scanner satellite imagery and pigment-productivity algorithms applied to the satellite imagery. Because the shipboard sampling program conducted over the past decade in Washington coastal waters was limited in its temporal and spatial coverage, it is not possible to determine interannual patterns of variability in primary production. However, from available ship-board productivity data and from analysis of more recently available satellite color imagery of the spatial extent of the highly productive regions, we estimate that an average of 11.5 metric tons of carbon are annually fixed in the waters overlying the shelf and slope off Washington.

3.2 METHODS

Coastal Zone Color Scanner (CZCS) and Advanced Very High Resolution Radiometer (AVHRR) satellite data were obtained as Level 1 radiance data tapes. The data were processed using the University of Miami algorithms for sea surface temperature and ocean color (multiple scattering algorithm; Gordon and Castaño, 1987; Gordon *et al.*, 1988). The Level 3 satellite data products shown in this Chapter are earth-navigated images of sea surface temperature in degrees Celsius and of pigment concentrations in mg chlorophyll *a* plus phaeopigment m^{-3} (Fig 3.1).

Productivity data were obtained from 17 cruises conducted off the Washington coast between 1974 and 1982. During these cruises a total of 139 productivity stations were occupied. The stations were concentrated between 46° and 48° N and extended from the shore (124°W) seaward to approximately 127° W. The stations located in waters shallower than 200 m are designated as shelf stations in this chapter; those located in waters with depths between 200 and 2,000 m as slope stations; those located in waters deeper than 2,000 m as oceanic stations.

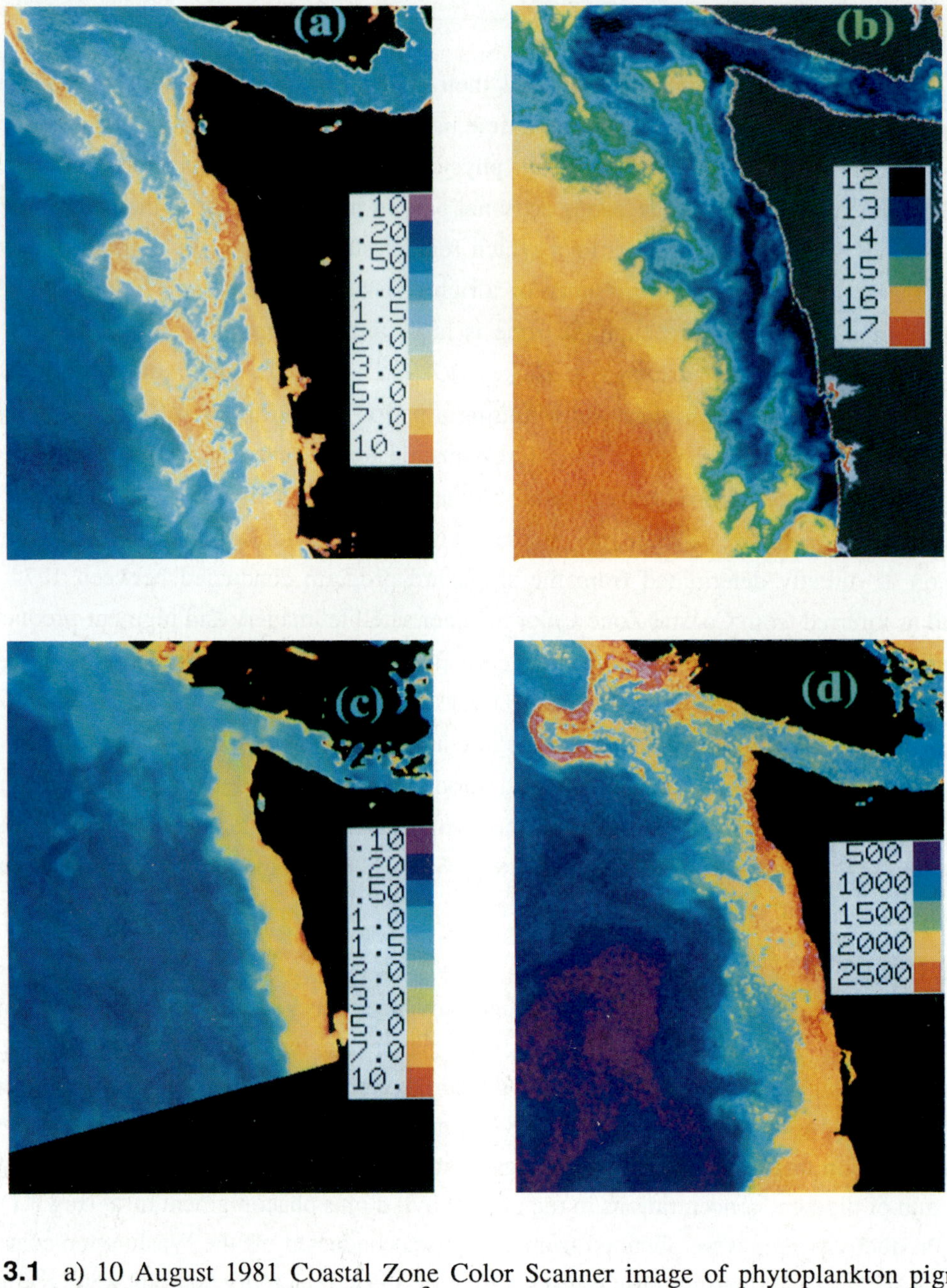

Fig. 3.1 a) 10 August 1981 Coastal Zone Color Scanner image of phytoplankton pigments (mg chlorophyll *a* plus phaeopigments) m $^{-3}$. A color bar indicates concentration. b) 10 August 1981 Advanced Very High Resolution Radiometer image of sea surface temperature (°C). A color bar indicates temperature. c) 17 March 1981 Coastal Zone Color Scanner image of phytoplankton pigments; units as in 3.1a d) Map of primary production for 6 September 1981, derived from Coastal Zone Color Scanner pigment concentration and the pigment-productivity algorithm from Figure 3.11 (mg C m $^{-2}$ d $^{-1}$).

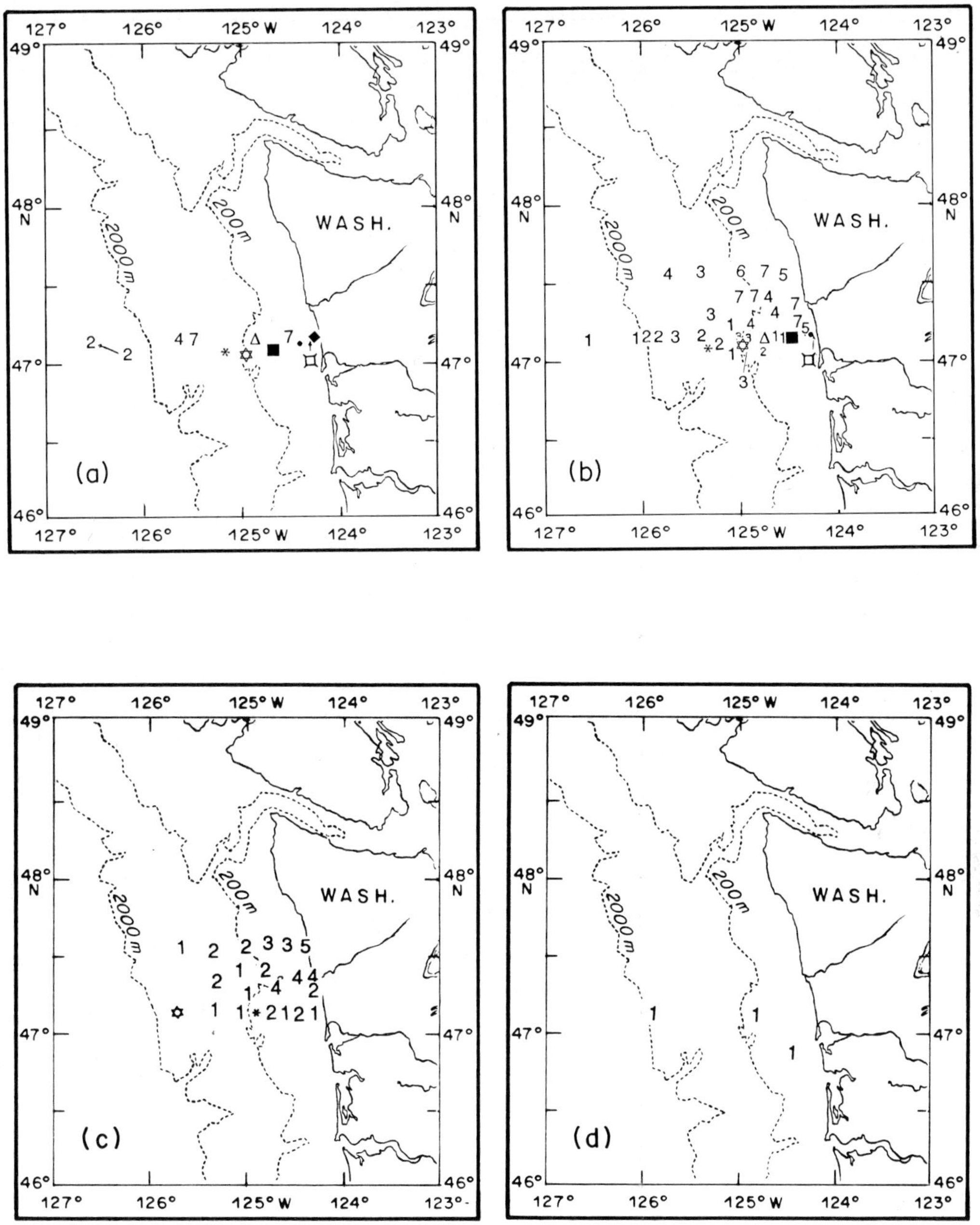

Fig. 3.2 Seasonal maps of primary production: (a) spring, (b) summer, (c) autumn and (d) winter. The numbers indicate ranges in measured rates of primary production (mg C m^{-2} d^{-1}); 1 = 0-500; 2 = 501-1,000; 3 = 1,001-1,500; 4 = 1,501-2,000; 5 = 2,001-2,500; 6 = 2,501-3,000; 7 = > 3,000. Symbols represent locations at which multiple stations were occupied between 1974 and 1982. The symbols are defined on the next page.

Fig. 3.2 Continued.

Season	Symbol	Number of Stations at a Location With Following Values ($mg\ C\ m^{-2}\ ^{-1}$) 0-500	501-1,000	1,001-1,500	1,501-2,000	2,001-2,500	2,501-3,000	>3,000
Spring	*			2	1			
Spring	✡		2	3				
Spring	Δ			2	1			
Spring	■					1		4
Spring	•					1		6
Spring	¤		1		1	1		3
Spring	◆		1		1		1	2
Summer	*	3	1					
Summer	✡		4	1	1			
Summer	Δ	1	2	3	1			
Summer	■		3	2	1	1		1
Summer	•		1	1	2	2	2	1
Summer	¤			1	3		1	2
Autumn	*	1						
Autumn	✡	1	1					

Table 3.1 Temporal and spatial locations of primary productivity stations occupied between 1974 and 1982.

Cruise	Month	Year	Number of Productivity Stations Shelf	Slope	Oceanic
TT091	July	1974	2		
YA199	January	1975	1		
TT096	March	1975			1
TT100	July	1975	1	1	1
TT105	October	1975	2		
TT111	September	1976	2		1
TT112	January	1977	2	1	
TT116	April	1977	2	1	1
TT122	August	1977	1	1	1
CY378	March	1978	2		
CY678	June	1978	3	1	
CY978	September	1978	1		
TT135	October	1978	4		
TT140	July	1979	10	9	
W8009	September	1980	11	8	
TT160	August	1981	26	14	
W8206	June	1982	25	3	

These designations agree reasonably well with the biomass gradients observed in CZCS and AVHRR images (Fig. 3.1). The temporal and spatial distributions of stations exhibit strong biases in sampling density (Fig. 3.2; Table 3.1). Spring and summer months were more highly sampled than autumn and winter; shelf waters, more highly sampled than slope or ocean waters.

Primary production was measured according to Steemann Nielsen (1952). Rates were calculated from incorporation of ^{14}C-carbon by particles collected on 0.45 μm Millipore filters; all samples were HCl-fumed and analyzed with a Packard TriCarb liquid scintillation counter. No dark-bottle corrections were applied. Waters from six depths were incubated on deck with neutral density filters to simulate the *in situ* irradiance field. The depths sampled corresponded to the following irradiances, given as percentages of the surface irradiance: 1-2.5%, 5-7%, 10-15%, 25-35%, 45-55% and 95-100%. The specific irradiance percentage varied among cruises but not within a cruise.

Between 1974 and 1982, four different incubation periods were employed: (1) sunrise to sunset (i.e, daylight incubation); (2) sunrise to local apparent noon; (3) local apparent noon to sunset; and (4) 24 h (i.e., complete daylight plus night-time incubation). The ^{14}C-incorporation data derived from the second, third and fourth types of incubations were standardized to the first incubation type, i.e., the daylight incubation (from sunrise to sunset). The two different types of half-day incubations were normalized to full daylight incubations by scaling the integrated radiation for the half-day incubation period to the total sunrise-to-sunset radiation. Primary production from the 24-h incubations were standardized to the daylight incubations by adding back the carbon respired during the night (see below).

The respiratory loss factor was determined by comparing stations with simultaneous daylight and 24-h incubations. An average night-time respiration or loss of the ^{14}C-carbon incorporated during daylight was estimated to be 20% from comparison made at six stations; this value is not inconsistent with the respiration coefficients reported by others (cf. Parsons *et al.*, 1974; Vezina and Platt, 1988). No significant difference in the respiration factor was found among seasons or with depth.

Data collected during cruises conducted prior to 1974 were not included in the analyses reported in this chapter. The methods used to measure ^{14}C underwent significant changes in the early 1970's (e.g., geiger counting *vs.* liquid scintillation counting of ^{14}C samples; Anderson and Zeutschel, 1970). A trend of uniformly low rates of production normalized to chlorophyll *a* was observed for the pre-1974 data set in contrast to the post-1974 data sets (Bolger and Perry, 1987), suggesting a problem with the earlier data set (Fig. 3.3). In the last decade other investigators working in diverse regions of the ocean have also reported systematically higher rates of production in contrast to measurements taken in the same areas in the 1950's and 1960's (cf. Eppley, 1982; Chavez and Barber, 1987). Although we cannot rule out the possibility of decadal, long-term changes in rates of primary production, the trend evident in Figure 3.3 suggests that problems exist with the earlier data from the Washington and Oregon coast. Hence these data are not included in any other figure or table presented in this chapter. Summaries of the productivity data for years prior to 1974 are given in Table 3.2 to allow comparison to earlier observations (cf. Anderson, 1964, 1972).

Samples for pigment analyses were taken from the same water bottles as the productivity samples. Chlorophyll *a* and phaeopigment concentrations were determined by fluorometric procedures (Lorenzen, 1967). Daily shortwave radiation incident to the sea surface, as well as

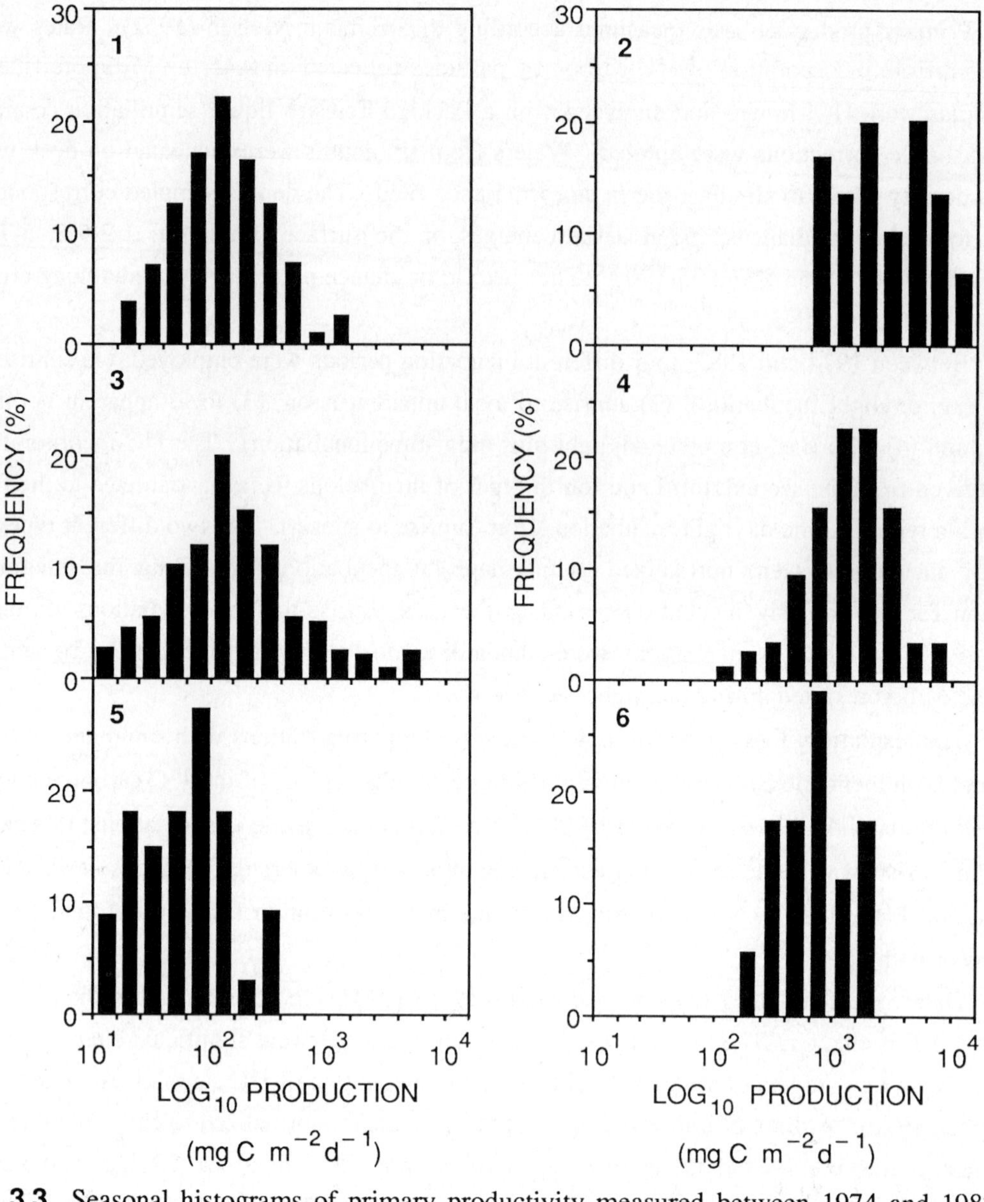

Fig. 3.3 Seasonal histograms of primary productivity measured between 1974 and 1982 in comparison with rates measured prior to 1974: (1) spring pre-74; (2) spring post-74; (3) summer pre-74; (4) summer post-74; (5) autumn pre-74; and (6) autumn post-74.

the total shortwave radiation incident to the productivity samples on deck during simulated *in situ* incubations, was monitored with a Belfort pyranometer. Photosynthetically active radiation, as μ moles photons m $^{-2}$ s $^{-1}$, was estimated as 50% of the shortwave radiation (Baker and Frouin, 1987). Attenuation coefficients for computing the percentage of surface radiation were derived from Secchi disc lowerings; the depth of the euphotic zone, calculated as 2.8 times the Secchi depth, was taken as the depth where irradiance was 1% of surface irradiance.

3.3 PATTERNS OF PHYTOPLANKTON DISTRIBUTION

Satellite images of phytoplankton pigment biomass and sea surface temperature were derived from CZCS and AVHRR sensors for 10 August 1981 (Figs. 3.1a and b). During the

Table 3.2 Primary productivity data collected prior to 1974 (mg C m $^{-2}$ d $^{-1}$).

Season	Shelf	Slope	Oceanic
Spring			
Median	207	121	105
Mean	300	190	144
C.V.%	84%	107%	106%
Summer			
Median	421	175	82
Mean	949	328	186
C.V.%	144%	137%	324%
Autumn			
Median	44	43	46
Mean	60	55	57
C.V.%	91%	60%	91%
Winter			
Median	71	137	80
Mean	77	179	124
C.V.%	62%	75%	86%

spring-summer period, CZCS images from other days and years are similar to the August 1981 image in Figure 3.1a with respect to the general pattern of phytoplankton pigment distribution and relative concentration of pigment; hence, this image is used to point out some of the important features for the Washington coastal zone during the upwelling seasons. First, phytoplankton biomass is high. In waters overlying the shelf, the pigment concentrations in Figure 3.1a ranged between 1.0 mg chlorophyll *a* plus phaeophytin m $^{-3}$ to 11.0 mg m $^{-3}$, while the range for slope waters was from 0.3 to 8.5 mg m $^{-3}$. In contrast, the pigment concentration in offshore waters (e.g., 140 km from the coast), averaged 0.3 mg m $^{-3}$. Second, features similar to the squirts or jets observed off California (c.f. Kelly, 1985) exist and may be important in transporting phytoplankton-rich coastal water into the more oligotrophic oceanic ecosystem offshore. Third, the distribution of phytoplankton pigments is patchy or non-uniform in both the longshore and the cross-shelf directions (Fig. 3.4a and b). The spatial features evident in the CZCS image are also apparent in the AVHRR image, indicating that mesoscale three-dimensional physical circulation plays a dominant role in the observed distribution pattern of phytoplankton in these waters. Similarity of features in co-located CZCS and AVHRR imagery have been noted in other regions (e.g., in California waters, Abbott and Zion, 1985). Fourth, the scales of patchiness evident in the phytoplankton distribution indicate that a serious bias will exist if data collected from ship transects is extrapolated to the entire region.

A CZCS image from late winter is given in Figure 3.1c for 17 March 1981, as a contrast to the spring-summer upwelling period. Although fewer images are available for the late autumn -winter period due to low solar angle and increased cloud cover, the available images do show agreement with the general pattern derived from shipboard chlorophyll *a* data as presented by Landry *et al.* (1989, Chapter 1). The important features of the winter distribution include a

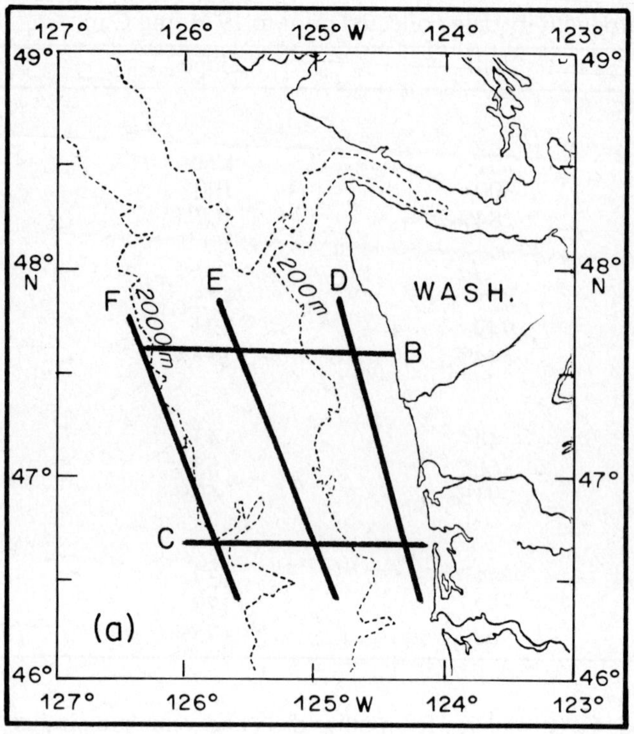

Fig. 3.4a Location of subsamples of pigment concentrations from a 10 August 1981 Coastal Zone Color Scanner image.

generally lower phytoplankton abundance in shelf, slope and oceanic regions, with a restriction of the higher phytoplankton abundances to an area immediately adjacent to the coast.

3.4 ENVIRONMENTAL PARAMETERS

The two primary environmental parameters that regulate the abundance and productivity of phytoplankton in the global ocean are light and nutrients, primarily nitrogen. In this mid-latitude upwelling region, the dominant seasonal and spatial patterns of phytoplankton biomass can similarly be explained by the annual patterns of the flux of solar radiation to the sea surface and the flux of upwelled nitrate into euphotic waters. In addition in this region, the longshore and cross-shelf currents play an important role in generating and maintaining the phytoplankton patchiness evident in the CZCS images. The role of advection in maintaining chlorophyll and nutrient distributions is treated explicitly by Hermann *et al.* (1989, Chapter 6), who used a two-dimensional array of current measurements to separate advective contributions from non-conservative changes observed along a transect across the mid-Washington shelf during two summer periods.

The total daily flux of solar radiation to the top of the atmosphere has a strong seasonal signal at mid-latitudes. The generalized curve in Figure 3.5 represents the solar radiation reaching the top of the atmosphere. Variable cloud cover will affect the magnitude of radiation reaching the sea surface. Additionally, daylength varies from a minimum of 8.2 h at the winter solstice to a maximum of 16.1 h at the summer solstice. It should be noted that the total hours of

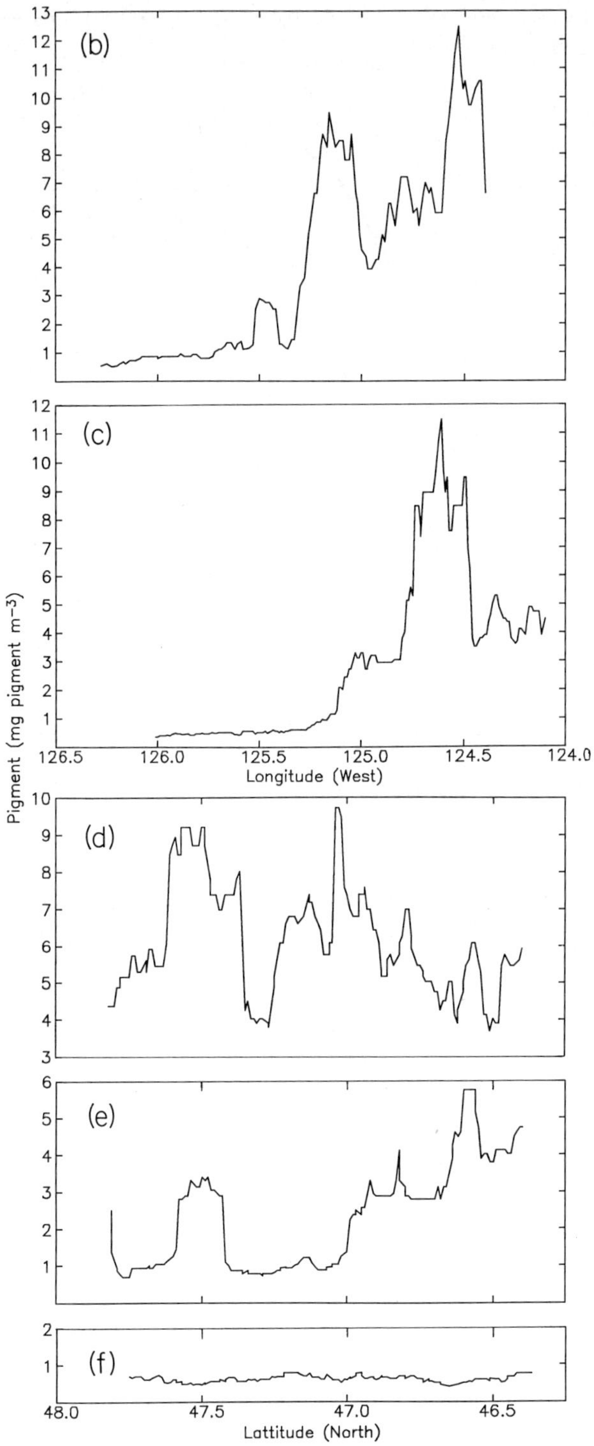

Fig. 3.4b The sampling lines for pigment concentrations from a 10 August 1981 Coastal Zone Color Scanner image: (b) northern and (c) southern cross-shelf samples and (d) shelf, (e) slope and (f) ocean alongshore samples. Locations are shown in Figure 3.4a.

available light, as well as the total incident photosynthetically available radiation, are important in determining the daily phytoplankton division rate (Vaulot and Chisholm, 1987); hence, the

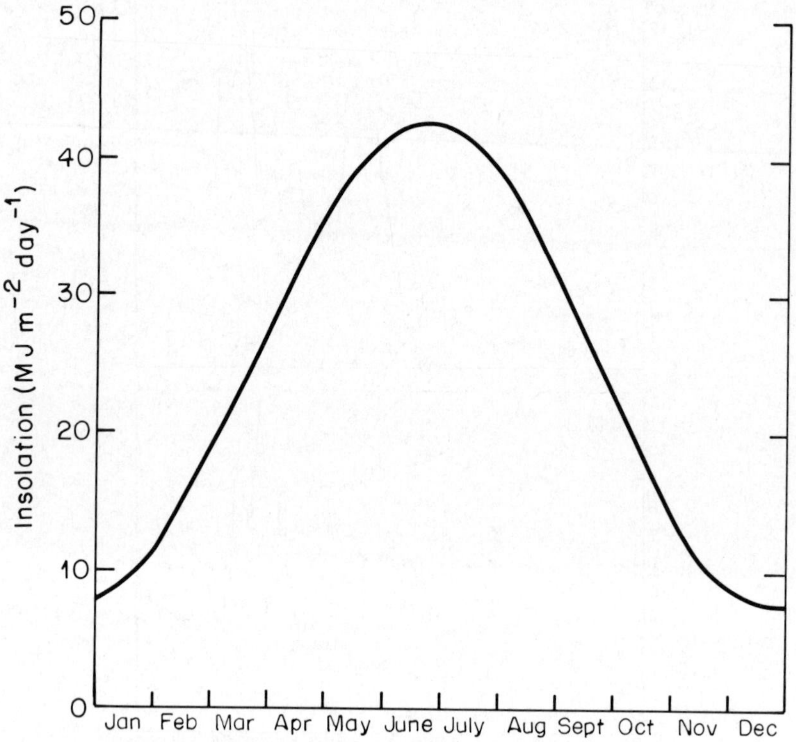

Fig. 3.5 Generalized annual cycle of solar shortwave radiation (285-2,000 nm) reaching the sea-surface at 48° N. Redrawn from Kirk (1983).

importance of considering daylength as an independent regulating parameter. During the spring and summer upwelling season, cloud cover due to regional storms or coastal weather attenuate radiation reaching the earth's surface by as much as 25 to 75% (DOE, 1980). Although systematic records of cloud cover are not available for years of the field program, inspection of available records indicate that the reduction in solar radiation due to clouds is episodic, with a time scale of days to weeks.

The influx of new nutrients into euphotic waters also varies seasonally, in response to the intensity and duration of upwelling winds and the depth of the nutricline in the upwelling-source waters. The seasonal nutrient patterns as a function of depth and distance offshore are described by Landry *et al.* (1989, Chapter 1). It should be noted that the persistence of significant concentrations of nitrate in the surface waters during the winter is primarily a result of the reduced phytoplankton growth rates; these low growth rates, in turn, are a consequence of the low flux of solar radiation to the sea surface and deep winter mixing.

Although upwelling can occur in any season in association with northerly winds, it predominates in the spring and summer seasons. Upwelling has a strong effect on temperature and nutrient distributions over the shelf as deeper cold, nutrient-rich water is moved up the slope onto the shelf and into the surface layers. The seasonal pattern of winds in the Pacific Northwest, with their predominantly northerly component in the spring-summer season and southerly component in the autumn-winter, are well described by Hickey (1989, Chapter 2). The most persistent upwelling occurs in the spring and early summer, with its persistence declining as the summer advances, as described by Hermann *et al.* (1989, Chapter 6). Although considerable nitrogen regeneration within euphotic waters does occur as a result of zooplankton and microzooplankton metabolism, the flux of nitrate from upwelling source waters is the prime factor responsible for the high phytoplankton biomass and rates of primary production, particularly new production (*sensu* Eppley and Peterson, 1979 and as discussed by Dortch and Postel, 1989, Chapter 4). While the availability of photosynthetically utilizable radiation determines the actual daily *in situ* rate of production, it is the rate of renewal of dissolved and available nitrogen that is ultimately responsible for the high primary production in upwelling regimes.

3.5 PATTERNS OF PRIMARY PRODUCTION

Rates of primary production in the Washington coastal region vary spatially and temporally; although the cruise sampling design was inadequate to accurately resolve the scales, a general pattern can be extracted from the seasonally-averaged ship data. Seasonal maps of observed daily rates of carbon fixation, integrated for the water column, are given in Figure 3.2 for cruises made between 1974 and 1982. These maps also illustrate the variation in sampling density among seasons and regions; e.g., no winter productivity stations were occupied on the slope and only two were occupied on the shelf and one in the oceanic region; no oceanic data were collected in the autumn (see also Table 3.1).

The data are also summarized by season as the median, mean and coefficient of variation about the mean according to location (i.e., shelf, slope and oceanic regimes) in Figure 3.6 During the winter between 1974 and 1982, only two productivity stations were occupied in shelf waters and one in oceanic waters. The winter mean for the shelf waters was 114 mg C m^{-2} d^{-1}, with a C.V.% of 37%. The single value for the oceanic station was 446 mg C m^{-2} d^{-1}; this high oceanic production rate cannot be explained with the available data.

To compute a regional annual average, winter slope production was assumed to be equal to winter shelf production; autumn oceanic production was assumed to be equal to autumn slope production. The resulting annually averaged rates of primary production are: 646 g C m^{-2} y^{-1} in waters overlying the shelf, 294 g C m^{-2} y^{-1} in waters overlying the slope and 229 g C m^{-2} y^{-1} in oceanic waters.

Several trends in rates of primary production are evident in the maps and histograms given in Figures 3.2 and 3.6. First, rates of primary production are highest during the spring in all regions. Second, productivity was highest in waters overlying the shelf, intermediate in slope waters and lowest in oceanic waters during all seasons (see note on paucity of winter data, above). Third, the skewed or non-normal distribution of the productivity data for some regions and seasons (cf. Figs. 3.3 and 3.7a for post-1974 data) dictate the use of medians for calculating seasonal and annual averages (non-parametric statistics do not have an underlying assumption of normality). Even so, the available data from the spring and summer months are dominated by the August 1981 (Summer) and June 1982 (Spring) cruises – both periods of high upwelling (Hermann *et al.*, 1989, Chapter 6). Fourth, low rates of production could be observed even during the most productive period (e.g., in shelf waters during the summer, Figs. 3.2b and 3.7b).

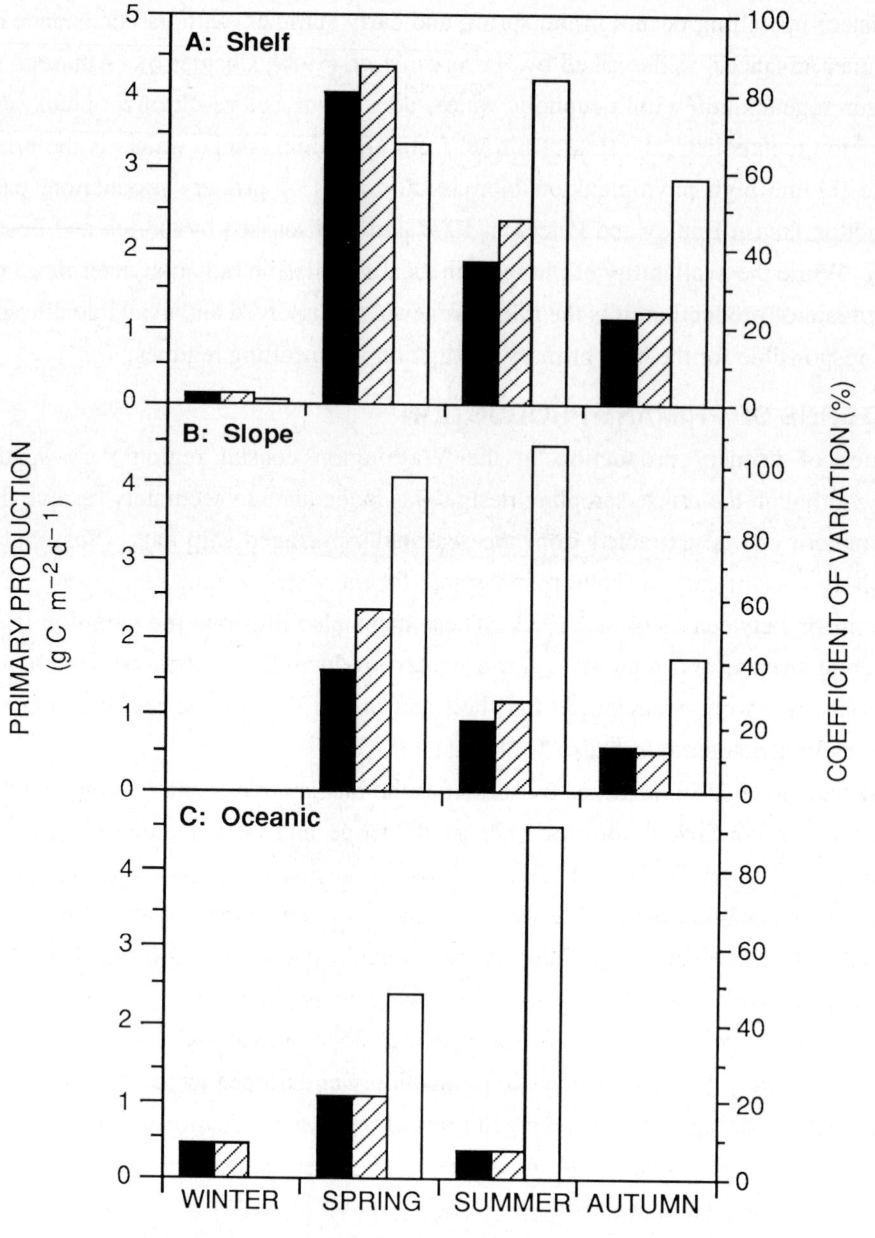

Fig. 3.6 Seasonal histograms of medians, means and coefficient of variation as percent of the mean for the (a) shelf, (b) slope and (c) oceanic regions. Solid bars are medians; diagonal bars are means; white bars are % coefficient of variation of the mean.

For all regions, the variability in primary production was generally greatest during the summer months (note the high CV% for summer season in shelf, slope and ocean regions). In waters overlying the shelf, the high variance is most likely a consequence of the more intermittent nature of the upwelling during the summer (Hermann *et al*, 1989, Chapter 6); the extremes in summer production can be related to upwelling and downwelling events. Based on hydrographic data, an upwelling event occurred during TT140 on 20-21 July 1979 (Fig. 3.8a). The median rate of primary production in shelf waters during this cruise was 3.20 g C m^{-2} d^{-1}. In contrast,

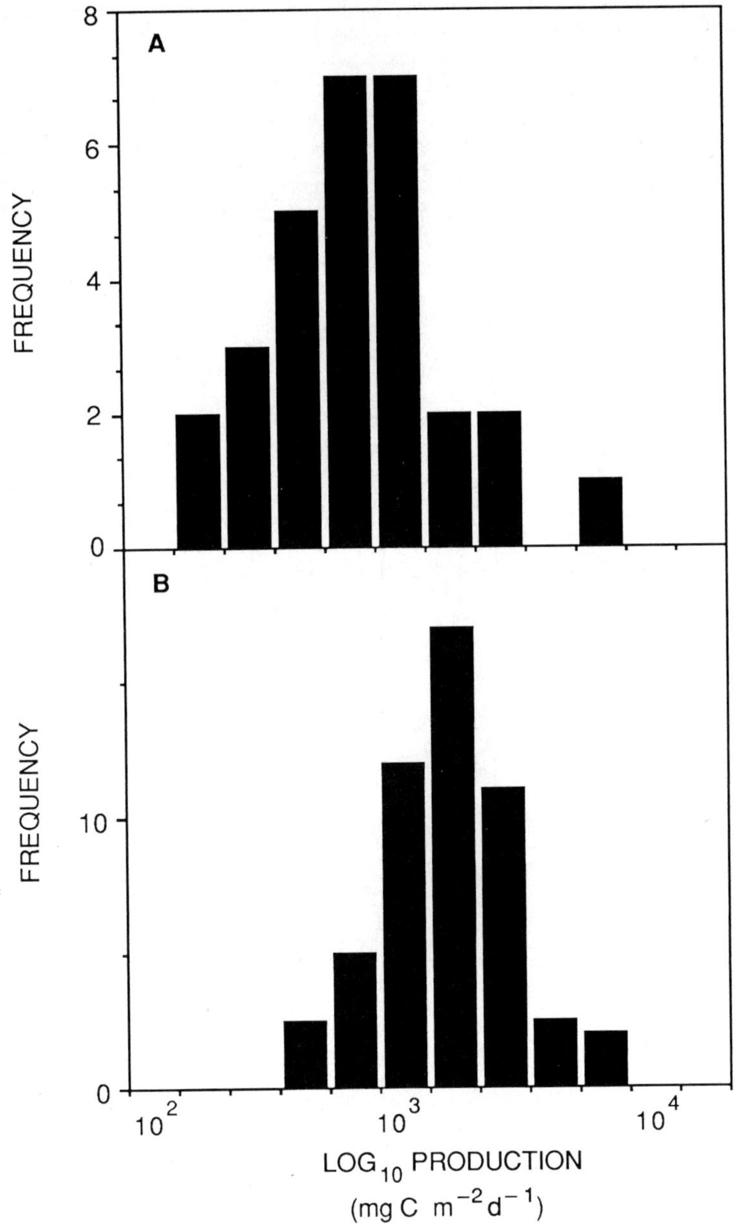

Fig. 3.7 Histograms of $\log_{10}$ summer rates of primary production in waters over the (a) slope and (b) shelf.

a downwelling event was observed on 26-27 August 1981 during TT160 (Fig. 3.8b); the corresponding median production rate was 1.39 g C m^{-2} d^{-1}. Similar responses to upwelling and relaxation events have been observed in other eastern boundary current regions (Barber and Smith, 1980), including the system off Oregon (Small and Menzies, 1981). The upwelling or downwelling conditions prevailing during a specific cruise or portion of a cruise bias the seasonal averages of primary production in Washington waters, particularly for the spring and summer seasons.

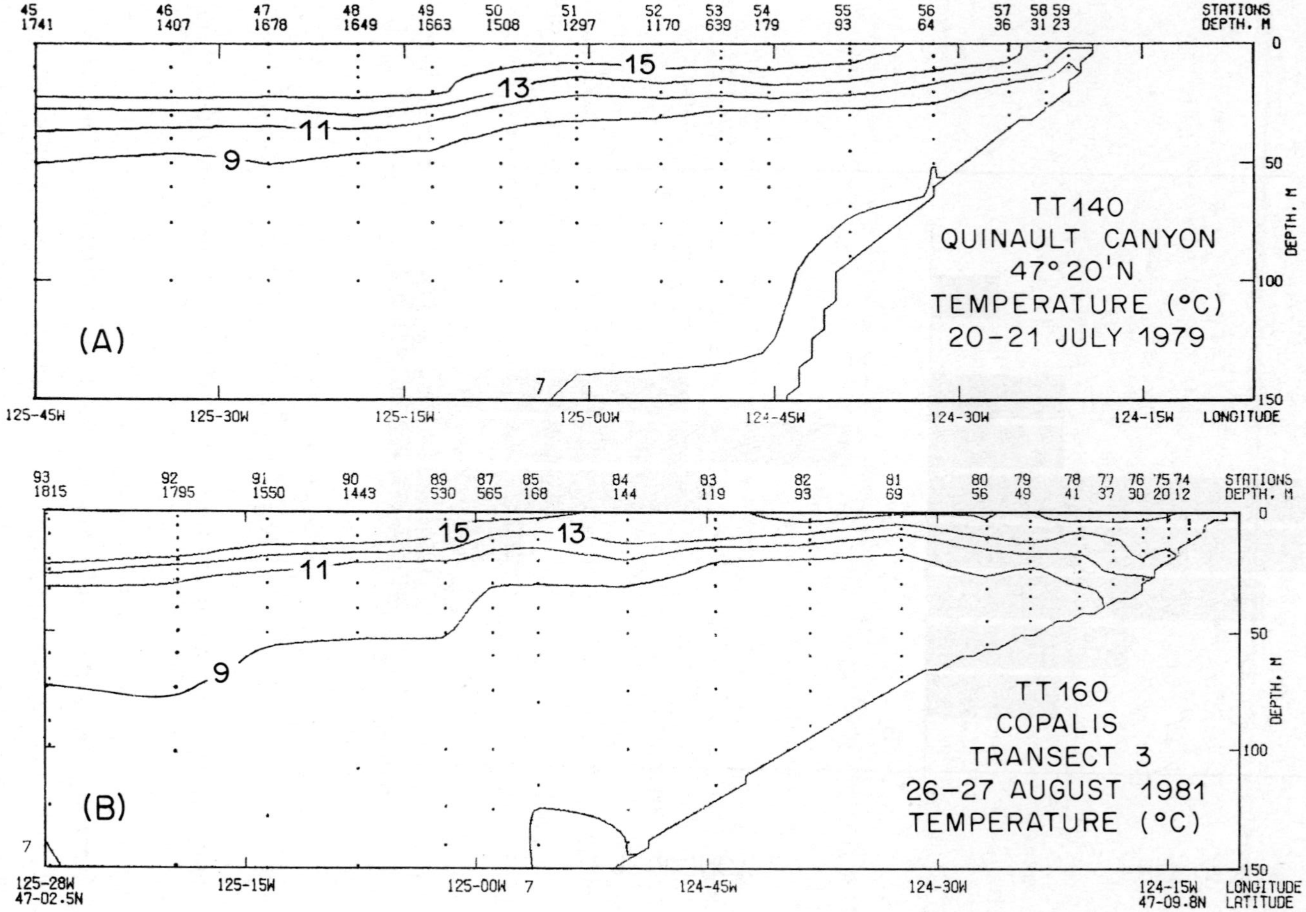

Fig. 3.8 Temperature sections from (a) TT140 and (b) TT160.

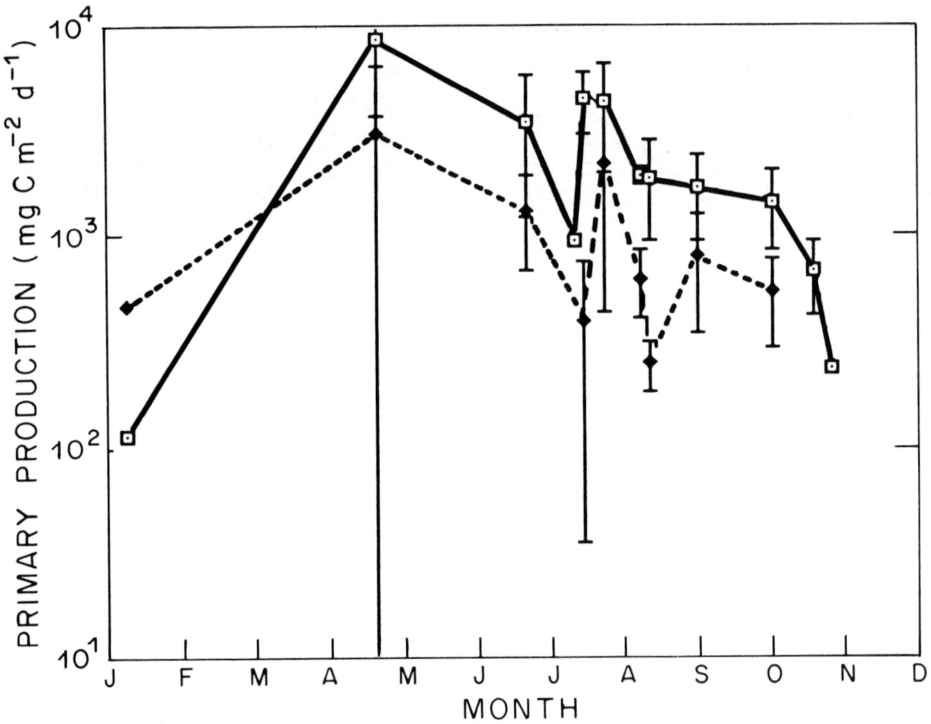

Fig. 3.9 Monthly averages of daily rates of integrated water column primary production, with standard deviations, for 17 cruises; open squares are shelf stations; closed diamonds are combined slope and oceanic stations.

The variability in primary production in slope and oceanic waters during summer is greater than that over the shelf. This increased variability may be a consequence of offshore transport of highly productive coastal water by squirts (see high CV% in Figs. 3.6b and c). For example, the high CV% observed in the oceanic region during the summer, relative to the spring and autumn (Fig. 3.6c), could be the result of episodic injection of highly productive coastal water by squirts into the oligotrophic offshore waters (see features in the CZCS image in Fig. 3.1a). A second example is the skewed distribution of primary production in the slope data (Fig. 3.7a) in contrast to the log normal distribution for the shelf waters (Fig. 3.7b). The deviation of the primary production in slope waters from a log normal distribution may relate to transport of richer coastal water by the squirts, or alternatively may be related to wind-induced upwelling at the shelf break.

3.6 REGULATION OF PRIMARY PRODUCTION

One of the fundamental causes of variability in total water column rates of production is the variability in the abundance of phytoplankton. A comparison between Figures 3.9 and 3.10 illustrates this point: on an annual scale, total water column production varied by almost two orders of magnitude, while productivity normalized to phytoplankton-pigment biomass varied by less than one order of magnitude. The causes of variability in the rate of primary production per unit biomass will include any physiological parameter that affects rates of photosynthetic reduction of carbon dioxide per unit phytoplankton-pigment biomass. The later encompasses

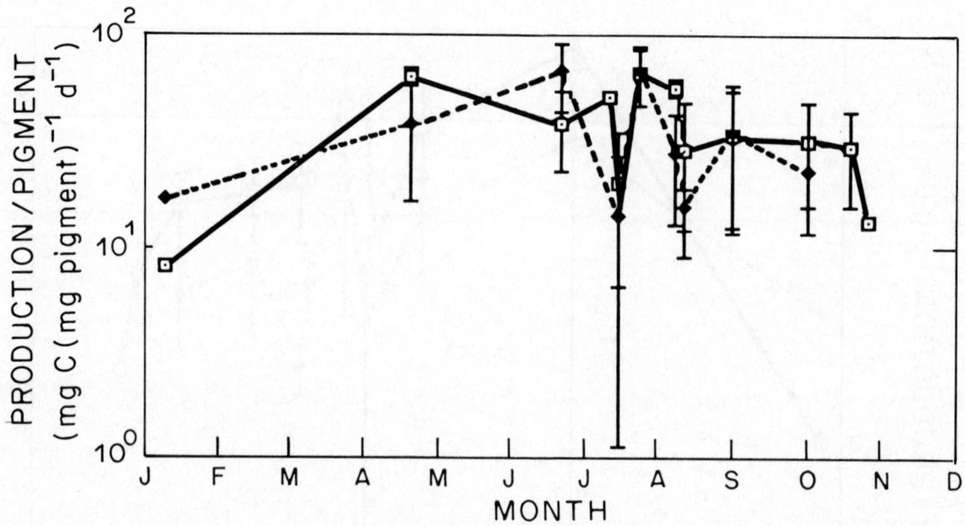

Fig. 3.10 Monthly averages of daily rates of integrated water column primary production normalized to chlorophyll *a*, with standard deviations, for 17 cruises; open squares are shelf stations; closed diamonds are combined slope and oceanic stations.

variation in photosynthetically available solar radiation (Perry *et al.*, 1981), nitrogen availability (Welschmeyer and Lorenzen, 1981), temperature (Eppley, 1972) and vertical mixing (Frost, 1987). As discussed earlier, solar radiation, nutrient flux, water temperature and mixing vary within a season, as well as seasonally and interannually.

The distinction between biomass *vs.* physiological control of the variability in rates of primary production is further illustrated by Figures 3.11-3.13. The relationship between water column primary production and surface phytoplankton-pigment biomass in Figure 3.11 was developed to investigate the utility of using remotely-sensed pigment concentrations to predict primary production from space (Eppley *et al.*, 1985; Perry, 1986; Bolger and Perry, unpublished manuscript). For the present application, Figure 3.11 illustrates that biomass variability can account for much, but not all, of the variability in primary production. The relationship between primary production and biomass (Fig. 3.11) can be applied to a CZCS image to generate a map of primary production (Fig. 3.1d). This map can be regarded as an example of the future potential of satellites for estimating primary production in spatially complex areas.

Additional variation can be accounted by the available radiant energy; a photosynthesis *vs.* irradiance curve determined for a specific water sample illustrates the non-linear exponential nature of the light response of photosynthesis (Talbot *et al.*, 1985; Fig. 3.12). The regulation of primary production by irradiance in mid-latitude upwelling waters is also evident when rates of production normalized to phytoplankton-pigment biomass are summarized for discrete water samples as a function of the incident daily irradiance at depth (cf. summer data, Fig. 3.13; note that the Y-axis is log-base 2). Such a strong relationship between production and irradiance is not observed at lower latitudes, where irradiance plays a less direct role in regulating primary production (e.g., in the Eastern Tropical Pacific, Banse and Yong, unpublished manuscript). The scatter in the data about the non-linear exponentially fitted curve is primarily a function of the

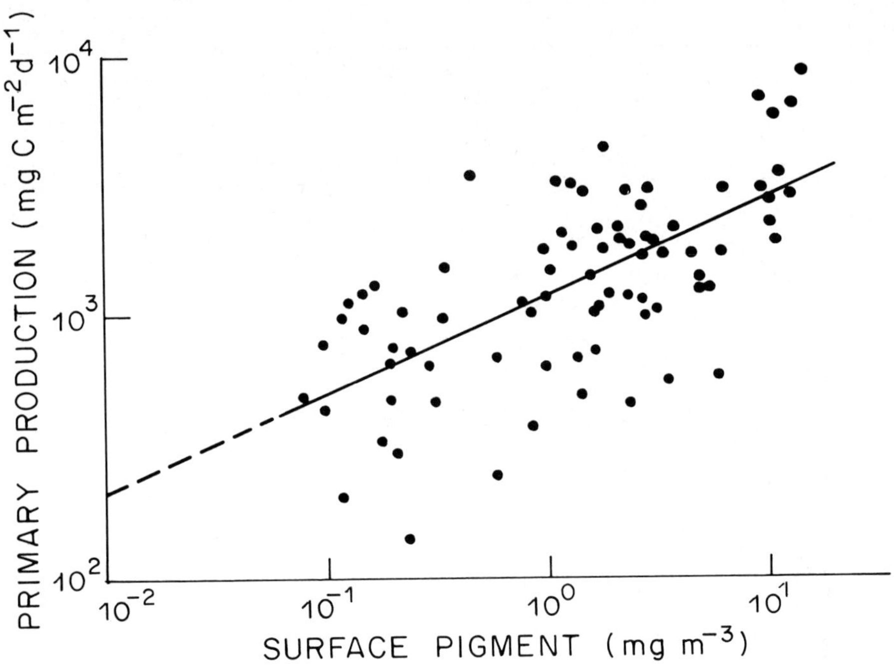

Fig. 3.11 Rates of integrated water column primary production as a function of surface chlorophyll *a* concentrations; summer data. From Bolger and Perry (unpublished manuscript).

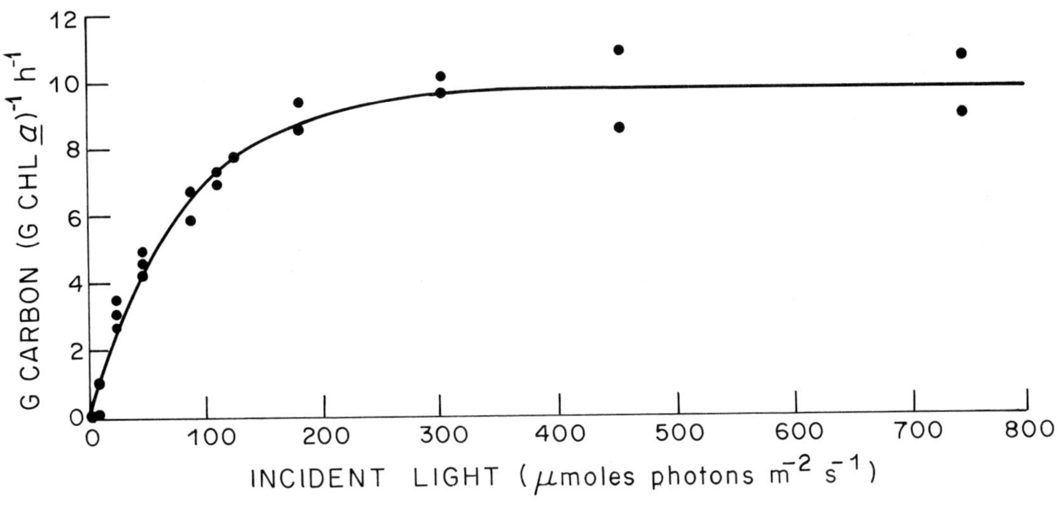

Fig. 3.12 Photosynthetic rate as a function of irradiance for a water sample taken from 2 m on TT160 in slope waters.

availability of nitrogen. The relationship shown in Figure 3.13 is an important summary of the regulation of primary production by solar radiation in this mid-latitude upwelling region.

3.7 FUTURE DIRECTIONS

The CZCS and AVHRR images provide insight into the spatial variability in the distribution of phytoplankton biomass and the complex physical circulation patterns in Washington

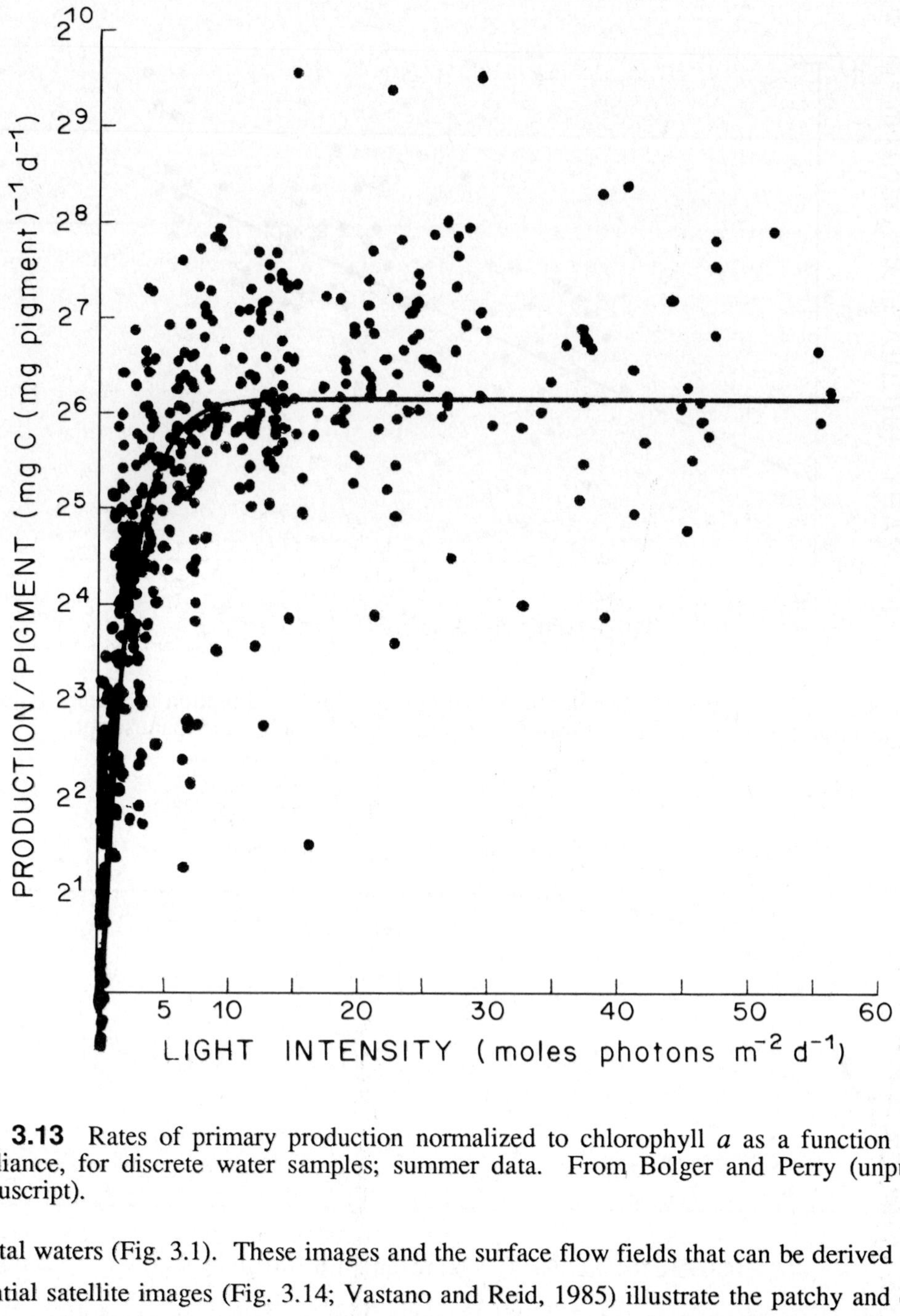

Fig. 3.13 Rates of primary production normalized to chlorophyll *a* as a function of daily irradiance, for discrete water samples; summer data. From Bolger and Perry (unpublished manuscript).

coastal waters (Fig. 3.1). These images and the surface flow fields that can be derived from sequential satellite images (Fig. 3.14; Vastano and Reid, 1985) illustrate the patchy and dynamic nature of this ecosystem; these images also illustrate the biases inherent in standard shipboard sampling designs. Future cruises in coastal upwelling systems would benefit from satellite or low-flying aircraft remote sensing of the sea surface temperature and color fields.

Additionally, satellite imagery can be used to determine what component of the day-to-day variability in pigment concentration can be attributed to physical advection as opposed to biological processes (Perry *et al.*, unpublished manuscript). The flow fields depicted in Figure 3.14, derived from two sequential AVHRR images from 10 and 11 August 1981 can be

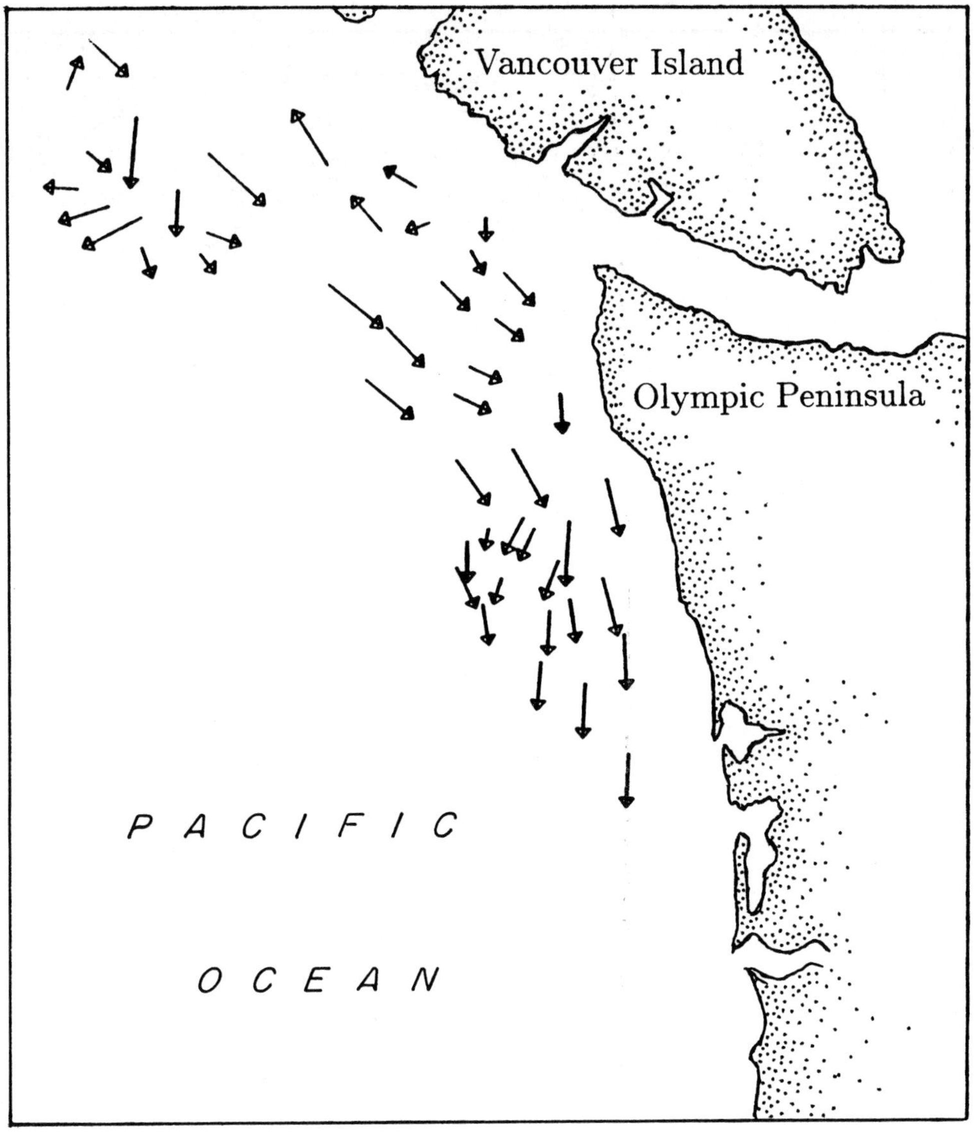

Fig. 3.14 Map of surface velocity vectors from between 10 and 11 August 1981, derived from a sequential pair of Advanced Very High Resolution Radiometer images. From Perry *et al.* (unpublished manuscript).

applied to sequential CZCS images from the same period to correct for the advective displacement of phytoplankton pigments. By combining pigment-primary productivity algorithms with sequential satellite images of ocean color and surface flow fields, it is possible to compute what the expected phytoplankton pigment concentrations would be on the second day of any two-day image sequence in the absence of zooplankton grazing. The actual pigment concentrations on the second day are almost always less than the expected concentrations, as a consequence of zooplankton grazing. A map of the carbon lost to zooplankton grazing or by other mechanisms between 10 and 11 August 1981 is shown in Figure 3.15. The future combined use of satellite and shipboard observations should improve our information on the annual and interannual

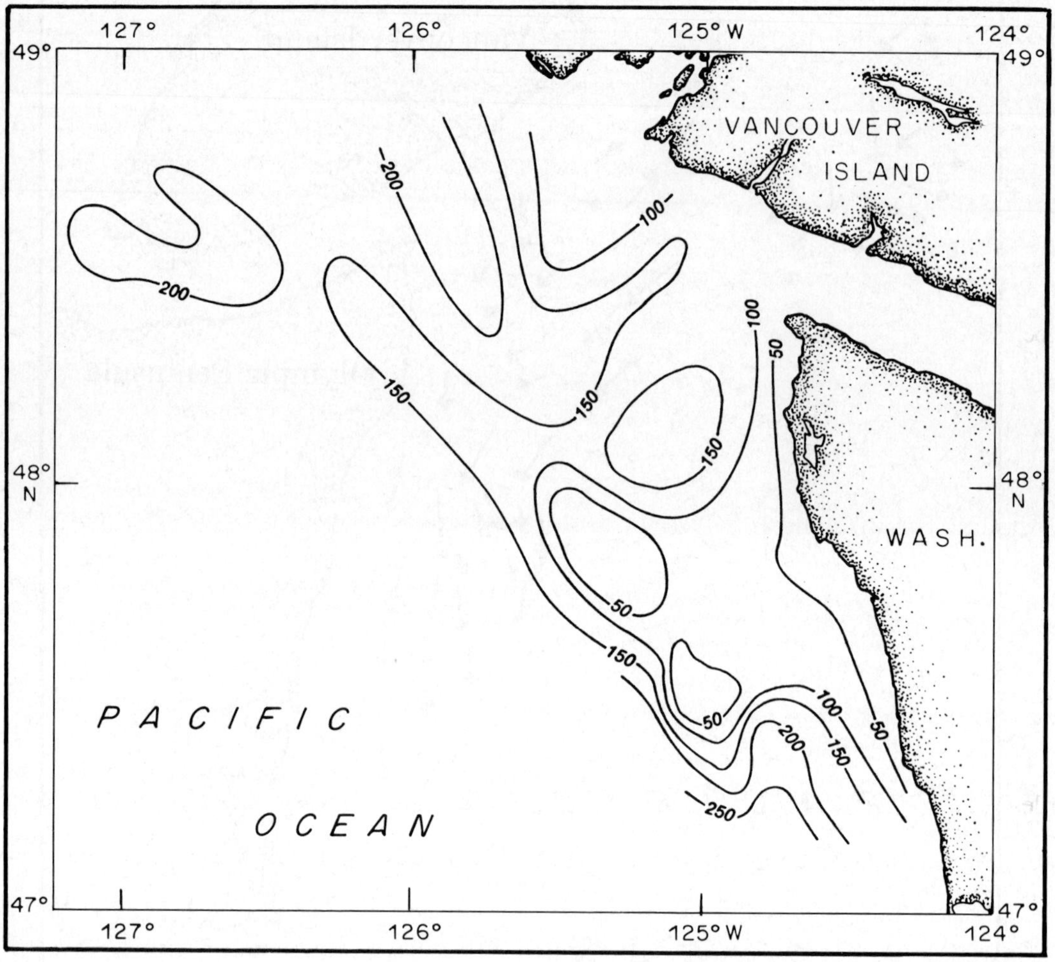

Fig. 3.15 Map of phytoplankton carbon removed from surface waters between 10 and 11 August 1981 (mg C m^{-3} d^{-1}). From Perry *et al.* (unpublished manuscript).

variability in the spatial distribution of phytoplankton, their primary production and the flux of carbon from euphotic waters.

ACKNOWLEDGEMENTS

DOE contract EV-76-S-062225-TA-26 supported collection of field data; NASA grant NAGW-889 supported satellite data processing. This is a component of Contribution No. 1784 from the School of Oceanography, University of Washington.

REFERENCES

Abbott, M.R. and P.M. Zion. 1985. Satellite observations of phytoplankton variability during an upwelling event. *Cont. Shelf Res.*, 4: 661-680.

Anderson, G.C. 1964. The seasonal and geographic distribution of primary productivity off the Washington and Oregon coasts. *Limnol. Oceanogr.*, 9: 284-302.

Anderson, G.C. 1972. Aspects of marine phytoplankton studies near the Columbia River, with special reference to a subsurface chlorophyll maximum. Pages 219-240 in: Pruter, A.T. and D.L. Alverson (eds.), *The Columbia River Estuary and Adjacent Ocean Waters: Bioenvironmental Studies*. Univ. of Washington Press, Seattle, Wa.

Anderson, G.C. and R.P. Zeutschel. 1970. Release of dissolved organic matter by marine phytoplankton in coastal and offshore areas of the northeast Pacific Ocean. *Limnol. Oceanogr.*, 15: 402-407.

Baker, K.S. and R. Frouin. 1987. Relation between synthetically available radiation and total insolation at the ocean surface under clear skies. *Limnol. Oceanogr.*, 32: 1370-1372.

Barber, R.T. and R.L. Smith. 1980. Coastal upwelling ecosystems. Pages 31-68 in: Longhurst, A. (ed.), *Analysis of Marine Ecosystem.* Academic Press, London.

Bolger, J. and M.J. Perry. 1987. Primary productivity values for the Washington coast between 1961 and 1982. Pages 2-5 in: J. Marra (ed.), *Biowatt News*. Columbia Univ., Palisades, N.Y.

Chavez, F.P. and R.T. Barber. 1987. An estimate of new production in the equatorial Pacific. *Deep-Sea Res.,* 34: 1229-1243.

Dortch, Q. and J. R. Postel. 1989. Phytoplankton and nitrogen intereactions. Pages 139-174 in: M.L. Landry and B.M. Hickey (eds.), *Coastal Oceanography of Washington and Oregon.* Elsevier, Amsterdam, Neth.

Eppley, R.W. 1972. Temperature and phytoplankton growth in the sea. *Fish. Bull.,* 70: 1063-1085.

Eppley, R.W. 1982. The PRPOOS program. *EOS, Trans. Am. Geophys. Union,* 63: 522-523.

Eppley, R.W. and B.J. Peterson. 1979. Particulate organic matter flux and planktonic new production in the deep ocean. *Nature,* 282: 677-680.

Eppley, R.W., E. Stewart, M.R. Abbott and U. Heyman. 1985. Estimating ocean primary production from satellite chlorophyll: introduction to regional differences and statistics for the Southern California Bight. *J. Plankton Res.*, 7: 57-70.

Frost, B.W. 1987. Grazing control of phytoplankton stock in the open subarctic Pacific Ocean: a model assessing the role of mesozooplankton, particularly the large calanoid copepods *Neocalanus* spp. *Mar. Ecol. Progr. Ser.,* 39: 49-68.

Gordon, H.R. and D.J. Castaño. 1987. Coastal Zone Color Scanner atmospheric correction algorithm: multiple scattering effects. *Appl. Opt.,* 26: 2111-2122.

Gordon, H.R., J.W. Brown and R.H. Evans. 1988. Exact Rayleigh scattering calculations for use with the Nimbus-7 Coastal Zone Color Scanner. *Appl. Opt.,* 27: 862-871.

Hermann, A. J., B.M. Hickey, M.L. Landry and D.F. Winter. 1989. Coastal upwelling dynamics. Pages 211-254 in: M.L. Landry and B.M. Hickey (eds.), *Coastal Oceanography of Washington and Oregon.* Elsevier, Amsterdam, Neth.

Hickey, B. M. 1989. Patterns and process of circulation over the shelf and Slope. Pages 41-116 in: M.L. Landry and B.M. Hickey (eds.), *Coastal Oceanography of Washington and Oregon.* Elsevier, Amsterdam, Neth.

Kelly, K. 1985. The influence of winds and topography on the sea surface temperature patterns over the northern California slope. *J. Geophys. Res.*, 90: 11,783-11,798.

Kirk, J.T. 1983. *Light and photosynthesis in aquatic ecosystems.* Cambridge Univ. Press, Cambridge, Mass. and New York, N.Y., 357 pp.

Landry, M.R., J. R. Postel, W.K. Peterson and J. Newman. 1989. Broad scale patterns in the distribution of hydrographic variables. Pages 1-40 in: M.L. Landry and B.M. Hickey (eds.), *Coastal Oceanography of Washington and Oregon.* Elsevier, Amsterdam, Neth.

Lorenzen, C.J. 1967. Determination of chlorophyll and phaeopigments: spectrophotometric equations. *Limnol. Oceanogr.*, 12: 343-346.

Parsons, T.R., M. Takahashi and B. Hargrave. 1984. *Biological Oceanographic Processes*, 3rd ed. Pergamon Press, N. Y.

Perry, M.J. 1986. Assessing marine primary production from space. *BioScience*, 36: 461-467.

Perry, M.J., M.C. Talbot and R.S. Alberte. 1981. Photoadaptation in marine phytoplankton: response of the photosynthetic unit. *Mar. Biol.,* 62: 91-101.

Small, L.F. and D.W. Menzies. 1981. Patterns of primary productivity and biomass in a coastal upwelling region. *Deep-Sea Res.*, 28: 123-149.

Steeman Nielsen, E. 1952. The use of radioactive carbon (C^{14}) for measuring organic production in the sea. *J. Cons. Int. Explor. Mer.*, 18: 117-140.

Talbot, M.C., D.S. Thoreson and M.J. Perry. 1985. Photosynthesis vs. light intensity measurements: a miniaturized incubator. ONR Tech. Rep., Univ. of Washington, Seattle, Wa.

Vastano, A.C. and R.O. Reid. 1985. Sea surface topography estimation with infrared satellite imagery. *J. Atmos. Oceanogr. Tech.*, 2: 393-400.

Vaulot, D. and S.W. Chisholm. 1987. A simple model of the growth of phytoplankton populations in light/dark cycles. *J. Plankton Res.,* 9: 345-366.

Vezina, A.F. and T. Platt. 1988. Food web dynamics in the ocean: I. Best estimates of flow networks using inverse methods. *Mar. Ecol. Prog. Ser.,* 42: 269-287.

Welschmeyer, N.A. and C.J. Lorenzen. 1981. Chlorophyll-specific photosynthesis and quantum efficiency at subsaturating light intensities. *J. Phycol.,* 17: 283-293.

CHAPTER 4

PHYTOPLANKTON - NITROGEN INTERACTIONS

Q. Dortch[1] *and J. R. Postel*[2]
[1] *Louisiana Universities Marine Consortium, Chauvin, LA 70344*
[2] *School of Oceanography, University of Washington, Seattle, WA 98195*

4.1 INTRODUCTION

In coastal areas, especially those where upwelling occurs, the availability of suitable dissolved nitrogen sources for phytoplankton growth is postulated to strongly influence phytoplankton productivity (e.g., Ryther and Dunstan, 1971; Dugdale, 1976; MacIsaac *et al.*, 1985). The importance and nature of the coupling between nitrogen supply and productivity is currently a highly debated topic obscured by lack of methods for direct measurement and of understanding of the processes involved. From the distribution of variables off the Washington and Oregon coasts described in Landry *et al.* (1989, Chapter 1) an inverse correlation between changes in nitrate concentration and concentrations of dissolved oxygen and chlorophyll occurs frequently, suggesting that the two variables are closely linked via phytoplankton nitrogen uptake.

In order to characterize nitrogen uptake by phytoplankton off the Washington coast, ^{15}N uptake was measured at selected stations from 1971-79 (Table 4.1). Three types of information were obtained during these studies: 1) experimental data on light, nitrogen concentration and size dependence of nitrogen uptake; 2) the space and time distribution of nitrate, ammonium and urea uptake; and 3) the relationship between primary production and nitrogen uptake. While the results can be examined in many different ways, the most important question to be addressed is the role of nitrogen supply in regulating primary production.

4.2 METHODS

The historical data presented here were collected using currently-available, standard methods (e.g, Dugdale and Goering, 1967; Eppley *et al.*, 1973; McCarthy *et al.*, 1977; MacIsaac *et al.*, 1985). General views on the physiology and methodology of nitrogen uptake have changed markedly in recent years, due particularly to the influence of McCarthy and Goldman (1979). The ^{15}N uptake methodology used in this study will be described in some detail and then critically evaluated in the light of present knowledge.

4.2.1 Description

Water samples were collected by standard hydrocast and dispensed into 2-liter Pyrex bottles to which ^{15}N-labeled ammonium (99 atom % enrichment), nitrate (95% enrichment), nitrite (95% enrichment), or urea (99.2% enrichment) was added. Unless otherwise noted, bottles were incubated on deck at surface seawater temperatures and screened to *in situ* light intensities with neutral density screens. Saturating nitrogen concentrations were used when uptake was measured as a function of light level; incubations ranged in length from 4.5 to

TABLE 4.1 Types (*) of measurements made on each cruise. All stations were on Copalis Line unless noted.

Cruise	Dates	Season[a]	^{15}N Uptake		^{14}C and ^{15}N Uptake[c]	Indicators of N deficiency[d]
			Ecophysiological Experiments[b]	Space/Time Distribution		
TT082[e]	July 1973	Summer	*			*
TT091	July 1974	Summer	*	*		*
TT096	Mar-Apr 1975	Spring		*		*
TT100	July 1975	Summer	*	*		*
TT105	Oct 1975	Fall/Winter		*		
TT111	Sept 1976	Summer	*	*	*	
TT112	Jan 1977	Fall/Winter	*	*	*	
TT116	April 1977	Spring	*	*	*	
TT122	Aug 1977	Summer	*	*	*	
CY678	Mar-Apr 1978	Spring	*	*	*	
CY978	Sept 1978	Summer		*		
TT140[f]	July-Aug 1979	Summer		*	*	
CY082	June 1982	Spring		*	*	*

[a]Characterized according to criteria outlined in Landry *et al.* (1989, Chapter 1) and Section 4.3.2. of this chapter

[b]Light, nitrogen concentration or size dependence experiments

[c]Cruises where ^{15}N and ^{14}C uptake were measured at the same stations and depths over comparable incubation times

[d]Enhancement of ^{14}C uptake in the dark by ammonium (Morris *et al.*, 1971) or biochemical indicators of nitrogen deficiency (Dortch *et al.*, 1984, 1985)

[e]Data collected on Newport, OR line

12.15 h during the daylight period. For experiments in which uptake rate was measured as a function of substrate concentration, incubations ranged from 4 to 10.5 h and only one nitrogen source was varied at a time. In experiments to determine the inhibition of nitrate uptake by ammonium the amount of unlabeled ammonium was varied while the $^{15}NO_3^-$ concentration was constant and saturating (> 5 μM).

Greatest effort went into determining temporal and spatial variations in nitrogen uptake in 3 locales off Copalis Head, Washington: the shelf, shelf break and offshore (Table 4.1). For routine measurements (depth profiles) of ^{15}N uptake, the amount of labeled nitrate or ammonium added was a compromise to minimize several problems. Enough ^{15}N had to be added so that ^{15}N enrichment in the particulate matter was measurable and to avoid nitrogen depletion during the incubation. At low ambient nitrogen concentrations even a small addition of ^{15}N considerably increased the concentration, perhaps stimulating uptake and making the rate of uptake vary over time. Saturating concentrations of urea were always added. Incubations varied in length from 4 to 28 h, but most were greater than 18 h.

Following incubation, samples were filtered onto precombusted Reeve Angel 945 AH or Whatman GF/F filters. For size fractionation studies, the samples were first passed through a 20 μm Nitex prefilter. The material on the 20 μm Nitex was then washed onto a second glass filter. The filters were dried at 60° C and stored over dessicant at room temperature. In the laboratory, the ^{15}N samples were analyzed for both particulate nitrogen and ^{15}N content using the method of Günther *et al.* (1966). Following combustion in sealed glass ampules, the amount of N_2 produced from each sample was measured in a calibrated expansion chamber and then directed into a Uthe Technology International Model 100C Precision Mass Analyzer to determine the fraction of ^{15}N in the assay gas.

The uptake rates were calculated as follows:

$$V = \frac{C_p - C_{nat}}{C_d t} \tag{4.1}$$

where C_p is the final atom % enrichment of the particulate matter, C_{nat} is the natural ^{15}N enrichment of the particulate matter, C_d is the initial ^{15}N enrichment of the dissolved nitrogen compound, t is the incubation time (h) and V is the specific uptake rate (h^{-1}) for each nitrogen source as designated in Table 4.2. The substrate transport rate, ρ (μmol N l^{-1} h^{-1}) is related to the specific uptake rate and the final concentration of particulate nitrogen, PN, by

$$\rho = VPN \tag{4.2}$$

Symbols for transport rates of each nitrogen source are designated in Table 4.2. According to Collos (1987), Equation 4.2 introduces a small but constant error unless the initial and final particulate nitrogen are approximately the same.

Replicate ^{15}N incubations were made on numerous cruises. Although there is considerable variability, a coefficient of variation (standard deviation/mean) of 25% is generally applicable (Table 4.3).

Most of the other methods for data used herein are standard oceanographic methods. ^{14}C uptake was measured as reported by Perry *et al.* (1989, Chapter 3). ^{14}C and ^{15}N uptake were compared only for those experiments where simulated *in situ* incubations were conducted simultaneously. Nutrient concentrations were determined by AutoAnalyzer according to Friederich and Whitledge (1972). Urea was measured by the method of Newell *et al.* (1967), which gave rather constant, low values. More recent data on urea distributions (e.g., Kaufman *et al.*, 1983; Kristiansen, 1983; Harrison *et al.*, 1985) indicate that higher and more variable concentrations are usually measured in coastal areas, suggesting a problem with this urea method and any calculations based on urea concentrations. Thus, a relative preference index (RPI) for urea (McCarthy *et al.*, 1977) cannot be calculated and C_d in Equation 4.1 may be in error, although the use of saturating urea concentrations may reduce the magnitude of the error.

TABLE 4.2 Abbreviations and units.

Abbreviation	Description	Units
$V_{NO_3^-}$	Specific nitrate uptake rate (Eq. 4.1)	h^{-1}
$V_{NH_4^+}$	Specific ammonium uptake rate (Eq. 4.1)	h^{-1}
$V_{NO_2^-}$	Specific nitrite uptake rate (Eq. 4.1)	h^{-1}
V_u	Specific urea uptake rate (Eq. 4.1)	h^{-1}
V_N	Totala nitrogen specific uptake rate (Eq.4.2)	h^{-1}
$\rho_{NO_3^-}$	Nitrate transport rate	$\mu mol\ l^{-1}\ h^{-1}$
$\rho_{NH_4^+}$	Ammonium transport rate	$\mu mol\ l^{-1}\ h^{-1}$
$\rho_{NO_2^-}$	Urea transport rate	$\mu mol\ l^{-1}\ h^{-1}$
ρ_u	Urea transport rate (Eq. 4.2)	$\mu mol\ l^{-1}\ h^{-1}$
ρ_N	Total[a] nitrogen transport rate (Eq. 4.2)	$\mu mol\ l^{-1}\ h^{-1}$
ρ_c	Carbon transport rate	$\mu mol\ l^{-1}\ h^{-1}$
PN	Particulate nitrogen concentration	μM-N
Chl	Chlorophyll *a* concentration	$\mu g\ l^{-1}$
NO_3^-	Nitrate concentration	μM
NH_4^+	Ammonium concentration	μM
U	Urea concentration	μM-N
I_0	Light intensity at surface	$ly\ d^{-1}$
$\%I_0$	Light intensity at depth relative to surface	%
I	Approximate absolute light level ($I_0 \times \%I_0$)	$ly\ d^{-1}$
z	Depth	m
$RPI_{NO_3^-}$	Relative preference index for nitrate[b]	none
Σ IN	Intracellular NO_3^- + NH_4^+ + free amino acid-N concentration	μM-N
AA/Pr	Free amino acid-N/protein-N	% mole N

[a]sum of nitrate, ammonium and urea uptake

[b]from footnote h, Table 4.10

4.2.2 Evaluation

Two major criticisms of the methods can be made: 1) the methodology, particularly for the incubations, is now dated and 2) the sampling scheme is not adequate to fully address the dynamic aspects of the system. However, similar flaws will be found in any data set collected over a similarly long period. Despite these problems, this is a unique data set because of its size and the number of other available measurements. A constant methodology was maintained throughout so the data are comparable. Thus, although the data must be interpreted with caution, they can still illuminate fundamental questions.

TABLE 4.3 Mean coefficient of variation (Standard deviation/mean x 100), calculated from n experiments (3-12 replicates/experiment).

	Coefficient of Variation (%)			
	V	ρ	PN	n
NO_3^-	16.51	22.10	22.26	31
NH_4^+	24.92	35.56	26.43	19

In early work with ^{15}N, uptake rates were assumed to be constant over time, so that the length of an incubation was important only to the extent that the ^{15}N-labeled compound was not depleted and the dark period was either consistently included or excluded. Recent studies, summarized in Collos (1983), have shown that as phytoplankton become nitrogen deficient, the responses of uptake rates to substrate additions are quite variable. In general, ammonium uptake rates are enhanced initially but decrease, whereas nitrate uptake rates are often very low but increase. Considerable species variation is observed in both. Off the Washington coast the surface layer is often depleted of nitrogen and the addition of even small amounts of ^{15}N-labeled substrate represents a large increase in ambient concentration. These populations are also the most likely to be nitrogen deficient and thus, temporal changes in uptake rates can be expected. Furthermore, up to 60% of the total nitrogen (ambient plus added) may be used during some of the longer incubations, so the uptake rate can decrease due to substrate depletion. Thus, whether the ^{15}N uptake rates are over or underestimates of the *in situ* rate depends on the nutrient in question, the nitrogen deficiency of the population, the initial substrate concentration and the length of the incubation. In addition, these factors may introduce variability into the data which is difficult to explain.

During $^{15}NH_4^+$ uptake experiments, zooplankton excretion and bacterial remineralization dilutes $^{15}NH_4^+$ with $^{14}NH_4^+$ (Glibert *et al.*, 1982; Garside and Glibert, 1984). Our calculations of ammonium uptake use only the initial enrichment of the dissolved ammonium C_d in Equation 4.1. Thus, the calculated ammonium uptake rate is underestimated (Glibert *et al.*, 1982; Garside and Glibert, 1984) and cannot be corrected because the time course of isotopic dilution is not known.

Several lesser problems must also be considered. Time course measurements of ^{15}N incorporation into particulate matter sometimes indicate a significant loss of ^{15}N, possibly as dissolved organic nitrogen, starting about 6 h after beginning the incubation (Laws, 1984; Price *et al.*, 1985; Dortch and Postel, unpublished). There can also be a significant loss of particulate nitrogen during an incubation which does not necessarily correlate with the loss in ^{15}N from the cell (Dortch and Postel, unpublished). Both of these little understood processes can affect the calculation and interpretation of uptake rates.

Finally, the spatial and temporal coverage of the data is rather patchy. Most sampling was conducted in spring and summer (Table 4.1). In addition, the stations form groups along the Copalis line: nearshore on the shelf with water depths between 20 and 150 m; at the shelf break in water depths of 200 m to 1,700 m; and offshore with water depths of 2,000 m to 2,500 m. A fixed, patchy sampling program such as this does not provide sufficient spatial resolution for such a dynamic region. In particular, unusual events, such as upwelling or brief winter/spring blooms may be underrepresented. Finally, when the data were compiled to examine the variation in the nitrogen uptake with time, location and other environmental variables, average values rather than depth-integrated values were used, in part, because a similar approach was used in Landry *et al.* (1989, Chapter 1) and because integrated values would not be comparable due to limited numbers of points in each profile, missing data and different depths sampled on different cruises.

4.3 RESULTS AND DISCUSSION

4.3.1 Experimental Measurements

Nitrogen Uptake as a Function of Concentration

When phytoplankton are exposed to increasing concentrations of a nitrogen compound, the uptake rate can be described hypothetically by a Michaelis-Menten-type equation,

$$V = V_{max} \frac{S}{k_s + S} \tag{4.3}$$

where V is the uptake rate, V_{max} is the maximum possible rate, S is the concentration of the nitrogen substrate of interest and k_s is the half saturation constant (Dugdale, 1967). Comparison of k_s and V_{max} between species in culture or natural populations can indicate their ability to utilize nitrogen under different nutrient regimes (Eppley *et al.*, 1969; MacIsaac and Dugdale, 1969). Although it is now known that the physiological response of phytoplankton to changing nutrient conditions is much more complex (Collos, 1983; Goldman and Glibert, 1983; Wheeler, 1983), this simpler approach may still provide information about how phytoplankton along the Washington coast are adapted to their environment.

Eighty-nine kinetics experiments were conducted on 12 cruises from 1973 to 1978 during all seasons. Each experiment represented from 4 to 24 measurements of uptake, but usually 5 to 6, at different nitrogen concentrations. Depending on the nitrogen source, visual inspection of the data clearly showed that between 0 and 62% of the experiments (Table 4.4) could not be described by Equation 4.3. In many cases of non-Michaelis-Menten kinetics (18 experiments, all but one involving nitrate or ammonium uptake), there was an inverse relationship between uptake rate and concentration. This phenomenon was first described by MacIsaac and Dugdale (1969), but the causes are still unknown since there is no correlation with any environmental parameter and no methodological problems can readily explain it. In other cases the uptake rate did not vary at all with nitrogen concentration, but this occurred primarily when ambient nitrogen concentrations were already saturating.

Excluding the experiments with non-Michaelis-Menten kinetics, the mean k_s for different nitrogen sources were ordered nitrite > urea > ammonium > nitrate. The rank-order of the mean V_{max} were ammonium > nitrate > urea > nitrite (Table 4.4). Eppley *et al.* (1969) proposed that a

TABLE 4.4 Effect of nitrogen concentration on nitrogen uptake rate. The k_s (μM) and V_{max} (h^{-1}) were calculated for each experiment in which the data fit Equation (4.3) using a Wolfe linearization (Dortch, 1980). The mean ± 1 standard error (number of experiments) are also included.

	Nitrogen Source			
	NO_3^-	NH_4^+	NO_2^-	Urea
Number of Experiments	29	30	5	25
% Non-Michaelis -Menten Kinetics	62	40	0	24
$k_s \pm$ SE (n)	0.50 ± 0.108 (11)	0.71 ± 0.122 (18)	1.9 ± 0.87 (5)	0.78 ± 0.167 (19)
$V_{max} \pm$ SE (n)	0.0058 ±0.00230 (11)	0.00682 ± 0.000930 (18)	0.0033 ± 0.00109 (5)	0.0046 ± 0.00106 (19)

low k_s would be associated with a low V_{max} and a high V_{max} would result in a high k_s. When this was tested with this large number of kinetics experiments, k_s and V_{max} were not significantly related for nitrate, ammonium, or urea uptake, but for nitrite the k_s was positively correlated with V_{max} ($r^2 = 0.9059$). The mean k_s values for nitrate and ammonium uptake were typical of those observed in coastal areas, as summarized in Goldman and Glibert (1983). Fewer k_s values are available for urea uptake, but in both phytoplankton cultures (Syrett and Leftley, 1976; Healey, 1977; Rees and Syrett, 1979) and coastal areas (Kristiansen, 1983), values are quite similar to those measured off the Washington coast. Although it was expected that the k_s and V_{max} values would be influenced by environmental variables, there was no variation in k_s or V_{max} with season, distance offshore, light level or biomass except for ammonium uptake. For ammonium, V_{max} was significantly greater ($p < 0.05$, one tailed t-test) for populations on the shelf than for those offshore. Further, the V_{max} for ammonium uptake was highly correlated with biomass, expressed either as particulate nitrogen or chlorophyll. The mean k_s for the summer was twice that for the fall/winter/spring ($p < 0.05$, one tailed t-test).

Nitrogen Uptake as a Function of Light Level

Uptake can be related to light level by a rectangular hyperbola similar to Equation 4.3, with a k_{LT} and V_{max} analogous to the k_s and V_{max} for uptake as a function of nitrogen concentration (MacIsaac and Dugdale, 1972). However, if there is significant dark N uptake or photoinhibition, then the rectangular hyperbola may not be appropriate because it only describes a small fraction of the data. In general, ammonium is thought to be more easily taken up in the dark than nitrate (MacIsaac *et al.*, 1974; Conway and Whitledge, 1979; Nelson and Conway, 1979; Price *et al.*, 1985). Furthermore dark uptake is thought to increase in nitrogen-deficient phytoplankton (Malone *et al.*, 1975; Caperon and Ziemann, 1976; Harrison, 1976; Healey, 1977; Rees and Syrett, 1979; Fisher *et al.*, 1982; Horrigan and McCarthy, 1982; Kanda *et al.*, 1985; Price *et al.*, 1985). Photoinhibition may be more typical in oligotrophic waters (MacIsaac and

TABLE 4.5 Effect of light on nitrogen uptake.

	Nitrogen Source			
	NO_3^-	NH_4^+	NO_2^-	Urea
Number of Experiments	9	4	2	3
Dark uptake (%)[a]	21 ± 7.1	43 ± 13.0	23	38.4 ± 2.78
Photoinhibited uptake (%)[b]	84 ± 7.1	84 ± 10.2	87	97.6 ± 2.43
Light level maximum uptake rate ($\%I_0$)[c]	46 ± 11.9 (6)	25 ± 0.0 (3)	25 (2)	10 ± 0.0 (3)

[a]Mean (dark rate/maximum rate) x 100 ± standard error

[b]Mean (rate at 100% I_0/maximum rate) x 100 ± standard error

[c]Mean relative light level resulting in maximum uptake rate ± standard error (number of experiments). Number of experiments in the fourth row is less than in the first row because some experiments did not show saturated uptake rate

Dugdale, 1972), but is also observed in coastal waters (Garside, 1981; Fisher *et al.*, 1982; Glibert *et al.*, 1982). The preconditioning *in situ* light level is also a factor (MacIsaac and Dugdale, 1972; Conway and Whitledge, 1979).

Eighteen experiments on five cruises from 1976 to 1978 were conducted to determine the effect of light on nitrogen uptake; half examined nitrate uptake. Dark uptake was measurable in all experiments and photoinhibition was apparent in half the experiments. Thus, only a small part of each experiment can be described by a rectangular hyperbola and four experiments show no resemblance to a Michaelis-Menten relationship. Consequently, only the dark uptake, the light level at which the uptake rate is first saturated and the degree of photoinhibition are given (Table 4.5). The high rates of dark uptake and the low light levels at which uptake rates are maximal suggest that ammonium and urea uptake are the least light dependent and nitrate and nitrite uptake are the most light dependent. However, ammonium and nitrate uptake are photoinhibited to approximately the same degree. For the most part, the data agree with predictions from earlier studies, but the relative lack of light dependence for urea uptake (2 cruises, both in summer) was unexpected, suggesting that either phytoplankton took up urea in the dark or non-photosynthetic organisms were utilizing urea.

Nitrate was the only nutrient for which there were enough experiments to examine factors affecting the interaction of light and nutrient uptake. Dark uptake was not significantly related to relative light level (% incident light, $\%I_0$) or the approximate absolute light level ($\%I_0 \times I_0$; not corrected for incubation length or time of day) from which the water was originally taken, depth, season, or total ambient nitrogen concentration. The latter was examined as a rough indicator of nitrogen sufficiency although particulate C/N ratios or ratios of C/N uptakewould have been more appropriate if they had been available. Similarly, no significant relation was found between photoinhibition and the previous light history of the sample.

TABLE 4.6 Effect of ammonium on nitrate uptake rate (relative $V_{NO_3^-} = V_{NO_3^-}$ at lowest NH_4^+ concentration/$V_{NO_3^-}$ at 1 μM x 100) on a summer cruise (TT122). One μM ammonium is the concentration above which it is generally accepted that nitrate uptake is inhibited (Dugdale, 1976).

Station	Light (%I_0)	Size	Rel. $V_{NO_3^-}$
13	3%	Total	28%
	1%	Total	0%
23	> 10%	> 20 μm	37%
		< 20 μm	54%
	< 10%	> 20 μm	100%
		< 20 μm	100%

Ammonium Inhibition of Nitrate Uptake

Ammonium may inhibit nitrate uptake, but the magnitude of the inhibition varies with nutrient prehistory, species identity and perhaps other environmental factors such as light (see Blasco and Conway, 1982; Dortch and Conway, 1984; and references therein). Several experiments were conducted in which increasing concentrations of unlabeled ammonium were added to incubations with a constant, saturating addition of $^{15}NO_3^-$. The effect was measured at several light levels at two stations. In high light, a typical pattern of high $V_{NO_3^-}$ at low ammonium concentrations decreasing to low constant values at concentrations above 1 μM was observed. Thus, the data are summarized by calculating nitrate uptake rate at 1 μM ammonium as a percent of the uninhibited uptake rate (Table 4.6). At low light, uptake rates were low and variable with no clear dependence on ammonium concentration. The question of ammonium inhibition of nitrate uptake will be in considered further in Section 4.3.3.

Nitrogen Uptake by Organisms of Different Size

Malone (1980) postulated that large cells (net plankton) should dominate nitrogen uptake in the spring, in part because their high uptake rates and storage capacities are advantageous when ambient nitrogen concentrations, especially nitrate, are high. Conversely, when nitrogen concentrations are low, small phytoplankton, which have a low k_s for uptake (Eppley *et al.* 1969), would grow chiefly at the expense of regenerated nitrogen. Studies of nitrogen uptake by different size classes in coastal areas support the hypothesis that net plankton dominate nitrogen uptake during winter/spring blooms (Glibert *et al.*, 1982; Furnas, 1983) and in coastal eutrophic regions in comparison with oceanic and oligotrophic regions (Probyn, 1985). Conflicting results were obtained when the ratio of nitrate uptake to total uptake was examined in different size classes (Furnas, 1983; Probyn, 1985).

The differences in nitrogen uptake between net (> 20 μm) and nano- (< 20 μm) plankton was examined in a series of four cruises (two summer, one winter, one spring) in 1976 and 1977. Most emphasis was placed on measuring nitrate, ammonium and urea uptake at different depths from representative stations, resulting in 90 pairs of uptake measurements. Additional experiments examined the size dependence of k_s and V_{max} for uptake of the four nitrogen

TABLE 4.7 Seasonal variation in nitrogen transport rates (ρ) and particulate nitrogen concentrations (PN) by net plankton (> 20 μm) as a percentage of ρ and PN for all size classes. Percentages are expressed as mean ± 1 standard error (number of measurements).

	Spring	Other Seasons
$\rho_{NO_3^-}$	87.9 ± 1.08 (4)	52.0 ± 4.04 (27)
$\rho_{NH_4^+}$	85.4 ± 1.34 (4)	31.0 ± 4.0 (25)
ρ_u	86.3 ± 1.90 (4)	19.0 ± 3.6 (27)
PN	86.5 ± 0.83 (12)	52.6 ± 1.34 (79)

sources (April, 1977), the effect of size on interactions between nitrate and ammonium uptake (August, 1977) and the differences in light/dark uptake ratios for different sizes.

During fall, winter and summer the net plankton comprised approximately half the particulate nitrogen (Table 4.7), but as expected for a coastal area (Malone, 1980) their proportion of the biomass was even greater in the spring. Uptake of all three nitrogen sources was significantly (one tailed t-test, $p < 0.025$) greater for the larger size class in the spring than in the other three seasons, but even in those seasons, nitrate uptake was dominated by the net plankton whereas ammonium and urea uptakes were dominated by the nanoplankton. All of the data were analyzed to determine if there were variations in the size dependence of uptake with depth or distance offshore, but neither factor appeared to play an important role.

On a spring cruise (TT116) the concentration dependence of nitrate, ammonium, nitrite and urea uptake was determined simultaneously for the net and nanoplankton. The net plankton/nanoplankton (> 20 μm/< 20 μm) ratio for k_s (3.7 ± 0.58; mean ± standard error, n = 6) and V_{max} (9.8 ± 1.87; n = 6) were both significantly greater than one (one tailed t-test; $p < 0.005$), supporting the earlier hypothesis of Eppley *et al.* (1969) that k_s should vary with size and partially explaining the dominance of net plankton in utilizing any nitrogen source in the spring (Malone, 1980). It is somewhat puzzling that a similar dominance is not also seen in the winter.

The inhibition of nitrate uptake by ammonium for net and nanoplankton was examined as described in the preceding section (Table 4.6). At higher light levels, it appears that nanoplankton nitrate uptake was less affected by ammonium than net plankton nitrate uptake because 54% of the uninhibited nitrate uptake remained in the nanoplankton at 1 μM ammonium whereas only 37% remained in the net plankton.

One small experiment comparing light and dark nitrate and ammonium uptake for the two size fractions was conducted on a summer cruise (TT122). The ratios of light to dark nitrate transport rates were about the same for the two size classes, but nanoplankton had an ammonium light/dark uptake ratio of 2.3, whereas net plankton had a ratio of 1.6.

4.3.2 Variation of Nitrogen Uptake in Time and Space

During most of the cruises, depth profiles of nitrate, ammonium and urea uptake were measured along the Copalis Line. The stations were not distributed evenly along the line so that it is not possible to construct sections of mean uptake rates, as in Chapters 1 and 3 for other variables. However, since the stations were clustered into three groups, the average rates were calculated for each group by season in depth increments of 0-5, 6-10, 11-15, 16-20, 21-30, 31-40 and 40-60 m in a manner analogous to that in Landry *et al.* (1989, Chapter 1) (Fig. 4.1-4.5). Despite the large number of uptake measurements involved (477), the distribution of the data is uneven; the least data are for the offshore stations and during winter and fall. In addition to the data from the Copalis Line, uptake data is also available for a summer cruise to examine the effect of the Quinault Canyon on productivity (TT140). These uptake rates are given separately (Fig. 4.6) and represent individual data points, not averages, for one station on the Copalis Line and three others to the north of it and into the Quinault Canyon.

The uptake data demonstrate the dynamic nature of phytoplankton processes along the Washington coast and the difficulty of studying such processes with occasional cruises and stations along a single transect. In other areas the shape of the depth profiles of nitrogen uptake, especially for nitrate, has been used as evidence concerning the nature of the nutrient regime. High uptake rates near the surface, decreasing with depth are typical of areas with plentiful nitrogen whereas low uptake rates at the surface but a subsurface maximum in or just above the nitracline are typical of areas where nitrogen is limiting (MacIsaac and Dugdale, 1969; Harrison *et al.*, 1983; Cochlan, 1986). For the Washington coast, within-cruise variability in uptake profiles is as great as between seasons. Thus, the pattern of the averaged profiles does not show any consistent trend (Fig. 4.1-4.5), although, for example, the $\rho_{NO_3^-}$ maximum in the summer on the shelf (Fig. 4.1) is exactly what would be expected.

Hydrographic, nutrient and chlorophyll data show that sporadic upwelling occurs in the late spring and summer (Landry *et al.*, 1989, Chapter 1). Except for TT140, where weak upwelling was suggested in the hydrographic and nutrient sections, none of these cruises and stations appeared to represent the extraordinarily high nitrogen uptake rates which should develop subsequent to an upwelling event (Wroblewski, 1977). MacIsaac *et al.* (1985) showed that such events are difficult to monitor with the sampling strategy used in this study. Nevertheless, nitrogen uptake rates measured along the Washington coast were within the range of those observed in other upwelling and very productive coastal areas (e.g., MacIsaac and Dugdale, 1969, 1972; MacIsaac *et al.*, 1974, 1985; Conway and Whitledge, 1979; Nelson and Conway, 1979; Glibert *et al.*, 1982; Paasche and Kristiansen, 1982; Furnas, 1983; Harrison *et al.*, 1983; Kaufman *et al.*, 1983; Kristiansen, 1983; Price *et al.*, 1985; Probyn, 1985; Cochlan, 1986).

With a few notable exceptions, there is remarkably little variation in the magnitude of the uptake rates with season or distance offshore, two major factors expected to affect uptake rates (Fig. 4.1-4.5). In fall/winter, the total nitrogen transport rates (Fig. 4.4) tend to be lower than in spring and summer although this trend is more obvious in the profiles of urea uptake (Fig. 4.3) than for nitrate (Fig. 4.1) or ammonium (Fig. 4.2) uptake rates. In the spring, elevated nitrate, ammonium and urea uptake occur at the shelfbreak stations. Part of the increase can be accounted for by higher biomass since the difference is less evident in specific uptake rates (Fig. 4.5).

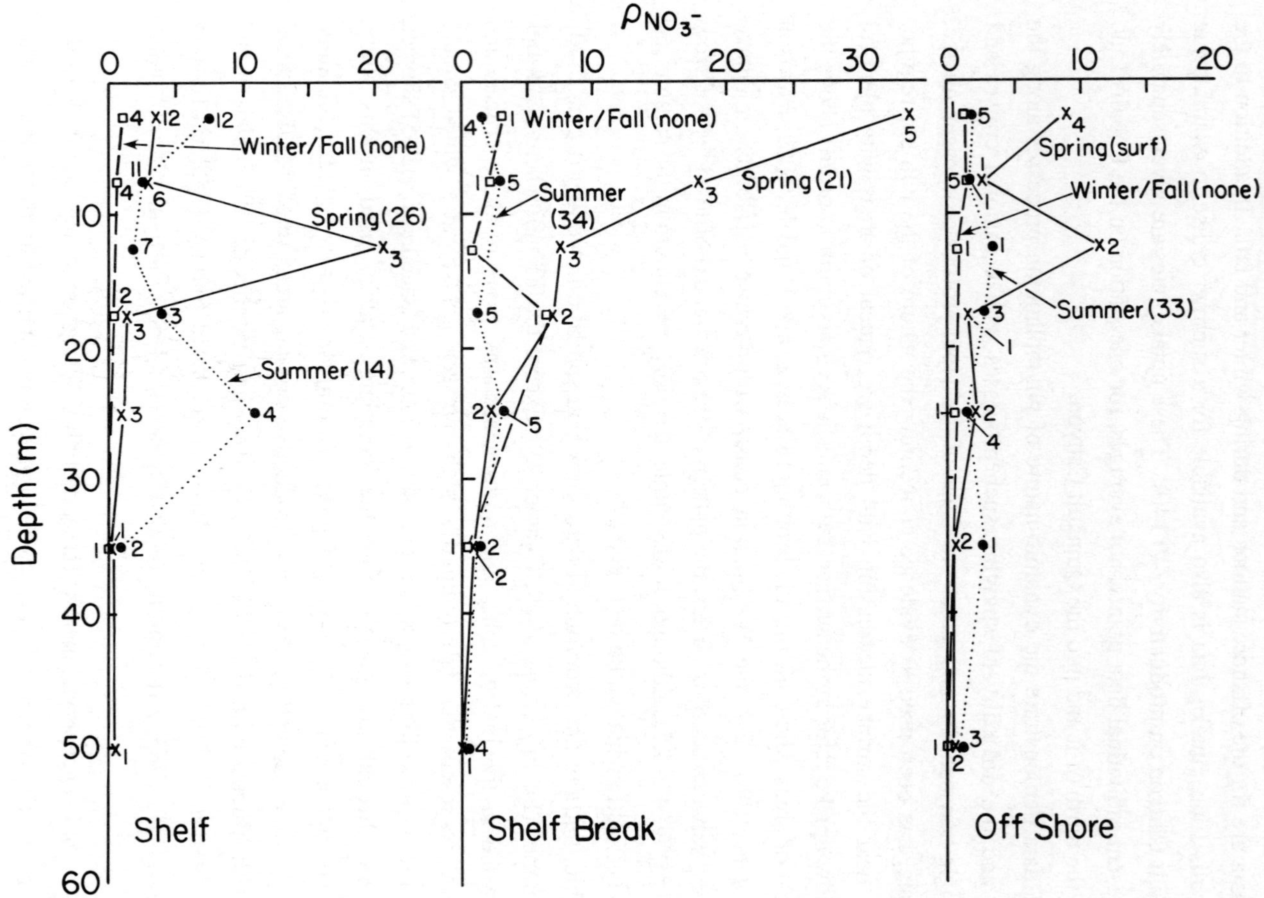

Fig. 4.1 Nitrate transport rates ($\rho_{NO_3^-}$ x 10 ^{3}) averaged over time and space as described in the text, as a function of season and distance offshore on the Copalis transect. The small numbers adjacent to each mean are the number of individual points, and the average depth of the nitracline (m) follows the season.

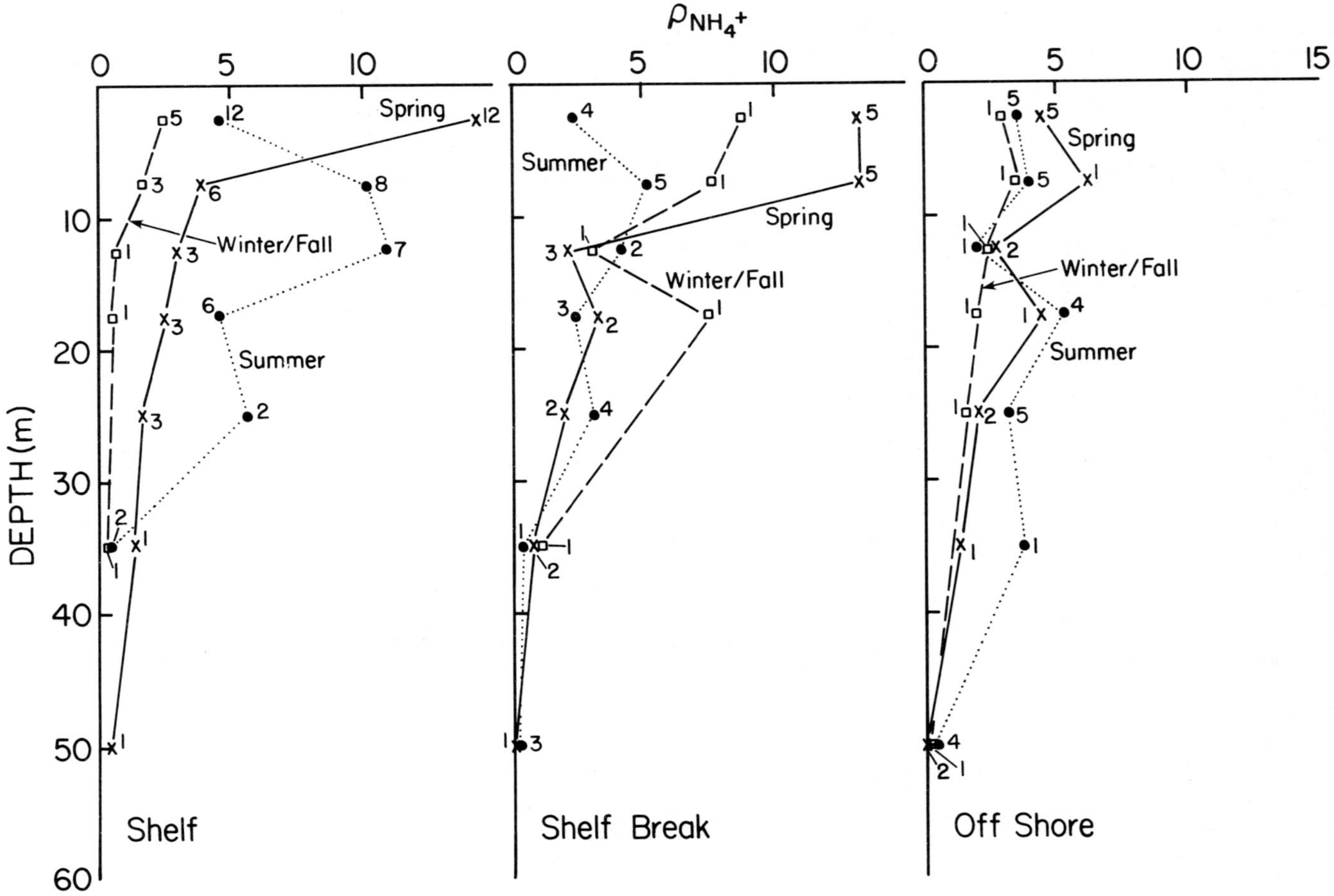

Fig. 4.2 Ammonium transport ($\rho_{NH_4^+}$), as in Figure 4.1

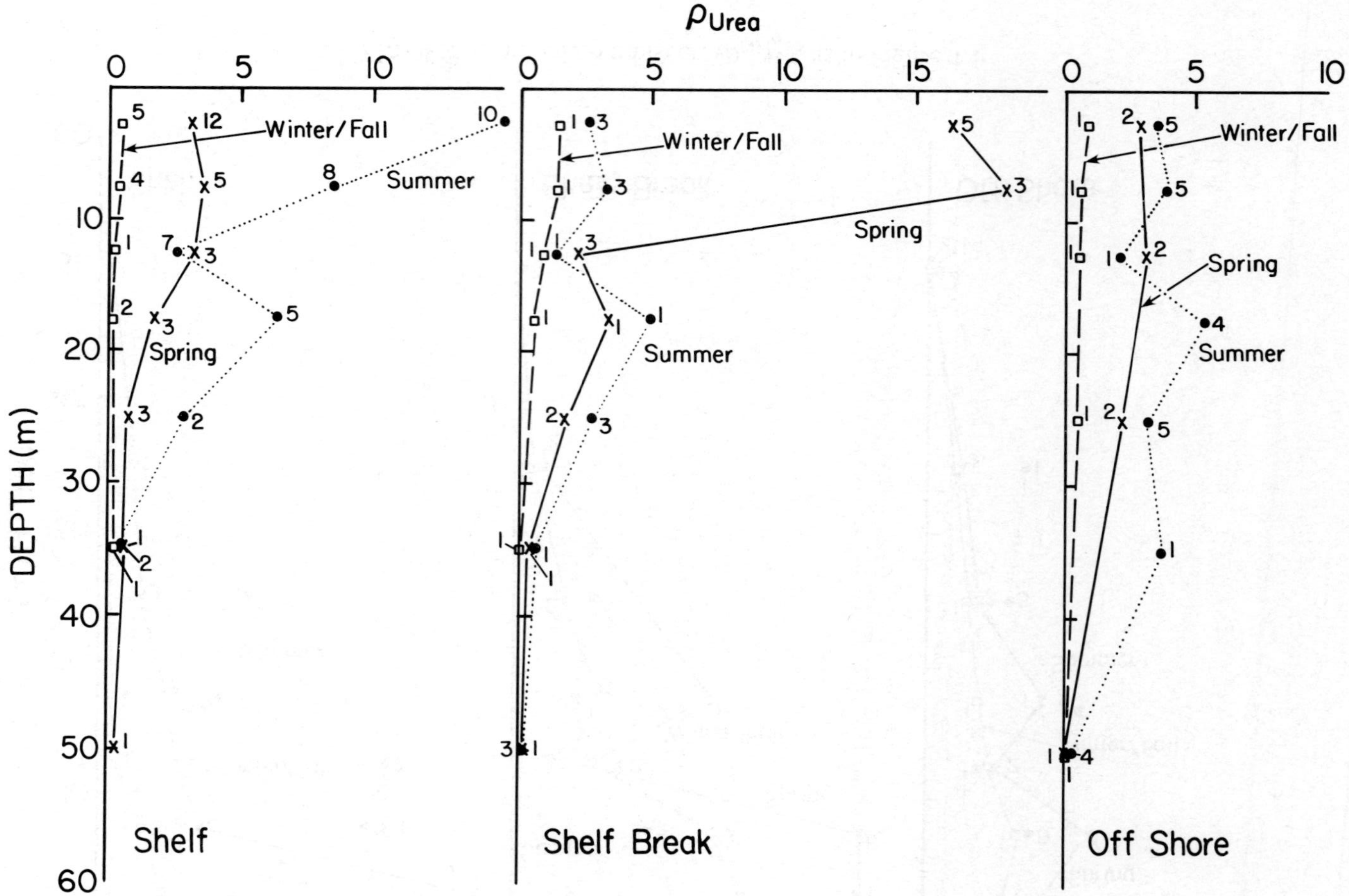

Fig. 4.3 Urea transport (ρ_u), as in Figure 4.1

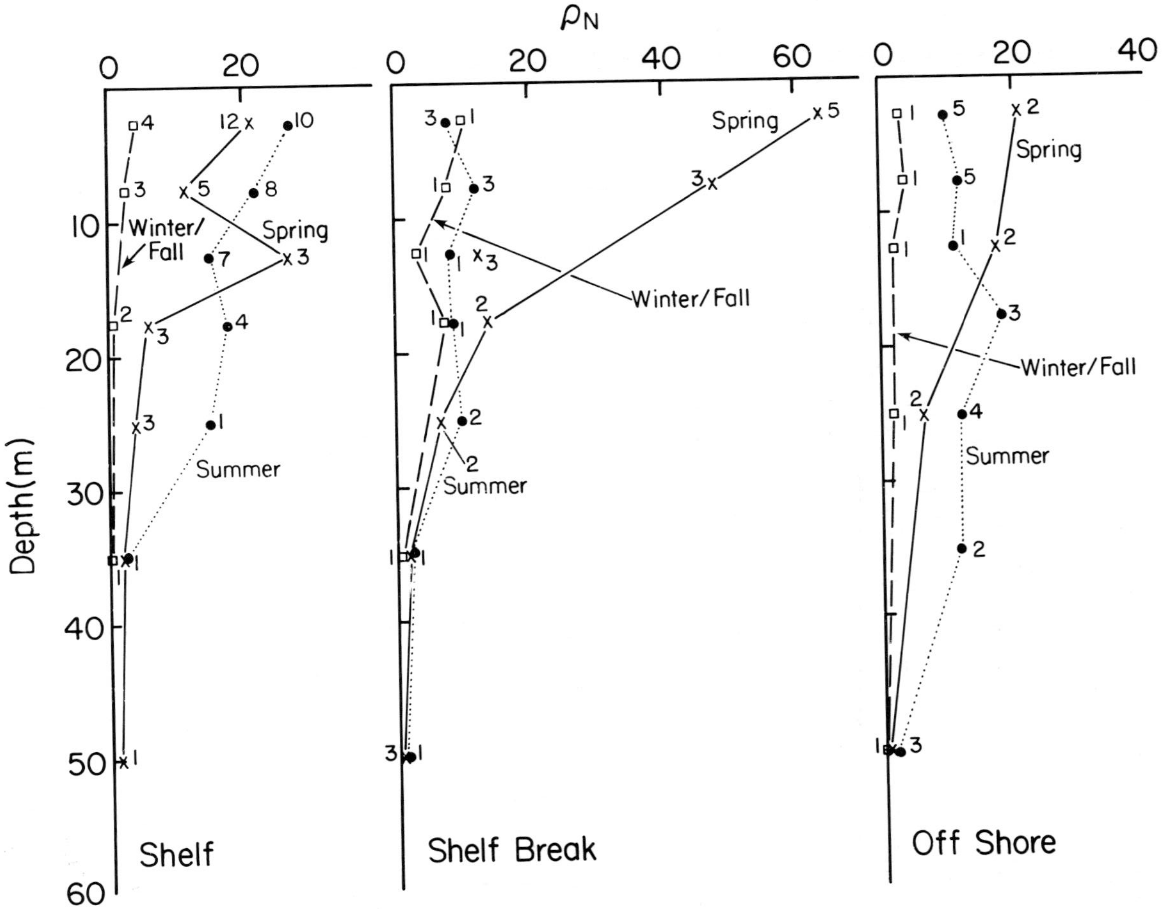

Fig. 4.4 Total nitrogen transport (ρ_N), as in Figure 4.1 Plots include only data where uptake rates of three nitrogen sources were measured on samples from the same depth.

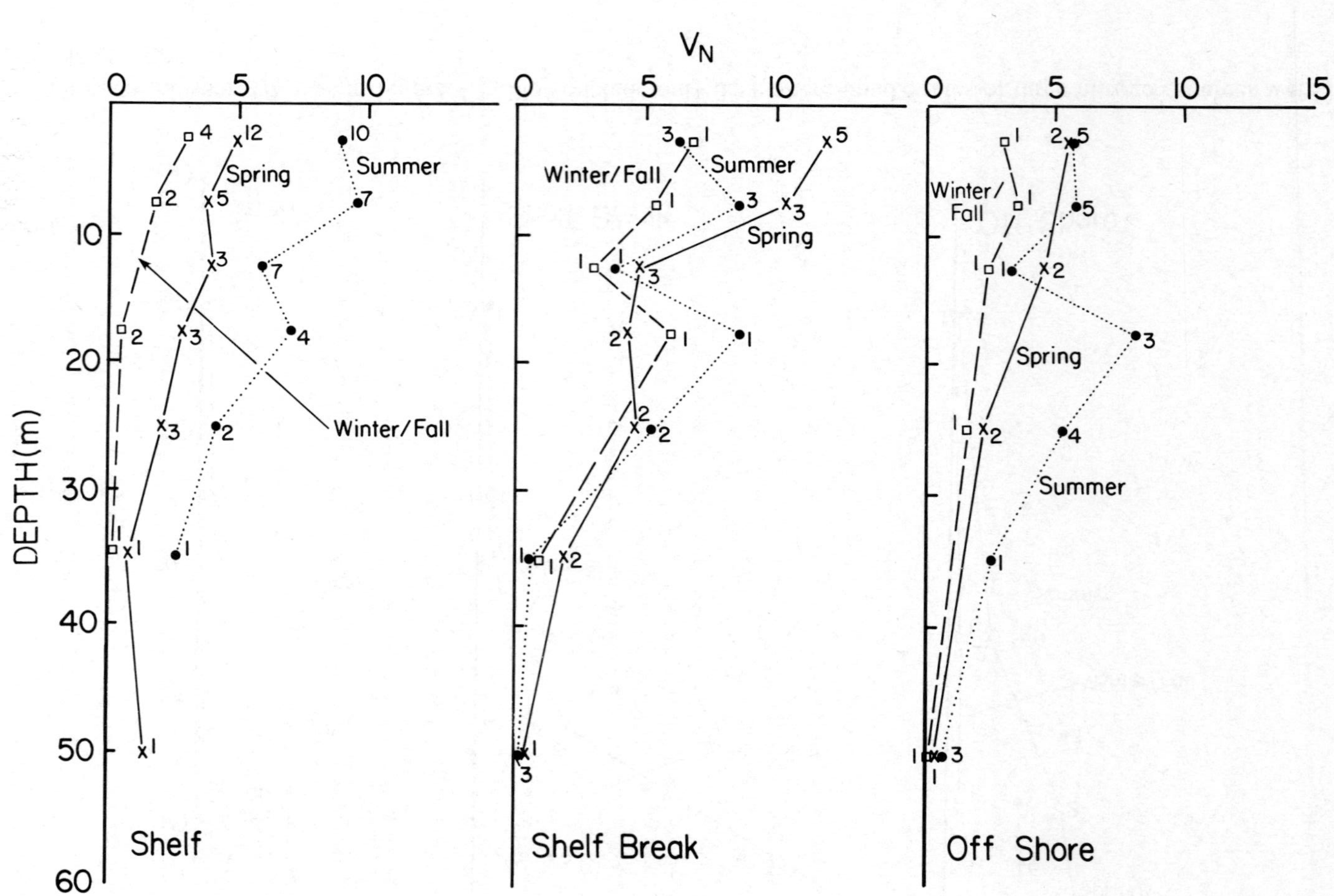

Fig. 4.5 Total nitrogen specific uptake rate ($V_N \times 10^3$), as in Figure 4.1 Plots include only data where uptake rates of three nitrogen sources were measured.

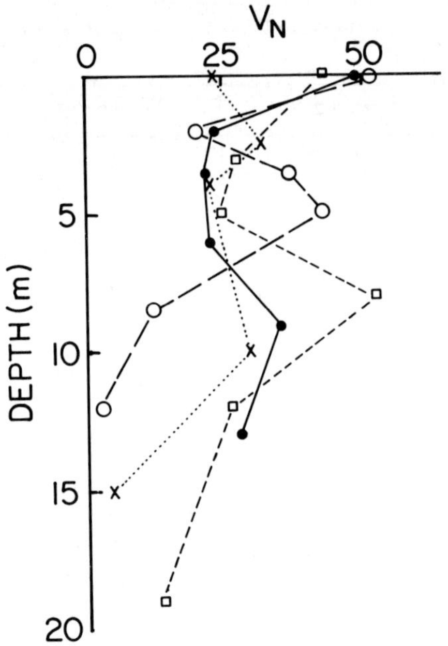

Fig. 4.6 Total nitrogen specific uptake rate (V_N x 10^3), at four stations during cruise TT140: ○, shelf on Copalis Line; ●, shelf, north of Copalis Line and south of Quinault Canyon; ✖, shelf break, north of Copalis Line and south of Quinault Canyon; □, Quinault Canyon, > 1500 m. Total nitrogen uptake includes nitrate, ammonium, and urea uptakes.

Furnas (1983) also commented on the lack of seasonal variation of N uptake in Narragansett Bay. Usually, however, nitrate uptake in coastal areas, is high in the spring and sometimes summer and ammonium and urea uptake rates are high in the summer.

The remarkable lack of seasonal or spatial variation in mean uptake rates may result from the high variability in individual uptake rates measured for the same season, depth interval and location. When the coefficient of variation (standard deviation/mean x 100) is calculated for each mean value in Figures 4.1-4.3, the mean of all coefficients is 92% (n = 80), which is considerably larger than the coefficient of variation of replicate samples reported in Table 4.3 (~25%). The extreme variation in the data is demonstrated by two examples. Table 4.8 contrasts uptake rates and other environmental variables for the shelfbreak stations in the spring. While values are generally high, there is an order of magnitude difference in $\rho_{NO_3^-}$ between stations. If the data are normalized to biomass, as chlorophyll (Table 4.8) or particulate nitrogen (Fig. 4.5), the differences decrease. The high stations are clearly associated with higher ambient nitrogen concentrations and higher chlorophyll values. The second example is the extremely high specific uptake rates obtained on TT140 (Fig. 4.6), caused primarily by elevated nitrate and urea uptake rates. Even on this cruise, ambient nitrogen concentrations in most of the euphotic zone were less than 1 μM. Besides the different location and recent upwelling, these results were obtained with ~4-h incubations, which are considerably shorter than any other incubations in this study. The degree to which shorter incubations should affect the apparent uptake rates is complex. By excluding the dark period, all rates, but especially nitrate uptake rates, should be elevated. This should be particularly so if the phytoplankton are nitrogen-sufficient, as suggested by the low, but still measurable, ambient nitrogen concentrations and very low C/N uptake ratios (Table 4.9). In general shorter incubations result in higher ammonium uptake rates (Goldman and Glibert, 1983) even when phytoplankton are nitrogen-sufficient and growing on nitrate (Horrigan and McCarthy, 1982). However, ammonium uptake rates were not appreciably higher

TABLE 4.8 Comparison of mean[a] uptake rates and enviromental variables for low and high stations from the shelfbreak during spring, using data from surface to 10% light depth. Refer to Table 4.2 for abbreviations and units.

Cruise - Station	TT096	TT116-13	CY378-9	CY678-8
Date	03/31/75	04/19/77	03/31/78	06/10/78
I_0	362	557	NA	444
Depths 10% I_0	10	25	22	38
ρNO_3^-	0.044 $\pm$ 0.0122 (4)	0.020 $\pm$ 0.0042 (3)	0.0040 $\pm$ 0.00154 (4)	0.00164 $\pm$ 0.000207 (4)
ρNH_4^+	0.024 $\pm$ 0.0050 (4)	0.00354 $\pm$ 0.000122 (3)	0.0021 $\pm$ 0.00184 (4)	0.00153 $\pm$ 0.000184 (4)
ρ_u	0.030 $\pm$ 0.0041 (4)	0.00344 $\pm$ 0.000095 (3)	0.00204 $\pm$ 0.000060 (4)	0.00112 $\pm$0.000059 (4)
Incubation Length (h)	22.9-24.1	19	24	23.9-24.5
NO_3^-	1.88 $\pm$ 0.44 (4)	2.51 $\pm$ 0.023 (3)	0.010 $\pm$ 0.0063 (3)	0.3 $\pm$ 0.32 (4)
Nitracline depth[b]	None	None	> 40 m	23-38 m
NH_4^+	0.6 $\pm$ 0.35 (4)	NA	0.020 $\pm$ 0.0150	0.000 $\pm$ 0.0050(4)
Chlorophyll *a*	4.31 $\pm$ 0.223 (3)	1.530 $\pm$ 0.0098 (3)	0.32 $\pm$ 0.096 (4)	0.16 $\pm$ 0.03 7(4)
ρ_C/ρ_N (by mole)	NA	1.61 $\pm$ 0.57 (3)	15.0 $\pm$ 4.0 (4)	20.0 $\pm$ 4.9 (4)
ρNO_3^-/ Chl *a*	0.009 $\pm$ 0.0056 (3)	0.0134 $\pm$ 0.00274 (3)	0.016 $\pm$ 0.0085(4)	0.0115$\pm$.00262(4)
$\rho NO_3^-/\rho_N$(%)	43.0 $\pm$ 4.7 (4)	73.0 $\pm$ 4.0 (3)	44.0 $\pm$ 8.2 (4)	39.0 $\pm$ 5.8 (4)

[a] $\pm$ 1 standard error (number of measurements)
[b]Depth interval over which nitrate increases from the limit of detection to > 1 μM
[c]($\rho NO_3^-/\rho_N$) x 100

than usual. Thus, it seems possible that the apparent high uptake rates reflect an enhanced uptake ability by the populations encountered on this cruise.

Although these two examples support the generally accepted hypothesis that high nitrogen transport rates occur when ambient nutrient and chlorophyll concentrations are high, a more detailed quantitative analysis can be made using the entire data set. Pearson correlation coefficients were calculated for 33 variables using the Statistical Package for Social Scientists

TABLE 4.9 Significant correlations (two tailed t-test, $p \leq 0.05$) between uptake rates (ρ) and environmental variables based on Pearson correlation coefficients (r). Abbreviations as in Table 4.2.

Season	r	$\rho_{NO_3^-}$	$\rho_{NH_4^+}$	ρ_{urea}
Fall/winter	$r \geq 0.5$	I, Chl, V_N, $V_{NH_4^+}$ $\rho_{NH_4^+}$, ρ_u, ρ_c	I, Chl, V_N, $\rho_{NO_3^-}$ ρ_u, ρ_c	I, Chl, V_N, $\rho_{NO_3^-}$ $\rho_{NH_4^+}$, ρ_c
	$0 < r < 0.5$			
	$r > 0$		Z	
Spring	$r \geq 0.5$	Chl, $V_{NO_3^-}$		Chl, V_N, $\rho_{NO_3^-}$ ρ_c,,ρ_N
	$0 < r < 0.5$	V_N, V_u		$V_{NO_3^-}$, $V_{NH_4^+}$, V_u Z, NO_3^-
	$r < 0$			
Summer	$r \geq 0.5$	Chl, $V_{NO_3^-}$, V_N , $\rho_{NH_4^+}$, ρ_c	Chl, $V_{NO_3^-}$, $V_{NH_4^+}$, V_N, V_u, $\rho_{NO_3^-}$, ρ_u	I, U
	$0 < r < 0.5$	V_u, ρ_u	ρ_c	%I_0, Chl, $V_{NO_3^-}$, $\rho_{NO_3^-}$
	$r < 0$	Z	Z	Z, NH_4^+
All	$r \leq 0.5$	$V_{NO_3^-}$		I, V_u, V_N
	$0 < r < 0.5$	I, Chl, V_u, V_N, ρ_u		%I_0, $\rho_{NO_3^-}$
	$r < 0$	Z		Z

(SPSS). The variables to which uptake rates show significant correlation ($p > 0.05$) are listed in Table 4.10. The results are interesting both for what is and what is not, correlated. As expected, nitrogen transport rates (ρ) often appear to be correlated with biomass expressed as chlorophyll. They are also often correlated with the specific uptake rate (V) for the same nitrogen compound and the transport (ρ) and specific uptake (V) rate of both total nitrogen and individual nitrogen compounds. No correlation is apparent between the concentration of a specific nitrogen compound and its uptake rate, except for urea in the summer. This contradicts the previous conclusion that very high nitrogen uptake rates appear to be measured when ambient nitrogen concentrations are high. In fact Wroblewski (1977) and MacIsaac *et al.* (1985) showed that the highest nitrogen uptake rates were observed as ambient nitrogen becomes depleted after an input

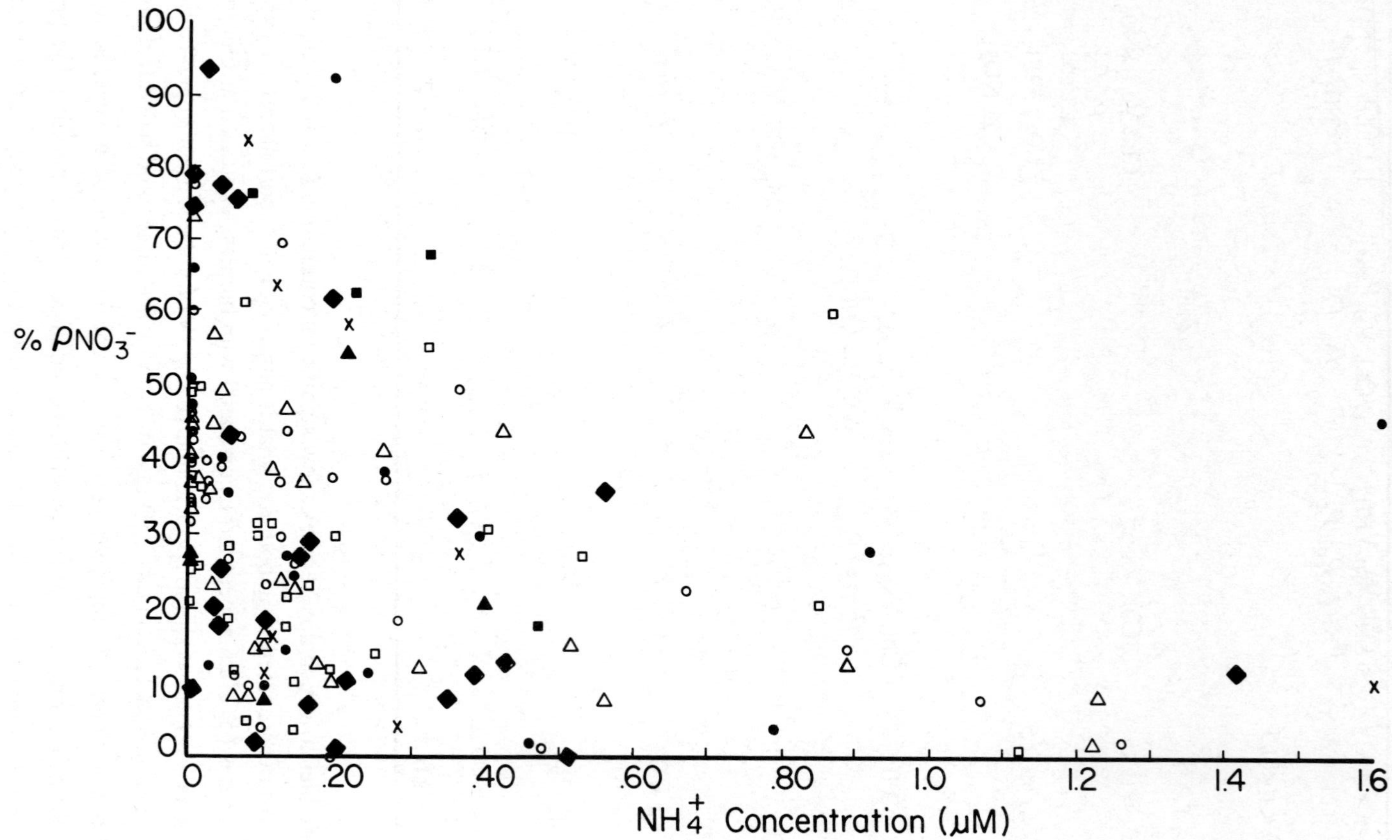

Fig. 4.7 Nitrate uptake relative to total nitrogen uptake, $\%\rho_{NO_3^-} = [\rho_{NO_3^-} / (\rho_{NO_3^-} + \rho_{NH_4^+} + \rho_u) \times 100]$; f according to Eppley and Peterson (1979), as a function of ambient ammonium concentration and light level (% I_0): ○, 80-100%; △, 43-51%; □, 23-30%; ✖, 15%; ●, 9-10%; ▲, 3-5%; ■, 2%; ◆, 1%. Plots include data from Copalis Line and TT140 for which uptake rates of nitrate, ammonium, and urea were measured.

of nitrate. Correlations demonstrating inhibition of nitrate uptake by ammonium are also lacking, although urea uptake may be inhibited by nitrate in the spring and by ammonium in the summer. Light levels are important for uptake of all nitrogen compounds in winter but less so in other seasons. Urea uptake appears to be more light dependent than either nitrate or ammonium uptake, in contrast to the results from light kinetics experiments in which urea uptake was the least light dependent. The absence of expected correlations could be real, or could be related to 1) the difficulty of assessing processes in a very dynamic system from data with limited spatial and temporal coverage or 2) the inappropriate use of Pearson correlation coefficients which assume a linear relationship between variables.

In summary, nitrogen uptake rates can be quite high off the Washington coast. The highest values are associated with periods of high biomass, usually in the spring bloom or in the summer during upwelling, which may be underrepresented in this data set. Within any season or for any location, the rates are quite variable, resulting in very little difference in mean rates between seasons or with distance offshore and reflecting the very dynamic nature of the processes affecting phytoplankton growth on the coast.

4.3.3 Utilization of Different Nitrogen Sources

Since Dugdale and Goering (1967) first distinguished between production occurring at the expense of 'new' nitrogen and that fueled by "regenerated" nitrogen, assessing the relative importance of nitrate uptake on the one hand *vs.* ammonium and urea uptake on the other has been the subject of many field studies (e.g., MacIsaac and Dugdale, 1969, 1972; MacIsaac *et al.*, 1974; Conway, 1977; McCarthy *et al.*, 1977; Conway and Whitledge, 1979; Nelson and Conway, 1979; Olson, 1980; Garside, 1981; Blasco and Conway, 1982; Glibert *et al.*, 1982; Paasche and Kristiansen, 1982; Furnas, 1983; Harrison *et al.*, 1983, 1985; Kaufman *et al.*, 1983; Kristiansen, 1983; Ronner *et al.*, 1983; Kanda *et al.*, 1985; Price *et al.*, 1985; Probyn, 1985; Cochlan, 1986). As will be seen in the next section, these relative rates of uptake of new and regenerated nitrogen can be related to the magnitude of primary production and ecosystem dynamics (Dugdale and Goering, 1967; Dugdale, 1976; Eppley *et al.*, 1979; Eppley and Peterson, 1979; Eppley, 1981). Relative rates can be expressed in a number of ways, each of which may have different ecological or physiological interpretations.

Rates of transport of specific nitrogen sources relative to total nitrogen transport rates are quite variable (Table 4.9, Figs. 4.7 and 4.8), although on the whole regenerated nitrogen accounts for 64% of total nitrogen uptake $[(\rho_{NH_4^+} + \rho_u)/\rho_N] \times 100$. Nitrate was most important as an nitrogen source in the fall/winter and spring in water on the shelf and at the shelf break (Table 4.9), which is consistent with other nitrogen uptake studies showing high nitrate uptake rates during spring blooms. The $\%\rho_{NH_4^+}$ was highest on the shelf in all seasons and elsewhere during the summer. The high $\%\rho_{NH_4^+}$ was considerably higher than would have been predicted from ammonium excretion studies in the same area (Jawed, 1973). Some of the highest ρ_u and $\%\rho_u$ ever reported were observed during these cruises, especially in the summer on TT140. Not only does regenerated nitrogen tend to be more important when the surface waters are stratified

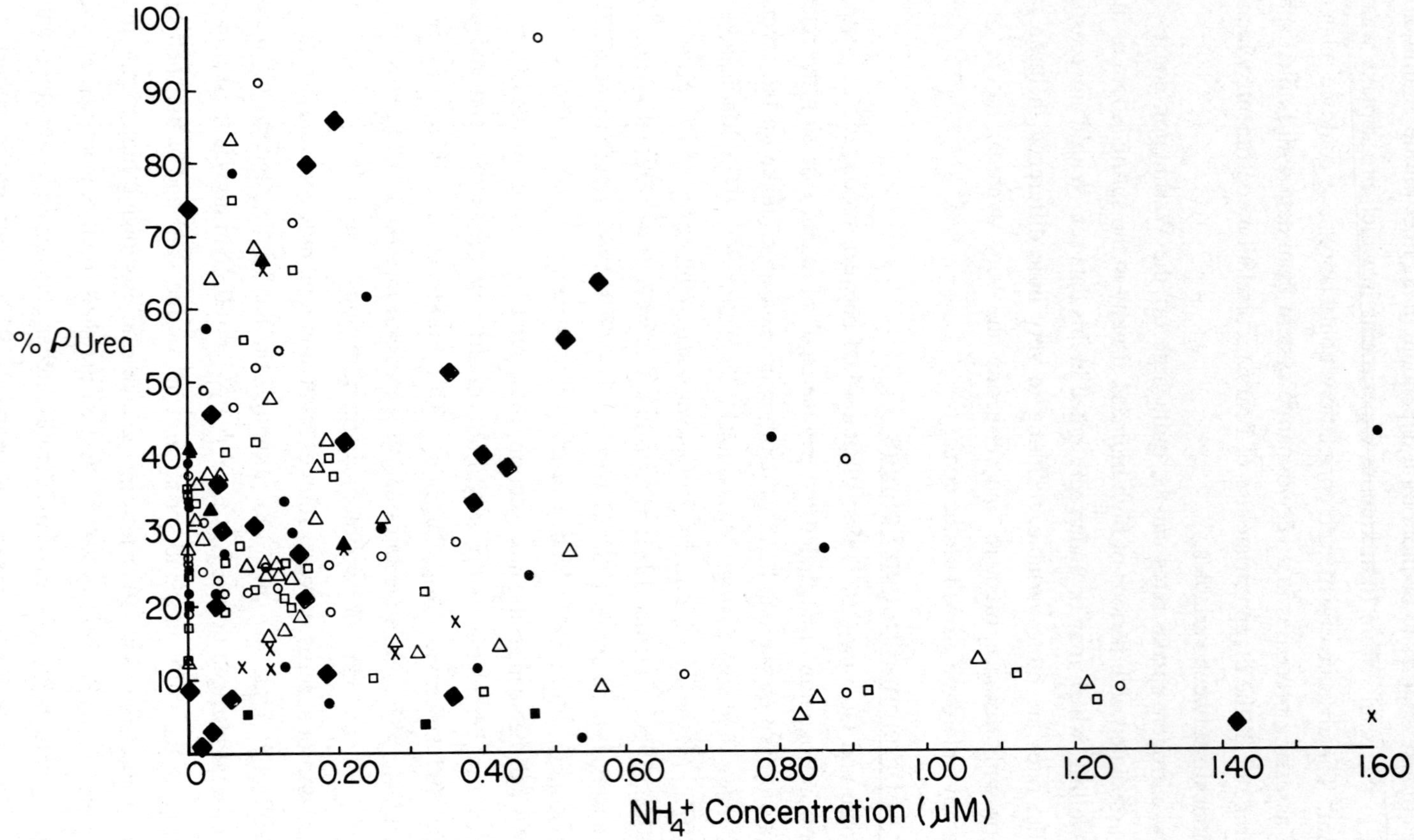

Fig. 4.8 Urea uptake relative to total nitrogen uptake, $\%\rho_{\text{urea}} = [\rho_{\text{urea}} / (\rho_{NO_3^-} + \rho_{NH_4^+} + \rho_u) \times 100]$, as a function of ambient ammonium concentration and light level, as in Figure 4.7.

TABLE 4.10 Relative transport rates (%ρ)[a,e,f,g] of different nitrogen sources, relative preference index for nitrate ($RPI_{NO_3^-}$)[a,h], ammonium to nitrate transport ratios ($\rho_{NH_4^+}/\rho_{NO_3^-}$)[a] and carbon to nitrogen transport ratios (ρ_C/ρ_N) a as a function of season and location.

	Shelf				Shelf Break				Offshore				All
	Fall/Winter	Spring	Summer[b]	TT140[c]	Fall/Winter	Spring	Summer[b]	TT140[c]	Fall/Winter	Spring	Summer[b]	TT140[c]	
% $\rho_{NO_3^-}$(f)[e]	23	32	21	38	53	45	28	23	45	56	21	33	31.5
	3.4	3.5	3.7	7.5	13.3	5.3	4.8	6.2	8.2	7.7	3.4	8.0	1.72
	(10)	(28)	(30)	(12)	(5)	(19)	(13)	(5)	(5)	(7)	(22)	(6)	(162)
% $\rho_{NH_4^+}$ [f]	63	41	45	17.1	32	27	41	25	35	21	58	18	39.4
	3.7	3.7	5.5	2.64	9.8	3.2	4.0	5.7	5.5	3.4	3.4	3.3	1.80
	(10)	(28)	(30)	(12)	(5)	(19)	(13)	(5)	(5)	(7)	(22)	(6)	(162)
% ρ_u [g]	13.9	26.3	34	45	15	26.8	30	52	20	23	21.9	49	29.1
	2.51	1.87	5.7	6.0	3.9	2.84	4.4	4.9	4.0	4.5	2.86	7.8	1.57
	(10)	(28)	(30)	(12)	(5)	(19)	(13)	(5)	(5)	(7)	(22)	(6)	(162)
$RPI_{NO_3^-}$ [h]	0.30	1.0	0.97	0.77	0.64	1.2	0.57	0.59	0.57	0.52	0.65	0.69	0.74
	0.045	0.39	0.161	0.082	0.137	0.54	0.069	0.108	0.076	0.076	0.184	0.145	0.069
	(10)	(12)	(26)	(10)	(5)	(10)	(15)	(6)	(5)	(12)	(17)	(6)	(134)
$\rho_{NH_4^+}/\rho_{NO_3^-}$	17	3.4	11	0.98	1.0	1.8	2.2	1.7	0.94	1.8	4.9	2.7	4.9
	14.5	1.32	4.2	0.279	0.36	0.94	0.43	0.74	0.250	0.66	1.65	2.07	1.17
	(10)	(29)	(33)	(12)	(5)	(19)	(22)	(6)	(5)	(12)	(21)	(6)	(180)
ρ_c/ρ_N [i]	9	20.6	19	7.0	6.6	17.0	9.5	3.7	15.4	21.1	13.1	6.3	14.4
	3.4	2.05	3.4	1.51	2.38	2.10	2.60	1.08	1.92	1.56	2.14	1.72	0.90
	(6)	(28)	(16)	(12)	(5)	(14)	(9)	(5)	(5)	(7)	(17)	(6)	(130)

[a]Mean followed by standard error and number of measurements (n)
[b]Data from Copalis Line
[c]Data from Quinalt Canyon Experiment
[d]In Quinalt Canyon, depth ~1500 m
[e]$[\rho_{NO_3^-}/(\rho_{NO_3^-} + \rho_{NH_4^+} + \rho_u)] \times 100$
[f]$[\rho_{NH_4^+}/(\rho_{NH_4^+} + \rho_{NO_3^-} + \rho_u)] \times 100$
[g]$[\rho_u/(\rho_u + \rho_{NO_3^-} + \rho_{NH_4^+})] \times 100$
[h]$[\rho_{NO_3^-}/(\rho_{NO_3^-} + \rho_{NH_4^+})]/[NO_3^-/(NO_3^- + NH_4^+)]$
[i]Includes only data with nitrate, ammonium, and urea uptake measurements and ρ_c measured under identical conditions.

and N depleted, but ammonium uptake rates and possibly urea uptake rates, appear to be somewhat temperature dependent (Fisher *et al.*, 1982; Glibert *et al.*, 1982; Paasche and Kristiansen, 1982; Kristiansen, 1983; Kanda *et al.*, 1985) which may explain the higher $\%\rho_{NH_4^+}$ in the summer.

Considering that this is a productive coastal area with sporadic upwelling, the low relative nitrate uptake rates are surprising. Because of the methodological problems mentioned in the methods, both ammonium and urea uptake rates may be underestimated, so that the dependence on regenerated N may actually be underestimated. However, the data collection may have been biased towards areas and times when regenerated nitrogen was important. Additionally, many of the measurements were taken in shallow water on the shelf where benthic regeneration may have been a major factor contributing to low relative nitrate uptake in all seasons.

The relative preference index (RPI) was first used by McCarthy *et al.* (1977) to express uptake of a specific nutrient relative to its ambient concentration (Table 4.9, footnote h). In previous studies, even when $\rho_{NO_3^-}$ and $\%\rho_{NO_3^-}$ was high, the $RPI_{NO_3^-}$ was often less than one, indicating that nitrate was taken up less rapidly than would be expected from its availability (e.g., McCarthy *et al.*, 1977; Glibert *et al.*, 1982; Paasche and Kristiansen, 1982; Furnas, 1983; Kaufman *et al.*, 1983; Harrison *et al.*, 1985; Probyn, 1985; Cochlan, 1986). Similarly, the $RPI_{NO_3^-}$ was less than one for the results reported here. The exceptions, when the RPI was greater than one in the spring, are consistent with results from other studies. However, the high value for the shelf in the summer (0.97) is hard to explain in the light of the high $\%\rho_{NH_4^+}$ (45%) and $\%\rho_u$ (34%) for the same period. Urea uptake rates were not included in calculations of the $RPI_{NO_3^-}$. The RPI for urea was not calculated because of uncertainty about urea concentrations. Furthermore, $RPI_{NO_3^-}$ data are biased because the ratio cannot be calculated if ambient nitrate concentrations are zero, as frequently occurs.

Low nitrate uptake rates are often explained by inhibition by ammonium, especially when the $RPI_{NO_3^-}$ is low. Inhibition may be greatest when ammonium concentrations exceed a threshold value. Ecophysiological experiments, described previously, showed significant inhibition of nitrate by added ammonium (Table 4.6). When $\%\rho_{NO_3^-}$ is plotted as a function of ambient ammonium concentrations (Fig. 4.7) from all depth profile data, it is evident that high $\%\rho_{NO_3^-}$ is not observed at high ammonium concentrations. However, low ammonium concentrations do not necessarily result in high $\%\rho_{NO_3^-}$; therefore, some other factor(s) must be important. Since ammonium uptake and assimilation may require less energy than nitrate uptake, light levels are included in Figure 4.7 to determine if they can account for low $\%\rho_{NO_3^-}$ at low ammonium concentrations. No such relationship is apparent. Further, there does not appear to be any dependence on absolute light levels or depth (data not shown). The relationship between nitrate uptake and ambient ammonium concentration may be strongly influenced by the history of exposure to ammonium. In fact, it may not be so much the concentration of ammonium as its flux, both in the past and the present, which controls nitrate uptake.

Like nitrate uptake, urea uptake is thought to be inhibited by ammonium (McCarthy, 1972; Healey, 1977; Rees and Syrett, 1979; Kristiansen, 1983). Plots of $\%\rho_u$ against ambient ammonium concentrations (Fig. 4.8) show again that $\%\rho_u$ tends to be low at high ammonium, but varies over a wide range at low ammonium. The variation appears to be independent of light.

An alternate approach is to calculate the ratio of ammonium to nitrate transport rates (Table 4.9) as a measure of how many times an atom of nitrogen recycles in the euphotic zone before being exported from the upper water column (Eppley, 1981; Harrison *et al.*, 1983). As expected, high values were observed on the shelf and in the summer. The mean value of five seems high for a coastal area, particularly considering the mean values for each season and location (Table 4.9). However, as in the Middle Atlantic Bight (Harrison *et al.*, 1983), the data are not normally distributed and only 12% of the values are greater than the mean. The median value of approximately 2 seems more appropriate and is similar to the results of Harrison *et al.* (1983).

In conclusion, nitrate accounted for a surprisingly small proportion of the nitrogen uptake by phytoplankton in this productive area. This result might be explained in part by underrepresentation of times and places when nitrate would be important, such as late winter and spring blooms and upwelling periods. Inhibition of nitrate uptake by ammonium may be a factor, but ammonium supply rates may be more important than actual concentrations. Urea could also supply a large portion of the phytoplankton nitrogen requirement. Finally, however, environmental variables measured during these studies do not appear to explain the large variation in relative uptake of different nitrogen sources, or in fact, the large variation in many other aspects of the data in this study. If the data are broken down by cruise and not season or location as in Table 4.9, there is as much variation between cruises in one season, or stations within one cruise, as between seasons. This suggests that preconditioning and/or species composition may be a critical factor in how nitrogen is utilized.

4.3.4 Assessment of Nitrogen Limitation

As described in Landry *et al.* (1989, Chapter 1), near-surface nitrogen concentrations decline rapidly in early spring and, except for sporadic upwelling, remain low throughout the spring and summer. Thus, the average nitracline depths in Figure 4.1 are quite deep in those seasons. While it is unwise to conclude that phytoplankton are nitrogen limited from low ambient nitrogen concentrations because of the possibility of rapid recycling, it is clear that N is not plentiful. The degree of nitrogen deficiency can be assessed by a variety techniques.

The results of field experiments do not support severe nitrogen limitation in any season. Ratios of light to dark nitrate uptake are 8.8 in the summer and 8.2 in the fall/winter/ spring and are, thus, quite high in comparison with ratios of 1-2 in severely nitrogen-deficient phytoplankton. Eppley *et al.* (1969) and MacIsaac and Dugdale (1969, 1972) proposed that the k_s for uptake is lower in phytoplankton from oligotrophic areas. If that is the case, the lower k_s for ammonium uptake in the spring/fall/winter suggests that nitrogen is more limiting then, but the k_s

for uptake of nitrate and urea do not show a similar pattern. Moreover, the nutrient data itself does not support such a hypothesis.

Another, perhaps more useful indicator because of the larger amount of data, are ρ_C to ρ_N ratios (Table 4.9). Values near or below the Redfield ratio (6.6 by atom) indicate nitrogen sufficiency and larger values indicate nitrogen deficiency. Ratios of ρ_C/ρ_N are quite variable. For TT140, a summer cruise with weak upwelling and extraordinarily high ρ_C and ρ_N, the low ρ_C/ρ_N ratios clearly indicate nitrogen sufficiency. Otherwise the ratios range from values near Redfield to very high values, with the highest values usually in the spring and summer on the shelf. These values are typical of those obtained in other studies (summarized in Fisher *et al.*, 1982; Harrison *et al.*, 1983, 1985; Kanda *et al.*, 1985; Cochlan, 1986). Fisher *et al.* (1982) suggested that high values would result if incubations were too short and substantial nitrogen uptake occurred at night, as for nitrogen-deficient phytoplankton. Most incubations in this study were approximately 24 h and there was no correlation between ρ_C/ρ_N and the length of the incubation. The ρ_C/ρ_N ratio is not an unequivocal indicator of nitrogen deficiency because it is also influenced by light level (Slawyk, 1979; Olson, 1980; Harrison *et al.*, 1982; Cochlan, 1986). Off the Washington coast, the ratio was significantly (one tailed t-test, $p < 0.005$) lower for the populations at light levels < 5% I_0 (10.4 ± 1.91, mean ± 1 standard error, n = 20) than for populations at light levels > 5% I_0 (17.8 ± 1.10, n = 86).

On four cruises (Table 4.1), ^{14}C uptake was measured in the dark with and without added ammonium (Peterson and Postel, unpublished) as described in Morris *et al.* (1971). In nitrogen-deficient phytoplankton, the presence of ammonium enhances ^{14}C uptake in the dark by at least a factor of two (Morris *et al.*, 1971; Yentsch *et al.*, 1977; Goldman and Dennett, 1983). The ratios of dark ^{14}C uptake with and without added ammonium were remarkably constant at 1.18 ± 0.064 (mean ± 1 standard error, n = 32) for stations on the Copalis Line and 1.09 ± 0.046 (n = 49) for stations on the Newport Line. In both cases the ratios are significantly greater than one (one tailed t-test, $p < 0.05$), but they are considerably less than two, which would indicate nitrogen deficiency. Most of these data were acquired when the euphotic zone was very nitrogen depleted, particularly on TT082, yet the data show no serious nitrogen deficiency.

During a much later cruise (CY082; Table 4.1) biochemical indicators of nitrogen deficiency were measured at the surface on a quick transit along the Copalis Line (Table 4.11). Nearshore nitrate concentrations were high, but decreased offshore. In contrast, maximum chlorophyll concentrations were observed at low nitrate concentrations. The spatial separation between maximum nitrate and chlorophyll concentrations is typical of upwelling areas and may represent a time lag between nitrate uptake and phytoplankton growth (Wroblewski, 1977; Small and Menzies, 1981; MacIsaac *et al.*, 1985). The most nitrogen-deficient populations would be anticipated nearshore, before nitrate uptake has occurred and further offshore, after growth has depleted internal nitrogen reserves. In Table 4.11 intracellular nitrate plus ammonium plus free amino acid concentrations (ΣIN; Dortch *et al.*, 1984, 1985) were always measurable, suggesting that the phytoplankton were not nitrogen starved. Low values of ΣIN offshore do, however, indicate possible nitrogen deficiency. Similarly, free amino acid-nitrogen/protein nitrogen

TABLE 4.11 Biochemical indicators of nitrogen deficiency at the surface measured from nearshore to the shelf break along the Copalis Line on June 11, 1982 (Postel *et al.*, 1984).

Station	Water Depth	NO_3^-	Σ IN	AA/Pr	Chl
13	9	21.28	0.30	9.5	0.95
15	27	3.53	0.36	11.6	1.20
17	41	0.34	0.56	14.8	5.12
19	54	0.01	0.74	23.2	4.90
21	93	0.02	1.41	28.2	12.53
23	147	0.02	0.24	11.0	0.39
25	880	0.06	0.07	8.6	0.17

(AA/Pr) ratios greater than 10 (Dortch *et al.*, 1984, 1985) indicate nitrogen sufficiency at most stations, whereas AA/Pr values of 9.5 and 8.6 suggest slight nitrogen deficiency nearshore and again at the station furthest offshore. In a seasonal study in Dabob Bay, Washington (Dortch *et al.*, 1985), which is often studied as an analog of the open ocean (Shuman, 1978; Copping, 1982), biochemical indicators showed that plankton were never extremely nitrogen deficient, even after the euphotic zone had been nitrogen depleted for months. Thus, it appears from biochemical indicators that in Washington coastal waters plankton are either nitrogen sufficient or at most slightly nitrogen deficient.

In summary, the different approaches to determining the degree of nitrogen deficiency provide conflicting results. Certainly, ambient nitrogen concentrations are frequently quite low and a major part of the phytoplankton nitrogen requirement comes from recycled nitrogen, but absolute uptake rates are generally high. The relatively high ρ_c/ρ_N ratios argue for nitrogen limitation whereas the lack of enhancement of dark ^{14}C uptake by ammonium and the biochemical indicators suggest nitrogen sufficiency. Although the N uptake data can be faulted on methodological grounds, the other approaches still yield contradictory results about the degree of nitrogen limitation, as might be expected in a dynamic system. Perhaps the best way to describe this system is as being, on the average, poised between nitrogen sufficiency and deficiency. Because it is a dynamic system, individual stations or cruises can deviate considerably from the mean state. Nitrogen availability does play a critical role, even if the phytoplankton are not strictly nitrogen deficient, because of the delicate balance between supply and demand.

4.3.5 Relationship Between Nitrogen and Productivity

When nitrogen supply potentially limits phytoplankton growth, productivity should be closely linked with nitrogen uptake (e.g., Dugdale and Goering, 1967; Eppley and Peterson, 1979; Eppley *et al.*, 1979). On the Washington coast in fall, winter and spring, ρ_c is significantly ($p < 0.001$) correlated with the $\%\rho_{NO_3^-}$ at light levels above 5% incident radiation (Fig. 4.9A). However, to obtain this correlation, it was necessary to exclude the data from Station 21 on cruise TT116 (April 1977), which had the highest chlorophyll values in the entire data set and nutrient concentrations and light levels more typical of summer conditions. At low

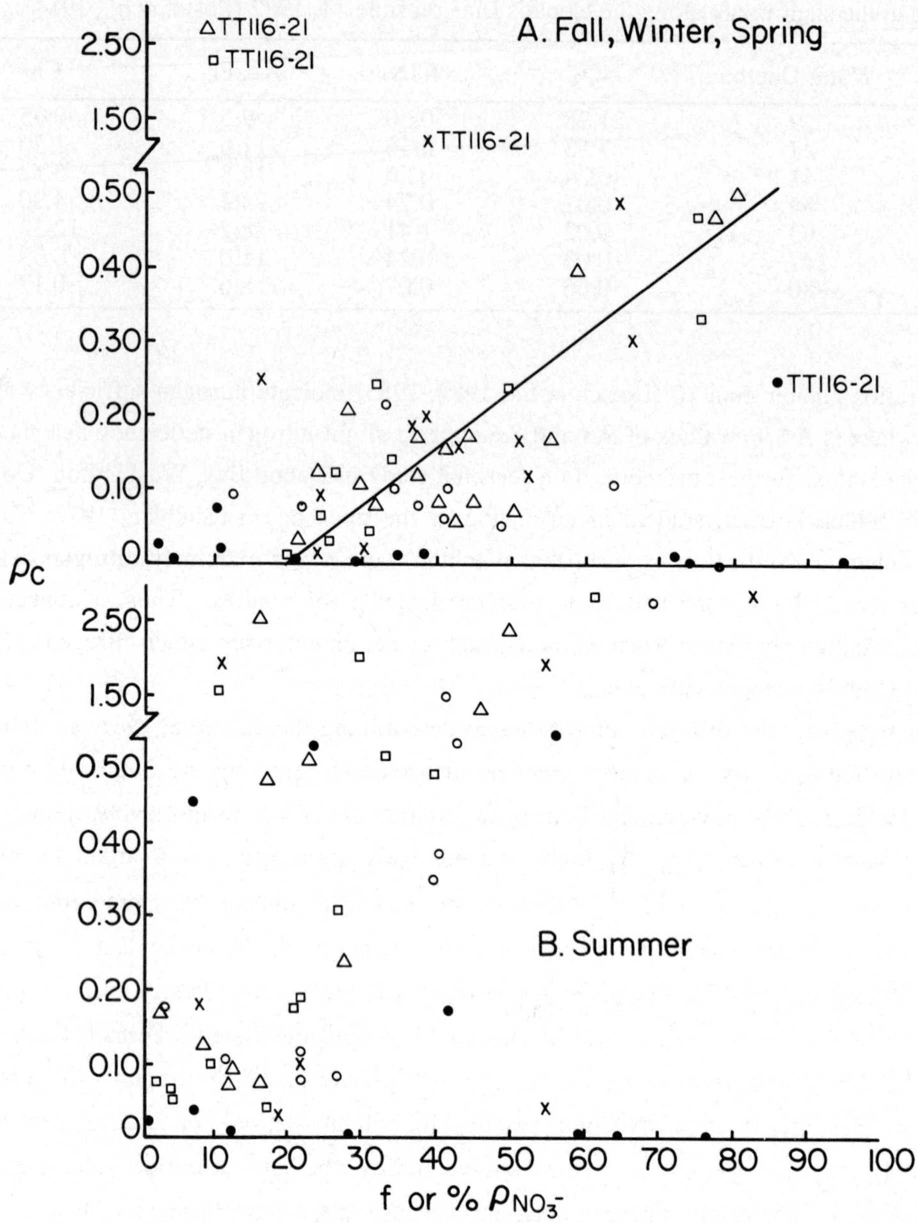

Fig. 4.9 Carbon transport vs. $\%\rho_{NO_3^-}$ (see Fig. 4.7) at different relative light levels: ○, 80-100% surface incident light; △, 43-51%, □, 23-30%; ✖, 9-15%,●, 1-5%. In A, $\rho_c = 0.00787f - 0.1559$ ($n = 48$; $r^2 = 0.52$) from geometric mean regression analysis (Laws and Archie, 1981), excluding station TT116-21 (see text) and all depths with relative light levels < 5% I_0. Note scale change on y axis. Plot includes all data from Copalis Line and TT140 as described in footnote i, Table 4.10.

light levels (< 5% I_0), ρ_c appears to be independent of $\%\rho_{NO_3^-}$ but is quite low. A similarly significant, but inverse relationship (not shown) is obtained between ρ_c and $\%\rho_{NH_4^+}$ for the same data set. This result is contrary to what would be expected from the observations of Eppley *et al.* (1979) that $\rho_{NH_4^+}$ is dependent on $\rho_{NO_3^-}$, implying that ρ_c should vary with $\rho_{NH_4^+}$ as well. Off the Washington coast, however, $\rho_{NH_4^+}$ and $\rho_{NO_3^-}$ were not significantly correlated. In the summer there is no correlation between ρ_c and $\%\rho_{NO_3^-}$ (Fig. 4.9B) or $\%\rho_{NH_4^+}$ (not shown).

A more appropriate test of the relationship between nitrogen uptake or supply and productivity is to estimate productivity from either nitrogen uptake or supply and compare it with actual values of annual ^{14}C primary production, which Perry *et al.* (1989, Chapter 3) estimate, based on data from 1974 to 1982, to be 646 g C m $^{-2}$ y $^{-1}$ for the shelf region and 294 g C m $^{-2}$ y $^{-1}$ for the slope. Primary production can be calculated from the $\%\rho_{NO_3^-}$ or f ratio using the empirical relationship derived by Eppley and Peterson (1979). From the mean f ratio for all data from the Washington coast (Table 4.9) and actual ρ_u data rather than the standard correction of Eppley and Peterson (1979), an annual productivity of 126 g C m $^{-2}$ y $^{-1}$ is predicted. This value seems low both for a temperate coastal area with intermittent upwelling and especially compared with the measured productivity. Eppley and Peterson (1979) suggest that the relationship between $\%\rho_{NO_3^-}$ (f) and productivity does not apply when the water depth is less than 200 m because of the importance of benthic regeneration. However, if the calculation is repeated using only the average $\%\rho_{NO_3^-}$ for the offshore stations along the Washington coast where the water depth exceeds 200 m, the estimated productivity is still only 127 g C m $^{-2}$ y $^{-1}$. The problem may be the difficulty in adequately sampling periods of high productivity in the late winter/spring bloom or during upwelling events when nitrate is a dominant nitrogen source.

An alternate approach for estimating production on the shelf which circumvents this problem takes into account the sporadic nitrate supply. New production using nitrate can be calculated for each season, assuming that different supply mechanisms dominate. From May to September, nitrate is supplied by upwelling. If the upwelling index is 0.20 m ^{3} s $^{-1}$ m $^{-1}$ (Bakun, 1973, 1975, pers. comm.), the nitrate concentration at depth is 15 mg-at m $^{-3}$, the width of the shelf is 7 x 10 ^{4} m and the ρ_c/ρ_N is 14.4 (from Table 4.9), then the summer new production can be estimated to be 98.7 g C m $^{-2}$. During the spring bloom, new production can be calculated from nitrate disappearance as 34.6 g C m $^{-2}$, assuming that the initial nitrate concentration is 8 mg-at m $^{-3}$, the euphotic zone is 25-m deep and the ρ_c/ρ_N is 14.4. In the fall and winter, production along the Washington and Oregon coasts is equal to 25% of the spring production (Anderson, 1964). The new production estimated for that period is 8.6 g C m $^{-2}$. After summing the seasonal data, the annual new production is 142 g C m $^{-2}$ y $^{-1}$. Using the observed $\%\rho_{NO_3^-}$ (Table 4.9), the total annual production is calculated to be 450 g C m $^{-2}$ y $^{-1}$. However, this value may represent an upper limit because $\rho_{NO_3^-}$ may be low. Hence, $\%\rho_{NO_3^-}$ could be underestimated and ρ_c/ρ_N, overestimated. If instead, the highest mean value of $\%\rho_{NO_3^-}$ given in Eppley and Peterson (1979), 50%, and a ρ_c/ρ_N of 6.625, the Redfield ratio, are used in this calculation, the annual production is 131 g C m $^{-2}$ y $^{-1}$. The variation in estimates of total production from nitrogen uptake and supply data demonstrates the difficulty of relating productivity and nitrogen, even in an area where it seems very likely that the two must be related.

4.4 CONCLUSIONS

Nitrate, ammonium and urea uptakes were measured during every season on a series of cruises along the Washington coast over the course of eight years in an effort to determine how productivity is related to nitrogen supply. As in many data sets collected over a long period, some of the methods, especially the ^{15}N-incubation procedure, are outdated and may lead to under- or over- estimates of uptake rates, depending on a number of factors. Further, the sampling scheme, in which measurements were made at three general locations on one line perpendicular to the coast on selected cruises, may not have been adequate to fully describe such a dynamic system. However, this is one of the largest ^{15}N data sets available and can provide useful information.

Experiments were conducted to characterize the nitrogen uptake of phytoplankton populations. The k_s for uptake were typical of coastal phytoplankton and were higher in net (> 20 μm) than in nanoplankton (< 20 μm). Uptake rates as a function of nitrogen concentration frequently did not show typical kinetics. Moreover, uptake rates as a function of light level almost never showed typical kinetics because both dark uptake and photoinhibition were observed. Nitrate and nitrite uptake were the most light dependent.

Depth profiles of uptake rates showed enormous variation in pattern, magnitude and relative utilization of specific nitrogen sources within cruises and between cruises, such that it was often difficult to relate uptake to obvious environmental variables. The variation was much greater than that for replicate samples, so it must reflect the dynamic nature of the ecosystem.

Despite the variation, some trends were apparent. Uptake rates tended to be lower in the fall/winter than at other times and were usually correlated with biomass levels but not nutrient concentrations. Overall, regenerated nitrogen (ammonium + urea) accounted for most of the nitrogen uptake, but nitrate was relatively more important in the spring and ammonium in the summer. The preference for regenerated nitrogen could not be adequately explained by either inhibition of nitrate uptake by ammonium or low light levels. Thus, as is typical of coastal areas every atom of nitrogen appeared to be recycled about twice before it was transported below the euphotic zone.

Conflicting results with several indicators of nitrogen deficiency lead to the hypothesis that phytoplankton off the Washington coast are poised precariously between nitrogen sufficiency and nitrogen deficiency. In such a state a small change in nitrogen supply or an environmental factor which affects the ability to utilize nitrogen could alter nitrogen uptake rates and productivity considerably. Thus, the variability in the data and the difficulty relating rates to environmental conditions may be explained, particularly when time lags between cause and effect are considered. Since we hypothesized that phytoplankton growth was dependent on nitrogen supply, if not strictly nitrogen limited, it was possible to calculate primary production either from nitrogen uptake or nitrogen supply. The estimates range from 126 to 450 gC m^{-2} y^{-1} depending on assumptions made during the calculations. Such a wide range, especially when compared with actual values of 650 g C m^{-2} y^{-1} from ^{14}C uptake (Perry *et al.*, 1989, Chapter 3), demonstrates the difficulty of determining the exact role of nitrogen in regulating primary production off the Washington coast.

ACKNOWLEDGEMENTS

This work was supported by the Atomic Energy Commision and the Department of Energy under contracts AT-(45-1)-2225-TA26, AT-(45-1)-1725-TA26, E-76-S-06-2225-TA26, DE-AT06-76-EV-75026 and DE-AT06-79-EV-10047. This is a component of Contribution No. 1784 from the School of Oceanography, University of Washington.

REFERENCES

Anderson, G.C. 1964. The seasonal and geographic distributions of primary productivity off the Washington and Oregon coasts. *Limnol. Oceanogr.*, 9: 284-302.

Bakun, A. 1973. Coastal Upwelling Indices, West Coast of North America, 1946-71. NOAA Tech. Rep. NMFS SSRF #671, 103 pp.

Bakun, A. 1975. Daily and Weekly Upwelling Indices, West Coast of North America, 1967-73. NOAA Tech. Rep. NMFS SSFR #693, 114 pp.

Blasco, D. and H.L. Conway. 1982. Effect of ammonium on the regulation of nitrate assimilation in natural phytoplankton populations. *J. Exp. Mar. Biol. Ecol.*, 61: 157-168.

Caperon, R. and D.A. Ziemann. 1976. Synergistic effects of nitrate and ammonium ion on the growth and uptake kinetics of *Monochrysis lutheri* in continuous culture. *Mar. Biol.*, 36: 73-84.

Cochlan, W.P. 1986. Seasonal study of uptake and regeneration of nitrogen on the Scotian Shelf. *Cont. Shelf Res.*, 5: 555-577.

Collos, Y. 1983. Transient situations in nitrate assimilation by marine diatoms. 4. Non-linear phenomena and the estimation of the maximum uptake rate. *J. Plankton Res.,* 5: 677-691.

Collos, Y. 1987. Calculations of ^{15}N uptake rate by phytoplankton assimilating one or several nitrogen sources. *Int. J. Radiat. Appl. Instrum. Part A, Appl. Radiat. Isot.,* 38: 275-282.

Conway, H.L. 1977. Interactions of inorganic nitrogen in the uptake and assimilation by marine phytoplankton. *Mar. Biol.*, 39: 221-232.

Conway, H.L. and T.E. Whitledge. 1979. Nitrogen uptake during a spring bloom in the New York Bight. *J. Mar. Res.*, 37: 657-668.

Copping, A.E. 1982. The distribution and passage of organic matter in the marine food web, using nitrogen as a tracer. Ph.D. Diss., Dept. of Oceanogr., Univ. of Washington, Seattle, Wa., 155 pp.

Dortch, Q. 1980. Nitrate and ammonium assimilation in three marine diatoms. Ph.D. Diss., Dept. of Oceanogr., Univ. of Washington, Seattle, Wa., 299 pp.

Dortch, Q., J.R. Clayton, Jr., S.S. Thoresen and S.I. Ahmed. 1984. Species differences in accumulation of nitrogen pools in phytoplankton. *Mar. Biol.*, 81: 237-250.

Dortch, Q., J.R. Clayton, Jr., S.S. Thoresen, J.S. Cleveland, S.L. Bressler and S.I. Ahmed. 1985. Nitrogen storage and use of biochemical indices to assess nitrogen deficiency and growth rate in natural plankton. *J. Mar. Res.,* 43: 437-464.

Dortch Q. and H.L. Conway. 1984. Interactions between nitrate and ammonium uptake: variations with growth rate, nitrogen source and species. *Mar. Biol.,* 79: 151-164.

Dugdale, R.C. 1967. Nutrient limitation in the sea: dynamics, identification and significance. *Limnol. Oceanogr.*, 12: 685-695.

Dugdale, R.C. 1976. Nutrient cycles. Pages 141-172 in: D.H. Cushing and J.J. Walsh (eds.), *The Ecology of the Sea.* Blackwell, London.

Dugdale, R.C. and J.J. Goering. 1967. Uptake of new and regenerated forms of nitrogen in primary productivity. *Limnol. Oceanogr.*, 12: 196-206.

Eppley, R.W. 1981. Autotrophic production of particulate matter. Pages 343-361 in: A.R. Longhurst (ed.), *Analysis of Marine Ecosystems.* Academic Press, London.

Eppley, R.W. and B.J. Peterson. 1979. Particulate organic matter flux and planktonic new production in the deep ocean. *Nature*, 282: 677-680.

Eppley, R.W., E.H. Renger and W.G. Harrison. 1979. Nitrate and phytoplankton production in southern California coastal waters. *Limnol. Oceanogr.*, 24: 483-494.

Eppley, R.W., E.H. Renger, E.L. Venrick and M.M. Mullin. 1973. A study of plankton dynamics and nutrient cycling in the central gyre of the N. Pacific Ocean. *Limnol. Oceanogr.,* 18: 534-551.

Eppley, R.W., J.N. Rogers and J.J. McCarthy. 1969. Half-saturation constants for uptake of nitrate and ammonium by marine pytoplankton. *Limnol. Oceanogr.*, 14: 912-919.

Fisher, T.R., P.R. Carlson and R.T. Barber. 1982. Carbon and nitrogen primary productivity in three North Carolina estuaries. *Est. Coast. Shelf Sci.,* 15: 621-644.

Friederich, G.E. and T.E. Whitledge. 1972. Autoanalyzer procedures for nutrients. Pages 38-55 in: S.P. Pavlou (ed.), *Phytoplankton Growth Dynamics.* Tech. Ser. 1, Chemostat methodology and chemical analyses. Spec. Rep., Dept. of Oceanogr., Univ. of Washington, Seattle, Wa.

Furnas, M.J. 1983. Nitrogen dynamics in lower Narragansett Bay, Rhode Island. *J. Plankton Res.*, 5: 657-676.

Garside, C. 1981. Nitrate and ammonium uptake in the apex of the New York Bight. *Limnol. Oceanogr.,* 26: 731-739.

Garside, C. and P.M. Glibert. 1984. Computer modeling of ^{15}N uptake and remineralization experiments. *Limnol. Oceanogr.*, 29: 199-204.

Glibert, P.M., F. Lipschultz, J.J. McCarthy and M.A. Altabet. 1982. Isotope dilution models of uptake and remineralization of ammonium by marine plankton. *Limnol. Oceanogr.,* 27: 639-650.

Glibert, P.M., J.C. Goldman and E.J. Carpenter. 1982. Seasonal variations in the utilization of ammonium and nitrate by phytoplankton in Vineyard Sound, Massachusetts, USA. *Mar. Biol.*, 70: 237-249.

Goldman, J.C. and M.R. Dennett. 1983. Effect of nitrogen source on short-term light and dark CO_2 uptake by a marine diatom. *Mar. Biol.*, 76: 7-15.

Goldman, J.C. and P.M. Glibert. 1983. Kinetics of inorganic nitrogen uptake by photoplankton. Pages 233-274 in: E.J. Carpenter and D.G. Capore (eds.), *Nitrogen in the Marine Environment.* Academic Press, London.

Günther, H., H.G. Floss and H. Simon. 1966. A simple procedure for ^{15}N determination. *Analyt. Chem.,* 218: 401-408.

Harrison, W.G. 1976. Nitrate metabolism of the red tide dinoflagellate *Gonyaulax polyedra* Stein. *J. Exp. Mar. Biol. Ecol.,* 21: 199-209.

Harrison, W.G., D. Douglas, P. Falkowski, G. Rowe and J. Vidal. 1983. Summer nutrient dynamics of the Middle Atlantic Bight: nitrogen uptake and regeneration. *J. Plank. Res.*, 5: 539-556.

Harrison, W.G., E.J.H. Head, R.J. Conover, A.R. Longhurst and D.D. Sameoto. 1985. The distribution and metabolism of urea in the eastern Canadian Arctic. *Deep-Sea Res.*, 32: 23-42.

Harrison, W.G., T. Platt and B. Irwin. 1982. Primary production and nutrient assimilation by natural phytoplankton populations of the Eastern Canadian Arctic. *Can. J. Fish Aquat. Sci.,* 39: 335-345.

Healey, F.P. 1977. Ammonium and urea uptake by some freshwater algae. *Can. J. Bot.*, 55: 61-69.

Horrigan, S.G. and J.J. McCarthy. 1982. Phytoplankton uptake of ammonium and urea during growth on oxidized forms of nitrogen. *J. Plankton Res.*, 4: 379-389.

Jawed, M. 1973. Ammonia excretion by zooplankton and its significance to primary productivity during summer. *Mar. Biol.*, 23: 115-120.

Kanda, J., T. Saino and A. Hattori. 1985. Nitrogen uptake by natural populations of phytoplankton and primary production in the Pacific Ocean: regional uptake capacity. *Limnol. Oceanogr.*, 30: 987-999.

Kaufman, Z.G., J.S. Lively and E.J. Carpenter. 1983. Uptake of nitrogenous nutrients by phytoplankton in a barrier island estuary: Great South Bay, N.Y. *Est. Coast. Shelf Sci.*, 17: 483-493.

Kristiansen, S. 1983. Urea as a nitrogen source for the phytoplankton in the Oslo fjord. *Mar. Biol.,* 74: 17-24.

Landry, M.R., J.R. Postel, W.K. Peterson and J. Newman. 1989. Broad-scale patterns in the distribution if hydrographic variables. Pages 1-40 in: M.R. Landry and B.M. Hickey (eds.), *The Coastal Oceanography of Washington and Oregon.* Elsevier, Amsterdam, Neth.

Laws, E. 1984. Isotope dilution models and the mystery of the vanishing ^{15}N. *Limnol. Oceanogr.*, 29: 379-386.

Laws, E. and J.W. Archie. 1981. Appropriate use of regression analysis in marine biology. *Mar. Biol.,* 65: 13-16.

MacIsaac, J.J. and R.C. Dugdale. 1969. The kinetics of nitrate and ammonia uptake by natural populations of marine phytoplankton. *Deep-Sea Res.*, 16: 45-57.

MacIsaac, J.J. and R.C. Dugdale. 1972. Interactions of light and inorganic nitrogen in controlling nitrogen uptake at sea. *Deep-Sea Res.*, 19: 209-232.

MacIsaac, J.J., R.C. Dugdale, R.T. Barber, D. Blasco and T.T. Packard. 1985. Primary production cycle in an upwelling center. *Deep-Sea Res.,* 32: 503-529.

MacIsaac, J.J., R.C. Dugdale and G. Slawyk. 1974. Nitrogen uptake in the northwest Africa Upwelling Area: results from the Cineca-Charcot II cruise. *Tethys*, 6: 69-76.

Malone, T.C. 1980. Algal size. Pages 209-232 in: I. Morris (ed.), *The Physiological Ecology of Phytoplankton.* Blackwell, London.

Malone, T.C., C. Garside, K.C. Haines and D.A. Roels. 1975. Nitrate uptake and growth of *Chaetoceros* sp. in large outdoor continuous cultures. *Limnol. Oceanogr.*, 20: 9-19.

McCarthy, J.J. 1972. The uptake of urea by natural population of marine phytoplankton. *Limnol. Oceanogr.*, 17: 738-748.

McCarthy, J.J. and J.C. Goldman. 1979. Nitrogenous nutrition of marine phytoplankton in nutrient-depleted waters. *Science*, 203: 670-672.

McCarthy, J.J., W.R. Taylor and J.L. Taft. 1977. Nitrogenous nutrition of the phytoplankton in the Chesapeake Bay. I. Nutrient availability and phytoplankton preferences. *Limnol. Oceanogr.,* 22: 996-1011.

Morris, I., C.M. Yentsch and C.S. Yentsch. 1971. Physiological state with respect to N of phytoplankton from low nutrient subtropical water as measured by effect of NH_4^+ on dark CO_2 fixation. *Limnol. Oceanogr.*, 16: 859-868.

Nelson, D.M. and H.L. Conway. 1979. Effects of the light regime on nutrient assimilation by phytoplankton in the Baja California and northwest Africa upwelling system. *J. Mar .Res.,* 37: 301-318.

Newell, B.S., B. Morgan and J. Cundy. 1967. The determination of urea in seawater. *J. Mar. Res.*, 25: 201-202.

Olson, R.J. 1980. Nitrate and ammonium uptake in Antarctic waters. *Limnol. Oceanogr.*, 25: 1064-1074.

Paasche, E. and S. Kristiansen. 1982. Nitrogen nutrition of the phytoplankton in the Oslo fjord. *Est. Coast. Shelf Sci.*, 14: 237-249.

Perry, M.J., J.P. Bolger and D.C. English. 1989. Primary production. Pages 117-138 in: M.R. Landry and B.M. Hickey (eds.), *The Coastal Oceanography of Washington and Oregon.* Elsevier, Amsterdam, Neth.

Postel, J.R., W.K. Peterson and Q. Dortch. 1984. Data report of the DOE sponsored Northwest Marine Sciences Group, June 1982 cruise. DOE Report DE/EV/10047-8, Univ. of Washington, Dept. of Oceanogr., Seattle, Wa., 591 pp.

Price, N.M., W.P. Cochlan and P.J. Harrison. 1985. Time course of uptake of inorganic and organic nitrogen by phytoplankton in the Strait of Georgia: comparison of frontal and stratified communities. *Mar. Ecol. Prog. Ser.*, 27: 39-53.

Probyn, T.A. 1985. Nitrogen uptake by size-fractionated phytoplankton populations in the southern Benguela upwelling system. *Mar. Ecol. Prog. Ser.*, 22: 249-256.

Rees, T.A.V. and P.J. Syrett. 1979. The uptake of urea by the diatom. *Phaeodactylum. New Phytol.*, 82: 169-178.

Ronner, U., F. Sorensson and O. Holm-Hansen. 1983. Nitrogen assimilation by phytoplankton in the Scotia Sea. *Polar Biol.,* 2: 137-147.

Ryther, J.H. and W.M. Dunstan. 1971. N, P and eutrophication in the coastal environment. *Science*, 171: 1008-1013.

Shuman, F.R. 1978. The fate of phytoplankton chlorophyll in the euphotic zone-Washington coastal waters. Ph.D. Diss., Dept. of Oceanogr., Univ. of Washington, Seattle, Wa., 238 pp.

Small, L.F. and D.W. Menzies. 1981. Patterns of primary productivity and biomass in a coastal upwelling region. *Deep-Sea Res.*, 28: 123-149.

Slawyk, G. 1979. ^{13}C and ^{15}N uptake by phytoplankton in the Antarctic upwelling area: results from the Antiprod I cruise in the Indian Ocean sector. *Aust. J. Mar. Freshwater Res.*, 30: 431-438.

Syrett, P.J. and J.W. Leftley. 1976. NO_3^- and urea assimilation by algae. Pages 221-234 in: N. Sutherland (ed.), *Perspectives in Experimental Biology*. Vol. 2. Pergamon Press, LOndon.

Wheeler, P.A. 1983. Phytoplankton nitrogen metabolism. Pages 309-346 in: E.J. Carpenter and D.G. Capone (eds.), *Nitrogen in the marine environment.* Academic Press, N. Y..

Wroblewski, J.S. 1977. A model of phytoplankton plume formation during variable Oregon upwelling. *J. Mar. Res.*, 35: 357-394.

Yentsch, C.M., C.S. Yentsch and L.R. Strube. 1977. Variations in ammonium enhancement and indication of nitrogen deficiency in New England coastal phytoplankton. *J. Mar. Res.*, 35: 539-555.

CHAPTER 5

ABUNDANCE, DISTRIBUTION, AND GRAZING IMPACT OF ZOOPLANKTON ON THE WASHINGTON SHELF

M.R. Landry[1] *and C.J. Lorenzen*[2]
[1] *Department of Oceanography and Hawaii Institute of Geophysics, University of Hawaii, Honolulu, HI 96822*
[2] *School of Oceanography, University of Washington, Seattle, WA 98195*

5.1 INTRODUCTION

The aim of this chapter is to provide a first-order estimate of the grazing capability of zooplankton populations on the Washington shelf during summer. Our approach is to first review and update information on the abundance, distribution and biomass of important grazer populations in the Oregon and Washington coastal environment. Second, we derive estimates of biomass-specific grazing potential for each taxonomic grouping based on organism size and feeding and growth rates in laboratory and field studies. Grazing rate estimates and population biomass distributions are then combined to provide cross-shelf estimates of community grazing potential relative to primary production during field experiments in August 1981 and June 1982. Lastly, as a test of the grazing rate calculations, predicted rates and relationships are compared to grazing and vertical flux rate measurements on the Washington mid-shelf during a field experiment in August 1983.

5.2 ABUNDANCES AND DISTRIBUTIONS OF GRAZER POPULATIONS

5.2.1 Data Sets

The abundances and distributions of zooplankton in coastal waters of the U.S. Pacific Northwest are best known from studies off central Oregon conducted from the late 1960s to the mid-1970s (Table 5.1). These studies emphasized copepod populations retained by nets with relatively coarse meshes (> 200 μm). Other zooplankton taxa are less well studied. Chester's (1978a,b) studies are the only systematic assessments of protozoan populations on the Washington/Oregon shelf. Investigations of salp abundances by Hubbard and Pearcy (1971) and euphausiid distributions by Smiles and Pearcy (1971) and Alton and Blackburn (1972) provide a limited data set for determining mean abundances of larger suspension-feeding consumers.

Regarding the better studied copepods off Oregon, Frolander *et al.* (1973) described the seasonal cycles of dominant near-shore species in Yaquina Bay from net samples taken weekly at two stations for seven years (1963 -1970). Zimmerman (1972) and Johnson (1981) extended the Yaquina Bay time series, respectively, with weekly sampling at five stations from 1969 to 1971 and with biweekly sampling from June to November, 1972 to 1974. Peterson and Miller (1976) described the zooplankton community off Newport, Oregon from biweekly collections from June 1969 to August 1972. This study concentrated on a transect of four nearshore stations (2 to 18-km offshore) in the upwelling zone, but from February 1971 to August 1972 the transect

TABLE 5.1 Selected studies of zooplankton abundance and distribution in the Washington-Oregon coastal ecosystem. Locations refer to Washington (Wa) and Oregon (Or) shelf (Sh) and slope (Sl) regions and/or to specific sites, e.g., cross-shelf transects off Newport, Ore. and Copalis, Wa. **Sta/Smpl.** refers to numbers of different sampling locations and total number of samples collected. **Gear** indicates type of collection (**O** = integrated oblique, **V** = integrated vertical, **H** = horizontal or depth stratified), type of sampling gear (**IKMT** = Isaacs-Kidd Midwater Trawl, **CB** = Clarke-Bumpus sampler, **Bongo** = double ring) and size of retention mesh.

Month/Year	Location	Sta./Smpl.		Gear	Reference
Jan'61-Oct'62	Or&Wa-Sh,Sl	128	434	H; 12-cm CB, 110μm	Peterson (1972)
Jul'61-Jun'63	Or-Sh, Sl	21	639	OV; IKMT 13mm; 1-m 570μm	Hubbard and Pearcy (1971)
Jun'63-Jul'67	Or-Sh,Sl, Newport	4	199	OV; 1-m ring, 570μm	Smiles and Pearcy (1971)
Jan'63-Sept'70	Or-Yaquina Bay	2	498	O; 12-cm CB, 240μm	Frolander *et al.* (1973)
Jun'69-Aug'72	Or-Sh, Newport	12	499	O; 20-cm Bongo, 240μm	Peterson and Miller (1976; 1977)
Jul-Aug'67	Wa-Sh	3	43	H; IKMT, 3.2 mm	Alton and Blackburn (1972)
Jun-Nov,'72-'74	Or-Yaquina Bay	5	650	O; 12-cm CB, 112μm	Johnson (1981)
Jul-Sept'73; Aug'74	Or-Sh	20	250	H; 12-cm CB & pump, 120μm	Peterson *et al.* (1979)
Jul'74	Wa-Sh,Sl, Copalis	7	38	H; Van Dorn bottle	Chester (1978a)
Feb'76-Oct'77	Wa-Sh, Juan de Fuca	3	36	Surface; Niskin bottle	Chester (1978b)
Jul-Aug'77	Or&Wa-Sh	19	301	H; Pump, 64μm	Peterson (1979)
Aug'77	Wa-Sh,Sl	6	12	V; 50-cm ring, 73μm	Postel *et al.* (1980)
Jul-Aug'79	Wa-Sh,Sl	14	16	V; 50-cm ring, 216μm	Postel *et al.* (1980)
Sept'80	Wa-Sh,Sl	19	38	V; 50-cm Bongo, 60 & 216μm	Postel *et al.* (1982)
Aug-Sept'81	Wa-Sh, Copalis	8	74	V; 50-cm Bongo, 60 & 216μm	Landry (unpubl.)
Jun'82	Wa-Sh, Copalis	8	399	H; Pump, 52μm	Landry (unpubl.)
Aug'83	Wa-Sh, Copalis	1	99	H; Pump, 52μm	Landry (unpubl.)

was extended by eight additional stations to 100-km offshore. The cross-shelf depth distribution of copepods in relation to upwelling activity was investigated by Peterson *et al.* (1979) from net and pump samples collected in July-September 1973 and August 1974. Peterson (1980) reported additional pump profiles from six transects off 45° N from July and August 1977 and seven transect lines from central Oregon to Vancouver Island, Canada (late July 1977).

Distributions and abundances of meso- and macrozooplankton for the Washington coast are known only from Peterson's (1972) and Jawed's (1973) seasonal studies which emphasized offshore populations and the distinction between the Columbia River plume and ambient seawater. Jawed's (1973; see also Hameedi, 1974) extensive analysis of samples from 1963, for example, includes only one summer net haul from a station outside of the plume. Studies comparable to the detailed investigations of shelf populations conducted off Oregon were not initiated until the late 1970s (e.g., Postel *et al.*, 1980, 1982; Landry, unpublished). Results of these studies are presented here for the first time.

Sampling stations for August 1977, July-August 1979 and September 1980 in Table 5.1 were distributed over the central Washington shelf and slope regions from 47° 07' N (Copalis) to 47° 32' N. Subsequent studies (August-September 1981 and June 1982) involved repeated sampling of an 8-station transect from the near-shore to the mid-shelf off Copalis Head. These latter studies were components of multi-disciplinary investigations of upwelling effects on the Washington inner shelf. Measurements were also made of current and wind velocities, temperature, salinity, dissolved nutrients and oxygen, chlorophyll, phytoplankton species and various experimental rate determinations (^{14}C-uptake, ^{15}N-nitrate uptake and ^{15}N-ammonium uptake and oxidation rates).

For the 1981 study, eight transect stations shoreward of the 70-m isobath were sampled five times at 3-4 d intervals. Vertically-integrated samples were taken with a modified bongo net frame with high-porosity 60- and 216-μm mesh nets. The sampler was designed such that the paired net mouths faced each other in a 'closed' position during lowering to depth, approximately 2-m above the bottom. When the wire angle was vertical, the net frame was opened by messenger and the nets were pulled through the water at 30 m min $^{-1}$. Samples were preserved in buffered 4% Formalin.

For the 1982 study, depth-stratified samples were collected using a high-capacity (4,80 l min $^{-1}$) pump with an 8-cm diameter intake hose. Sampling was conducted on the 8-station Copalis transect on seven dates at a frequency of 2 to 3 d. Standard sampling depths were 5-m intervals from the surface to 30-m and 10-m intervals from 30 to 60 m or near the bottom. The pump hose was allowed to flush for one minute before sampling at the desired depth. Pump effluent was concentrated in 52-μm nets. Transect sampling was started in early morning (0700-0800) and usually completed in the early afternoon (1400-1600).

Samples collected on the August 1983 cruise were part of a diel study of the relationships among zooplankton distributions, grazing and particle flux. Depth-stratified pump samples were taken over a 48-h period at an 80-m deep, mid-shelf station 30-km off Copalis. Pump samples were taken at approximately 0100, 0600, 1000, 1600 and 2200 at 5-m intervals to 30-m and 10-m intervals to 50 m. Vertical hauls with a 1-m, 500-μm mesh net were also taken to provide estimates of euphausiid densities and to correct for pump collection efficiency of larger copepodids and adults of *Calanus*. The zooplankton sampling station was positioned in the center of four moored sediment-trap arrays, each with two traps at 50 m. One array was 1.6 km directly shoreward of the main station, one 1.6 km directly seaward and the remaining stations were at similar distances to the north and south of the main station. At the central station, an 8-trap array was deployed, sampled and redeployed at 5 to 7-h intervals to obtain time and depth profiles of particulate fluxes. Single traps were positioned at 10, 15, 20, 30, 40, 50, 60 and 75 m. Standard deployment periods were 0900-1400, 1430-1900, 1930-0200, 0230-0800.

At selected stations during the above studies, primary production rates were determined from ^{14}C-uptake of phytoplankton sampled from depths corresponding to 95, 50, 39, 15, 5 and 1% of surface light irradiance. Generally, replicate light bottles and one dark bottle from each depth, screened to the appropriate light irradiance, were incubated on deck for 24-h in surface-water cooled baths. During the August 1983 study, deck incubations for ^{14}C-uptake were conducted for 5 to 7-h intervals corresponding to the trap deployment times. In addition, *in situ* ^{14}C-production rates were measured from bottles incubated 24-h on a separate float attached to the drifting sediment trap array. As part of the standard sampling procedure, water samples from each of the primary production stations were preserved with buffered 2% formalin for identification of phytoplankton and ciliate populations using an inverted microscope.

5.2.2 Planktonic Copepods

Abundances and distributions of zooplankton in the Oregon shelf region vary seasonally, interannually and with distance from shore (Peterson and Miller, 1975, 1976, 1977; Peterson *et al.*, 1979). Seasonal distributions are influenced by the direction of prevailing currents (Frolander *et al.*, 1973). Two copepod species with southern affinities, *Paracalanus parvus* and *Ctenocalanus vanus*, are important during winter months but are usually absent during the summer unless advected into the region by anomalous events (Peterson and Miller, 1977). Five species with northern affinities, *Acartia clausii*, *A. longiremis*, *Calanus marshallae*, *Oithona similis* and *Pseudocalanus* spp., are present throughout the year and dominate during summer upwelling months. Interannual variability in distributional patterns and relative abundance levels are related to year-to-year differences in mean upwelling intensity as well as global-scale processes such as El Niño events (Peterson and Miller, 1975; Miller *et al.*, 1983).

Mean abundances of adult copepods overlying the central Oregon shelf off Newport (Fig. 5.1; data from Peterson and Miller, 1976) show characteristic cross-shelf patterns in distribution. Upwelling is most intense in the nearshore environment within 15 km of the coast. The effective shelf break occurs at about 40-km offshore. *Acartia clausii* is most abundant shoreward of the upwelling front. Population densities are highest at shallow-water coastal stations and in adjacent estuaries (e.g., Frolander *et al.*, 1973). Peterson and Miller (1976) reported a maximum nearshore concentration of about 4 x 10 ^{5} adult *A. clausii* m $^{-2}$ at their

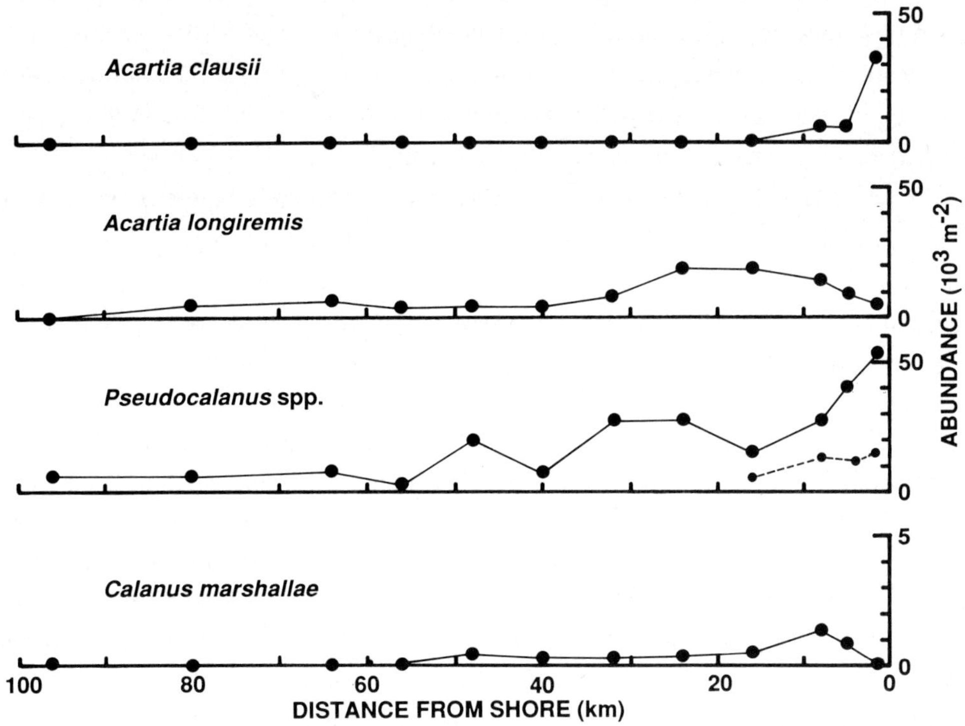

Fig. 5.1 Cross-shelf abundances of dominant planktonic copepods off Newport, Ore. during summer months (June through September). Mean data from Peterson and Miller (1976). Dashed line for *Pseudocalanus* figure indicates median abundance.

innermost sampling station in June 1970 (1.8 km from shore, 18-m bottom depth). Mean summer densities are about an order of magnitude lower.

Acartia longiremis is more broadly distributed across the shelf and in oceanic waters beyond, but densities are highest overlying the mid-shelf region. Maximum summer density of adults in this region approaches 80,000 copepods m $^{-2}$ (September 1969; August 1971 and 1972; and July 1972; Peterson and Miller, 1976).

Peterson and Miller's (1975, 1976, 1977) *Pseudocalanus* spp. refers to a suite of closely-related species of which *P. mimas* is the predominate species in the open coastal region of Washington and Oregon (B. Frost, pers. comm.). The genus *Pseudocalanus* numerically dominates the zooplankton across the Oregon shelf (Fig. 5.1). Population densities of *Pseudocalanus* spp. are maximum, on average, in the nearshore open coast environment, although abundances drop off rapidly in shallower coastal estuaries relative to true shallow estuarine copepods such as *Acartia clausii* and *A. californiensis* (Frolander *et al.*, 1973; Johnson, 1981). A maximum density of 8 x 10 ^{5} *Pseudocalanus* spp. adults m $^{-2}$ was observed in July 1969 in the Newport time-series (Peterson and Miller, 1976).

Abundances of *Calanus marshallae* adults are low in the shallow coastal stations where *Acartia clausii* and *Pseudocalanus* spp. have their highest population densities but peak within the 15-km upwelling zone (Fig. 5.1). Intermediate abundances occur across the mid- and outer shelf regions during the summer months. A high density of 13,000 *C. marshallae* adults m^{-2} was observed off Newport in August 1969 (Peterson and Miller, 1976). Again, this is about an order of magnitude greater than the mean summer density.

The population distribution of *Oithona similis* is not plotted in Figure 5.3 because the net mesh size used by Peterson and Miller (1976) was not fine enough to sample the species quantitatively. Distribution of *Centropages abdominalis*, the sixth most abundant copepod in the summer plankton off Oregon, is also not presented since the species is not common off Washington except in years of anomalous southern transport (e.g., the 1983 El Niño).

Peterson *et al.* (1979) speculated that zooplankton populations may be more broadly distributed across the Washington shelf relative to the Oregon shelf because the Columbia River plume does not act as a seaward barrier to cross-shelf transport off the Washington coast during summer. Distributions of planktonic copepods off Washington support this view (Fig. 5.2). Highest densities of *Acartia clausii*, for instance, are not found directly adjacent to the Washington coast as they are off Oregon. A near-shore maximum was observed in the June 1982 transect data, but adult population densities increased seaward to the mid-shelf during the August-September 1981 study. Presumably the latter pattern reflected upwelling-related, cross-shelf advection of surface water from the near-shore containing high numbers of *A. clausii*.

Maximum densities of *Acartia longiremis* adults occur over the mid-shelf region of Washington approximately 20 to 40 km from shore. *Pseudocalanus* spp. densities peak beyond 10-km offshore and remain high over the mid-shelf. Maximum densities of *Calanus marshallae* are observed shoreward of 10-km off Oregon. However, the copepod is virtually absent from the plankton at a similar distance off Washington. This difference is explained in part by the shallower slope of the Washington shelf. The bottom depth 10 km off Copalis is only about 35 m. *Calanus* is more abundant off Oregon seaward of the 40-m isobath.

Peterson *et al.* (1979) observed that the distribution of *Oithona similis* off Oregon did not demonstrate a clear coastal affinity. The population density of *Oithona* increases gradually across the Washington shelf, peaking offshore (Fig. 5.2).

Although we cannot properly compare abundance levels from the available zooplankton data from Washington and Oregon since they were collected in different years and with different sampling gears, it seems that summer copepod densities off Washington may exceed those off Oregon. This is certainly the case for total shelf populations, since regions of high densities are more broadly distributed across the Washington shelf. However, population densities per unit of sea surface area may also be greater off Washington. It is notable, for example, that high densities of the dominant copepod, *Pseudocalanus* spp., were found 10 to 30-km offshore on all of the Washington cruises. In this region of the shelf, the median observed density of *Pseudocalanus* adults was over 70,000 m^{-2}, about five times the summer median density off Oregon shoreward of 10 km, the region of greatest abundance. Over the course of their four-summer study, Peterson and Miller (1976) observed *Pseudocalanus* densities in excess of

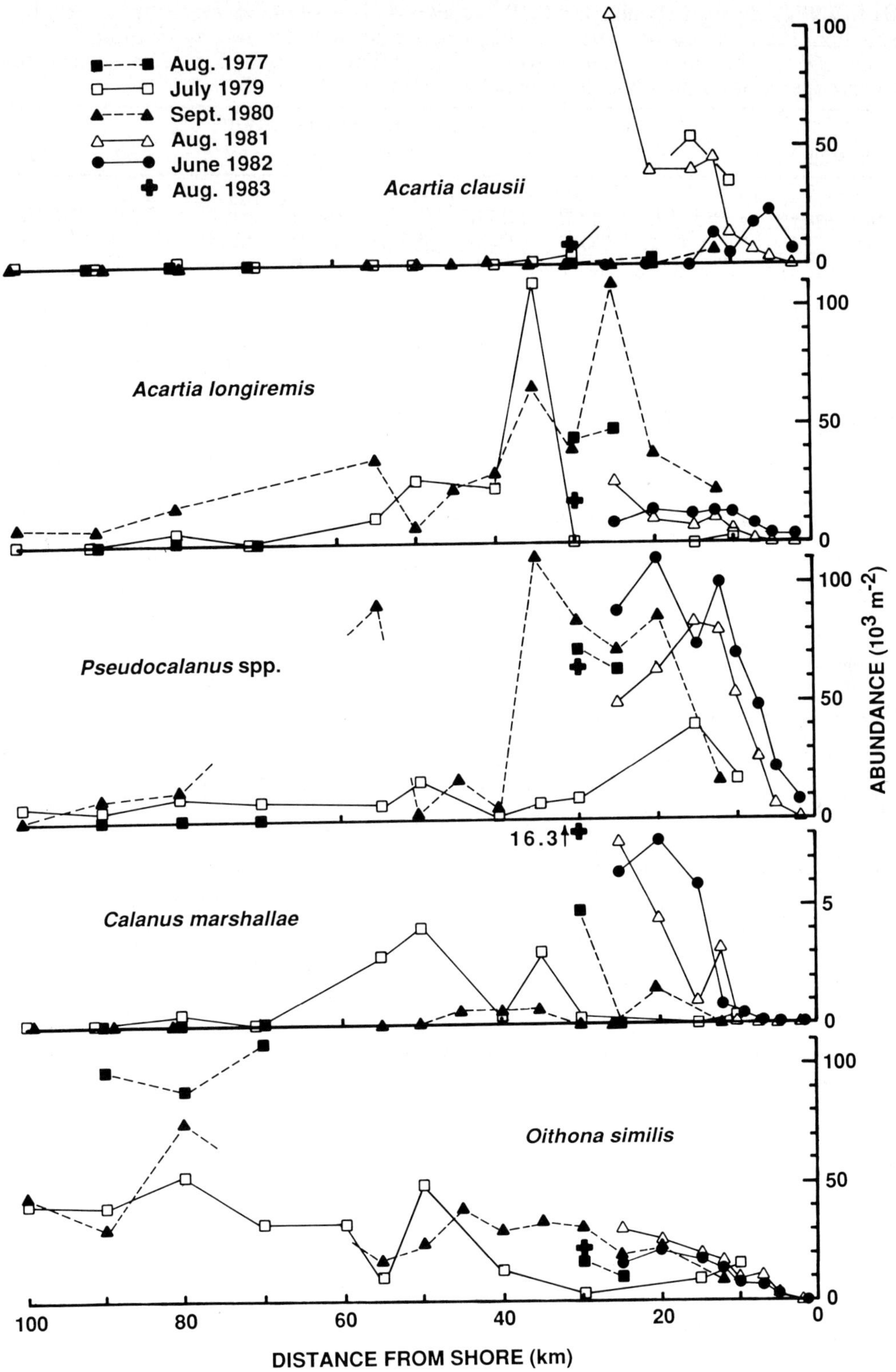

Fig. 5.2 Cross-shelf abundances of dominant planktonic copepods off Copalis, Wa. during summer months (June through September).

TABLE 5.2 Cross-shelf abundances (x10 ^{3} copepods m $^{-2}$) of dominant copepod species on the Washington shelf in August 1981. Abundance estimates are means computed from five transects off Copalis, Wa. sampled over a 17-d period. **n**, **c**, and **a** refer, respectively, to naupliar, copepodid and adult developmental stages.

Species/stage		Distance Offshore (km) 2	5	7	10	12	15	20	25
Pseudocalanus	n	371.0	543.0	505.0	616.0	350.0	600.0	606.0	348.0
	c	41.4	167.0	204.0	350.0	337.0	362.0	367.0	157.0
	a	0.5	7.7	26.8	53.6	79.8	82.7	63.8	49.8
C. marshallae	n	9.0	11.3	41.6	36.0	131.9	80.4	83.1	8.9
	c	0.0	0.0	0.0	17.7	9.6	4.6	4.1	5.1
	a	0.0	0.0	0.0	0.0	3.2	1.0	4.1	7.7
A. clausii	n	86.4	102.0	93.8	84.5	100.0	116.0	93.6	73.5
	c	3.1	18.5	11.5	36.1	51.7	58.0	28.5	43.6
	a	0.0	2.8	1.0	14.2	46.7	39.7	39.7	111.0
A. longiremis	n	15.5	69.4	46.6	43.1	48.2	51.1	43.7	70.9
	c	0.1	6.7	0.6	4.4	43.1	16.4	10.7	14.7
	a	0.0	0.5	1.3	4.1	12.8	7.7	10.4	26.2
O. similis	c	3.0	7.5	14.7	28.9	24.9	35.3	23.4	34.6
	a	1.0	4.4	11.5	9.8	16.9	19.9	25.5	30.7

70,000 adults m $^{-2}$ in only 14% of their samples at the two near-shore stations off Newport. Similar to the case for *Pseudocalanus*, the median population density of *A. longiremis* between 20 and 40-km off the Washington coast, about 30,000 adults m $^{-2}$, is approximately three times greater than the median abundance observed at the 16-km station off Newport, again the region of greatest abundance. Higher summer zooplankton densities off Washington could be related to the higher mean concentrations of surface chlorophyll observed over the Washington mid-shelf during summer (Landry *et al.*, 1989, Chapter 1). It could also mean that predators of planktonic copepods (planktivorous fishes and medusae) could be more active or abundant off central Oregon.

In the recent sampling work on the Washington shelf (Table 5.1), we used fine-mesh nets capable of retaining even the smallest developmental stages of the dominant copepod species. This was not the case for the earlier work off Oregon. Consequently, the distributions of developmental stages in the two regions cannot be adequately compared.

Tables 5.2 (August 1981) and 5.3 (June 1982) summarize the available information on cross-shelf stage structure of dominant copepod species on the Washington innershelf. Observed abundances, cross-shelf distributions and population structures of the dominant copepods are quite comparable in the two data sets. For the most part, nauplii numerically dominate the plankton at all stations although copepodid stages and adults become increasingly more important with distance offshore. Differences in the spatial distributions of nauplii and adult copepods are particularly apparent for *Calanus*. These differences are consistent with onshore advection of early developmental stages (eggs or nauplii) and/or offshore advection during development

TABLE 5.3 Cross-shelf abundances (x10 ^{3} copepods m $^{-2}$) of dominant copepod species on the Washington shelf in June 1982. Abundance estimates are means computed from seven transects off Copalis, Wa. sampled over a 14-d period. **n, c** and **a** refer, respectively, to naupliar, copepodid and adult developmental stages. Naupliar stages and early copepodids of *Acartia* species were combined.

		Distance Offshore (km)							
Species/stage		2	5	7	10	12	15	20	25
Pseudocalanus	n	249.0	569.0	789.0	1090.0	1340.0	1380.0	1220.0	989.0
	c	139.0	375.0	448.0	349.0	227.0	259.0	135.0	142.0
	a	8.9	22.0	48.3	69.4	99.0	77.3	110.0	85.1
C. marshallae	n	13.8	32.9	64.1	89.9	37.8	60.1	99.0	48.3
	c	2.8	14.8	21.4	38.3	36.6	22.8	30.3	16.8
	a	0.0	0.0	0.2	0.5	0.9	5.9	7.8	6.4
Acartia species	n	113.0	159.0	184.0	258.0	192.0	275.0	350.0	305.0
C1-C3	c	25.4	35.7	28.1	24.9	10.5	9.0	12.4	8.9
A. clausii	c	7.3	13.0	8.4	3.2	2.5	0.6	0.1	0.3
	a	7.3	20.4	17.3	15.8	3.3	0.3	0.1	0.0
A. longiremis	c	2.6	5.6	6.6	7.8	4.6	5.2	6.3	3.5
	a	2.2	3.6	9.1	13.0	10.8	12.1	14.0	9.0

through successive life-history stages (Peterson *et al.*, 1979; Wroblewski, 1980, 1982). Collections from the mid-shelf of Washington in August 1983 also yield similar abundance estimates (see Table 5.13). The consistency in these data from different years, different summer months and different hydrographic conditions (e.g., June 1982 sampling occurred during a major upwelling event; August 1983 sampling was during a major El Niño year) suggest that the typical structure of the summer zooplankton community is adequately represented by these collections.

The cross-shelf structure of zooplankton biomass on the Washington shelf was determined from population counts by assigning each developmental stage of each dominant copepod species a carbon estimate. Estimates for adult copepods were derived from published values for specimens of comparable size to the mean body length of animals in the Washington plankton. Biomass estimates for developmental stages were derived from published length-weight relationships or from of the juvenile:adult weight ratios multiplied times the appropriate estimate of adult weight. Where conversions were required, we used a carbon:dry weight ratio of 0.4. Biomass estimates and reference sources are given in Table 5.4.

Cross-shelf biomass distributions clearly identify *Pseudocalanus* spp. as the dominant copepod on the Washington shelf (Tables 5.5 and 5.6). *Pseudocalanus* accounted for an average of 64.5% of the copepod biomass for the eight stations in the August 1981 study and 77.1% in June 1982. At individual sites, *Pseudocalanus* represented up to 84% of the copepod biomass. *Calanus* increased in relative importance with distance from shore. At the near-shore stations in August 1981, for example, *Calanus* (mostly naupliar stages) accounted for only 1-2% of the zooplankton biomass. During the same study, *Calanus* represented 40% of the biomass near the

TABLE 5.4 Carbon values used to convert population abundance estimates to biomass.

Stage	*A. clausii*	*A. longiremis*	*Pseudocalanus*	*C. marshallae*	*O. similis*
N1-N2	0.05	0.05	0.08	0.22	--
N3	0.09	0.09	0.18	0.28	--
N4	0.14	0.14	0.27	0.48	--
N5	0.20	0.20	0.40	0.76	0.04
N6	0.32	0.32	0.64	1.1	--
C1	0.28	0.28	0.72	2.1	--
C2	0.48	0.48	1.2	4.4	--
C3	0.76	0.80	2.0	12.0	0.20
C4	1.2	1.5	3.4	29.0	--
C5	1.8	2.2	6.0	74.0	--
Female	3.0	4.0	8.0	112.0	0.80
Male	2.1	2.8	6.0	92.0	--

A. clausii – from Landry (1978); female body length = 1,060 μm
A. longiremis – from Landry (1978); female body length = 1,240 μm
Pseudocalanus spp.– from Vidal (1980a), Davis (1984); female body length = 1,600 μm
C. marshallae – from Vidal (1980a), Peterson (1980), Mullin and Brooks (1970); female length = 3,400 μm
O. similis – estimated body volume relative to *Pseudocalanus*; nauplii, copepodid and adults combined

mid-shelf, exceeding the contribution of *Pseudocalanus* at that station. Combined, *Pseudo-calanus* and *Calanus* generally represent 80% of the copepod biomass shoreward of the Washington mid-shelf. If contribution to grazing follows contribution to biomass, then feeding rates of these two species should figure prominently in the community grazing impact on the shelf. In contrast, despite its numerical abundance, *Oithona* contributes so little to the zooplankton biomass on the inner shelf (generally < 1%) that its grazing impact can be safely ignored. *Acartia* species are intermediate in abundance. Together, they typically represent 10 to 20% of the copepod biomass on the inner shelf.

For simplicity, zooplankton ecologists sometimes assume that the ecological impact of a given species is adequately represented by the production and feeding rates of adults. This would not be the case for copepods on the Washington inner shelf where naupliar and copepodid stages account for the majority of biomass (Tables 5.4 and 5.6). Even with the conservative assumption that grazing contribution is proportional to biomass (i.e., no rate enhancement for smaller stages), any reasonable assessment of community grazing impact must consider juvenile stages of the dominant copepods.

5.2.3 Euphausiids

Euphausiids dominate the Oregon shelf zooplankton in collections made with larger, coarse-meshed nets (Hebard, 1966; Laurs, 1967). They are also the major food items in the diets of size-selective, planktivorous fish, such as salmonids (Peterson *et al.*, 1982; Brodeur, 1986). Two species are important in the shelf ecosystem. *Thysanoessa spinifera* is most abundant in shallow waters shoreward of the mid-shelf (Peterson and Miller, 1975). *Euphausia pacifica* is maximally abundant over the mid- to outer shelf; larger size-classes are found rarely in oceanic waters overlying the slope (Smiles and Pearcy, 1971). A third species, *Nyctiphanes simplex*,

TABLE 5.5 Cross-shelf distributions of biomass (mg C m $^{-2}$) for dominant copepod species collected on the Washington shelf in August 1981.

		Distance Offshore (km)							
Species/stage		2	5	7	10	12	15	20	25
Pseudocalanus	n	101.0	158.0	137.0	168.0	95.0	163.0	165.0	95.0
	c	83.0	335.0	408.0	701.0	675.0	724.0	734.0	313.0
	a	3.8	58.0	204.0	407.0	606.0	629.0	485.0	378.0
C. marshallae	n	3.6	4.5	17.0	14.0	53.0	32.0	33.0	3.6
	c	0.0	0.0	0.0	213.0	115.0	55.0	49.0	61.0
	a	0.0	0.0	0.0	0.0	358.0	114.0	456.0	861.0
A. clausii	n	15.0	18.0	17.0	15.0	18.0	21.0	17.0	13.0
	c	3.6	21.0	13.0	42.0	60.0	67.0	34.0	51.0
	a	0.0	9.4	23.0	47.0	155.0	132.0	132.0	367.0
A. longiremis	n	3.1	14.0	9.3	8.6	9.6	10.0	8.8	14.0
	c	0.1	9.3	0.9	6.2	60.0	23.0	15.0	21.0
	a	0.0	2.1	5.1	16.0	51.0	31.0	42.0	105.0
O. similis		1.4	5.0	12.0	14.0	19.0	23.0	25.0	31.0
Total (g C m $^{-2}$)		0.21	0.63	0.84	1.65	2.28	2.02	2.20	2.31

appears to be advected into the shelf system in low numbers during El Niño events (Brodeur, 1986).

Smiles and Pearcy's (1971) four-year study of *Euphausia pacifica* off Oregon represents the most extensive data set for assessing abundances of euphausiids on the Pacific Northwest shelf. Over the course of the study, a maximum density of 27 *E. pacifica* m $^{-3}$ was observed. Data from two mid-shelf stations yield a mean population density of 2 euphausiids m $^{-3}$, about 140 animals m $^{-2}$ assuming a mean water depth of 120 m. Alton and Blackburn (1972) did not rigorously measure the sampling volumes filtered during their brief study of euphausiid populations on the Washington shelf. However, from the mean ship-speed of their IKMT tows, assuming 100% net efficiency, their nighttime collections at three depths on July 19-20, 1967 yield population estimates of 40 *E. pacifica* m $^{-2}$ and 92 *T. spinifera* m $^{-2}$. These are minimum estimates since the large mesh size of their nets (3.2 mm) did not allow for quantitative collection of the smallest and most abundant, juvenile stages. Peterson and Miller (1975) occasionally found densities in excess of 50 *T. spinifera* m $^{-3}$ at shallow stations on the inner shelf off Oregon. They suggested that population densities of *T. spinifera* on the inner shelf might exceed mid-shelf populations of *E. pacifica* by about an order of magnitude.

Smiles and Pearcy (1971) provide the only published information on size-structure of euphausiid populations on the Washington-Oregon shelf. Table 5.7 summarizes their mean size-specific abundance estimates for the Oregon mid-shelf. Euphausiid biomass was estimated from the mean size structure and the total length-carbon relationship determined by Ross (1982a). This analysis yields a mean biomass estimate of 176 mg C m $^{-2}$ for euphausiid populations on the mid-shelf (100-m water depth). On average, the biomass is approximately

TABLE 5.6 Cross-shelf distributions of biomass (mg C m $^{-2}$) for dominant copepod species collected on the Washington shelf in June 1982.

		Distance Offshore (km)							
Species/stage		2	5	7	10	12	15	20	25
Pseudocalanus	n	62.0	142.0	197.0	283.0	362.0	373.0	329.0	267.0
	c	250.0	750.0	896.0	695.0	477.0	544.0	338.0	355.0
	a	69.0	169.0	372.0	534.0	772.0	603.0	847.0	663.0
C. marshallae	n	4.8	16.0	29.0	40.0	17.0	27.0	46.0	23.0
	c	17.0	92.0	143.0	237.0	239.0	207.0	301.0	184.0
	a	0.0	0.0	11.0	11.0	21.0	119.0	162.0	130.0
Acartia spp.	n	15.2	22.0	25.0	35.5	26.0	37.2	47.3	41.1
	c	28.3	47.0	39.0	31.6	16.1	15.0	18.2	11.4
A. clausii	a	19.0	53.0	45.0	41.0	8.6	0.8	0.3	0.0
A. longiremis	a	7.5	12.0	31.0	44.0	37.0	41.0	41.0	31.0
O. similis		2.0	5.2	11.0	9.9	17.9	19.9	21.5	23.3
Total (g C m $^{-2}$)		0.50	1.31	1.80	1.96	1.99	1.99	2.15	1.73

evenly distributed among size classes from 8 to 22 mm in total length. However, the distribution at a given point in time should vary with seasonal development of the population (Smiles and Pearcy, 1971).

Vertical net hauls (75m-surface, 1-m ring net, 500-μm mesh) taken at night during the August 1983 field experiment on the Washington mid-shelf gave population estimates of 1,740 euphausiids m $^{-2}$, about one order of magnitude greater than the mean population density of *Euphausia pacifica* off Oregon (Smiles and Pearcy, 1971). The size structure of this mixed assemblage of *E. pacifica* and *T. spinifera* was virtually identical to the mean composition off Oregon. For example, the < 10mm, 10-20 mm and > 20 mm size classes accounted, respectively, for 65.5, 31.5 and 3% of the total population off Washington and 72.1, 25.6 and 2.4% in the Smiles and Pearcy (1971) composite. Given this similarity, we used the mean size structure in Table 5.7 to correct for sampling bias in August 1981 plankton collections. Since samples during this study were collected during daytime hours with small nets, few euphausiids > 10 mm were captured. To account for inadequate sampling of the larger organisms, we multiplied observed abundances of < 10 mm animals by 1.4. Biomass estimates in Table 5.8 were based on these corrected densities and a mean carbon content of 0.87 mg C euphausiid $^{-1}$ from Table 5.7.

Reliable estimates of euphausiid population abundances could not be ascertained from the daytime pump samples during the June 1982 study. Vertical hauls with the 0.5-m bongo net at two intermediate stations (12 and 15-km offshore in Table 5.8) in the sampling transect gave estimates that were substantially less than observed in August 1981 but comparable to the mean densites observed by Smiles and Pearcy (1971). We corrected observed densities of < 10 mm euphausiids to total population estimates as above. These abundance estimates were then

TABLE 5.7 Size-structure of *Euphausia pacifica* on the Oregon mid-shelf. Mean abundance estimates from Smiles and Pearcy (1971). Carbon estimates from Ross (1982a).

Total Length (mm)	Abundance (Number/10 ^{3} m ^{3})	Carbon (μg C/animal)	Carbon (μg C/m ^{3})
3	30	11	0.3
4	158	28	4.4
6	355	100	35.5
8	316	245	77.4
9	316	363	114.7
10	282	502	141.6
11	251	680	170.7
12	100	896	89.6
13	45	1155	52.0
14	56	1462	82.0
15	40	1819	74.0
16	56	2233	125.0
17	38	2707	102.8
18	56	3245	181.7
19	35	3853	134.9
20	25	4534	113.4
21	32	5293	169.4
22	14	6135	85.9
	2.02 individuals m $^{-3}$		1.76 mg m $^{-3}$

extrapolated to other stations on the cross-shelf transect (Table 5.8) using the observed pattern of relative abundance from August 1981.

5.2.4 Pelagic Tunicates

Hubbard and Pearcy (1971) identified six salp species in 634 samples collected over a four-year period (1961-1964) off Oregon. *Salpa fusiformis* and *Iasis zonaria*, the two most abundant species, were encountered, respectively, in only 19% and 12% of the mid-water trawl samples, each sample representing a mean volume of 13.5 x 10 ^{3} m ^{3}. The other species, *Pagea confoederata*, *Thetys vagina*, *Thalia democratica* and *Helicosalpa virgula* occurred in less than 2% of the samples. Mean densities of 2 x 10 $^{-3}$ *S. fusiformis* and 7 x 10 $^{-5}$ *I. zonaria* m $^{-3}$ were reported. The mean summer-averaged abundance of *S. fusiformis* on the Oregon shelf off Newport was 1.2 x 10 $^{-4}$ salps m $^{-3}$. Peterson and Miller (1975) reported no salps during their 1969-1972 study off Newport. Even allowing for the possibility of substantial underestimation of the salp population due to inappropriate mesh size or net destruction, we conclude that salps are not generally abundant enough to contribute substantially to grazing impact on the shelf.

Appendicularians (*Oikopleura* sp.) are the preferred prey of larval English sole and their abundance during the early feeding period in winter is strongly linked to the recruitment success of this fish species (Gadomski and Boehlert, 1984). Maximum abundances of appendicularians occurred in autumn and winter samples off Newport; they were found in only 27% of the summer plankton samples (Peterson and Miller, 1975). The presence of *Oikopleura* sp. was significantly related to waters of reduced salinity suggesting an affinity to offshore waters

TABLE 5.8 Estimates of euphausiid biomass (mg C m $^{-2}$) from plankton collections on the Washington shelf during August 1981, June 1982 and August 1983.

	Distance Offshore (km)								
	2	5	7	10	12	15	20	25	30
August, 1981	0.0	3.5	8.7	52.2	399.0	387.0	302.0	451.0	
June, 1982	0.0	0.6	1.5	9.3	42.6	97.4	54.0	80.0	
August, 1983									1510.0

influenced by the Columbia River plume. Peterson and Miller's (1975) results provide a mean population estimate of 160 *Oikopleura* m $^{-2}$ for the Newport transect stations during summer.

Population densities of *Oikopleura* sp. on the Washington shelf do not display the summer consistency noted for dominant copepods. Significant populations were not evident in samples collected in June 1982, July 1979 and September 1980. However, the mean inner-shelf abundance for the eight transect stations in the August 1981 study was in excess of 34,000 *Oikopleura* m $^{-2}$. Maximum population abundances of 1.7 x 10 ^{5} m $^{-2}$ were observed in the nearshore stations. In August 1977 samples, mid-shelf densities of *Oikopleura* sp. exceeded 10 ^{5} m $^{-2}$. Mackas and Sefton (1982) reported a mean density of 2,900 larvaceans m $^{-2}$ for samples collected during July and August, 1979 and June and August, 1980 off Vancouver Island, Canada. This may substantially underestimate the actual mean population abundances due to the coarse mesh (233 μm) of plankton nets used in the study. The available evidence thus points to the possibility that Oregon and Washington shelf waters differ in the extent to which larvacean populations develop during summer months. This could be due to greater summer upwelling intensity on the Oregon shelf (Landry *et al.*, 1989, Chapter 1) as larvaceans are rare in high salinity water (> 33.5 ppt) associated with upwelling (Peterson and Miller, 1975).

The size structure of the larvacean population was not measured in our samples. From King's (1981) observations of *Oikopleura dioica* populations in CEPEX enclosures in Saanich Inlet, British Columbia, we assume a modal trunk length of 0.5 mm. Larvaceans of this size have a carbon content of 0.9 μg C (King *et al.*, 1980). Consequently, the mean population density of 34,000 *Oikopleura* m $^{-2}$ in August 1981 corresponds to a biomass of 30.6 mg C m $^{-2}$.

5.2.5 Ciliated Protozoans

Although recent studies emphasize the importance of protozoan micro- and nanozooplankton as primary consumers in marine systems, they are the most poorly studied group of grazers in Oregon and Washington coastal waters. Chester (1978a) observed a mean ciliate density of about 2,000 l $^{-1}$ in the upper 20 m of three shelf stations sampled off Copalis, Washington in July 1974. A density of about 1,500 cells l $^{-1}$ occurred at the shelf break. Comparable densities were found at the depth of the chlorophyll maximum (40-50 m) in offshore, oceanic waters, where near-surface densities were reduced (500-1,000 cells l $^{-1}$). Postel (1975) also noted 3 to 6 times surface ciliate densities in the deep chlorophyll maximum in offshore waters off Oregon. Densities as high as 10 ^{4} cells l $^{-1}$ were found during the summer in surface waters of the Juan

TABLE 5.9 Estimates of ciliate biomass (mg C m $^{-2}$) from plankton collections on the Washington shelf during August 1981 (Postel *et al.*, 1983), June 1982 and August 1983. Estimates are based on average densities in the depth range of 0 to 30 m (or bottom) and a mean carbon content of 1 ng C per ciliate (Chester, 1978a).

	Distance Offshore (km)					
	2	7	10	15	20	30
August, 1981	48.0		64.5		103.0	
June, 1982	85.0	172.0		191.0		333.0
August, 1983						318.0

de Fuca Strait (Chester, 1978b). Loricate ciliates (tintinnids) generally comprised only 5% of the ciliate community on the Washington shelf; oligotrichs dominated (Chester, 1978a). Chester's (1978a) biomass estimates for ciliate populations in near-surface waters (0 -20 m) of the Washington shelf range from 0.94 to 3.87 μg C l^{-1}. The mean estimate of 2.1 μg C l^{-1} corresponds to about 42 mg ciliate C m $^{-2}$ in the euphotic zone.

Ciliate biomass estimates in Table 5.9 were determined from measured densities in the depth range of 0 to 30 m (or to the bottom for shallow stations) and Chester's (1978a) average carbon content of 1 ng C ciliate $^{-1}$. These biomass estimates are higher than Chester's, particularly at the mid-shelf station where observed densities are on the order of 10 ^{4} ciliates l^{-1}. However, the densities are in the range, 0.3 to 2.8 x 10 ^{4} ciliates l^{-1}, reported by Landry and Hassett (1982) from samples collected in October 1980, and preserved in Lugol's fixative. Fixatives are a concern in interpreting ciliate abundances since formalin, by present standards, is not considered optimal for preserving delicate oligotrichs (D. Gifford, pers. comm.). Our samples were preserved with formalin; hence, our abundance estimates are conservative.

5.3 FEEDING RATE ESTIMATES

5.3.1 Growth and Feeding Studies

With the exception of ciliates, the dominant groups of suspension-feeding zooplankton on the Washington-Oregon shelf have been the focus of considerable experimental study in the laboratory and the field (Table 5.10). *Acartia* species have been studied with an emphasis on the dynamics of field populations (e.g., Landry, 1978; Johnson, 1981). *Calanus* and *Pseudocalanus* species have been preferred subjects for detailed laboratory investigations of feeding, growth and energetics (e.g., Frost, 1972; Vidal, 1980a, b, c). The feeding, growth and development of *Calanus marshallae* is known almost exclusively from the work of Peterson (1980, 1986) off Oregon although Landry and Hassett (1985) reported diel patterns in gut fluorescence and digestive enzyme activities in field work on the Washington shelf. However, many of the characteristics of *C. marshallae* are similar to those of the closely-related *C. pacificus* which dominates in Puget Sound, Washington and on the California shelf. Therefore, extrapolation from the latter species' extensive data base may be justified. *Paracalanus parvus*, characteristic of the late summer, early autumn zooplankton in Puget Sound and the autumn-winter

TABLE 5.10 Studies of the feeding, growth and population dynamics of important zooplankton species on the Washington-Oregon coast.

Species	Parameters	References
Acartia californiensis	Growth, Population Dynamics	Johnson (1981)
Acartia clausii	Growth, Population Dynamics	Landry (1978)
	Feeding	Donaghay (1980)
Acartia longiremis	Population Dynamics	Beers *et al.* (1977)
Calanus marshallae	Feeding, Growth, Population Dynamics	Peterson (1980, 1986)
Calanus pacificus	Feeding	Frost (1972, 1977); Runge (1980); Hassett and Landry (1983)
	Growth	Vidal (1980a,c); Runge (1984, 1985)
Paracalanus parvus	Feeding, Growth	Bartram (1980); Checkley, (1980a,b)
	Population Dynamics	Grice *et al.* (1977); Sonntag and Parslow (1981)
Pseudocalanus spp.	Feeding	Frost (1980); Paffenhofer and Harris (1976)
	Growth	Vidal (1980 a,b); Corkett and McLaren (1978)
	Population Dynamics	Beers *et al.* (1977); Ohman (1983)
Euphausia pacifica	Feeding, Growth	Ross (1981, 1982 a,b)
	Feeding	Ohman (1984)
	Population Dynamics	Smiles and Pearcy (1971)
Oikopleura dioica	Feeding	King *et al.* (1980)
	Feeding, Growth	Paffenhofer (1975)
	Growth, Population Dynamics	King (1981, 1982)

zooplankton off the outer-coast of Washington and Oregon, is best known from Checkley's (1980a, b) work off California.

Ross (1981, 1982a, b) documented feeding, growth and energetic parameters for all size classes of *Euphausia pacifica*. Comparable studies have not been conducted with the common near-shore euphausiid *Thysanoessa spinifera*. King's (1981) investigations of *Oikopleura dioica* represent the primary data base for larvaceans in the Washington coastal region although Paffenhöfer's (1975) results from the North Sea are also applicable.

5.3.2 Maximum Ingestion *vs.* Maximum Clearance

Ignoring behaviors at very low food concentration (e.g., Frost, 1975), the relationship between food concentration and ingestion of suspension-feeding zooplankton can be thought of as having two parts. Above a given concentration, ingestion is maximal and independent of food level (Frost, 1972). Below this concentration, water is cleared of food at a constant maximal rate such that ingestion increases approximately linearly with increasing food. Therefore, in computing the grazing impact of zooplankton populations, we must apply maximum ingestion rate, maximum clearance rate, or both rates depending on the ambient food level relative to the concentration that saturates ingestion.

The issue of food limitation among zooplankton, marine copepods in particular, has been widely debated during the past decade with no emergent consensus (e.g., Conover, 1978; Mayzaud and Poulet, 1978; McLaren, 1978; Checkley, 1980b; Durbin *et al.*, 1983; Huntley and Boyd, 1984; Runge, 1985; Ohman, 1985). However, it seems likely that different species encounter different degrees of food limitation at different locations and at different times of the year (Frost, 1985). Among the important zooplankton in the coastal plankton off Washington, *Pseudocalanus* is probably the least affected by food abundance (Ohman, 1985). This, most likely, explains its numerical and biomass dominance in spite of its low population growth potential relative to co-occurring copepod species (e.g., about 3 - 4 eggs female $^{-1}$ d $^{-1}$ *vs*. 24 eggs female $^{-1}$ d $^{-1}$ for *Calanus marshallae*). In Dabob Bay, Washington, generation times, prosome lengths and fecundities of *Pseudocalanus* suggest that the copepod is almost always operating under food concentrations supporting maximum growth, hence maximum ingestion rates (Ohman, 1985). Maximum growth of *Pseudocalanus* occurs at phytoplankton densities above 50 μg C l $^{-1}$ at 8° C and 150 μg C l $^{-1}$ at 15.5° C (Vidal, 1980a). Assuming a carbon to chlorophyll ratio of 50, mean carbon concentrations exceeding 150 μg C l $^{-1}$ are typical of the inner shelf, near-surface waters off Washington and Oregon from July to September (Landry *et al.*, 1989, Chapter 1, Fig. 1.24). Higher densities, exceeding 300 μg C l $^{-1}$, occur on average during June.

From similarities in cohort development times relative to maximum, temperature-dependent laboratory rates and comparable sizes of laboratory and field-raised stages up to C4, Peterson (1980) argued that *Calanus marshallae* may not be food-limited off Oregon. In laboratory studies, both *C. marshallae* and *C. pacificus* approach food satiation at similar concentrations of the diatom *Thalassiosira weissflogii*, about 300 μg C l $^{-1}$ (Frost, 1972; Peterson, 1986). However, *T. weissflogii* is relatively small (11-12 μm diameter) and food satiation should occur at lower concentrations for larger diatom cells, or diatom chains characteristic of upwelling zones (e.g., Frost, 1972). For larger cells (e.g., *T. angstii*, 64 μm) maximum growth rate of *C. pacificus* is achieved at food concentrations of about 160 μg C l $^{-1}$ at 12 to 15.5° C and 120 μg C l $^{-1}$ at 8° C (Vidal, 1980a). The application of these laboratory results to the field is complicated by the fact that the copepod may feed disproportionately in local patches of phytoplankton higher than mean levels and that the diet probably includes prey other than phytoplankton (e.g., Landry, 1981). Therefore, it is possible that *C. marshallae* could meet its maximum food requirements at phytoplankton densities typical of the Washington-Oregon coast during summer. Nonetheless, this would be a radical assumption in view of the fact that Runge (1985) and Frost (1985) clearly demonstrate chronic food limitation, as measured by fecundity, for *C. pacificus* in Puget Sound except during periods of major phytoplankton blooms.

Consequently, we assume that *Calanus* populations on the Washington shelf are more likely to be near food satiation (maximum ingestion rates apply) when water temperatures are cold and phytoplankton densities maximum (e.g., in June) and food-limited (maximum clearance rates apply) in late summer.

Acartia populations are probably also food-limited during the late summer, although probably to a lesser extent than *Calanus*. Landry (1978) observed that prosome length, development times, growth rates and fecundity of a lagoon population of *A. clausii* were depressed during summer months when water temperatures were elevated and phytoplankton was dominated by small flagellates. However, these indices were generally in the range of optimal laboratory values during colder spring months when diatoms were abundant. Durbin *et al.* (1983) reported that fecundity and condition factor of *A. tonsa* in Narragansett Bay, Rhode Island were suboptimal below chlorophyll concentrations of 6 to 8 μg Chl *a* l^{-1} (about 800 μg C l^{-1} given their observed C:Chl ratio of 45 and a background detritus concentration of 450 μg C l^{-1}). Based on these observations, we would expect maximum ingestion rates in early summer (June) and reduced feeding in late summer (August-September).

In view of the arguments for food-limited growth of populations of *Calanus* and *Acartia* on the Washington shelf, one would also expect that even larger suspension-feeders, the euphausiids, would demonstrate suboptimal growth under ambient food conditions. However, Smiles and Pearcy (1971) remarked on the rapid growth rates of *Euphausia pacifica* inferred from seasonal changes in the size-structure of the population off Newport, Ore.. Mean growth rates were 0.065 mm total length d^{-1}, but growth rates of 0.095 mm d^{-1} were noted for smaller, juvenile stages. From the length-carbon relationship of Ross (1982b), a growth rate of 0.065 mm d^{-1} corresponds to a growth increment of 29 μg C d^{-1} for a 16-mm euphausiid or about 1.3% body C d^{-1}. Similarly, a growth increment of 0.095 mm d^{-1} corresponds to a carbon growth of 15.5 μg d^{-1} for a 10-mm euphausiid or about 3.1% body C d^{-1}. In comparison, allometric growth equations at 12°C indicate maximum growth rates of 1.2% and 2.4% body C d^{-1}, respectively, for euphausiids of these sizes (Ross, 1982b). Thus, the available evidence points to optimal or near-optimal growth conditions for the euphausiid population. We advance this notion cautiously as heavy predatory impact on smaller size-classes of euphausiids (e.g., 8-9 mm; Peterson *et al.*, 1982) could give the appearance of greater growth rates than actually occurred.

The link between food concentration and maximal growth and ingestion of larvacean and protozoan populations is difficult to assess. The increased relative abundance of these forms in regions and times of lower chlorophyll concentrations suggest that their maximum needs are met at lower food densities than most co-occurring planktonic crustaceans. However, the larger phytoplankton cells and diatom chains characteristic of upwelling systems may not be generally available to these suspension-feeders or may interfere with their feeding mechanisms. Paffenhöfer (1975) suggested that food did not limit *Oikopleura dioica* populations in the North Sea and King (1981) observed rapid blooms of *O. dioica* in Saanich Inlet, British Columbia when contained populations were removed from predators. King (1981, 1982) estimated that

Oikopleura could grow at maximum rates, one biomass doubling d $^{-1}$ at 14° C, when concentrations of particles smaller than 10 μm exceeded 60 μg C l $^{-1}$.

5.3.3 Copepod Feeding Rates

Pseudocalanus spp.

Grazing estimates of *Pseudocalanus* spp. are made with the assumption that maximum ingestion rates are always applicable during summer months in the Washington shelf environment. Harris *et al.* (1982) estimated maximum ingestion rates of approximately 50% body C d $^{-1}$ for *Pseudocalanus* spp. females in CEPEX enclosures in Saanich Inlet, British Columbia. Frost's (1980) laboratory grazing data for female *Pseudocalanus* sp. are maximum at 38% body C d $^{-1}$.

From body mass values in Table 5.4 and 15° C developmental times in Landry (1983), we estimate mean growth rates of 26% body C d $^{-1}$ for naupliar stages of *Pseudocalanus* sp. and 30% C d $^{-1}$ for early copepodids. Vidal (1980) determined maximum growth rates of about 25.5% C d $^{-1}$ for early copepodids at 15.5° C, 22% C d $^{-1}$ at 12° C and 15.5% C d $^{-1}$ at 8° C. Therefore, even a relatively high growth efficiency of 35% would require maximum ingestion rates ranging from 44% body C d $^{-1}$ at 8° C to 73% body C d $^{-1}$ at 15.5° C. Paffenhöfer and Harris (1976) determined that weight-specific ingestion rates for *Pseudocalanus elongatus* were constant for most developmental stages, but typically declined for late copepodids and adults. Their maximum sustained ingestion rate was 148% body weight d $^{-1}$, but their estimated growth efficiencies were correspondingly low (12-17% d $^{-1}$; Harris and Paffenhöfer, 1976).

Based on the preceding analysis, we adopt the view that maximum ingestion rates of about 50% body C d $^{-1}$ are appropriate for *Pseudocalanus* spp. populations at typical June temperatures in the upper 20 m of the water column (9-10° C). A maximum ingestion rate of 60% is conservatively assumed for typical August temperatures (14-15° C). These rates apply to all developmental stages except adults. Vidal (1980a) found that maximum growth rates for late copepodids were about 70% of the rates for smaller stages. Therefore, we assume maximum ingestion rates for adult *Pseudocalanus* equal 70% of the rates for juvenile stages, that is 35% body C d $^{-1}$ at 9-10° C and 42% body C d $^{-1}$ at 14-15° C. These values are in good agreement with the experimental results of Frost (1980) and Harris *et al.* (1982).

Calanus marshallae

Both food-sufficiency and food-limitation are considered in evaluating the grazing impact of *Calanus marshallae*. Peterson (1980) observed that *C. marshallae* attains a maximum ingestion rate of 31.5% body C d $^{-1}$ (11.7 x 10 ^{3} cells copepod $^{-1}$ h $^{-1}$) at food densities of the diatom *Thalassiosira weissflogii* above 3,700 cells ml $^{-1}$. *C. pacificus* achieves a maximum ingestion rate of 39% body C d $^{-1}$ (12.1 x 10 ^{3} cells copepod $^{-1}$ h $^{-1}$) at a comparable cell density (Frost, 1972). The difference in ingestion rates as % body C d $^{-1}$ derives from the fact that *C. marshallae* is slightly larger than *C. pacificus*. We adopt the former estimate (31.5% body C d $^{-1}$) as the maximum ingestion rate for *C. marshallae* females and use growth relationships among developmental stages to scale maximum ingestion rates to size. It appears from comparisons of the maximum fecundities of adult females that *C. pacificus* (13% body C d^{-1} at 8° C and 21% C d^{-1} at 15° C; Runge, 1984) has a substantially higher growth potential than *C. marshallae* (6.4% body C d^{-1} at 10-12° C; Peterson, 1980) at optimal food densities. Peterson (1980) computed a gross growth efficiency of egg production (egg production/ingestion) of 22% for *C. marshallae.*

From body mass estimates in Table 5.4 and stage-specific developmental rates in Peterson (1986), maximum growth rates of *C. marshallae* nauplii (N3-N6) are estimated to be about 11% body C d^{-1} at 10° C. Similarly, development times imply mean maximum growth rates of 12.5% body C d^{-1} at 10° C and 21% body C d^{-1} at 15° C for copepodid stages (C1-C5). At 10° C, therefore, growth rates of *C. marshallae* juveniles exceed maximal egg production rates of adult females by about a factor of 1.8. Comparing Runge's (1984) egg production rates to Vidal's (1980a) growth rates for early copepodids (C2 and C3), the factor for *C. pacificus* ranges from 1.4 at 8° C to 1.8 at 15.5° C. We assume an intermediate factor of 1.6 and similar gross growth efficiencies for all developmental stages. Consequently, maximum ingestion rates for *C. marshallae* at 10° C (June) are 31.5% body C d^{-1} for adults and 50% body C d^{-1} for immature stages. We do not correct adult rates for the higher August temperature because their daily migration takes them through a wide range of temperature. However, most of the immature stages experience elevated temperature in later summer because they reside predominately in the upper 20 m of the water column. Maximum ingestion rates of nauplii and copepodids, therefore, are assumed to be elevated by 20% (i.e., 60 % body C d^{-1}) during August. This conservative estimate is identical to the temperature enhancement we use for *Pseudocalanus* spp. Huntley and Boyd's (1984) general regressions of clearance rate *vs.* zooplankton body size suggest a three-fold temperature enhancement effect for 5-10 μg C copepodids at 14.4° C relative to 9.8° C. Vidal's (1980a) growth rate estimates for *C. pacificus* C2 and C3 indicate a temperature enhancement factor of 2.2 from 8 to 15.5° C and 1.7 from 8 to 12° C.

Peterson's (1980) functional response curve for *C. marshallae* females provides an estimate of maximum clearance rate of 75 ml copepod d^{-1} at 10°C, about 0.67 ml μg C d^{-1}. Clearance rates for juvenile stages can be scaled relative to this estimate using the size-dependent, regression relationships at 9.8° C in Huntley and Boyd (1984). Accordingly, nauplii of mean size 0.48 μg C have maximum clearance rates of 1.3% of adult females (i.e., about 1 ml nauplii d^{-1} or 2 ml μg C d^{-1}) and copepodids of mean size 12 μg C have maximum clearance rates of 17% of adult females (i.e., about 12.8 ml copepodid d^{-1} or 1.07 ml μg C d^{-1}). To determine grazing potential of *Calanus* developmental stages from these weight-specific clearance rates, we first determined food levels from mean euphotic zone chlorophyll concentration and a C:Chl ratio of 50. We used maximum ingestion rates where food abundance exceeded the concentrations implying satiation.

Acartia species

Whether conditions are optimal for temperature-dependent growth in the laboratory or suboptimal under field conditions, growth of *Acartia* species can generally be expressed in terms of a single growth coefficient applicable to all developmental stages (e.g., Miller *et al.*, 1977; Landry, 1978). Consequently, if different stages have comparable growth efficiencies, they should also have similar weight-specific ingestion rates. Temperature-dependent maximum growth rates for *A. clausii* vary from 14% body C d^{-1} at 10° C to 24% d^{-1} at 15° C (Landry, 1978). Grazing rates for *A. clausii* females correspond to a maximum ingestion of 83% body C d^{-1} (Donaghay, 1980). This implies a growth efficiency of 29% relative to maximum growth rate at 15° C.

Landry (1978) observed that growth rates of *A. clausii* under field conditions were generally comparable to maximum rates in the laboratory at ambient temperatures in the range of 9 to 14° C. Substantial departures occurred at warmer temperatures. The mean growth rate of six field generations developing at temperatures in the range of 8 to 15° C was 18% body C d^{-1}. We therefore assume that *Acartia* species can achieve maximum growth rates during the period of high food concentration in June. With a 29% growth efficiency and 18% growth d^{-1}, the calculated maximum ingestion rate is about 60% body C d^{-1}. We assumed this rate applies throughout the summer months.

5.3.4 Euphausiid Feeding Rates

The grazing impact of euphausiids can be estimated from the mean size-structure of the shelf populations (Table 5.7) and size-specific ingestion rates determined by Ross (1982a). Over the size range of juvenile and adult *E. pacifica*, daily ingestion at 12° C varies from 32 to 18% body C d^{-1} (Ross 1982a). For euphausiids between 9 and 20 mm in length, carbon-specific ingestion varies only between 24 and 19% body C d^{-1}. We assume that an intermediate estimate of 21% body C d^{-1} is appropriate for late summer (August) conditions when euphausiids migrate daily between 8° C and 14° C strata. During early summer (June), an 8° C temperature is be more appropriate. Ross's (1982a) weight-specific ingestion rates at 8° C are uniformly 62% of the 12° C rates. We, therefore, compute the grazing potential of euphausiids in June 1982 by applying a biomass-specific grazing rate of 13 % body C d^{-1} to the population biomass estimate.

5.3.5 Appendicularian Feeding Rates

Both Paffenhöfer (1975) and King (1981) noted that all size classes of *Oikopleura dioica* can double their biomass daily at food concentrations less than 100 μg C l^{-1} and temperatures in the range of 13-14° C. King (1981) suggested weight-specific ingestion rates on the order of 200 to 300% body C d^{-1}. A 0.5-mm larvacean (0.9 μg C) feeding at a clearance rate of 26 ml animal $^{-1}$ d^{-1} (Paffenhöfer, 1975) could achieve a maximum ingestion rate of 250% body C d^{-1} at a food concentration of 87 μg C l^{-1}. We assume that the larvacean population on the Washington shelf in August is growing at the rate of one doubling d^{-1} with a weight-specific ingestion rate of 250% body C d^{-1}.

5.3.6 Protozoan Feeding Rates

Phytoplankton densities on the Washington shelf are sufficiently high that we might expect near maximum growth and ingestion rates for ciliated protozoans in near-surface waters for most of the summer. Heinbokel (1978) and Verity (1985) found that high growth rates of tintinnids could be achieved at food densities on the order of 100 μg C l^{-1}. Verity (1985) determined maximum ingestion rates for the tintinnid, *Tintinnopsis vasculum*, of approximately 170% body C d^{-1} at 10° C and 230% body C d^{-1} at 15° C. We adopt these rate estimates for ciliate populations on the Washington shelf. However, *T. vasculum* is a relatively large cell (8.9 ng C cell $^{-1}$) compared to the estimated mean size of protozoans off Washington (1.0 ng C cell $^{-1}$; Chester, 1978a). Given the demonstrated effect of cell size on growth rate (e.g., Heinbokel, 1978), it is likely that we conservatively estimate the grazing impact of ciliate populations on the shelf.

5.4 COMMUNITY GRAZING IMPACT

Estimates of zooplankton community grazing impact on the Washington shelf can be derived from observed abundances of suspension-feeding taxa during the 1981 and 1982 field experiments and previous assumptions regarding their feeding rates. Accordingly, calculated carbon demand for August 1981 ranges from about 300 mg m^{-2} d^{-1} in the shallow, near-shore environment to 1,300 mg m^{-2} d^{-1} near mid-shelf (Table 5.11). Primary production, measured as ^{14}C-uptake, substantially exceeds grazing potential in the region shoreward of 10 km, an area of high nutrient levels due to upwelling and relatively low standing stock of zooplankton. This excess near-shore production is advected offshore by upwelling where it encounters a region of high exploitation near the mid-shelf. Two mechanisms probably account for this zone of elevated grazing impact. Abundance levels of smaller, surface-dwelling taxa develop with increasing phytoplankton biomass as near-shore water is advected offshore. In addition, the greater water depth over the mid-shelf region allows larger, migrating species to occur there.

Population abundance estimates for August 1981 indicate that copepods, particularly *Pseudocalanus* spp., account for the largest fraction of grazing potential shoreward of the mid-shelf (Fig. 5.3). Protozoans and larvaceans are relatively more important in the near-shore environment where they account, respectively, for maxima of 36.6% and 25.6% of the calculated grazer impact. A quasi-balance must exist between production and grazing farther offshore where surface nutrients are exhausted, photosynthetic production is reduced and phytoplankton are distributed with a pronounced subsurface maximum (e.g., Jamart *et al.*, 1977). Protozoans may increase in relative importance as grazers in this offshore, oceanic region since their abundance levels in the subsurface chlorophyll maxima can be comparable to those in near-shore surface waters (e.g., Chester, 1978a) while primary production rates are much reduced.

The distinction between grazing of metazoans and protozoans in Table 5.11 does not directly translate into a division of grazing by macro- and microzooplankton. A substantial fraction of the metazoan grazers fit either the formal (i.e., they pass through a 200-μm mesh screen; Dussart, 1965) or functional (i.e., they produce slowly settling fecal debris; Welschmeyer

TABLE 5.11 Estimates of zooplankton grazing impact and primary production (mg C m^{-2} d^{-1}) for August 1981. Grazing estimates are based on cross-shelf biomass distributions. Ciliate biomass estimates are from mean abundance in the depth range of 0-30 m at productivity stations only. Other ciliate estimates are interpolated.

		Distance Offshore (km)							
Species/stage		2	5	7	10	12	15	20	25
Pseudocalanus	n	61	95	82	101	57	98	99	57
	c	50	201	245	421	405	434	440	188
	a	2	24	86	171	254	264	204	159
C. marshallae	n	2	3	10	7	22	11	10	1
	c	0	0	0	53	25	10	8	9
	a	0	0	0	0	49	13	44	83
Acartia species	n	11	20	16	15	17	19	16	17
	c	2	19	9	30	74	56	30	45
	a	0	7	17	39	128	101	108	293
Euphausiids		0	1	2	11	84	81	63	96
Oikopleura		61	177	186	77	54	39	20	6
Ciliates		110	123	135	148	178	207	237	237
Total Impact		299	670	788	1073	1347	1333	1279	1197
^{14}C-uptake		1949			2083			1045	

and Lorenzen, 1985) definitions of microzooplankton. Realistically, all of the copepod nauplii, a large fraction of the copepodid stages of small copepod species (say, 50% of *Acartia* and *Pseudocalanus* copepodids) and possibly most of the *Oikopleura* function as microzooplankton. If we include these groups with the protozoa, the computed microzooplankton grazing for August 1981 ranges from 91% of total grazing at the shallowest station to 36% at mid-shelf (Fig. 5.3). In the only published study of microzooplankton grazing on the Washington coast, Landry and Hassett (1982) measured daily grazing rates of combined protozoan and copepod nauplii equivalent to 6 to 24% of phytoplankton standing stock and 17 to 52% of primary production.

Estimates of zooplankton grazing for June 1982 range from 400 to 1,600 mg C m^{-2} d^{-1} (Table 5.12). These rates are slightly higher than estimates for August 1981, but cross-shelf distributions are similar. Also as in August 1981, copepods dominate among the various grazer taxa and *Pseudocalanus* dominates among the copepods. *Pseudocalanus, Calanus* and ciliates account, on average, for more than 93% of the estimated grazing impact at the 8 shelf stations in June 1982. Neither larvacean nor euphausiid populations were large enough to make significant contributions. Estimated microzooplankton grazing, grouping nauplii and small copepodids with ciliates, ranged from 68% for the near-shore station to 42% of the total grazer impact for the 20-km station (Fig. 5.4).

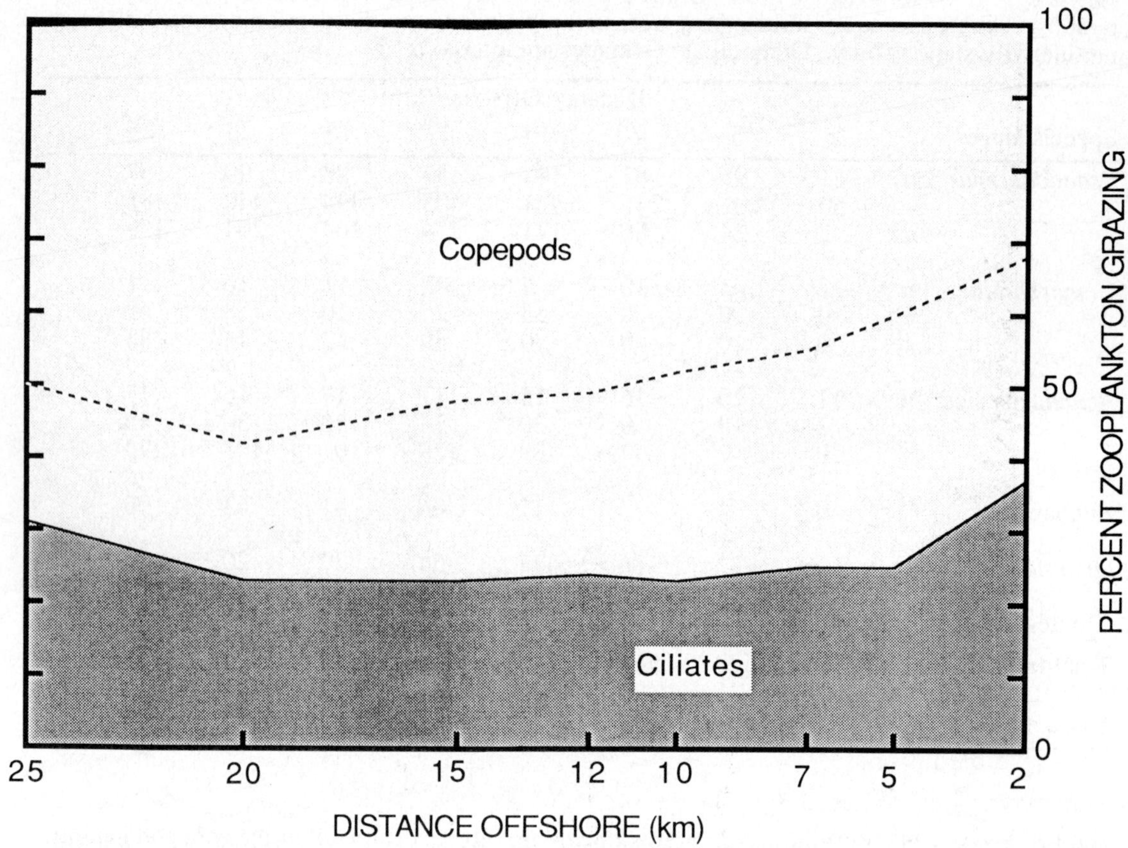

Fig. 5.3 Distribution of total grazing within the zooplankton community as a function of distance from shore for August 1981 (Table 5.11). Dashed line hypothetically divides functional microzooplankton (below) from functional macrozooplankton (above) as described in text.

In contrast to the observations for August 1981, grazing estimates for June 1982 do not balance or exceed observed primary production rates anywhere on the inner shelf transect. Production exceeds grazing by at least a factor of 2.5 at each of the stations where both are estimated. Chlorophyll concentrations and primary production rates in June 1982 were the highest ever measured for the Washington mid-shelf. They are clearly atypical of mean concentrations and rates during summer. Nonetheless, they dramatically illustrate that sustained periods of high production may periodically occur in response to major episodes of upwelling. Such periods can swamp the grazing potential of zooplankton stocks across much of the shelf. We cannot presently account for the fate of this excess production. However, extremely high standing stocks of phytoplankton, mainly large, chain-forming species, remained tightly confined to the upper 20-m of the water-column during the 1982 study, providing no evidence for mass sinking of ungrazed cells. Production rates beyond mid-shelf decreased to less than 1,000 mg C m^{-2} d^{-1} during the June study. Therefore, grazing may have exceeded production on the outer-shelf during this period (see also Hermann *et al.*, 1989, Chapter 6).

TABLE 5.12 Estimates of zooplankton grazing impact and primary production (mg C $m^{-2} d^{-1}$) for June 1982. Grazing estimates are based on cross-shelf biomass distributions. Ciliate biomass estimates are from mean abundance in the depth range of 0-30 m at productivity stations only. Other ciliate estimates are interpolated.

		Distance Offshore (km)							
Species/stage		2	5	7	10	12	15	20	25
Pseudocalanus	n	31	71	98	142	181	186	164	134
	c	125	375	448	348	238	272	169	178
	a	24	59	130	187	270	211	296	232
C. marshallae	n	2	8	14	20	8	14	23	12
	c	8	46	109	184	209	182	263	147
	a	0	0	7	14	28	180	240	190
Acartia species	n	9	14	15	21	15	22	29	25
	c	18	29	24	19	10	9	11	8
	a	16	39	45	51	28	25	25	19
Euphausiids		0	0	0	1	6	13	7	10
Ciliates		144	218	292	303	314	325	372	420
Total Impact		377	859	1182	1290	1307	1439	1599	1375
^{14}C-uptake		2930		3720			4920		3520

A large fraction of primary production sinks unutilized to the seafloor in shelf regions where phytoplankton bloom cycles are too rapid for the limited response capabilities of zooplankton populations (e.g., Smetacek, 1985; Laws *et al.*, 1988). In other regions, the northeast coast of the United States, for example, export of excess shelf production has been hypothesized to be an important source of organic matter input to the deep sea (Walsh *et al.*, 1981, 1988). In contrast, physical and biological characteristics of the Washington coastal ecosystem provide a setting in which the vast majority of shelf production can be retained and utilized locally. Intermittent upwelling circulation minimizes transport loss from the shelf region during summer. The mean shelf residence time for inert particles in the mixed layer is about 30 d (Hickey, 1989, Chapter 2). From observed mean values of depth-integrated standing stock (about 50 mg Chl m^{-2}; Landry *et al.*, 1989, Chapter 1) and production rate (about 1,000 mg C $m^{-2} d^{-1}$; Perry *et. al.*, 1989, Chapter 3) during summer and an assumed C: Chl ratio of 50, the mean turnover time of phytoplankton is much shorter, only 2 to 3 d. Furthermore, high phytoplankton concentrations, stimulated by intermittent upwelling, extend over a relatively long period, thereby, reducing the lag response of zooplankton populations to new production events. Since zooplankton populations are not subjected to a dramatically seasonal, "feast or famine" food environment, they can integrate over periods of relatively more and relatively less food, maintaining large stocks capable of consuming high average levels of phytoplankton production. In addition, zooplankton can exploit the two-layer upwelling circulation directly (vertical migration) or indirectly (reproductive products) to reduce their rate of washout from the shelf ecosystem (e.g., Peterson *et al.*, 1979; Wroblewski, 1982).

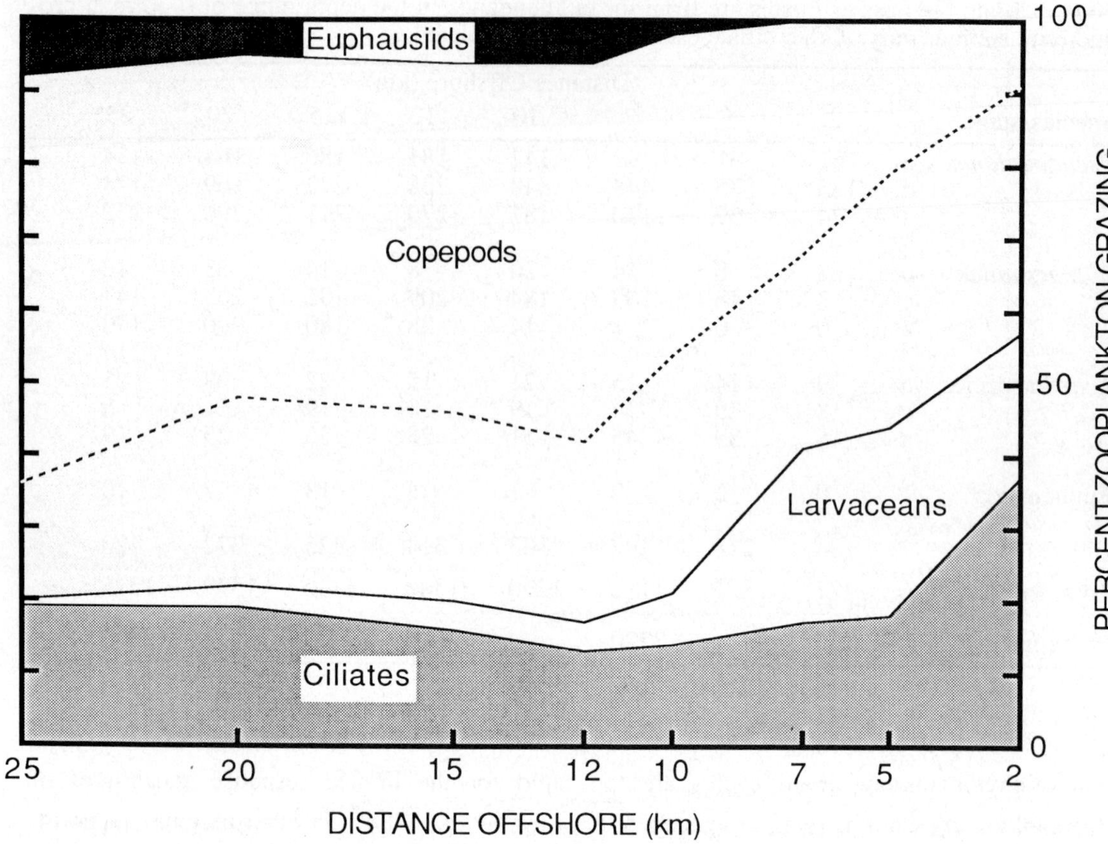

Fig. 5.4 Distribution of total grazing within the zooplankton community as a function of distance from shore for June 1982 (Table 5.12). Dashed line hypothetically divides functional microzooplankton (below) from functional macrozooplankton (above).

5.5 GRAZING AND PARTICLE FLUX

Table 5.13 gives abundance, biomass and grazing estimates for dominant zooplankton taxa on the Washington mid-shelf in August 1983. Despite the fact that 1983 was a major El Niño year with little coastal upwelling, zooplankton biomass, in general, was higher than mid-shelf estimates for the 1981 and 1982 studies. In particular, densities of larger organisms, *Calanus* adults and euphausiids, were notably elevated. This unexpected observation could be fortuitous, or it could be due to El Niño effects on planktivorous fishes (e.g., reduced abundances or feeding rates). Whatever the reason, computed grazing rates based on zooplankton biomass are higher than estimates from either August 1981 or June 1982 while primary production rates are lower. If these estimates are correct, then zooplankton grazing at this time exceeded primary production by a factor of 2.5 (i.e., 1,880 *vs*. 749 mg C m^{-2} d^{-1}).

Gut fluorescence and particle flux measurements made during the August 1983 study provide independent assessments of grazing rates to compare with the rates predicted from zooplankton biomass. Gut fluorescence measurements for *Calanus* females (mean of 38 determinations during the 2-d study) averaged 9.2 ng pigment copepod $^{-1}$. Gut evacuation rates, deter-

TABLE 5.13 Abundance (10^3 m $^{-2}$), biomass (mg C m $^{-2}$) and grazing (mg C m $^{-2}$ d $^{-1}$) estimates for zooplankton populations on the Washington mid-shelf during August 1983.

Taxa		Abundance	Biomass	Grazing
Pseudocalanus	n	377.0	69	41
	c	129.0	406	244
	a	64.8	497	209
C. marshallae	n	12.3	6	2
	c	9.3	266	38
	a	16.3	1679	149
Acartia species	n	499.0	40	24
	c	39.4	37	22
A. clausii	a	7.3	17	10
A. longiremis	a	16.9	52	31
O. similis	n	153.0	6	
	c	59.3	12	
	a	22.2	18	
Euphausiids		1.7	1510	317
Ciliates			318	795
Total			4933	1882
^{14}C-uptake				749

mined from the time course of gut pigment decrease in filtered water, averaged 0.04 min $^{-1}$. If feeding occurred during a 12-h day, then each *Calanus* female produced fecal material containing 0.264 μg phaeopigment d $^{-1}$. Assuming a carbon: chlorophyll ratio of 50, the corresponding estimate of carbon ingestion is 13.2 μg C *Calanus* $^{-1}$ d $^{-1}$, or about 11.8% body C d $^{-1}$. Our computed grazing rates for *Calanus* females were 8.9% body C d $^{-1}$. Similar gut pigment analyses for adult euphausiids > 20 mm in length give pigment grazing estimates of 1.5 μg phaeopigment (mg euphausiid C) $^{-1}$ d $^{-1}$ and carbon ingestion rates of 7.6% body C d $^{-1}$. These latter estimates are substantially lower than the 21% used to estimate grazing rates. However, non-pigmented food may contribute substantially to the euphausiid nutrition (e.g., Ohman, 1984).

Pigment budget estimates of phytoplankton loss rates were computed according to the method of Welschmeyer and Lorenzen (1985) except that a factor of 1.51 was not used to convert phaeopigments to molar chlorophyll equivalents (Conover *et al.*, 1986). Macrozooplankton grazing rate, measured as phaeopigment flux in sediment traps below the euphotic zone, was 0.188 d $^{-1}$ (43.9% of total phytoplankton loss). Phytoplankton loss rate due to microzooplankton grazing, calculated from standing stocks of phaeopigments in the euphotic zone and light-dependent degradation, was 0.224 d $^{-1}$ (52.3% of total). Cell sinking, from chlorophyll flux to the traps, equaled 0.016 d $^{-1}$ (3.7% of total). These pigment budget estimates of grazing losses due to functionally-defined micro- and macrozooplankton correspond closely to estimates based on community biomass structure, 52.9% microzooplankton and 47.1% macrozooplankton.

There is also good agreement between between the phaeopigment fluxes measured in sediment traps below the euphotic zone in August 1983 and independent gut fluorescence estimates of pigment flux for the larger zooplankton taxa. Pigment flux for the four traps deployed for three days at 50 m averaged 8 mg phaeopigment m^{-2} d^{-1}. The mean flux from the short-term trap array was 11 mg m^{-2} d^{-1}. In comparison, combined pigment flux based on gut fluorescence and gut clearance data for the large zooplankton was 7.1 mg m^{-2} d^{-1}. Of this, adult female *Calanus*, *Calanus* stage C5 and euphausiids accounted, respectively, for 4.3, 0.5 and 2.3 mg phaeopigment m^{-2} d^{-1}. According to our calculations, these large taxa accounted for only 21% of community grazing pressure but at least 65% of pigment flux. This is consistent with the view that larger zooplankton taxa contribute disproportionately to the production of fecal pellets which sink rapidly through the water column (e.g., Hofmann *et al.*, 1981.).

Pigment budget-derived estimates of grazing losses due to micro- and macrozooplankton can be converted to carbon equivalents from the depth-integrated standing stock of chlorophyll (59.5 mg m^{-2}) and a carbon: chlorophyll ratio of 50. Accordingly, the depth-integrated phytoplankton standing stock is equivalent to 2.98 g C m^{-2} d^{-1} and the combined grazing loss rate (0.41 d^{-1}) corresponds to 1.97 g C m^{-2} d^{-1}. This carbon loss rate is remarkably similar to the estimate of 1.88 g C m^{-2} d^{-1} based on zooplankton biomass. Measured carbon flux averaged 268 mg C m^{-2} d^{-1} for the three-day trap deployment and 423 mg C m^{-2} d^{-1} for the short term traps. These flux rates are 14 and 22% of estimated grazing, respectively. If the standard 70% assimilation efficiency for carbon is reasonable, these flux measurements imply that the carbon fixed by phytoplankton passes through more than one but less than two trophic steps, on average, before it sinks below the euphotic zone. This is consistent with an equal division of grazing impact between macro- and microzooplankton.

The good agreement and internal consistency in calculated and experimentally-derived estimates of zooplankton feeding and particle flux in the August 1983 study, do not demonstrate that all aspects of the calculated rates are correct. The grazing contribution of euphausiids, for example, may have been substantially overestimated. More significantly, the abundance estimates, carbon contents and carbon-specific ingestion rates used in computing protozoan grazing impact are all open to question. Nonetheless, the agreement does lend some credibility to the computed rates. Given the behavioral considerations that typically confound the measurement and interpretation of zooplankton feeding under field conditions, it is perhaps notable that these computations are based on simple extrapolations of laboratory results.

5.6. SUMMARY

A large standing stock of zooplankton, in excess of 2 g C m^{-2}, resides on the Washington shelf during summer in a zone extending from about 10-km offshore to the mid-shelf (30-km offshore in the Copalis area) and probably beyond. The region of high abundance is more broadly distributed than the zone of peak zooplankton abundance off Oregon and it may also contain higher densities of organisms on an areal basis. Calanoid copepods, particularly *Pseudocalanus* spp., dominate the zooplankton community of the inner shelf in terms of biomass and grazing impacts, where the latter are computed from the biomass structure and laboratory growth and feeding studies. Ciliated protozoans and larvaceans are relatively more important in the shallow, near-shore environment, as are the smaller developmental stages of individual copepod species. *Calanus marshallae* and euphausiids increase in relative importance in the mid-shelf region. Community grazing estimates for August 1981 and June 1982 show zooplankton carbon demand increasing from 300-400 mg C m^{-2} d^{-1} in the near-shore environment to 1,300-1,600 mg C m^{-2} d^{-1} approaching the mid-shelf. In both field studies, primary production in the near-shore region occurred significantly in excess of local zooplankton grazing capability. Excess carbon production was also apparent at all locations shoreward of the mid-shelf during June 1982. However, this intense upwelling period was atypical of the shelf system during summer. During periods of more representative summer production levels, such as August 1981 and 1983, calculated zooplankton grazing potential in the mid-shelf region exceeded local phytoplankton production. In general, surface circulation associated with summer upwelling carries phytoplankton produced in excess of grazing in the near shore environment to a mid-shelf region of high grazing loss. Zooplankton grazing on the mid-shelf appears to account for virtually all of the phytoplankton lost in this region. There is no evidence for substantial loss due to mass sinking.

Grazing by microzooplankton dominates (70 to 90%) in the shallow, near-shore environment. About half of microzooplankton grazing can be directly attributable to protozoans; the remainder is due to smaller developmental stages of metazoans. Macrozooplankton increase in relative importance toward the mid-shelf where they represents about half of the community grazing impact. Since macrozooplankton are functionally distinguished from microzooplankton by their production of rapidly sinking fecal debris, the mid-shelf region is characterized by enhanced sinking of particulate organic matter from the euphotic zone. In contrast, microzooplankton dominance in the inshore environment results in more efficient recycling of nutrients in the euphotic zone. The rain of fecal debris produced by large planktonic grazers near mid-shelf must play an important role in the scavenging and transport of particle-reactive chemicals from their source in deep water advected onto the shelf by upwelling to their sink in the mid-shelf silt deposit (Carpenter and Peterson, 1989, Chapter 10).

Grazing rates predicted from the zooplankton biomass structure in August 1983 were remarkably consistent with rates from gut fluorescence and sediment trap experiments conducted at the same time. This is a good indication that knowledge of grazing rates of the dominant zooplankton group, copepods, is reasonably well constrained by laboratory studies. In future investigations, it would be desirable to obtain better estimates of euphausiid population densities, growth and grazing rates since these animals contribute disproportionately to the vertical transport of surface particulates to the sea floor and to the diets of important planktivorous fishes such as salmonids. Appendicularians can be significant consumers of phytoplankton on the Washington shelf. Moreover, their presence in high numbers is linked to the recruitment success of English sole. Therefore, some effort to understand the factors controlling temporal and spatial variations in population densities of appendicularians would also be worthwhile.

The high degree of uncertainty associated with protozoan abundance and feeding estimates is clearly a major gap in our attempt to reconcile grazing and production rates on the Washington shelf. However, it is symptomatic of a much broader need to incorporate microbial processes and interactions into food-web studies in this region. Over the past decade, the classical food-chain paradigm of diatom-copepod-fish has been assimilated into a more complex paradigm which emphasizes the role of smaller size components--bacteria, nanoplankton and protozoans–in energy utilization and cycling at the base of the food-web (e.g., Pomeroy, 1974; Williams, 1981; Azam *et al.*, 1983). This microbial community has been virtually ignored in plankton studies off Washington and Oregon to date.

ACKNOWLEDGEMENTS

We gratefully acknowledge the outstanding efforts of V. Fagerness, J. Lehner-Fournier, W. Peterson, J. Downs, R. Hassett and B. Monger who assisted with various aspects of sample collection, microscopical enumeration and data analysis. The text greatly benefitted from the comments of C.B. Miller and K. Banse. This work was supported by the Department of Energy under contract DE-AT06-76-EV-75026. Contribution No.1784 from the School of Oceanography, University of Washington.

REFERENCES

Alton, M.S. and C.J. Blackburn. 1972. Diel changes in the vertical distribution of the euphausiids, *Thysanoessa spinifera* Holmes and *Euphausia pacifica* Hansen, in coastal waters of Washington. *Calif. Fish Game*, 58: 179-190.

Azam, F., T. Fenchel, J.G. Field, J.S. Gray, L.A. Meyer-Reil and F. Thingstad. 1983. The ecological role of water-column microbes in the sea. *Mar. Ecol. Prog. Ser.*, 10: 257-263.

Bartram, W.C. 1980. Experimental development of a model for the feeding of neritic copepods on phytoplankton. *J. Plank. Res.*, 3: 25-51.

Beers, J.R., M.R. Reeve and G.D. Grice. 1977. Controlled ecosystem pollution experiment: Effect of mercury on enclosed water columns. IV. Zooplankton population dynamics and production. *Mar. Sci. Comm.*, 3: 355-394.

Brodeur, R.D. 1986. Northward displacement of the euphausiid *Nyctiphanes simplex* Hansen to Oregon and Washington waters following the El Niño event of 1982-83. *J. Crustacean Biol.*, 6: 686-692.

Carpenter R. and M.L. Peterson. 1989. Chemical cycling in Washington's coastal zone. Pages: 366-510 in: M. R. Landry and B. M. Hickey (eds.), *Coastal Oceanography of Washington and Oregon*. Elsevier, Amsterdam, Neth..

Checkley, D.M., Jr. 1980a. The egg production of a marine planktonic copepod in relation to its food supply: Laboratory studies. *Limnol. Oceanogr.*, 25: 430-446.

Checkley, D.M., Jr. 1980b. Food limitation of egg production by a marine, planktonic copepod in the sea off southern California. *Limnol. Oceanogr.*, 25: 991-998.

Chester, A.J. 1978a. Microzooplankton relative to a subsurface chlorophyll maximum layer. *Mar. Sci. Comm.* 4: 275-292.

Chester, A.J. 1978b. Microzooplankton in the surface waters of the Strait of Juan de Fuca. NOAA Technical Rep. ERL 403-PMEL 30, 26 pp.

Conover, R.J. 1978. Feeding interactions in the pelagic zone. *Rapp. P.-v. Reun. Cons. Int. Explor. Mer* 173: 66-76.

Conover, R.J., R. Durvasula, S. Roy and R. Wang. 1986. Probable loss of chlorophyll-derived pigments during passage through the gut of zooplankton and some of the consequences. *Limnol. Oceanogr.*, 31: 878-887.

Corkett, C.J. and I.A. McLaren. 1978. The biology of *Pseudocalanus*. *Adv. Mar. Biol.*, 15: 1-231.

Davis, C.S. 1984. Predatory control of copepod seasonal cycles on Georges Bank. *Mar. Biol.*, 82: 31-40.

Donaghay, P. 1980. Grazing interactions in the marine environment. Pages 4-240 in: W. C. Kerfoot (ed.), *Evolution and Ecology of Zooplankton Communities*. Univ. Press, New England.

Durbin, E.G., A.G. Durbin, T.J. Smayda and P.G. Verity. 1983. Food limitation of production by adult *Acartia tonsa* in Narragansett Bay, Rhode Island. *Limnol. Oceanogr.*, 28: 1199-1213.

Dussart, B.M. 1965. Les différentes catégories de plancton. *Hydrobiologia.*, 26: 72-74.

Frolander, H.F., C.B. Miller, M.J. Flynn, S.C. Myers and S.T. Zimmerman. 1973. Seasonal cycles of abundance in zooplankton populations of Yaquina Bay, Oregon. *Mar. Biol.*, 21: 277-288.

Frost, B.W. 1972. Effects of size and concentration of food particles on the feeding behavior of the marine planktonic copepod *Calanus pacificus*. *Limnol. Oceanogr.*, 17: 805-815.

Frost, B.W. 1975. A threshold feeding behavior in *Calanus pacificus*. *Limnol. Oceanogr.*, 20: 263-266.

Frost, B.W. 1977. Feeding behavior of *Calanus pacificus* in mixtures of food particles. *Limnol. Oceanogr.*, 22: 472-492.

Frost, B.W. 1980. The inadequacy of body size as an indicator of niches in the zooplankton, pages 742-753 on: W. C. Kerfoot (ed.), *Evolution and Ecology of Zooplankton Communities*. Univ. Press, New England.

Frost, B.W. 1985. Food limitation of the planktonic marine copepods *Calanus pacificus* and *Pseudocalanus* sp. in a temperate fjord. *Arch. Hydrobiol. Beih. Ergebn. Limnol.*, 21: 1-13.

Gadomski, D.M. and G.W. Boehlert. 1984. Feeding ecology of pelagic larvae of English sole *Parophrys vetulus* and butter sole *Isopsetta isolepis* off the Oregon coast. *Mar. Ecol. Prog. Ser.*, 20: 1-12.

Grice, D.G., M.R. Reeve, P. Koeller and D.W. Menzel. 1977. The use of large volume (1,300 m^3) transparent enclosed sea-surface water columns in the study of stress on plankton ecosystems. *Helgol. Wiss. Meeresunters.*, 30: 118-133.

Hameedi, M.J. 1974. Quantitative studies of phytoplankton and zooplankton and their interrelationships off Washington and Oregon. Ph.D. Diss., Dept. of Oceanogr., Univ. of Washington, Seattle, Wa., 288 pp.

Harris, R.P., G.A. Paffenhöfer. 1976. The effect of food concentration of cumulative ingestion and growth efficiency of two small marine planktonic copepods. *J. Mar. Biol. Ass. U.K.*, 56: 875-888.

Harris, R.P., M.R. Reeve, G.D. Grice, G.T. Evans, V.R. Gibson, J.R. Beers and B.K. Sullivan. 1982. Trophic Interactions and production processes in natural zooplankton communities in enclosed water columns. Pages 353-387 in: G.D. Grice and M. R. Reeve (eds.), *Marine Mesocosms: Biological and Chemical Research in Experimental Ecosystems.* Springer-Verlag, New York, N.Y..

Hassett, R.P. and M.R. Landry. 1983. Effects of food-level acclimation on digestive enzyme activity and feeding behavior of *Calanus pacificus*. *Mar. Biol.*, 75: 47-55.

Hebard, J.F. 1966. Distribution of Euphausiacea and Copepoda off Oregon in relation to oceanographic conditions. Ph.D. Diss., Oregon St. Univ., Corvallis, Ore., 85 pp.

Heinbokel, J.F. 1978. Studies on the functional role of tintinnids in the Southern California Bight. I. Grazing and growth rates in laboratory cultures. *Mar. Biol.*, 47: 177-189.

Hermann, A.J., B.M. Hickey, M.R. Landry and D.F. Winter. 1989. Costal upwelling dynamics. Pages: 211-254 in: M. R. Landry and B. M. Hickey (eds.), *Coastal Oceanography of Washington and Oregon.* Elsevier, Amsterdam, Neth..

Hickey, B.M. 1989. Patterns and processes of circulation over the shelf and slope. Pages: 41-116 in: M. R. Landry and B. M. Hickey (eds.), *Coastal Oceanography of Washington and Oregon.* Elsevier, Neth..

Hofmann, E.E., J.M. Klinck and G.-A. Paffenhöfer. 1981. Concentrations and vertical fluxes of zooplankton fecal pellets on a continental shelf. *Mar. Biol.*, 61: 327-335.

Hubbard, L.T., Jr. and W.G. Pearcy. 1971. Geographic distribution and relative abundance of Salpidae off the Oregon coast. *J. Fish. Res. Bd. Canada*, 28: 1831-1836.

Huntley, M. and C. Boyd. 1984. Food-limited growth of marine zooplankton. *Am. Nat.*, 124: 455-478.

Jamart, B.M., D.F. Winter, K. Banse, G.C. Anderson and R.K. Lam. 1977. A theoretical study of phytoplankton growth and nutrient distribution in the Pacific Ocean off the northwestern U.S. coast. *Deep-Sea Res.*, 24: 753-773.

Jawed, M. 1973. Numerical abundance of dominant copepods from the northeast Pacific Ocean: Columbia River effluent area, 1963. Dept. of Oceanogr., Univ. of Washington, Seattle, Wa., Tech. Rep. 286 (Ref: M73-40), 106 pp.

Johnson, J.K. 1981. Population dynamics and cohort persistence of *Acartia californiensis* (Copepoda: Calanoida) in Yaquina Bay, Oregon. Ph.D. Diss., Oregon St. Univ., Corvallis, Ore., 305 pp.

King, K.R. 1981. The quantitative natural history of *Oikopleura dioica* (Urochordata: Larvacea) in the laboratory and in enclosed water columns. Ph.D. Diss., School of Oceanogr., Univ. of Washington, Seattle, Wa., 152 pp.

King, K.R. 1982. The population biology of the larvacean *Oikopleura dioica* in enclosed water columns. Pages 341-351 in: G.D. Grice and M.R. Reeve (eds.), *Marine Mesocosms: Biological and Chemical Research in Experimental Ecosystems.* Springer-Verlag, New York, N.Y..

King, K.R., J.T. Hollibaugh and F. Azam. 1980. Predator-prey interactions between the larvacean *Oikopleura dioica* and bacterioplankton in enclosed water columns. *Mar. Biol.*, 56: 49-57.

Landry, M.R. 1978. Population dynamics and production of a planktonic marine copepod, *Acartia clausii*, in a small temperate lagoon on San Juan Island, Washington. *Int. Revue ges. Hydrobiol.*, 63: 77-119.

Landry, M.R. 1981. Switching between herbivory and carnivory by the planktonic marine copepod *Calanus pacificus*. *Mar. Biol.*, 65: 77-82.

Landry, M.R. 1983. The development of marine calanoid copepods with comment on the isochronal rule. *Limnol. Oceanogr.*, 28: 614-624.

Landry, M.R. and R. P. Hassett. 1982. Estimating the grazing impact of marine micro-zooplankton. *Mar. Biol.*, 67: 283-288.

Landry, M.R. and R.P. Hassett. 1985. Time scales in behavioral, biochemical and energetic adaptations to food-limiting conditions by a marine copepod. *Arch. Hydrobiol. Beih. Ergebn. Limnol.*, 21: 209-221.

Landry, M.R., J.R. Postel, W.K. Peterson and J. Newman. 1989. Broad-scale distributional patterns of hydrographic variables on the Washington/Oregon shelf. Pages 1-41 in: M. R. Landry and B. M. Hickey (eds.). *Coastal Oceanography of Washington and Oregon.* Elsevier, Amsterdam, Neth..

Laurs, R.M. 1967. Coastal upwelling and the ecology of lower trophic levels. Ph. D. Diss., Oregon St. Univ., Corvallis, Ore., 121 pp.

Laws, E.A., P.K. Bienfang, D.A. Ziemann and L.D. Conquest. 1988. Phytoplankton population dynamics and the fate of production during the spring bloom in Auke Bay, Alaska. *Limnol. Oceanogr.*, 33: 57-65.

Mackas, D.L. and H.A. Sefton. 1982. Plankton species assemblages off southern Vancouver Island: Geographic pattern and temporal variability. *J. Mar. Res.*, 40: 1173-1200.

Mayzaud, P. and S.A. Poulet. 1978. The importance of the time factor in the response of zooplankton to varying concentrations of naturally occurring particulate matter. *Limnol. Oceanogr.*, 23: 1144-1154.

McLaren, I.A. 1978. Generation lengths of some temperate marine copepods: Estimation, prediction and implications. *J. Fish. Res. Bd. Canada,* 35: 1330-1342.

Miller, C.B., H.P. Batchelder, R.D. Brodeur and W.G. Pearcy. 1983. Response of the zooplankton and icthyoplankton off Oregon to the El Niño event of 1983. Pages 185-187 in: W. S. Wooster and D. L. Fluharty (eds.), *El Niño North. Niño Effects in the Eastern Subarctic Pacific Ocean.* Washington Sea Grant, Seattle, Wa..

Miller, C.B., J.K. Johnson and D.R. Heinle. 1977. Growth rules in the marine copepod genus *Acartia. Limnol. Oceanogr.*, 22: 326-335.

Mullin, M.M. and E.R. Brooks. 1970. Production of the planktonic copepod, *Calanus helgolandicus*. Pages 185-187 in: J. D. H. Strickland (ed.), *The ecology of the plankton off La Jolla, California, in the period April through September, 1967.* Univ. of Calif. Press, Berkeley, Calif..

Ohman, M.D. 1983. The effects of predation and resource limitation on the copepod *Pseudocalanus* sp. in Dabob Bay, a temperate fjord. Ph.D. Diss., School of Oceanogr., Univ. of Washington, Seattle, Wa.

Ohman, M.D. 1984. Omnivory by *Euphausia pacifica*: the role of copepod prey. *Mar. Ecol. Progr. Ser.*, 19: 125-131.

Ohman, M.D. 1985. Resource-satiated population growth of the copepod *Pseudocalanus* sp. *Arch. Hydrobiol. Beih. Ergebn. Limnol.*, 21: 15-32.

Paffenhöfer, G.A. 1975. On the biology of Appendicularia of the southeastern North Sea. 10th *European Symp. Mar. Biol., Belgium*, 2: 437-455.

Paffenhöfer, G.A. and R.P. Harris. 1976. Feeding, growth and reproduction of the marine planktonic copepod *Pseudocalanus elongatus* Boeck. *J. Mar. Biol. Ass. U.K.*, 56: 327-344.

Perry, M.J., J. Bolger and D.C. English. 1989. Primary production in Washington coastal waters. Pages 117-138 in: M. R. Landry and B. M. Hickey (eds.). *Coastal Oceanography of Washington and Oregon.* Elsevier, Neth..

Peterson, W.K. 1972. Distribution of pelagic Copepoda off the coasts of Washington and Oregon during 1961 and 1962. Pages 313-343 in: A. T. Pruter and D. L. Alverson (eds.). *The Columbia River Estuary and Adjacent Ocean Waters: Bioenvironmental Studies.* Univ. of Washington Press, Seattle,Wa..

Peterson, W.T. 1980. Life history and ecology of *Calanus marshallae* Frost in the Oregon upwelling zone. Ph.D. Diss., Oregon St. Univ., Corvallis, Ore., 199 pp.

Peterson, W.T. 1986. Development, growth and survivorship of the copepod *Calanus marshallae* in the laboratory. *Mar. Ecol. Prog. Ser.*, 29: 61-72.

Peterson, W.T., R.D. Brodeur and W.G. Pearcy. 1982. Food habits of juvenile salmon in the Oregon coastal zone, June 1979. *Fish. Bull., U.S.*, 80: 841-851.

Peterson, W.T. and C.B. Miller. 1975. Year-to-year variations in the planktology of the Oregon upwelling zone. *Fish. Bull., U.S.*, 73: 642-653.

Peterson, W.T. and C.B. Miller. 1976. Zooplankton along the continental shelf off Newport, Oregon, 1969-1972: Distribution, abundance, seasonal cycle and year-to-year variations. Oregon St. Univ., Sea Grant Pub. ORESU-T-76-002, 111 pp.

Peterson, W.T. and C.B. Miller. 1977. Seasonal cycle of zooplankton abundance and species composition along the central Oregon coast. *Fish. Bull.*, 75: 717-724.

Peterson, W.T., C.B. Miller and A. Hutchinson. 1979. Zonation and maintenance of copepod populations in the Oregon upwelling zone. *Deep-Sea Res.*, 26: 467-494.

Pomeroy, L.R. 1974. The ocean's food web, a changing paradigm. *BioScience*, 24: 499-504.

Postel, J.R. 1975. Comparisons of phytoplankton populations in surface waters and in the deep chlorophyll maximum layer from a section off the Oregon coast. M.S. Thesis, School of Oceanogr., Univ. of Washington, Seattle, Wa., 104 pp.

Postel, J.R., W.K. Peterson, J.C. Kitchen and D.W. Menzies. 1980. Data report of the DOE-sponsored Northwest Marine Sciences Group July-August 1979 cruise. School of Oceanogr., Univ. of Washington, Ref. A80-37, 453 pp.

Postel, J.R., W.K. Peterson, J.C. Kitchen and D.W. Menzies. 1982. Data report of the DOE-sponsored Northwest Marine Sciences Group September-October 1980 cruise. School of Oceanogr., Univ. of Washington, Ref. A82-1, 498 pp.

Postel, J.R., W.K. Peterson, J.C. Kitchen and D.W. Menzies. 1983. Data report of the DOE-sponsored Northwest Marine Sciences Group August-September 1981 cruise. School of Oceanogr., Univ. of Washington, Ref. A83-2, 568 pp.

Ross, R.M. 1981. Laboratory culture and development of *Euphausia pacifica*. *Limnol. Oceanogr.*, 26: 235-246.

Ross, R.M. 1982a. Energetics of *Euphausia pacifica*. I. Effects of body carbon and nitrogen and temperature on measured and predicted production. *Mar. Biol.*, 68: 1-13.

Ross, R.M. 1982b. Energetics of *Euphausia pacifica*. II. Complete carbon and nitrogen budgets at 8° and 12° C throughout the life span. *Mar. Biol.*, 68: 15-23.

Runge, J.A. 1980. Effects of hunger and season on the feeding behavior of *Calanus pacificus*. *Limnol. Oceanogr.*, 25: 134-145.

Runge, J.A. 1984. Egg production of the marine, planktonic copepod, *Calanus pacificus* Brodsky: Laboratory observations. *J. Exp. Mar. Biol. Ecol.*, 74: 53-66.

Runge, J.A. 1985. Relationship of egg production of *Calanus pacificus* to seasonal changes in phytoplankton availability in Puget Sound, Washington. *Limnol. Oceanogr.*, 30: 382-396.

Smetacek, V. 1985. Role of sinking in diatom life history cycles: Ecological, evolutionary and geological significance. *Mar. Biol.*, 85: 239-251.

Smiles, M.C., Jr. and W.G. Pearcy. 1971. Size structure and growth rate of *Euphausia pacifica* off the Oregon coast. *Fish. Bull., U.S.*, 69: 79-86.

Sonntag, N.C. and J. Parslow. 1981. Technique of systems identification applied to estimating copepod production. *J. Plank. Res.*, 3: 461-473.

Verity, P.G. 1985. Grazing, respiration, excretion and growth rates of tintinnids. *Limnol. Oceanogr.*, 30: 1268-1282.

Vidal, J. 1980a. Physioecology of zooplankton. I. Effects of phytoplankton concentration, temperature and body size on the growth rate of *Calanus pacificus* and *Pseudocalanus* sp. *Mar. Biol.*, 56: 111-134.

Vidal, J. 1980b. Physioecology of zooplankton. II. Effects of phytoplankton concentration, temperature and body size on the development and molting rates of *Calanus pacificus* and *Pseudocalanus* sp. *Mar. Biol.*, 56: 135-146.

Vidal, J. 1980c. Physioecology of zooplankton. III. Effects of phytoplankton concentration, temperature and body size on the metabolic rate of *Calanus pacificus*. *Mar. Biol.*, 56: 195-202.

Walsh, J.J., D. A. Dieterle and M.B. Meyers. 1988. A simulation analysis of the fate of phytoplankton within the Mid-Atlantic Bight. *Cont. Shelf. Res.*, 8: 757-787.

Walsh, J.J., G.T. Rowe, R.L. Iverson and C.P. McRoy. 1981. Biological export of shelf carbon is a sink of the global $C0_2$ cycle. *Nature,* 291: 196-201.

Welschmeyer, N.A. and C.J. Lorenzen. 1985. Pigment budgets: Zooplankton grazing and phytoplankton growth in a temperate fjord and the Central Pacific Gyre. *Limnol. Oceanogr.*, 30: 1-21.

Williams, P.B. LeB. 1981. Incorporation of microheterotrophic processes into the classical paradigm of the planktonic food web. *Kieler Meeresforsch., Suppl.*, 5: 1-28.

Wroblewski, J.S. 1980. A simulation of the distribution of *Acartia clausi* during Oregon upwelling, August 1973. *J. Plank. Res.*, 2: 43-68.

Wroblewski, J.S. 1982. Interaction of currents and vertical migration in maintaining *Calanus marshallae* in the Oregon upwelling zone–a simulation. *Deep-Sea Res.*, 29: 665-686.

Zimmerman, S.T. 1972. Seasonal succession of zooplankton populations in two dissimilar marine embayments on the Oregon coast. Ph. D. Diss., Oregon St. Univ., Corvallis, Ore.

CHAPTER 6

COASTAL UPWELLING DYNAMICS

A. J. Hermann [1]*, B. M. Hickey* [1]*, M. R. Landry* [2] *and D. F. Winter* [3]
[1] *School of Oceanography, University of Washington, Seattle, WA 98195*
[2] *Dept of Oceanography and Hawaii Institute of Geophysics, University of Hawaii, Honolulu, HI 96822*
[3] *Dept. of Engineering, Univ. of Redlands, Redlands, CA 92373*

6.1 INTRODUCTION

Although it is generally accepted that physical variability drives biological fluctuations on many scales (e.g., Platt *et al.*, 1975), efforts to quantitatively assess the physical contribution to variance in biological parameters are just beginning. Small and Menzies (1981) have explored the interaction of physical and biological variability in the coastal upwelling region off Oregon. They found mesoscale patterns of primary production to be highly dependent on the upwelling state (e.g., upwelling *vs.* relaxation), with a response time of less than one day. In the same region, Peterson *et al.* (1979) concluded that copepod species interact with the circulation pattern in such a way as to maintain their populations within particular domains in the upwelling velocity field. Wroblewski (1980), in a modeling study, suggested that the observed pattern could be due to the combined effects of intermittent upwelling and mortality on the copepod *Acartia clausii*. Much of this earlier work was hampered by a lack of physical measurements nearshore and in the upper 20 m of the water column. Recent studies of upwelling centers have included near-surface measurements of both currents and biological dynamics and have documented the 'aging' of upwelled water with drogue studies (Brink *et al.*, 1981; MacIsaac *et al.*, 1985; Wilkerson and Dugdale, 1987). However, the three-dimensional nature of upwelling in such areas has prevented a clear determination of the role of physical and biological processes in maintaining nutrient and plankton distributions.

This chapter describes the results of two experiments explicitly designed to elucidate the physical control of mesoscale (~10 km) biomass and nutrient patterns in a region where the upwelling processes may be quasi-two-dimensional. Although the study region (the mid-Washington shelf; Fig. 6.1) exhibits significant alongshore gradients in physical and biological properties, it is not nearly as three-dimensional as persistent upwelling centers near large promontories, such as near Point Conception, California (Brink, 1983). The approach used is to distinguish conservative physical change from nonconservative biological change by first estimating the conservative flux using simultaneous measurements of currents, salt, chlorophyll and nutrients. Nonconservative changes are then inferred by comparing predicted conservative changes with observed changes in dissolved and suspended constituents. The relative contributions of individual conservative processes (e.g., longshore advection, cross-shelf advection, mixing) to the observed constituent variance are compared with each other and with the contribution of nonconservative processes under various physical environments.

The first experiment (henceforth termed 'late summer') took place during a period of intermittent upwelling (fluctuations on a 2-3 d time scale) in late August, 1981. The second

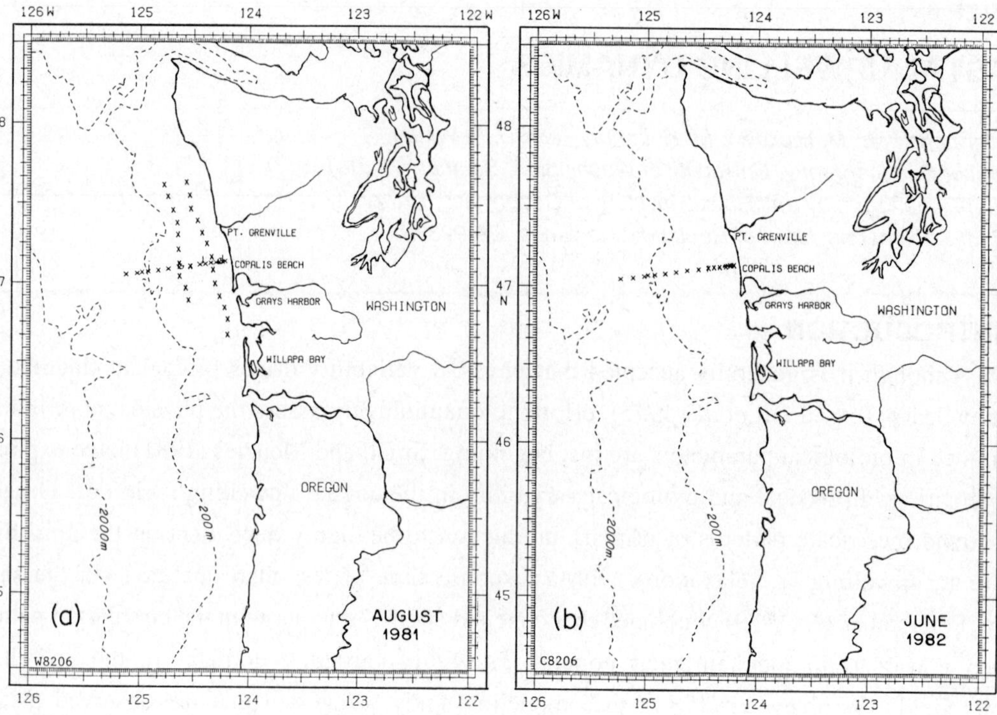

Fig. 6.1 Location of CTD/nutrient/pigment stations for (a) late summer (August-September 1981) and (b) early summer (June 1982) experiments.

experiment ('early summer') occurred during a period of persistent upwelling in June, 1982. Differences in the structure of the upwelling ecosystem under intermittent *vs.* persistent circulation regimes will arise both directly, from differences in patterns of advection and indirectly, from the biological responses to intermittent *vs.* persistent supplies of nutrients, phytoplankton stocks, etc.

The next section of this chapter (Section 6.2) describes the physical, chemical and biological data sets used in the analysis. The physical setting and biological patterns for the two experiments are discussed in Section 6.3. Models of the advective-diffusive control of biological parameters are developed in Section 6.4 and quantitative results for the various physical settings are compared in Section 6.5.

6.2 THE DATA SET

The late summer experiment involved five current meter moorings maintained from August 17 to September 7, 1981 at stations with bottom depths of 20 ('near-shore'), 50, 80 ('mid-shelf'), 130 and 185 m ('outer shelf') (Fig. 6.2a). At the 20 and 80-m stations, currents in the upper water column were measured with Vector Measuring Current Meters (VMCMs) suspended beneath a surface toroid buoy. All other moorings were of the subsurface taut wire type, with Aanderaa current meters at various depths. Surface winds were measured by VMCMs mounted on rigid 4-m towers on the surface toroids at mid-shelf (both experiments) and on the

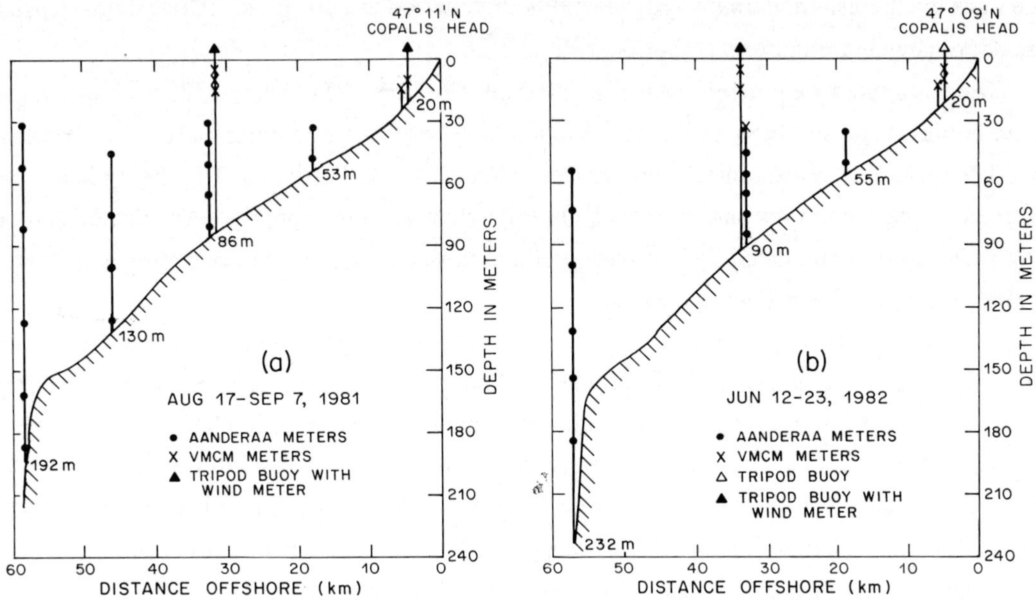

Fig. 6.2 Configuration of current meter moorings situated along the cross-shelf transects illustrated in Figure 6.1 for the late summer (a) and early summer (b) experiments.

inner shelf (late summer only). The sampling intervals were 4 min for VMCMs and 10 min for Aanderaa current meters.

In the early summer experiment, current meters were maintained from June 7 to June 24, 1982. Mooring locations were similar to those in the first experiment except that the 130-m station was omitted (Fig. 6.2b). Additional current meters were added to the remaining four moorings to improve vertical resolution.

Current and temperature records were processed by removing outliers, applying a binary filter to remove instrument noise and finally using a 25-h running mean filter (half power point of 0.43 cpd) to remove tidal and inertial oscillations. Although such a filter exhibits side lobes, the running mean was used to retain the first and last days' records, which would have been lost if a more efficient filter (e.g., a symetrical cosine type) had been used. Subsequently, current vectors from this subtidal data set were rotated into a local frame of reference for each mooring, to define cross-shelf (u) and alongshore (v) velocities.

Time-dependent changes in water properties (T, S, σ_t) were monitored with CTD profiles at ~15 stations along a cross-shelf section off Copalis Beach at approximately three-day intervals. A total of six sections were obtained for the late summer experiment and seven were obtained for the early summer experiment (see Figs. 6.1a and b). Alongshore variations were estimated for the late summer experiment by sampling at 20-km intervals along the 40-m isobath from 60 km north to 60 km south of the Copalis line during a period of relatively steady winds. A second alongshore transect, parallel to the first, followed the 100-m isobath. The 60-km station line represents the distance over which particles and nutrients could be advected in three days by alongshore currents typical of this time period and location. Alongshore transects were not obtained during the early summer experiment.

Water samples were taken from discrete depths on the CTD upcast with rosette-mounted Niskin bottles. The selection of sampling depths was based on hydrographic information from the CTD profiles. Subsamples were drawn from each of 12 depths for determination of dissolved oxygen, nutrients and pigments, chlorophyll *a* and phaeophytin. At selected stations along each transect, simulated ^{14}C primary production was measured for six depths representing the 100, 50, 30, 15, 5 and 1% light levels.

During the early summer experiment, a high capacity (480 l min $^{-1}$) pump system was used to sample the abundances and vertical distributions of zooplankton populations at eight stations shoreward of mid-shelf. One-minute pump samples were collected at 5-m depth intervals at each station. Nets of 52 μm Nitex mesh were used to retain even the smallest developmental stages in the pump discharge. Zooplankton counts (metazoans > 52 μm) were converted to total biomass using (= 2.5 x carbon) dry weight estimates for each species and developmental stage (see Landry and Lorenzen, 1989, Chapter 5).

6.3 ENVIRONMENTAL SETTING

Intermittent vs. steady upwelling

Although the magnitudes of upwelling favorable winds were similar during the early summer (June) and late summer (August) experiments, the persistence of the winds differed markedly. During early summer, winds were equatorward over the two-week experiment except for a 1/2 day relaxation event which appears only as a weakening of equatorward winds in the subtidal data set (Fig. 6.3a). In contrast, the wind reversed between poleward and equatorward on scales of 2-3 d during the late summer experiment (Fig. 6.3b). Consequently, we will denote the early summer experiment as one of 'persistent' upwelling and the late summer experiment as one of 'intermittent' upwelling (or intermittent downwelling).

Both the fluctuating and the mean current structures during the intermittent upwelling period differed from those of the persistent upwelling period. The seasonal mean alongshore current structure changes dramatically between June and September (see Hickey, 1989, Chapter 2, Fig. 2.12). During June, currents over the shelf are almost uniformly strong and equatorward and a coastal jet occurs in the surface layers at mid-shelf. In September, equatorward flow occurs only in a thin (5-10 m) surface layer, except on the outer shelf. Alongshore current fluctuations also differed between early and late summer. The amplitude of fluctuations was largest on the inner shelf during late summer, whereas it was largest at mid-shelf during early summer (see Hickey, 1989, Chapter 2, Fig. 2.21).

Cross-shelf currents also had a different structure in the early and late summer periods. In general, mean flow corresponded to a single-cell upwelling pattern in early summer and a single-cell downwelling pattern in late summer (Fig. 6.4a). In early summer, onshore flow at mid-shelf locations was strongest at mid-water-column depths, rather than in the bottom

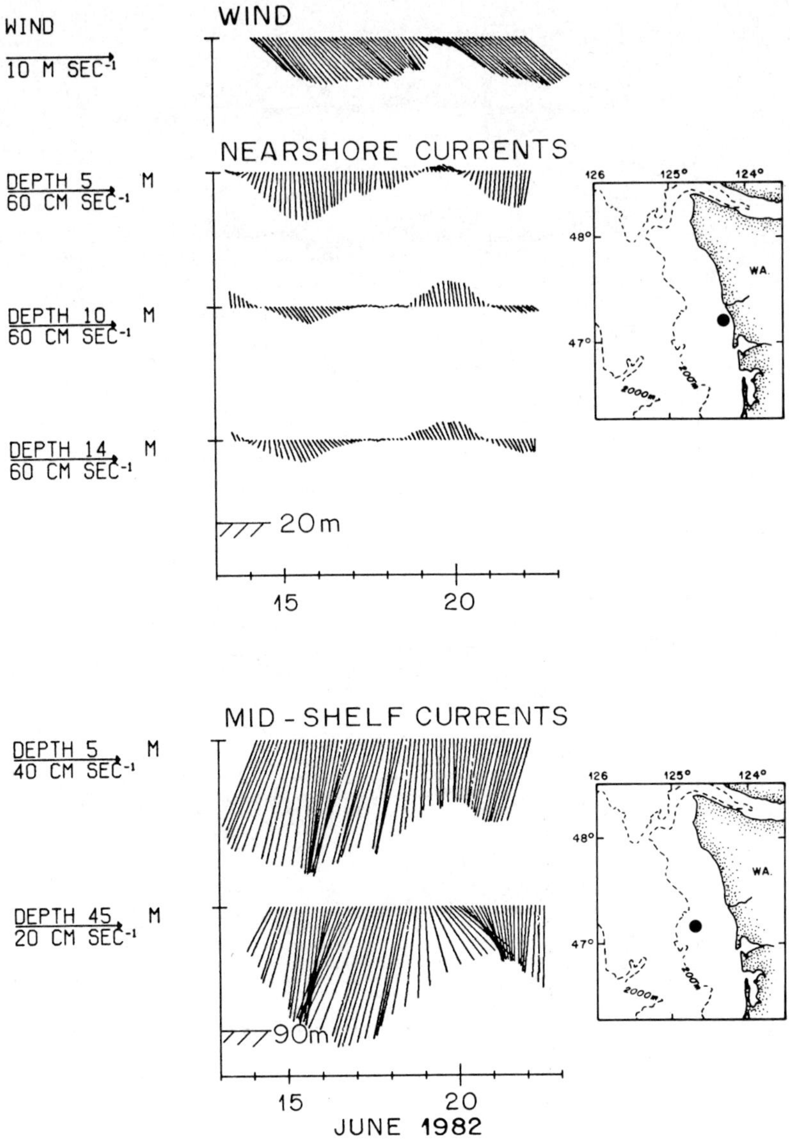

Fig. 6.3a Time series of winds and representative currents at near-shore (~5-km offshore) and mid-shelf (~30-km offshore) locations for the early summer experiment. All vectors are shown unrotated. Coordinate systems chosen for current series were: −10°, +10° and 0° from true N for the near-shore, mid-shelf and outer shelf locations, respectively.

boundary layer. The spatial structure of the fluctuations in cross-shelf currents also differed dramatically between the two periods (Fig. 6.4b). The dominant cross-shelf eigenmode for the late summer, intermittent upwelling period exhibits a shallow surface layer of strong cross-shelf fluctuations, which is not evident in the dominant eigenmode for the early summer, persistent upwelling period. The second cross-shelf eigenmode for the late summer has a mid-depth maximum in amplitude, with no zero crossing in the mid-shelf or near-shore region. The second

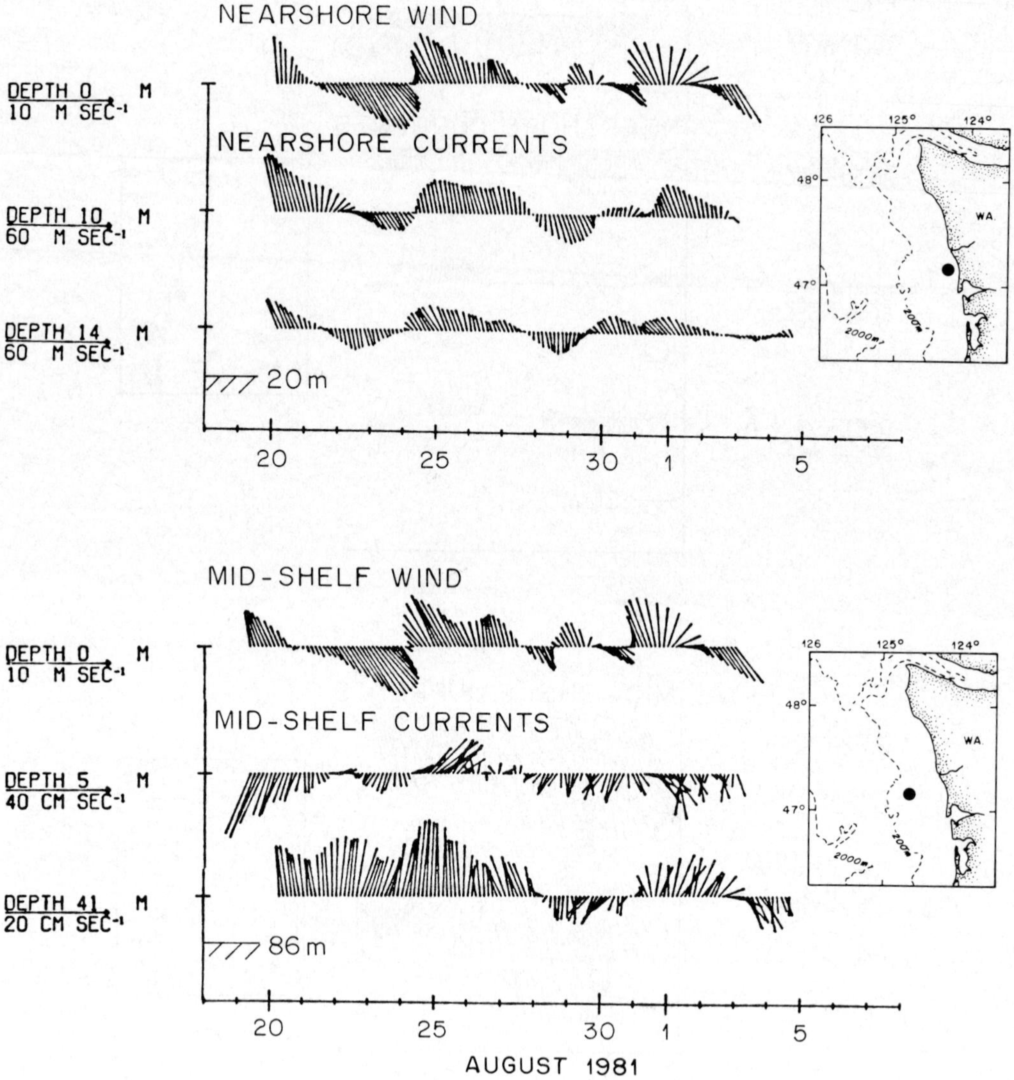

Fig. 6.3b Time series of winds and representative currents at nearshore (~5-km offshore) and mid-shelf (~30-km offshore) locations for the late summer experiment. All vectors are shown unrotated. Coordinate systems chosen for current series were: –10°, 0° and 0° from true N for the near-shore, mid-shelf and outer shelf locations, respectively.

cross-shelf eigenmode for the early summer period exhibits considerable structure with depth, with a zero crossing at mid-depth all across the shelf.

Alongshore current fluctuations during late summer appeared to be related to local wind stress. Equatorward flow was associated with equatorward wind stress and poleward flow with poleward wind stress events (Fig. 6.3b). Cross-shelf current fluctuations also exhibited some correspondence with local wind stress. A comparison of alongshore wind stress for the late

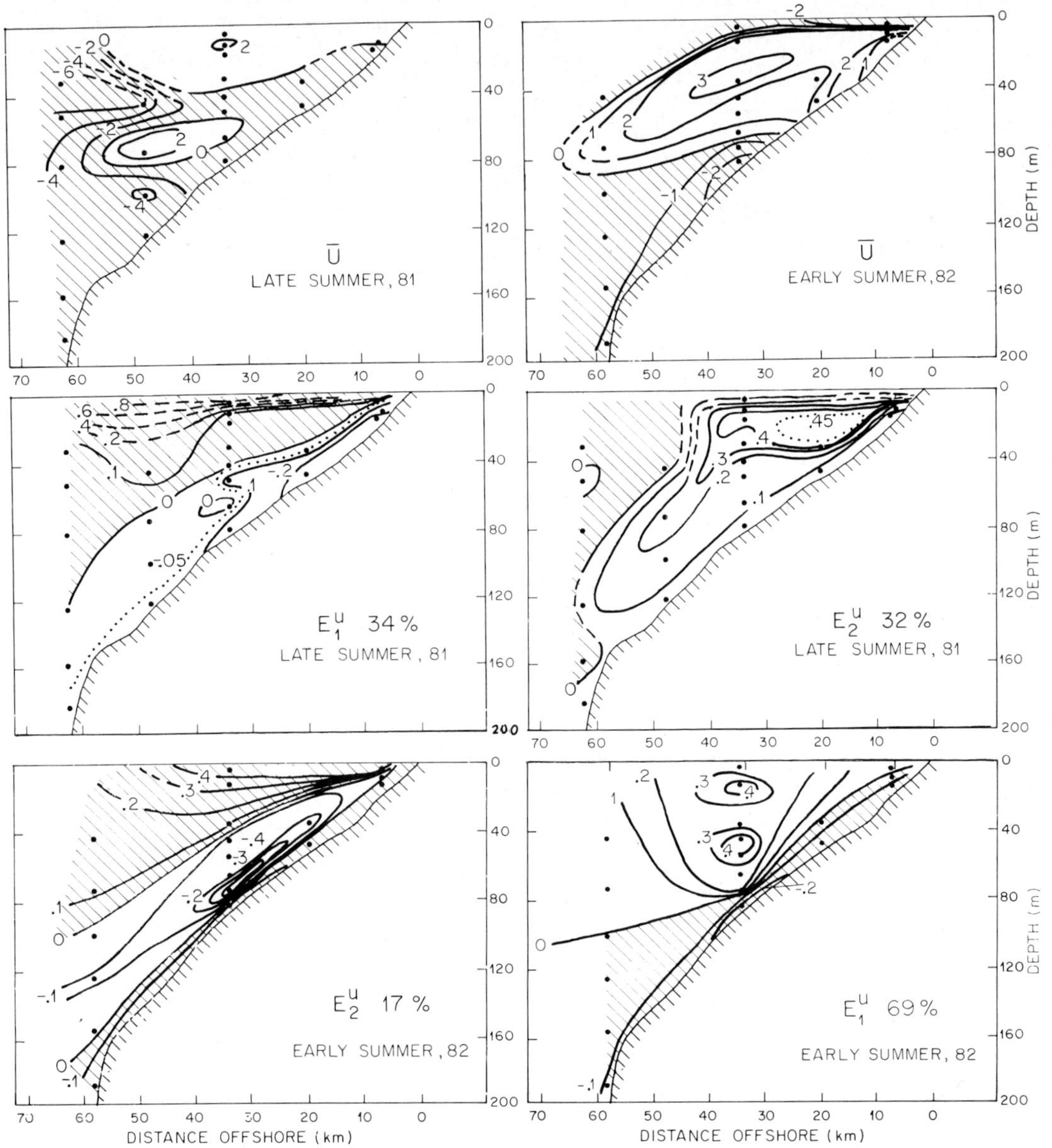

Fig. 6.4 (a) Mean cross-shelf velocities ($\overline{u}$) for the late summer and early summer experiments. Positive (negative) values designate onshore (offshore) flow in cm s $^{-1}$. (b) The amplitude of the first (E_1^u) and second (E_2^u) emperical eigenmodes of cross-shelf velocity for each experiment. When multiplied by the corresponding time series in Figure 6.5, these yield the time series of currents in each mode.

summer period with the amplitude time series for the first eigenfunction of cross-shelf current fluctuations (Fig. 6.5a) suggests locally wind-forced upwelling/downwelling for that period.

Alongshore current fluctuations during early summer are also related, although less strongly, to the local winds (Fig 6.3a). The second cross-shelf eigenmode is weakly related to the wind (Fig. 6.5b), but the variance accounted for by this eigenmode is only half that of the late summer period (Fig (6.4). This result may reflect the greater persistence of local winds during

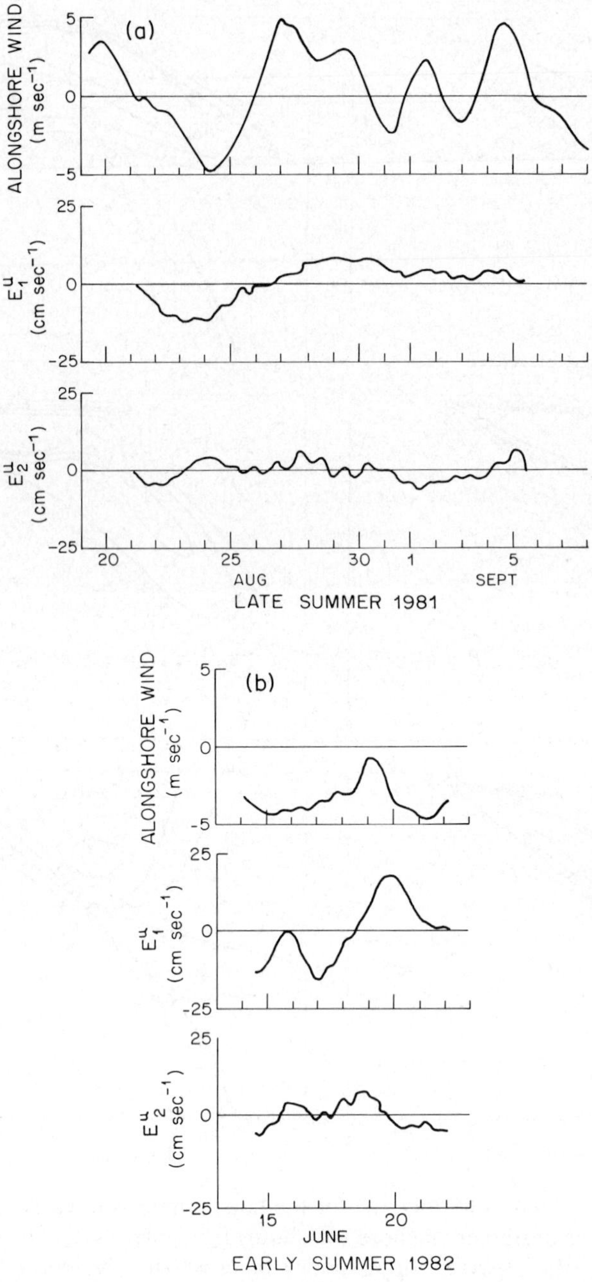

Fig. 6.5 The alongshore component of wind stress and the amplitude of the first and second eigenmodes of cross-shelf velocity (*u*) for (a) late summer 1981 and (b) early summer 1982 experiments.

the early summer period. With steady local winds, much of the variance in local currents could be attributable to eddy activity or non-local wind forcing (e.g., coastal-trapped waves forced to the south of the study area), even though the mean currents could be forced by the steady local and mean nonlocal winds. Indeed, the spatial pattern of the mean cross-shelf current structure during the early summer, persistent upwelling period (Fig. 6.4a) bears a similarity to that of the second cross-shelf eigenmode for the late summer, intermittent upwelling period. This eigenmode exhibits no strong correlation with local winds and may represent remotely-forced changes. Freely propagating, remotely forced coastal-trapped waves are likely to be particularly important in the Pacific Northwest during early summer, when local winds are relatively light and winds off California are at their seasonal maximum. The reader is referred to Hickey (1989, Chapter 2) for a complete discussion of the importance of propagating waves to the shelf circulation in the U.S. Pacific Northwest.

Examples of cross-shelf flow patterns have been delineated in objectively contoured isotach maps (Fig. 6.6) (see Section 6.4 for a discussion of the objective interpolation technique). Mid-shelf density profiles for each of these examples are also shown. On June 16, 1982, winds were equatorward (i.e., upwelling favorable; $v = -5$ m s $^{-1}$). Cross-shelf currents at 1200 PST (Pacific Standard Time; 1200 PST = 1900 GMT) were onshore in a bottom layer of ~10-m depth and offshore elsewhere. Maximum offshore flow occurred near the surface and in the pycnocline. This flow was stronger at mid- and outer shelf locations than over the inner shelf. On August 23, 1981, winds were southward and comparable in speed to those than on June 16, 1982. However, the structure of cross-shelf flow was markedly different than during the early summer. Offshore flow was weak (~4 cm s $^{-1}$) and confined to a very shallow (5-10-m thick) surface layer. Onshore flow was a maximum at mid-depth rather than in the bottom boundary layer. The sharper pycnocline may account for the thiner layer of offshore flow during the August 23, 1981, upwelling event, as compared with the more gradual density structure and thicker layer of offshore flow during the June 16, 1982, upwelling event. A typical downwelling event occurred on August 25, 1981 (Fig. 6.6c). During this event, onshore flow extended from the surface to a depth of 40 m at mid-shelf and to ~20 m on the inner shelf. Offshore flow was a maximum in the bottom boundary layer.

A purely two-dimensional, cross-shelf mass balance was not generally observed during upwelling or downwelling events in either the early summer or late summer periods; i.e., the cross-shelf velocity pattern was divergent at any instant. In the upwelling cases shown in Figure 6.6a and b, cross-shelf flow within the geostrophic interior exceeded cross-shelf flow at either the surface or the bottom boundary layer. This result is consistent with observations for upwelling areas off Oregon (Huyer, 1983) and underscores the significance of alongshore structure of the velocity field, e.g., the importance of coastal-trapped waves in this region (Battisti and Hickey, 1984; Allen and Denbo, 1984).

Qualitative changes in property fields during upwelling events are readily distinguished. For example, the upwelling events delineated in Figures 6.6a,b caused significant changes in the isopleth structure for density, temperature, salinity, chlorophyll, nitrate and silicate (see Figs. 6.7a,b). All of the property fields rose upward near the shore following both of the upwelling events, suggesting vertical advection. However, the upwarping of salinity, nitrate and silicate (but not temperature) isopleths at ~10-km offshore during the late summer experiment may have been due in part to alongshore, rather than cross-shelf, advection of a dramatic low salinity lens

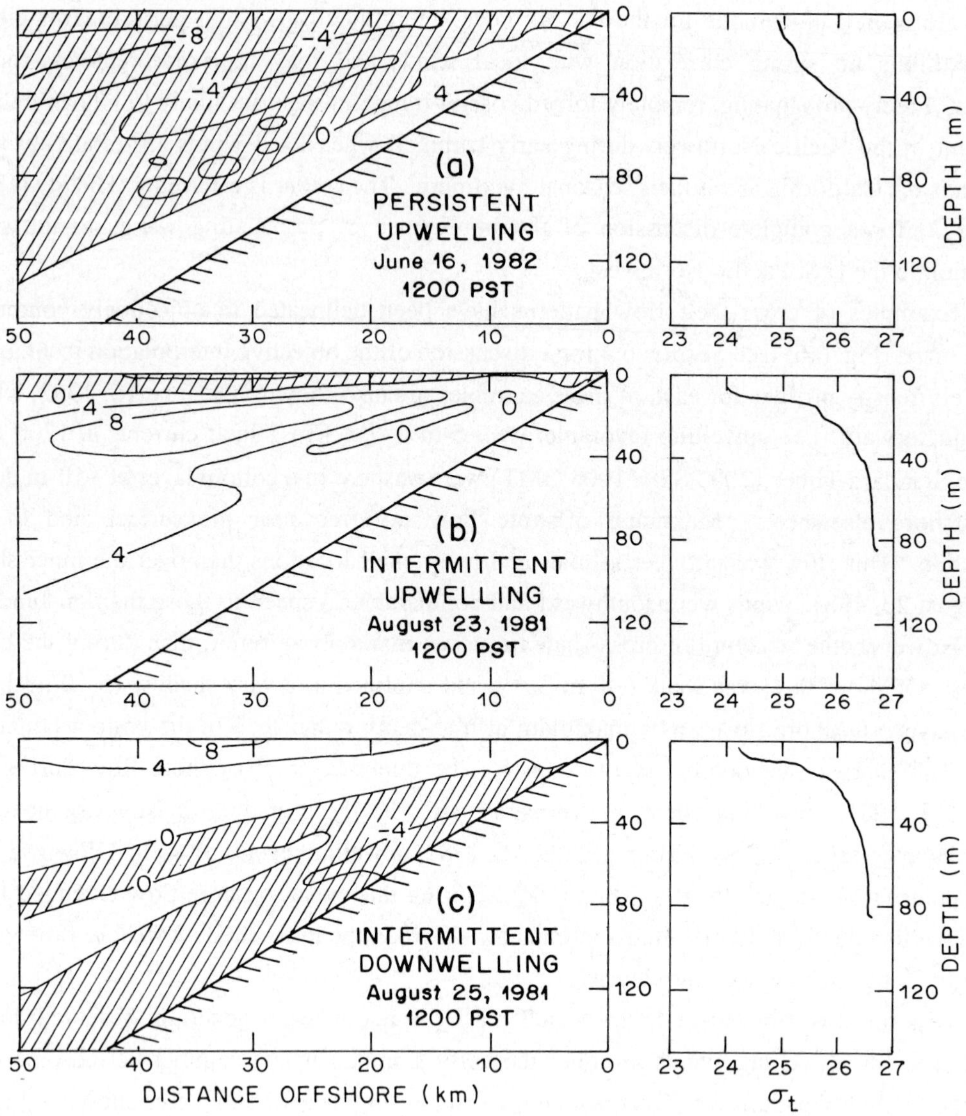

Fig. 6.6 Objectively-contoured isopleth maps of cross-shelf velocities (*u*) during early summer and late summer experiments illustrating (a) 'persistent upwelling' on June 16, 1981; (b) 'intermittent upwelling' on August 23, 1981; (c) 'downwelling' on August 25, 1981. Positive (negative) values designate onshore (offshore) flow in cm s^{-1}. Simultaneous mid-shelf (~30-km offshore) density profiles are shown for each map.

observed in an alongshore transect during that experiment (Fig. 6.7c). This lens was oblong in shape, with approximate dimensions of ~20 km cross-shelf, ~100-km alongshore and ~10-m depth. The absence of a strong temperature anomaly in this lens is consistent with the absence of a strong upwarping of temperature isopleths at the 10-km offshore location.

Estimates of total zooplankton biomass obtained just before and just after the 16-18 June 1982 upwelling event are shown in Figure 6.8. A core of high zooplankton density (~300 mg d.w. m^{-3}) was observed at ~10-m depth and 10-km offshore on June 16. The core

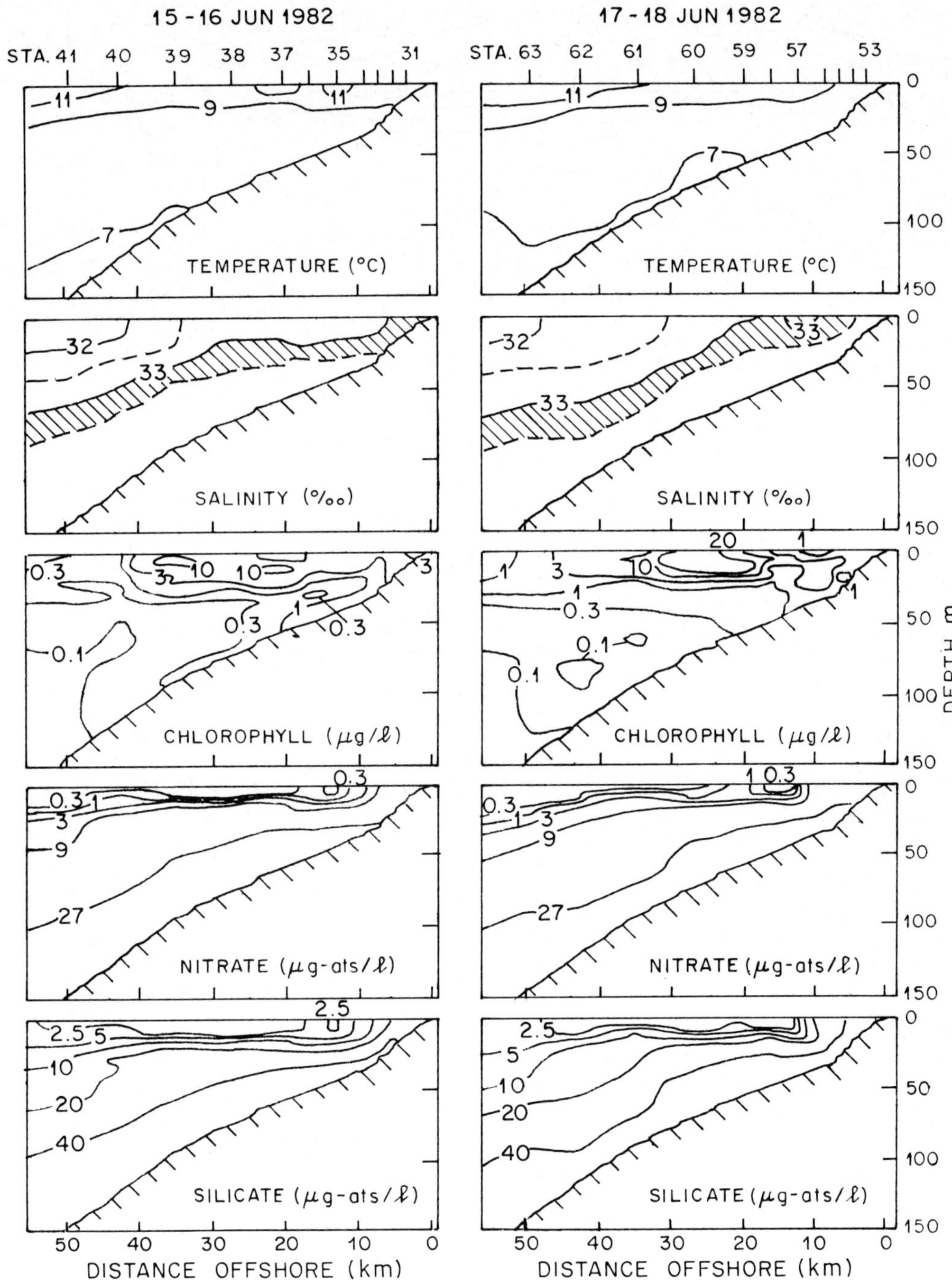

Fig. 6.7a Temperature, salinity, chlorophyll, nitrate, and silicate isopleths before and after the 'persistent upwelling' event in June 1982.

shoaled and moved onshore by June 18, consistent with the observed upwelling of other properties during this event.

The three periods of 21-24 August 1981, 24-27 August 1981 and 16-18 June 1982 will be emphasized in the ensuing analysis. They will be referred to as 'intermittent upwelling', 'intermittent downwelling' and 'persistent upwelling' events, respectively.

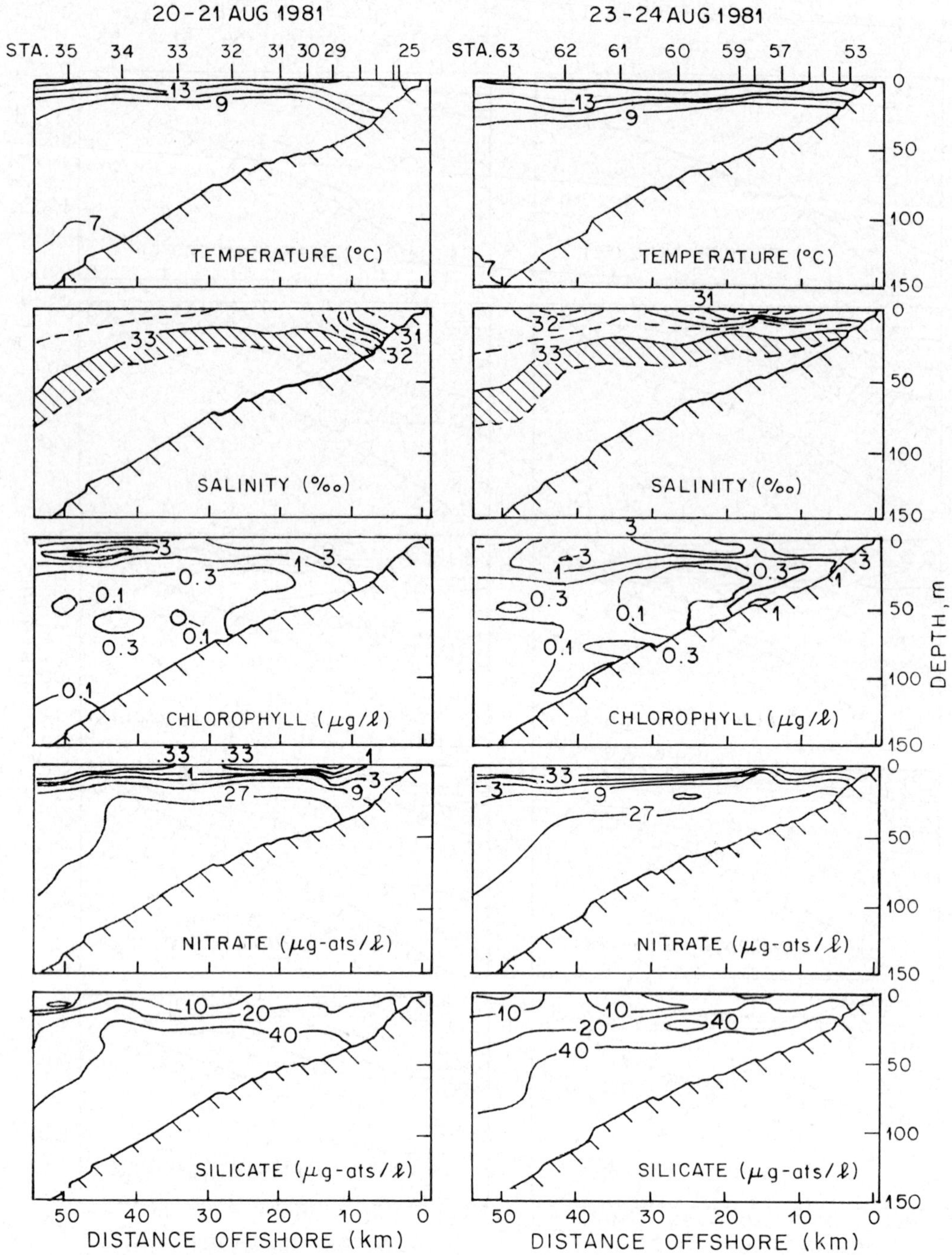

Fig. 6.7b Temperature, salinity, chlorophyll, nitrate, and silicate isopleths before and after the 'intermittent upwelling' event in August 1981.

6.4 ANALYSIS

6.4.1 Philosophy

Changes in dissolved constituents at any point (neglecting changes in the density of material elements) must be due to advection of constituents past that point, lateral and vertical

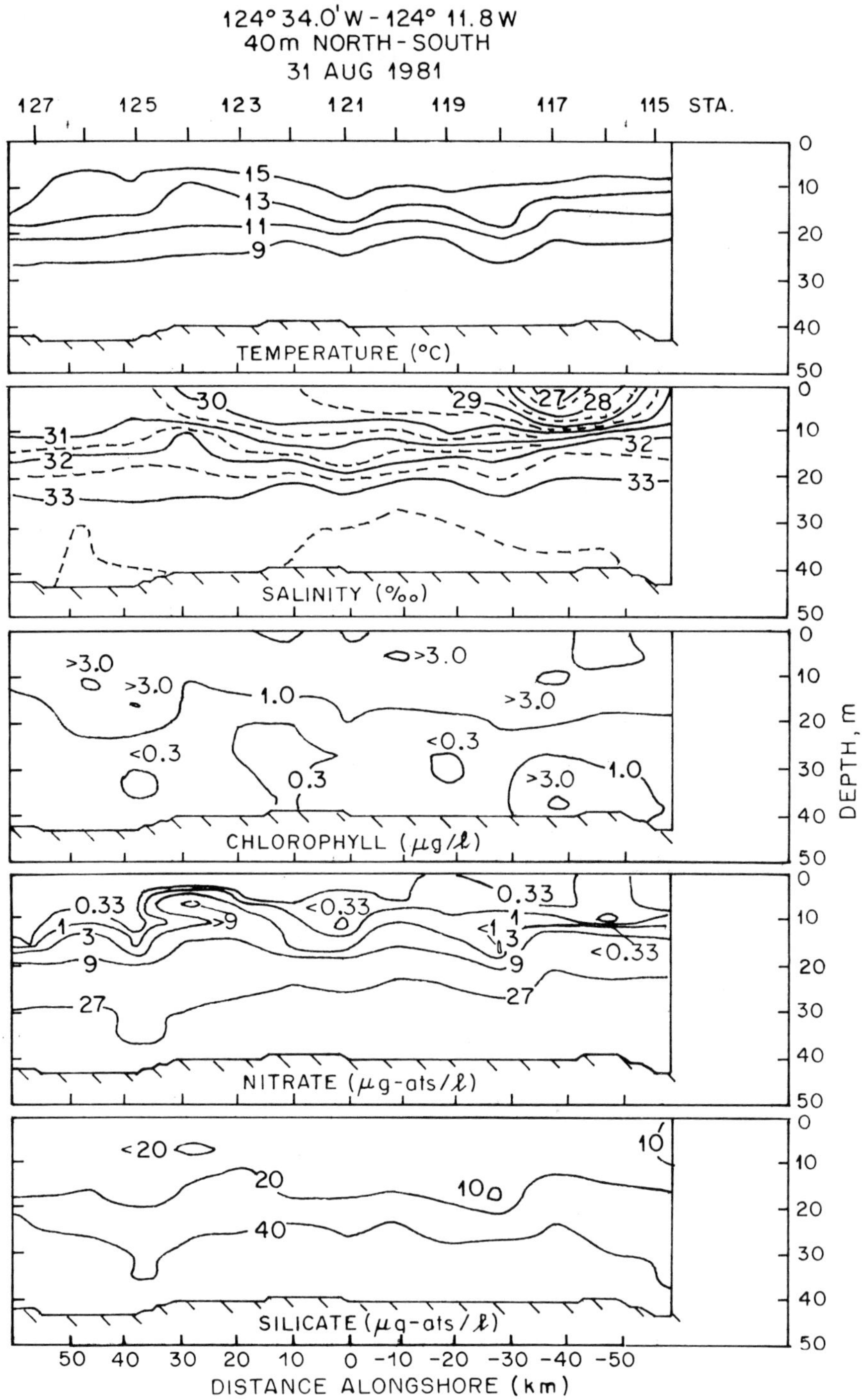

Fig. 6.7c Temperature, salinity, chlorophyll, nitrate, and silicate isopleth structure along the 40-m isobath during the late summer experiment (August 31, 1981).

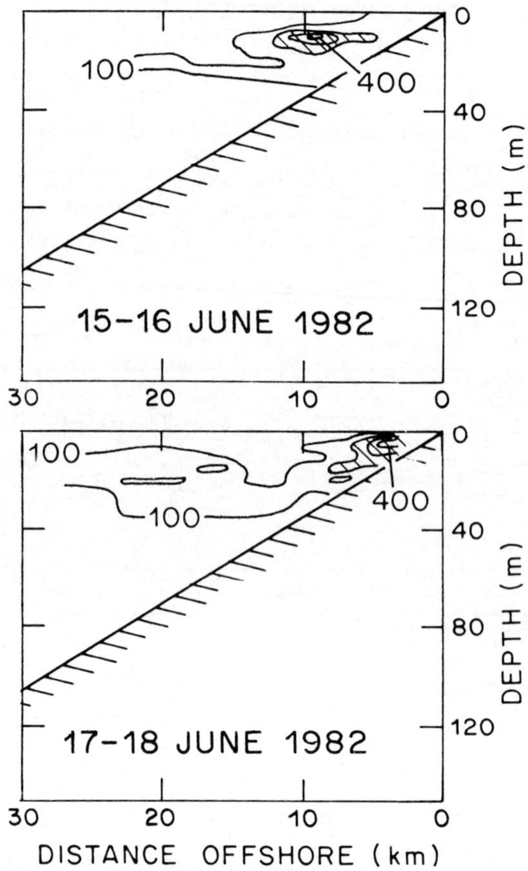

Fig. 6.8 Distributions of total zooplankton biomass (dry weight of metazoans > 60 μm, all species and life stages) before and after the 'persistent upwelling' event on June 16, 1982.

diffusion to or from adjacent parcels and nonconservative processes occurring at the point. An equation summarizing these processes is:

$$\frac{\partial\gamma}{\partial t} = -u\frac{\partial\gamma}{\partial x} - v\frac{\partial\gamma}{\partial y} - w\frac{\partial\gamma}{\partial z} + \frac{\partial}{\partial x}\left[K_H\frac{\partial\gamma}{\partial x}\right] + \frac{\partial}{\partial y}\left[K_H\frac{\partial\gamma}{\partial y}\right] + \frac{\partial}{\partial z}\left[K_V\frac{\partial\gamma}{\partial z}\right] \quad (6.1)$$

$$+ \textit{Nonconservative Processes}\ .$$

where γ represents the value of some dissolved property which varies in space and time, x,y,z are the onshore (cross-shelf), alongshore and upward directions, u,v,w represent the cross-shelf, alongshore and upward water velocities and K_H, K_V refer to eddy diffusivities in horizontal and vertical directions, respectively. For particulate materials, a sinking term, $-\partial\,(w_s\gamma\,)/\partial z$, where w_s is the settling velocity, must also be added.

Investigation of advective-diffusive and nonconservative changes ultimately requires the use of a finite difference analog of Equation 6.1 with sufficient temporal and spatial resolution to ensure the stability and accuracy of the solution procedure. Estimates of u,v,w are required at each grid point and throughout each period where the terms in Equation 6.1 are evaluated. Estimates of selected properties, γ, were interpolated from the 3-km horizontal x 5-m vertical sampling grid to initialize the process. The nonconservative term was then estimated by comparing observed changes in concentration at a point with changes predicted by integrating only the advective (e.g., $u\,\partial\gamma/\partial x$) and diffusive [e.g., $\partial(k_v\,\partial\gamma/\partial z)/\partial z$] terms in the equation. The integration was accomplished via a finite difference model that incorporates the observed velocity and property fields (see Section 6.4.2). The residuals between predicted, advective-diffusive changes and observed changes in concentration at a point approximately represent the sum of all nonconservative processes which occurred within the water parcel whose path (streakline) terminates at that point for the interval under study. Some inadvertent smoothing of the residual field is expected, insofar as this technique cannot account for diffusion of material produced during the integration interval. Still, residuals calculated in the Eulerian framework may be viewed as approximate Lagrangian integrals of nonconservative behavior for each water parcel. The residual values for chlorophyll and phaeopigments also include the net influx minus efflux of pigmented particles sinking through the water. However, passive sinking should not constitute a large bias for chlorophyll on ~3-d time scales.

6.4.2 The Velocity Field

Objective interpolation techniques were used to obtain reasonable estimates of u and v at each point in the cross-shelf (x-z) plane (the Copalis line) passing through the current meter moorings of the two experiments. The technique as applied here considers the value of a scalar quantity at a particular time at an unmeasured location to be the weighted average of the observations made nearby at that same time (Bretherton *et al.*, 1976). The weights assigned to each observation depend upon the expected covariance of the scalar at the measured location with its value at the unmeasured location and the covariance of the scalar between each pair of measurement locations. In practice, objective interpolation requires choosing a function to fix the covariance relationships. The precise form of this function depends both on the observed covariances among time series as well as *a priori* biases regarding the covariance structure of the property of interest. Since the scale of the physics governing cross-shelf and alongshore velocities is expected to differ, the two velocity components were treated independently in the analysis.

One natural way to represent the velocity field in a coastal boundary area with a gently sloping bottom is in terms of cylindrical coordinates, where the axis of the 'cylinder' coincides with the coastline (Fig. 6.9). Intuitively, equatorward Ekman flux at the surface of the coastal ocean should be correlated with onshore, poleward flux at depth. A negative covariance is thus expected for angular separations corresponding to the water column depth at some radial distance from the coast. Similarly, Ekman flux should be coherent in the cross-shelf (radial) direction, in response to upwelling-favorable winds which are larger in scale than the shelf width. These *a priori* considerations suggest a covariance function for cross-shelf (u) and alongshore (v) velocities of the form:

$$F(\Delta r,\Delta\theta) = \exp[-k_1(\Delta r)^2 - k_2(\Delta\theta)^2]\cos(k_3\Delta r)\cos(k_4\Delta\theta) \qquad (6.2)$$

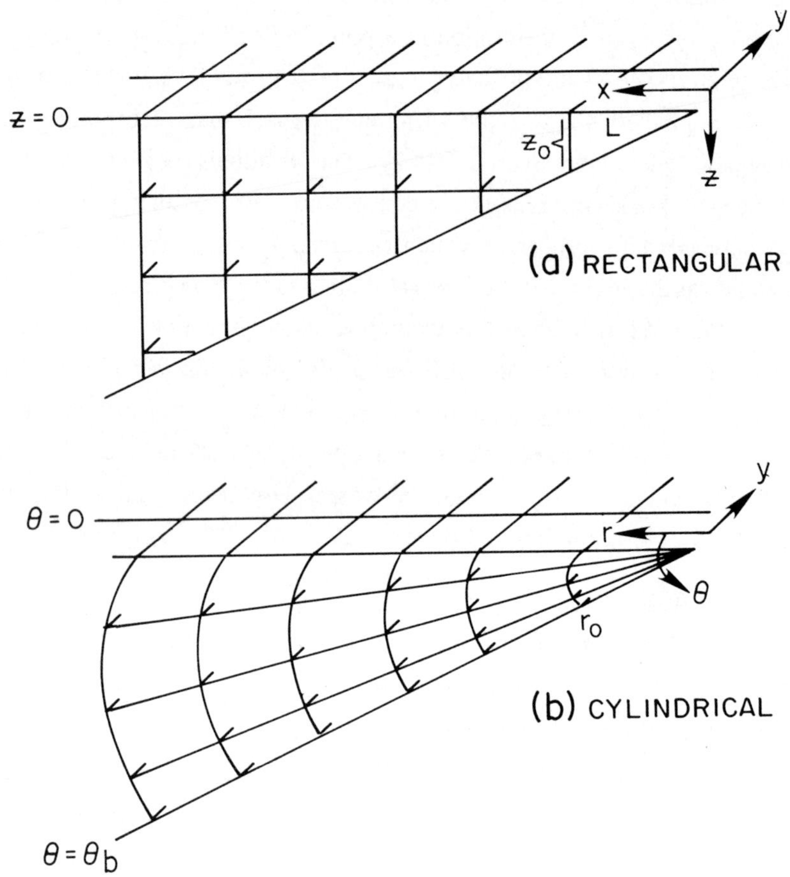

Fig. 6.9 Comparison of (a) rectangular and (b) cylindrical coordinate systems for the coastal regime. $\theta = \theta_b$ roughly corresponds to the bottom boundary of the continental shelf. The diagrams indicate cross-shelf (x) and vertical (z) directions for the rectangular case, and radial (r) and azimuthal (θ) directions for the cylindrical case.

where Δr and $\Delta\theta$ represent the radial and angular separation, respectively, between two locations. The coefficients k_1 through k_4 were chosen to yield a reasonable fit of the function to observed covariances between pairs of current meter observations. Chosen values for k_1 and k_3 correspond to an e-folding scale of 25 km in the radial direction for both u and v and a cross-over to negative covariance at radial separations of 100 km. Values for k_2 and k_4 correspond to an e-folding scale of 0.2 times the water column depth for u (1.0 times for v), with a crossover to negative covariance at 0.5 times the water column depth for both u and v. Observed covariances between current meter pairs are compared with the functional form (Eq. 6.2) in Figure 6.10. Comparisons are made separately as a function of Δr and $\Delta\theta$, utilizing those pairs with small $\Delta\theta$ and Δr respectively. The deviation of the observed covariance from the functional value for many of the pairs reflects the difficulty of accounting for the many length scales for u and v in the coastal environment with any simple functional form. Such difficulties ultimately necessitated using only the current meters within 40 km of the coast (< 100-m depth) to obtain

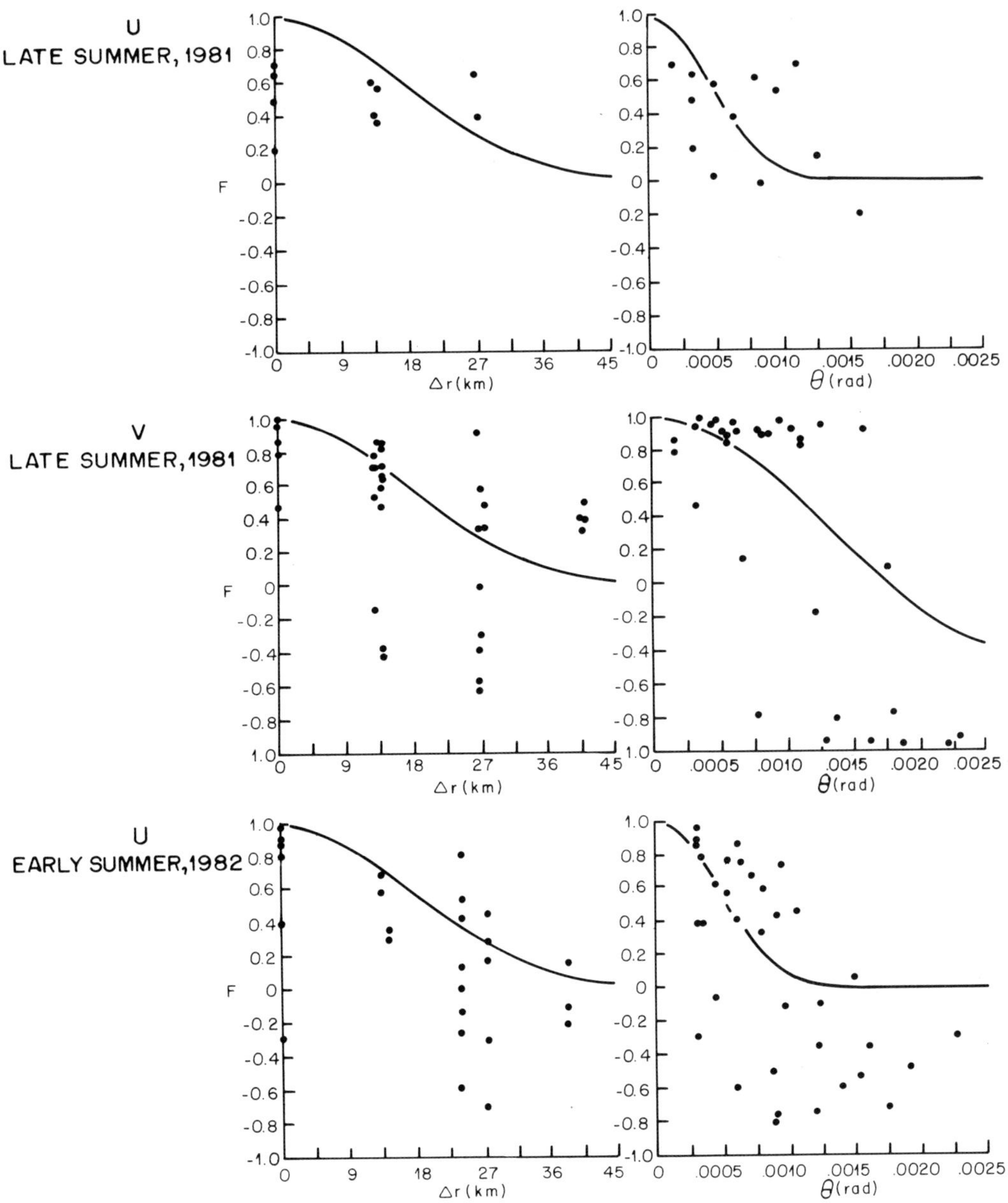

Fig. 6.10. Observed covariance among current meter pairs (solid dots) as a function of radial (Δr) and angular ($\Delta\theta$) separation, compared with the covariance function F (Eq. 6.2; solid lines) used in the objective interpolation.

reasonable maps of cross-shelf velocities in the near-shore region for the late summer period. The full current meter set was used for alongshore velocities in the late summer period and for cross-shelf velocities in the early summer period. Despite these difficulties, the objective interpolation technique was considered superior to cruder methods (e.g., linear interpolation) for our irregularly spaced current meter array. All methods of interpolation involve implicit assumptions about correlation scales; the objective scheme used here has the advantage of making these assumptions explicit.

Note that the values of k_1 and k_3 used here are different than those determined by Denman and Freeland (1985) for a region off Vancouver Island, where they assumed an isotropic velocity field in *u,v*. For an isotropic covariance function, they determined scales of ~50 km (e-folding) and ~35 km (crossover). Note also that while the isotropic form of the covariance function can imply unrealistic negative spectra, our chosen form of Equation 6.2 cannot, for any choice of k_1-k_4.

Maps of probable error in *u* and *v*, expressed as a fraction of their respective temporal variances (spatially-averaged) were computed for the current meter arrays of the early and late summer experiments, using the chosen forms of the covariance function (Fig. 6.11). Smallest probable error for estimates of *u* resulted at mid-shelf during both experiments, where the sampling arrays for currents were especially dense. Largest probable error resulted near the surface in the near-shore zone, especially during the late summer experiment. Lower values of probable error were estimated for *v*, due to the stronger covariance in θ of the chosen covariance function.

After utilizing the objective interpolation algorithm, minor adjustments were made to the series for *u* and *v* to ensure: 1) zero tangential flow at the bottom (the 'no-slip' condition), linearly varying both cross-shelf and alongshore velocity from zero at the bottom to the objectively obtained value at 5 m above the bottom; 2) zero cross-shelf flow at 2-km offshore, linearly varying cross-shelf velocity from zero at 2-km offshore to the objectively obtained value at 5-km offshore; 3) zero alongshore flow at the coast, linearly varying alongshore velocity from zero at the coast to the objective value at 5-km offshore. The 2-km offshore position, where the depth is ~6 m, was chosen as a convenient coastward boundary. Mitchum and Clark (1986) concluded that a condition of no net flux of water across the shelf can be reasonably imposed at a depth corresponding to ~3 times the Ekman depth, which occurs seaward of the 2-km offshore position for realistic values of vertical eddy viscosity. As with the choice of a covariance function, these adjustments reflect *a priori* biases concerning the flow field. Furthermore, they yielded a better fit to observed salinity data near the bottom and coastline, than was obtainable using objective interpolation alone.

Once estimates of alongshore and cross-shelf velocity had been obtained, the vertical velocity was estimated at each point. Correct estimates of *w* are especially desirable because our technique is intrinsically biased for those properties with strong vertical gradients. For such properties, small changes in the estimates of vertical velocity will effect large changes in the residuals between observed and predicted fields. Consider the cases of nitrate and chlorophyll. The former exhibits a strong increase in concentration with depth, whereas the latter exhibits a strong decrease (at least below 20 m; e.g., Fig. 6.7). Hence a large overestimate of upward vertical velocity would naturally produce a negative residual for nitrate and a positive residual for chlorophyll.

$$k^* = \frac{K_h}{L^2} = \frac{K_v}{z_0^2} = 4.0 \times 10^{-6} \text{s}^{-1} \tag{6.10}$$

We defined horizontal velocity u^* in units of u_0 and vertical velocity w^* in units of w_0, where $w_0 = u_0 z_0 / L$ from the incompressibility condition. With u_0 set at 0.016 m s $^{-1}$, a representative near-surface speed under typical conditions, we also have the relation

$$\frac{u_0}{L} = \frac{w_0}{z_0} = 1.6 \times 10^{-5} \text{s}^{-1} = 4k^* \tag{6.11}$$

The non-dimensional velocity components can be expressed in terms of the scaled cylindrical coordinate system as q_r and q_θ from the relations

$$u^* = u_0\, u = u_0\, (\cos\theta\, q_r - \sin\theta\, q_\theta) \tag{6.12}$$

$$w^* = w_0\, w = w_0\, (\sin\theta\, q_r + \cos\theta\, q_\theta)$$

Combining these definitions with the assumptions regarding turbulent mixing, the governing equations in the scaled cylindrical system take on a simpler form. For example, the conservation of chlorophyll is expressed as

$$\frac{\partial Chl}{\partial t} = k^* \nabla^2 Chl - 4k^* \tilde{q} \cdot \tilde{\nabla}\, Chl - v \frac{\partial Chl}{\partial y} - Sink. + Prod. - Resp. - Graz. \tag{6.13}$$

where Sink., Prod., Resp. and Graz. represent net loss by sinking, gross production, respiration and grazing, respectively. In this equation, the '∇' operator is expressed in cylindrical coordinates and $\hat{q} = q_r \hat{r} + q_\theta \hat{\theta}$. A similar equation holds for the conservation of salt:

$$\frac{\partial S}{\partial t} = k^* \nabla^2 S - 4k^* \tilde{q} \cdot \tilde{\nabla} S - v \frac{\partial S}{\partial y} \tag{6.14}$$

Alongshore advection is added to the model for the late summer experiment by: 1) using the cylindrical grid to interpolate parameter values measured in the two alongshore transects to a three-dimensional field, 110 km in alongshore extent, centered on the Copalis transect and 2) allowing the alongshore profile of interpolated values at each offshore location and depth to advect up or down the coast at the velocity predicted by objective interpolation for each time step. The alongshore advective term, $-v\, \partial\gamma / \partial y$, is calculated as the rate of change in property between time steps due to this advection of each 'frozen' alongshore profile at a particular offshore location and depth. At time scales such that the 110-km alongshore profile is advected beyond either endpoint, the advective term is set equal to zero; hence the model underestimates advection at the high end. This deficiency applies mainly in the offshore zone, where alongshore currents frequently carry water parcels more than 55 km up or down the coast in a single day. Since no alongshore CTD transects were made during the early summer experiment, no alongshore advection estimates could be made for this period. Alongshore diffusion estimates could not be made for either experiment.

The governing equations for advective-diffusive change are solved only in the domain of $r_0 < r \leq 28r_0$, $0 < \theta \leq \pi/4$, subject to appropriate initial and boundary conditions. With the velocity field such that $q_r = 0$ at the coastward boundary, we have

$$\frac{\partial Chl}{\partial r} = \frac{\partial S}{\partial r} = 0 \qquad on \qquad r = r_0 \tag{6.15}$$

At the sea surface and at the bottom, $q_\theta = 0$ and the flux of salt, phytoplankton and nitrogen are all presumed negligible. For advective-diffusive transport, we can then write

$$\frac{1}{r}\frac{\partial Chl}{\partial \theta} = \frac{1}{r}\frac{\partial N}{\partial \theta} = \frac{1}{r}\frac{\partial S}{\partial \theta} = 0 \tag{6.16}$$

Appropriate seaward boundary conditions are not immediately obvious. A workable procedure is to prescribe the concentration at the seaward boundary, $Chl_b(\theta)$, using the observed values at the beginning (t_0) and end (t_1)of the simulated period and interpolating between those boundary conditions for intermediate times; that is,

$$Chl(28r_0,\theta,t) = Chl_b(\theta,t) = Chl_b(\theta,t_0) + (t - t_0)\,\frac{Chl_b(\theta,t_1) - Chl_b(\theta,t_0)}{(t_1 - t_0)} \tag{6.17}$$

By means of a second coordinate transformation, the conservative equations can be written in a form convenient for finite difference solution in a rectangular domain. In particular, let

$$\xi = ln\,(r/r_0)$$

The chlorophyll equation can then be written in the form

$$\frac{\partial Chl}{\partial t} = f_1\left(\frac{\partial^2 Chl}{\partial \xi^2} + \frac{\partial^2 Chl}{\partial \theta^2}\right) - f_2\frac{\partial Chl}{\partial \xi} - f_3\frac{\partial Chl}{\partial \theta} - v\,\frac{\partial Chl}{\partial y} \tag{6.18}$$

In this expression,

$$f_1 = (k^*/r_0^{\,2})\exp(-2\xi)$$

$$f_2 = \left[\,4k^* q_r + (w^*_s \sin\theta/z_0)\right] r_0^{-1}\exp(-\xi)$$

$$f_3 = \left[\,4k^* q_b + (w^*_s \cos\theta/z_0)\right] r_0^{-1}\exp(-\xi)$$

The equations are solved in the transformed domain $0 < \xi < \ln 28$, $0 \leq \theta \leq \pi/4$, subject to the transformed boundary conditions.

The conservation equations were approximated by finite difference with derivative forms of second order accuracy. The increment $\delta\theta$ was chosen small enough so that the depth increment ranged from 0.5 m near the coast to 2.5 m at the seaward edge of the domain. The integration in time was carried out by the Alternating-Direction Implicit method with $\delta t = 0.5$ h, the choice of the time step being a compromise between accuracy and economy.

6.5 RESULTS

6.5.1 Advective/Diffusive Changes

Salt

Since salinity is a conservative property of the water column, "goodness of fit" of the model-predicted salinity to the observed salinity field can be used as a test of the advective-diffusive model. Comparison of salinities predicted by the model with observed salinity along the Copalis section for the persistent upwelling (~2 d), intermittent upwelling (~3 d) and intermittent downwelling (~3 d) events suggests that the model is reasonably successful for all three events (Fig. 6.13). Note that the large spatial variance in observed salinity fields during the latter two events effects a higher degree of misfit with the model when misfit (observed minus predicted values, henceforth called the "residual") is expressed in ppt salinity.

During the persistent upwelling event, isopleths of salinity in the near-shore zone were observed to shoal and move offshore. Less offshore movement and even some onshore movement at depth was evident at mid-shelf. The advective-diffusive model was able to reproduce much of the offshore motion in the near-shore zone, although the degree of shoaling was underestimated. A large surface pool of water with < 33 ppt salinity is also predicted at 10-30-km offshore; the observed pool at 10-15-km offshore is much smaller. The residual map indicates that the best near-surface fit occurs at mid-shelf and well offshore. Largest positive residuals (underestimation of salinity by as much as 0.3 ppt) occur at ~15-km offshore and inshore of ~8 km. Largest negative residuals (overestimation of salinity by as much as 0.6 ppt) occur over the outer shelf below the surface layer.

During the intermittent upwelling event, a core of low salinity (< 32 ppt) water shoaled and spread out from within 10 km of the coast to between 30 km and the coast. The model reproduced most of this offshore spreading, though both spreading and shoaling of, e.g., the 32 ppt isopleth were somewhat underestimated. Small regions of salinity > 34 ppt predicted at depth below the pycnocline were not apparent in the observations. Largest negative residuals occurred near the surface at ~20-km offshore, where the model overestimated the salinity by ~1 ppt; largest positive residuals occurred within 10 km of the shore, where the model underestimated salinity by ~1 ppt.

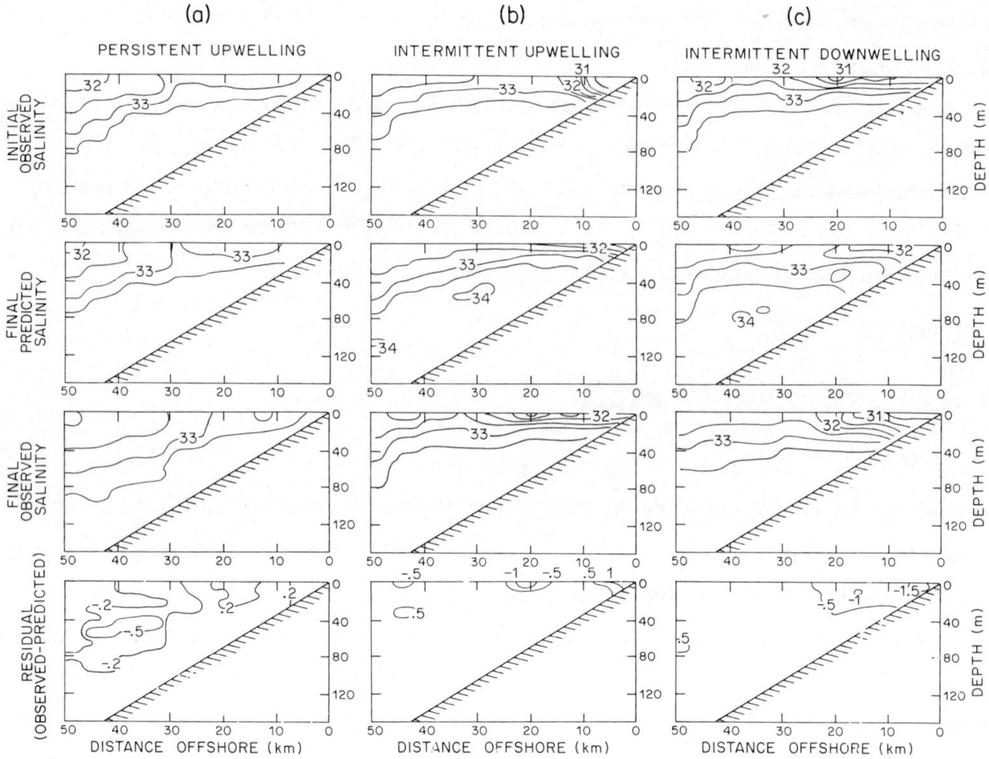

Fig. 6.13. Predicted and observed salinities ($^o/_{oo}$) for (a) 'persistent upwelling', (b) 'intermittent upwelling', and (c) 'intermittent downwelling' events. The initial salinity as well as the residual between predicted and observed salinity are also shown.

During the downwelling event, which immediately followed the intermittent upwelling event, the core of low salinity water returned to a deeper, nearshore location. The model reproduced qualitatively this near-shore, descending motion, but overestimated the salinity values (i.e., negative residuals) by as much as 1.5 ppt in the nearshore area. Near-surface salinity offshore was slightly underestimated.

A summary of the mean squared residual values for each period compared with mean squared observed change (observed final minus observed initial values) is given in Table 6.1. The ratio of these two variance statistics for salinity provides an objective measure of the relative goodness-of-fit of the model for the three periods. We note that the mean squared values are derived using all grid points in the model's stretched radial coordinate system (described in Section 6.4). Thus, the near-shore area, where the density of gridpoints is highest, is weighted more heavily in the statistics than offshore regions. Of the three representative periods, the intermittent upwelling event exhibits by far the lowest global ratio of residual to observed change and hence the best fit.

The relative contribution of the various advective-diffusive terms in Equation 6.14 (i.e., the salt conservation equation in radial coordinates) is delineated in Figure 6.14. For a perfect match of model to data, the observed change would precisely equal the total advective-diffusive flux at each location. The intermittent upwelling case approaches this ideal most closely. The

TABLE 6.1 Comparison of mean squared (i.e., for all grid points) residual salinity *vs*. observed change in salinity for each of the three study periods.

	Σ (residual) ^{2}	Σ (Obs. Change) ^{2}	Ratio
Persistent upwelling	0.071	0.094	0.76
Intermittent upwelling	0.403	1.35	0.30
Downwelling	1.15	1.25	0.92

breakdown of total advective-diffusive flux into its constituents reveals that radial advection is the dominant term in each case (at the surface radial advection is equivalent to cross-shelf advection; at the bottom it represents advection parallel to the bottom topography, which includes both cross-shelf and vertical components). Eddy diffusion (cross-shelf plus vertical, i.e., radial plus azimuthal) is the next largest term. Alongshore advection is significant in the intermittent upwelling case but not in the intermittent downwelling case. Alongshore advection was not measured for the persistent upwelling case.

Specifically, radial advection supplies as much as a 0.2 ppt increase in near-surface salinity both nearshore and offshore in the persistent upwelling case (δt = 48 h). At depth nearshore, eddy diffusion depletes much of the advective increase, resulting in a negligible total advective-diffusive change at this location. In a similar fashion, radial advection accounts for a more than 7 ppt increase nearshore during the intermittent upwelling period, whereas eddy diffusion acts as a sink for as much as 4 ppt at that location. The net result is a total advective-diffusive increase of no more than ~2 ppt nearshore. Alongshore advection depletes salinity by as much as 1 ppt in the upper layers at a distance of 10-km offshore in this case and azimuthal advection makes a small contribution to the very nearshore increase. For the intermittent downwelling case, radial advection and diffusion do not compensate one another; instead both act to decrease salinity in the region within 5 km of the coastline and to generally increase salinity farther seaward. Alongshore and azimuthal advection are negligible (everywhere less than ± 0.5 ppt) for this period.

Chlorophyll a

Maps of the total predicted advective-diffusive change for chlorophyll *a* during the three representative periods, as well as isopleth maps of individual contributions to the total change, are given in Figure 6.15 (note contour spacing is irregular for clarity, since chlorophyll *a* tends to be log-normally distributed). The maps of total change indicate that much of the variance in the observations can be reproduced by the physical model. For the persistent upwelling period, a loss of > 3 $\mu g\ l^{-1}$ of chlorophyll *a* in the surface layers in the region shoreward of 15 km can be partially attributed to the advective-diffusive loss of > 1 $\mu g\ l^{-1}$ at that location. Beyond 40-km offshore, the observed increase of ~3 $\mu g\ l^{-1}$ chlorophyll *a* is overpredicted by the total advective-diffusive increase of > 3 $\mu g\ l^{-1}$. At mid-shelf, observed change (> 20 $\mu g\ l^{-1}$) far exceeds model-predicted change (~1 $\mu g\ l^{-1}$). For the intermittent upwelling period, observed and model-predicted changes are much more comparable; for example, observed loss of chlorophyll

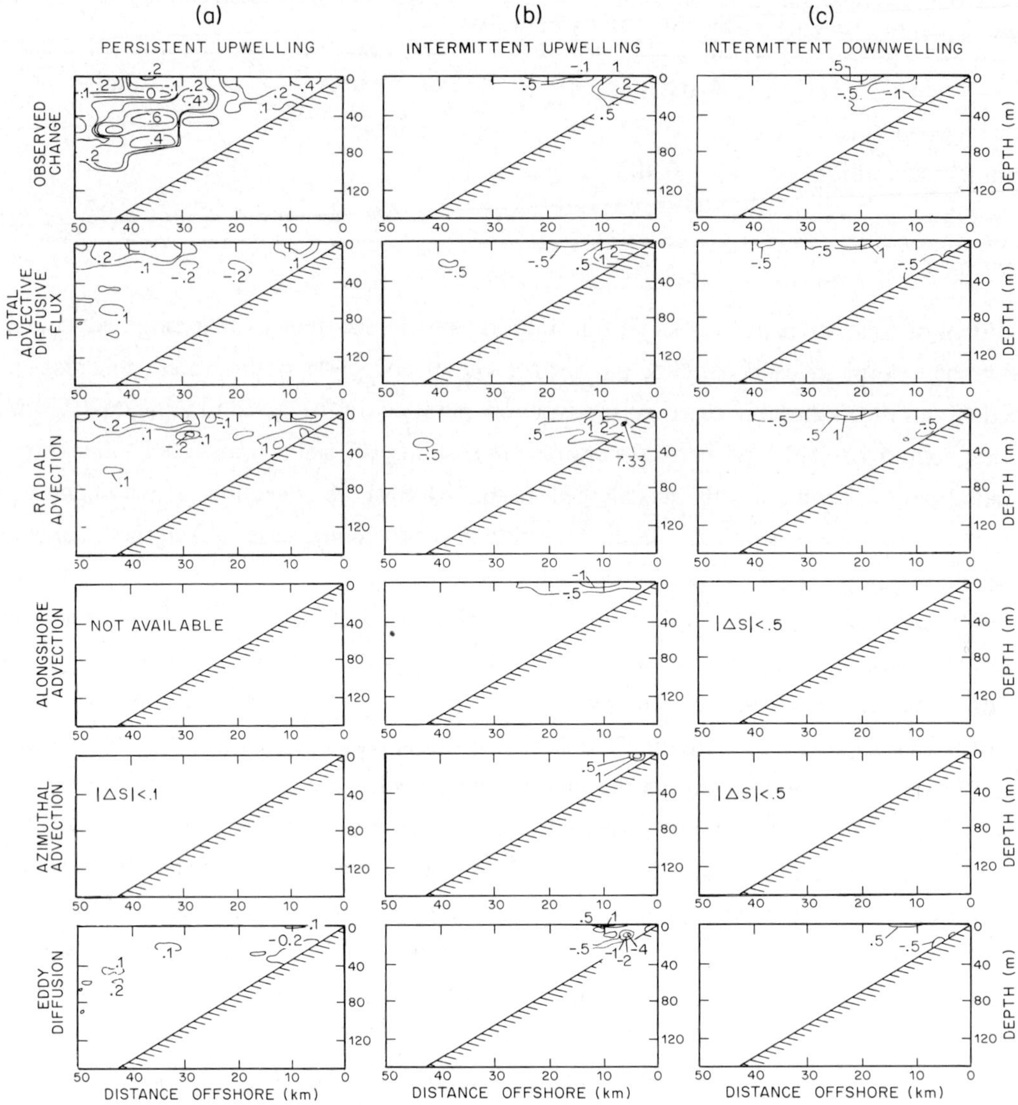

Fig. 6.14. Predicted advective-diffusive change and observed change in salinity, for (a) 'persistent upwelling', (b) 'intermittent upwelling', and (c) 'intermittent downwelling' events. Shown are (from top to bottom): observed change, and changes due to total advection-diffusion, radial advection, alongshore advection, azimuthal advection, and eddy diffusion.

a nearshore is of the same order of magnitude (2-4 $\mu g\, l^{-1}$) as model-predicted change. Also, over the outer shelf at ~10-m depth, observed and predicted loss are both ~3 $\mu g\, l^{-1}$. The intermittent downwelling period also exhibits a reasonably good match between regions of observed and predicted change. Both observation and prediction show near-shore increases of > 3 $\mu g\, l^{-1}$, while over the outer shelf and at mid-shelf both exhibit losses of 1-2 $\mu g\, l^{-1}$ chlorophyll *a*.

As with salt, radial advection accounts for the largest portion (as much as 100%) of the advective-diffusive change in each period for all locations. However, eddy diffusion contributes very significantly where chlorophyll concentrations are high (see Figs. 6.7a,b), exceeding

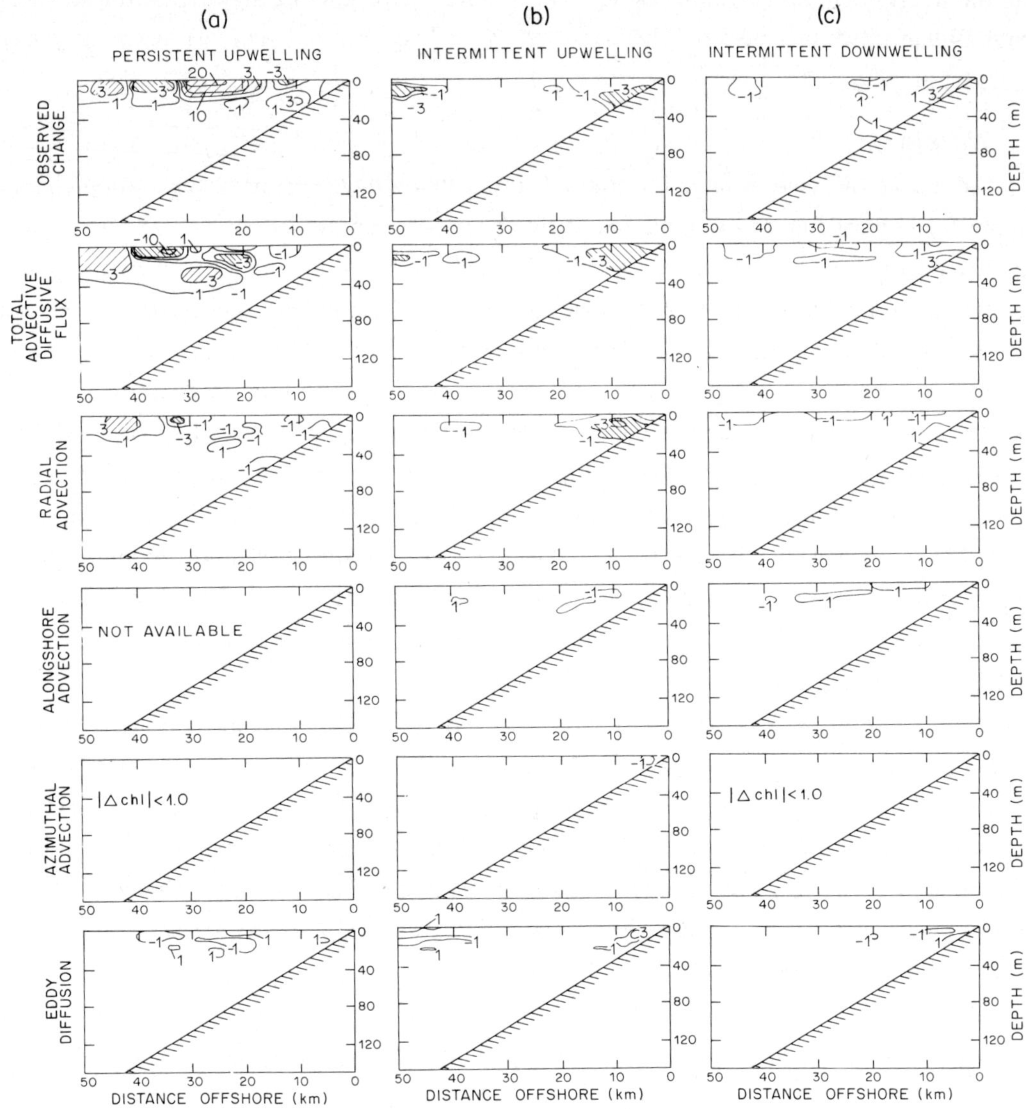

Fig. 6.15. Predicted advective-diffusive change in chlorophyll (μg l $^{-1}$), compared with observed change in the concentration field, for (a) 'persistent upwelling', (b) 'intermittent upwelling', and (c) 'intermittent downwelling' events. Shown are (from top to bottom): observed change, and changes due to total advection-diffusion, radial advection, alongshore advection, azimuthal advection, and eddy diffusion.

advective change near the shelf break during the intermittent upwelling case (50-km offshore). For each period, most diffusive change is negative; during the persistent upwelling period the diffusive model predicts a loss of between 1 and 3 μg l $^{-1}$ at mid-shelf. However, during the intermittent upwelling period a net increase in chlorophyll *a* of > 3 μg l $^{-1}$ via diffusive flux is predicted below 5 m in the region shoreward of 10 km. Some compensation of advective and diffusive flux is observed in near-shore areas, as occurred also for salinity. Alongshore advection makes significant contributions (on the order of 1 μg l $^{-1}$) to near-shore, mid-depth losses in chlorophyll *a* during intermittent upwelling and to gains in a broad near-surface area during the intermittent downwelling period. Azimuthal advection is insignificant for all except the intermittent upwelling period, where it serves as a loss term (~1 μg l $^{-1}$) in the very near-shore (< 5-km) zone.

Nitrate

Advective-diffusive change for nitrate is compared with observed concentration changes in Figure 6.16. Large areas of significant differences between predicted and observed change occur, especially in the offshore region during the intermittent downwelling period. Nevertheless, a qualitative match is achieved for the intermittent upwelling period both nearshore and offshore and for the very near-shore area (< 5 km) during the intermittent downwelling and persistent upwelling periods. For persistent upwelling, the observed increase exceeds modeled increase by as much as 5 μM nearshore. Also, an observed decrease of 5 μM over the outer shelf is not predicted by the model. For the intermittent upwelling period, observed decrease is nearly matched by model-predicted decrease in the offshore zone (both ~10-20 μM), while observed increase (~10 μM) is considerably smaller than the model-predicted increase (as much as 20 μM) nearshore of 20 km. In the downwelling period, both model and observations exhibit a decrease of 10 μM in the near-shore, near-bottom region. A decrease of ~5 μM near the surface at ~15-km offshore is not predicted by the model.

As for the other variables, radial advection makes the largest contributions to the total advective-diffusive change; in the persistent upwelling case, it is the only significant contributor (all others < 5 μM). During the intermittent upwelling period, a partial compensation between radial advective increase (as much as 40 μM) and diffusive decrease (as much as 20 μM) occurs nearshore, as was the case for salinity. Alongshore advection contributes as much as 10 μM over the outer shelf and ~5 μM in a broad region at depth, while azimuthal advection is only significant (~5 μM) shoreward of 5 km. Alongshore advection is a significant contributor to the predicted increase during the downwelling event on the outer shelf (though observed change was far greater). Both radial advection and eddy diffusion effect a loss in the near-shore area.

Ammonium

As with nitrate, model-predicted ammonium and observed changes exhibit significant differences in many areas (Fig. 6.17). Observed change during the persistent upwelling period is generally greater than the predicted change. Although both exhibit an increase of 1-2 μM in the near-shore zone, an observed increase of > 2 μM at depth ~20-km offshore is not predicted by the model. For intermittent upwelling, both observed and predicted maps exhibit increases of ~1 μM near the surface at ~12-km offshore and decreases of 1-2 μM at depth. The predicted increase of 2-4 μM at 20-m depth offshore is greater than the observed decrease (~1 μM) at that location. Other increases and decreases in the observed map for intermittent upwelling are not reproduced by the model. Observed near-shore decreases of 1-2 μM at 20-m depth in the downwelling period are generally predicted by the model, while an observed increase of > 2 μM at 40 m is not. Over the outer shelf, observed and predicted patterns do not generally match. Radial advection is the only significant contributor to modeled change for the persistent

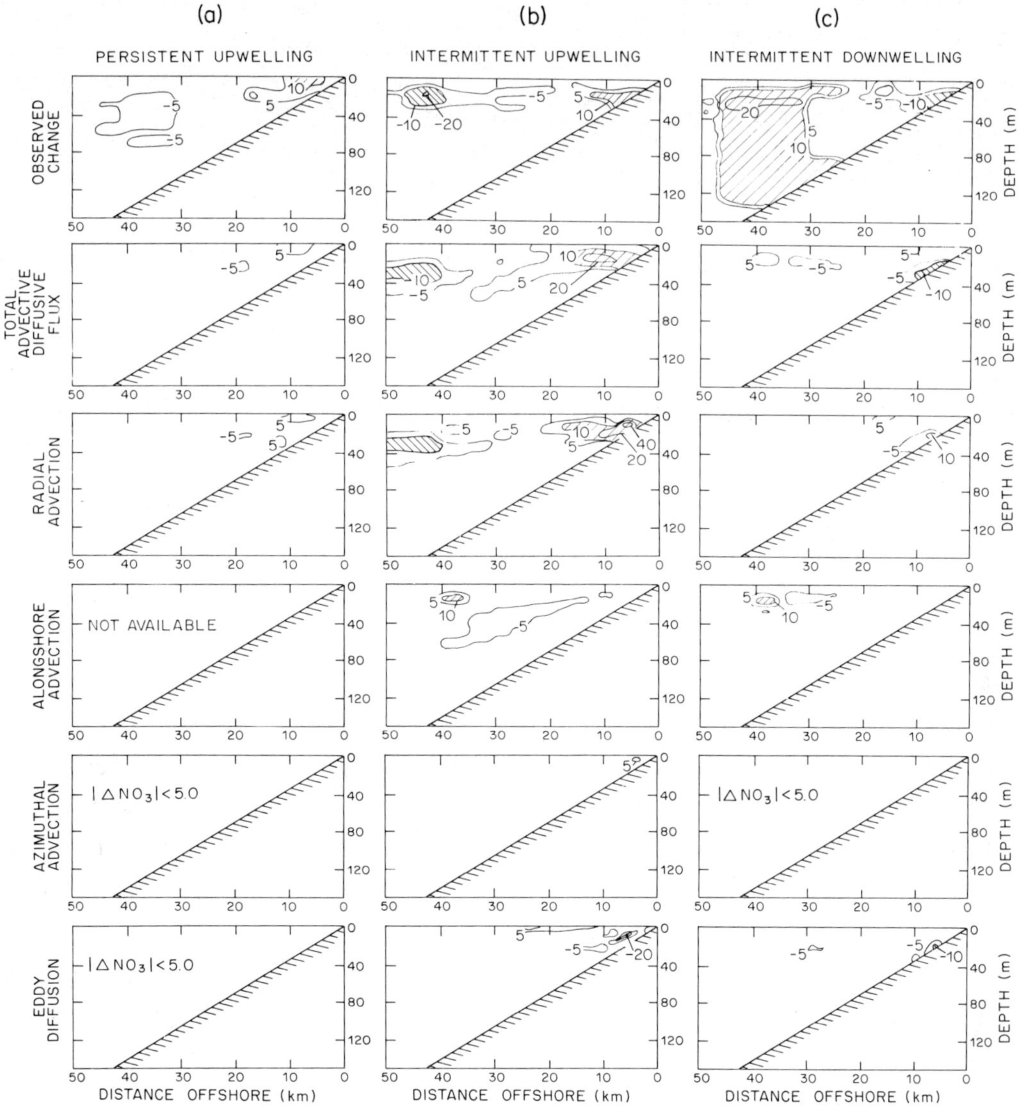

Fig. 6.16. Predicted advective-diffusive change in nitrate (μM), compared with observed change in the concentration field, for (a) 'persistent upwelling', (b) 'intermittent upwelling', and (c) 'intermittent downwelling' events. Shown are (from top to bottom): observed change, and changes due to total advection-diffusion, radial advection, alongshore advection, azimuthal advection, and eddy diffusion.

upwelling case. However, both radial and alongshore advection, as well as eddy diffusion, are strong contributors to modeled change during intermittent upwelling and downwelling periods. Indeed, alongshore advection effects a larger share of the total predicted change for ammonium than was the case for any other modeled variable. For the downwelling case in particular, alongshore advection accounts for nearly all of the predicted loss nearshore at mid-depth, shoreward of 20 km. The pattern for eddy diffusive change is irregular and acts as both source and sink (as much as 2 μM) for ammonium. Azimuthal advection is negligible (i.e., < 1 μM) for all time periods.

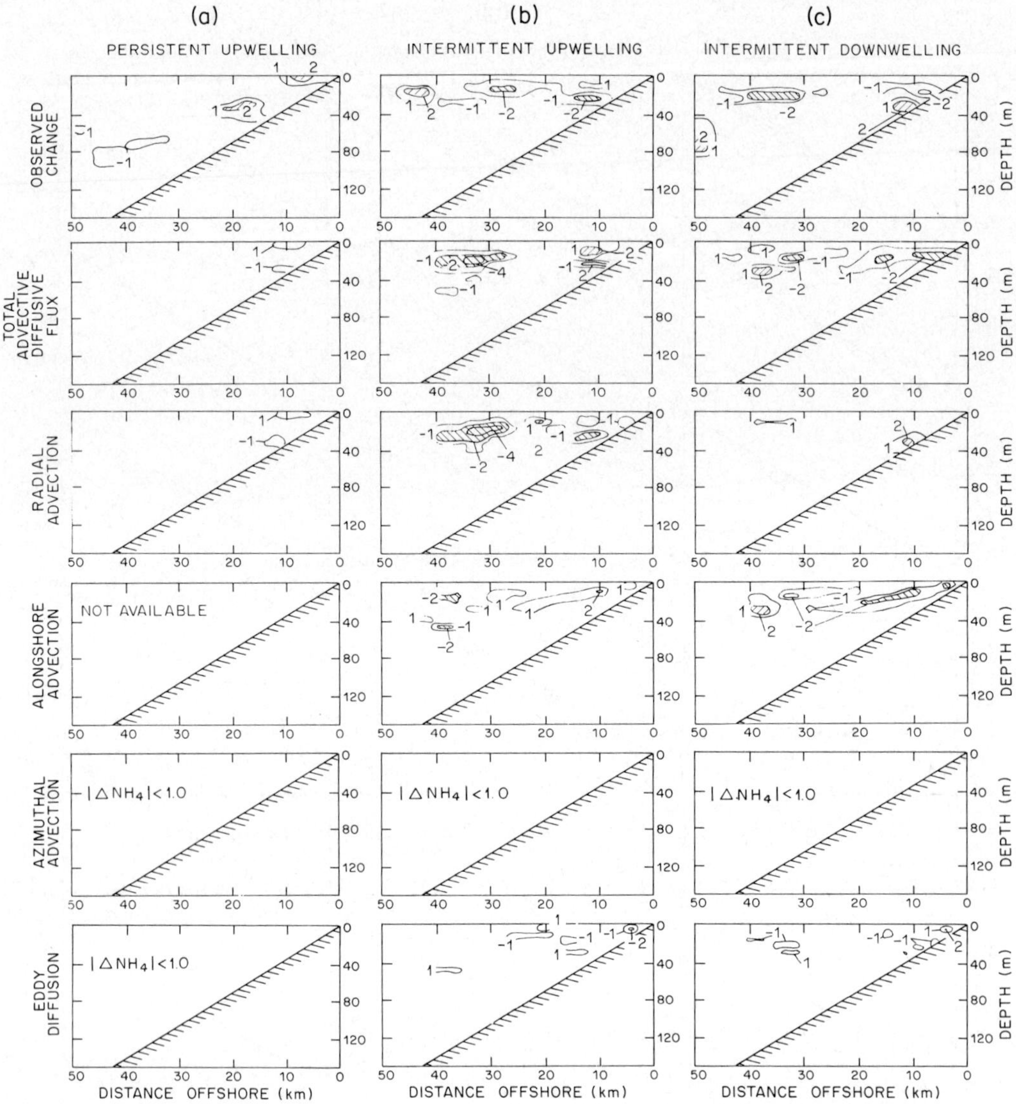

Fig. 6.17. Predicted advective-diffusive change in ammonium (μM), compared with observed change in the concentration field, for (a) 'persistent upwelling', (b) 'intermittent upwelling', and (c) 'intermittent downwelling' events. Shown are (from top to bottom): observed change, and changes due to total advection-diffusion, radial advection, alongshore advection, azimuthal advection, and eddy diffusion.

Clearly, many areas exist where chlorophyll and nutrients exhibit changes which cannot be attributed to the advective-diffusive transport as modeled in this study. While a portion of the difference can be attributed to the inadequacy of any advective-diffusive model (especially one which cannot entirely resolve alongshore advection), much of the difference will be due to nonconservative changes resulting from biologically mediated processes. Hence, as described in the Methods section, we provide a qualitative description (e.g., spatial patterns) of nonconservative behavior in the next section by comparing predicted with observed change for the nonconservative variables.

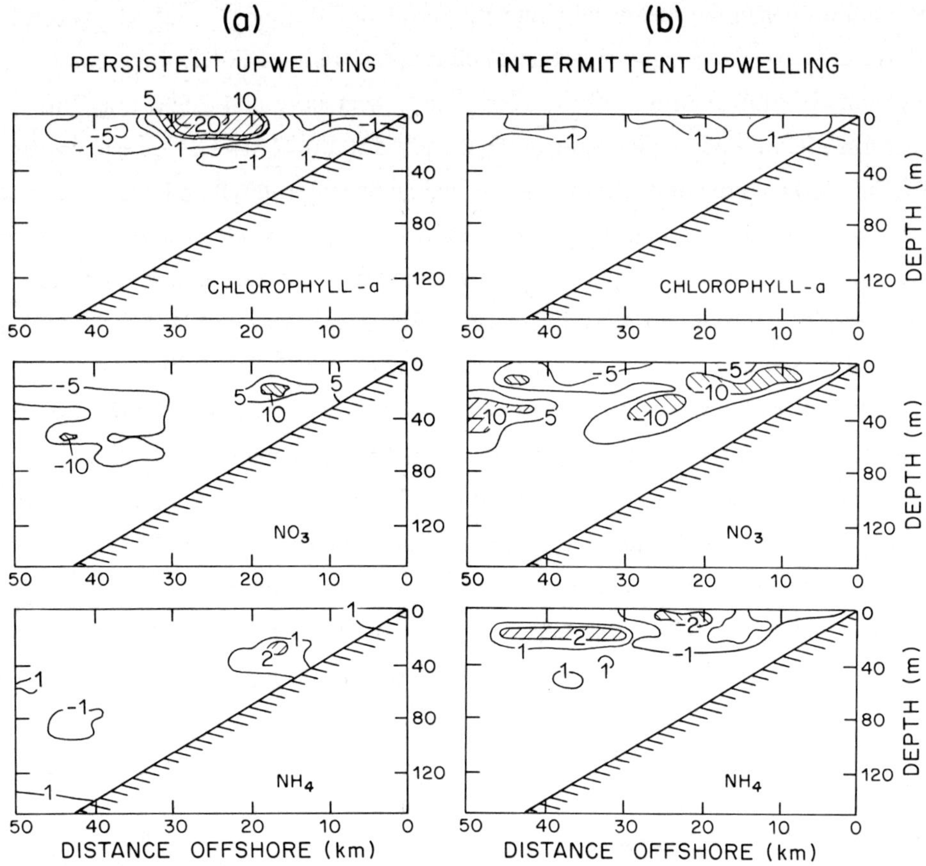

Fig. 6.18 Residual maps (observed minus predicted concentrations) for chlorophyll ($\mu g\ l^{-1}$), nitrate (μM), and ammonium (μM) after (a) 'persistent upwelling' and (b) 'intermittent upwelling'.

6.5.2 <u>Nonconservative changes</u>

Maps of inferred nonconservative change during the persistent and intermittent upwelling periods were obtained by subtracting the predicted concentration field from the observed field at the end of each period (note that this is identical to subtracting the predicted advective-diffusive change from the observed change in each of Figs. 6.15, 6.16 and 6.17). As with salinity, the difference is referred to as the "residual" for each variable (Fig. 6.18).

During the persistent upwelling period, extremely high values of new chlorophyll (> 20 $\mu g\ l^{-1}$) are observed at the mid-shelf location. A nonconservative decrease of 1-5 $\mu g\ l^{-1}$ is revealed seaward of that region, with slight decreases beneath and shoreward of the region. A small increase in chlorophyll is also observed just above the bottom at ~10-km offshore. The residual map for nitrate bears little resemblance to the map for chlorophyll during this period. Nitrate shows increases (5-10 μM) especially at ~17-km offshore and decreases (5-10 μM) near the shelf break at depth. The residuals for ammonium are largest (> 2 μM) in the lower half of the water column at 15-20-km offshore, with small scattered areas of slight increases and decreases both shoreward and seaward of that core. The calculated nonconservative change in chlorophyll is lower than the measured value for production during this period. When both are expressed as mg C m^{-2} d^{-1} using a carbon-to-chlorophyll ratio of 50: 1, the predicted nonconservative change is equivalent to 250 mg C m^{-2} d^{-1} nearshore and 2,500 mg C m^{-2} d^{-1} at mid-shelf, while ^{14}C measurements yielded 2,200 and 4,150 mg C m^{-2} d^{-1}, respectively.

The inferred patterns of nonconservative change for chlorophyll and nutrients are considerably different for the intermittent upwelling period. Apparent production of new chlorophyll with a near-shore maximum (1-3 μg l^{-1}) is accompanied by a decrease in both nitrate (5-10 μM) and ammonium (1-2 μM). An apparent decrease in chlorophyll near the shelf break (1-3 μg l^{-1}) is accompanied by some decrease in nitrate (~5 μM) but especially by an increase in ammonium (1-3 μM). Some nitrate changes appear to be unrelated to those of chlorophyll; in particular, a depletion at depth 25-km offshore and an increase at depth near the shelf break. A comparison of nonconservative estimates for chlorophyll with the ^{14}C measured in-situ net production again shows the net nonconservative rate of change in chlorophyll to be consistently lower than the measured net production rate. The predicted nonconservative change in chlorophyll for the photic zone is equivalent to 330 mg C m^{-2} d^{-1} nearshore and 85 mg C m^{-2} d^{-1} at mid-shelf. The ^{14}C net production measurements indicate 3,000 and 1,600 mg C m^{-2} d^{-1} at those locations.

6.6 DISCUSSION

6.6.1 Net Production and Grazing

The nonconservative residual calculated for chlorophyll and nutrients in this study includes errors resulting from scales of motion and processes such as alongshore advection and diffusion that are not resolved adequately by our technique. Nonetheless, the residual is a useful statistic for examining the gross features of the ecosystem; e.g., spatial patterns of nonconservative change.

The apparent nonconservative behavior for chlorophyll or nutrients can be viewed as a holistic property of the upwelling ecosystem. It represents the sum of all processes, productive and consumptive, which influence the concentration of a property within a water parcel. Minas *et al.* (1986) also investigated such time-integrated nonconservative behavior for several upwelling ecosystems. Whereas our technique attempts to specifically account for advective-diffusive and nonconservative change, their method utilized an idealization of one-celled upwelling, together with the mixing line technique (Broenkow, 1965), to infer nonconservative change. The nonconservative behavior of chlorophyll so obtained in their study was designated 'net community production', a term we shall employ here.

In this study, large differences were found between the net production rates measured by ^{14}C and the net community growth rates inferred from nonconservative changes in chlorophyll. During the persistent upwelling period in June 1982, for instance, the difference between net production (4,150 mg C m^{-2} d^{-1}) and net growth (2,500 mg C m^{-2} d^{-1}) was 1,650 mg C m $^{-2}$ d $^{-1}$ at the mid-shelf. During the period of intermittent upwelling in August 1981, the disparity was 1,515 mg C m $^{-2}$ d $^{-1}$. Differences on this order are in close agreement with the calculated levels of zooplankton grazing in the mid-shelf area during theses two periods (Landry and Lorenzen, 1989, Chapter 5). In contrast, the large differences between net production and net growth for the near-shore region are not consistent with zooplankton grazing. The seaward edge of the chlorophyll plume in upwelling systems is expected to be a region of high grazing (e.g., off Peru, Boyd and Smith, 1983). Substantial grazing activity on the outer shelf of Washington is indicated by negative net community growth in this area during both upwelling experiments. However, since zooplankton sampling did not extend to the outer shelf on these cruises, we have no independent assessment of zooplankton grazing for this region.

Although the periods studied are not directly comparable, the overall patterns of ^{14}C measured production in this study were consistent with the ^{14}C measurements of Small and Menzies (1981) for a location south of the Columbia River plume. They found a single core of high primary production under strong, 'steady-state' upwelling in early summer. For weaker 'steady-state' upwelling, they found largest primary production and biomass further inshore, consistent with our results for the intermittent upwelling case. Hence our intermittent upwelling case resembles their weak upwelling example more closely than their strong upwelling example. Net community production estimates cannot be obtained from their data, since maps of chlorophyll *a* were not presented for successive days.

6.6.2 Nutrient Sinks and Sources

One striking inference from our analysis is that a large nonconservative increase in chlorophyll occurs at mid-shelf during persistent upwelling without an accompanying depletion of nitrate or ammonium. Measurements of the internal nitrogen (i.e., nitrogen within particulate matter) to chlorophyll ratio suggest that during this period the phytoplankton stocks were primarily utilizing internally-stored nutrients at the mid-shelf location, rather than drawing nutrients from the water column (Dortch and Postel, submitted). Total internal nitrogen was approximated using internal organic nitrogen and amino acids plus particulate protein concentrations (expressed as μM-N). The internal nitrogen to chlorophyll ratios obtained at 1-m depth along a cross-shelf transect on June 11 (4 d prior to the persistent upwelling event) exhibit a steady decline in the offshore direction out to the region of maximal chlorophyll, from 3.19 μM-N: μg chlorophyll within 5 km of the shoreline to 0.49 at 30-km offshore (Table 6.2). A similar, but shorter, transect at 10 m on June 13 exhibited a similar trend (Dortch, unpublished data). For a persistent upwelling circulation pattern, where surface waters are ultimately advected offshore, this decline suggests production of new chlorophyll without a large assimilation of external nitrogen. Indeed, the concentration of internal nitrogen itself increases only by ~3 μM between the near-shore and mid-shelf (maximal chlorophyll) locations. The water parcel trajectory maps for the persistent upwelling period indicate an offshore travel of only 4 km over that two-day period. Values in Table 6.2 for the mid-shelf area suggest that

TABLE 6.2 Internal nitrogen to chlorophyll ratio measured at 1-m depth in a cross-shelf transect for the study area on June 11, 1982 (Dortch and Postel, submitted), and at 10-m depth on June 13, 1982 (in parentheses) (Dortch, unpublished data).

Distance Offshore (km)	Chlorophyll *a* ($\mu g\ l^{-1}$)	Internal nitrogen:chlorophyll *a* (μM-N:μg-chl)	Internal nitrogen (μM-N)
3.2	0.95	3.19	3.03
6.9	1.2 (1.45)	2.73 (1.85)	3.28 (2.68)
12.5	5.12 (2.68)	0.80 (1.26)	4.08 (3.38)
20.5	4.9 (8.12)	0.77 (0.44)	3.76 (3.58)
35.4	12.53	0.49	6.19
49.9	0.39	5.95	2.32
60.9	0.17	4.54	0.77

phytoplankton within a water parcel traveling from 28 to 32-km offshore would have only assimilated ~0.8 μM dissolved nitrogen, the remainder of nitrogen requirements being supplied from preexisting internal supplies. Hence negligible loss of dissolved nitrogen would be expected at the mid-shelf location, as shown in the residual maps of Figure 6.18.

A nonconservative increase in ammonium (> 1 μM) was predicted during persistent upwelling for the area within 10 km of the coast, where zooplankton density (and, presumably, grazing activity) was highest (compare Figs. 6.8 and 6.18). The ammonium released through grazing (~0.2-0.5 $\mu M\ d^{-1}$ for the observed density of ~300 mg d.w. m^{-3}; e.g. Jawed, 1973, Miller and Landry, 1984) could serve as a nutrient source for newly upwelled phytoplankton at this location. Tight coupling of this release and that from protozoans, with assimilation by phytoplankton in the near-shore region ('recycled' production; Dugdale and Goering, 1967) could account for the absence of any near-shore negative ammonium residual. Some of the assimilated nitrogen could still come from nitrate (i.e., 'new' production) for newly upwelled cells. Perhaps a portion of this requirement is met gradually during onshore flow at depth.

Barber and Smith (1981) have presented an upwelling scenario in which nitrogen in sinking particles, remineralized by zooplankters or bacteria, is returned to the near-shore zone as ammonium and urea by onshore flow at depth. This helps maintain higher levels of productivity in the upwelling ecosystem by allowing enhanced recycled production. In this study, the large (> 2 μM) positive residual for ammonium near the bottom at 10-25-km offshore during the persistent upwelling period suggests such 'nutrient-trapping'.

Residual maps for the intermittent upwelling event suggest that nutrient coupling was weaker than in the persistent upwelling event. Large regions of both nitrate and ammonium depletion, accompanied by increases in chlorophyll, were observed nearshore. Apparently both new and recycled nutrients were being utilized, although it is difficult to quantify the relative amounts of each using the residual maps because of the possibility of other sources and sinks. In addition, urea measurements are lacking. Typical values for the ratio of new to total nitrogen-based recycled production, i.e., the 'f-ratio' or 'Eppley ratio', are in the range of 0.5 to 0.8 for upwelling systems (Codispoti, 1984). At the very least, the residual nutrient maps indicate that assimilation of ammonium exceeded remineralization in the zone shoreward of 30 km, whereas remineralization (presumably by zooplankton grazing, to be consistent with the chlorophyll loss), predominated seaward. Considering the shallow depth of the layer of offshore flow during this period (see Fig. 6.6), such remineralization constitutes nutrient-trapping, as the released ammonium would be carried shoreward. The residual maps suggest that new (nitrate supplied) production occurred across the entire shelf, with a maximum within 25 km of the shore.

The predicted net increases in nitrate during both the persistent and intermittent upwelling events cannot be ascribed to nitrification, since measured nitrification rates in this same upwelling system (Ward *et al.*, 1984) suggested a maximum rate of 0.035 μM d $^{-1}$. Rather, such increases probably result from the inadequacy of our estimation technique for nitrate, especially seaward of 30 km, where concentrations exhibit significant alongshore structure and the bottom topography is more irregular (see Fig. 6.1). These factors make alongshore advection difficult to resolve, since alongshore gradients of concentrations and current velocities are strong in this region. Furthermore, any nonconservative depletion of nitrate at depth would be difficult to identify by our technique, since small changes in the advective flux of the large concentrations at depth could easily overwhelm the nonconservative signal.

6.6.3 Impact of Circulation on Flux Balances

The relative significance of conservative *vs.* nonconservative fluxes in a marine ecosystem strongly depends on the nature of the flow field. O'Brien and Wroblewski (1973) presented scaling arguments to quantify the relative importance of advection and biologically mediated change in determining the spatial configuration of plankton biomass. For the Peruvian upwelling region, they concluded that these two factors should play an equally important role. Here we have utilized an advective-diffusive model to quantify these factors more precisely, for the upwelling region off Washington. Given an absolutely steady upwelling circulation, it might be possible for the upwelling ecosystem to achieve a steady-state configuration where conservative and nonconservative fluxes balance at all locations so as to produce a stationary concentration pattern for each nonconservative variable. Obviously this was not the case for our study; considerable observed change, especially in chlorophyll, occurred between the beginning and end of even the 'persistent' upwelling period. For an approximate steady-state to be achieved, the circulation pattern would need to persist for at least as long as it takes an upwelled water parcel to be advected out to the shelf break. For a typical cross-shelf velocity of 4 cm s $^{-1}$ (~4 km d $^{-1}$), this time requirement for our system is in excess of 10 d. Wroblewski (1977), in his modeling study of the Oregon upwelling system, likewise did not find a steady-state pattern of chlorophyll and net production after 10 d of steady, upwelling-favorable winds. The salinity isopleths for June 11 and June 13 and unfiltered winds for June 13-14 (not shown) suggest that a persistent upwelling situation was not in effect prior to June 13; hence a period of only 5 d was available for physical/biological spin-up prior to the end of the persistent upwelling period (June 16-18) considered in our study. The large observed change in this period thus does not preclude the possibility of steady-state biological patterns when physical forcing is very persistent.

The intermittency of the upwelling circulation pattern prior to (and during) the August 21-24, 1981, period itself suggests that advective change will be important for any variable with spatial gradients. The effect of physical intermittency on nonconservative processes within the ecosystem is less clear. Wroblewski (1977) predicted similar spatial features for variable upwelling *vs.* steady upwelling for the daily values of primary production (increased with upwelling-favorable winds, with ~1 d time lag). In part such effects depend on the pattern of return flow, which supplies newly upwelled water to the system. For the persistent upwelling case, offshore flow extends further down in the water column and return flow at mid-shelf appears to come from a deeper layer (> 60 m) than in the intermittent upwelling period (10-30 m; see Fig. 6.6). Phytoplankton in deeper water arriving near the surface would need to adjust to the new, favorable light conditions, but would not in general be nutrient-starved. MacIsaac *et al.* (1985) showed how cells newly upwelled near the coast at a persistent upwelling center may experience a period of physiological 'shift up' to higher growth rates as they are advected offshore. Initially, specific growth rates and nutrient depletion rates are slow relative to their offshore values. Hence we could expect physical fluxes of nitrate in the near-shore region during persistent upwelling to predominate over nonconservative loss. The lack of any negative residual for nitrate in the near-shore zone tends to support this concept, although the strong gradients (and hence strong advective flux) of nitrate in this region makes any firm conclusion difficult. Conversely, for the intermittent upwelling period, offshore flow is limited to a very thin (~5 m) layer at the surface and return flow is concentrated just beneath the surface layer (Fig. 6.6). Phytoplankton in that return flow, exposed to higher light and lower nutrient levels than at depth, would be nutrient-starved when exposed to the most favorable light conditions after upwelling. The intense demand for nutrients would be reflected in a large loss of nitrate and possibly ammonium (depending on regeneration rates). Again, the residual map supports this scenario. Because of nutrient starvation, the initial growth rates of the near-surface phytoplankton would be slow. This allows the predominance of conservative fluxes in changing the concentration of chlorophyll at a fixed location: the advective-diffusive map for chlorophyll in this period can account for as much as 100% of the observed change (Fig. 6.15) of 1-3 $\mu g\ l^{-1}$ of chlorophyll.

As mentioned in the previous section, the existence of onshore flow near the surface during the intermittent upwelling event could result in a significant trapping of nutrients within the system, as newly regenerated ammonium is returned to the near-shore zone. Trapping within the observed persistent upwelling circulation regime could be expected to occur at a deeper location than in the intermittent case, since regenerating particles need to fall further before reaching the layer of return flow. The residual maps for ammonium indeed suggest shallow regeneration and onshore flux during the intermittent upwelling period, with deeper regeneration and onshore flux during the persistent upwelling period. Klein and Steele (1985), by modeling the effects of advection and diffusion on a simple pelagic ecosystem, found that advection without diffusion could spatially separate zooplankton and regenerated nutrients from phytoplankton, reducing the level of primary production below that of a model with no advection. Diffusion served to lessen this spatial segregation of the pelagic food chain and hence increase primary production above the advection-only level. In our upwelling system, nutrient trapping plays a supplemental role to such diffusive enhancement, probably raising the levels of recycled production in the intermittent upwelling case, while lowering the level of net community production as zooplankters are provided greater access to fixed carbon.

6.7 SUMMARY

We have used objectively-mapped velocities in an advective-diffusive model to determine the physical contributions to variance in chlorophyll, nitrate and ammonium for two environmental situations – persistent and intermittent upwelling. Fortuitously, winds were of comparable magnitude during the two upwelling experiments, allowing the effects of persistence to be assessed. Salt was used to test model performance. Results indicated that the model could account for as much as 70% of the variance in this conservative tracer over 2-3 d periods, despite incomplete information for flux contributions by alongshore advection.

Significant differences between persistent and intermittent upwelling regimes were observed. Strong advective-diffusive control of the chlorophyll distribution was demonstrated for the intermittent upwelling (late summer; August 1982) situation. Conversely, the advective-diffusive model accounts for little of the observed change in chlorophyll during a period of persistent (early summer; June 1981) upwelling. Relative contributions of the various advective-diffusive terms were a function of cross-shelf location for all upwelling conditions. For example, during the intermittent upwelling period, both radial (cross-shelf and vertical) advection and eddy diffusion were significant shoreward of 10 km, whereas only the diffusive term was significant seaward of 40 km. Advective-diffusive change did not generally balance nonconservative change for chlorophyll during any of the 2-3 d periods studied. Hence, on this time scale, the system was unsteady for all sets of conditions.

Gross features of the nonconservative behavior of chlorophyll and nitrogen were inferred from the residual between advective-diffusive model results and observed concentration changes. These features include: 1) greater net depletion of nitrogen (nitrate or ammonium) during the intermittent upwelling event than during the persistent upwelling event (> 5 μM over a 3-d interval *vs.* < 1 μM over a 2-d interval, respectively). The shallower return flow during the intermittent upwelling event may account for this difference, having supplied the surface offshore flow with cells which were more nutrient starved than in the persistent upwelling case. 2) Regeneration of ammonium at 20-m depth over the mid- and outer shelf during the intermittent upwelling event. 3) Greatest net depletion of ammonium at mid-shelf and greatest net depletion of nitrate on the inner shelf during the intermittent upwelling event. Shallow offshore and onshore flow appears to have enhanced the use of regenerated nutrients across the shelf, as newly regenerated ammonium at ~20-m depth on the outer shelf was carried inshore. 4) Regeneration of ammonium (> 1 μM over 2 d), partially by zooplankton, within 10 km of the shore during the persistent upwelling event.

Despite the limitations imposed by the data sets, our coupling of physical and biological measurements has led to significant results. The reasonable success of our technique in the absence of extensive alongshore information was possible because the Pacific Northwest shelf has less alongshore topographic variation than other areas of coastal upwelling. In particular, this shelf has few promontories and is wide enough to inhibit the penetration of the eastern boundary current system and associated eddies. Future experiments should be carefully designed to include the significant contributions of alongshore advection to property distributions, given the three-dimensional variation of currents, nutrients and pigments in most coastal environments (especially upwelling centers). Implementation of such experiments is likely to include pigment monitoring drogues, as well as the integration of remotely-sensed temperature and color imagery with *in situ* measurements from a three-dimensional sampling array.

ACKNOWLEDGEMENTS

A large number of people were involved in the collection of the chlorophyll, phytoplankton, zooplankton, nutrient, temperature and salinity data used in this analysis. We would especially like to thank V.L. Fagerness, J.M. Lehner-Fournier, K. Krogslund, W.K. Peterson and J.R. Postel for their efforts. The text benefited greatly from the comments of E.E. Hofmann and K. Banse. Discussions with Q. Dortch were especially helpful in our interpretation of the nitrogen data. Both field work and analysis were supported by the Department of Energy under contract DE-AT06-76EV-71025 to B. Hickey. This is a component of Contribution No. 1784 from the School of Oceanography, University of Washington.

REFERENCES

Allen, J.S. and D.W. Denbo. 1984. Statistical characteristics of the large-scale response of coastal sea level to atmospheric forcing. *J. Phys. Oceanogr.,* 14: 1079-1094.

Barber, R.T. and R.L. Smith. 1981. Coastal upwelling ecosystems. Pages 31-68 in: A.R. Longhurst (ed.), *Analysis of Marine Ecosystems*. Academic Press, N. Y.

Battisti, D.S. and B.M. Hickey. 1984. Application of remote wind-forced coastal-trapped wave theory to the Oregon and Washington coasts. *J. Phys. Oceanogr.*, 14: 887-903.

Boyd, C.M. and S.L. Smith. 1983. Plankton, upwelling and coastally trapped waves off Peru. *Deep-Sea Res.*, 30: 723-742.

Bretherton, F.P., R.E. Davis and C.B. Fandry. 1976. A technique for objective analysis and design of oceanographic experiments applied to MODE-73. *Deep-Sea Res.*, 23: 559-582.

Brink, K.H., B.H. Jones, J. Van Leer, C.N.K. Mooers, D. Stuart, M.K. Stevenson, R.C. Dugdale and G. Heburn. 1981. Physical and biological structure and variability in an upwelling center off Peru near 15°S during March 1977. Pages 473-495 in: F.A. Richards (ed.), *Coastal Upwelling Research.* American Geophysical Union, Washington, D.C.

Brink, K.H. 1983. The near-surface dynamics of coastal upwelling. *Prog. Oceanogr.*, 12: 223-257.

Broenkow, W.W. 1965. The distribution of nutrients in the Costa Rica Dome in the eastern tropical Pacific Ocean. *Limnol. Oceanogr.*, 10: 40-52.

Codispoti, L.A. 1984. Nitrogen in upwelling systems. Pages 512-564 in: Carpenter, E.J. and D.G. Capone (eds.), *Nitrogen in the Marine Environment.* Academic Press, N. Y.

Denman, K.L. and H.J. Freeland. 1985. Correlation scales, objective mapping and a statistical test of geostrophy over the continental shelf. *J. Mar. Res.*, 43: 517-539.

Dortch, F.Q. and J.R. Postel. Submitted. Biochemical indicators of phytoplankton N utilization during upwelling off the Washington coast. *Limnol. Oceanogr.*

Dugdale, R.C. and J.J. Goering. 1967. Uptake of new and regenerated forms of nitrogen in primary productivity. *Limnol. Oceanogr.*, 12: 196-206.

Hickey, B. M. 1989. Patterns and process of circulation over the shelf and slope. Pages 41-116 in: M.L. Landry and B.M. Hickey (eds.), *Coastal Oceanography of Washington and Oregon.* Elsevier, Amsterdam, Neth..

Holloway, G. 1984. Effects of velocity fluctuations on vertical distributions of phytoplankton. *J. Mar. Res.*, 42: 559-571.

Huyer, A. 1983. Coastal upwelling in the California Current system. *Prog. Oceanogr.*, 12: 259-284.

Jamart, B.M., D.F. Winter, K. Banse, G.C. Anderson and R.K. Lam. 1977. A theoretical study of phytoplankton growth and nutrient distribution in the Pacific Ocean off the northwestern U.S. Coast. *Deep-Sea Res.*, 24: 753-773.

Jawed, M. 1973. Ammonia excretion by zooplankton and its significance to primary production during summer. *J. Mar. Biol.*, 23: 115-120.

Klein, P. and J.H. Steele. 1985. Some physical factors affecting ecosystems. *J. Mar. Res.*, 43: 337-350.

Landry, M.L. and C.J. Lorenzen. 1989. Utilization and Transformation of primary production by zooplankton. Pages 175-210 in: M.L. Landry and B.M. Hickey (eds.), *Coastal Oceanography of Washington and Oregon.* Elsevier, Amsterdam, Neth.

Landry, M.L., J.R. Postel, W.K. Peterson and J. Newman. 1989. Broad-scale patterns in the distribution of hydrographic variables. Pages 1-40 in: M.L. Landry and B.M. Hickey (eds.), *Coastal Oceanography of Washington and Oregon.* Elsevier, Amsterdam, Neth.

MacIsaac, J.J., R.C. Dugdale, R.T. Barber, D. Blasco and T.T. Packard. 1985. Primary production cycle in an upwelling center. *Deep-Sea Res.*, 32: 503-529.

Miller, C.A. and M.R. Landry. 1984. Ingestion-independent rates of ammonium excretion by the copepod *Calanus pacificus*. *J. Mar. Biol.*, 78: 265-270.

Minas, H.J., M. Minas and T.T. Packard. 1986. Productivity in upwelling areas deduced from hydrographic and chemical fields. *Limnol. Oceanogr.*, 31: 1182-1206.

Mitchum, G.T. and A.J. Clarke. 1986. The frictional nearshore response to forcing by synoptic scale winds. *J. Phys. Oceanogr.*, 16: 934-946.

O'Brien, J.J. 1970. Alternative solutions to the classical vertical velocity problem. *J. Appl. Meteorol.*, 9: 197-203.

O'Brien, J.J. and J.S. Wroblewski. 1973. On advection in phytoplankton models. *J. Theor. Biol.*, 38: 197-202.

Peterson, W.T., C.B. Miller and A. Hutchinson. 1979. Zonation and maintenance of copepod populations in the Oregon upwelling zone. *Deep-Sea Res.*, 26A: 467-494.

Platt, T., K.L. Denman and A.D. Jassby. 1975. Modeling the productivity of phytoplankton. Pages 807-856 in: Goldberg, E.D. (ed.), *The Sea.* Vol. 6 (Modeling). John Wiley, N.Y.

Small, L.F. and D.W. Menzies. 1981. Patterns of primary productivity and biomass in a coastal upwelling region. *Deep-Sea Res.*, 28: 123-149.

Smith, W.O., G.W. Heburn, R.T. Barber and J.J. O'Brien. 1983. Regulation of phytoplankton communities by physical processes in upwelling ecosystems. *J. Mar. Res.*, 41: 539-556.

Ward, B.B., M.C. Talbot and M.J. Perry. 1984. Contributions of phytoplankton and nitrifying bacteria to ammonium and nitrite dynamics in coastal waters. *Cont. Shelf Res.*, 3: 383-398.

Wilkerson, F.P. and R.C. Dugdale. 1987. The use of large shipboard barrels and drifters to study the effects of coastal upwelling on phytoplankton dynamics. *Limnol. Oceanogr.*, 32: 368-382.

Wroblewski, J.S. 1980. A simulation of the distribution of *Acartia calusi* during Oregon upwelling, August 1973. *J. Plank. Res.*, 2: 43-68.

Wroblewski, J.S. 1977. A model of phytoplankton plume formation during variable Oregon upwelling. *J. Mar. Res.*, 35: 357-394.

Young, W.R., P.B. Rhines and C.J.R. Garrett. 1982. Shear-flow dispersion, internal waves and horizontal mixing in the ocean. *J. Phys. Oceanogr.*, 12: 515-527.

CHAPTER 7

SEASONAL DYNAMICS OF SUSPENDED PARTICULATE MATTER

L.F. Small, H. Pak, D.M. Nelson, and C.S. Weimer
College of Oceanography, Oregon State University, Corvallis, OR 97331

7.1 INTRODUCTION

As in other coastal areas, the distributions and characteristic properties of suspended particulate matter (SPM) off the northwest coast of the United States are subject to great variability in time and space. Particles are introduced into the coastal ocean by the wind and by rivers and streams, and they are generated *in situ* by biological production. Once in the coastal water column they are transported by currents and gravitational settling, they are resuspended by wave action and near-bottom turbulence, they are consumed by organisms, and they change their characteristic properties by dissolution, fragmentation, and aggregation.

Identification of continuity and of repeatable patterns of SPM in Oregon-Washington coastal waters, and thus understanding the seasonal dynamics of SPM, is made particularly difficult by seasonal changes in the regional wind patterns. These changes in turn bring about seasonal changes in the position and intensity of the Columbia River discharge off the coast (Barnes *et al*, 1972; Hickey, 1989, Chapter 2)), and in the presence and intensity of coastal upwelling (Bakun, 1975; Huyer, 1977, 1983; Hermann *et al.*, 1989, Chapter 6; Landry *et al.*, 1989, Chapter 1).

In winter (generally mid-October through mid-March) the prevailing winds are from the south and southwest about 70% of the time (Barnes *et al.*, 1972; Pittock *et al.*, 1982) and storms are frequent. The general consequence of this winter wind pattern is a mixed surface layer of 25 to 50 m over the Oregon outer shelf (Huyer, 1977) and 20 to 40 m over the Washington outer shelf (Hickey, 1989, Chapter 2). Near-surface current flow is generally to the north and downwelling is often evident at the coast. The effluent of the Columbia River is directed northward immediately adjacent to the Washington coast, and there is a mean northward flow of bottom water over most of the shelf. The northward movement of bottom water along the Washington coast apparently persists throughout much of the year (Morse *et al.*, 1968; Hickey, 1979; Hickey, 1989, Chapter 2).

In summer (generally June through mid-September), prevailing winds are from the north and northwest about 80% of the time (Barnes *et al.*, 1972; Huyer, 1977; Pittock *et al.*, 1982). The major outflow from the Columbia River is in June, and the large tongue of relatively warm, fresh water is directed southward at the surface off the Oregon coast; thus, it is not usually present off Washington in the summer (Barnes *et al.*, 1972; Landry *et al.*, 1989, Chapter 1). At the same time a northward undercurrent develops, so that maximum shear in the water column is eventually attained at 20 to 30-m depths. Upwelling of relatively cold, saline water from > 60-m depths also begins immediately adjacent to the Oregon and Washington coasts in response to the prevailing northerly winds. Such complicated hydrography leads to complex and changeable frontal patterns and vertical distributions of properties.

Transition periods occur in spring and fall, when wind and current patterns shift from winter to summer conditions, or summer to winter conditions, respectively. It also should be noted that the general conditions above are subject to unseasonable events such as severe storms or periods of southerly winds in summer, or periods of calm, sunny weather in winter. The coastal water column responds quickly to these forcings (Small and Menzies, 1981; Hickey, 1989, Chapter 2). Finally, global events such as El Niño affect Oregon-Washington shelf waters principally by skewing regional climate and coastal hydrographic measurements away from long-term means (Landry *et al.*, 1989, Chapter 1; Hickey, 1989, Chapter 2). It is in the above complex hydrographic setting that we have been attempting to elucidate the seasonal dynamics of suspended particulate matter.

7.2 METHODS AND DATA BASE

We have drawn heavily on data from 15 short-term cruises carried out over the central Oregon continental shelf approximately each month from June 1980 to October 1981. On each cruise we occupied eight stations along a track normal to the local trend of the isobaths, with the seaward station located approximately 28 km off the coast at the shelf break and a distance between stations of about 3.7 km. At all stations we obtained CTD and beam attenuation profiles from the surface to within about 2 m of the bottom. At the 24-km station we also took water samples at nine depths, surface to near-bottom, for dissolved nutrients and for size and compositional analysis of the SPM (Table 7.1). In addition to the monthly short-duration cruises, we have relied heavily on selected long cruises in November 1977 off central Oregon, and in July 1979 and September-October 1980 off south and central Washington. These cruises allowed us to examine the connections between the SPM fields over the continental shelf and slope in winter and summer, and to do time-course studies on board ship. To a lesser extent we examined data from long cruises over the shelf and slope off Washington in October 1978 and August-September 1981, and utilized historical cruise data back to 1960 for trends and anomalies in certain hydrographic, optical, and chlorophyll distributions. It should be emphasized here that details and rates of change of distributions cannot be estimated from these data sets.

7.3 SEASONAL CHANGES OVER THE CONTINENTAL SHELF

7.3.1 Total Suspended Particulate Matter in Relation to Temperature and Nutrient Fields

The basic seasonal changes in water-column structure and SPM distribution can be demonstrated using onshore-offshore sections of the beam attenuation coefficient and temperature at different times (Figs. 7.1 and 7.2). The beam attenuation coefficient is defined as: $c = -(1/r)\ln(I_r/I_0)$, where r is the pathlength (m) of a light beam in a transmissometer, and I_r and I_0 are the light energy at the receiver and the source, respectively. During spring and summer a three-layered pattern of attenuation emerges over the Oregon shelf: a surface turbid layer, a mid-depth layer of relatively clear water and a bottom nepheloid layer. The clearest water registers beam attenuation coefficients less than 0.5 m^{-1}, and this mid-depth layer tends to appear quickly over the shelf in March and disappear in October or November. This clear layer appears

Table 7.1 Measurements taken on 15 monthly one-day cruises off central Oregon, from June 1980 to October 1981. Temperature, salinity and beam attenuation were measured at each of eight stations along a transect normal to the bathymetry, beginning inshore at 44° 48' N. Other measurements were from water bottle samples at nine depths at one station over 185 m total water depth (44° 48' N, 124° 30' W).

Analysis	Method	References
Temperature	CTD	---
Salinity	CTD	---
Sigma-t	CTD	---
Nutrient concentrations [NO_3^-, NO_2^-, $HPO_4^=$, Si $(OH)_4$]	Automated colorimetric analyses	Atlas *et al.* (1971)
Beam attenuation	Beam transmissometry	Bartz *et al.* (1978)
Particle volume, size distribution	Coulter counting using 2400 channel analyzer	Kitchen *et al.* (1975); Kitchen (1978)
POC, PON	Pyrolysis and gas chromatography (CHN analyzer)	----
Chlorophyll, phaeophytin	Fluorometry on extracted samples	Strickland and Parsons (1972)
Total particulate silica	Hydrofluoric acid digestion, spectrophotometry	Eggiman *et al.* (1976)
Biogenic silica	Hot alkaline digestion, spectrophotometry	Paasche (1973); Krausse *et al.* (1983)
Bacterial abundance	Epifluorescence microscopy	Hobbie *et al.* (1977)

all along the northwestern U.S. coast from early spring through fall, and the relative clarity of the layer is nearly uniform in different areas and different years. Anomalous conditions occur on smaller scales. For example, in December 1980 a small region of clear (< 0.5 m^{-1}) water appeared (Fig. 7.1). Such events likely result from wind reversals and subsequent movement of clear water towards or away from the coast over some depth interval. Such conditions appear to override the prevailing summer or winter condition but are transitory (Small and Menzies, 1981; Hickey *et al.*, 1986). The general persistence of the clear mid-layer over the shelf in summer and the disappearance of that layer in winter is most easily seen over the outer shelf (Fig. 7.3a). The layer tends to dissipate in the coastal region, which leaves the water column turbid very near-shore at most times of year (Fig. 7.1). This near-shore turbid condition is most pronounced in March and April as a result of increased phytoplankton growth in conjunction with vertical mixing which continues into the spring period (Figs. 7.1 and 7.2). Greater stratification of SPM and temperature and greater penetration of clear water into the nearshore region are apparent from May through October. The thin, turbid surface layer which develops strongly in summer

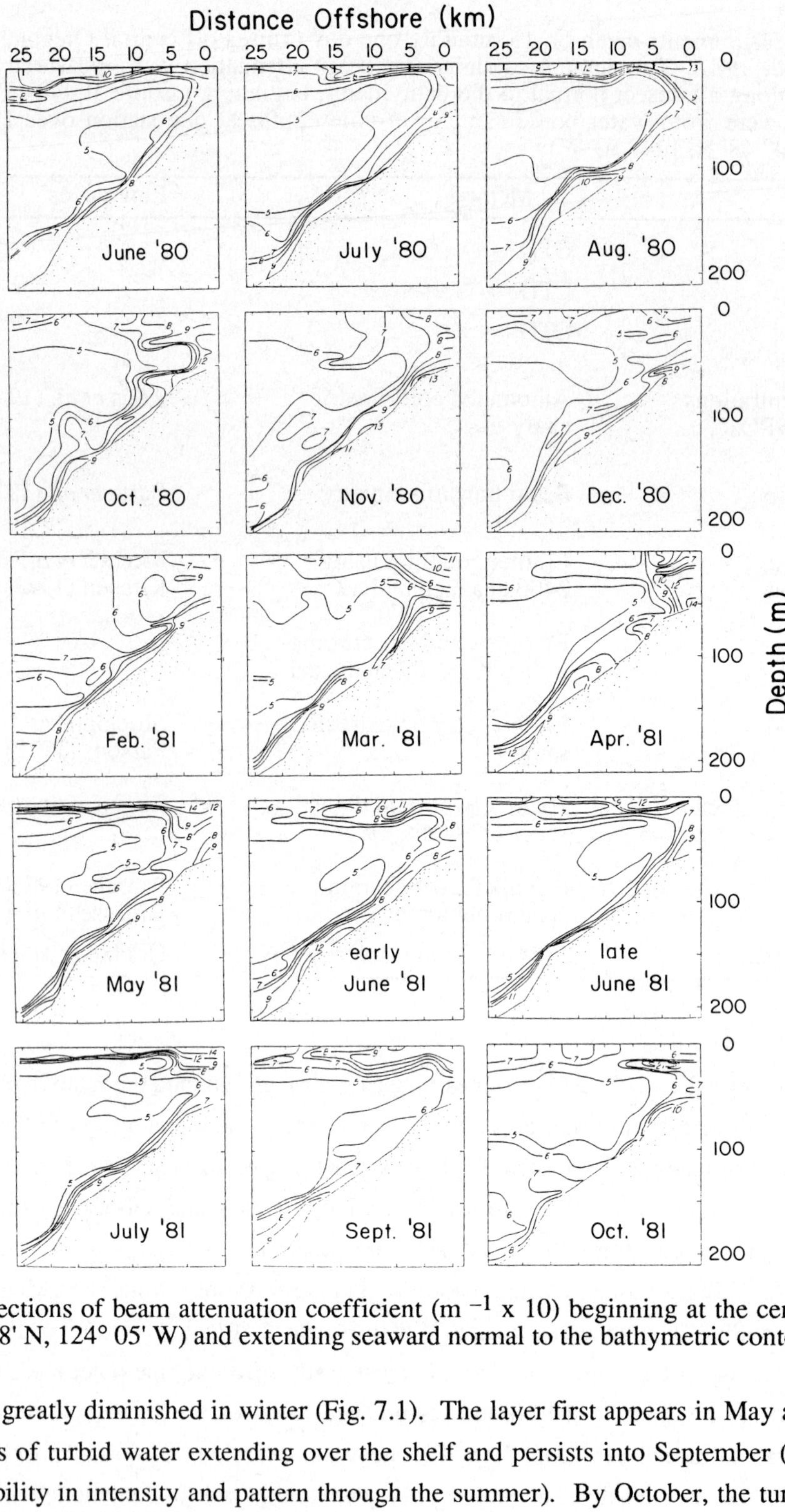

Fig. 7.1 Sections of beam attenuation coefficient (m $^{-1}$ x 10) beginning at the central Oregon coast (44° 48' N, 124° 05' W) and extending seaward normal to the bathymetric contours.

is absent or greatly diminished in winter (Fig. 7.1). The layer first appears in May as a well-developed lens of turbid water extending over the shelf and persists into September (though with much variability in intensity and pattern through the summer). By October, the turbid water in the surface layer has receded to coastal waters so that the pattern generally resembles that in March and April. From November through February, the upper water column is usually well mixed with no distinct surface boundary layer in SPM.

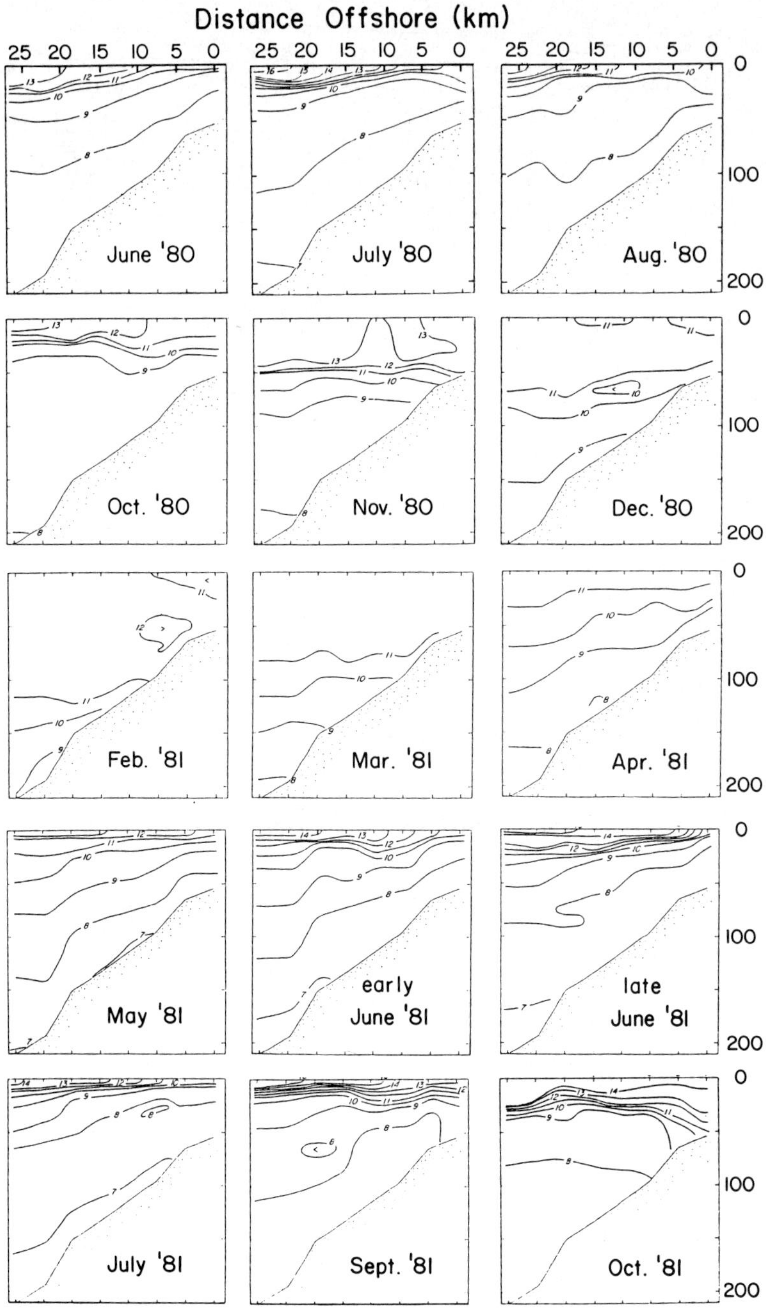

Fig. 7.2. Sections of temperature (°C) along the same transect as in Figure 7.1.

The seasonal changes in SPM in the upper water column over the shelf, including the appearance and disappearance of the surface turbid layer and mid-depth clear layer, are governed by the seasonally changing meteorological and hydrographic conditions over the shelf. In spring the increasing input of solar radiation begins to warm and stabilize the upper layer, while the prevailing northerly winds and/or the longshore sea level slope act to intensify the longshore flow in a southward direction and the surface zonal flow in an offshore direction (Hickey, 1989,

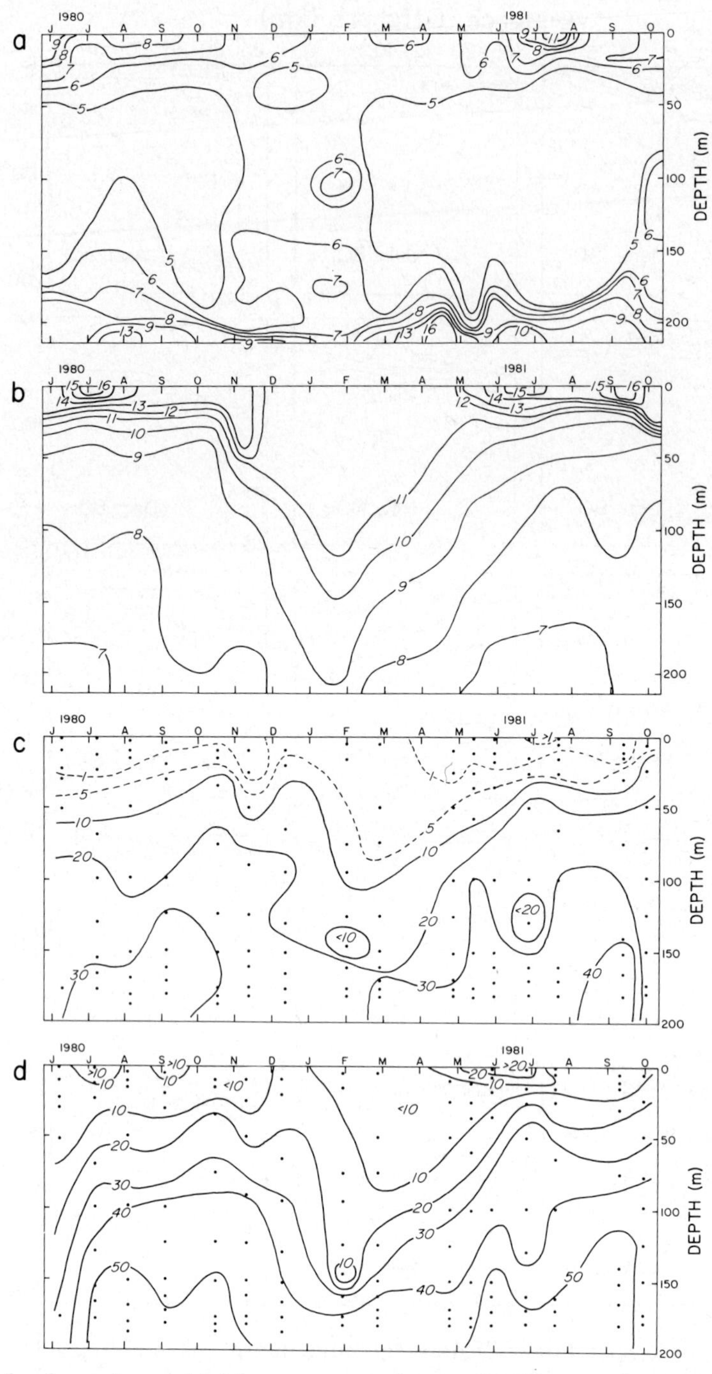

Fig. 7.3 Time-depth sections of (a) beam attenuation coefficient (m^{-1} x 10); (b) temperature (°C); (c) dissolved nitrate (μM) and (d) silicic acid (μM) at a station 24-km off the central Oregon coast (44° 56' N) along the same transect as in Figures 7.1 and 7.2.

Chapter 2). The result of this circulation is that the turbid surface water inshore is moved southward along the coast and seaward at the surface, and cold, nutrient-rich water is brought onto the shelf where it upwells near the coast (Figs. 7.1 to 7.3). Surface turbidity begins to increase at this time of year as upwelled nutrients stimulate phytoplankton growth.

In late May and June, the thin surface lens of warm, nutrient-poor Columbia River plume water off Oregon, coupled with increased surface heating, establishes the sharp thermocline which restricts the surface turbid layer to the upper 10 to 20 m and to varying distances offshore (Figs. 7.1 to 7.3). At the same time, upwelling has intensified, so that for much of this period water of < 10° C and > 25 μM nitrate reaches the surface adjacent to the coast. This pattern of cold nutrient-rich surface water near-shore and warm, nutrient-impoverished surface water off-shore results in steep horizontal temperature and nutrient gradients at the surface (Mooers *et al.*, 1976; Small and Ramberg, 1971; Small and Menzies, 1981). Such gradients, and their temporal variations, can cause complex patterns of SPM distribution. For example, a particle maximum can develop just under the nutrient-depleted surface water offshore of the temperature and nutrient gradient. Such a subsurface particle maximum often appears to be contiguous with the near-shore surface particle maximum when deduced from beam attenuation distributions (for example, early June 1981, Figure 7.1), but the two particle maxima are often different in their composition. Kitchen *et al.* (1978) have shown from analysis of particle size distributions that the subsurface maximum often is composed of phytoplankton cells smaller than those found in the surface particle maximum adjacent to the coast. The subsurface maximum therefore is not necessarily a seaward extension of the surface particle maximum, but rather might be the inherent condition of the more offshore region unaffected by the upwelling and nutrient enrichment adjacent to the coast. In apparently rarer instances, subsurface particle maxima are indeed extensions of surface maxima (Pak *et al.*, 1970; Small and Menzies, 1981); however, these extensions appear less as long, thin seaward protrusions in or just below the seasonal thermocline, and more as short, broad nodes more steeply inclined along shoaling isopycnal surfaces near the coast. Other variations occur as well, making cause-effect relationships exceedingly difficult to identify with certainty.

With the Columbia River plume displaced offshore and with weak, upwelling-favorable winds (which is frequently the case in late summer off Oregon), the turbid surface layer can continue offshore as a relatively unbroken feature (August 1980 and July 1981, Figure 7.1). Should the winds change direction from northerly (upwelling-favorable) to southerly (upwelling-unfavorable) the SPM distribution undergoes a rapid transition and SPM begins to accumulate adjacent to the coast (Small and Menzies, 1981). Return to northerly winds again readjusts the distribution. Wind reversals to the southerly direction occur frequently enough throughout the summer, and are often of long-enough duration (1 to 10 days), to complicate distributional analyses of SPM through the season on the basis of only a few sampling dates (Small and Menzies, 1981).

With the onset of southerly winds in the late fall, the seasonal thermocline begins to erode and the Columbia River plume is diverted northward along the Washington coast. Continued southerly winds, intensifying in strength, generate convergence of surface water against the coast, with the ultimate effect that the water column is more deeply mixed over the shelf. The fall-to-winter transition coincides with the disappearance of the warm surface turbid layer and mid-depth clear layer and with the establishment of a more uniform distribution of SPM through the water column (Figs. 7.1 to 7.3).

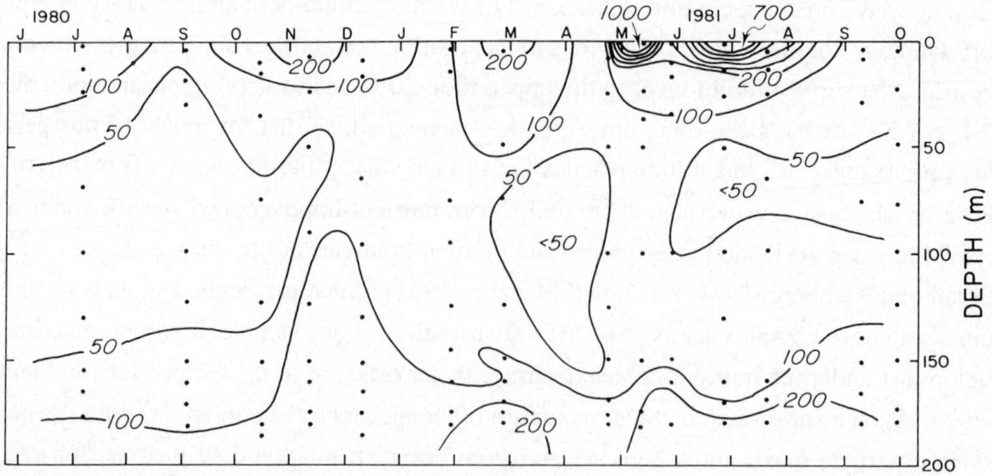

Fig. 7.4 Time-depth section of total suspended particle volume (nl l^{-1}) at the 24-km station off central Oregon.

Unlike the surface and mid-depth layers, the bottom nepheloid layer (BNL) on the shelf shows a more subtle seasonality with respect to total SPM (Fig. 7.1, and see Pak and Zaneveld, 1977). The layer tends to be less well defined along the outer shelf in winter (Figs. 7.1 and 7.3a), but remains fairly compact over the mid-shelf at this time of year (Fig. 7.1). There is northward transport of the BNL in winter off Washington, as deduced from depositional patterns of terrigenous sediments derived from the Columbia River (Gross, 1966; Morse *et al.*, 1968; Duncan *et al.*, 1970; Karlin, 1980; Kachel and Smith, 1989, Chapter 8). Off central Oregon, however, the picture is not so clear. In spring and summer the BNL tends to have a greater vertical extent along the outer shelf than inshore and a summer plume of SPM, contiguous with the shelf BNL, is observed at intermediate depths over the continental slope (see later). A detailed analysis of the processes that form the BNL over the shelf is beyond the scope of this chapter; however, we may accept the general view that the BNL is maintained by turbulence associated with the bottom boundary layer. Variations in the vertical extent of the BNL with distance from the coast within any one season reflect variations in bottom boundary dynamics not specifically addressed here.

7.3.2 Characteristics and Composition of Suspended Particulate Matter

The total volume concentration of suspended particles is a direct measure of total SPM in a water parcel and is best estimated from the area under a particle size distribution curve (Kitchen, 1978). In most cases total volume concentrations bear a close relationship to the beam attenuation data, thus allowing the use of the indirect but more easily obtained attenuation data to estimate suspended particle concentrations. Such a relationship exists in the water column over the northwestern U.S. shelf (Kitchen, 1978). Therefore, the seasonal distribution of total suspended particle volume concentration over the mid-shelf region off central Oregon (Fig. 7.4) bears fairly close resemblance to the seasonal distribution of attenuation values (Fig. 7.3a). The clear layer is identified in Figure 7.4, as in the attenuation data (Fig. 7.3a), though some

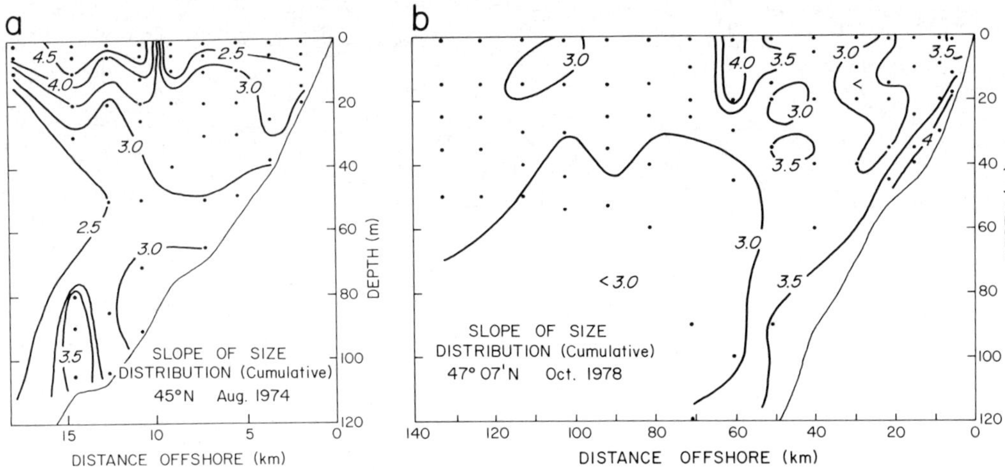

Fig. 7.5 Slopes of the cumulative particle size distributions along transects at (a) 45° 00' N in August; and (b) 47° 07' N in October.

discrepancies are noted. For example, the surface turbid layer and corresponding bottom nepheloid layer in late spring and summer show quite different particle volume concentrations (Fig. 7.4), but these differences are not reflected in the attenuation data (Fig. 7.3a). Such information indicates that the composition of particles in the surface layer is very different from that in the BNL at this time of year. In mid-winter, the particle volume concentrations are reasonably similar throughout the water column (Fig. 7.4), in agreement with winter attenuation profiles.

The use of total particle volume for sensitive characterizations of particle fields is limited by the fact that different particle size distributions and/or particle compositions can yield similar total volumes. Characterization of the particle size distribution (PSD) itself is thus a suitable means to describe a collection of particles further. The number of particles or the particulate volume is plotted as a function of some size parameter (often the particle diameter or spherical equivalent diameter for non-spherical particles), and this PSD usually is described well by either a power law (Bader, 1970) or an exponential function (Zaneveld and Pak, 1973). The fit of the cumulative size distribution (CSD) to the power law is even better, over a small particle size range. Thus, $N = N_0 D^{-s}$ where s is the regression slope determined from a plot of the logarithm of the particle number concentration (N) greater than the particle diameter (D) *vs.* the logarithm of the particle diameter. Higher slopes indicate a relatively greater percentage of small particles in the sample. In Figure 7.5a, low slopes (< 3.0) of the cumulative size distributions for inshore surface waters in summer correspond with waters that have recently been upwelled, while higher slopes (> 4.0) for surface waters farther offshore correspond with waters that are more depleted in nutrients. The inshore cumulative size distributions result from intense diatom blooms that can only be supported by nutrient-rich water, while the offshore distributions are produced by communities of smaller phytoplankton which may be better able to live in low nutrient environments (Kitchen *et al.*, 1978). CSD slopes in the BNL in both summer and winter tend to increase with increasing turbidity, probably because the more easily resuspended small particles

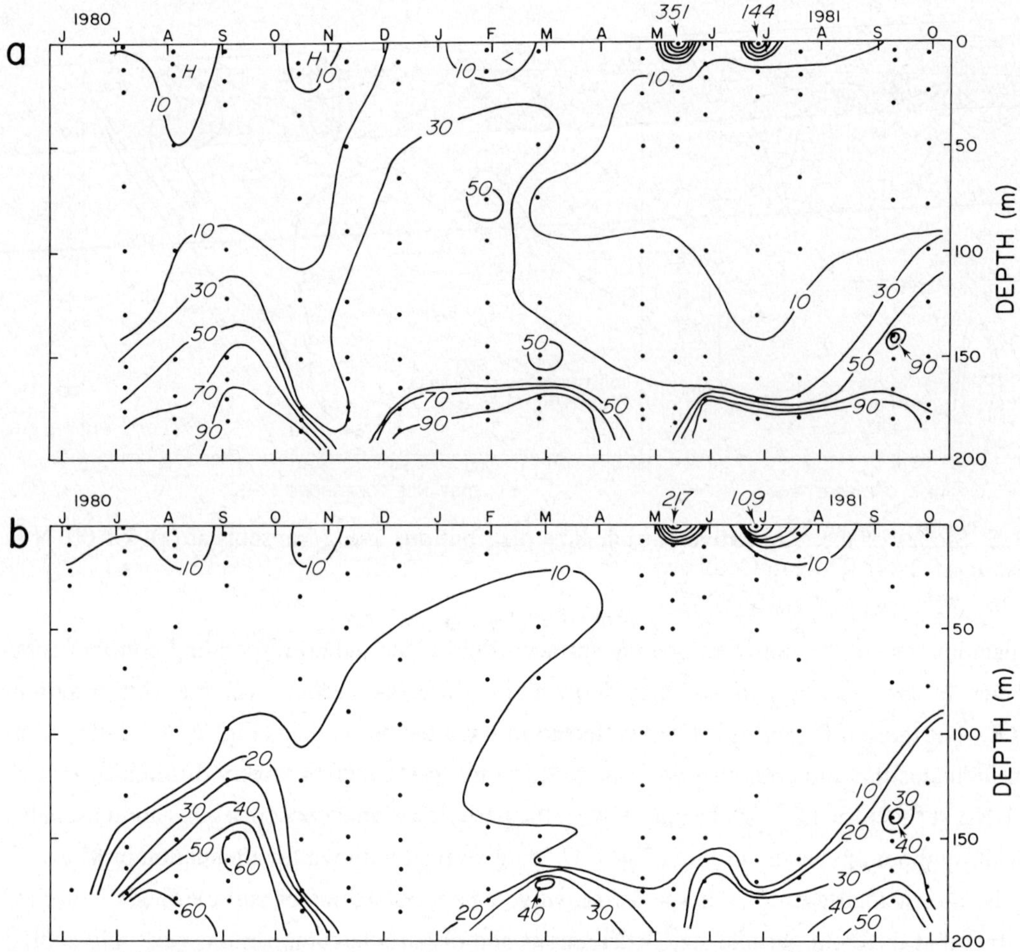

Fig. 7.6 Time-depth sections of (a) total particulate silica (μg Si l^{-1}); and (b) biogenic silica (μg Si l^{-1}) at the 24-km station off central Oregon.

are the ones responsible for increasing turbidity. In fall and winter surface waters, most CSD slopes are generally only slightly higher (< 3.0) than those characteristic of mid-depth layers (Fig. 7.5b).

The silica content of SPM is composed of both crystalline lithogenic phases and biologically-derived amorphous silica from the frustules and skeletons of diatoms, silicoflagellates, radiolarians and other organisms. These can be distinguished analytically because amorphous silica dissolves much more rapidly than crystalline silicates in hot alkaline solution (Krausse *et al.*, 1983). Total particulate silica concentrations are highest in the bottom nepheloid layer during all seasons, but substantial concentrations are found throughout the water column in winter (Fig. 7.6a). It is likely that these enhanced winter concentrations result from particles settling out of the surface mixed layer plus some resuspension from the BNL. High total silica concentrations at the surface in spring and early summer undoubtedly are associated with blooms of diatoms, as indicated by the correspondingly high concentrations of biogenic silica (Fig. 7.6b). One might expect that particulate silica in the BNL would be mainly lithogenic in character, but this seems not to be the case, at least in summer (Fig. 7.6). Usually over half the total particu-

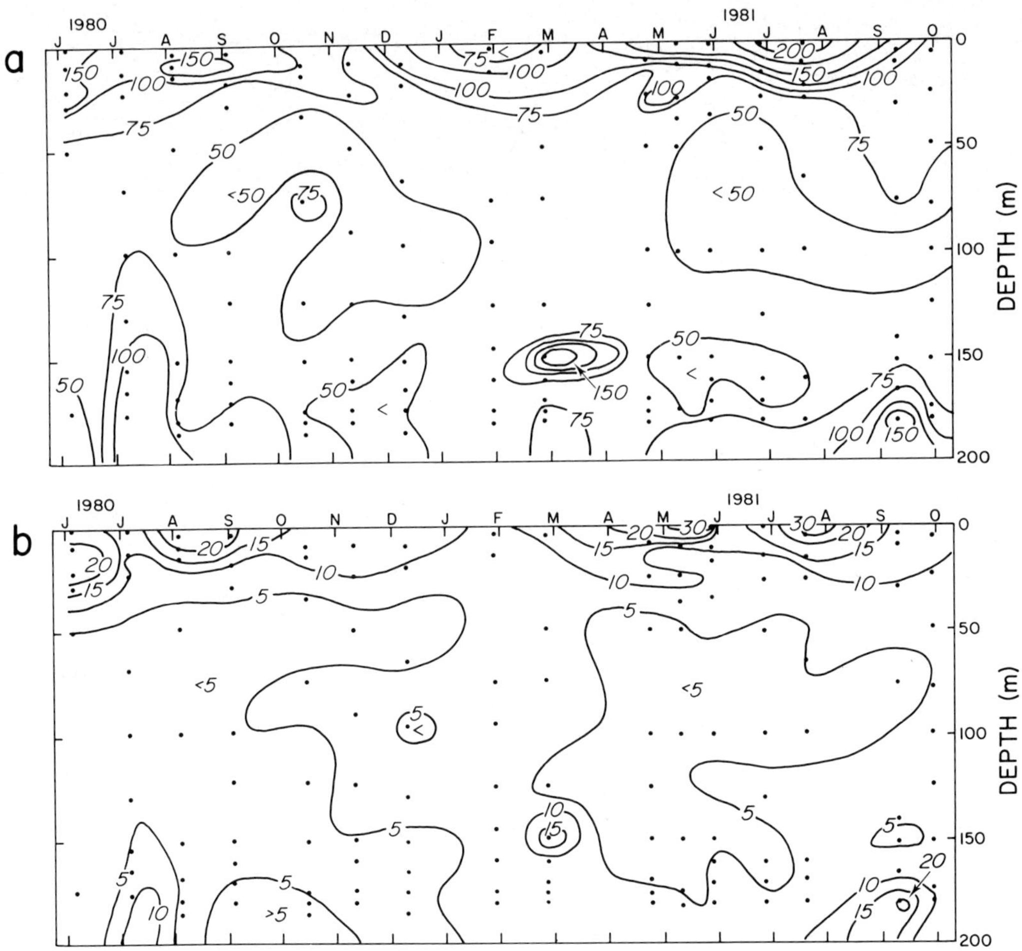

Fig. 7.7 Time-depth sections of (a) particulate organic carbon ($\mu g\ l^{-1}$); and (b) particulate organic nitrogen ($\mu g\ l^{-1}$) at the 24-km station off central Oregon.

late silica in the BNL in summer is biogenic, even though a clear intermediate-water layer separates the top productive layer from the BNL. In winter, the biogenic fraction of total particulate silica is less than in summer, with biogenic silica in the BNL comprising < 30% of the total.

The three-layered structure of the summer water column is also apparent in the vertical distribution of particulate organic carbon and nitrogen (POC and PON) (Fig. 7.7). The PON series particularly seems coherent with respect to the summer structure and also reflects the disappearance of the three-layered structure in winter. PON concentrations therefore may reflect increases and decreases in recently produced organic matter better than POC concentrations. POC often contains significant quantities of relict and detrital carbon from many sources.

Time series of particulate chlorophyll and phaeophytin (Fig. 7.8) show the expected high but variable concentrations at the surface, likely brought about by phytoplankton bloom and non-bloom events of unknown spatial extent. In general, the highest concentrations of chlorophyll *a* correspond to low concentrations of phaeophytin *a* and vice versa. Almost no active chlorophyll *a* is present in the bottom nepheloid layer at any time, but phaeophytin levels there

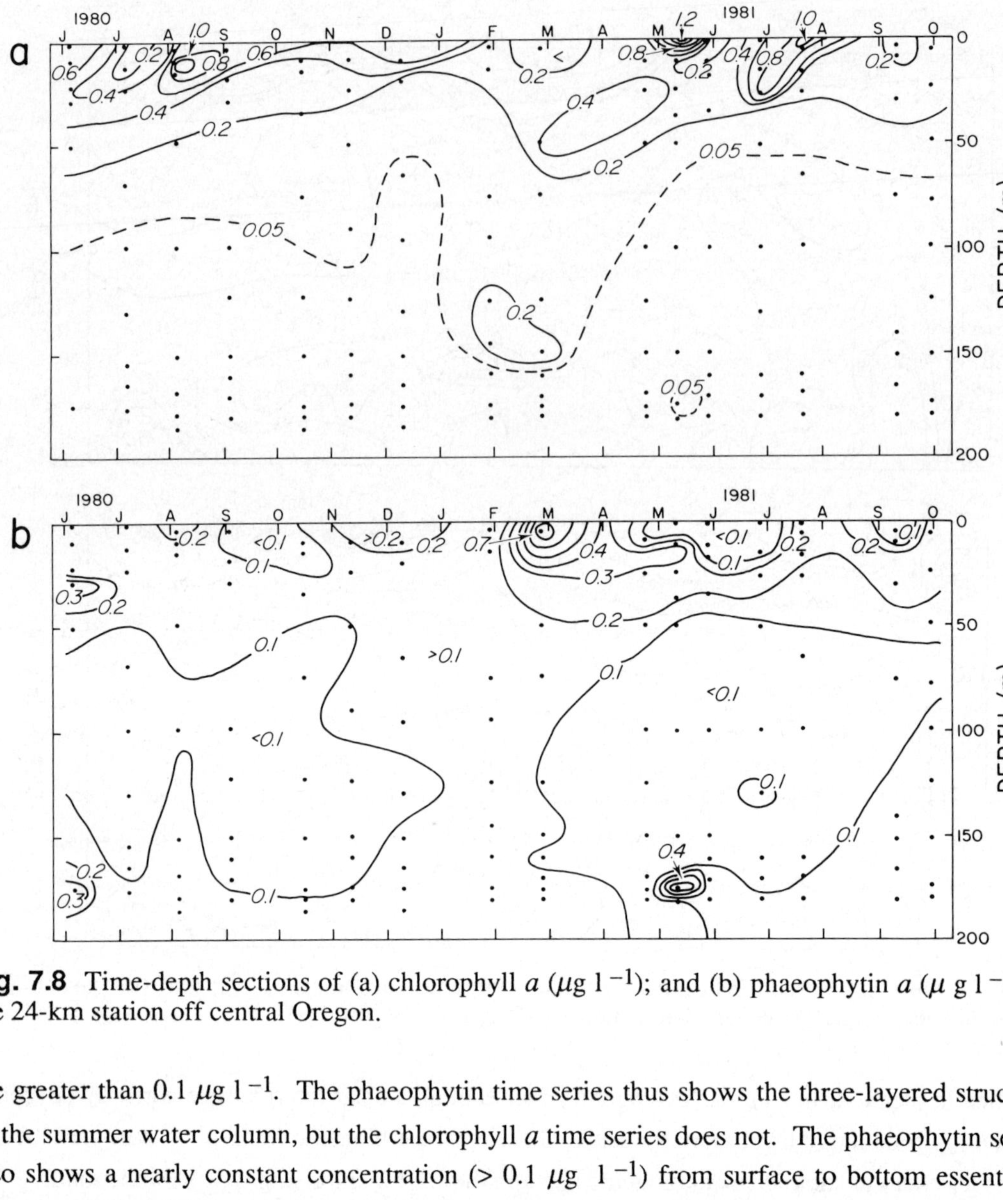

Fig. 7.8 Time-depth sections of (a) chlorophyll *a* (μg l^{-1}); and (b) phaeophytin *a* (μ g l^{-1}) at the 24-km station off central Oregon.

are greater than 0.1 μg l^{-1}. The phaeophytin time series thus shows the three-layered structure of the summer water column, but the chlorophyll *a* time series does not. The phaeophytin series also shows a nearly constant concentration (> 0.1 μg l^{-1}) from surface to bottom essentially from December through February (Fig. 7.8b). The concentration of chlorophyll *a* > 0.2 μg l^{-1} between 125 and 150 m in February and early March (Fig. 7.8a) likely is not a persistent feature of late winter waters over the shelf, as it does not appear in long-term mean plots (Landry *et al.*, 1989, Chapter 1).

The BNL in summer is basically non-chlorophyllous but is high in biogenic silica and almost always has POC : PON ratios exceeding 10 (often greatly exceeding 10). Up to 80% of the BNL can be composed of biogenic debris whose source must have been, at one time or another, the surface waters. The presence of the mid-depth clear layer indicates that differential cross-shelf advection between the surface layer and the water below (Hickey, 1989, Chapter 2) prevents the continuous direct settling of small particles from surface to bottom over the shelf in summer. Repackaging of small particles into zooplankton fecal pellets of a size and density which sink rapidly through the clear layer to the BNL is one possible mechanism for delivering surface materials quickly to the bottom (Fowler and Small, 1972; Small *et al.*, 1979, Bruland and Silver, 1981). Most rapidly sinking pellets have slightly elevated POC : PON ratios at the time of release from the zooplankton (Small *et al.*, 1983) and further reworking in the BNL or in the bottom sediments might continue to increase the POC : PON ratios. In addition, fecal pellets tend to preserve biogenic silica in their matrices, at least over the short time required to sink through water columns over continental shelves (Turner 1977, 1984; Honjo and Roman, 1978). Finally, chlorophyll is converted to phaeopigments via the processes of zooplankton grazing and voiding of fecal pellets (Shuman and Lorenzen, 1975; Welschmeyer and Lorenzen, 1985), which is consistent with the appearance of phaeophytin, but not chlorophyll, in the BNL over the Oregon shelf. Because the larger fecal pellets which settle rapidly are also relatively rare (McCave, 1975), it is presumed they can sink through the mid-depth clear layer without disintegration and without perceptibly contributing to the turbidity of this layer. Very small fecal pellets tend to be recycled in upper waters (Bishop *et al.*, 1977; Paffenhofer and Knowles, 1979; Hoffman *et al.*, 1981; Small *et al.*, 1987) and thus would not participate in rapid removal of material to the bottom.

Another mechanism for delivery of surface materials quickly to sediments has been proposed recently by Smetacek (1985). Increased mucous secretion by diatoms during blooms leads to aggregate formation and enhanced sinking. These sticky aggregates scavenge other particles (including fecal pellets) during descent, which further accelerates the sinking rate. Mass diatom sedimentation through the clear layer to the shelf bottom after blooms would be sporadic and easily missed by infrequent and short-term shipboard sampling. Lack of chlorophyll *a* signatures in the BNL could mean either that mass sedimentation was not an important vertical transport mechanism off Oregon, or that the chlorophyll *a* had been degraded between sampling periods.

Movement of organic-rich SPM onto the shelf along the bottom from offshore is unlikely during summer, as the near-bottom flow is directed offshore on the mid- and outer shelf (Huyer *et al.*, 1978; see Fig. 2.29 in Hickey, 1989, Chapter 2). However, sporadic movement from the innermost shelf regions in the offshore direction along the bottom is a distinct possibility. The three-layered structure of the water column is lost close to shore, even in summer (Fig. 7.1), so that the near-shore SPM field could act, at least intermittently, as a source of particles for the BNL over the complete shelf width. This is another possible mechanism for delivery of surface SPM to the BNL without invoking sedimentation through the mid-depth clear layer.

7.3.3 Biological Potential of Waters from the Mid-Depth Layer and Bottom Nepheloid Layer

Seasonal changes in SPM in near-surface waters over the Oregon/Washington shelf are partly the result of changes in phytoplankton productivity through the year. Primary productivity in the Oregon/Washington coastal region is well documented in previous chapters (Landry *et al.*, 1989, Chapter 1; Perry *et al.*, 1989, Chapter 3) and in the literature (Anderson, 1964, 1972;

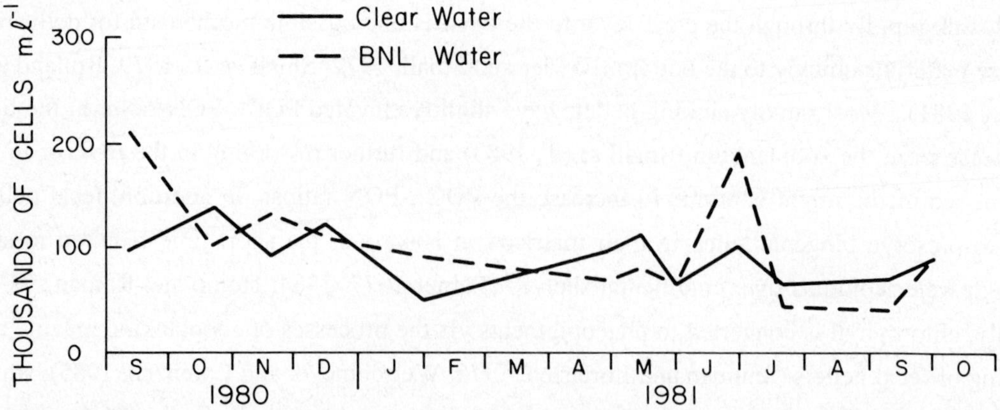

Fig. 7.9 Concentrations of bacterial cells in mid-depth clear water and in bottom nepheloid layer (BNL) water at different times of year.

Small *et al.*, 1972a,b; Small and Menzies, 1981). What is less well studied is the seasonal potential of the mid-depth clear layer and BNL to act as sources of microfloral seed stock for the lighted surface layer. In 1980 and 1981, sterile bag techniques were used to enumerate bacterial cells, determine viable phytoplanktonic material, and evaluate phytoplankton growth potential in these layers throughout the year.

Sterile bag samples were filtered through 0.2-μ m Nuclepore filters and the filters examined using epifluorescent microscopic techniques. Concentrations of bacteria over the shelf ranged from 0.5 to 2.1 x 10^5 cells ml $^{-1}$, with little sustained difference among clear-water samples in summer, mixed-layer samples in winter and BNL samples the year around (Fig. 7.9). No distinction was made between free-living and particle-attached bacteria, so it could not be determined whether relatively high concentrations of free-living bacteria in the clear water compensated relatively high concentrations of attached forms in the BNL. Jacq and Prieur (1985) found that particle-bound bacteria were absent below the thermocline in the Quessant estuary and outfall area in France. The below-thermocline region might be comparable to the mid-depth clear layer over the Oregon shelf. In mixed waters in the Quessant system (perhaps similar to the winter condition off Oregon), particle-bound bacteria were present throughout the water column, with maxima near the surface and bottom. Distributions of free-living and particle-bound bacteria were complex in frontal areas of the Quessant system. Unfortunately, metabolic activities of neither the Quessant nor the Oregon bacterial populations were assessed, so there is no way of knowing whether activity differences occurred between clear-water populations and those in either surface or BNL waters.

Sterile bag samples from the mid-water and bottom nepheloid layers were also incubated under continuous irradiance at 58 μ E m $^{-1}$ sec $^{-1}$ at 15° C, without nutrient enrichment (near-bottom waters were consistently rich in nutrients; see Figure 7.3). Sequential fluorometric readings were made to ascertain phytoplankton growth. When growth occurred, it was observed after 5 to 27 d of incubation. With one exception, growth occurred in the months of June, July, August and September in two different years, in both the BNL and the clear layer (Fig. 7.10). With one exception, no growth was observed in mid-depth water from October through May. In

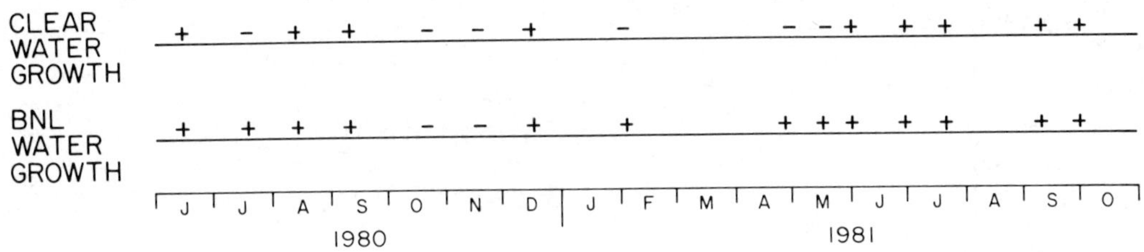

Fig. 7.10 Growth of phytoplankton from inoculations of mid-depth clear water and BNL water into cell-free water under given conditions (see text). + ,– indicate that growth or no growth was observed, respectively, after 27-d incubation.

contrast, the BNL contained viable algal cells capable of growth at all times except October and November. Thus, the summer months provide particles with growth potential at all subsurface depths, while the winter and spring months provide growth potential almost entirely in the BNL. In all cases, the predominant cells that grew were small, centric diatoms. The lack of growth potential in the BNL in October and November is unexplained.

That no growth, or very little growth, occurred in either the clear water or nepheloid samples in late April and mid May 1981 (Fig. 7.10) was unexpected. It is important to stress that all sterile bag samples were collected and incubated in such a way that no contamination from other water depths or from the laboratory occurred. Thus, the clear layer in summer, the surface mixed layer in winter, and the bottom nepheloid layer at all times of year contain some viable algal cells and/or spores, with most of the growth potential occurring during summer months.

With four exceptions, it appeared that once growth was observed the rates were fairly constant between 0.45 and 0.95 doubling d $^{-1}$ (mean = 0.67 doubling d $^{-1}$), regardless of the origin of the seed stock (BNL or clear layer) or of the time that elapsed before growth became measurable (Fig. 7.11). These growth rates were close to the 1.0 doubling d $^{-1}$ reported by Hollibaugh *et al.* (1981) for diatom resting spores incubated under conditions similar to ours. Of the four higher rates measured (1.28 to 1.76 doublings d $^{-1}$), three were in late summer and one in spring, and three were from the BNL and one from the mid-depth clear layer; thus, no meaningful pattern emerged to help explain the higher rates.

7.3.4. Correlations of Properties on the Shelf

That striking seasonal changes in the SPM field and associated attributes occur over the shelf is unquestionable (Figs. 7.1, 7.3a, 7.4, 7.6, 7.7 and 7.8). That patterns of seasonal change of SPM properties are often visually similar further suggests that the properties remain correlated to some degree throughout the annual cycle. Regression analyses among different pairs of selected variables were done at different progressive depth intervals over the outer shelf off Oregon (24-km station with 183-m total water depth), in order to examine to what degree pairs of properties covaried through the water column and through all seasons.

The variables examined were beam attenuation and concentrations of particle volume, chlorophyll, POC and PON (Table 7.2). Depths < 30 m corresponded approximately to the depths representing the surface turbid layer in summer, while depths < 50 m encompassed the maximum mixed layer depth in winter (Huyer, 1977). Depths < 150 m roughly included those

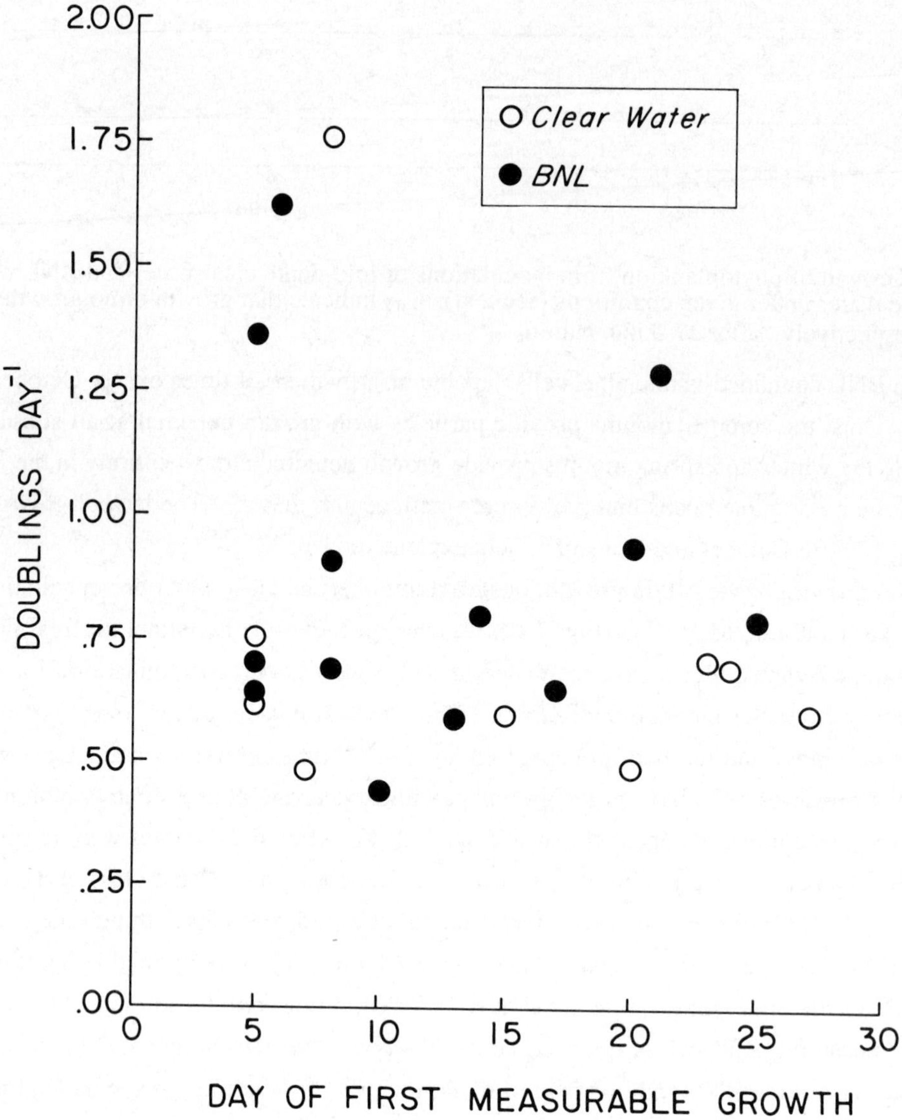

Fig. 7.11 Growth rate of phytoplankton seed stock as a function of the first day of measureable growth from natural inocula.

from both the turbid and mid-depth clear layers in summer. Evaluation of all depths included the BNL in addition to the rest of the water column.

Correlation coefficients were always high ($r > 0.75$ and usually > 0.80) for most pairs of variables if the BNL is excluded (Table 7.2). Thus, the beam attenuation coefficient was always highly correlated with total particle volume, POC and PON in surface and mid-depth clear waters regardless of season. Chlorophyll *a* content of particles was the only attribute measured that consistently showed lower correlation with the other variables in surface and mid-depth layers. A substantial amount of the total particulate volume during all seasons was non-chlorophyllous but was still composed of organic carbon and nitrogen. Hence, higher correlations among these variables (and with beam attenuation) is to be expected regardless of season. Furthermore, the

Table 7.2 Correlation coefficients (r) for pairs of selected attributes through different depth horizons over the outer shelf off central Oregon. ATN = beam attenuation coefficient, VOL = particle volume, CHL = chlorophyll *a*, POC = particulate organic carbon, PON = particulate organic nitrogen.

	Depth Horizon (m)	ATN	VOL	CHL	POC	PON
ATN	< 30		0.83	0.60	0.80	0.85
	< 50		0.84	0.64	0.83	0.87
	< 50		0.78	0.65	0.75	0.82
	All		0.68	0.23	0.52	0.57
VOL	< 30			0.44	0.81	0.76
	< 50			0.48	0.81	0.77
	< 150			0.55	0.78	0.75
	All			0.42	0.65	0.64
CHL	< 30				0.47	0.55
	< 50				0.54	0.61
	< 150				0.68	0.71
	All				0.58	0.62
POC	< 30					0.85
	< 50					0.87
	< 150					0.86
	All					0.83

chlorophyll content of chlorophyll-containing particles is dependent on the light field. In outer shelf surface waters in summer light levels are often high and chlorophyll can be bleached. Lower in the euphotic zone the chlorophyll content of particles can increase without concomitant change in particle volume and without measurable change in POC or PON content.

It is interesting that the correlation coefficients between chlorophyll and the other variables apparently increased with depth in the water column (Table 7.2). For example, the correlation between chlorophyll and POC was weak in the upper 30 m of water (r = 0.47), strengthened through the top 50 m (r = 0.54), and was highest over a 150-m water column (r = 0.68). Such a pattern indicates that although concentrations of both chlorophyll and POC were often high in surface waters (Figs. 7.7a and 7.8a), they were not always in phase. Relatively high concentrations of POC often yielded relatively low chlorophyll concentrations (in winter, for example), and similar concentrations of chlorophyll often were associated with different concentrations of POC. Conversely, in mid-depth waters (50-150 m) the relationship between the small concentrations of chlorophyll and POC must have held throughout the year because the correlation coefficients measurably increased with each increment of added depth.

When data from the BNL are included in the different regression analyses, the correlation coefficients nearly always decrease (Table 7.2). For example, the correlation coefficient for the relationship between beam attenuation coefficient and total particle volume dropped from an average of 0.82 in the top 150 m to 0.68 when BNL data were included in the analysis. The relationship between beam attenuation and chlorophyll *a* off the southern Washington coast in September 1980 (Fig. 7.12) shows, as expected, that the two attributes do not covary at all in the

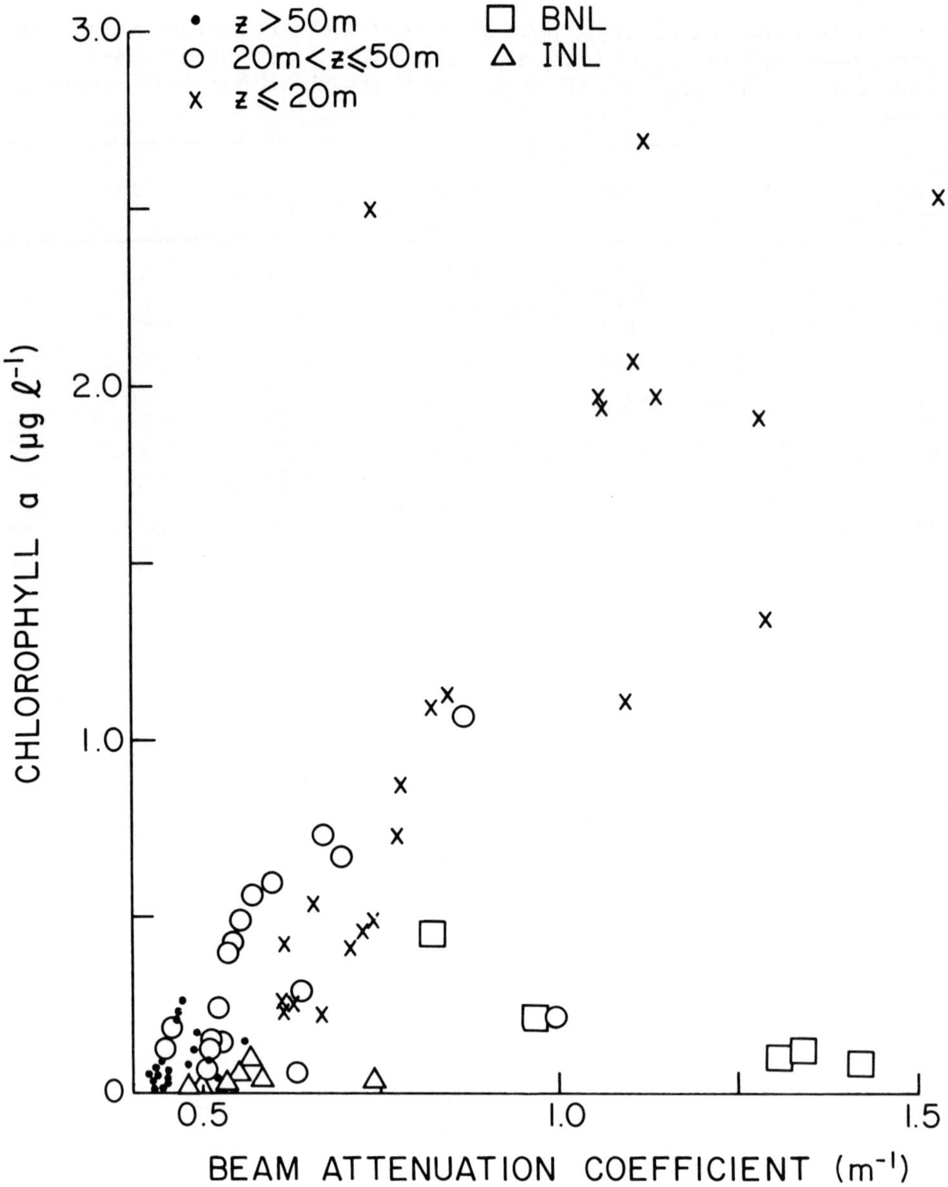

Fig. 7.12 Chlorophyll *a* as a function of beam attenuation coefficient at different depth strata off the southern Washington coast in September, showing the lack of a relationship in the BNL and intermediate nepheloid layer (INL).

BNL. Poor correlation between these two variables in the intermediate nepheloid layer (see later) is also shown along with the typical point scatter through the remainder of the water column during any given cruise period. The only relationship that remains high when data from the BNL are combined with water-column data is the correlation between POC and PON (Table 7.2).

The reduced correlation between beam attenuation and particle volume in the BNL indicates a significant change in the particle size distribution between the BNL and the remainder of the water column. The proportion of biogenic material in the BNL also is less than that in the upper water column. For example, even though there is more biogenic silica in the BNL than in surface and mid-depth waters during much of the year (Fig. 7.6b), the fraction of total particulate silica that is biogenic is lower on average in the BNL than in overlying waters (compare Figure 7.6a with Figure 7.6b). Because lithogenic particles generally have higher densities, they must be smaller in size than the biogenic particles to remain in suspension in the BNL. The relationship between the slope of the particle size distribution and beam attenuation shows the clear distinction between the BNL and the overlying waters (Fig. 7.13). It is obvious that a relatively large number of small particles define the BNL (high beam attenuation and high PSD slope), whereas fewer, larger particles are present in the water column (low beam attenuation and low PSD slope). Note that particles in both surface and mid-depth layers are of similar size (similar PSD slopes), even though surface water on average contains more particles (higher attenuation coefficient). Analysis of particles in the intermediate nepheloid layer (INL) over the continental slope indicates that they are strongly related to shelf BNL particles in size and number (Fig. 7.13).

7.4 CONTINUITY OF SHELF SPM DISTRIBUTIONS OVER THE CONTINENTAL SLOPE

Distributions of SPM and other attributes over the continental slope aid in understanding the seasonal shelf dynamics. In July off the southern Washington coast (47° 07' N), for example, the mid-depth clear layer over the shelf is seen in beam attenuation sections as an intrusion of clear intermediate-depth slope water (Fig. 7.14). Baker and Hickey (1986) and Hickey (1989, Chapter 2) show similar distributions for the fall and winter months. The clear layer over the slope actually is given definition by the seaward extension of both the shelf surface turbid layer and the BNL. The seaward extension of the shelf BNL over the slope has been termed the intermediate nepheloid layer, or INL (Pak and Zaneveld, 1978; Pak *et al.*, 1980). In addition to being observed in the attenuation data, the INL often can be well defined by the total silica content of the SPM (Fig. 7.15a), and to a lesser extent by the distributions of biogenic silica (Fig. 7.15b) and organic carbon (not illustrated). The lack of chlorophyll in this layer (Fig. 7.12), the relatively high mineral content relative to organic content (Fig. 7.15), and the similarity in particle size distribution to the BNL (Fig. 7.13) leaves little doubt that the INL is formed by seaward transport of particles from the shelf BNL. Particle size distributions in the INL and in the clearer water just above it further suggest that the INL is not formed principally by particles settling from above and accumulating. Off central Oregon in November 1977, for example, the particle size distribution in the INL differed significantly from that in the clear water for particle sizes of < 15-μm diameter (Fig. 7.16). The steeper slope for INL particles indicates a substantially greater number of particles < 15 μm in diameter in that layer. There were about equal (and relatively low) concentrations of particles > 15 μm in the INL and clear water (Fig. 7.16). INLs thus occur primarily as a result of the persistence in the water column of the fine-particle fraction from the source sediments of the shelf BNL.

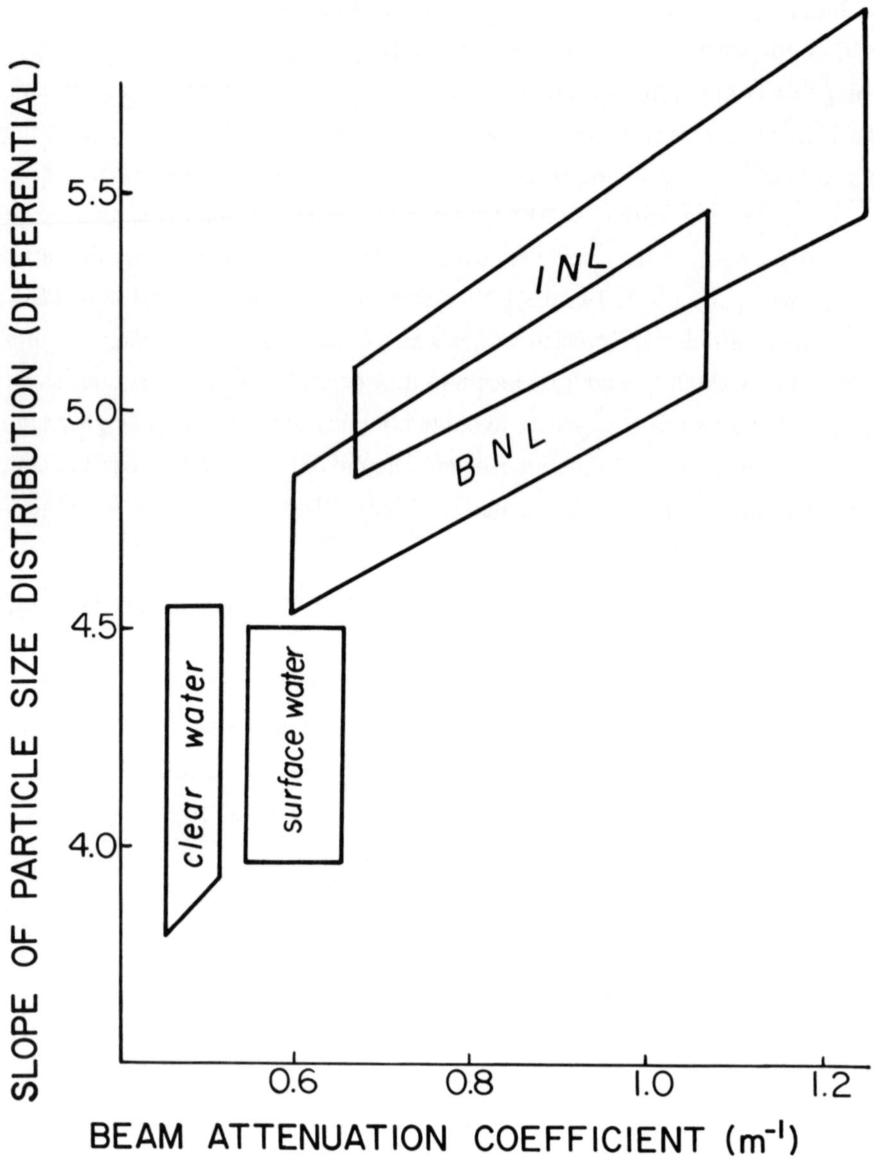

Fig. 7.13 Slope of the particle size distribution as a function of beam attenuation coefficient, showing general functional response zones for mid-depth clear water, surface water, the BNL and the INL.

Relatively particle-free water is found below the INL over the continental slope, and a BNL is always present just above the slope bottom (Fig. 7.14). The above pattern occurs all along the Oregon-Washington shelf-slope region, though the seaward extent of the INL can vary (compare Fig. 7.14 and 7.17a, for example, and see Hickey, 1989, Chapter 2,). The onshore-offshore extent, thickness, and particle concentration of the INL at any given time are functions of 1) flow direction and speed over the outer shelf and slope, 2) the timing and intensity of resuspension of bottom sediments on the shelf, 3) particle settling, upwelling and downwelling events in relation to flow over the slope, and 4) the shelf-slope bathymetry (whether the region shows

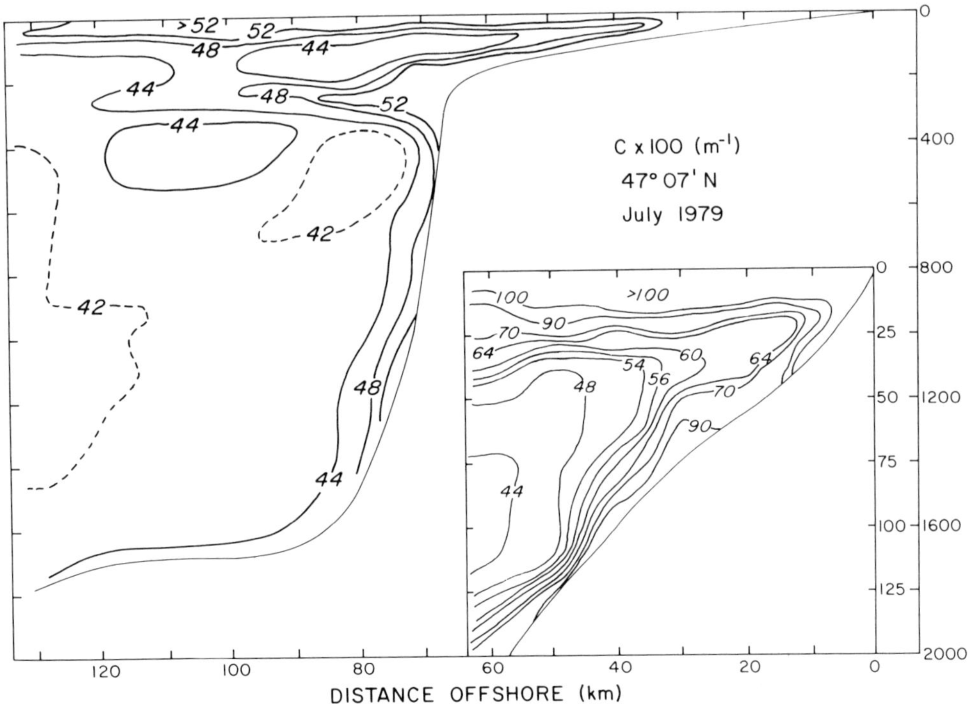

Fig. 7.14 Beam attenuation coefficient (m^{-1} x 100) over the southern Washington continental slope and continental shelf (inset) in July, showing the INL over the slope.

relatively smooth bathymetry or is cut by canyons, for example) (Hickey *et al.*, 1986). Baker and Hickey (1986) illustrate that the seaward extension and particle concentration in observed INLs off southern Washington were positively and significantly correlated with the magnitude of resuspension events on the adjacent shelf and with short-term (about 10 d) temporal variation of deposition rates on the slope.

There is some evidence to suggest that the INLs themselves might record successive resuspension events that occurred earlier on the shelf. For example, within the general matrix and along the zonal axis of some INLs are alternating cores of relatively more particle-rich and particle-poor waters (Figs. 7.14 and 7.17a). If the shelf BNL were steadily supplying particles to the INL one would expect a continuous gradient of SPM concentration from the shelf break seaward. Plots of total particulate silica (Fig. 7.15a) and biogenic silica (Fig. 7.15b) in July 1979 confirm the existence of an INL core at that time. A region of oxygen depletion coincides with the core of the September 1980 INL (Fig. 7.17b), suggesting significant heterotrophic activity in this core. No such O_2 depletion exists in the July 1979 core, however (not illustrated). No particulate silica or carbon data were available for the September 1980 INL core.

The appearance of cores of more concentrated SPM within the INL matrix does not necessarily indicate that the cores are derived from evenly spaced pulses of material off the shelf. In fact, great variability can occur in both the timing of resuspension events and in the SPM concentration within each cloud of resuspended material. Hickey (1989, Chapter 2) shows a week-long time series of particle concentrations 5 m above the bottom at the shelf edge off

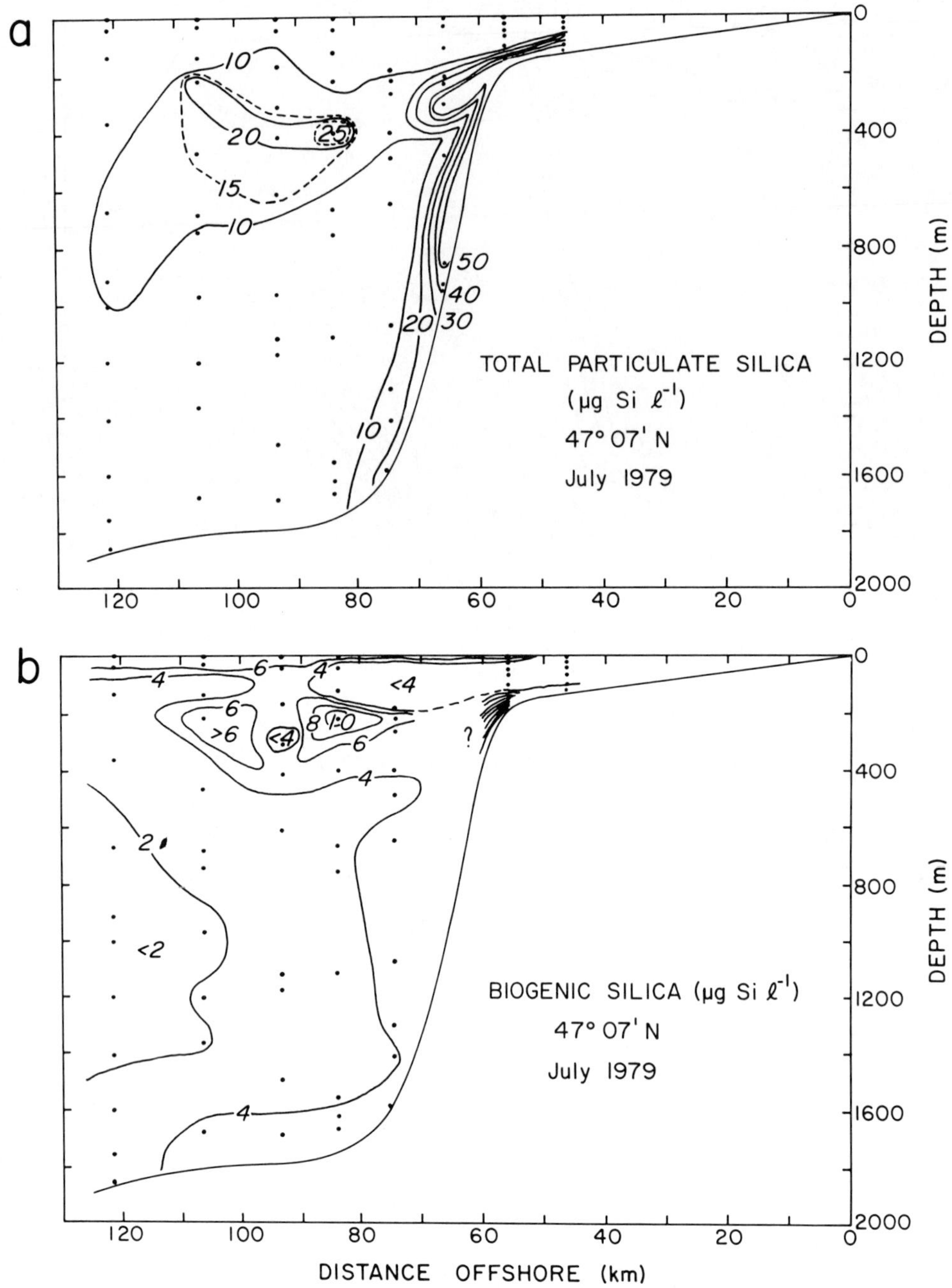

Fig. 7.15 (a) Total particulate silica and (b) biogenic silica defining the INL over the southern Washington continental slope.

southern Washington. The striking feature of that series is the order-of-magnitude changes that occur in a matter of hours.

Exceedingly large short-term variability can of course mask interannual variability in any region that is sampled only sporadically. For example, particle volume concentration in the BNL near the central Oregon shelf edge (180-m depth, 24 km from shore) in November 1977 exceeded 500 nl l^{-1} (Fig. 7.18a), while in November 1980 at the same depth and distance from

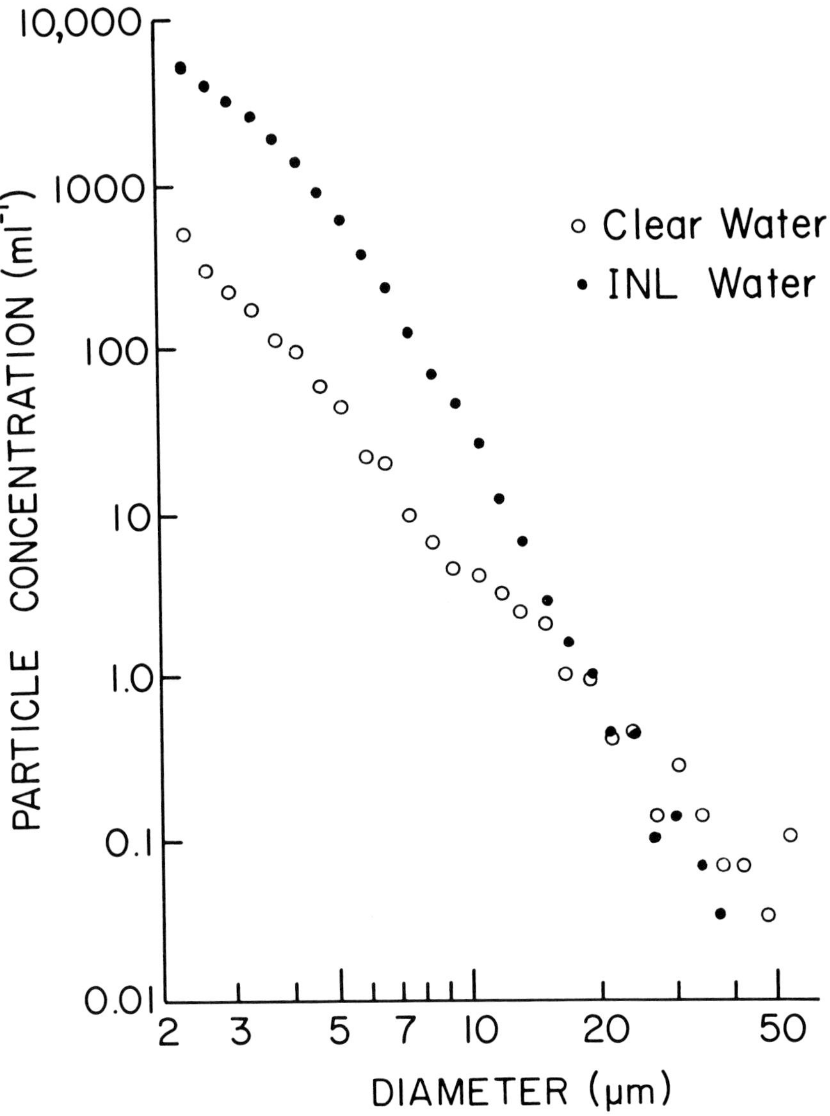

Fig. 7.16 Particle concentration (number ml $^{-1}$) as a function of particle diameter (mm), showing the separation of mid-depth clear water and INL water for particles < 15-μm diameter.

shore the particle volume concentration did not reach 200 nl l $^{-1}$ (Fig. 7.4). Similarly, total particulate silica was >100 μg l $^{-1}$ in November 1977 (Fig. 7.18b), but only 30 μg l $^{-1}$ in November 1980 (Fig. 7.6a). It is tempting to classify these differences as interannual differences, perhaps related to the different intensities and durations of summer particle production in surface waters (Small and Menzies, 1981) as well as to different turbulent interactions in the bottom waters which generate resuspension of sediments (Hickey *et al.*, 1986). There are no long-term time series of data through several summer-fall-winter periods to test this, however, and year-to-year spot comparisons in highly patchy surface and bottom waters are inadequate for the purpose. The differences could just as easily have been the result of short-term variability.

Interannual variability might be addressed from the standpoint of compositional differences of the SPM in addition to concentration differences. Presumably SPM composition in the

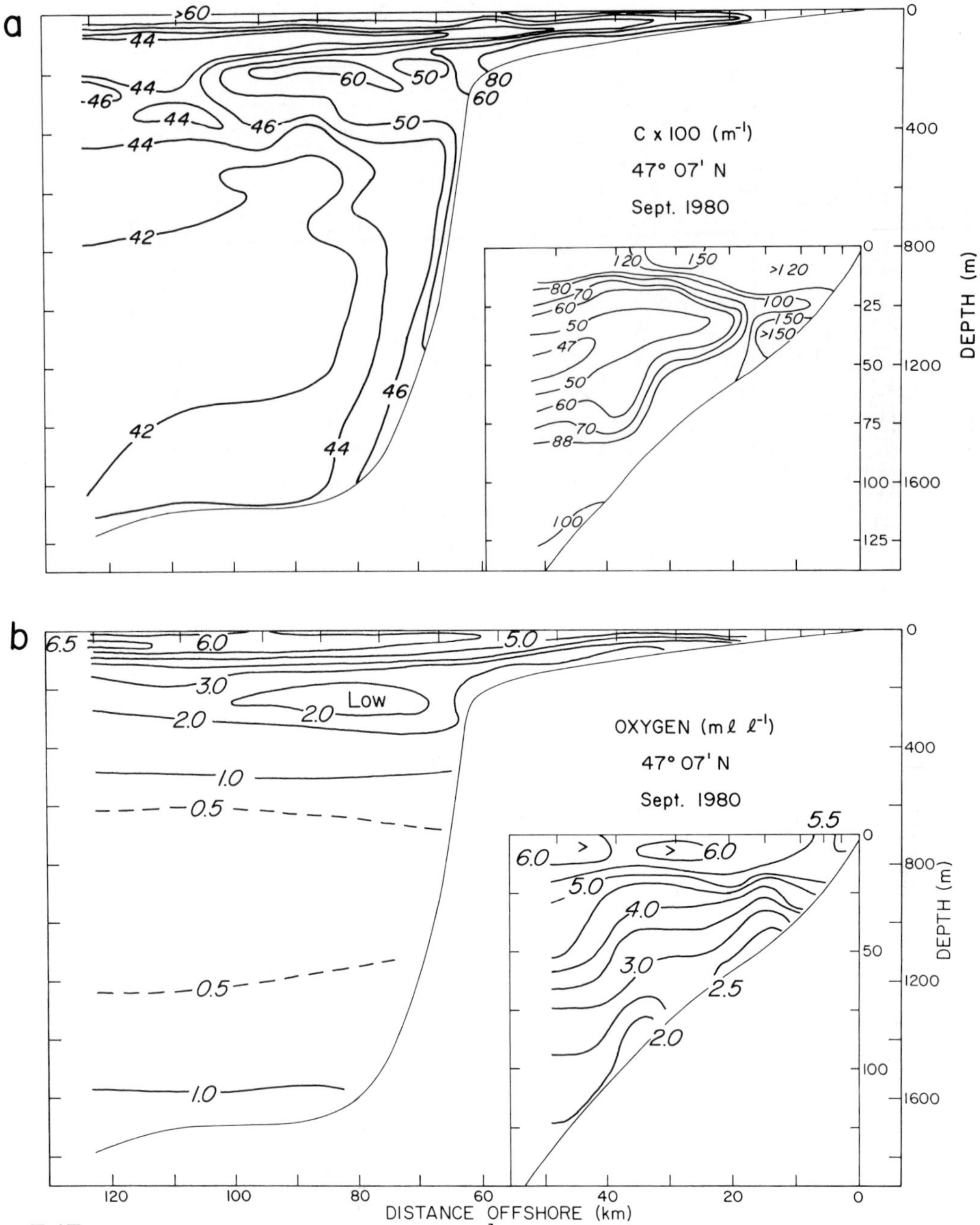

Fig. 7.17 (a) Beam attenuation coefficient (m^{-1} x 100) and (b) dissolved oxygen concentration (ml l^{-1}) over the southern Washington continental slope and continental shelf (inset) in September, showing the INL and O_2 depletion in the core of the INL.

BNL would be more conservative than SPM concentration in the BNL during any given season of different years, so that large differences in composition might signal fundamental, long-term BNL differences which would then be reflected in interannual compositional differences in the INL. The silica and carbon fractions of total BNL SPM can be roughly calculated for November 1977 and November 1980. If the specific gravity of the BNL particles is assumed to be 2.0 for purposes of initial comparison, then the 500 nl l^{-1} particle volume contour in Figure 7.18a

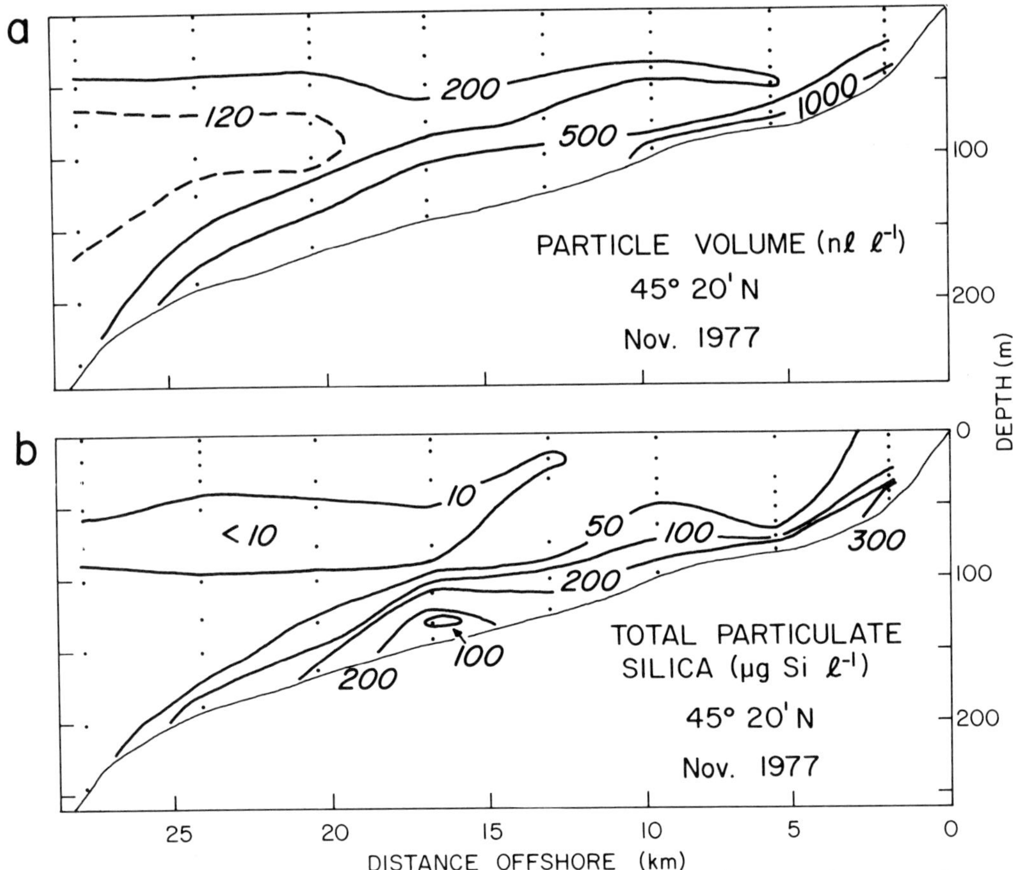

Fig. 7.18 (a) Total suspended particle volume and (b) total particulate silica over the central Oregon continental shelf in early winter.

equates to about 1000 μg l $^{-1}$ of particle mass. The 500 nl l $^{-1}$ contour coincides reasonably well with the 100 μg Si l $^{-1}$ contour in Figure 7.18b. Because 100 μg Si l $^{-1}$ is approximately equivalent to 200 μg SiO_2 l $^{-1}$, the siliceous fraction of the BNL in November 1977 is 200/1,000 = 0.20. Over the outer shelf, 1,000 μg l $^{-1}$ of particle mass coincides roughly with 100 μg POC l $^{-1}$ (not illustrated), to yield an organic carbon fraction of about 0.10 for the BNL in November 1977 (over the inner shelf the fraction of organic carbon in the BNL approaches 0.20). In November 1980 over the outer shelf, the siliceous fraction was 60/ 264 = 0.23 (computed from actual data points) if the specific gravity of BNL SPM was again assumed to be 2.0 and SiO_2, concentrations are double the Si concentrations. This November 1980 fraction was similar to the 0.20 value for November 1977. The organic carbon fraction was 40/264 = 0.15 in November 1980, a value intermediate between those of the outer shelf (0.10) and the inner shelf (0.20) in November 1977. These comparisons suggest that there were no remarkable compositional differences between the BNL SPM of the two Novembers, and thus no interannual variability in composition for the early winter. The INLs generated from these November BNLs would be expected to have similar SPM compositions if not similar SPM concentration.

There may be fundamental differences in the seasonal INLs off Oregon and Washington, irrespective of great short-term variability in both INLs and their source BNLs in both regions.

Winter INLs off Washington appear to have greater particle concentrations and extend farther offshore than winter INLs off Oregon (Pak *et al.*, 1980). The majority of the particle load of the Columbia River is deposited on the Washington shelf rather than the Oregon shelf, where it forms a thick mid-shelf silt deposit (Nittrouer, 1978). This material, when resuspended by moderate or strong currents or wave action, becomes much of the source material for INLs off Washington (Hickey *et al.*, 1986). In addition, the general east-west orientation of the shelf-slope canyons off Washington relative to the prevailing northward currents in winter tends to more sharply define the INLs over the slope. Resuspension events on the shelf plus canyon "focusing" thus allow formation of relatively robust INLs that can extend fairly great distances offshore (Hickey *et al.*, 1986). The Oregon Coast has no canyons south of the Astoria Canyon (this canyon is just offshore of the Columbia River), and there is essentially no sedimentary contribution to the Oregon shelf from the Columbia River. Other Oregon coastal streams are small. Thus, there is no comparable source of SPM off Oregon as there is off Washington, and no large irregularities in the bathymetry to direct INLs well over the slope.

In summer, the Columbia River effluent is moved southward by the prevailing surface currents, but the significant offshore component of these currents tends to direct the river outfall seaward off the Oregon shelf except in the immediate proximity of the river mouth. Consequently, the Oregon shelf probably only rarely receives the accumulation of sediments typical of the Washington shelf, and thus seldom has the source material to extend INLs very far seaward. Even in summer, the INLs off Washington can be prominent (Figs. 7.14 and 7.15), as much source material remains on the shelf to be resuspended under appropriate current or wave events.

7.5 SUMMARY AND CONCLUSIONS

Seasonal changes in the prevailing wind directions and speeds over the continental shelf result in complex seasonal shelf circulation patterns which in turn yield changing seasonal distributions of SPM. The three-layered structure of the shelf SPM which persists through spring and summer gives way to a deeper mixed layer and more uniform SPM distribution as the prevailing northwesterly winds yield to winter southerlies.

In summer the surface turbid layer is characterized by a relatively large fraction of biogenic, chlorophyllous particles. The bottom nepheloid layer contains a relatively greater proportion of finer particles than the surface layer, and the total near-bottom particle field is relatively enriched in mineral silica and almost devoid of chlorophyll. The relatively few particles in the mid-depth clear layer have a particle size distribution and general composition most similar to particles in the surface turbid layer, though these mid-depth particles do not simply settle out from the surface layer. The zonal component of surface layer currents is offshore at this time of year while that of clear-layer currents is onshore. The sources of surface-layer particles in summer are thus mainly the shallow, near-coast areas and *in situ* growth of phytoplankton stimulated by nutrients upwelled into the well-lit surface waters. The source of the mid-depth clear layer likely is particle-impoverished offshore water. The smallest particles (< 15-μm diameter) from the shelf BNL are moved out over the continental slope at this time of year to form the INL. The

BNL may pulse particles into the INL in response to resuspension events in the BNL and to disruption and resumption of prevailing circulation patterns over the shelf in summer. Similar particle size distributions and chemical compositions in the INL and shelf BNL (relative to the surface layer and mid-depth clear layer) further indicate that the shelf BNL is the main particle source for the INL.

The winter water column over the shelf shows relatively uniform vertical distributions of beam attenuation coefficients, total particle concentrations, biogenic silica, PON and phaeophytin *a*. Chlorophyll *a* and POC concentrations tend to remain higher at the surface, but there are indications of patchy distributions through the winter water column. Total particulate silica concentrations remain higher near the bottom in winter, mainly a function of the vertical distributions of mineral silica. INLs form from shelf BNL material in winter much as they do in summer.

Regardless of time of year, bottom nepheloid water and mid-depth water apparently contain comparable concentrations of bacterial cells as deduced by epifluorescence counting of stained cells. However, the BNL apparently can be a ready source of viable phytoplanktonic cells or spores at all times of year, while mid-depth water rarely contains a viable seed stock from fall through early spring. Once growth is established from seed stock in either BNL or mid-depth water, the growth rate tends to be constant (range between 0.45 and 0.95 doubling d^{-1} at 58 mE m^{-2} s^{-1} constant light and 15° C).

Regardless of time of year, beam attenuation and concentrations of particle volume, POC and PON were always well correlated (usually $r > 0.80$) through depth horizons < 30 m, < 50 m and < 150 m off the central Oregon outer shelf. These correlations usually were lowered markedly when data from the BNL were included in the calculations, indicating a significant difference between most particle characteristics in the BNL and those in the remainder of the water column. Only the correlation between POC and PON remained high ($r = 0.83$) when the BNL data were included. Chlorophyll *a* concentrations in the < 30-m and < 50-m depth horizons did not correlate particularly well with the other attributes examined. Such concentrations were often patchy and not tightly in phase with POC, PON, etc. Conversely, in mid-depth waters (50 to 150 m), the relationships between the small concentrations of chlorophyll and the small concentrations of POC, PON and particle volume must be robust, because all correlation coefficients in the < 150-m depth horizon increased markedly over those coefficients in either the < 30-m or < 50-m layers.

Lastly, we note that there appears to be fundamental differences in the BNLs, and hence in the INLs, off Washington and Oregon. The INLs off Washington appear to have greater particle concentrations and extend farther offshore, than those off Oregon. This is a likely result of 1) the much greater deposit of fine-grained material on the Washington shelf available for resuspension in the BNL and 2) the presence of several canyons intersecting the Washington shelf, which help to more sharply define INLs over the slope.

ACKNOWLEDGEMENTS

Data collection for this chapter was carried out under U.S. Department of Energy contracts EY-76-S-06-2227, Task #29 and DE-AT06-76EV70029. This is a component of Contribution No. 1784 from the School of Oceanography, University of Washington.

REFERENCES

Anderson, G.C. 1964. The seasonal and geographic distribution of primary productivity off the Washington and Oregon coasts. *Limnol. Oceanogr.*, 9:284-302.

Anderson, G.C. 1972. Aspects of marine phytoplankton studies near the Columbia River, with special reference to a subsurface chlorophyll maximum. Pages 219-240 in: A. T. Pruter and D. L. Alverson (eds.), *The Columbia River Estuary and adjacent ocean waters. Bioenvironmental Studies*, Univ. of Washington Press, Seattle, Wa.

Atlas, E.L., L.I. Gordon, S.W. Hager and P.L. Park. 1971. A practical manual for use of the Technicon Auto Analyzer in seawater nutrient analyses; revised. Tech. Rep. 215, Ref. 71-11, Oregon St. Univ., School of Oceanogr., 48 pp.

Bader, H. 1970. The hyperbolic distribution of particle sizes. *J. Geophys. Res.*, 75: 2823-2830.

Baker, E.T. and B.M. Hickey. 1986. Contemporary sedimentation processes in and around an active West Coast submarine canyon. *Mar. Geol.*, 71: 15-34.

Bakun, A. 1975. Daily and weekly upwelling indices, west coast of North America, 1967-1073. NOAA Tech. Rep., NMFS SSRF-693.

Barnes, C.A., A.C. Duxbury and B.A. Morse. 1972. Circulation and selected properties of the Columbia River effluent at sea. Pages 41-80 in: A.T. Pruter and D.L. Alverson (eds), *The Columbia River Estuary and adjacent ocean waters. Bioenvironrmental studies.* Univ. of Washington Press, Seattle, Wa..

Bartz, R., J.R.V. Zaneveld and H. Pak. 1978. A transmissometer for profiling and moored observations in water. *Proc. Soc. Photo-Opt. Instrum. Eng.*, 160: 102-108.

Bishop, J.K., J.M. Edmond, D.R. Ketten, M.P. Bacon and W.B. Silker. 1977. The chemistry, biology and vertical flux of particulate matter from the upper 400 m of the equatorial Atlantic Ocean. *Deep-Sea Res.*, 24: 511-548.

Bruland, K.W. and M.W. Silver. 1981. Sinking rates of fecal pellets from gelatinous zooplankton (salps, pteropods, doliolids). *Mar. Biol.*, 63: 295-300.

Duncan, J.R., L.D. Kulm and G.B. Griggs. 1970. Clay-mineral composition of Late Pleistocene and Holocene sediments of Cascadia Basin, Northern Pacific Ocean. *J. Geol.*, 78: 213-221.

Eggiman, D.E., F.T. Manheim and P.R. Betzer. 1976. Dissolution and analysis of amorphous silica in marine sediments. *J. Sediment. Petrol.*, 50: 215-225.

Fowler, S.W. and L.F. Small. 1972. Sinking rates of euphausiid fecal pellets. *Limnol. Oceanogr.*, 17: 293-296.

Gross, M.G. 1966. Distribution of radioactive marine sediment derived from the Columbia River. *J. Geophys. Res.*, 71: 2017-2021.

Hermann A.J., B. M. Hickey, A.J. ,M.R. Landry and D.F. Winter. 1989. Coastal upwelling dynamics. Pages 211-254 in: M.R. Landry and B. M. Hickey (eds.), *The Coastal Oceanography of Washington and Oregon*. Elsevier, Amsterdam, Neth.

Hickey, B.M. 1979. The California Current System- hypothesis and facts. *Prog. Oceanog.*, 8: 191-279.

Hickey, B.M. 1989. Patterns and process of circulation over the shelf and slope. Pages 41-116 in: M.R. Landry and B. M. Hickey (eds.), The Coastal Oceanography of Washington and Oregon. Elsevier, Amsterdam, Neth.

Hickey, B.M., E. Baker and N. Kachel. 1986. Suspended particle movement in and around Quinault submarine canyon. *Mar. Geol.*, 71: 35-83.

Hobbie, J.H., R.J. Daley and S. Jasper. 1977. Use of Nuclepore filters for counting bacteria by epifluorescence microscopy. *Appl. Environ. Microbiol.*, 33: 1225-1228.

Hofmann, E.E., J. M. Klinck and G.A. Paffenhofer. 1981. Concentrations and vertical fluxes of fecal pellets on a continental shelf. *Mar. Biol.*, 61: 327-335.

Hollibaugh, J.T., D.L.R. Seibert and W.H. Thomas. 1981. Observations on the survival and germination of resting spores of three Chaetoceros (Bacillariophyceae) species. *J. Phycol.*, 17: 1-9.

Honjo, S. and M.R. Roman. 1978. Marine copepod fecal pellets: production, preservation and sedimentation. *J. Mar. Res.*, 36: 45-57.

Huyer, A. 1977. Seasonal variation in temperature, salinity and density over the continental shelf off Oregon. *Limnol. Oceanogr.*, 22: 442-453.

Huyer. A., R.L. Smith and E.J.C. Sobey. 1978. Seasonal differences in low-frequency current fluctuations over the Oregon continental shelf. *J. Geophy. Res.*, 83: 3495-3505.

Jacq, E. and D. Prieur. 1985. Comparative distribution of free-living and particle-bound bacteria in the Quessant frontal system. *Oceanis,* 11: 223-234 (In French, Eng. abstract).

Kachel, N.B. and J.D. Smith. 1989. Sediment transport and deposition. Pages 287-348 in: M.R. Landry and B. M. Hickey (eds.), *The Coastal Oceanography of Washington and Oregon*. Elsevier, Amsterdam, Neth.

Karlin, R. 1980. Sediment sources and clay mineral distribution off the Oregon coast. *J. Sediment. Petrol.*, 50: 543-560.

Kitchen, J.C. 1978. Particle size distributions and the vertical distribution of suspended matter in the upwelling region off Oregon. M.S. Thesis, Oregon St. Univ., Corvallis, Ore., 118 pp.

Kitchen, J.C., D.W. Menzies, H. Pak and J.R.V. Zaneveld. 1975. Particle size distributions in a region of coastal upwelling analyzed by characteristic vectors. *Limnol. Oceanogr.*, 20: 775-783.

Kitchen, J.C., J.R.V. Zaneveld and H. Pak. 1978. The vertical structure and size distributions of suspended particles off Oregon during the upwelling season. *Deep-Sea Res.*, 25: 453-468.

Krausse, G.L., C.L. Schelske and C.O. Davis. 1983. Comparison of three wet-alkaline methods of digestion of biogenic silica in water. *Freshwater Bio.*, 13: 1-9.

Landry, M.R., J. Postel, W.K. Peterson and J. Newman. 1989. Broad-scale patterns in the distribution of hydrographic variables. Pages 1-40 in: M.R. Landry and B. M. Hickey (eds.), *The Coastal Oceanography of Washington and Oregon*. Elsevier, Amsterdam, Neth.

McCave, I.N. 1975. Vertical flux of particles in the ocean. *Deep-Sea Res.*, 22: 491-502.

Mooers, C.N.K., C.A. Collins and R.L. Smith. 1976. The dynamic structure of the frontal zone in the coastal upwelling region off Oregon. *J. Phys. Oceanogr.*, 6: 3-21.

Morse, B.A., M.G. Gross and C.A. Barnes. 1968. Movement of seabed drifters near the Columbia River. J. Waterways Harbors Div., *Am. Soc. Civil Eng.*, 94: 93-103.

Nittrouer, C.A. 1978. The process of detrital sediment accumulation in a continental shelf environment: an examination of the Washington shelf. Ph.D. Diss., Univ. of Washington, Seattle, Wa., 243 pp.

Paasche, E. 1973. Silicon and the ecology of marine plankton diatoms. I. Thalassiosira pseudonana (Cyclotella nana) grown in a chemostat with silicate as limiting nutrient. *Mar. Biol.*, 19: 117-126.

Paffenhofer, G.A. and S.C. Knowles. 1979. Ecological implications of fecal pellet size, production and consumption by copepods. *J. Mar. Res.*, 37: 35-49.

Pak, H., G.F. Beardsley, Jr. and R.L. Smith. 1970. An optical and hydrographic study of a temperature inversion off Oregon during upwelling. *J. Geophys. Res.*, 75: 629-636.

Pak, H. and J.R.V. Zaneveld. 1977. Bottom nepheloid layers and bottom mixed layers observed on the continental shelf off Oregon. *J. Geophys. Res.*, 82: 3921-3931.

Pak, H. and J.R.V. Zaneveld. 1978. Intermediate nepheloid layers observed over the continental margins off Oregon. *Proc. Soc. Photo-Opt. Instrum. Eng.*, 160: 9-17.

Pak, H., J.R.V. Zaneveld and J. Kitchen. 1980. Intermediate nepheloid layers observed off Oregon and Washington. *J. Geophys. Res.*, 85: 6697-6708.

Perry, M.J. and J.P. Bolger and D. English. 1989. Primary production. Pages 117-138 in: M.R. Landry and B. M. Hickey (eds.), The Coastal Oceanography of Washington and Oregon. Elsevier, Amsterdam, Neth.

Pittock, H.L., W.E. Gilbert, A. Huyer and R.L. Smith. 1982. Observations of sea level, wind and atmospheric pressure at Newport, Oregon 1967-1980. Data Rep. 98, Ref. 82-12, College of Oceanogr., Oregon St. Univ., Corvallis, Ore.

Shuman, F.R. and C.J. Lorenzen. 1975. Quantitative degradation of chlorophyll by a marine herbivore. *Limnol. Oceanogr.*, 20: 580-586.

Small, L.F., H. Curl, Jr. and W.A. Glooschenko. 1972a. Estimates of primary production off Oregon using an improved chlorophyll-light technique. *J. Fish Res. Bd. Canada*, 29: 1261-1267.

Small, L.F., H. Curl, Jr. and W.A. Glooschenko. 1972b. Effects of solar radiation and upwelling on daily primary production off Oregon. *J. Fish. Res. Bd. Canada*, 29: 1269-1275.

Small, L.F., S.W. Fowler, S.A. Moore and J. LaRosa. 1983. Dissolved and fecal pellet carbon and nitrogen release by zooplankton in tropical waters. *Deep-Sea Res.*, 30: 1199-1220.

Small, L.F., S.W. Fowler and M.Y. Unlu. 1979. Sinking rates of natural copepod fecal pellets. *Mar. Biol.*, 51: 233-241.

Small, L.F., G.A. Knauer and M.D. Tuel. 1987. The role of sinking fecal pellets in stratified euphotic zones. *Deep-Sea Res.*, 34: 1705-1712.

Small, L.F. and D. W. Menzies. 1981. Patterns of primary productivity and biomass in a coastal upwelling region. *Deep-Sea Res.*, 28: 123-149.

Small, L.F. and D. A. Ramberg. 1971. Chlorophyll *a,* carbon, and nitrogen in particles from a unique coastal environment. Pages 475-492 in: J.D. Costlow (ed.), Fertility of the sea, Vol. 2. Gordon and Breach.

Smetacek, V.S. 1985. Role of sinking in diatom life-history cycles: ecological, evolutionary and geological significance. *Mar. Biol.*, 84: 239-251.

Strickland, J.D.H. and T.R. Parsons. 1972. A practical handbook of seawater analysis. *Fish. Res. Bd. Canada Bull.* 167 (2nd ed.)., 310 pp.

Turner, J.T. 1977. Sinking rates of fecal pellets from the marine copepod *Pontella meadii. Mar. Ecol. Prog. Ser.*, 40: 249-259.

Turner, J.T. 1984. Zooplankton feeding ecology: contents of fecal pellets of the copepods *Eucalanus pileatus* and *Paraclanus quasimodo* from continental shelf waters of the Gulf of Mexico. *Mar. Ecol. Prog. Ser.*, 15: 27-46.

Welschmeyer, N.A. and C.J. Lorenzen. 1985. Chlorophyll budgets: Zooplankton grazing and phytoplankton growth in a temperate fjord and the Central Pacific Gyres. *Limnol. Oceanogr.*, 30: 1-21.

Zaneveld, J.R.V. and H. Pak. 1973. Method for the determination of the index of refraction of particles suspended in the ocean. *J. Opt. Soc. Amer.*, 63: 321-324.

CHAPTER 8

SEDIMENT TRANSPORT AND DEPOSITION ON THE WASHINGTON CONTINENTAL SHELF

N. B. Kachel[1] and J. D. Smith[2]
[1]*School of Oceanography, University of Washington, Seattle, WA 98195*
[2]*Geophysics Program, University of Washington, Seattle, WA 98195*

8.1 INTRODUCTION

Geological and physical oceanographers during the past twenty years have collected a large body of data relevant to the description of the sediments and sediment transport processes that occur on the Washington continental shelf. These include studies of the currents (Smith and Hopkins, 1972; Smith *et al.*, 1976; Hickey, 1981; Huyer *et al.*, 1978; Hickey, 1984; Battisti and Hickey, 1984; in particular, Hickey, 1989, Chapter 2) and of the wave climate (Larsen and Fenton, 1974; Sternberg and Larsen, 1975) on the shelf, as well as investigations of the spatial variation of sediment texture (McManus, 1972), composition (White, 1970a) and accumulation rates (Nittrouer and Sternberg, 1981). Theoretically sound fluid mechanical models of the structure of the benthic boundary layer over this and other continental shelves (Smith and Long, 1976; Long, 1981) have been developed, as have general theories for the non-linear interaction of the boundary layers associated with waves and currents (Smith, 1977; Grant and Madsen, 1979). Finally, these latter models have been coupled to the sediment transport theories of Smith and McLean (1977a,b) and to a procedure for calculating sediment bed armoring in order to create a numerical model of sediment erosion, transport and deposition on the Washington continental shelf (Kachel, 1980; Kachel and Smith, 1986).

Observations of the resuspension of bottom sediments at mid- to outer-shelf depths during winter storms have been made on the Washington continental shelf by Sternberg and Larsen (1976) and Hickey *et al.* (1986), off northern California (Drake and Cacchione, 1985, 1986; Caccione and Drake, 1982; and Cacchione *et al.*, 1982, 1987) and elsewhere by Butman (1987), Drake *et al.* (1980) and Drake and Cacchione (1986). Drake and Cacchione (1986) observed that substantial increases in suspended sediment that occurred during the winter months on the northern California shelf and were primarily caused by the coincidence of larger waves (with periods >12 s) from distant storms and strong currents associated with local storms. Cacchione *et al.* (1987) concluded that "the resultant high bottom stress due to the combined effects of waves and currents is a major factor in controlling the distribution of bottom sediments on the central portion of the northern California shelf."

There are two reasons for using numerical models to study continental shelf sediment transport mechanisms. First, the system of interest is very complex due to interactions among wave and current boundary layers, stratification of the water column induced by suspended sediment and composition of the bed, so the use of mathematical models is necessary to evaluate the specific contributions of the various component processes. Second, the major sediment transport events are driven by conditions too infrequent and too severe to be investigated effectively by standard observational techniques. Sediment transport is highly non-linear with respect to current velocity and wave amplitude; therefore, the infrequent extreme events arising from a combination of strong storm-driven circulation and large swell are the conditions that mark the stratigraphic record of a continental shelf such as the one bordering the state of Washington (Kachel and Smith, 1986). These extreme events erase the accumulated effect of smaller storms that occur in the intervening years.

In this chapter, we compile the current and wave data available for the southern Washington continental shelf in order to characterize typical and extreme sediment transporting events in terms of their, magnitudes, durations and their frequencies of occurrence. We then use this information both as input to a sediment transport model and as an aid to understanding the relative importance of transport processes on this shelf. Our model, as presented here, includes the feedback interactions between waves, currents, suspended sediment-induced density stratification and the size composition of the sediment bed that are necessary to predict the structure of the benthic boundary layer, the depths of erosion and reworking of sediment, the thicknesses and compositional changes in the reworked sediment and the paths and the rates of sediment transported as bedload and as suspended load. The model is first used to investigate the effect of resuspended sediment on the shear velocities associated with the waves and currents in interacting boundary layers. Then the predicted structure of the combined benthic boundary layer is presented for an infrequent storm event coinciding with the arrival of large swells. These results are then compared with those from four other sets of input conditions. Next, the climatologic conditions for a coincidence of waves and currents, expected to occur once every 9.6 y, are input to the model and the results are presented for a set of model runs at a series of locations along and across the continental shelf. These latter results establish patterns of erosion, reworking, deposition, transport and sediment redistribution on the Washington continental shelf during the physical oceanographic events that appear to have the greatest geological impact. Finally, the divergence of the sediment transport is used in combination with the expected frequency of occurrence of wave-current events to calculate the average annual deposition and the thickness of the Holocene mid-shelf silt deposit produced from Columbia River-derived sediment.

8.2 DESCRIPTION OF THE WASHINGTON CONTINENTAL SHELF

8.2.1 Geologic Setting

The geologic setting of the continental margin of the Pacific Northwest is typical of that of other convergent plate margins, that is, it is relatively narrow and adjacent to a mountainous coast. The oceanic crust offshore of this region is part of the Juan de Fuca Plate, which is moving eastward from the Juan de Fuca and Gorda ridges and obliquely subducting under the North American plate. Pleistocene thrust-folding and uplift of Cascadia Basin sediments has caused the formation of a series of en-echelon anticlinal ridges with intervening sediment-filled basins. Many of these have been incorporated locally into the structure of the continental slope

(Barnard, 1978; von Huene and Kulm, 1973). The continental shelf adjacent to Washington and Oregon varies in width between 25 and 60 km, with a shelf break depth of 150 to 200 m (Byrne, 1963; McManus, 1972). This shelf is cut by a series of submarine canyons that head at a depth of approximately 130 m and reduce the shelf width to 15-30 km. According to Barnard (1978) during the last glacial epoch and prior to 10,500 B.P., all the submarine canyons incising the upper slope were probably active conduits for turbidity currents. In contrast, during the early postglacial period most of the coarse material transported by small coastal streams was probably trapped on the shelf, thus forming a transgressive deposit. The Columbia River, which had previously emptied into Astoria Canyon, probably has had a greater percentage of its load distributed northward to Willapa Canyon since sea level began to rise. According to Barnard (1978) substantial turbidity current activity occurred in Willapa Canyon until 6,750 B.P., but after sea level attained approximately its present position (6,750 y B.P.), most of the canyons that incise the slope off Washington and Oregon have been filling mainly with hemipelagic sediment. Quinault Canyon, however, has acted as a funnel for sediment moving northward along the continental shelf. In this system fine sediment accumulates for about 500 y between flushings by turbidity currents (Barnard, 1978).

The bathymetry of the Washington shelf (Fig. 8.1) can be generally described as the smooth topography of a sediment wedge with contour lines trending approximately N 15° W and diverging systematically to the north by roughly 10 degrees (McManus, 1972). Exceptions to the pattern of smooth topography, however, do occur. They are: on the outer shelf just north of Astoria Canyon where hummocks give the region a local relief of up to 10 m; on the mid-shelf region off Willapa Bay, which has submerged pinnacles 20-m high; and on the inner shelf north of Grays Harbor, which has many small islands, stacks and submerged outcrops (Nittrouer, 1978). The change in inner shelf topography reflects that of the adjacent coastal physiography varying from a coastal plain in the south to a region of active geologic uplift in the north. The latter is characterized by seacliffs, headlands, pocket beaches and small local rivers. Off Oregon, submarine banks form prominent shoals on the outer portion of the shelf. Rock outcrops dot the inner shelf between Coos Bay and Cape Blanco. Small swales or valleys occur behind the banks (Kulm *et al.*, 1975), but substantial sediment accumulation is rare.

Studies of shallow seismic stratigraphy on the Washington shelf (Bennett and Grim, 1968; Nittrouer and Sternberg, 1981) delineate numerous piercement structures that have been compared to the Miocene siltstone melanges intruded into Pliocene rocks on the Olympic Peninsula. On the shelf most of these structures have been covered by Holocene sediments, but the resistant cores of some structures (typically located mid-shelf near 46° 40') protrude as pinnacles up to 20-m above the surrounding sea floor. The gently folded siltstone bedrock exposed in some places near the shelf break has been identified by Nittrouer (1978) and by Byrne (1963) as Pliocene in age. This interpretation is corroborated by data from core 176 taken by the Deep Sea Drilling Project (Kulm and von Huene, 1973) at a depth of 193 m, just south of Astoria Canyon. This core shows major periods of uplift and erosion ranging in age from Eocene to Miocene, covered by a Pliocene shale which, in turn, is overlain by a Pleistocene

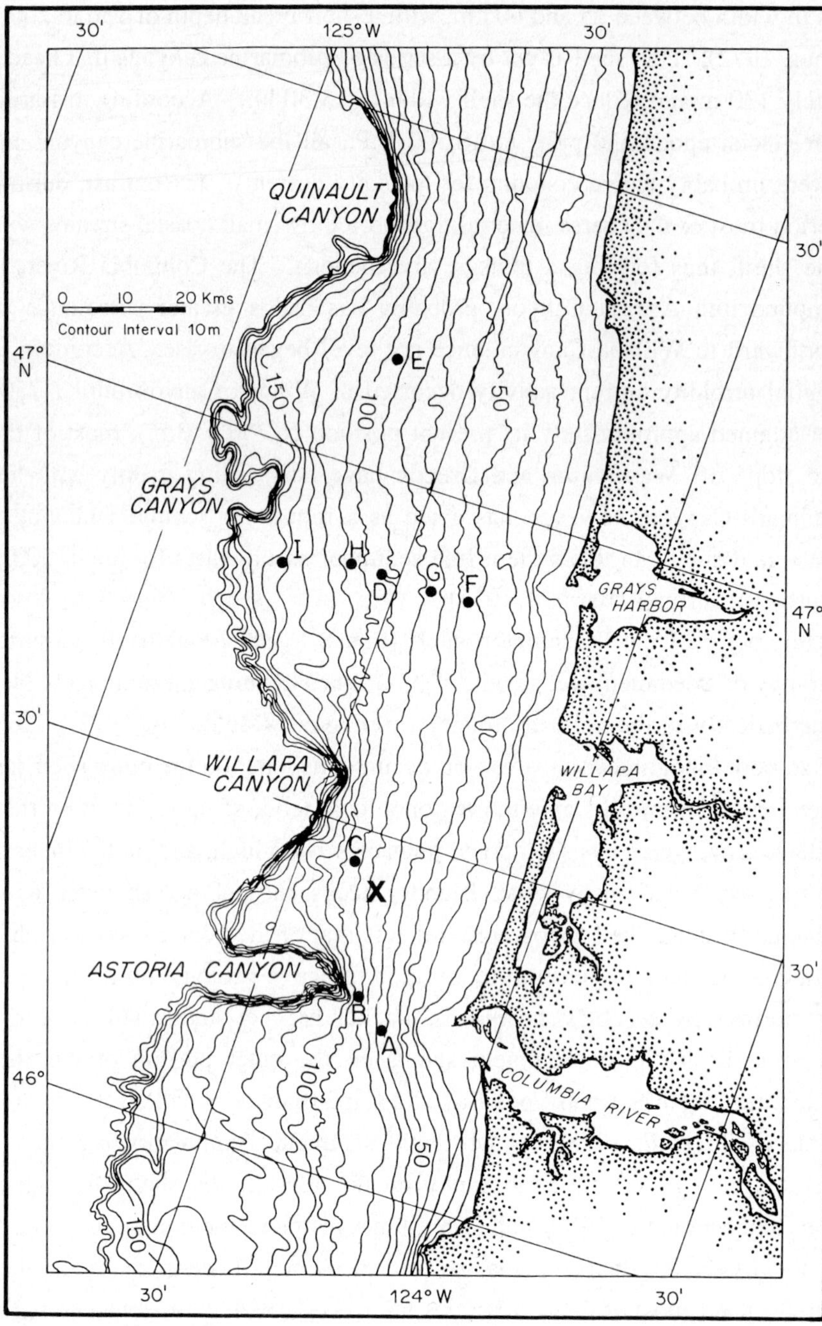

Fig. 8.1 Bathymetry of the Washington continental shelf with 10-m contour interval. Lettered dots mark locations of model runs.

regressive and transgressive sequences composed of silts and coarser sediments. Based on gravity cores and shallow seismic records taken north of the Columbia River, Nittrouer (1978) and Nittrouer and Sternberg (1981) describe the shallow stratigraphy in this region in terms of a

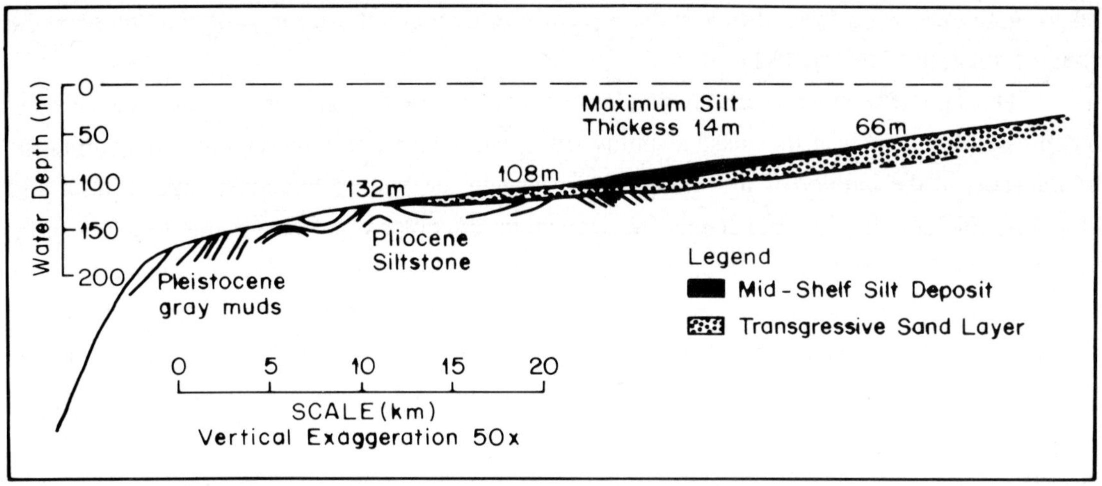

Fig. 8.2 Shallow seismic stratigraphy across the shelf on line from F to I in Figure 8.1 which coincides with Nittrouer's experimental transect. After Nittrouer (1978).

classic transgressive sequence in which a basal sand unit represents the inner shelf environment migrating eastward as sea level rose. Figure 8.2 shows this sand overlain at mid-shelf depths by a lens-shaped, mid-shelf silt deposit with a maximum thickness of 14 m on a line west of Grays Harbor. This layer probably did not begin to form until nearly 7,000 y ago, when sea level reached its present elevation and modern shelf circulation patterns were established.

8.2.2 Circulation

The circulation on the Washington continental shelf has been described by Barnes *et al.* (1972), Smith and Hopkins (1972), Hickey (1981, 1984), Battisti and Hickey (1984) and Werner and Hickey (1983). The most complete description is given by Hickey (1989, Chapter 2). The seasonal pattern that dominates the shelf circulation develops in response to large scale atmospheric flow. During winter, low pressure cells originating in the Gulf of Alaska cause a change in the dominant wind direction from northerly during fair weather to southerly or south-westerly during storms. This response of the seasonal circulation to atmospheric forcing is addressed at length in Chapter 2. During summer, a southward, alongshelf wind stress and an opposing northward pressure gradient force result in a southward near-surface flow and a northward undercurrent over the shelf and slope. During winter, both the wind stress and pressure gradient reverse directions, producing northward flow over the shelf and in the upper water column over the slope and, at times, a southward undercurrent over the slope. During spring, both the wind stress and alongshelf pressure gradient force are southward, resulting in maximum equatorward flow. Hickey's observations illustrate that as much as 70-90% of the variance in the alongshelf flow over the mid-shelf at this time of the year, 50% of the variance in the flow on the outer shelf and most of the fluctuations in the coastal sea level can be explained as first mode coastal trapped waves. These waves are generated primarily off northern California, particularly during late spring and summer when coastal wind stress increases southward from Washington to California. In contrast, motion of the shelf water during the winter is dominated by the response to local wind stress, because the local winds are much more energetic than those to the south (Battisti and Hickey, 1984).

During fair weather, northerly winds cause an offshore Ekman transport near the surface, a depressed sea level near the coast, a southward geostrophic current and an onshore component of transport in the interior of the water column and in the bottom boundary layer (Smith and Hopkins, 1972; Hickey, 1989, Chapter 2). Upwelling of cooler water occurs due to the onshore bottom flow. This produces a strong pycnocline in the mid- to outer shelf region and results in a baroclinic component of the alongshelf flow that opposes the barotropic component, hence, causing a strong vertical shear. Weak northward flow occurs near the sea bed due to the presence of the northward undercurrent. Bottom stresses under these conditions, however, are almost always too low to cause sediment transport. During winter storms, southerly winds force an onshore Ekman transport in the surface layer which results in a northward geostrophic current. The northward interior geostrophic flow drives a bottom boundary layer with net offshore near-bottom transport. In the absence of a strong undercurrent opposing the barotropic flow, bottom stresses are much higher than during the summer. The offshore turning in the bottom Ekman layer during winter is illustrated by the vector diagrams in Figures 8.3a and b (after Smith and Long, 1976).

For the purposes of constructing a theoretical model of typical or extreme sediment transport events in an area, it is necessary to compile information on the observed frequency, intensity and duration of events likely to move sediment, as well as on the pattern of vertical and cross-shelf variations in the flow. These data are used to construct forcing functions for the sediment transport model which are: the geostrophic (interior) velocity and the bottom wave period and orbital velocity (a function of wave amplitude, period and water depth). The pattern of the cross-shelf variation of the flow during a period with a large storm event is illustrated in Figure 8.3. At both the mid-shelf and offshore sites, during the periods of northward, storm-driven currents, the offshore turning of the velocity vectors increases toward the bottom. The magnitude of the interior current at the 170-m deep outer-shelf site is roughly two-thirds that at the 78-m deep, mid-shelf site. This result is correlated statistically by the EOF (orthogonal eigenfunction) analysis shown by Hickey (1989, Fig. 2.21, Chapter 2) for two other mid-winter data sets: January, 1972 and February, 1973. During those periods the interior velocities at the offshore site are 37 and 84%, respectively, of those at the mid-shelf site. Therefore, in constructing the model at a series of sites across the shelf, the geostrophic velocity at any site was linearly interpolated as a function of depth from that at the mid-shelf location using the approximation that the offshore site velocity is two-thirds that of the mid-shelf.

In order to establish the frequency of occurrence of storm-generated currents capable of resuspending and transporting sediment along the shelf, we used current meter data collected from 1967-1982 under the direction of J.D. Smith or B.M. Hickey (Hopkins, 1971; Smith and Hopkins, 1972; Smith and Long, 1976; Smith *et al.*, 1976; Hickey, 1981; 1984; Hickey *et al.*, 1986; Hickey, 1989, Chapter 2). The temporal coverage of these data in time is presented as Table 1 (Hickey, 1989, Chapter 2). Before 1971, data were collected using Braincon current

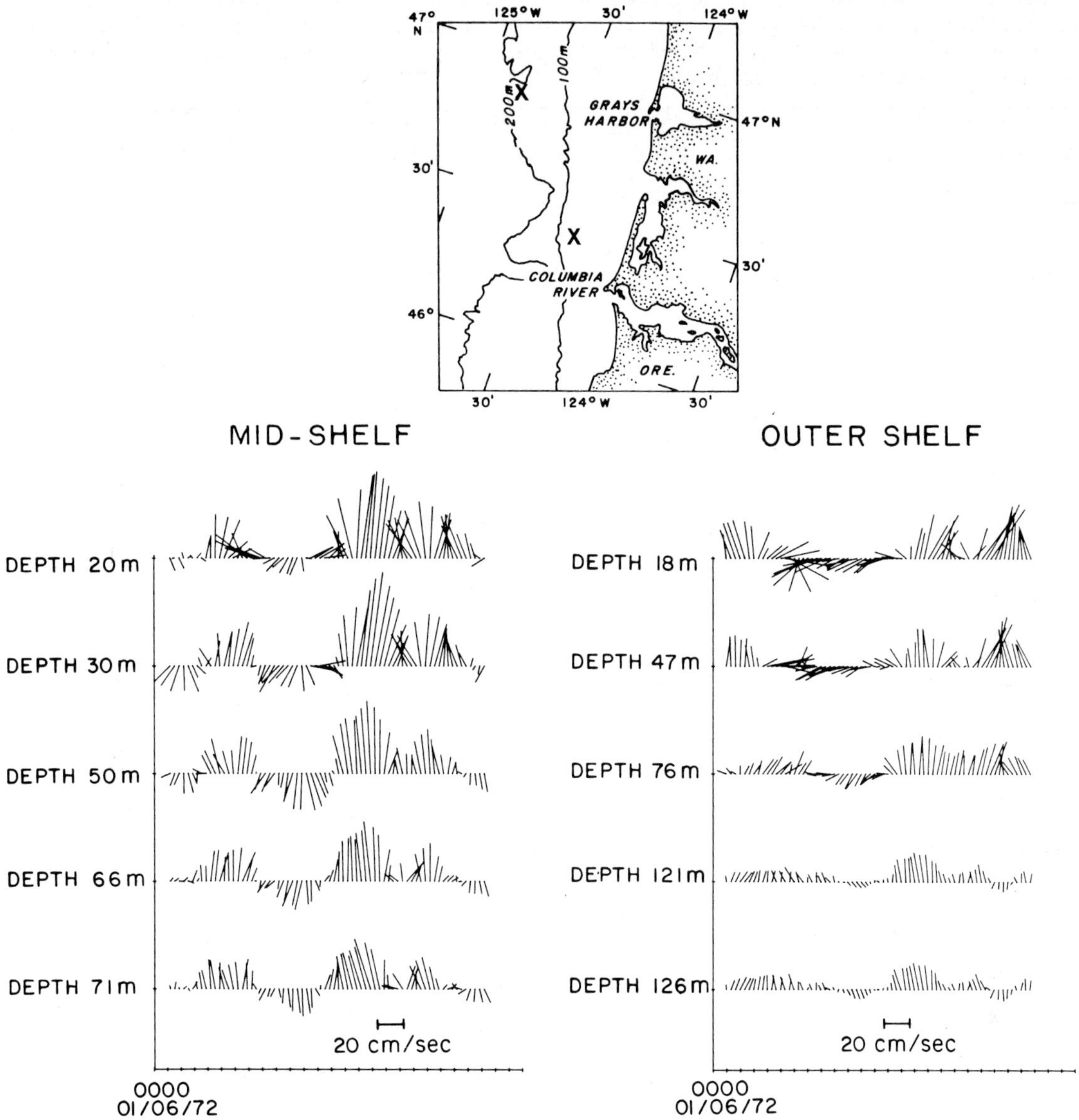

Fig. 8.3 Vector velocities (after Smith *et al.*, 1976) at a mid- and an outer shelf site on the Washington continental shelf during January 1972. Upward-pointing vectors indicate northward flow along the isobaths whereas downward-pointing vectors indicate southward flow along the isobaths. Note that at extended periods of quasi-steady northward flow, the mid- to bottom depth flow is relatively lower at the outer shelf site than it is at the mid-shelf. The locations for the data are marked on the insert.

meters. Subsequently, Aanderaa type current meters suspended from taut-wire moorings were used. Pumping by oscillatory currents due to waves is a potential problem with Aanderaa current meters. However, as shown by Hammond and Collins (1979), differences between indicated and actual unidirectional currents are small when the magnitudes of the wave orbital velocity is smaller than that of the current. This is the case during the time intervals of interest in this

TABLE 8.1 Distribution (h) of current events longer than 12 h with magnitude > 0.50 m s $^{-1}$ at mid-depth or > 0.40 m s $^{-1}$ for 21,179 h of observations made on the Washington continental shelf, 1967-1982. Numbers in parentheses are number of hours for which mid-depth currents were ≥ 0.80 m s $^{-1}$.

Year	OCT	NOV	DEC	JAN	FEB	MAR	Time (h) Sampled by Data
1967-68	0	23 (1)	64	0	--	--	2381
1968-69	--	--	23	0	0	--	1395
1970-71	--	--	--	--	0	0	345
1971-72	0	12	--	--	--	--	707
1972-73	--	--	--	0	0	0	1764
1973-74	0	9	22	59 (7)	--	--	2256
1974-75	--	--	--	0	21	--	1764
1975-76	--	--	--	0	0	0	1966
1977-78	0	17	27 (2)	17	--	--	3417
1980-81	0	0	0	0	--	--	2772
1981-82	27	38 (12)	--	--	--	--	2412
1967-82	27	90 (13)	136 (2)	76(7)	21	0	21179
Average Probability of Occurrence	0.0013	0.0042	0.0064	0.0036	0.0010	0	Total 0.0165

chapter. Additional support for the lack of significant wave pumping, at least on the outer shelf, is provided by an anaylsis of two winters (8 months) of data on the outer shelf off Washington (Hickey, *et al.*, 1986). The authors found that current maxima are not correlated with the occurrence of large near-bottom wave orbital velocities.

To define the characteristics of storm-driven current events on the Washington shelf, hourly averages of the current meter records from 1967-1982 were scanned. It was found that the typical duration of such events is 2-5 d, although it is not uncommon for two or more such events to occur in rapid succession. We define large storm events as those during which mid-depth currents exceeded 0.50 m s $^{-1}$, or with near-bottom (within 6 m of the bed) currents exceeded 0.40 m s $^{-1}$ lasting for 12 h or longer. Table 8.1 presents the distribution of 18 large observed events during the fall and winter months. No such events were observed during the months from April to September. The data collected over the years span a total of 21,179 h (882 d or 4.8 y). Of the fall and winter records, procured between October 1 and March 31, currents exceeding 0.40 and 0.80 m s $^{-1}$ occurred with probabilities of 0.017 and 0.001, respectively. Typically, large storms occurred between the months of November and January while less frequent or less severe storms occurred throughout the other fall and winter months. According to these data an 'average' year will have 72 h of mid-shelf, mid-depth currents greater than

0.50 m s $^{-1}$ and 4 h with currents exceeding 0.80 m s $^{-1}$. It should be noted, however, that the data suggest considerable interannual variation in the distribution of such currents.

8.2.3 Wave Climate

The continental shelf off the Pacific Northwest is subjected to relatively severe waves, producing orbital velocities that occasionally are capable of moving sandy and silty sediments even at the outer edge of the shelf. Sternberg and Larsen (1976) estimated that wave-induced oscillatory currents exceeded the threshold speed for sediment approximately 53 d y $^{-1}$ at 80-m depth in 1973, compared to an estimated 22 d y $^{-1}$ for low frequency currents that year. However, the interaction of the wave and current boundary layers enhances the boundary shear stress associated with each (Smith, 1977; Grant and Madsen, 1979), so that sediments eroded from the bed are diffused higher into the water column than they would be in the absence of waves. At these greater distances from the sea bed the current velocities are higher. The role of waves in shelf sediment transport, therefore, is not only to resuspend bottom sediments for subsequent transport alongshelf and offshelf by currents, but also to increase the rate of sediment transport by the currents. The importance of longer period (16-20 s) waves and swell for sediment transport is that their orbital velocities decay more slowly with depth than those for shorter period waves. For wave amplitudes typical on this continental shelf, it is primarily these longer period waves that cause bottom shear stresses which exceed the threshold of sediment motion at mid- to outer shelf.

There are three major data sources for information on the wave climate on the Washington shelf. These are: (1) three years of hindcast data by National Marine Consultants (NMC) (1961); (2) pressure data collected for 205 d on the top of Cobb Seamount by Larsen and Fenton (1974); and (3) wave-rider buoy data taken off Grays Harbor (Coastal Data Information Program, 1981-1984). Each of these sources provides somewhat different information. Hindcast data is derived from synoptic weather charts by taking wind speed, fetch and duration data to predict the wave spectra produced and the effect that time and distance traveled will have on that spectra at a point of interest some distance away from the generation area. The method used is that of Pierson *et al.* (1955). The usefulness of this type of data is that it gives valuable insight into the meteorological conditions that generate those longer period waves that are able to move sediment at continental shelf depths, as well as providing an estimate of their frequency of occurrence. The pressure data of Larsen and Fenton (1974) provide a continuous eight-month record of the actual wave field at a point 400 km west of the Washington shelf. Sternberg (1986), after reviewing evidence from other wave data sources, concludes that the Cobb Seamount observations were taken during a relatively mild year. The waverider buoy data analyzed by the Coastal Data Information Program provides energy spectra 3-4 times a day at a site on the Washington shelf, but technical breakdowns during the data collection period 1981-1984 biased the record against severe wave conditions.

National Marine Consultants (1961) analyzed deep-water wave conditions for three years (1956-58) on the Washington and Oregon coast to determine the meteorological situations that produced anomalously large waves. They found that the most severe wave conditions are caused by extratropical cyclones that intensify in the eastern Pacific near Japan and move onto the U.S. Northwest coast. Storm winds ahead of warm fronts blow from the south and generate waves with significant heights up to 7 m, while winds associated with cold fronts blow from WSW to NW and can generate waves of 8-10 m significant height. Storms generated in the vicinity of Japan, which then move into the Gulf of Alaska, produce waves in their southwest sector that can be characterized on the Washington coast as long period swell (T = 14-18 s) with heights less than 3 m. When a storm path follows a more southerly course, California experiences its most severe wave conditions, while only low amplitude, west to southwest swell conditions occur on the Washington coast. Intense storms developing in the Western Pacific serve as another source of swell on the Washington coast during the winter months. These storms generate waves that travel a great circle route to arrive on the Northwest coast as very low amplitude, long-period swell from the west-northwest. In addition to these conditions, Snodgrass *et al.* (1966) suggest that swell from storms in the Antarctic may occasionally be large enough to move sediment at depths comparable to those on the middle of the Washington continental shelf, but the impact of such events is reduced because they occur principally during the summer months when currents are weakest. The usefulness of the NMC data is not so much for the significant heights as for the information provided about the distribution of directions during wave events.

The data of Larsen and Fenton (1974) also show the changes in the wave spectrum during a sediment-transporting event. Their data show that a typical history for an event off the Washington shelf begins with longer period waves (17.5 to 23.0 s) of low amplitude followed by successively shorter period waves which first increase, then decrease in amplitude. The typical duration of a sediment-transporting event due to these waves on the mid-continental shelf is 24-32 h, although wave conditions capable of moving sediment for periods up to 80 h in duration were recorded. These longer episodes occur during a series of storms in the Pacific when new waves arrive from an offshore storm before sediment transport has ceased from prior events.

In order to obtain a more detailed picture of the wave climate on this shelf in terms of the frequency of occurrence of joint distributions of period and amplitude, as well as period and direction, two sets of records were re-analyzed. In Tables 8.2 to 8.4 only those combinations of period and amplitude (or height) are included that are capable of producing bottom orbital velocities in 100-m depth (mid-shelf) exceeding Shields' (1936) threshold criterion. For the most easily moved sediment fraction, the very fine sands, found on the mid-shelf off Washington, we calculate that the Shields' criterion for the threshold of sediment motion, $u_{*cr} = 0.011$ cm s^{-1}, is met when waves with periods in the range of 8-16 s have bottom orbital velocities $u_0 = 0.07$ to 0.08 m s^{-1}. This is consistent with the empirically derived criteria of Sternberg and Larsen (1975). To exceed the threshold criterion in water approximately 100-m deep, waves with periods of 10-12 s must have heights of approximately 3 m, 12-14 s waves must be 1.8 m in height, 14-16 s waves must be 1.2 m in height and 18-20 s waves must be approximately 0.6 m high.

TABLE 8.2 Joint percent frequency distribution of predicted waves with sufficient significant period and height to move sediment at 100-m depth on the Washington continental shelf. Distributions were derived from hindcast data of National Marine Consultants (1961) for the years 1956-1958. The numbers in italics are associated with conditions in which $u_{0cr} < u_0 < 0.30$ m s^{-1}, those in normal style with 0.30 m s^{-1} $< u_0 <$ 0.60 m s^{-1} and those in bold style with $u_0 > 0.60$ m s^{-1}.

Significant Height, H_s (m)	Significant Period (s) 10-12	12-14	14-16	16-18	18+	Total	No. of Hours with H_s Exceeding
1.0-2.0		*4.20*	*3.38*	*1.04*	*0.04*	8.66	215
2.0-3.0	*2.33*	*1.65*	*0.94*	0.17	0.01	5.10	1391
3.0-4.0	*3.71*	*1.24*	*0.55*	0.08	0.02	5.60	945
4.0-5.0	*1.98*	*0.59*	0.33	0.04	**0.00**	2.94	279
5.0-6.0	*1.26*	0.34	0.07	**0.08**	**0.01**	1.76	197
6.0-7.0	*0.02*	0.17	0.12			0.31	43
7.0-8.0		0.14				0.14	16
8.0-9.0		0.04				0.04	4
Total	9.30	8.37	5.39	1.41	0.08	24.5	
h y^{-1}	815	733	472	123	7	2150	

The first data set analyzed, that from National Marine Consultants (1961), represents the results of three years of wave-hindcast information computed every six hours in accordance with the statistical methods of Pierson *et al.* (1955). The synoptic weather maps that were used were from the years 1956, 1957 and 1958, years chosen by National Marine Consultants (NMC) because they were consecutive, yet significantly different from each other in terms of storm frequency. In the present analysis the assumption is made that sediment moves if an observed or predicted wave has a particular significant height and period associated with a boundary shear stress at 100-m depth exceeding Shield's threshold for sediment motion. It should be noted that even if the waves are able to move the sediment present, these particles will not be transported a net distance over time unless some unidirectional current is present. The resulting joint distribution of significant height and period of the NMC predicted waves is shown in Table 8.2. The areas of the table in normal and bold type delineate those conditions associated with bottom orbital velocities (u_0) at 100-m depth greater than 0.30 m s^{-1} and 0.60 m s^{-1}, respectively. These data suggest that during an average year the threshold condition for sediment movement is exceeded by the near-bottom flow due to waves alone 24.6% of the time (or 90 d y^{-1}) on the mid-continental shelf; orbital velocities exceed 0.30 m s^{-1} for 2.1% of the time (7.7 d y^{-1}) and exceed 60 m s^{-1} for 0.1% of the time (0.4 d y^{-1}). In the most common case, the significant height of those waves moving sediment is 1-2 m with an associated wave period of 12-16 s. The

TABLE 8.3 Joint percent frequency distribution of direction and period of predicted waves and swell able to move sediment at 100-m depth on Washington shelf, derived from hindcast data of National Marine Consultants (1961) for the years 1956-1958.

Significant Period (s)	Direction NNW	NW	WNW	W	WSW	SW	SSW	S	SSE
10-12	--	0.50	1.51	1.62	0.55	0.77	0.74	0.89	0.37
12-14	0.07	0.68	1.96	2.38	0.84	0.48	0.90	0.91	0.16
14-16	0.02	0.33	1.77	2.36	0.40	0.35	0.16	--	--
16-18	--	0.02	0.51	0.98	0.26	0.08	--	--	--
18+	--	--	0.06	0.08	--	--	--	--	--
Total	0.09	1.53	5.81	7.42	2.09	1.68	1.80	1.80	0.53

predominant directions calculated for the sediment transporting waves in deep water are between WNW and WSW with due west as the modal direction (Table 8.3).

The second, and best, wave data set available is that of Larsen and Fenton (1974), which consists of 30 min of 10 kHz pressure data collected every eight hours for 205 d by a bottom mounted sensor on Cobb Seamount, a site 465 km west of the shelf break off southern Washington. Larsen and Fenton calculated spectra and wave statistics for each of the 2,039 point records. Sternberg and Larsen (1975) found excellent correlation between this wave record and pressure fluctuations measured at an 80-m deep continental shelf site that was occupied for 17 d in March 1973. They concluded that the wave conditions recorded over Cobb Seamount were responsible for the pressure fluctuations measured 9-10 h later on the Washington shelf. Sternberg and Larsen (1976) used the Cobb record to calculate bottom orbital velocity, u_0, at a depth of 75 m, from which they decided that their threshold criterion for sediment movement was exceeded approximately 53 d, with almost all movement confined to the months between November and March. Here we have re-analyzed Larsen and Fenton's (1974) data to yield the frequency of occurrence of sediment transporting waves at the 100-m depth in terms of joint distribution by amplitude and period. Each spectral record was inspected to determine if any frequency band had sufficient amplitude to produce bottom orbital velocities exceeding the threshold for initial motion. If more than one frequency band exceeded this criterion, the dominant one was recorded (Table 8.4). The results show that during the 1972-73 season threshold conditions were exceeded 13.8% of the year (50 d y $^{-1}$) at the 100-m depth while u_0 exceeded 0.30 m s $^{-1}$ for 4.9% of the year (17.9 d) and u_0 exceeded 0.60 m s $^{-1}$ for 0.08% of the year (3 d). The difference between the Table 8.4 results and those presented in Table 8.2 is explained in part by inter-annual variations in the wave climate as suggested by Sternberg and Larsen (1976), but also certainly reflects the difference in the type of calculations made on the different data sets. The periods associated with the predicted significant heights (the NMC data) do not

TABLE 8.4 Joint distribution of annual percent frequency by period and amplitude of waves able to move sediment at 100-m depth on the Washington continental shelf, derived from spectral data of Larsen and Fenton (1974). The numbers in italics are associated with conditions in which $u_{0_{cr}} < u_0 < 0.30$ m s^{-1}, those in normal style with 0.30 m s$^{-1} < u_0 < 0.60$ m s^{-1}, and those in bold style with $u_0 > 0.60$ m s^{-1}.

Amplitude (m)	Dominant Period (s) 10-12	12-14	14-16	16-18	18-20	20-22	22+	Total	Cumulative % of Time with Amplitude Exceeding
0-0.5							*0.09*	0.09	13.68
0.5-1.0		*0.27*	*0.37*	*1.10*	*0.36*	0.37	0.09	2.45	13.59
1.0-1.5		*0.27*	*1.37*	*2.10*	*1.10*	0.37	0.18	5.39	11.14
1.5-2.0		*0.37*	0.73	1.83	0.37	0.18		3.48	5.75
2.0-2.5			0.27	0.82	**0.37**	**0.09**		1.55	2.27
2.5-3.0			0.09	**0.09**	**0.18**			0.36	0.72
3.0-3.5				**0.27**	**0.09**			0.36	0.36
Total	0.0	0.91	2.83	6.21	2.47	1.01	0.36	13.78	
h y^{-1}	0	80	248	544	216	88	32	1208	
Hours with Period Exceeding		1208	1128	880	336	120	32		

indicate the presence in the spectra of any longer period components which can be present and have sufficient energy to move sediment in deeper water; therefore, the data in Table 8.2 under-estimate the occurrence of the longer period swell that produce the higher bottom orbital velocities. This is indicated by the modal period of 12-14 s in Table 8.2 compared to the modal value of 16-18 s in Table 8.4.

A third source of wave data on this shelf is from the Coastal Data Information Program, 1981-1984. These data indicate sediment movement could have occurred 7 to 11.6% of the time (found by the same technique used to analyze Larsen and Fenton's data) during the October-March period over three years (1981-1984); however, instrument failures were frequent (30-60% of the time) and usually occurred during the severe wave conditions of greatest interest. Therefore it appears that the Larsen and Fenton (1974) data set and thus Table 8.4 provides the most accurate and complete information currently available for the period and amplitude of waves on the Washington shelf. Their statement that the winter of 1972-73 was less severe than usual is corroborated by the current meter records in Table 8.1. As the wave rider buoy system becomes reliably functional during more extreme conditions, it should provide an excellent source of information on the wave climate.

Conclusions from this analysis of available wave data are: 1) the dominant direction of approach for waves able to resuspend mid-shelf sediment are west to west-northwest [from the

TABLE 8.5 Joint probability distribution that conditions of waves and storm currents exceeding listed values will occur for at least 12 consecutive hours at 100-m depth on the Washington continental shelf between October 1 and March 31. Only those occurrences are included in which currents exceeded 0.50 m s $^{-1}$ for 12 h or more. At other velocities the total number of hours of observed currents were used in the calculation of frequency of occurrence. Values presented indicate the expected number of such periods in a single year.

$\|u_0\|$ (m s $^{-1}$)	$\|u_g\|_{max}$ (m s $^{-1}$) 0	> 0.50*	> 0.60	> 0.70	> 0.80
0		5.94	3.28	1.16	0.374
$u_{0_{cr}}$	49.3	0.860	0.432	0.115	0.050
0.30 (T=16 s)	33.1	0.540	0.299	0.104	0.035
0.60 (T=16 s)	9.61	0.097	0.054	0.019	0.006

National Marine Consultants' (1961) data]; and 2) the bottom orbital velocity in 100 m of water exceeds threshold criteria for 50 d y $^{-1}$(13.8%) and exceeds 0.30 d y $^{-1}$ (0.08%) [from Larsen and Fenton's (1974) data]. These are the values used in calculating the joint probabilities of occurrence of various combinations of waves and storm currents in the model and the discussion that follow.

8.2.4 Joint Occurrences of Waves and Currents

Modeling experiments have demonstrated that sediment transport on the Washington continental shelf is dominated by events in which large wind-driven currents occur at the same time as long period (≥ 16 s) waves (Kachel and Smith, 1986). The frequency of occurrence of such joint events can be estimated from the independent expected frequencies of occurrence of individual wave and current events. Evidence that the wave and current occurrences are independent stems from consideration of the pattern of wave events observed in the Cobb Seamount data (Larsen and Fenton, 1974). Long-period forerunner waves or swell (17-22 s) are followed by successively shorter period waves in the 12-16 s band. Since the long period waves cross the shelf far ahead of the storm that generated them, the concurrence of large swell and currents may be considered statistically independent. The joint probability of occurrence of waves and currents of selected sizes, calculated as $p_{wc} = p_w p_c$, is presented in Table 8.5. In tallying occurrences of currents of magnitude 0.50 m s $^{-1}$, only those events are included in which this value was met or exceeded for a duration of 12 h or more as discussed previously. Owing to the enhancement of the boundary shear stress by wave-current interactions and the use of threshold conditions to limit the wave and current data sets individually, not all of the sediment-transporting events have been identified in the available data sets, but the ones that have been missed are weak and contribute little to the overall sediment flux at a given locality. They also are small compared to the limitations of the data sets. Events of less than 12 h duration have been omitted because the currents over such a short period have limited ability to transport large volumes of resuspended sediment. These small events are too short to develop the thick boundary layers that allow large volumes of sediment to be diffused and transported high into the water column. Therefore, they are of much less importance in the sediment transport history than the larger events.

8.2.5 Sediment Transport

The major source of sediment to the Washington shelf is the Columbia River, which has an estimated average annual sediment discharge of 0.5 to 2.1 x 10^{10} Kg y^{-1} (van Winkle, 1914a,b; Judson and Ritter, 1964; Haushild *et al.*, 1966; Simonstad *et al.*, 1984). Karlin (1980) analyzed the contributions of sediment from sources from Cape Mendocino north to the Strait of Juan de Fuca and found that the Columbia River accounts for only 21.1% of the total sediment load to this region, although it discharges 72.3% of the water. Nevertheless, it is responsible for over 90% of the sediment discharged onto the Washington shelf.

The general surface sediment pattern on the Washington shelf displays a near-shore sand unit extending north along the coast out to a depth of 50 m, the depth at which typical waves in this region are first capable of stirring sediment with sufficient intensity to prevent significant deposition of silt. Seaward of the near-shore sand is a mid-shelf silt deposit that trends north-northwestwardly from the mouth of the Columbia River. It is partially intersected by Quinault Canyon. Within this unit sediment becomes progressively finer with distance from the river. The existence of measurable accumulation rates and the occurrence of clay mineralogy characteristic of a Columbia River source (Nittrouer, 1978) make it apparent that modern clays and silts are being deposited on most of this shelf. Sandy sediments occupy the outer shelf at depths generally deeper than 130 m. The cross-shelf progression of sediment characteristics was illustrated in Figure 8.2 and by the histograms of grain size in Figure 8.4. Nittrouer (1978) and Nittrouer *et al.* (1979), using Pb-210 analysis of sediment cores, found that a zone of high-accumulation rates extends from the Columbia River to Quinault Canyon with an axis corresponding to that of the mid-shelf deposit (Fig. 8.5). Accumulation rates decrease with increasing distance from the river from a maximum value of 1.7 kg m^{-2} y^{-1} near the mouth of Astoria Canyon to value less than 0.20 kg m^{-2} y^{-1} at the shelf break. On the promontory just north of Astoria Canyon and at other sites along the outer continental shelf, at depths greater than 150 m the sediment contains a relict sand component identified by coarser sediment textures (McManus, 1972; Roberts, 1974) and by the presence of both authigenic and reworked glauconite (White, 1970a). The results of investigations of heavy minerals and of gravel deposits (Venkatarathnam and McManus, 1973) indicate that north of Grays Harbor the Columbia River becomes less important as a source of near-shore sand relative to sediment derived locally from the coast.

When Smith and Hopkins (1972) reviewed the size analysis data for sediment on the southern Washington shelf presented by Kelley and McManus (1969; 1970) and McManus (1972), they found that the size distributions are polymodal. At almost all stations across the shelf they found sand modes at 2.5ϕ and 3.0-3.5ϕ and a broad silt mode. Nittrouer's (1978) data

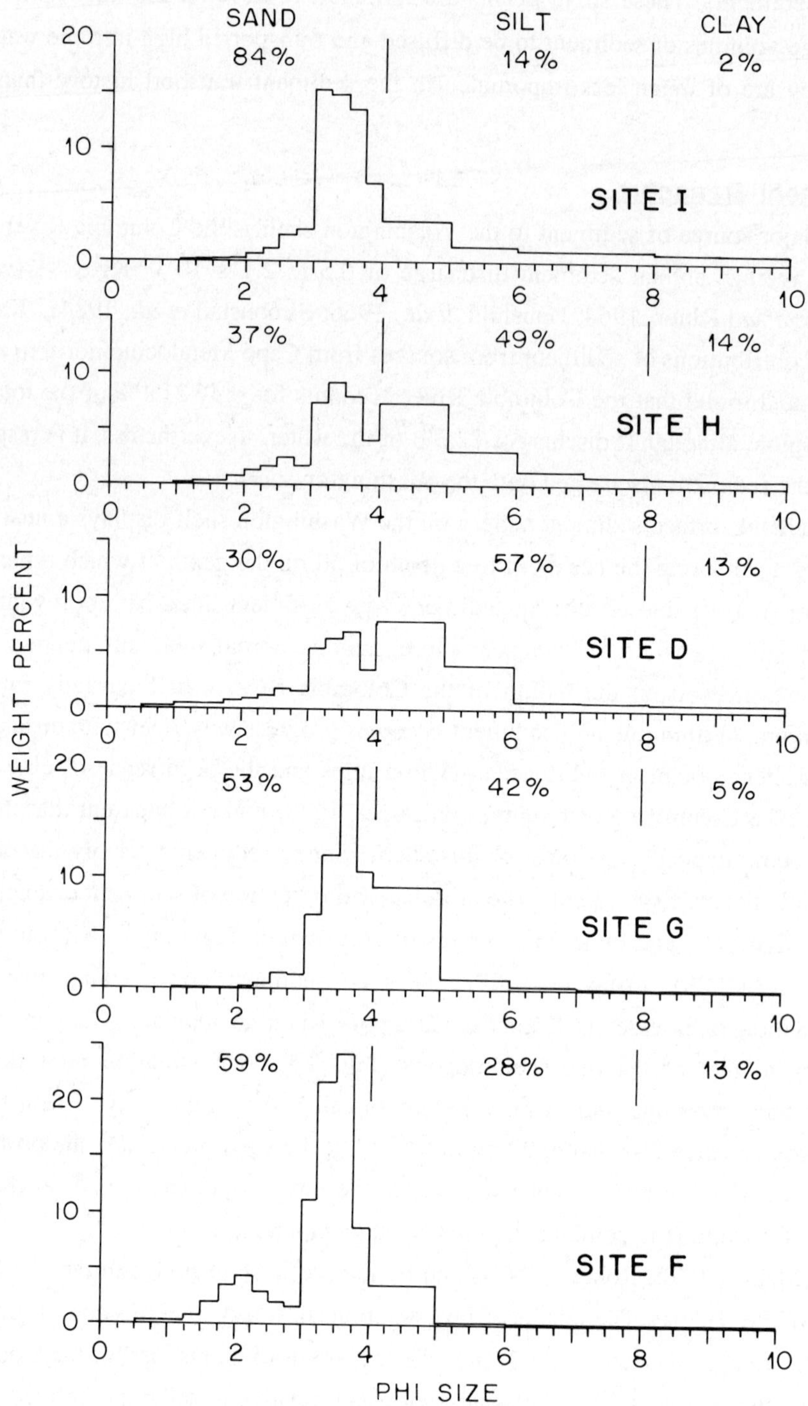

Fig. 8.4 Histograms of grain size observed in surface sediment samples from the cross-shelf transect in Figure 8.1. The data represent average values of sediment distribution for one to six samples taken at each site during August 1976. The data were presented by Nittrouer (1978).

presented in part in Figure 8.4 confirms the polymodality of sediments. The presence of multiple modes indicates that more than one mechanism of sediment transport acts on this shelf.

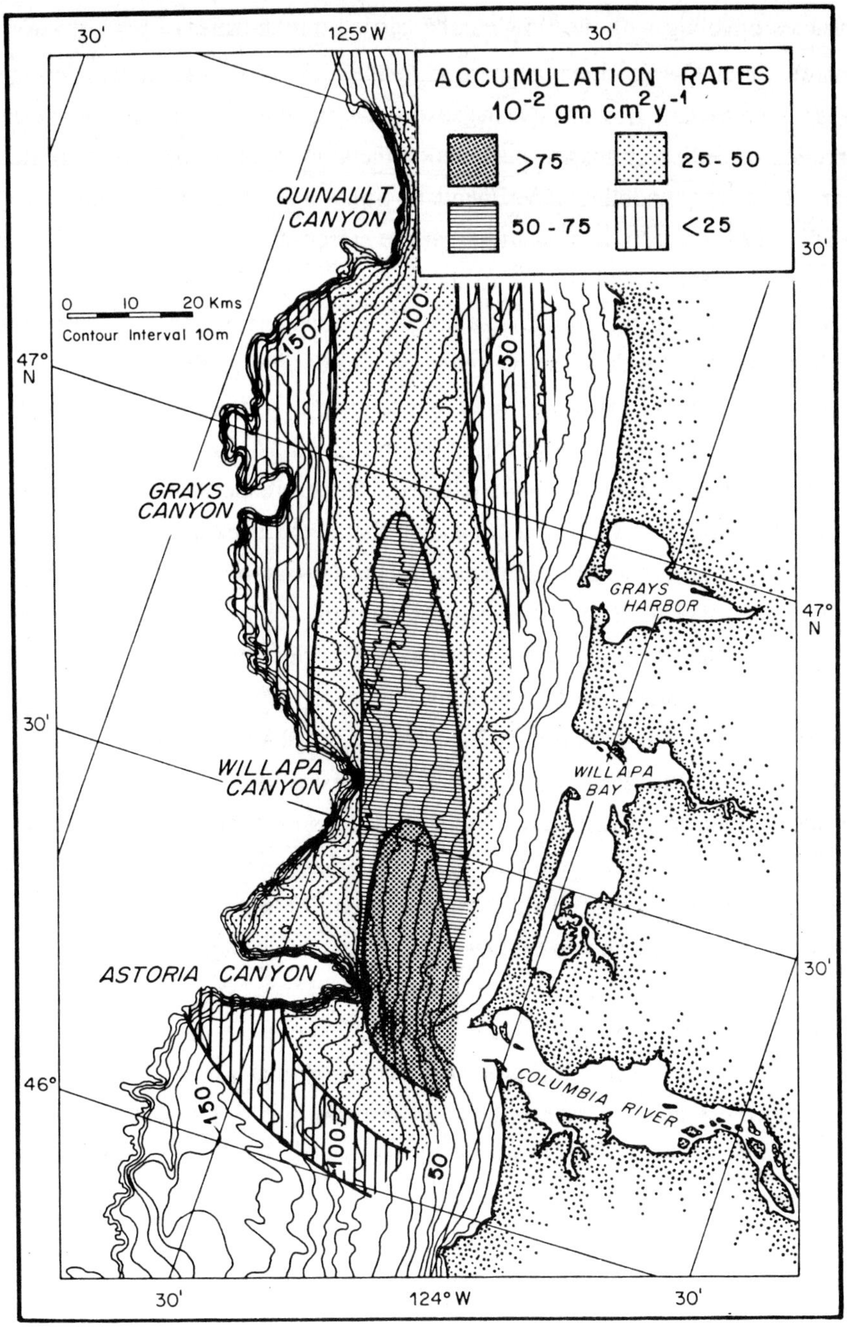

Fig. 8. 5 Accumulation rates for the Washington continental shelf as derived by Nittrouer (1978) using Pb-210 geochronology.

Kachel (1980) found that for the range of values of boundary shear stress likely to occur on the Washington shelf the sands with sizes larger than 3.0ϕ will travel as bedload and the silts will be carried in suspension and that intermediate-sized sediments in the 3.5-4.0ϕ (0.07 mm) mode are carried either as bedload or in suspension within a few meters of the bed. This means that suspended particles traveling with the flow can be carried on the order of tens of kilometers per year, while particles in the bedload may be carried on the order of tens of meters per year.

Although a northward flowing quasi-geostrophic current is produced by storm-induced set up, the direction of bedload transport depends on the amount of turning in the bottom Ekman boundary layer. Results of modeling experiments by Smith and Long (1976), Kachel (1980) and Kachel and Smith (1986) clearly demonstrate that the effect of increasing the density stratification by salinity or suspended sediment is to increase the offshore turning in the near-bottom flow. They found that during the course of a sediment transporting storm, stratification of the water column increases rapidly as sediment goes into suspension; as the event proceeds, mixing in the bottom Ekman boundary layer continues, yet the supply of fine material available for resuspension from the bed is reduced by armoring, so the density stratification in the Ekman layer breaks down and the amount of its turning is reduced. Therefore, sediment transported later in a storm is directed in a more longshore direction than that occurring during the initial periods of higher density stratification. In addition, finer sediments are capable of transported by the lower boundary shear stresses typical of the latter stages of storms. Thus they diffuse further from the sea bed as the bottom boundary layer grows in height, so these sediments are found farther to the north on the Washington continental shelf than are the coarser fractions that are transported only near the peaks of storms. The directions of the time-averaged transport, predicted by Kachel (1980) and Kachel and Smith (1986) are 23°-28° northwest of the isobaths for the bedload and 9-11° for the suspended load. Kachel and Smith (1986) also reanalyzed the sediment size data presented by Nittrouer (1978) and found that when the silt mode is removed, the aerial distribution of the dominant remaining sand mode is that depicted in Figure 8.6. The predicted paths of the bedload and suspended load shown on the sand and silt mode maps (Figs. 8.6 and 8.7) illustrates that the model results compare favorably with the observed orientation of the modal-size patterns. The differences in these paths illustrate a divergence in the mechanisms of transport of the sand and silt fractions. The net transport of the fine sediments is via suspension and therefore they have a trajectory that is closer to the along-isobath direction of the geostrophic interior flow than the trajectory of the sand fractions, which are carried primarily as bedload. The explanation for the existence of coarser sands on the promontories north of Astoria and Grays canyons that have been previously described as relict (White, 1970a), appears to be that these areas are cut off from supplies of modern bedload sediment. The continuity of the remainder of the sand mode patterns from the inner to the outer continental shelf suggest that with few exceptions the modal sands are present in response to modern transport processes, rather than to paleo-oceanographic processes during the Pleistocene transgression.

8.3 A SEDIMENT TRANSPORT MODEL FOR A STRATIFIED FLOW IN COMBINED WAVE-CURRENT EVENTS: THEORY

The major physical phenomena affecting the near-bottom transport of non-cohesive sediment on the mid to outer continental shelf that need to be included in a model of sediment transport are as follows: (1) the interaction between the surface gravity waves and the relatively low frequency currents of the Ekman boundary layer, that in combination are able to erode and

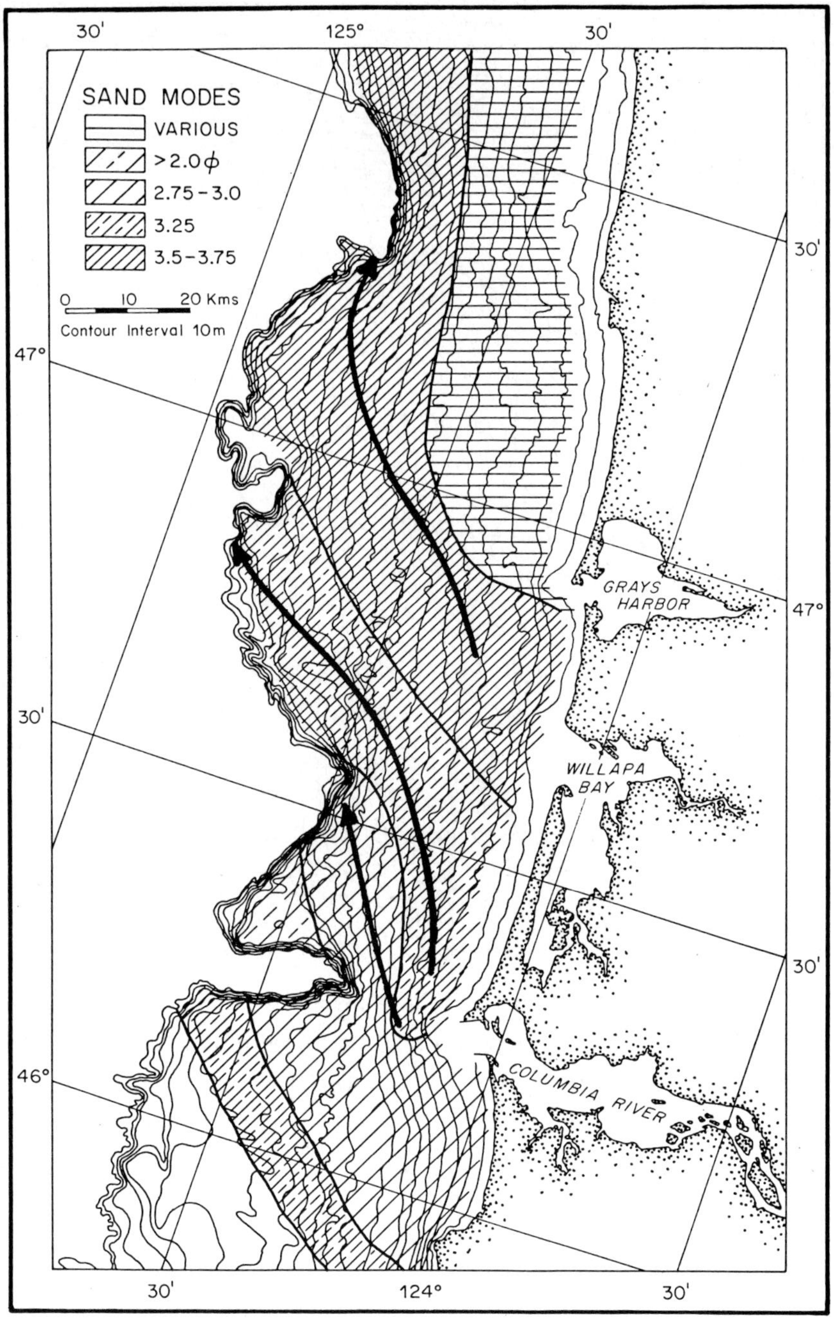

Fig. 8.6 Distribution of sand modes on the Washington continental shelf (from data of Nittrouer, 1978; Roberts, 1974). Superimposed are typical predicted paths for grains carried as bedload. These arrows begin at arbitrary locations and follow paths 25° to the west of the direction of the isobaths. From Kachel and Smith (1986).

to transport the bottom sediments, (2) the interaction between the resuspended sediments and the flow, such that the resuspended sediments increase the density gradient within the fluid and thus inhibit the transfer of momentum within the boundary layer flow and (3) the interaction between

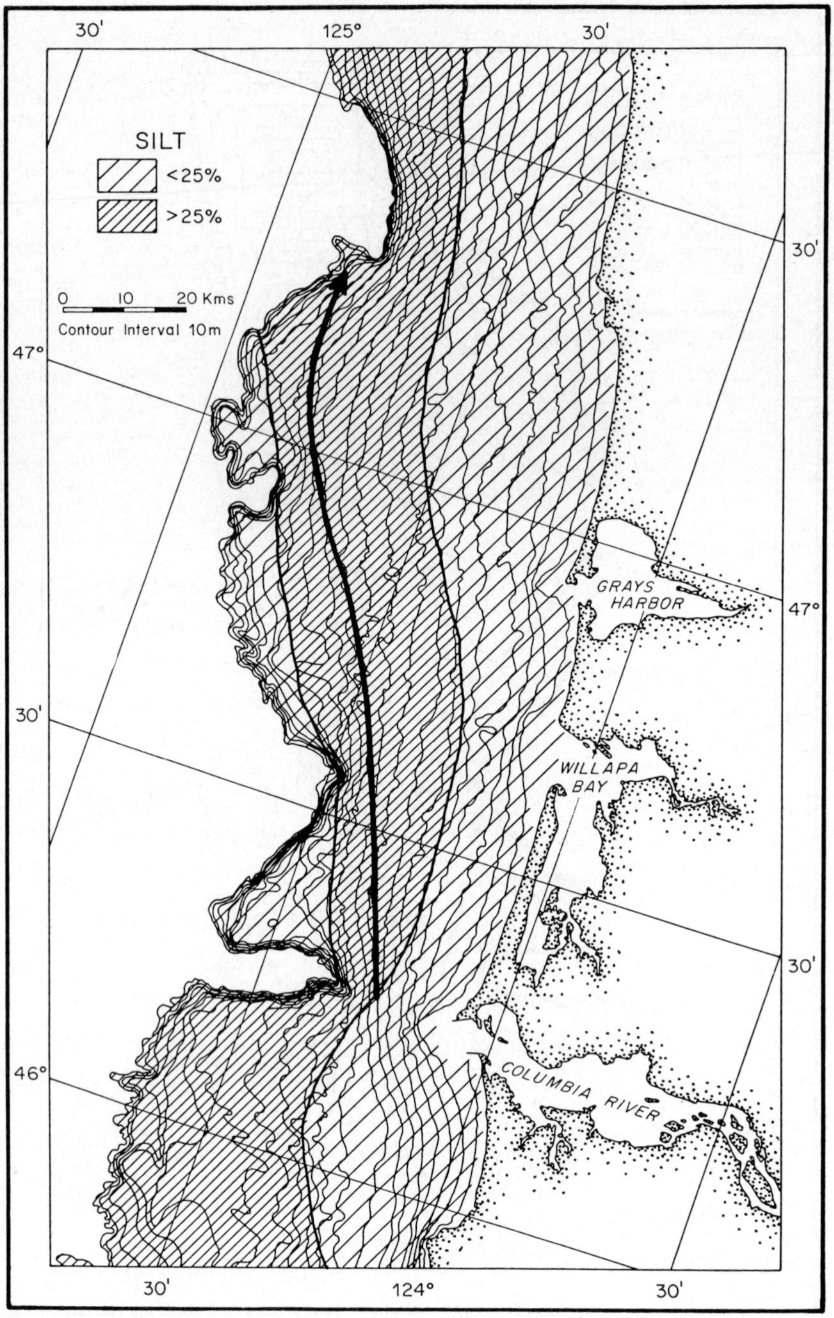

Fig. 8.7 Distribution of the silt mode on the Washington continental shelf. Superimposed is a typical predicted path of fine sediment carried in suspension during modeled events. The arrow begins at an arbitrary point and follows a course 10° to the west of the direction of the isobaths. From Kachel and Smith (1986).

a heterogeneous sediment bed and the flow, as finer, more mobile grains are removed from the bed and transported as bedload or in suspension, while other grains remain on the bed, thus armoring it from further erosion.

The non-linear interactions of wave and current boundary layers were first characterized by Smith (1977) and by Grant and Madsen (1979), subsequently improved and verified by Wiberg and Smith (1983), Grant *et al.* (1983, 1984) and Glenn (1983) and recently reviewed and summarized by Grant and Madsen (1986) and Glenn and Grant (1987). The importance of the wave-current interaction is that the length scales of the wave boundary layer are short compared with those of the current boundary layer, so the boundary shear stresses associated with the waves are much larger than those for the current. The instantaneous direction of the bottom boundary shear stress is the vector sum of that for the wave and that for the current. The eddy viscosity, hence the turbulent diffusion of momentum away from the bed, is enhanced in the region of the wave boundary layer, which leads to higher values of both the wave and current shear stresses than would exist for either component alone.

The boundary shear stress associated with a given flow velocity at some height above the bed is affected by the roughness of the bed. If no sediment is moving and the sediment bed is flat, the roughness length scale is related to grain diameter (Nikuradse, 1933). When sediment is put into motion, sandy sediments form ripples that increase the bottom roughness of the bed and reduce the boundary shear stress by adding a form drag to the drag created by the grains (Smith and McLean, 1977a; Grant and Madsen, 1982). The extraction of momentum from the flow caused by movement of grains in the bedload layer also increases the effective roughness of the bed (Smith and McLean, 1977a; Dietrich, 1982). As the flow intensifies the ripples will grow until so many bedload grains are being lifted into temporary suspension at the ripple crest that the ripples wash out. Once the ripples wash out the effect of sediment being in suspension inside the wave boundary layer is greater on the shear stress associated with the thin wave boundary layer, which is on the order of a few centimeters thick, than it is on the shear stress associated with the currents for which the boundary layer thickness is on the order of tens of meters. The effect of those sediments thrown high into suspension is to increase the density stratification of the fluid in the current boundary layer, thus reducing the downward diffusion of momentum from the current (Smith and McLean, 1977a; Kachel, 1980; Kachel and Smith, 1986; Glenn and Grant, 1987). As the finer sediment grains in the bed are resuspended, the size composition of the sediment remaining either on the bed or in the bedload becomes coarser. Eventually a sufficiently thick layer of grains, depleted of fines, is formed on top of the bed to armor the sediments below it from further erosion and to limit the volume of sediment available for transport by the flow (Kachel, 1980; Kachel and Smith, 1986).

The theoretical model presented here to describe sediment transport during storm events on the Washington shelf is based on Smith and Long's (1976) characterization of the storm-driven geostrophic flow as well as the development of a time-dependent model that accurately predicts the velocity and stress fields in a slowly-varying, stably-stratified planetary boundary layer (Long and Smith, 1979; Long, 1981). Included are the extremely important wave-current interactions discussed qualitatively above. This information is combined to yield the basic velocity and shear stress equations for the model. The sediment transport aspects of the model are based on Smith and McLean's (1977a,b) derivation of the suspended load transport field as well as their characterization of enhanced bed roughness due to the sediment transport process. Bedload transport is computed using Yalin's (1963) equation as described by Smith (1977). This set of equations for sediment transport in a benthic boundary layer under the influence of interacting waves and currents is coupled to a model (Kachel, 1980; Kachel and Smith, 1986) that accounts for changes in the composition of a heterogeneous sediment bed due to the differences in responses of various components (e.g., size classes) to such flows. The main purpose of this paper is to present this model and to apply it to a set of locations on the Washington shelf in order to understand the erosion and deposition patterns in this region.

The modeled variables include: $\tilde{u}$, current velocity; $\tilde{\tau}$, shear stress; K_m and K_s, eddy diffusion coefficients for momentum and mass; s salinity; C_j, seven size classes of sediment; ρ, water density; $\tilde{u}_0$, wave orbital velocity; and $\tilde{\tau}_w$, wave induced shear stress. The initial conditions for the model are: water depth, sediment composition of the bed and the pre-storm density field. The sediment inputs are: D, the grain diameter, w_s, the settling velocity and τ_{cr}, the critical shear stress parameter (Shields, 1936; Miller *et al.*, 1977). It is assumed that initially there is no sediment in suspension. The forcing functions for the time-dependent boundary layer models are: idealized functions of $\tilde{u}_g$, geostrophic velocity and $\tilde{u}_0$, wave orbital velocity, as functions of time during the modeled event, as well as the period of the waves.

8.3.1 Velocity and Shear Stress Due to Currents

For shelf sediment transport models the bottom boundary layer is of primary importance, therefore it is advantageous to force such models using estimated interior geostrophic currents rather than surface stresses calculated from poorly known wind velocities. Available flow information indicates that the geostrophic velocity during storms of moderate to long duration is oriented parallel to the isobaths (Smith and Hopkins, 1972; Smith *et al*, 1976; Kachel and Smith, 1986; Hickey, 1989, Chapter 2). In the steady state condition, where the flow follows the isobaths and the mass balance is two-dimensional, the onshore surface Ekman transport is balanced by offshore transport in the bottom boundary layer. Therefore, the net cross-shelf transport integrated from the bottom at z_0 to the surface, S, is

$$Q_x = \int_{z_0}^{S} u_x \, dx = 0 \tag{8.1}$$

in a coordinate system defined with the y-axis positive northward and parallel to the isobaths and the x-axis positive in the onshore direction. We note that two-dimensional mass balance has not generally been observed in the coastal ocean (Hickey, 1989, Chapter 2). However, with the exception of the vicinity of submarine canyons, in the Pacific Northwest where coastal promontories are relatively rare, the mass balance may be close to two-dimensional during winter storms when local wind forcing of the coastal currents predominates.

The lower limit of integration for the bottom boundary layer is the bottom roughness length, z_0, defined as the height above the bed at which $\tilde{u} = 0$. Over geometrically smooth beds z_0 can be determined from Nikuradse's (1933) diagram as $z_0 = z_n = f\,(\mathrm{Re})$, where $\mathrm{Re} = (u_* k_s / \nu)$ is the roughness Reynolds number, k_s is the scale length for the sediment grains composing the bed, u_* is the shear velocity (see Eq. 8.14) and ν is the viscosity. When sediment is moving there is an additional loss of momentum from the near-bed flow. Smith and McLean (1977a and b) and Dietrich (1982) have related z_0, under sediment transporting conditions, to the thickness of the bedload layer. The model uses the method presented by Smith and McLean (1977b) and Kachel (1980) to relate z_0 to the concentration-weighted average of the two expressions for a mixed bedload and suspended load over a heterogeneous sediment bed. A more modern approach to the problem of calculating z_0 is provided by Wiberg and Ruben (in press) for bedload transport and by Gelfenbaum and Smith (1986) for suspended load transport. Using Carstens *et al.'s* (1969) experiments, Grant and Madsen (1982) parameterized the effect of ripples on relative roughness as $k_b = 2.77(H^2/\lambda)$ where H is ripple height, λ is ripple length and H/λ is the ripple index. The presence of ripples increases the scale length, k_s, used in establishing a roughness Reynolds number for the calculation of z_0 in the present model. Glenn (1983) assessed the relative contributions to z_0 of pressure drag on bed forms and sediment transport and found that for a fine sand bed the roughness lengths associated with drag on bed forms and with sediment transport are of the same order. Over silt beds, sediment transport effects dominate z_0 while over beds of medium sand the z_0 associated with bedforms is most important.

The model uses the planetary boundary layer algorithm of Long (1981). The Reynolds-averaged momentum equations for a horizontally uniform flow in a time dependent, stably-stratified, planetary boundary layer are

$$\frac{\partial u}{\partial t} - fv = -\frac{1}{\rho}\frac{\partial p}{\partial x} + \frac{\partial}{\partial z}K_m\frac{\partial u}{\partial z} \tag{8.2}$$

and

$$\frac{\partial v}{\partial t} + fu = -\frac{1}{\rho}\frac{\partial p}{\partial y} + \frac{\partial}{\partial z}K_m\frac{\partial v}{\partial z} \tag{8.3}$$

where an eddy coefficient closure scheme is used to model the vertical turbulent fluxes. The equation for the distribution of mass is

$$\frac{\partial \rho}{\partial t} = \frac{\partial}{\partial z}K_\rho\frac{\partial \rho}{\partial z} \tag{8.4}$$

Above the bottom boundary layer the pressure gradient is given by an inviscid balance. If the flow is additionally hydrostatic and barotropic, the velocity equations can be written

$$\frac{\partial u_\infty}{\partial t} + fv_\infty = -\frac{1}{\rho}\frac{\partial p}{\partial x} \tag{8.5}$$

$$\frac{\partial v_\infty}{\partial t} - fu_\infty = -\frac{1}{\rho}\frac{\partial p}{\partial y} \tag{8.6}$$

When Equations 8.5 and 8.6 are substituted into Equations 8.2 and 8.3 and the latter are combined using complex notation, the result is

$$\frac{\partial}{\partial t}(\tilde{u} - \tilde{u}_\infty) + if(\tilde{u} - \tilde{u}_\infty) = \frac{\partial}{\partial z} K_m \frac{\partial \tilde{u}}{\partial z} \tag{8.7}$$

in which $\tilde{u} = u + iv$ represents a complex velocity. The boundary conditions on this expression are that as $z \to z_\infty$, $\tilde{u} \to \tilde{u}_\infty = \tilde{u}_g$ (the geostrophic velocity discussed in Eq. 8.1) and that $z \to z_0$, $\tilde{u} \to 0$. The shear stress, $\tilde{\tau}$, associated with a velocity is found from the relationship

$$\tilde{\tau} = K_m \frac{\partial \tilde{u}}{\partial z} \tag{8.8}$$

8.3.2 Momentum and Shear Stress Due to Combined Waves and Currents

The wave-current interaction equations used here follow the development presented by Wiberg and Smith (1983) who refined the model first suggested by Smith (1977) and later by Grant and Madsen (1979). Under high-frequency surface, gravity waves the momentum equation for the boundary layer is

$$\frac{\partial \tilde{u}_w}{\partial t} = \frac{\partial \tilde{U}_0}{\partial t} + \frac{\partial}{\partial z} K_m \frac{\partial \tilde{u}_w}{\partial z} \tag{8.9}$$

where $\tilde{U}_0$ is the maximum oscillatory velocity in the inviscid region. With a sinusoidally-varying interior velocity field, the solution to (8.9) is found by postulating an orbital velocity of the form $\tilde{u}_\infty = \tilde{U}_0 R(e^{\,i\omega t})$ outside the boundary layer and

$$\tilde{u}_w = \tilde{U}_0 R\{e^{\,i\omega t}[1 - F(z)]\} \tag{8.10}$$

within the layer. Here R is an operator that yields the real part of the complex function and ω is the wave frequency. Substituting Equation 8.8 into 8.9 gives

$$i\omega F = \frac{\partial K_m}{\partial z}\frac{\partial F}{\partial z} + K_m \frac{\partial^2 F}{\partial z^2} \tag{8.11}$$

The boundary conditions on Equation 8.11 are set so that $F(z) \to 0$ as $z \to \infty$ and $F(z) = 1$ at $z = z_0$.

When a complicated depth dependence is chosen for K_m, it is necessary to solve Equation 8.11 for $F(z)$ numerically, then substitute the result into Equation 8.10 to obtain a velocity profile. From the wave velocity field, the boundary shear stress associated with the waves can be calculated employing Equation 8.8, such that

$$\tilde{\tau}_{b_w} = K_m \, (\partial \tilde{u}_w / \partial z) \tag{8.12}$$

Due to the highly non-linear relationship between the boundary shear stress and sediment transport, the maximum value of τ_{b_w} is the most appropriate choice with which to parameterize τ_{b_w} over the wave period. This choice follows from the assumption that turbulence generated by the peak velocity of a wave decays slowly relative to a wave period. Long (1981) demonstrated the validity of this assumption by comparing model predictions of profiles for the amplitudes of velocity components in an oscillating flow with Jonsson's (1963) data. The RMS difference between this theory and the observation was 4% of the interior flow speed. In sediment transport problems the non-linear relationship between sediment flux and boundary shear stress further emphasizes the need to choose approximations that give the greatest accuracy at the time of maximum boundary shear stress.

The shear velocity due to the waves is defined as

$$u_{*_w} = \left(\frac{\tau_{b_w}}{\rho} \right)^{1/2} \tag{8.13}$$

Similarly the shear velocity due to the current alone is defined as

$$u_{*_c} = \left(\frac{\tau_{b_c}}{\rho} \right)^{1/2} \tag{8.14}$$

The total boundary shear stress acting on sediment grains when waves and currents interact is obtained by vector addition of the average enhanced shear stress associated with the current and the maximum shear stress associated with the waves (Grant and Madsen, 1979), that is,

$$\tilde{\tau}_b = \tilde{\tau}_{b_c} + \tilde{\tau}_{b_w} \tag{8.15}$$

8.3.3 Salinity and Density Distribution

The importance of suspended sediment induced and ambient density stratification on the vertical transfer of mass and momentum, K_ρ and K_m, has been demonstrated by Smith and McLean (1977a), Long (1981) and Kachel (1980). Temperature, salinity (s) and the sum of the concentration of suspended sediment in all size classes, $\sum_j^N C_j$, make significant contributions to the density stratification in the water column. Although both the salinity and temperature profiles vary widely during the winter storm season, changes in salinity dominate the vertical density structure of the water column on the Washington shelf (Hickey, 1983). The generalized choice of this density structure at the beginning of a modeled storm is calculated from the salinity profile alone, but is representative of density profiles observed during such times by Hickey (1983). Kachel (1980) showed that the effect of resuspended sediments can be important in the lower 5-10 m of the flow, whereas salinity stratification affects the upper portion of the benthic boundary layer. The result is a reduced thickness and an increased the turning angle in the boundary layer. In a horizontally homogeneous fluid the time change in salinity distribution is given by

$$\frac{\partial s}{\partial t} = \frac{\partial s}{\partial z} K_s \frac{\partial s}{\partial z} \tag{8.16}$$

with boundary conditions $s \to s_\infty$ as $z \to \infty$ and $K_s\, \partial s/\partial z = 0$ at $z = 0$. The initial salinity profile chosen for the model runs varied linearly from 33.5 at the bed to 31.5 at the surface, these values being typical for the area of interest.

If the densities of the water and the sediment are ρ_w and ρ_s, respectively, and C_s is the concentration of suspended sediment, then the density of the sediment water mixture is

$$\rho = \rho_w + (\rho_s - \rho_w) C_s \tag{8.17}$$

In subsequent calculations ρ_s is taken to be 2.65 x 10^{3} kg m^{-3}, the density of quartz grains.

8.3.4 Suspended Sediment Distribution

The relevant equation for the distribution of suspended sediment is derived by Smith and Hopkins (1972), Smith (1977), Smith and McLean (1977a and b) and Kachel (1980) and Gelfenbaum and Smith (1986). The concentration of a single class, j, of suspended sediment, C_j, in horizontally uniform flow with more than one component is

$$\frac{\partial C_j}{\partial t} = w_j \left[C_j \frac{\partial}{\partial z} (1 - C_s) + (1 - C_s) \frac{\partial C_j}{\partial z} \right] + \frac{\partial K_s}{\partial z} \frac{\partial C_j}{\partial z} + K_s \frac{\partial^2 C_j}{\partial z^2} \tag{8.18}$$

where w_j is the settling velocity of the jth sediment class and K_s is the eddy diffusion coefficient for sediment. The term C_s indicates the sum of the sediment concentration for all classes and the $(1 - C_s) = C_w$ is the concentration of water. This term arises from the upward flow of water needed to conserve mass for this component as well. Glenn (1983) and Glenn and Grant (1987) used a somewhat different formulation for this equation. However, Wiberg and Smith (1983) showed their method produces similar resultant values to the one presented here under a wide range of conditions. The upper boundary condition for Equation 8.18 is $C_j \to 0$ as $z \to \infty$. When $\tau_b > \tau_{cr}$, sediment moves as bedload or suspended load depending on the value of w_j / u_*. The appropriate lower boundary condition requires specification of either the reference concentration C_{a_j} at z_0 or of the gradient $\partial C_s/\partial z$.

Smith and McLean's (1977a) empirical expression is utilized to find C_a at z_0,

$$C_{a_j} = \frac{i_j C_b \gamma_0 S_j}{1 + \gamma_0 S_j} \tag{8.19}$$

In this expression i_j is the fraction of sediment in settling velocity class w_j in the bed, C_b is the total concentration of sediment in the bed and S_j is the excess shear stress for class j. This is the value used in the model calculations presented here. In cases where $C_s << 1$, as it is here, the approximation $C_{a_j} = i_b C_b \gamma_0 S_j$ can be used instead of the full expression in Equation 8.19.

The choice for the value of γ_0 in Equation 8.19 is problematic and apparently is different for those authors using it to match field observations of suspended sediment concentration than to those making full, detailed boundary layer profile calculations. When using simplified models to predict suspended sediment concentration at some observed distance above the bed, investigators have used Equation 8.19 with estimates of boundary shear stress based on velocity profiles or on single point velocity measurements (and some assumed drag coefficient) made at a variety of distances up to several meters off the bottom. From these they calculate the reference layer concentration. An accurate knowledge of both the Rouse number (w_j / ku_*) and the frequency distribution of each size (or settling velocity) class within the suspended sediment sample is also required. Typically these values have not been procured in the field experiments used to calculate γ_0. Next a sediment profile of the form $C_z = C_a\ (z/z_a)\ \exp\ (-\ w_j / ku_*)$ is calculated. This formulation assumes a linear model for the eddy diffusion coefficient. By matching predicted with observed sediment concentrations at some height above the bed, field estimates for γ_0 have thus been derived. Drake and Cacchione (in press) found a range of values from 2 x 10 $^{-5}$ to 5 x 10 $^{-4}$ in their own data and those of others working on continental shelves and concluded that variations in γ_0 based on field observations are probably due to cohesive effects and armoring of heterogeneous sediment beds.

The value of γ_0 used by those who calculate non-linear profiles of the eddy diffusion coefficients for mass and momentum are approximately two orders of magnitude larger. Smith and McLean's (1977b) value of 2.4 x 10 $^{-3}$ was empirically derived from fitting such a model to velocity profiles and suspended sediment concentration observed 10 cm above a sand bed. Glenn (1983) used the laboratory results of Kalkanis (1964) and Abou-Seide (1965) for beds composed of fine to medium sand to obtain a value of of 0.003. From our own numerical experiments using Smith and McLean's (1977a) data set it appears that the choice of values for γ_0 is highly dependent on the values of the coefficients used in the formulation of the eddy diffusion coefficients (Eq. 8.22a and b) and, especially, on the value of von Karman's constant. Smith and McLean (1977a and b) used $k = 0.35$, while Glenn (1983) used 0.40. These authors used values for the coefficients in 8.22 of $\beta = \beta_p = 4.7$ and $\sigma = 0.74$ taken from Businger and Arya (1974). Wieringa (1980) reevaluated the data of Businger and Arya and set the values of the coefficients in the stratification correction at $\beta = 6.9$, $\beta_p = 9.2$ and $\sigma = 1.0$. Using Wieringa's values in conjunction with an exceptionally high value for von Karman's constant of 0.43 yields a value for γ_0 = 1.5 x 10 $^{-2}$ derived from Smith and McLean's (1977a,b) Columbia River data set. When Wieringa's values were used in conjunction with a more realistic value of 0.406 (Nowell and Long, 1983) a subsequent reanalysis of the Smith and McLean (1977a) data yielded a best fit value of γ_0 = 5.4 x 10 $^{-3}$. This latter value is used in this paper.

8.3.5 Eddy Diffusion Coefficient

The functional form of the neutral eddy diffusion coefficient K_0 used in turbulent boundary layer models by Kachel (1980), Long (1981), Wiberg and Smith (1983) and Kachel and Smith (1986) is based on previous work by Businger and Arya (1974). For a boundary layer characterized by a single length scale they suggest

$$K_0 = k u_* z e^{-3z/h} \tag{8.20}$$

where k is von Karman's constant (value = 0.406) and h is the length scale. For an Ekman boundary layer, this length scale is $h = h_c = \delta_E = u_{*c} / 6f$ where f is the Coriolis parameter and for an oscillatory flow the length scale is the thickness of the wave boundary $h_w = \delta_w = u_{*w}/6\omega$. Long (1981) successfully tested this formulation for K_0 against the data sets of Jonsson (1963) and Jonsson and Carlsen (1976) as well as others. Under storm conditions on the continental shelf, h_c is tens of meters while h_w is tens of millimeters.

When waves and currents interact, it is necessary to redefine the neutral eddy diffusion coefficient (Eq. 8.20) in order to include the turbulence produced in both boundary layers. Kachel and Smith (1986) and Wiberg and Smith (1983) suggest that the appropriate form is

$$K_0 = kz\left(u_{*c}^2 \, e^{-3z/h_c} + u_{*w}^2 \, e^{-3z/h_w} \right) \tag{8.21}$$

To model the effect of stratification on K_m and K_ρ, Long (1981) examined the scaling of terms in the turbulent kinetic energy and density variance equations for a quasi-steady case. The resultant forms for the eddy diffusion coefficients are

$$K_m = \frac{K_0}{1 + \beta\zeta} \tag{8.22a}$$

and

$$K_\rho = \frac{K_0}{\sigma + \beta_\rho \zeta} \tag{8.22b}$$

in which $\nu = 0.013 \text{ cm}^2 \text{ s}^{-1}$ ($1.3 \times 10^{-4} \text{ m s}^{-1}$), β, β_ρ and σ are stratification coefficients and ζ is the stratification scaling parameter. For computational ease Long (1981) set a lower limit for K_m equal to the molecular value, ν, at the top of the boundary layer. The values of β, β_ρ and σ used in the calculations were those of Wieringa (1980), namely 6.9, 9.2 and 1.0, respectively. Long (1981) defines the stratification scaling parameter ζ as

$$\zeta = \frac{\alpha Ri}{1 - \alpha\gamma Ri} \tag{8.23}$$

The term Ri is the gradient Richardson number, the ratio of turbulent energy loss due to stable density stratification to turbulent energy production by shear. $Ri = N^2/(\partial u/\partial z)^2$, where N^2 is the squared Brünt-Väisälä frequency, $(-g/\rho)(\partial\rho/\partial z)$, and $(\partial \tilde{u}/\partial z)$ is the shear. Near the bed (within approximately 5 m) N^2 varies with distance above the bottom primarily due to the decrease in $\partial C/\partial z$ with height but also to the breakdown in the salinity gradient as the boundary layer grows and mixing continues. The coefficient α varies from 1 to 0.74 depending on the value of the Richardson number, Ri; i.e.,

$$\alpha = \frac{K_\rho}{K_m} = \frac{1 + \beta\zeta}{\sigma + \beta_\rho \zeta} \tag{8.24}$$

8.3.6 Sediment Transport, Erosion and Deposition

After the boundary shear stress, the velocity and the sediment concentration are found, it is possible to calculate sediment transport rates and directions. As mentioned above, the Yalin (1963) equation is used to obtain the bedload transport rate, $\tilde{Q}_b$. Recently Wiberg (in press) has shown that this expression is the most accurate of the available semi-empirical bed load equations. The net bedload transport rate is assigned the direction of the boundary shear stress. Suspended load transport rate is defined as

$$\tilde{Q}_s = \int_{z_0}^{\infty} \tilde{q}_s \, dz = \int_{z_0}^{\infty} C \, \tilde{u} \, dz \tag{8.25}$$

Total load is the sum of the vector two, $\tilde{Q}_T = \tilde{Q}_b + \tilde{Q}_s$.

The equation for erosion or deposition of any class within the bed can be derived from mass conservation and is discussed by Smith (1977). The form of the expression for rate of change of elevation of the bed η is

$$\frac{\partial \eta}{\partial t} = -\frac{1}{C_b}\left[\frac{\partial}{\partial t}\int_{\eta}^{\infty} \sum_{j=1}^{N} C_j dz + \nabla\cdot\sum_{j=1}^{N}\tilde{Q}_j\right] \tag{8.26}$$

where N is the number of size, shape and specific gravity classes into which the bed sample is divided. Therefore, if we write $\eta = \sum_j^N \eta_j$ the expression in Equation 8.26 becomes

$$\frac{\partial \eta_j}{\partial t} = -\frac{1}{C_b}\left[\frac{\partial}{\partial t}\int_{\eta}^{\infty} C_j dz + \nabla\cdot\tilde{Q}_j\right] \tag{8.27}$$

The first term on the right hand side represents the change in the volume of sediment of a given class carried in suspension above a unit area of the bed; the second is the divergence of its transport rate. As a first approximation for the one-dimensional model it is assumed that the transport field is homogeneous. The model then is run for a series of shelf locations, after which it is possible to estimate the magnitude of the divergence term.

Kachel (1980) argued that on the continental shelf, suspended sediment transport is limited by the supply of sediment from the surface of the bed, not by the thickness of the boundary layer. Therefore, the use of a constant bottom composition is equivalent to providing an infinite supply of fine material to the suspended load. In order to model transport over a heterogeneous sediment bed it is important to consider armoring of the bed and the thickness of the layer of sediment that can interact with the flow.

8.3.7 Composition of the Bottom Sediments and Armoring of the Bed

The composition of the bed enters the model through the boundary condition on the sediment concentration profile. The reference level concentration given by Equation 8.19 depends directly on the frequency of occurrence of a particular grain size in the bed. The change in composition of the bed at the sediment-water interface during a specific time interval depends on the relative erosion or deposition of each size class present in the bed. The amount of material that goes into suspension depends on the accessibility to the flow of movable particles in the bed, that is, the volume of sediment interacting with the flow. In an extreme situation, a layer one to two grains thick of non-moving grains could totally armor the sediment surface, thereby preventing further erosion. Consequently, a critical problem in sediment transport modeling is to describe the bed composition as a function of time during a transition between a heterogeneous surface sediment and an armored surface layer depleted of mobile grains. If the sediment bed is flat and remains so, then the model proposed by Shi (1983) or Shi *et al.* (1985) is applicable. They used the observation that a layer of immobile or rolling grains with the thickness of two grain diameters is sufficient to block the removal of suspendible grains from the bed, that is, their mixing depth was $\alpha\mu = 2D_{max}$, where D_{max} is the maximum grain diameter in the bed. Kachel (1980; Kachel and Smith, 1986) pointed out that this lower plane bed condition is rare and not usually of major interest in sediment transport problems. The bed is unstable to the generation of bedforms and to include that effect Kachel (1980) developed a model that characterized the armoring of an initially heterogeneous sediment bed subjected to an accelerating flow. That model allows fine materials to go into suspension, while the coarser fractions (sand) moving as bedload quickly form ripples that migrate across the bed surface. Once all of the bed material can go into suspension (upper plane bed condition) no armor is possible, although under these conditions the rate of erosion of finer sizes is controlled in part by the rate of erosion of the coarser suspended material that is selectively concentrated at the sediment water interface. All of the material in the ripples will be exposed to the action of the flow during the time it takes for the ripple to move one full wavelength. If a triangular cross-section is assumed for the ripple with height, H and wavelength, λ, then in the time, t_λ, it takes the bedform to travel one wavelength, the average depth to which the bed has been effectively stirred is $\delta_m = H/2$. This is an upper limit on the average depth of mixing. In a computational time interval, Δt, less than t_λ, the ripple travels a distance $u_r\Delta t$, where u_r is the ripple migration rate. Thus the depth of mixing averaged over the ripple wavelength is on the average proportionally less

$$\delta_m = \frac{Hu_r}{2\lambda}\Delta t \tag{8.28}$$

Sternberg (1967) and Kachel and Sternberg (1971) presented data for the ripple velocity, u_r, as a function of, S, the excess shear stress $(\tau_b/\tau_{cr} - 1)$. A linear regression of the data resulted in a first order empirical fit of $u_r = \beta_H S$, where $\beta_H = 3 \times 10^{-5}$ m s^{-1}. Therefore, for S > 0, substituting for u_r in Equation 8.27 yields

$$\delta_m = \beta_H \frac{H}{2\lambda} S\,\Delta t \qquad \text{for } \Delta t \le t_\lambda \tag{8.29a}$$

and

$$\delta_m = \frac{H}{2} \qquad \text{for } \Delta t \ge t_\lambda \tag{8.29b}$$

The ripple wavelength λ is dependent on the grain diameter of the sediment and typically ranges from a hundred to over a few thousand grain diameters (e.g., $\lambda \approx 1{,}000\,D$ according to Yalin, 1964). The ripple index, λ/H, tends toward a value of 8 to 12, so, $\delta_m \approx 1.5 \times 10^{-6}\, S\,\Delta t$, for $\Delta t \ge t_\lambda$.

As shear stresses increase, more of the grains present on the bed are capable of being transported in suspension. As this happens, ripples wash out, or exceed the breakoff point defined by Grant and Madsen (1982, Fig. 1). Sediment transport is then considered to be in the state called upper plane bed regime with respect to ripple wavelengths. It is assumed here that during upper plane bed motion no less material is in contact with the flow than the maximum predicted by Equation 8.29. If erosion occurs, once upper plane bed conditions are reached, the effects of armoring are negligible. Therefore, the maximum value is used as a first approximation during such times. From the net amount of erosion or deposition and the depth disturbed by moving ripples, a new size frequency distribution F_j for the bed's surface can be found using a simple mixing algorithm

$$F_{j_1} = \frac{F_{j0}\,\delta_m + \Delta\eta_j}{\delta_m + \sum_j \Delta\eta_j} \tag{8.30}$$

where $\sum_j \Delta\eta_j$ is the net sum of erosion or deposition for all size classes during the interval Δt . During times of deceleration, δ_m decreases, deposition occurs and some material no longer interacts with the flow. In such a way a graded deposit forms on the bed. Kachel and Smith (1986) calculated erosion, deposition and changes in bed composition in order to predict the possible stratigraphic signature from various modeled sediment transporting events on the Washington continental shelf.

8.4 METHOD OF SOLUTION OF THE THEORETICAL MODEL

Although the relationships among the variables is complex, these equations probably represent the simplest realistic model for near bottom flow and sediment transport on a continental shelf. To solve the theoretical model numerically, a finite difference scheme with central differencing and unequally spaced elements in space and time was devised. A logarithmically changing increment size on the z axis is needed to minimize the sensitivity of the calculations to computational errors in the immediate vicinity of the bottom. Time steps vary from a few minutes to two hours, depending on the magnitude of the change in conditions during the time step. To solve the partial differential equations, a standard, implicit method, known as Thomas' algorithm (von Rosenburg, 1969; Ames, 1977; Carnahan *et al.*, 1969) was used.

8.4.1 Initial Conditions

The initial conditions for the model are: the size frequency distribution and the particle diameters, the settling velocities and critical shear stress values for seven classes of bottom sediments, as well as vertical profiles of current velocity, wave orbital velocity, salinity and eddy diffusivity coefficients for mass and of momentum. In all the cases discussed here the system is assumed to start from rest, so all these profiles except salinity are assumed to be zero throughout prior to $t = 0$.

The initial size frequency distribution of bottom sediments for all sites is presented in Table 8.6. The first sediment bed is composed of a sandy-silt characteristic of the mid-shelf region, while the second is a silty-sand typical of locations further offshore. The other sites are located along and across the shelf as shown in Figure 8.1. The size frequency distributions shown for site A to I were averaged from data presented by Nittrouer (1978) for samples collected in August and so represent the distribution before the first winter storms. The settling velocity and critical shear stress for each size class are also shown. The critical shear stress values used for the sands were derived using the modified version of Shield's diagram presented by Smith (1977). Threshold shear stress values for the fine sediments were inferred from observations of resuspension events made on the mid-shelf of Washington by Sternberg and Larsen (1976). During times of low wave activity, sediment resuspension was associated with a current of 0.26-0.30 m s^{-1} at one meter off the bed. If a logarithmic velocity profile and a geometrically smooth bed of silty sand are assumed these velocities correspond to τ_{cr} of 0.090 to 0.117 N m^{-2}. As mentioned above, the initial salinity profile used in each case decreases linearly from 33.5 to 31.5 parts per thousand from the bottom to the sea surface, a range typical of the Washington shelf during autumn and winter (Hickey, 1983). It is assumed that initially there is no sediment in suspension.

TABLE 8.6 Input composition of sediment size classes in bottom samples at stations on the Washington shelf. Values averaged over 1-3 samples per station analyzed by Nittrouer (1978). Grain diameter, D, settling velocity, w_s, and critical shear stress, τ_{cr}, for each size class modeled are also presented. Letter designation refer to sites in Figure 8.1, numbers in parentheses are Nittrouer's station numbers.

Phi-size class	< 2.0	2.0-2.5	2.5-3.0	3.0-3.5	3.5-4.0	4.0-6.0	> 6
D (mm)	0.30	0.21	0.15	0.11	0.07	0.032	0.016
w_j (cm s^{-1})	3.66	2.82	2.17	1.25	0.174	0.085	0.006
τ_{cr} (N m^{-2})	0.205	0.173	0.149	0.117	0.117	0.117	0.117
CASES							
SANDY-SILT	0.0130	0.0560	0.2600	0.1190	0.0950	0.2260	0.2430
SILTY-SAND	0.0180	0.0820	0.3800	0.1490	0.0570	0.1180	0.2060
A (81-82)	0.1488	0.1346	0.2942	0.1220	0.0279	0.1336	0.1391
B (83)	0.0498	0.1995	0.2654	0.1308	0.0356	0.1185	0.1938
C (79)	0.0378	0.1056	0.3610	0.1675	0.0296	0.1233	0.1701
D (25)	0.0239	0.0214	0.0356	0.1156	0.1153	0.4099	0.0654
E (44)	0.0120	0.0247	0.1247	0.0984	0.0826	0.4008	0.2567
F (23)	0.0633	0.0533	0.0366	0.3286	0.3344	0.1422	0.0416
G (24)	0.0054	0.0085	0.0302	0.1840	0.2065	0.4100	0.0654
H (27)	0.0326	0.0289	0.0449	0.1638	0.1134	0.3607	0.2603
I (28)	0.0084	0.0142	0.0541	0.3319	0.2299	0.1889	0.1726

8.4.2 Forcing Functions

The forcing functions for the model are: idealized relationships of geostrophic velocity, u_g, bottom wave orbital velocity, u_0, and wave period. In order to address the questions of geological interest that relate to the coincidence of extreme current and swell conditions, the forcing functions for u_g and u_0 were given a similar temporal structure. The shape of the input function for $|\tilde{u}_g|$ as a function of time will be discussed later.

The other forcing functions for the model are the maximum wave orbital velocity and period as functions of time. The wave period is chosen to be constant, but the amplitude varies. In order to reduce computation time, the problem of determining u_{*w} as a function of u_0, the bottom orbital velocity of a wave, is decoupled from the time-dependent problem of predicting the structure of the benthic boundary layer, sediment transport, erosion and deposition during wave-current events. This is allowed by the rapid adjustment of the near bottom flow and suspended sediment fields to changes in u_{*c} and u_{*w}, hence by the quasi-steady response of near bed flow conditions to changes in external forcing.

Solving u_{*w} as function of u_0

Before u_{*w} can be found as a function of the wave orbital velocity, a partial differential equation for $\tilde{\tau}$, the shear stress, must be derived. To do this, Equation 8.7 for the velocity profile is differentiated with respect to z and multiplied by K_m. Substitution of the relationship $\tilde{\tau} = K_m\,(\partial\tilde{u}/\partial z)$ into that result yields

$$\frac{\partial\tilde{\tau}}{\partial t} + if\,\tilde{\tau} = K_m \frac{\partial^2\tilde{\tau}}{\partial z^2} \tag{8.31}$$

To find u_{*w} in sediment-induced stratification conditions for given values of u_0, the wave orbital velocity just above the boundary layer and u_{100}, the current velocity at 1.0 m above the bed, the model equations can be used with a shooting technique. An initial estimate of u_{*c} is made from which first estimates of the profiles for K_m (Eq. 8.22a), for K_s (Eq. 8.22b) and for shear stress (Eq. 8.31) can be found. These permit computation of salinity (Eq. 8.16), suspended sediment (Eq. 8.18) and density (Eq. 8.17) profiles. Next, the profile of wave velocity (Eq. 8.10) and u_{*w} (Eq. 8.13) are calculated, from which better estimates of K_m and K_s can be made. Current velocity is then calculated by integration.

$$u_{100} = \int_{z_0}^{100} \frac{\tau}{K_m}\,dz \tag{8.32}$$

Comparison of this result with the known input value of u_{100} permits correction of the estimate for u_{*c}. This iterative procedure continues until stable values of u_{*c} and u_{*w} are found that yield the desired value of u_{100}.

For modeling combined wave-current events on the Washington shelf the angle between the direction of the waves and currents was chosen to be 60°, based on the observed NNW direction for the currents along the isobaths and our analysis of National Marine Consultants data (Table 8.3). Those data show that waves able to move mid- to outer shelf sediments predominantly arrive from the west to west-north-west direction (an angle of 50-75° to that of the currents). According to Figure 2 presented in Grant and Madsen (1979), even when wave and current velocities are nearly the same, the error in combined shear stress, τ_T, resulting from a choice of 60° instead of 90° would result in over-prediction of τ_c by only 10%.

Figure 8.8 shows the results of u_{*w} *versus* u_0 for calculations made for a variety of combinations of bottom compositions, current velocities, u_{100} and wave periods, as well as for two of sets calculations without sediment-induced density stratification. The non-linear effect of the presence of the current boundary layer on u_{*w}, the shear velocity associated with the waves, is illustrated by the two curves with no stratification correction. These calculations show an enhancement in the value of u_* of 15-20% when u_{100} increases from 0.10 to 0.40 m s $^{-1}$. By comparison, the curves for runs with stratification in the wave boundary layer show that for an example of $u_0 = u_{100} = 0.50$ m s $^{-1}$, u_{*w} is reduced an average of 55% and that the reduction

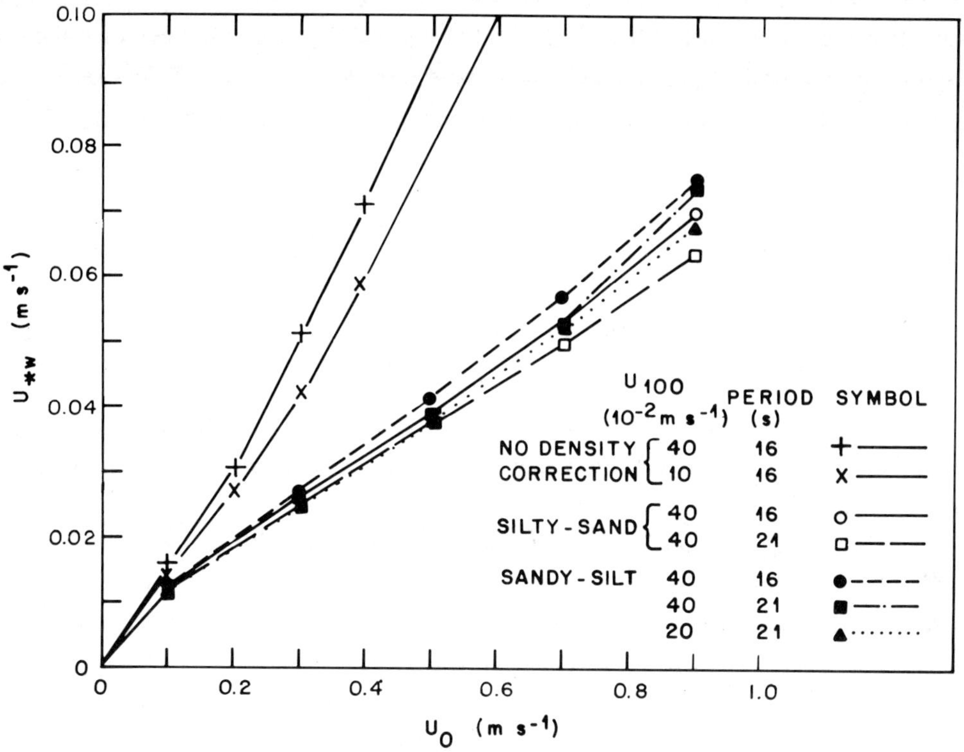

Fig. 8.8 Model results showing the relationship between u_{*w} and u_0, the bottom wave orbital velocity for cases with and without sediment induced stratification, for selected bottom sediment compositions, current velocities and wave periods. The current values are given in cm s^{-1}. The angle between the directions of the waves and currents is 60°. The size frequency distributions for the two bed types is given in Table 8.6.

caused by sediment-induced density stratification increases with wave orbital velocity. This lowering of u_{*w} values occurs because sediment-induced stratification reduces the eddy diffusive transfer of momentum to the bed with the wave boundary layer. Comparison of cases with periods of 16 and 21 s illustrate that the effect of lengthening the wave period is to decrease the shear velocity. The effect of grain size on the stratification and on u_{*w} is indicated by the differences between the curves for the two bed types, both composed primarily of fine sands and silts in different proportions. The effect of increasing the sand content of the bed is to increase the density stratification and its effects close to the bed, within the zone of the wave boundary layer, thus reducing the shear velocity due to waves, u_{*w}. The fine sediments that are carried higher into suspension have less impact in reducing u_{*w}, but can create stratification effects within the current boundary layer, thus reducing shear velocity due to the current (Kachel, 1980; Glenn, 1983; Kachel and Smith, 1986; Grant and Glenn, 1987).

To couple the wave-current interaction results to the time-dependent continental shelf model, polynomials were fitted to the curves shown in Figure 8.8 for u_* as a function of u_0. This was done for each of the bottom types and for the chosen wave period of 16 s, typical for swell events, with an assumption that u_{100}, the current 1.0 m off the bed is 0.50 m s $^{-1}$. Figure 8.8 illustrates that this simplification has least effect at low values of u_0. Yet even for the range of velocities for currents and waves modeled here the estimated error in u_{*w} is less than 10%. For waves with period T = 16 s, the following polynomial relationship holds for a sandy-silt bed:

$$u_{*w} = 0.5944 + 0.0622\,|\tilde{u}_0| \qquad (8.33a)$$

and for a silty-sand bed

$$u_{*w} = 0.5944 + 0.0622\,|\tilde{u}_0| - 8.4 \times 10^{-5}\,|\tilde{u}_0|^2 \qquad (8.33b)$$

The value of u_{*w} as a function of time is shown in Figure 8.9 for a modeled wave-current event with maximum bottom orbital velocity of 0.60 m s $^{-1}$ and period of 16 s over a sandy-silt bed.

8.4.3 Solution of the Model for Near-Bottom flow and Sediment Transport

When the model is used to calculate the near-bottom flow, sediment transport, erosion and deposition, the procedure is somewhat different than that described above. As introduced above, the geostrophic velocity, $\tilde{u}_g$, and bottom wave orbital velocity, $\tilde{u}_0$, are input as functions of time; however, $\tilde{u}_0$ is converted to u_{*w} using the polynomial expressions discussed above. Initial estimates for u_{*c} and z_0 are are used to make the first estimates of the profiles for K_m and K_s using Equations 8.22 and 8.23. The set of equations to be iterated includes those for the profiles of: current velocity (Eq. 8.7); salinity (Eq. 8.16); the volume concentration of suspended sediment in each of the size classes (Eq. 8.18); density (Eq. 8.17); wave orbital velocity (Eq. 8.9); eddy diffusion coefficients for mass and momentum (Eq. 8.22a and b); and τ_{cw}, the combined shear stress (Eq. 8.15). Iteration of these equations proceeds until stable values of $\tilde{\tau}$ are found at all levels. Subsequently, the rate of volume transport of sediment as bedload is found from the Yalin (1963) equation; the suspended load transport rate is found from Equation 8.25; the erosion or deposition, η, for each size class C_j is found from Equation 8.27; and the new distribution of size classes in the bed is found from Equation 8.30.

8.5 MODEL RESULTS AT A SINGLE SITE

The primary focus of this chapter is to model the impact of a large wave- current event at an array of locations, then to use these results to estimate the divergence of sediment transport on the shelf, hence, the net erosion or deposition during the storm (Eq. 8.25). The ultimate goal is to combine these results with the results of the analysis on the frequency of occurrence of such events in order to construct a sediment budget for the Washington continental shelf. However, before analyzing the results from the array of locations, it is necessary to explain the model results at a single site. From a geological point of view it is most interesting to model those extreme, infrequent events that can leave their mark on the stratigraphic record. Modeling such events also permits identification of the factors limiting transport and erosion in this system. Kachel and Smith (1986) showed that the coincidence of large swell (period ≥ 16 s) with storm

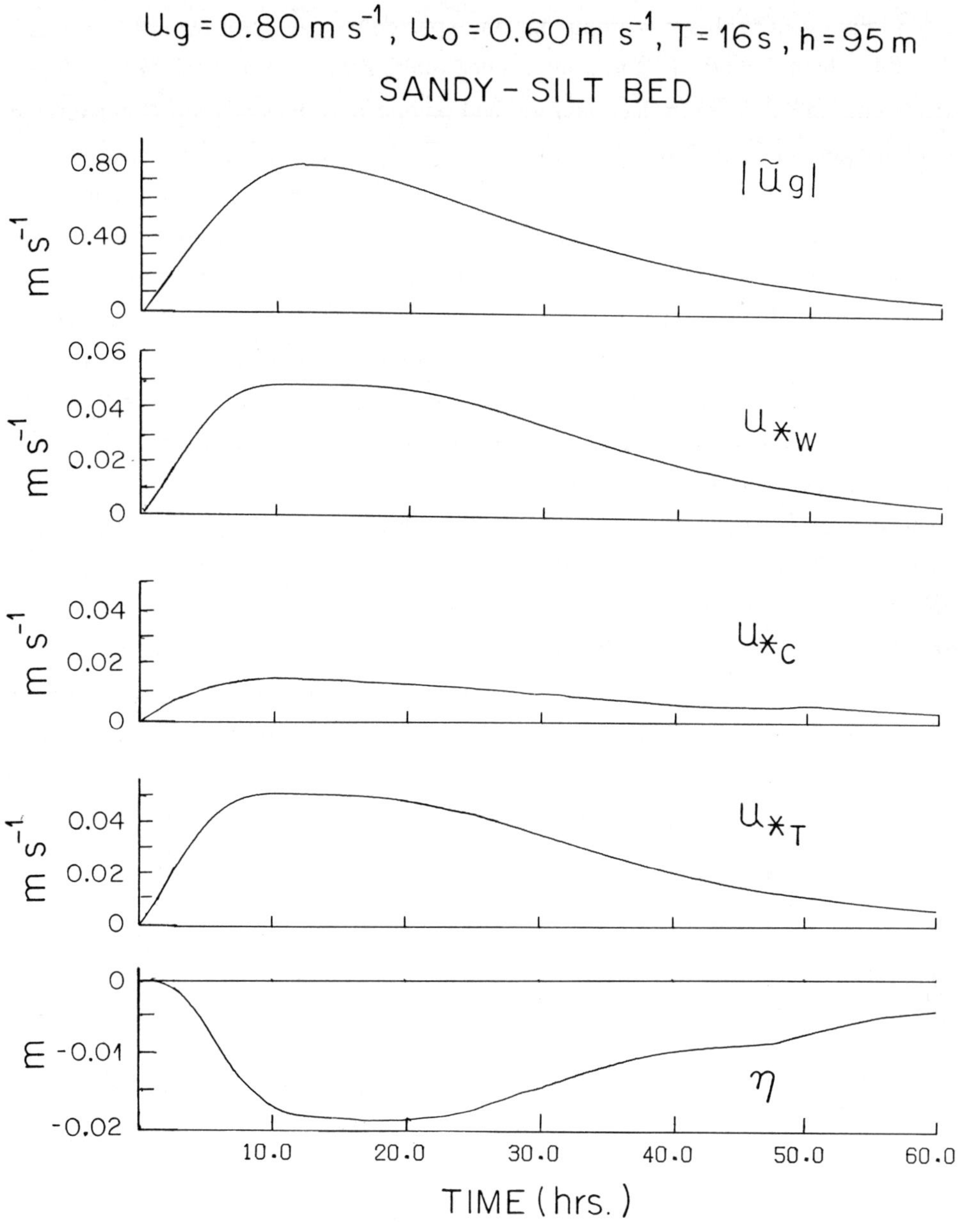

Fig. 8.9 Model input time-dependent functions of geostrophic velocity, $|\tilde{u}_g|$, calculated wave, current and total shear velocities and amount of erosion or deposition as reflected in the elevation of the bed, η.

driven currents produced almost an order of magnitude more sediment erosion and reworking than currents acting alone or in conjunction with shorter period (~12 s) locally generated storm waves. The amount of predicted erosion for the storm combined with the 16 s waves was 20-30 mm compared to 2-3 mm for the case with 12 s waves. Therefore, the event chosen to be modeled is one in which the maximum geostrophic current, $|\tilde{u}_g|$, reaches 0.80 m s $^{-1}$ coincident with the occurrence of waves of maximum orbital velocity, $|\tilde{u}_0|$ = 0.60 m s $^{-1}$ and period

T = 16 s. This event has $|\widetilde{u}_g| > 0.70$ m s $^{-1}$ and $|u_0| > 30$ m s $^{-1}$ for approximately 12 h. From Table 8.5 a joint event of this magnitude has an expected frequency of occurrence in any given year of 0.104, which is equivalent to a recurrence interval of once in every 9.6 y. The bottom depth at the chosen site is 95-m deep and the bed composition is sandy-silt characteristic of the mid-shelf silt deposit (Table 8.6).

The input for $|\widetilde{u}_g|$ as a function of time can be seen at the top of Figure 8.9 for the mid-shelf sandy-silt site with composition listed in Table 8.6 and a depth of 95 m. The input orbital velocity follows a similar curve to that shown for u_{*w},which is the result of applying Equation 8.33a to the input orbital velocity. The calculated values of u_{*c} and total shear velocity $u_{*T} = [(\tau_c + \tau_w)/\rho]^{1/2}$ are also presented, as is the predicted depth of erosion *versus* time for this case, which has no horizontal divergence in the sediment transport field included. The depth of erosion, multiplied by a unit area, represents the volume of sediment being carried in the bedload plus the suspended load above a unit area of bed.

The plot of u_{*T} illustrates that the contribution of shear velocity (or stress) associated with the waves dominates the total boundary shear velocity (or stress). At the bottom of Figure 8.9 is a plot of the cumulative erosion (–) or deposition (+) of the bed, which is depicted as the elevation (η) of the sediment as a function of time. It can be seen that the time of maximum erosion is at T ~ 18 h, which lags the time of maximum velocities and shear stresses by 8 h. The depth of maximum erosion of 22.4 mm coincides with a depth of an additional 24.4 mm of reworking, so the total depth of sediment affected by the modeled wave-current event is 46.6 mm.

8.5.1 Structure of the Bottom Boundary Layer

The interrelationship among some of the modeled variables can be seen in the time sequences of profiles of shear stress ($\widetilde{\tau}$), squared Brünt-Väisälä frequency (N^2) and suspended sediment concentration ($\sum_j^N C_j$) during the modeled wave-current event in Figure 8.10. The length scale for the Ekman boundary layer is defined as $\delta_B = u_{*c}/2f$, where f is the Coriolis parameter (1.06 x 10 $^{-4}$ s $^{-1}$ at 46° 50' N). The growth and decay of the boundary layer is best illustrated in the profiles of N^2, which are marked by a bump at the top of the Ekman layer, where density stratification increases and the turbulent mixing associated with the boundary approaches zero. The N^2 profiles reflect the structure of both the salinity and sediment concentration fields. The constant value for the upper portions of these profiles reflect the initial salinity gradient input to the model. The density gradient, hence N^2, increases throughout most of the boundary layer for the first 10-18 h, then decreases as coarser sediments settle out while the finer sediments continue to diffuse upward. The changing shape of the curves reflects the growth of and mixing in the bottom Ekman layer. By T = 36 h the concentration of sediments below z = 25 m is more uniform and the extent to which the boundary layer is well mixed can be seen in the reduced values of N^2. As $\widetilde{\tau}$, hence u_{*c} decreases, the length scale for the Ekman layer decreases, so that by T = 60 h the layer has been reduced to 15 m in thickness from its value of 32 m at T = 18 h.

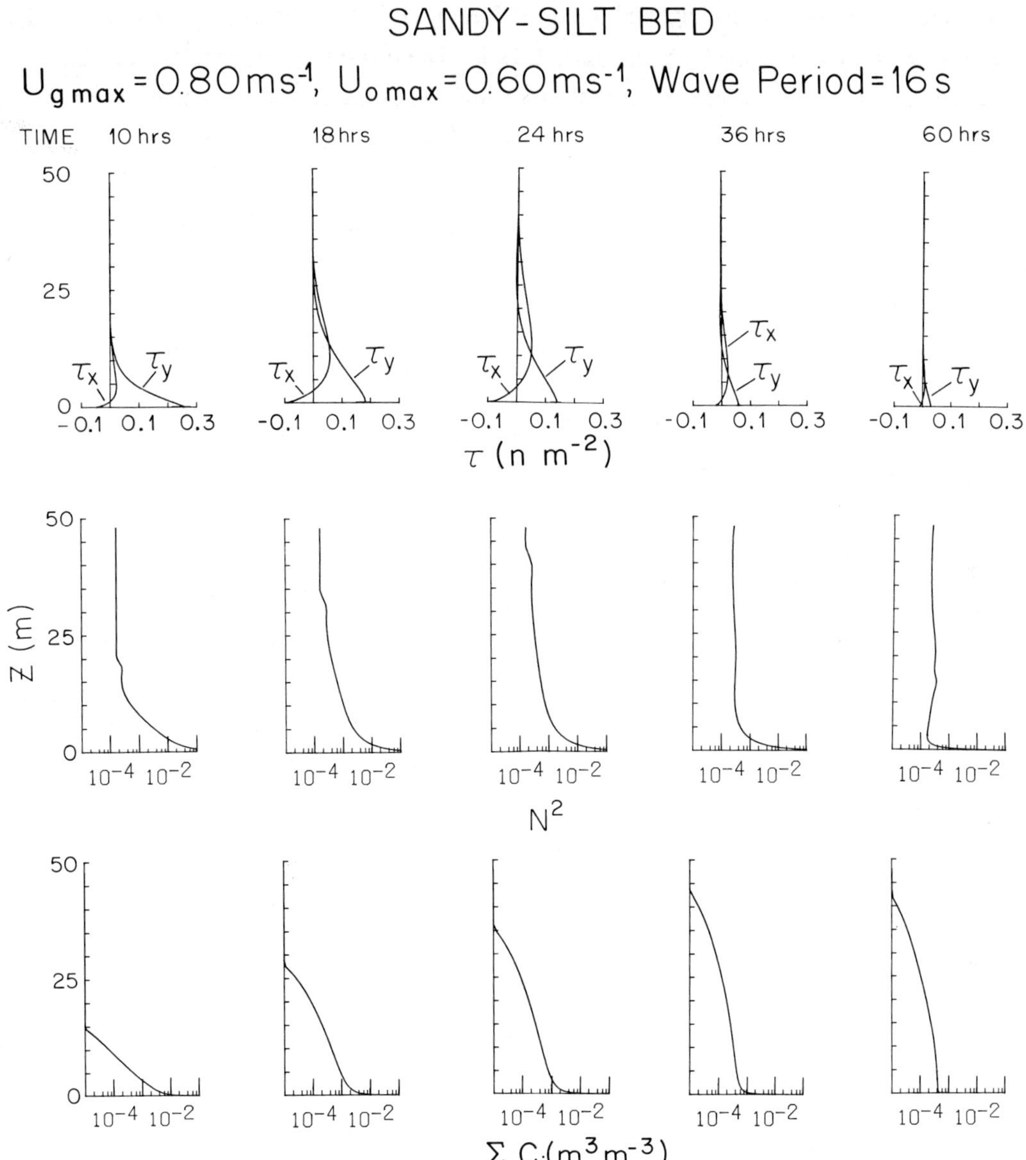

Fig. 8.10 Results of the wave-current-sediment interaction model depicting the growth and decay of the benthic boundary layer for the case with an initially sandy-silt bottom (see Table 8.6) and maximum values of $u_g = 0.80$ m s^{-1} and $u_0 = 0.60$ m s^{-1} for current and bottom wave orbital velocities, respectively. See Figure 8.9 for the form of the wave and current forcing functions. τ_x and τ_y represent the across- and along shelf components of the shear stress. The squared Brunt-Väisälä frequency, $N^2 = -(\rho/g)\ (\delta\rho/\delta z)$, parameterizes density stratification due to gradients in both suspended sediment and water (salinity). The bump in the N^2 profiles marks the top of the bottom Ekman boundary layer. Profiles of ΣC_j represent the sum over all seven size classes of the volume concentration of suspended sediment.

During the first 18 h the magnitude of the shear stress and the height of the Ekman layer increase. Simultaneously, there is an increase in the amount of turning between the direction of the geostrophic current (alongshelf) and the direction of the bottom boundary shear stress, that in this case is 28° offshore of the isobaths by T = 18 h. The turning is observed as a change in the relative magnitudes of the alongshelf and across-shelf components of the shear stress (τ_y and τ_x). As the storm abates the magnitudes of the velocity and the shear decrease, as does the stratification in the boundary layer, so that by T = 60 h the turning angle has been reduced to 18°. These calculations of the amounts of turning are greater than the 12-18° predicted by Smith and Long(1976) who included temperature and salinity but not suspended sediment effects on the density stratification.

The sequence of profiles of suspended sediment concentration summed over all size classes $\sum_j^N C_j$ illustrates the resuspension, upward diffusion and settling of sediment during passage of a storm. Once τ_b exceeds the threshold value for initiation of sediment movement, sediment begins to erode. Most of the sands are transported as bedload, or in suspension within a few centimeters of the bed, while the finest three classes in the model (diameter less than 3.5ϕ or 0.07 mm) are primarily carried in suspension. The distribution of sediment in the water column is a function of both the settling velocity and the diffusion coefficient for mass, K_ρ. During the early portion of the storm (10 to 18 h) upward diffusion of fine sediment dominates over settling so there is a continuous increase in both the volume of suspended material and the height off the bed to which it has been carried. After 19 to 20 h the volume in suspension begins to decline as coarser sediment settles; however, continued upward diffusion of fine sediment leads to a nearly uniform sediment concentration profile throughout the boundary layer.

8.5.2 Armoring of the Bed and Formation of a Graded Deposit

The impact of combined wave and current events on the stratigraphy of the sediment bed and the formation of storm deposits is discussed in detail by Kachel and Smith (1986). The effect of the interaction between the wave-current boundary layer of the flow and the bottom sediments is illustrated by the size frequency composition of the uppermost layer of the sediment bed. Figure 8.11 shows the progressive coarsening, or armoring, of the bed during the first 16 h as fine sediments erode and the successive deposition of fine fractions as the storm event wanes. If left undisturbed by reworking of organisms, the predicted stratigraphy from such an event would be that shown in Figure 8.12. At the bottom is the undisturbed zone that represents the original sandy silt bed. This is overlain by an armoring layer nearly depleted of finer sediment. As fine material settles and is deposited on the bed, it is incorporated into the layer moving as bedload. Although sediment is deposited between 18 and 47 h, τ_b is still greater than τ_{cr}, so bedload transport, ripple migration and mixing of sediment in the uppermost layer of the bed all continue to occur, yet some sediment becomes buried as the newly added sediment causes the ripples to climb. The composition of layers of sediment no longer in contact with the flow exhibit an upward fining. The frequency composition of the sediments deposited during arbitrarily long, successive time intervals during the waning of the storm event are also shown in Figure 8.12. After T = 48 h, the shear stress drops below the threshold for motion for all sizes

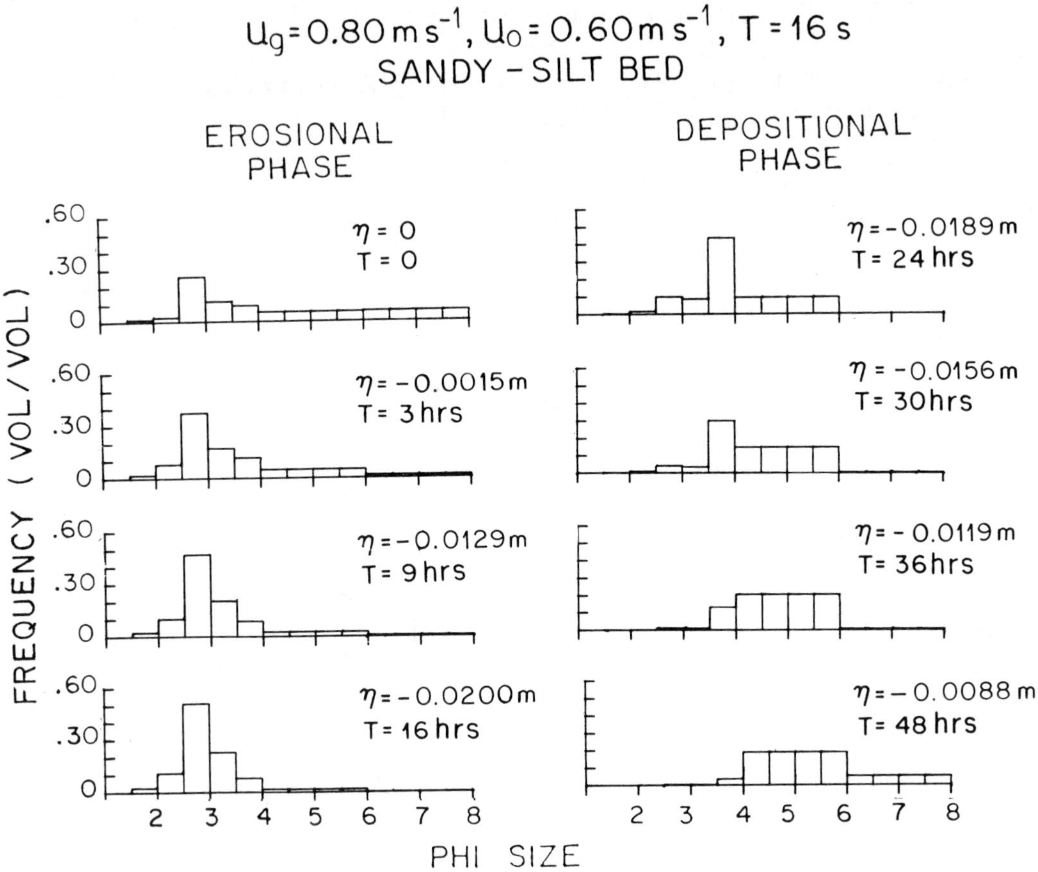

Fig. 8.11 Size-frequency distribution of bottom sediment at selected times. The bed elevation, η, at those times is indicated. During times of deposition the diagrams represent the net composition of sediment deposited during the preceding time interval.

present, so the top strata are composed of sediment deposited from suspension. Preservation in the geologic record of the deposit from any given event depends on the frequency and depth of erosion of transport events compared to accumulation rates and the presence or absence of benthic organisms able to mix this material down below the depth of erosion or physical reworking.

8.6 COMPARISON OF RESULTS OF SEVERAL CASES

Kachel and Smith (1986) present a detailed comparison of the results of the five modeled events shown in Figure 8.13 at T = 18 h, the time of peak erosion. The middle column shows the previously discussed case, including the profile of the eddy diffusion coefficient for momentum, K_m. The first three cases can be used to investigate the effects of changing the wave parameters. In the first case, the wave period has been reduced to 12 s and the maximum orbital velocity to 0.10 m s^{-1}. As was seen in Figure 8.8, at this low value of u_0, the effect of decreasing the wave period is small compared to that of increasing the amplitude of the orbital velocity. The effect of lowering the amplitude (or bottom wave orbital velocity) of waves is that much less sediment is

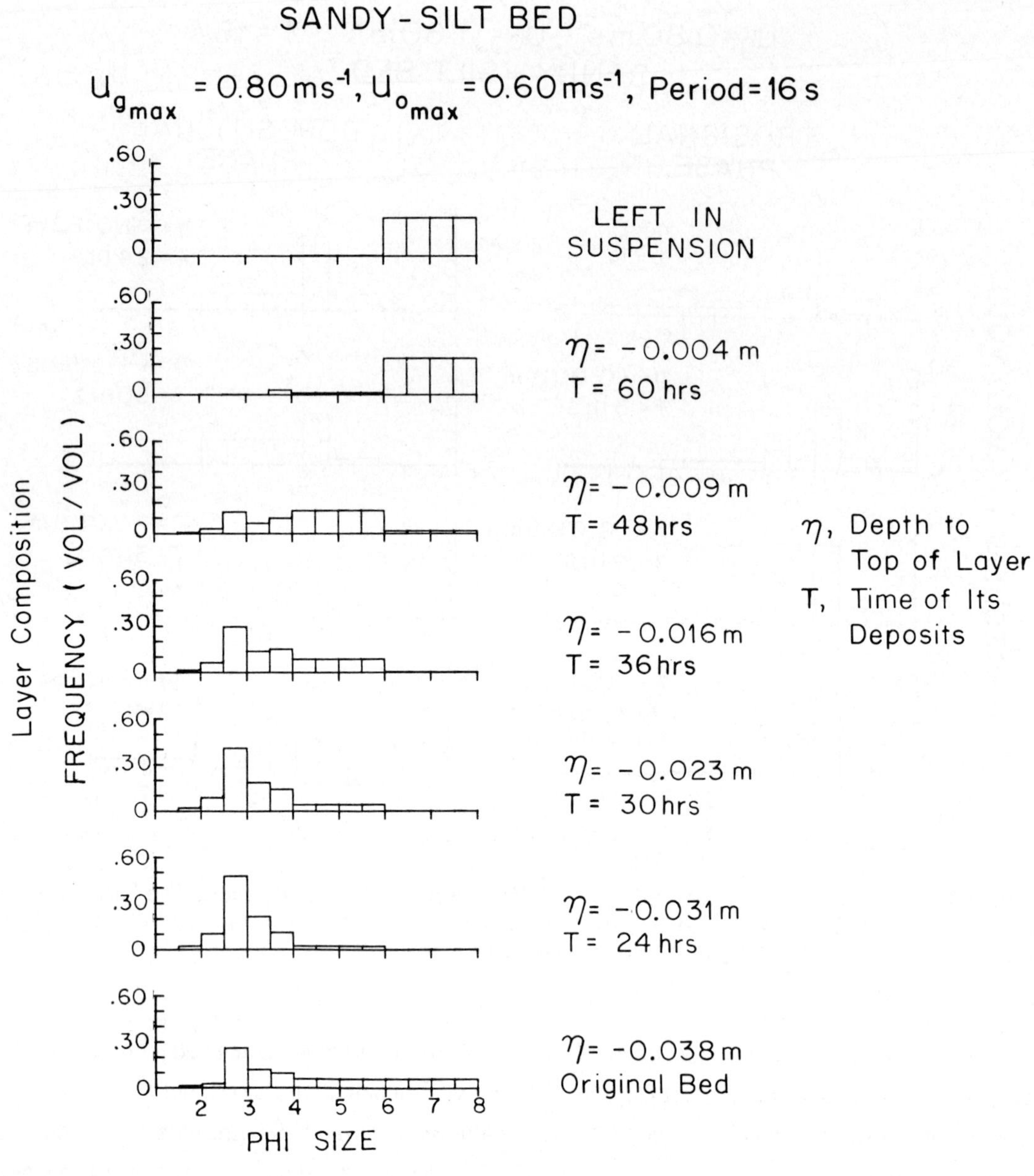

Fig. 8.12 Micro-stratigraphy predicted as a result of the modeled storm acting on a sandy-silt bottom. Times mark the end of the time interval of deposition of the layer of composition shown. Elevations (η) mark the depth in the bed at which a layer of that composition would be found. η is measured with reference to the ongoing bed elevation. Below $\eta = -0.038$ m is the original bed composition. It is overlain by an armoring layer of fine sand. As sediment is redeposited it enters the bedload until $\tau_b < \tau_{cr}$. Subsequently, the sediment remaining in suspension is deposited on top of any existing bedforms.

put into suspension, so density stratification is reduced. Therefore, turbulent eddy diffusion of momentum downward and of sediment upward is greater, shear stress values are higher through the benthic boundary layer and the rate of sediment transport, $\bar{q}_s$, is reduced throughout the benthic boundary layer.

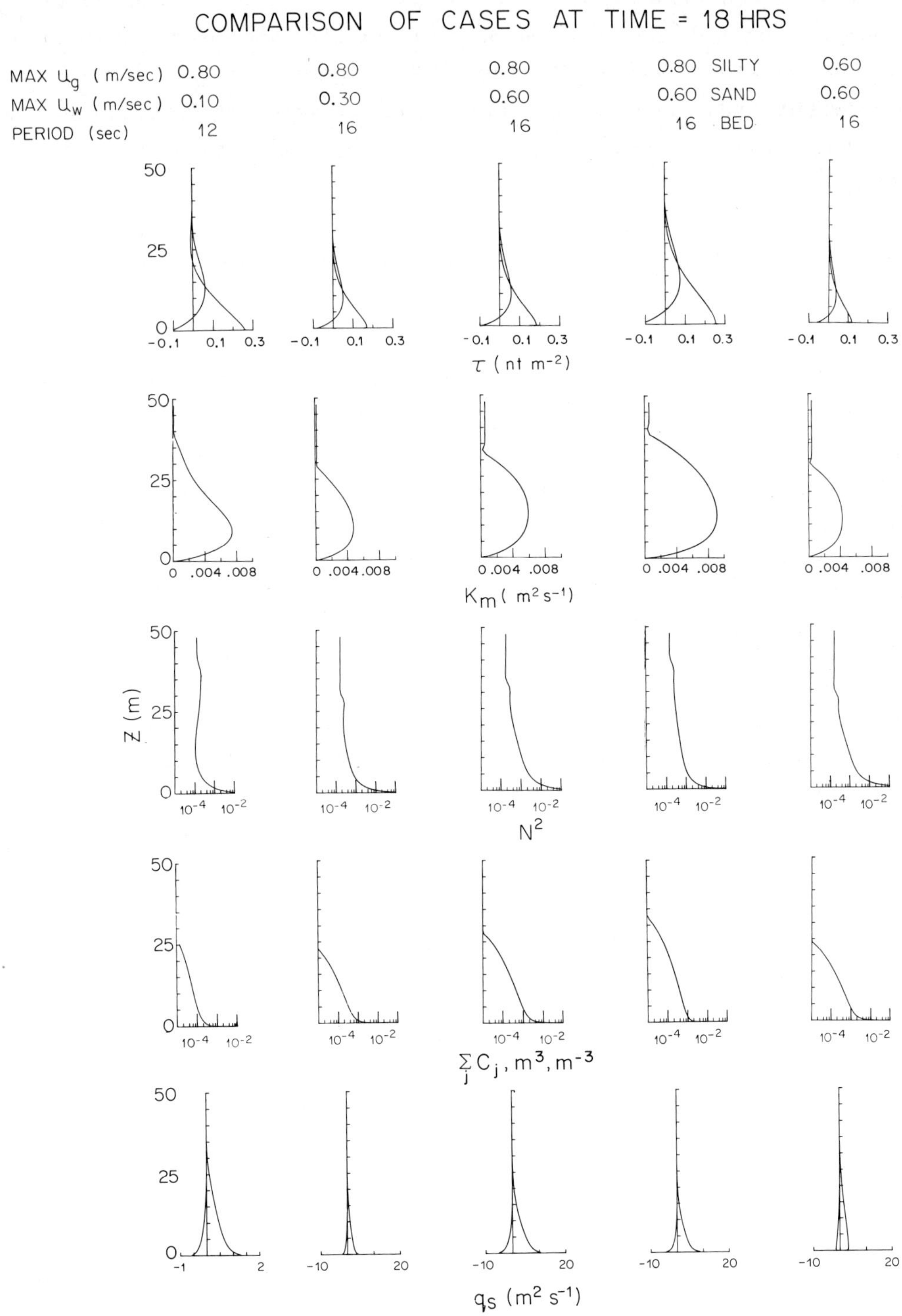

Fig. 8.13 Comparison of vertical profiles at 18 h of modeled time of shear stress, $\tilde{\tau}$, eddy diffusion coefficient for momentum, K_m, squared Brünt-Väisälä frequency, N^2, the sum of the concentration in all size classes, ΣC_j and the sediment transport rate, $\tilde{q}_s$. The case in column 1 has lower waves and lower wave period (12 s); case 2 has moderate swells, case 3 is that previously presented, case 4 has a sandier bed and case 5 has a smaller geostrophic current velocity. Note: the scale for $\tilde{q}_s$ in the first case is a factor of ten smaller than it is in the others.

As wave orbital velocity, u_0, increases in the first three cases, the magnitude of shear stress and the height of the boundary layer first decrease (the second case) and then increase slightly as the maximum orbital velocity of the waves continues to increase. Reduction of $\tilde{\tau}_c$, the shear stress due to the current, is caused by increased suspended sediment stratification in the water column due, in turn, to larger values of $\tilde{\tau}_w$ and $\tilde{\tau}_T$, which cause more sediment to be eroded from the bed.

The effect of reducing the amount of silt in the initial bed composition can be seen by comparing the case displayed in the the fourth column with that displayed in the third. As was seen in Figure 8.8, more sand being transported close to the bed reduces the shear velocity due to waves. Less fine sediment is available in the bed to be eroded and transported in suspension. Thus, the Ekman boundary layer is less stratified, the values of eddy diffusion coefficient and shear stress values are higher throughout and the scale length of the boundary layer is higher, but the total transport of sediment during the storm is reduced.

Insight into the importance of the current in a sediment transporting event involving interacting waves and currents can be gained by comparing the case in the middle column in Figure 8.12 with that in the last column, in which $|u_g|_{max}$ is reduced to 0.60 m s^{-1}. In the latter case, $\tilde{\tau}$ and K_m are reduced throughout the profile, the thickness of the Ekman layer is reduced, sediment is carried lower in the water column and the total sediment load is reduced. Comparison of the profiles for transport rate, $\tilde{q}_s$, illustrates that a reduction in the velocity, $|\tilde{u}_g|_{max}$, of 25% results in a reduction in the cumulative sediment transport, integrated over the thickness of the boundary layer and over the duration of the modeled storm to 38% of that occurring during the larger storm with the same bed and wave conditions. The model also predicts that during the event with a smaller current 0.024 m of sediment will be reworked, compared to 0.038 m during the larger event over a site with the same initial sediment composition.

Kachel (1980) and Kachel and Smith (1986) define average path lengths, $\tilde{D}_b$ or $\tilde{D}_s$, for particles in the bedload and suspended load, respectively, as

$$\tilde{D}_b = \int_{0h}^{60h} \frac{\tilde{Q}_b}{\eta + \delta_m}\, dt \quad \text{or} \quad \tilde{D}_s = \int_{0h}^{60h} \frac{\tilde{Q}_s}{\eta + \delta_m}\, dt \tag{8.34}$$

Where $\tilde{Q}_b$ and $\tilde{Q}_s$ have limits of L^2/T. For the range of cases modeled in Figure 8.13, path lengths of 4-85 m for the bedload compare with 5-52 km for the suspended load. This clearly illustrates the difference in net effect of these two sediment transport modes. Once eroded, particles carried in suspension travel quickly along and across the shelf compared to those in the bedload. Because the transport, $\tilde{Q}_s$, is the product of $\tilde{u}$ and $\sum_j C_j$ and because the concentration increases toward the bed, $\tilde{Q}_s$ is weighted most heavily in the direction of the lowest portions of the flow where the offshore turning is most pronounced. Note that the coarsest fractions of the bed, those subjected to bedload transport during the storm, move during the period when offshore turning is most intense. Fine fractions are transported in suspension higher in the flow and thus experience less turning. In addition, significant amounts of fine sediment continue in suspension as the net flow in the boundary layer aligns itself with the direction of the geostrophic current. Therefore, the direction of cumulative transport over the entire event is more offshore for the bedload than it is for the suspended load.

8.7 THE AREAL DISTRIBUTION OF EROSION AND DEPOSITION

In order to describe the pattern of sediment transport during a geologically interesting wave-current event on the Washington continental shelf, the model discussed above was applied to the array of locations along and across the shelf depicted in Figure 8.1. From the results of these runs, it is possible to estimate the divergence of the transport and thereby, the pattern of net erosion or deposition on the shelf (see Eq. 8.26). To this end the relatively infrequent (once every 9.6 y) combination of waves and currents used in the case previously discussed was run with the input conditions for each site shown in Table 8.7. The values of $u_{g_{max}}$ were empirically derived from current data taken during extended periods of northward flow (*e.g.*, Fig. 8.3). It was assumed during such times that the interior velocity, $\tilde{u}_g$, is inversely related to the water depth, h, such that $u_g / u_{g_0} = (0.67)\ (h_0/h)$, where u_{g_0} is the geostrophic velocity at a site of depth h_0. The rational for this choice was discussed previously in Section 8.2.2. The maximum bottom orbital velocity at any site was computed by linear wave theory from that at the 95-m site. The list of sites was divided into two sets in order to study across-shelf and alongshelf contributions to sediment dispersal. The alongshelf experiment was further divided into two phases in order to separate the counterbalancing effects of variations in bottom sediment silt content and the northward decrease in the geostrophic current related to the divergence of the mid-shelf isobaths.

8.7.1 Across-Shelf Effects

To test the effect of across-shelf changes in the size distribution of bottom sediments, the input forcing functions of geostrophic current and wave orbital velocity at the bottom were held constant while the size distribution frequency observed in the summer (Nittrouer, 1978) at each site was input (Table 8.6). Table 8.7 lists the input conditions for depth, current and wave conditions for the model run at each site. Sediment distribution is only suggested by the numbers for percent silt on Table 8.7; the actual frequency distribution input for each size class was presented in Table 8.6. The maximum calculated values for maximum shear velocity, u_{*max} and maximum depth of reworking $(\eta + \delta_m)_{max}$ are also shown. Plots of the across-shelf patterns of these variables are given in Figure 8.14. The cross-shelf variation in the values of u_{*max} seen in Table 8.7 and Figure 8.17 reflect not only the offshore reduction in $\tilde{u}_0$ and $\tilde{u}_g$, but also the reduction in u_* over the sandier beds found on the outer shelf. As was discussed above, sediment resuspension above sandier beds causes higher density stratification close to the bed than does resuspension of silts. Therefore, over the sandier bed, u_{*w} and thus u_{*T}, the total shear velocity are reduced. The depths of erosion and reworking predicted by running the one-dimensional model at five sites in the cross-shelf direction is shown in Table 8.7 and Figure 8.14. The predicted

TABLE 8.7 Model inputs and results for large wave-current events at several sites. Inputs for each case include bottom depth, maximum geostrophic velocity and maximum bottom wave orbital velocity. The waves at each site have heights of 5.8 m and periods of 16 s. Outputs listed are: the maximum combined shear velocities due to the combined wave and current velocities u_{*max}, the maximum depth of erosion η_{max}, and the combined depth of erosion and reworking of the sediment bed, $(\eta + \delta_m)$.

Site	Distance North of river (km)	Depth (m)	Silt (%)	Current $\lvert u_g \rvert_{max}$ (m s^{-1})	Waves $\lvert u_0 \rvert_{max}$ (m s^{-1})	u_{*max} (m s^{-1})	η_{max} (10^{-2} m s^{-1})	$(\eta + \delta_m)_{max}$ (10^{-2} m s^{-1})
Alongshelf								
A	−2	55	27.3	0.93	1.30	10.43	−0.96	−3.21
B	2	95	31.2	0.80	0.60	5.18	−0.99	−2.55
C	20	95	29.3	0.80	0.60	5.16	−1.23	−2.81
D	70	95	67.8	0.80	0.60	5.08	−2.02	−3.77
E	113	95	67.8	0.80	0.60	5.09	−1.66	−3.24
Across-shelf								
F	70	55	18.4	0.93	1.30	7.15	−1.96	−4.78
G	70	75	47.5	0.86	0.86	5.32	−1.84	−3.42
D	70	95	69.8	0.80	0.60	5.08	−2.02	−3.77
H	70	105	62.1	0.77	0.51	3.60	−1.10	−2.23
I	70	143	36.2	0.66	0.28	2.64	−0.37	−1.08
Effect of bathymetric divergence								
D'	70	95	69.81	0.60	0.60	4.03	−1.27	−2.39

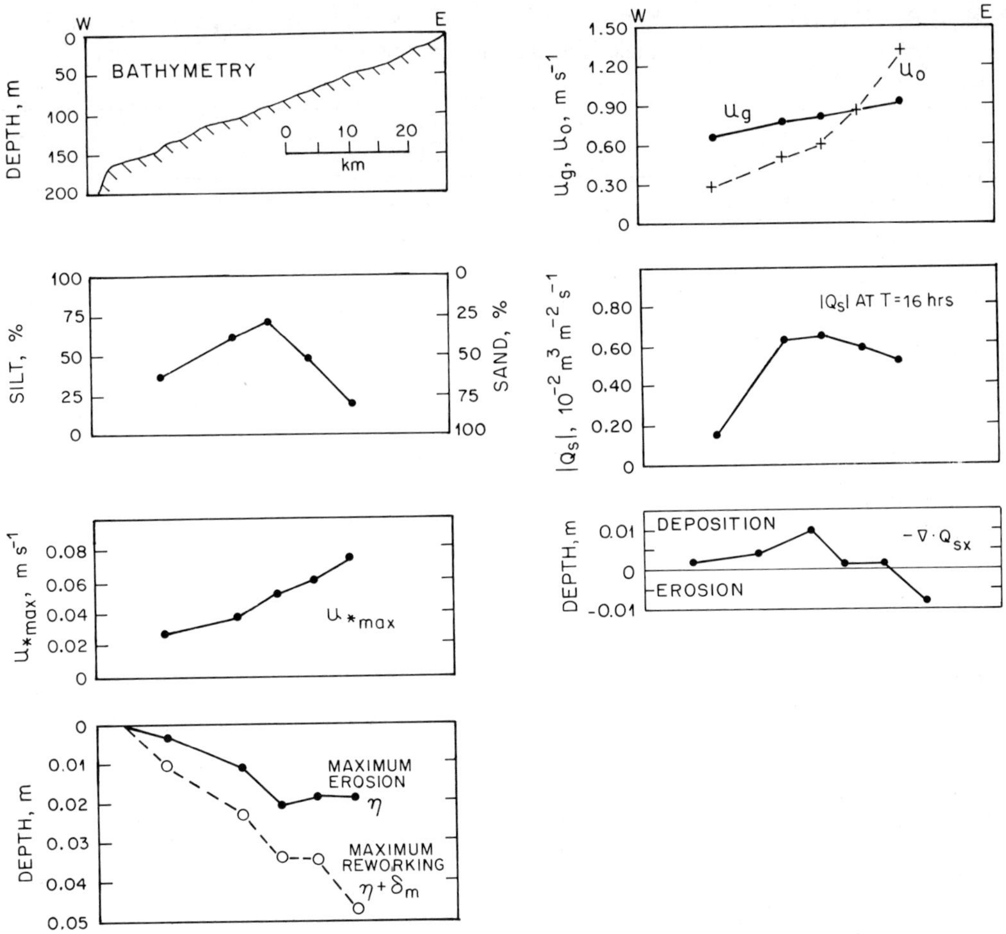

Fig. 8.14 The distribution across the shelf off Grays Harbor of the following model inputs: bathymetry, volume frequency distribution of silt and sand and maxima of $|\tilde{u}_g|$ and $|\tilde{u}_0|$ for constant wave amplitude and period across the shelf. Also shown are: the maximum depths of erosion (η) and reworking ($\eta + \delta_m$) predicted from the model runs at each site, as well as predicted peak transport ratio at each site. The across-shelf divergence of the sediment transport that causes net erosion or deposition of sediment (see Eq. 8.26) is shown in the lower right frame.

maximum thickness of bed material removed from the bottom during the storm is 0.004 to 0.020 m, while the average depth reworked is 0.01 to 0.05 m. Under the assumption of horizontal uniformity along the shelf, the material lifted into suspension eventually is deposited leaving a graded stratigraphic layer of thickness ($\eta + \delta_m$). From sites F to D (55 to 95-m depth) the almost level value of η_{max} is due to a balance between the effect of the offshore increase in silt content of the bed, which increases erosion and the decrease in u_* with decreasing current velocity, which decreases erosion. On the outer-shelf the decrease in u_* is the dominant effect so that the amount of erosion decreases in the offshore direction. The values for the depth of reworking ($\eta + \delta_m$) contain an additional grain size effect: as mean grain size on the bed increases, ripple height increases and therefore δ_m, the depth of mixing becomes greater. The pattern of the cross-shelf divergence of the sediment flux integrated over the storm is also shown in Figure

TABLE 8.8 Predicted across-shelf divergence of sediment transport, $Q_{s_x}\Delta t$, with units $m^3 m^{-1}$ storm $^{-1}$, for each size class during the modeled large storm/large swell event at locations shown in Figure 8.1.

	Site				
Size	I	H	D	G	F
> 2.0	−0.0018	−0.029	−0.056	−0.014	−0.17
2.0-2.5	−0.0027	−0.021	−0.039	−0.017	−0.12
2.5-3.0	−0.0090	−0.028	−0.056	−0.050	−0.068
3.0-3.5	−0.060	−0.13	−0.026	−0.37	−1.4
3.5-4.0	−4.2	−8.0	−16.0	−45.0	−80.0
4.0-6.0	−6.2	−36.0	−73.0	−70.0	−40.0
> 6.0	−5.4	−16.0	−24.0	−7.0	−9.4
Total	−16.0	−60.0	−110.0	−120.0	−130.0

8.14. Positive values indicate net removal of sediment from the inner shelf (depths < 75 m) and net negative values indicate deposition (negative) on the outer-shelf. Table 8.8 shows the same divergence data, but broken down into the contributions attributable to each grain size fraction. From these data it is apparent that between stations F and G, on the inner side of the mid-shelf silt deposit, both erosion of fine material (silts) and deposition of sands (especially the 3.5 to 4.0ϕ class) occur during the storm. Evidence for this phenomenon was observed by Nittrouer (1978), who sampled the surface sediment distribution at station F (Fig. 8.1) on six occasions between June 1976 and April 1977. He observed that the frequency of silt and clay is reduced from summer values of 18-21% to less than 1% in the winter. Concurrently, the median sediment diameter increased from 0.09 mm in summer to 0.29 mm in winter.

8.7.2 Alongshelf Effects

The alongshelf distributions of maximum erosion, reworking and peak transport rates are shown in Figure 8.15. The station A case was included in order to provide an input to the system at the southern boundary of the area near the Columbia River mouth. The shallower bottom depths in this area with their associated higher wave orbital velocities cause higher transport rates. This site is important because it is the source area for sediments moving on the mid-shelf of Washington, the area where sediment from the Columbia River is initially deposited from the plume during the non-storm conditions that prevail during spring runoff (Barnes *et al.*, 1972; Sternberg *et al.*, 1977; Baker *et al.*, 1980; Baker and Curl, 1981; Ridge and Carson, 1987). Figure 8.15 shows a high degree of correlation between the depth of erosion, or the peak rates of volume transport of sediment and the percentage of silt in the initial sediment bed. The higher the initial proportion of finer, suspendible material in the bed, the greater the possible erosion depth and the greater the cumulative sediment transport during the storm. This is a result of the armoring effect of the sand grains shielding the sediment bed from erosion and reducing the availability of fine suspendible material to the flow. Another effect of sediment size composition

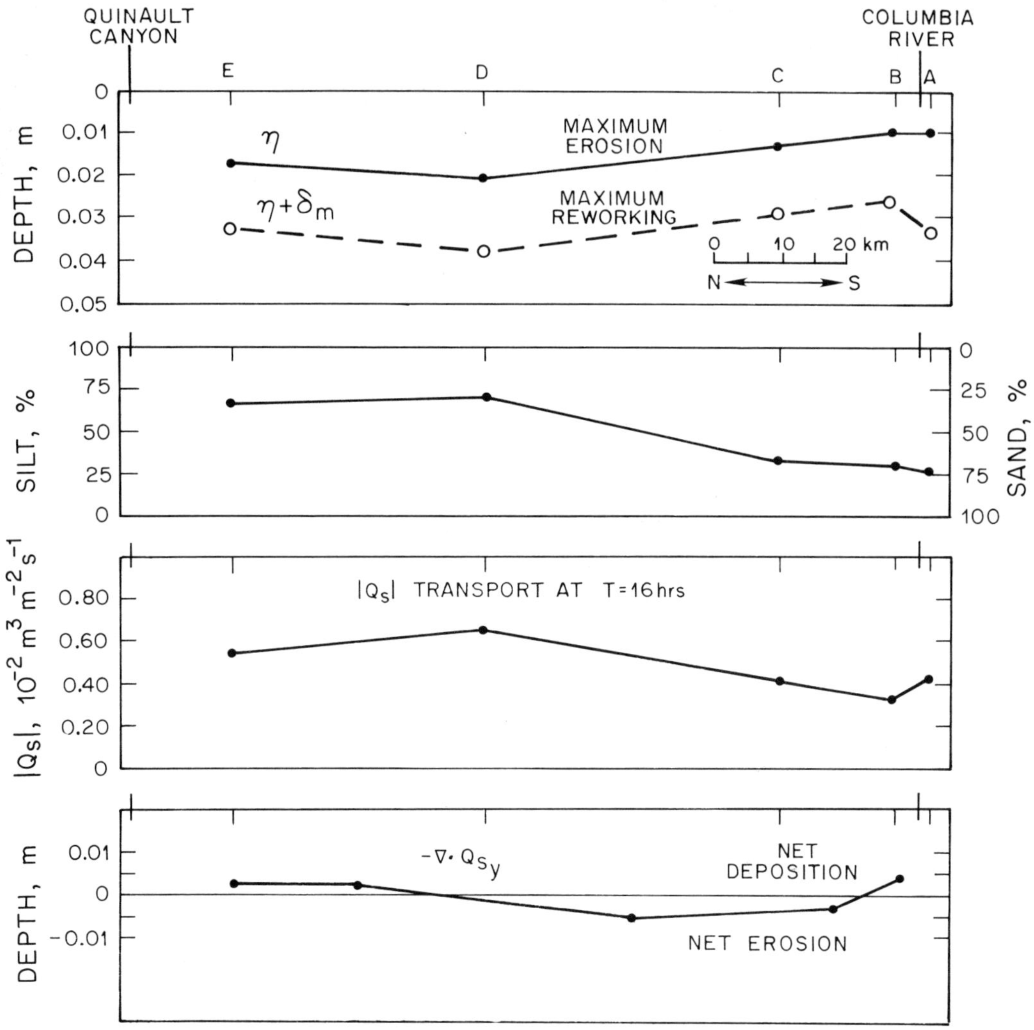

Fig. 8.15 Results of model runs at fine sites along the mid-shelf illustrates the alongshelf variations in maximum predicted depths of erosion and reworking, in sand and silt composition, in the peak rates of sediment transport (at ~16 h). The alongshelf divergence in the sediment transport shows areas of deposition near the Columbia River and just south of Quinault canyon. The alongshelf divergence of the sediment transport was calculated between pairs of sites from the total sediment loads, which are the values of the transport rate integrated over the boundary layer thickness and over the duration of the storm.

is seen at station A, where coarser grain sizes in the sand fraction lead to increased ripple height and in the volume of bed material from which fine material can be removed. The alongshelf divergence of the alongshelf component of sediment transport in Figure 8.15 shows an area of net erosion occupying most of the mid-shelf region between the Columbia River (site A) and Grays Harbor (Site D). Net deposition is indicated for a small area immediately off the Columbia River and in the zone north of Grays Harbor. When the typical assumption of a bed concentration of 0.6 is made, the maximum contribution to the net deposition calculated from the alongshelf divergence of the sediment transport on the mid-shelf for this storm is 0.006 m near the Columbia River and 0.004 m north of Grays Harbor. The maximum predicted erosional loss is –0.007 m in the vicinity of station C. These results indicate that effects of the northward fining of the sediment size composition are: to increase the volume of sediment being put into suspension northward along the mid-shelf; and to increase the depths of erosion and reworking. The alongshelf component of the divergence of the transport creates a net erosion between stations B and D and a net deposition north of D, during large, infrequent events such as the one modeled.

8.7.3 Effects of the Bathymetric Divergence

The divergence of the bathymetry on the Washington continental shelf in the area from 40-95 km north of the Columbia River mouth (the zone least affected by indentations of submarine canyons), as measured by the change in the distance between the 50 and the 100-m isobaths, is 0.12-km width per km length or an increase of 37.9% in width. Assuming a two-dimensional mass balance, divergence of the isobaths results in a decrease of the geostrophic velocity along the downstream section. A 25% decrease in $|u_g|$ would be expected at a downstream (northward) distance of 48 km, where the shelf is 33% wider. To examine the impact that this alongshore divergence in the geostrophic flow has on the volume of sediment transported, two cases were run with the same sediment bed, namely that found at site D and with maximum input 0.60 and 0.80 m s $^{-1}$, respectively. For the case with maximum current velocity $|u_*|_{max} = 0.80$ m s $^{-1}$, the alongshelf component of the cumulative sediment transport for the wave-current event equals 2.8 x 10 ^{2} m ^{3} m $^{-1}$, while for the case with $|u_*|_{max} = 0.60$ m s $^{-1}$ the cumulative sediment transport is 2.2 x 10 ^{2} m ^{3} m $^{-1}$. If it is assumed that (due to this divergence) the excess sediment is deposited evenly between these locations 48 km apart and that the volume concentration of sediment in the bed is 0.6, then the resulting deposit would have a thickness of 0.0018 m. To achieve the average annual accumulation rate of approximately 50 x 10 $^{-2}$ g cm $^{-3}$ y $^{-1}$ (5.0 kg m $^{-2}$ y $^{-1}$) measured by Nittrouer (1978; Nittrouer *et al.*, 1979) a net deposition of 0.003 m of sediment is needed each year. However, the present model is for the 9.6 y storm, so the amount of deposition possible from alongshelf bathymetric divergence effects is an order of magnitude too small.

8.7.4 Contribution of Large Events

A conclusion from the modeling results is that the large events that can leave their mark on the stratigraphic record by eroding and reworking the sediments as illustrated earlier, are not the only cause of redistribution of the fine sediments northward from the mouth of the Columbia River or of the net observed depositional pattern in Figure 8.5. The modeling predicts erosion on the southern portions of the mid-shelf silt deposit, as well as average sediment path lengths for the total load, $\tilde{D}_T$, on the order of 50 km for the case of a large storm coinciding with a large swell event. Because the source area of accumulation of the Columbia River fine sediments is limited in extent south of the river (see Fig. 8.5), events that move particles these predicted distances should be expected to deplete areas near the source. Therefore, more frequent, storm events with lower currents or shorter durations must contribute significantly to the dispersal of fine sediments northward on the mid-shelf. That this, in fact, does occur is evidenced by the observations of Ridge and Carson (1987) of the dispersal of Mount Saint Helens ash from October, 1980 to January, 1982. Between October, 1980 and 1981, Ridge and Carson (1987) calculated that five storms transported ash northward at rates of 14 to 38 km per storm, rates which are low compared to the distances reported by Kachel and Smith (1986). During the winter of 1980-81 Hickey *et al.* (1986) observed that most of the resuspension events that occurred on the outer- continental shelf near Quinault canyon could be attributed to the action of waves instead of currents, suggesting a reason for the lower rates that year. In contrast, during the period between October, 1981 and January, 1982 they observed several storms with bottom currents on the outer shelf capable of transporting sediment occurred, including the largest event recorded in Table 8.1 (see Fig. 2.50, Hickey, 1989; Chapter 2). Ridge and Carson (1987) found that during this period the ash content declined at half of their stations along the mid-shelf silt deposit and concluded that these storms had resuspended the ash and transported it beyond the area of their study (the Columbia River to the head of Quinault canyon). Therefore, these observations corroborate the model predictions of fine sediment removal during large events. In the context of the model presented in this paper, during the 1980-81 winter only small storm events occurred, whereas during the 1981-82 season there was a storm comparable in magnitude to the one discussed above.

8.7.5 Estimating the Contributions of Many Smaller Events

The large storm modeled in this paper can be used to assess the net impact of the many smaller storm events. The accumulation on the inner continental shelf off southern Washington of modern sands derived from the Columbia River is well documented (Kelley and McManus, 1969; 1970; McManus, 1972; Smith and Hopkins, 1972; Nittrouer, 1978). Let us assume that these inner shelf sands provide the source for the vast majority of the sand on the mid- and outer shelf and that they are moved to the mid- and outershelf only during large events. Between occurrences of large events, smaller events bring in sufficient silt to create the size distributions observed in summer. The calculated cross-shelf divergence of the sand transport can be used in conjunction with the observed frequency distribution of sand in sediment samples collected during the summer to calculate the net deposition of silts needed between large storms to produce the observed sediment distributions in Table 8.6 (or Fig. 8.4). The ratio of the volume of material in a chosen sand class to that in all of the silt classes (as determined by Nitrouer, 1978) is compared to the cross-shelf divergence of the chosen sand class during the 9.6 y event. To do this two further assumptions must be made: first, that the sediment distributions observed at the present time are approximately the same as have existed since sea level reached its present position roughly 7,000 years ago; second, that the inner shelf sediment source is constant enough along the shelf that a cross-shelf divergence calculation represents the divergence perpendicular to the path taken by the sands (~15° west of the isobaths for the 3.5ϕ class, to ~25° for the 2ϕ class).

There are several factors to consider in choosing the specific sand class that should be used in these calculations. First, the sand should be derived from the nearshore zone along the entire coast and not significantly from the silt deposit off the river mouth. Second, it must be present in large enough proportions that sample-to-sample variations do not contribute large errors in the calculations. Third, the sediment should be transported primarily in suspension during extreme wave-current events to reduce errors caused by small calculated divergences. The frequency histograms in Figure 8.4 and the patterns of distribution for the sand classes in Figure 8.6 show that all the sand classes are distributed continuously from the near-shore to the outer shelf thus they all satisfy the first criterion. Classes larger in size than 3.0ϕ fail the second criterion because of their low abundances in the samples. To choose between the remaining sand fractions (i.e. the 3.0-3.5ϕ, the 3.5-4.0ϕ or the combined 3.0-4.0ϕ classes), the modes of transport need to be discussed. When sediment is transported primarily as bedload or solely as suspended load, its volume discharge can be computed accurately; whereas, when sediment is moved in a mixed-mode (e.g. the Rouse number is near unity), the transport rates depend sensitively on the settling velocities. Also, even when the appropriate sediment transport parameters are known, computation of mixed mode sediment transport is substantially less accurate. In the case of interest, model results show that the 3.0-3.5ϕ sand is moved in the mixed mode, but that the 3.5-4.0ϕ sand is transported purely as suspended load and therefore is the appropriate size class to use for reference in the calculation. In addition, it is better to compare transport rates of sizes that are moved in the same mode.

The top panel in Table 8.9 presents calculations for η_c, the contribution of the cross-shelf divergence to the erosion or deposition of sediment (Eq. 8.26) for each size class for the modeled 9.6-y wave-current event. Three sets of calculations of the net deposition of silt between large events were made: for the inner to mid-shelf, for the mid- to outer shelf and for the whole mid-shelf section with depths of 55-143 m. The division of the shelf into two zones in addition to the whole shelf calculation was done to permit a better comparison with the observed cross-shelf seismic stratigraphy presented by Nittrouer (1978) and shown in Figure 8.2.

The calculations in the lower half of Table 8.9 use the ratio of the sediment in the 3.5-4.0ϕ and the 4.0-8.0ϕ size classes, as determined by Nittrouer (1978), in conjunction with the calculated cross-shelf divergence in the transport of material in the 3.5-4.0ϕ class from the 9.6-y event, to find the volume of silt deposited by smaller events during the interval between the 9.6-y extreme wave-current events. Between the inner and mid-shelf sites (F-D) during the 9.6-y event, divergence of the transport of the 3.5-4.0ϕ class accounts for a deposit of thickness $\eta_c = (-1/C_b\, \partial Q_s/\partial x\, \Delta t)$ of 0.0068 m. At the same time the silt fractions between these sites is removed. Between the occurrence of the 9.6-y event sufficient quantities of silt must be deposited between sites F and D to create the average frequency distribution observed in Nittrouer's (1978) samples, in which the 3.5-4.0ϕ sands comprise an average of only 14.9% (8-33%), not 99%, of the material present. The total thickness of material deposited during the 9.6-y recurrence

TABLE 8.9 Estimates of net deposition on the Washington shelf during the Holocene contributed by the cross-shelf divergence of the transport of very fine sand as predicted by the model run with an event with a 9.6 y recurrence interval. $\eta_{c_j} = -(1/\ C_b)\ (\Delta\ Q_{s_j}\ \Delta t\ /\Delta x)$ is the average thickness of sediment in size class deposited between pairs of sites shown in Figure 8.1.

Size Class(ϕ)	Inner to Mid-Shelf F–D ($m^3\ m^{-2}$)	Mid to Outer Shelf D–I ($m^3\ m^{-2}$)	Inner to Outer Shelf F–I ($m^3\ m^{-2}$)
> 2	1.2×10^{-6}	5.1×10^{-6}	8.5×10^{-6}
2.0-2.5	7.9×10^{-6}	1.7×10^{-6}	6.0×10^{-6}
2.5-3.0	1.1×10^{-6}	4.5×10^{-6}	2.9×10^{-6}
3.0-3.5	1.2×10^{-4}	1.9×10^{-5}	7.0×10^{-5}
3.5-4.0	6.7×10^{-3}	1.1×10^{-3}	3.8×10^{-3}
4.0-6.0	-3.5×10^{-3}	6.4×10^{-3}	1.8×10^{-3}
< 6	-1.6×10^{-3}	1.8×10^{-3}	2.0×10^{-3}
Total	1.7×10^{-3}	9.3×10^{-3}	7.7×10^{-3}
F_j(3.5-4.0ϕ) %	14.9	24.5	21.6
η_{c_j}(3.5-4.0ϕ) (m)	6.7×10^{-3}	1.1×10^{-3}	3.8×10^{-3}
$\eta_{c_{total}}$ (m)	2.5×10^{-2}	7.1×10^{-3}	1.6×10^{-2}
η_c (net between storms) (m)	2.3×10^{-3}	-2.2×10^{-3}	8.4×10^{-3}
Average Annual deposition ($m\ y^{-1}$)	2.7×10^{-3}	7.7×10^{-4}	1.8×10^{-3}
Deposition in 7,000 y (m)	18.5	5.4	12.4

interval can be found by setting $\delta_m = 0$, then rearranging Equation 8.30, so that it becomes $\sum_j^N \eta_{c_j} = \eta_{c_j}/F_j$. Using this relationship the net deposition over a 9.6 y interval would be 0.0250 m. Repetition of these calculations for the zone between the mid- and outer shelf yields a prediction of an average deposition of 0.0011 m of sand 3.5-4.0ϕ diameter in this outer zone. However, the predicted cross-shelf divergence of the silt fraction also causes deposition of more silts than are needed to account for the observed sediment distributions on the upper shelf. In the model results in Figure 8.15, the shelf at this latitude showed predicted erosional losses due to the alongshelf component of $\widetilde{Q}_s$ that would reduced this estimate. To establish an average annual deposition rate based on the cross-shelf divergence of the very fine sands, the deposit thicknesses established above must be divided by the 9.6-y recurrence interval. This yields an average annual deposition of 0.0027 m. Multiplication by the approximately 7,000 y of Holocene history results in a predicted average deposit thickness of 18.5 m over the 55-95-m depth range near 49°50' N (75 km north of the Columbia River). When similar calculations are made for the mid-to outer shelf the result is 5.4 m of deposition. Across the entire width of the mid-shelf silt deposit between sites F and I, 12.4 m of deposition is predicted. These predictions are somewhat high compared to the observed seismic stratigraphy for the same section, which indicates an average thickness of 7 m and a maximum value of 14 m of deposition at this latitude, but the measured thicknesses are certainly within the expected accuracy of the calculations.

8.8 SUMMARY

This chapter presented a model combining isobath-parallel geostrophic flow, a physically sound, time-dependent model of a stratified benthic boundary layer, a model of wave-current interaction (Smith, 1977; Grant and Madsen, 1979; 1982) and a model for armoring of a non-cohesive, heterogeneous sediment bed (Kachel and Smith, 1986). The integrated model was first used to investigate the effect of suspended sediment-induced density stratification in significantly reducing enhancement of boundary shear velocities when waves and currents interact. Then several sets of conditions, typical of extreme wave-current events on a continental shelf, were run to investigate the impact of sediment-induced stratification on the benthic boundary layer and on sediment transport during storm events. Finally, the model was run for nine locations along and across the southern Washington continental shelf for a set of conditions expected to occur once every 9.6 y. The results were then compared with observed patterns of erosion, deposition and accumulation in order to gain insight into the sediment transport processes taking place on this continental shelf.

The theory of wave-current interaction predicts that shear velocity due to waves is enhanced by the presence of a current and the velocity due to the current is increased in the presence of waves. However, when the effects of sediment resuspension are included in the density stratification of the water column and then used to correct the eddy diffusion coefficients K_m and K_s, not only are the enhancement effects greatly reduced, but the individual values for the shear velocities are also lowered. The effects of sediment-induced density stratification on u_{*w} is greater over beds with sandier compositions than over beds with a higher silt content. This is because sand grains are usually carried as bedload or in suspension close to the bed, which causes higher density stratification in the first few centimeters above the bed, the zone of the wave boundary layer. In contrast, silts are carried higher over the bed as suspended load, which can cause a density stratification within the current boundary layer.

In order to summarize available information regarding sediment transport events on the Washington continental shelf, current meter data collected between 1967 and 1982, as well as available gravity wave information, were compiled and analyzed. Records of 21,179 h of near-bottom or mid-depth current data from sites on the mid- to outer- continental shelf indicate that storm current conditions (where velocities exceed 0.4 m s $^{-1}$ near bottom for at least a 12-h period) occurred 1.7% of the year during fall and winter months and that near-bottom currents exceeding 0.80 m s $^{-1}$ occurred 0.1% of the time. Analysis of the wave spectra data of Larsen and Fenton (1974) by frequency band as well as by amplitude indicated that threshold conditions were exceeded at a depth of 100 m for 13.7% of the same time period (fall and winter), whereas bottom orbital velocities in excess of 0.60 m s $^{-1}$ occurred 1.7% of the time. Typical durations of sediment transporting conditions due to waves are on the order of 24 h. Data from hindcast predictions indicate that the predominant direction of approach for hindcast waves able to move sediment on the mid-shelf is WNW to WSW. The most common wave periods for mid-shelf transport events are 16-18 s. These are associated with swell events, rather than the 8-12 s periods associated with locally-generated wind waves.

Modeling results presented here and elsewhere (Kachel, 1980; Kachel and Smith, 1986) have demonstrated that the most geologically significant transport events in terms of erosion and reworking of the sediment bed and volumes or mass transported are those in which storm currents occur simultaneously with large swells. To assess the areal pattern of the impact that combined wave and current events have on the resuspension, transportation and deposition of bottom sediments on this shelf, a set of conditions expected to recur once every 9.6 y was input to the sediment transport model. Results from the model at one mid-shelf location illustrate the growth and structure of the benthic boundary layer. In particular, they show an offshore turning of ~25° relative to the assumed along-isobath direction of the geostrophic flow. This turning creates a divergence in transport paths between the sand-sized fractions carried in the bedload and those of the finer fractions carried in suspension with a net bearing of ~10° offshore. The predicted path length of an average particle in the bedload is 60-80 m (depending on bottom composition), while the average path of grains in the suspended load is on the order of 30-50 km. The predicted stratigraphic impact of such events is the creation of an upwardly fining, or graded, sediment layer on the order of 0.005-0.05-m thickness.

When the model for the 9.6-y wave-current event was applied at a set of sites along and across the shelf it was found that the increasing silt content to the north of the Columbia River mouth leads to an erosion along most of the shelf during this event. Divergence of the isobaths in the same region leads to net deposition on the order of 0.003-m per storm, almost an order of magnitude lower than the deposition rate needed to account for the observed thickness of the silt wedge on the southern portion of the shelf. Therefore the effect of extreme sediment transport events on this shelf is to remove fine material from the inner shelf and the southern portion of the shelf and to transport it seaward and northward in suspension. Such storms also transport very fine sands, primarily of the 3.5-4ϕ size range, away from their source along the inner shelf and across the shelf. The divergence of this cross-shelf sand transport leads to deposition of sands on the mid- to outer shelf. Sands account for an average of 35 to 45% of the sediment accumulating in the region. Using this information the net accumulation of sediment from this event plus the many smaller events that redistribute the silt alongshelf can be calculated. The effect of many small storms on this shelf is to transport silts to the north and to build up the silt deposit. When these results are extrapolated for the last 7,000 y, the approximate time since sea level attained its present elevation, the calculated thickness of the mid-shelf silt deposit is consistent with that observed by seismic stratigraphy methods.

The results presented here illustrate that, in spite of some uncertainties in empirical values used in sediment transport models, theories that include the complex interrelationships between waves, currents and sediments are useful for increasing our understanding of the patterns of sediment transport events on the continental shelf. When model results are combined with data for the frequency and severity of such events, the predicted annual sediment budget compares favorably with that derived from stratigraphic rate studies. As sediment transport models continue to be refined the models such as the ones presented here should become valuable tools for assessing characteristics of shallow marine environments, both ancient and modern, that are less well studied than the Washington continental shelf.

8.9 CONCLUSIONS

It is essential to include the important non-linear wave-current interactions and the effects of suspended-sediment-induced density stratification when calculating sediment transport on the continental shelf during large transport events. The events that leave their signature on the stratigraphic records of continental shelf deposits, such as the one off Washington, are those events during which the occurrence of large storm-driven currents coincide with the arrival of large swell events. The model results show that these large events not only remove silts from the shelf and transport them off the shelf, but also move sand from the nearshore sand zone out across the shelf. The role of the many smaller storms is to transport silts to the north from their initial place of deposition near the Columbia River mouth and to build up the mid-shelf silt deposit. These conclusions come from combining results of the sediment transport model with observations of the patterns of sediment distribution and transport.

ACKNOWLEDGMENTS

The majority of the work and described in this chapter were supported by the Energy Research and Development Administration under contract EV(45-1)-2225 TA25 and the Department of Energy under contract DE-AT06-76EV71025. Additional support was provided by the National Science Foundation under grants OCE 80-01712 and OCE 87-00659. This is a component of Contribution No. 1784 from the School of Oceanography, University of Washington.

REFERENCES

Abou-Seide, M.M. 1965. Bed load function due to wave action. Univ. of California, Berkeley, Hydraulic Engineering Lab. Rep. HEL-2-11, 78 pp.

Ames, W.F. 1977. Numerical methods for partial differential equations. Academic Press, New York, N.Y., San Francisco, Calif., 65 pp.

Baker, E.T. and H. Curl, Jr. 1981. Observations of the Columbia River plume after the eruption of Mount St. Helens. *Coastal Ocean Pollution News,* 1: 1-2.

Baker, E.T., R.A. Feely and H. Curl, Jr. 1980. Observations of the Columbia River Plume after eruption of Mount St. Helens. *Climatology News,* 2: 39-40.

Barnard, W.D. 1978. The Washington continental slope: Quaternary tectonics and sedimentation. *Mar. Geol.*, 27: 79-114.

Barnes, C.A., A.C. Duxbury and B.A. Morse. 1972. The circulation and selected properties of the Columbia River effluent at sea. Pages 41-80 in: D.L. Alverson and A.T. Pruter (eds.), *The Columbia River Estuary and Adjacent Ocean Regions: Bioenvironmental Studies.* Univ. of Washington Press, Seattle, WA.

Battisti, D.S. and B.M. Hickey. 1984. Application of remote wind forced coastal trapped wave theory to the Oregon and Washington coast. *J. Phys. Oceanogr.*, 14(5): 889-903.

Bennett, L.C. and M.S. Grim. 1968. Investigation of the continental shelf structure off the coast of Washington. *EOS, Trans. Am. Geophys. Union*, 49: 209.

Businger, J.A. and S.P.S. Arya. 1974. The height of the mixed layer in a stably stratified planetary boundary layer. *Adv. in Geophysics,* 18A: 74-92.

Butman, B. 1987. Physical processes causing surficial sediment movement. Pages 147-163 in: R.H. Backus and D.W. Bourne (eds.), *Georges Bank.* MIT Press.

Byrne, J.V. 1963. Geomorphology of the continental terrace off the northern coast of Oregon. *The ORE BIN*, 25(12): 201-207.

Cacchione, D.A. and D.E. Drake. 1982. Measurements of storm generated bottom stresses on a continental shelf. *J. Geophys. Res.*, 87: 1952-1960.

Cacchione, D.A., W.D. Grant, D.E. Drake and P. Wiberg. 1982. Velocity and bottom-stress measurements in the boundary layer, outer Norton Sound, Alaska. Pages 71-78 in C.H. Nelson and S.O. Nies (eds.), *The northeastern Bering Shelf: new perspectives on epicontinental shel processes and depositional products.* Geol. en Mijnbouw, Vol 61.

Cacchione, D.A., W.D. Grant, D.E. Drake and S.M. Glenn. 1987. Storm-dominated bottom boundary layer dynamics on the California continental shelf: measurements and predictions. *J. Geophys. Res.,* 92(C3): 1817-1827.

Carnahan, B., H.A. Luthes and J.O. Wilkes. 1969. Applied Numerical Methods. John Wiley and Sons, Inc., New York, N.Y., 604 pp.

Carstens, M.R., R.M. Neilson and H.D. Altinbilck. 1969. Bedforms generated in the laboratory under oscillatory flow: analytical and experimental study. Tech. Memo. 28. U.S. Army Corps of Engineers. Coastal Engineering Res. Center.

Coastal Data Information Program. 1981-1984. Monthly summary reports, Coastal Data Information Program, Nearshore Research Group, Institute of Marine Resources at Scripps Institution of Oceanography, State of California, Resource Agency, Dept. of Boating and Waterways.

Dietrich, W.A. 1982. Settling velocity of natural particles. *Water Resources Res.* 18: 1615-1626.

Drake, D.E. and D.A. Cacchione. 1985. Seasonal variation in sediment transport on the Russian River shelf, Calif. *Cont. Shelf Res.,* 4: 495-514.

Drake, D.E. and D.A. Cacchione. 1986. Field observations of bed shear stress on sediment resuspension on continental shelves, Alaska and California. *Cont. Shelf Res.,* 6: 415-429.

Drake, D.E. and D.A. Cacchione. In press. Estimates of the reference concentration (C_a) and resuspension coefficient (γ_0) from near-bottom observations on the California shelf. *Cont. Shelf Res.*

Drake, D.E., D.A. Cacchione, R.D. Muench and C.H. Nelson. 1980. Sediment transport in Norton Sound, Alaska. Mar. Geology, 36: 97-126.

Gelfenbaum, G. and J.D. Smith. 1986. Experimental evaluation of a generalized suspended sediment transport theory. Pages 133-144 in: R.J. Knight (ed.), *Shelf Sands and Sandstones. Can. Soc. Petrol. Geol. Mem.*, 11.

Glenn, S.M. 1983. A continental shelf bottom boundary layer model: the effects of waves, currents and a movable bed. Ph.D. Diss., Woods Hole Oceanogr. Inst., WHOI-83-6, 336 pp.

Glenn, S.M. and W.D. Grant. 1987. A suspended sediment stratification correction for combined wave and current flows. *J. Geophys. Res.*, 92(C8): 8244-8264.

Grant, W.D. and O.S. Madsen. 1979. Combined wave and current interaction with a rough bottom. *J. Geophys. Res.*, 84: 1779-1808.

Grant, W.D. and O.S. Madsen. 1982. Movable bed roughness in unsteady oscillatory flow. *J. Geophys. Res.*, 87(C1): 469-482.

Grant W.D. and O.S. Madsen. 1986. The Continental-shelf bottom boundary layer. *Ann. Rev. Fluid Mech.*, 18: 265-305

Grant, W.D., A.J. Williams III and S.M. Glenn. 1984. Bottom stress estimates and their prediction on the Northern California continental shelf during CODE-I: The importance of wave-current interaction. *J. Phys. Oceanogr.*, 14: 506-527.

Grant, W.D., A.J. Williams III, S.M. Glenn and D.A. Cacchione. 1983. High frequency bottom stress variability and its prediction in the CODE region. Tech. Rep. 83-19. Woods Hole Oceanogr. Inst. and CODE Tech. Rep. 15, 72 pp.

Hammond, T.M. and M.B. Collins. 1979. Flume studies of the responses of various current meter rotor/propellers to combinations of unidirectional and oscillatory flow. *Deutsche Hydrographische Zeitschrift, Jahrgang,* 32 (Sonderdruck): 40-58.

Haushild, W.L., R.W. Perkins, H.H. Sleutens, G.R. Dempster and J.L. Glenn. 1966. Radionuclide transport in the Pasco to Vancouver, Washington, reach of the Columbia River, July 1962 to September 1986. U.S. Geol. Survey, Open File Rep., Portland, Ore., 188 pp.

Hickey, B.M. 1981. Alongshore coherence on the Pacific Northwest continental shelf (January-April, 1985). *J. Phys. Oceanogr.*, 11(6): 822-835.

Hickey, B.M. 1983. Hydrographic observations over the Washington-Oregon continental shelf and slope during WISP (February-March, 1975). Univ. of Washington, Dept. of Oceanogr., Spec. Rep. No. 86, 296 pp.

Hickey, B.M. 1984. The fluctuating longshore pressure gradient on the Pacific Northwest shelf: A dynamical analysis. *J. Phys. Oceanogr.,* 14(2): 276-293.

Hickey, B.M., E.T. Baker and N.B. Kachel. 1986. Suspended particle movement in and around Quinault submarine canyon. *Mar. Geol.*, 71: 35-83.

Hickey, B. M. 1989. Patterns and process of circulation over the shelf and slope. Pages 41-116 in: M.L. Landry and B.M. Hickey (eds.), *Coastal Oceanography of Washington and Oregon*. Elsevier, Amsterdam, Neth.

Hopkins, T.S. 1971. On the circulation over the continental shelf off Washington. Ph.D. Diss., Dept. of Oceanography. Univ. of Washington, Seattle, Wa., 204 pp.

Huyer, A., R.L. Smith and E. Sobey. 1978. Seasonal differences in low-frequency current fluctuations over the Oregon continental shelf. *J. Geophys. Res.*, 83: 5076-5089.

Jonsson, I.G. 1963. Measurements on the turbulent wave boundary layer. *Proc. of the 10th Congr., Internatl. Ass. Hydraulic Res.*, London 1: 85-92.

Jonsson, I.G. and N.A. Carlsen. 1976. Experimental and theoretical investigations in an oscillatory turbulent boundary layer. *IANR J. of Hydraulic Res.*, 14(1): 45-60.

Judson, S. and R.F. Ritter. 1964. Rates of regional denudation in the United States. *J. Geophys. Res.*, 69: 3395-3401.

Kachel, N.B. 1980. A time-dependent model of sediment transport and strata formation on a continental shelf. Ph.D. Diss., Dept. of Oceanogr., Univ. of Washington, Seattle, Wa., 123 pp.

Kachel, N.B. and J.D. Smith. 1986. Geologic impact of sediment transporting events on the Washington continental shelf. Pages 145-162 in: R.J. Knight (ed.), *Shelf Sands and Sandstones. Can. Soc. Petrol. Geol. Mem.* 11.

Kachel, N.B. and R.W. Sternberg. 1971. Transport of bedload as ripples during an ebb current. *Mar. Geol.*, 19: 229-244.

Kalkanis, G. 1964. Transportation of bed material due to wave action. U.S. Army Corps of Engineers CERC Tech. Memo. No. 2, 38 pp.

Karlin, R. 1980. Sediment sources and clay mineral distributions off the Oregon coast. *J. Sedimentary Petrology*, 50(2): 543-560.

Kelly, J.C. and D.A. McManus. 1969. Optimizing sediment sampling plans. *Mar. Geol.*, 7: 465-476.

Kelley, J.C. and D.A. McManus. 1970. Hierarchical analysis of variance of shelf sediment texture. *J. Sed. Petrol.*, 40: 1335-1339.

Kulm, L.D., R.C. Roush, J.H. Harlett, R.H. Neudeck, D.M. Chambers and E.J. Runge. 1975. Oregon continental shelf sedimentation, interrelationships of facies distribution and sedimentary processes. *J. Geol.*, 83: 145-175.

Kulm, L.D. and R. Von Heine, *et al.* 1973. Initial reports of the Deep Sea Drilling Project, V.18, Washington (U.S. Government Printing Office): 213-231.

Larsen, L.A. and D. Fenton. 1974. Open ocean wave studies, vol. 2. Dept. of Oceanogr., Univ. of Washington, Seattle, Wa., Data Rep. 1-73, 560 pp.

Long, C.E. 1981. A simple mathematical model for time dependent stably stratified turbulent boundary layers. Ph.D. Diss., Dept. of Oceanogr, Univ. of Washington, Seattle, Wa., 170 pp.

Long, C.E. and J.D. Smith. 1979. A simple model of the time dependent, stably stratified, planetary boundary layer. DOE Rep. RLO-225-TA-25-27, Pollutant transport and sediment dispersal in the Washington-Oregon coastal zone. Dept. of Oceanogr., Univ. of Washington, Seattle, Wa., 51 pp.

McManus, D.A. 1972. Bottom topography and sediment texture near the Columbia River mouth. Pages 241-253 in: D.L. Alverson and A.T. Pruter (eds.), *The Columbia River Estuary and Adjacent Ocean Regions*. Univ. of Washington Press, Seattle, WA.

Miller, M.C., I.N. McCave and P.D. Komar. 1977. Threshold of sediment motion under unidirectional waves. *Sedimentology,* 24: 507-527.

National Marine Consultants. 1961. Wave statistics for three deep-water stations along the Oregon-Washington coast. National Marine Consultants, Santa Barbara, Calif., 16 pp.

Nikuradse, J. 1933. Laws of flow in rough pipes (Transl. from German, 1950). Natl. Advisory Comm. Aeronaut. Tech. Memo. No. 1292, pp: 1-62.

Nittrouer, C.A. 1978. The process of detrital sediment accumulation in continental shelf environment: an examination of the Washington shelf. Ph.D. Diss., Dept. of Oceanography, Univ. of Washington, Seattle, Wa., 243 pp.

Nittrouer, C.A. and R.W. Sternberg, 1981. The formation of sedimentary strata in an allochthonous shelf environment: the Washington continental shelf. *Mar. Geol.*, 42: 201-232.

Nittrouer, C.E., R.W. Sternberg, R. Carpenter and J.F. Bennett. 1979. The use of Pb-210 geochronology as a sedimentological tool: application to the Washington continental shelf. *Mar. Geol.,* 31: 297-316.

Nowell, A.R.M. and C.E. Long. 1983. An evaluation of von Karman's constant. In: B.M.Hickey (ed.), *Pollutant Transfer and Sediment Dispersal in the Washington-Oregon Coastal Zone.* DOE Rep. RLO-2225-TA25-64, 41 pp.

Pierson, W.J., G. Neumann and R.W. James. 1955. Practical methods for observing and forecasting ocean waves by means of wave spectra and statistics. H. Publ. 603, U.S. Navy Hydrogr. Office, Washington, D.C., 283 pp.

Ridge, M.J.H. and B. Carson. 1987. Sediment transport on the Washington continental shelf: estimates of dispersal rates from Mount St. Helens ash. *Cont. Shelf Res.,* 7: 759-772.

Roberts, R.W. 1974. Marine sedimentological data of the Washington continental shelf. Dept. of Oceanogr., Univ. of Washington, Seattle, Wa., Special Rep. 57, 129 pp.

Shi, N.C. 1983. Reverse sediment transport induced by amplitude modulated waves. Ph.D. Diss., Dept. of Oceanogr., Univ. of Washington, Seattle, Wa., 121 pp.

Shi, N.C., L.H. Larsen and J.P. Downing. 1985. Predicting suspended sediment concentration on continental shelves. *Mar. Geol.*, 62: 255-275.

Shields, A. 1936. Anwendung der Ahnlichkeitmekanik und Turbulenzforschung anf die Geschiebebewegung, Mitteil. Preuss. Versuchsanst. Wasser, Erd, Schiffsbau, Berlin, no. 26.

Simonstad, C.P., D.A. Jay, C.D. McIntyre, W. Nehlson, C. Sherwood and L. Small. 1984. The dynamics of the Columbia River Estuary ecosystem. Columbia River Estuary Data Development Program, Astoria, Ore., 2 vol.

Smith, J.D. 1977. Modeling of sediment transport on continental shelves. Pages 539-577 in: E.D. Goldberg, I.N. McCave, J.J. O'Brien and J.H. Steele (eds.), *TheSea.* Vol. 6. Wiley, New York, N.Y..

Smith, J.D., B. Hickey and J. Beck. 1976. Observations from moored current meters on the Washington continental shelf from February 1971 to February 1974. Dept. of Oceanogr., Univ. of Washington, Spec. Rep. No. 65, 346 pp.

Smith, J.D. and T.S. Hopkins. 1972. Sediment transport on the continental shelf off of Washington and Oregon in light of recent current measurements. Pages 143-180 in: D.J.P. Swift, D.B. Duane and O.H. Pilkey (eds.), *Shelf Sediment Transport.* Dowden, Hutchinson and Ross, Inc., Stroudsburg, Penn..

Smith, J.D. and C.E. Long. 1976. The effect of turning in the bottom boundary layer on continental shelf sediment transport. *Mem. Soc. Roy. Sci. Liege, 6e ser.*, 10: 369-396.

Smith, J.D. and S.R. McLean. 1977a. Spatially averaged flow over a wavy surface. *J. Geophys. Res.* 82: 1735-1746.

Smith, J.D. and S.R. McLean. 1977b. Boundary layer adjustments to bottom topography and suspended sediments. *Mem. Soc. Roy. Sci. Liege, 6e ser.*, 11: 123-151.

Snodgrass, F.E., G.W. Groves, K.F. Hasselmann, G.R. Miller, W.H. Munk and W.H. Powers. 1966. Propagation of ocean swell across the Pacific. *Phil. Trans. Roy. Soc. Lond., Ser. A,* 259: 431-497.

Sternberg, R.W. 1967. Measurements of sediment movement and ripple migration in a shallow marine environment. *Mar. Geol.*, 5: 195-205.

Sternberg, R.W. 1986. Transport and accumulation of river-derived sediments on the Washington continental shelf, U.S.A. *J. Geol. Soc., Lond.*, 143: 945-956

Sternberg, R.W., J.S. Creager, W. Glassley and J. Johnson. 1977. Aquatic disposal field investigations, Columbia River disposal site, Oregon, Appendix A: Investigation of the hydraulic regime and physical nature of bottom sedimentation. Tech. Rep. D-77-30, U.S. Army Engin. Waterways Expt. Station, Vicksburg, Miss., 327 pp.

Sternberg, R.W. and L.H. Larsen. 1975. Threshold of sediment movement by open ocean waves: observations. *Deep-Sea Res.*, 22: 299-309.

Sternberg, R.W. and L.H. Larsen. 1976. Frequency of sediment movement on the Washington continental shelf: a note. *Mar. Geol.*, 21: M37-M47.

Sternberg, R.W. and D.A. McManus. 1972. Implications of sediment dispersal from long-term, bottom-current measurements on the continental shelf of Washington. Pages 181-194 in: D.J.P. Swift, D.B. Duane and O.H. Pilkey (eds.), Shelf sediment transport. Dowden, Hutchinson and Ross, Stroudsberg, Penn..

van Winkle, W. 1914a. Quality of the surface waters of Washington. U.S. Geol. Survey, Water Supply Paper 363, 16 pp.

van Winkle, W. 1914b. Quality of the surface waters of Washington. U.S. Geol. Survey, Water Supply Paper 339, 105 pp.

Venkatarathnam, K. and D.A. McManus. 1973. Origin and distribution of sands and gravels on the northern continental shelf off Washington. *J. Sed. Petrol.*, 43: 799-811.

von Huene, R. and L.D. Kulm. 1973. Tectonic summary of Leg 18. Pages 961-976 in: L.D. Kulm, R. von Huene, *et al.* (eds.), *Initial Rep. of the Deep Sea Drilling Project.* Vol. 18. U.S. Government Printing Office, Washington, D.C.

von Rosenburg, D.V. 1969. Methods for the numerical solution of partial differential equations. American Elsevier Publ. Co. Inc., New York, N.Y., 128 pp.

Werner, F.E. and B.M.Hickey. 1983. The role of a longshore pressure gradient in Pacific Northwest coastal dynamics. *J. Phys. Oceanogr.*, 13(3): 395-410

White, S.M. 1970a. Mineralogy and geochemistry of continental shelf sediments off the Washington-Oregon coast. *J. Sed. Petrol.*, 40: 38-54.

White, S.J. 1970b. Plane bed thresholds of fine grained sediments. *Nature*, 228: 152-153.

Wiberg, P. and D. Ruben. In press. Bed roughness under sediment-transporting flows. *J. Geophys. Res.*

Wiberg, P. and J.D. Smith. 1983. A comparison of field data and theoretical models for wave-current interactions at the bed on the continental shelf. *Cont. Shelf Res.*, 2: 147-162.

Wieringa, J. 1980. A reevaluation of the Kansas mast influence on measurements of stress and cup anemometer. *Boundary Layer Meterol.*, 18: 222-250

Yalin, M. 1963. An expression for bed-load transportation. *J. Hydraulics Div., Proc. Am. Soc. Civil Engin.*, 89(HY3): 221-250.

Yalin, M. 1964. Geometrical properties of sand waves. *J. Hydraulics Div., Proc. Am. Soc. Civil Engin.*, 89(HY5): 105-119.

Yalin, M. 1972. *Mechanics of Sediment Transport.* Pergamon Press, New York, N.Y., 290 pp.

CHAPTER 9

BENTHOS AND ITS INTERACTION WITH BOTTOM BOUNDARY LAYER PROCESSES

P.A. Jumars and K. Banse
School of Oceanography, University of Washington, Seattle, WA 98195

9.1 INTRODUCTION

Work on the continental shelf benthos of the Pacific Northwest prior to the 1970s has been presented and summarized by Lie and Kisker (1970), Pamatmat (1973) and numerous chapters in Pruter and Alverson (1972). Since then, investigations in the area have been fragmentary and episodic. During these years, however, continental shelf benthos elsewhere within the United States has been studied intensively and extensively in preparation for oil exploration and recovery and basic research efforts in several other shelf settings around the globe have been continued or initiated. In addition, the bottom boundary layer environments of continental shelf benthos have become much better known through work on the Washington continental shelf and elsewhere. The study of organism-sediment relations has also matured and expanded to include fluid dynamics. It is our intent to use these various advances to place historical data from the U.S. Northwest continental shelf in a new context, underscoring gaps in existing information and consequent opportunities for future research. This examination can go little below the size range of the macrofauna, however, because the meiofauna and microbiota of the Northwest shelf are so little studied.

9.2 BENTHOS OF THE WASHINGTON SHELF

9.2.1 Spatial and Temporal Structure of Communities

Most of what is known about the standing crops and species compositions of the infaunal communities comes from a series of papers by Lie and his coworkers (Lie, 1969; Lie and Kelley, 1970; Lie and Kisker, 1970). The basic data set consists of 22 stations off the northwestern coast of Washington and 18 stations off the mid- and southwestern coast of Washington (Fig. 9.1). The fairly broadly distributed stations north of about 47° 50'N were occupied in May to August of 1967 and those further south, in July 1968. In most cases, 3 replicate, 0.2-m^2 van Veen grab samples were taken per site and sieved through a 1-mm mesh, with the retained animals identified to species, weighed and counted. For polychaetes the identification task was so large that only the species dominating in numbers or biomass were treated at the species level. The grab used was neither screened nor otherwise vented, so that epifauna probably were seriously undersampled. The stations fall primarily in what is now known (Kachel and Smith, 1989, Chapter 8) as the mid-shelf silt deposit and in the sands inshore of it, although a few are scattered seaward of the mid-shelf silts.

At the most gross level, but one entirely appropriate to the sparsity of samples, community classification yields three major groupings of stations on the basis of taxonomic composition

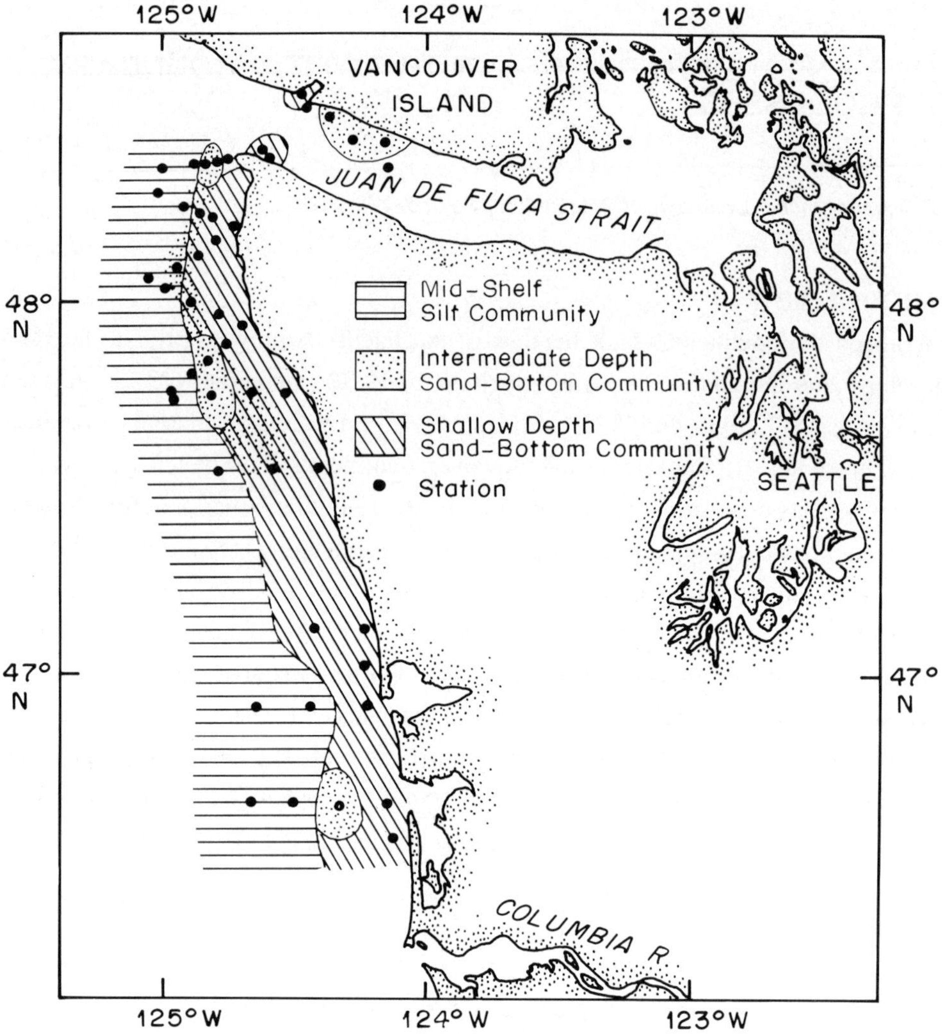

Fig. 9.1 Location of sampling and geographic distribution of the three offshore benthos communities. Modified from Lie and Kisker (1970).

(Lie and Kisker, 1970 and Fig. 9.1 herein). The community identified by them as a deep-water, mud-bottom community can on the basis of recent geological work be more accurately termed a mid-shelf silt community. It apparently extends (on the basis of two of Lie and Kisker's stations) some distance onto the muddy sands of the outer shelf. Immediately shoreward of this mid-shelf silt community and extending until the sediments reach approximately 80% sand is an intermediate-depth, muddy sand community. Still further landward, but offshore of the breakers, is a shallow-water sand community. Median ash-free dry weights (AFDW) of the macrofauna for the three communities at the stations north of 47° 50' (Fig. 9.1) are, respectively, 1.94, 1.98 and 1.17 g m^{-2} (based on Table 2 of Lie, 1969, with stations 7 and 20 omitted because they were classified differently in Lie and Kisker, 1970). Lie's (1969) biomass means for these three groups (2.27, 2.34 and 1.35 g m^{-2} AFDW), respectively, were influenced by a few stations with unusually high values. The same problem holds for the widely quoted means for all stations, of

3.06, 2.53 and 1.40 g m $^{-2}$ AFDW in Lie and Kisker (1970). Recognizing the difficulty of comparing and interpreting biomass data from studies employing different gears and sieves, Lie (1969) discussed only briefly that his means appeared to be low relative to other temperate sites (e.g., < 50 % of the infauna in Long Island Sound or off Santa Barbara, California, but comparable to the then-existing data from the Gulf of Alaska). A.G. Carey (Oregon State University, personal communication) using finer sieves and better vented sampling gear on the Oregon shelf has reported substantially higher standing crops. Without intercalibration it is only safe to make internal comparisons. Within Lie's samples, surface deposit feeders are important throughout and the relative abundance of suspension feeders decreases with water depth (Table 9.1). Unfortunately, the depth series does not extend much beyond the mid-shelf depths.

No surveys of comparable magnitude have been conducted since 1968 on the Washington shelf. More recent, scattered observations off Washington and off Oregon, however, confirm the general conclusions. In a transect off Newport, Oregon, Carey (1972, using dredges and a sieve of 0.42-mm aperture) also found depth zonation, with increasing importance of burrowing polychaetes and maximal standing crops toward the shelf break. Smethie *et al.* (1981) and Nittrouer and Sternberg (1981), analyzing only the polychaetes (from 0.35-mm-sieved subsamples out of box cores) taken at three sites in the mid-shelf silt deposit from off the Columbia River (about 46° 15' N) northward to off Gray's Harbor (about 46° 55' N), found a much greater prevalence of motile, subsurface deposit feeders (especially capitellid polychaetes) than observed by Lie and coworkers. It is not possible to attribute discrepancies among these studies to differences in sampling gear, station position, or time between the sampling programs.

Epibenthos of the Washington continental shelf has not been studied systematically, though it is probably qualitatively similar to that off Oregon (see Carey, 1972). In the latter region, mobile epifauna and the projecting soft body parts of infauna are very important items in the diets of commercially valuable fishes (e.g., Pearcy and Hancock, 1978; Hogue and Carey, 1982), especially in the youngest age classes.

No time series have been conducted to resolve patterns of benthic community change on the U.S. Pacific Northwest shelf. The evidence that substantial interannual changes may occur is of two kinds: data from a physically different environment in Puget Sound, with some overlap in faunal composition with the open shelf and data from physically similar but geographically remote environments with little faunistic overlap. Nichols (1975, 1985) has maintained a time series for over a two decades at a 203-m station (his Station 2) within the main basin of Puget Sound. This station (Station B of Lie, 1969) is similar in sedimentary and infaunal composition to the mid-shelf silt stations. Nichols (1985) has shown dramatic shifts in infaunal community dominance from one year or group of years to the next. Taking the other possible tack of looking at a time series from the best studied, exposed shelf area (at 55-m depth off Northumberland; Buchanan and Moore, 1986) – albeit one with a different fauna – leads to the same conclusion. Again, dramatic year-to-year shifts in the abundances of some populations were seen. Alone, these independent observations do not provide strong evidence for interannual changes in the

Table 9.1 Numerical (N) and biomass (B) dominants of shelf benthos, by community, numbers at right indicating numerical abundance, converted to m $^{-2}$. From Lie and Kisker (1970) (x = mean; Md = median). Dominants (*sensu* Lie and Kisker, 1970) are those with the highest ranks in mean abundance and biomass.

	Genus, species (feeding guild and major taxon)	x	Md
	Mid-Shelf Silt Community		
(N)	*Prionospio malmgreni* (surface deposit-feeding, spionid polychaete)	34	13
(N)	*Axinopsida serricata* (suspension-feeding, thyasirid bivalve)	30	25
(N,B)	*Sternaspis fossor* (subsurface deposit-feeding, sternaspid polychaete)	28	10
(N)	*Adontorhina cyclia* (suspension-feeding, thyasirid bivalve)	27	12
(N)	*Heterophoxus oculatus* (omnivorous, subsurface-feeding, phoxocephalid amphipod)	18	18
(N)	*Macoma carlottensis* (surface deposit-feeding, tellinid bivalve)	17	12
(N)	*Ninoe gemmea* (omnivorous, burrowing, lumbrinerid polychaete)	17	10
(B)	*Brisaster latifrons* (subsurface deposit-feeding, irregular urchin)	2	2
(B)	*Ophiura luetkeni* (omnivorous brittle star)	5	0
(B)	*Amphioplus* sp. (suspension[?]-feeding brittle star)	5	5
	Intermediate-Depth, Muddy Sand Community		
(N,B)	*Yoldia ensifera* (subsurface deposit-feeding, nuculanid bivalve)	76	3
(N,B)	*Paraphoxus variatus* (omnivorous, subsurface-feeding, phoxocephalid amphipod)	50	58
(N,B)	*Magelona* sp. (surface deposit-feeding, magelonid polychaete)	43	8
(N,B)	*Sternaspis fossor* (subsurface deposit-feeding, sternaspid polychaete)	39	13
(N,B)	*Nephtys* sp. (burrowing, carnivorous, nephtyid polychaete)	35	30
(N,B)	*Haploscoloplos elongatus* (subsurface deposit-feeding, orbiniid polychaete)	30	30
(N,B)	*Axinopsida serricata* (suspension-feeding, thyasirid bivalve)	26	17
(B)	*Macoma eliminata* (surface deposit-feeding, tellinid bivalve)	15	3
	Shallow-Water, Sand Community		
(N,B)	*Diastylopsis dawsoni* (suspension-feeding cumacean)	900	12
(N,B)	*Tellina salmonea* (surface deposit-feeding, tellinid bivalve)	613	0
(N,B)	*Owenia fusiformis* (surface deposit- and suspension-feeding, oweniid polychaete)	126	0
(N,B)	*Ampelisca macrocephala* (surface deposit- and suspension-feeding, ampeliscid amphipod)	73	47
(N,B)	*Paraphoxus obtusidens* (omnivorous, subsurface deposit -feeding, phoxocephalid amphipod)	34	8
(N,B)	*Macoma expansa* (surface deposit-feeding, tellinid bivalve)	33	0
(N,B)	*Eohaustorius washingtonensis* (burrowing, sand-browsing, haustoriid amphipod)	26	9
(B)	*Nephtys* sp. (burrowing, carnivorous, nephtyid polychaete)	19	13
(B)	*Chaetozone setosa* (surface and subsurface deposit-feeding, cirratulid polychaete)	23	10
(B)	*Siliqua patula* (suspension-feeding, solenid [razor] clam	20	0

benthos of the Washington continental shelf, but together they are more convincing. There are no obvious reasons why the shelf benthos should be any more stable than the benthos of Puget Sound. The episodic nature of large-magnitude storms on the shelf (Kachel and Smith, 1989, Chapter 8) suggests that even greater population fluctuations might be expected over the continental shelf.

It is perhaps worthwhile at this point to mention effects caused by El Niño. It is clear that both currents (see Hickey, 1989, Chapter 2) and water properties such as temperature, salinity and nutrients (see Landry *et al.*, 1989, Chapter 1) over the shelf in the Pacific Northwest have interannual fluctuations that are often related to El Niño events. Recruitment of shelf benthos certainly is likely to be affected too, both directly and through effects on predators of benthos (cf. the changed food ingested by planktonic larvae of sable fish, Grover and Olla, 1987). It is remarkable, though, how difficult it was to recognize change of abundance during the "century Niño" of 1982-1983 against the background, normal year-to-year variability in the intertidal benthos of the outer coast of Washington (Paine, 1986). Even when a change can be observed, the unresolved issue is to identify the mechanism. An obvious step is to correlate the series of some biotic parameter with any environmental data available over the same period. Temperature data are the most easily obtained. We caution against the temptation to leap from such correlations to assumptions that temperature is a root cause of population change (e.g., Pearson *et al.*, 1986; Buchanan and Moore, 1986). Many environmental variables, e.g., nutrient supply to phytoplankton (themselves food for planktotrophic larvae), correlate with temperature in upwelling regions and correlation is no better evidence of cause in time series of community structure than correlation with sediment type is of cause in spatial analyses of community structure.

Benthic rate information for the Washington shelf comes from measurements of oxygen consumption and calculations of fluxes from concentration gradients of a number of constituents. Pamatmat (1973) determined total oxygen consumption and community respiration in shipboard incubations of retrieved samples. He had tested for artifacts by comparing *in situ*-measured rates with core incubations down to 22-m depth in analog settings in Puget Sound and studying pressure effects up to 19.5 atm without finding any significant technique- or pressure-dependent differences (Pamatmat, 1971). Biological (community) oxygen consumption on the continental shelf, determined at about 8.5° C, ranged only between 1.8 and 6.9 ml m^{-2} h^{-1} in cores from seven stations in 25 to 115 m of water and showed no trend with depth over this limited range ($n = 7$, Spearman's $r = 0.08$, $p > 0.5$).

From conservative assumptions applied to their measurements of sulfate reduction at three stations of 85-m depth during August 1978, Christensen *et al.* (1987a) estimated that this process oxidized 3.1, 1.9 and 1.2 times as much carbon as did oxygen consumption from aerobic respiration and abiotic oxidation of reduced substances as previously measured by Pamatmat (1973) at three nearby sites; the ratio was 0.7 for a shelf-break station of 147-m depth. On the three shallow stations, the uppermost two or so centimeters presumably were metabolically most active (aerobic processes) but sulfate reduction integrated over 30-cm or more of depth in the sediment more than compensated for a low rate per unit volume of sediment. The equivalent release and flux into the overlying water column of ammonium should be high, enough to support several tens of percent of the primary production of the overlying water column where the nutrients are made available to the photic zone. Ammonium flux out of the sediment calculated from nutrient profiles, combined with considerations of radon fluxes at the same stations published by Smethie *et al.* (1981), however, amounted to only 19% of the expected rate. Instead, the seabed consumed bound nitrogen during August, owing to nitrification (oxidation of ammonium) in the sediment coupled with denitrification of nitrate fluxing in from the overlying bottom water. Christensen *et al.* (1987b) suggested that this coupling of nitrification and denitrification, with the resulting nitrogen balance, is common on continental shelves with aerated bottom water. Also noteworthy for the observations on the Washington shelf is that irrigation by burrowing animals approximately trebled the influx of nitrate (and, presumably, of other solutes) over that due to molecular vertical diffusion (Christensen *et al.*, 1987a). The calculated 'diffuse' (approximately three-dimensional, see Christensen *et al.*, 1984) irrigation coefficient changed materially with season so that one must expect large seasonality of nutrient fluxes; Christensen *et al.* (1984) had assumed a reduction by 60% for the cool season. Christensen *et al.* (1984) also noted for the same stations that animal irrigation affected principally the depths between 2 and 10 cm in the sediments.

9.2.2 Comparison with other shelves

In view of the meager local information, extensive comparison with other shelves would be unwise. In view of the limited information on shelves worldwide, it is impossible. The paucity of information on benthos can be exemplified by the complete lack of any papers dealing with it in a 1,200-page, two-volume collection on the sedimentary record of coastal upwelling (Suess and Thiede, 1983; Thiede and Suess, 1983). Two gross kinds of comparisons can be made, however, i.e., with other shelves in upwelling regions and with shelves from non-upwelling situations.

In the upwelling-region shelves of Peru and northwestern Africa, the abundance of aerobic benthos is linked to the rate of supply of organic material to the bottom *vs.* the rate of supply of oxygen. The latter in turn depends on bottom water concentrations of oxygen and advective-diffusive exchange. Including the Washington-Oregon shelf, all three regions are characterized by high rates of organic-matter flux to the bottom, but differ in the supply of oxygen by bottom currents (Barber and Smith, 1981), with the Washington continental shelf being intermediate in character. Standing stocks of aerobic benthos are high off Northwest Africa (Nichols and Rowe, 1977; Thiel, 1978 and 1982) because shelf currents prevent oxygen-demanding sedimentary accumulation yet provide high fluxes of particulate material. Consequently, suspension feeders are more prevalent than on the Washington shelf. Off Peru, shelf bottom currents are the weakest of the three regions and macrofauna is absent in this anoxic bottom (Rowe, 1971); much of the sediment-water interface is covered by filamentous bacteria (Gallardo, 1977).

Passive continental margins (or more specifically coasts with too little particulate material supply to yet have filled their estuaries since deglaciation, e.g., the U.S. east coast) show little accummulation of fine material in contrast to the Washington shelf. The shelf-estuary systems of the U.S. east coast are in early stages of geologic evolution and fine materials are mostly trapped in estuaries with the remainder being swept from the shelf in winter storms. There is no appreciable mid-shelf silt after winter storms. The benthos of shelves on the east coast of the U.S. is little studied, but densities are high and variable (Frankenberg and Leiper, 1977). Episodically, high abundance is supported by phenomena such as upwelling along the west wall of the Gulf Stream (Yoder *et al.*, 1983) and much of the variability is correlated with small- and medium-scale variation in topography (e.g., Schaffner and Boesch 1982). Following apparently universal generalizations about benthos, communities are zoned by depth and correlated with sediment type (Weston, 1988; Boesch and Bowen, in press). These generalizations hold for all shelf types classified by geological history and modern sedimentary regime (Curray, 1965; Swift, 1970): shelves with autochthonous (e.g., the east coast shelf), or allochthonous (e.g., the west coast shelf from northern California northward) and climax-graded (showing progressively finer sediments with increasing depth, e.g., south Texas, cf. Flint, 1981).

9.2.3 Organism-Sediment Relations on U.S. Pacific Northwest Shelves

Whether one is interested because of intellectual curiosity, because of importance to demersal shelf fisheries, or because of the potential for pollution impact on shelves, the interaction between shelf benthos and shelf sediments merits attention. Unravelling cause and effect in any system with tight feedbacks is a challenge. For example, productivity of the animals that demersal fishes eat may be tied to the frequency of sediment disturbance of the sediments (Rhoads *et al.*, 1978), which is difficult to quantify. Also, particulate or particle-adsorbed pollutants are most likely to enter the food web through deposit feeding.

Correlation of sediment type with community structure has in the past been interpreted as biological structure determined by geological structure. That interpretation is no longer tenable as the only possibility or even as the most likely one. As an example of the tight feedbacks in the organism-sediment system (see also papers in McCall and Tevesz, 1982), organisms can alter sedimentary structure by altering sediment erodibility (Jumars and Nowell, 1984). Therefore, part of the correlation between sedimentary structure and benthic community structure may involve a biological cause of a geological effect (see also Eckman *et al.*, 1981). DeFlaun and Mayer (1983) go so far as to suggest that bacteria cause clays to accumulate--rather than the other way around. On Pacific Northwest shelves prior to storms, organisms and their structures are likely to dominate surface roughness and therefore affect the fluid drag on the bottom within the mid-shelf silts (Cacchione *et al.*, 1983). Bioturbation and armoring of the bed by a lag layer of large grains (the so-called lag layer left behind from a heterogeneous mixture of grains when the finer ones are eroded) are now known to be strongly interactive (Butman, in preparation); during periods of high animal density and individual activity, disturbance can quickly eliminate lag layers. The sensitivity of dispersal of larvae of bottom animals to the physical characteristics of the larvae and to the local flow regime also has been recently documented (Eckman, 1983; Hannan, 1984; Butman, 1987). Thus, community structure and sediment type may correlate well in part because larvae of given species and sediments of given grain sizes have similar settling velocities rather than the grain size of the bed determining the larval choice. Finally, frequency and intensity of sediment transport determine food supply to deposit feeders (Miller *et al.*, 1984; Jumars and Self, 1986; Miller and Jumars, 1986; Miller and Sternberg, in press). Bottom boundary-layer fluid dynamics thus are likely to be strong determinants of both geological and biological structure.

Other water column-benthic interactions are also significant, in particular, the benthic regeneration of nutrients used by phytoplankton (e.g., Pilson, 1985) and the hatching of benthic resting stages of both phytoplankton (e.g., Anderson *et al.*, 1982) and zooplankton (e.g., Marcus, 1984). Benthic nutrient recycling as well as recruitment from benthic resting stages are favored by an upwelling circulation (Hermann *et al.*, 1989, Chapter 6), so that both processes are more likely to be important on the Washington shelf than in a non-upwelling setting. Sediment transport is likely to enhance both release of benthic nutrients (Fanning *et al.*, 1982) and hatching of benthic resting stages (Anderson and Wall, 1978). Evidence of benthic nutrient regeneration is present in nutrient profiles from the Washington shelf (Dortch and Postel, 1989, Chapter 4), but so few local measurements are available on either nutrient regeneration (Pamatmat, 1973) or recruitment of plankton from resting stages that one can do little more than assert their likely importance.

9.2.4 Feeding

Although local work on feeding, physical disturbance and recruitment of benthos is hardly better developed than studies of animal-sediment relations, enough pieces of the puzzle can be visualized from work in other locations and in the laboratory to discuss the outlines of a coherent picture. The benthos of the mid and outer-shelf off Washington is dominated by motile deposit feeders (Nittrouer and Sternberg, 1981, Figs. 6 and 7). On the inner shelf, suspension feeders also comprise an appreciable fraction of total fauna. If these fauna depend very directly on the flux of particulate organic matter to the bottom, either in the form of phytoplankton cells or fecal pellets (Landry *et al.*, 1989, Chapter 1), then one would expect feeding rates, pools of storage products, somatic growth and fitness to reflect and lag only slightly (by days or weeks) the pattern of primary production (Perry *et al.*,1989, Chapter 3) on this shelf. Major responses would occur during the summer upwelling season. In the extreme view, there should be little organic material of any food value left by the time the late fall and winter storm season arrives. We know this view to be wrong in its most extreme, however, because of the prevalence of subsurface deposit feeders in the mid-shelf silts (Nittrouer and Sternberg, 1981, Figs. 6 and 7).

A complicating factor in the rate of supply of food to sedentary benthos is the redistribution of sediments by physical transport processes. Transport can enhance the size of the pool of organic matter available to any sedentary animal (Miller *et al.*, 1984) and stimulate bacterial production in sediments (Findlay *et al.*, 1985). If animals depend for their food supply on this redistribution or stimulation, then one would expect to see positive faunal responses (accrued storage products, somatic growth and increased fitness) to transport events. As opposed to the intertidal zone (Miller, 1985; Jumars and Self, 1986; Miller and Sternberg, in press) redistribution of sediment via wave action is not a daily occurrence on the mid-shelf. Kachel and Smith (1989, Chapter 8) estimate that waves capable of moving sediments at 100-m depth off Washington occur 14% of the time and are strongly concentrated in the winter storm season. Also, feeding-rate depression in both surface (Miller and Jumars, 1986) and subsurface (Fuller *et al.*, in press) deposit feeders from shallow water has been observed when fecal material is allowed to accrue for a few hours. If this phenomenon extends to shelf faunas, then feeding rate enhancement may take place during and shortly after storms. Again, whether such rapid feeding bouts occur and whether they are important in the energetics of the individuals involved remain to be seen. A related issue is whether microbes and non-living organic coatings will be significantly abraded from sediments during transport events. Evidence from the intertidal zone, where transport is a daily occurrence, suggests that effects of storms on attached microbes are minimal (Miller, 1985 and in review). The infrequency of sediment transport at shelf depths may, however, make a larger fraction of the attached microbial flora susceptible to abrasive removal.

A particularly interesting occurrence that may help to evaluate the importance of sediment redistribution is the occasional arrival of Antarctic swells capable of moving bottom sediments during the (local) non-storm season. Because of the weak low frequency currents in this season (Kachel and Smith, 1989, Chapter 8), such events are ineffectual in moving bottom sediments over large scales. They certainly can redistribute sediments over the scale of tens or hundreds of individual feeding ambits (Miller *et al.*, 1984), however, and generate sufficient disturbance to affect microbial populations.

Animals experiencing daily- or tidally-repeating cycles of sediment transport often show switching from deposit to suspension feeding and back again in response to suspended loads (e.g., Taghon *et al.*, 1980; Dauer *et al.*, 1981; Muschenheim, 1987). The selective pressure to engage in such switching should be reduced in an environment where physical resuspension is much less frequent and predictable, such as during the spring and summer seasons in the Pacific Northwest. An added complication to feeding dynamics is that organisms would be expected to respond to the annual cycle of change in sediment grain size. At one site just landward of the Washington mid-shelf silt, for example, sediment composition varies from 20% silt and clay in summer to < 1% in winter, with median grain diameter shifting from 90 μm to 290 μm (Nittrouer, 1978).

Kachel and Smith (1989, Chapter 8) convincingly argue that redistribution routes and rates of transport of particles should vary with their settling velocities, more precisely with their Rouse numbers. Since organic content and sediment grain size are inversely correlated and since even fairly large organic particles or aggregates will have low settling velocities (Rouse numbers), then one might expect transport of food materials to follow isobaths and resuspended food particles to move on the order of 10-100 km per year northward (cf. Kachel and Smith, 1989, Chapter 8). Further, at the crudest level, organic matter tends to fall into one of two types, labile and refractory, characterized by different first-order rate constants for decomposition (Westrich and Berner, 1984). The effects of increased labile organic matter inputs from the Columbia River (due both to import and nutrient-caused enhancement of local phytoplankton production) in its immediate vicinity are readily observed as increased abundances and activities of organisms (Smethie *et al.*, 1981; Nittrouer and Sternberg, 1981). Sampling intensity northward along the mid-shelf silt has not been sufficient to resolve whether the northward-moving 'conveyor belt' of ever more refractory material also is responsible for a larger-scale gradient in organism abundance. However, there is some evidence for decreasing bioturbation rates on this broader alongshelf scale with distance northward from the Columbia river (Carpenter and Peterson, 1989, Chapter 10). Thus, the along-isobath transport of organic matter in the mid-shelf silts may display in the horizontal on very large scales, what would, in the absence of horizontal transport be limited to very small-scale vertical gradients.

9.2.5 Disturbance, Dispersal and Recruitment

Disturbances more frequent than the generation time must be accommodated via functional responses, while less frequent disturbances cause population fluctuations (Southwood, 1981). What makes the continental shelf benthos an interesting test case for such hypotheses is the lack of obvious abiotic disturbance other than sediment transport (here counting settling of organic matter from upwelling events as a biotic disturbance). In contrast to intertidal settings, the frequency, intensity and spatial scale of disturbance are quite readily predictable from time series measurements and transport models (Kachel and Smith, 1989, Chapter 8). What is virtually unknown, however, is organism response. The best time series of population abundance and size composition for shelf species come from deep sites in the comparatively protected environment of Puget Sound (Nichols, 1975, 1985), where sediment transport from wave- current interaction is lacking. Such a time series could be used as an experimental control for studies of resuspension effects on biota dwelling on the shelf.

Field data from the literature allow a spectrum of opinion regarding the importance of transport events as disturbance agents. Clearly from lifestyle and life history studies, faunas are adapted evolutionarily to the local transport regime (Purdy, 1964; Jumars and Fauchald, 1977; Maurer and Leathem, 1981). That extreme transport events do produce mortality of macrofauna nonetheless is undeniable (Lammens, 1967; Eagle, 1975; Rees *et al.*, 1976). The issue is further complicated by the fact that sediment transport also has an immigration or emigration role because animals as well as sediments are transported (Grant, 1981; Dobbs and Vozarik, 1983; Veer *et al.*, 1985). Consequently, apparent mortality (disappearance) due to sediment transport may be no more than lateral displacement, lateral displacement may or may not cause reduction in fitness or induce mortality and sediment transport may be an important term in the recovery (immigration) equation.

There are virtually no observations of organisms under realistically simulated conditions of sediment erosion, deposition and transport and the few that do exist (Nowell *et al.*, in press) place past experiments on burial effects in serious doubt. Most experiments have buried animals impulsively under a given depth of sediments in aquaria with comparatively still water, while natural burial by bedload is a grain-by-grain process, as is burial by sedimentation from suspension in a gradually waning transport event. Oxygen stress is less likely a problem under grain-by-grain deposition and natural fluid dynamics and animal digging rates are less likely to be exceeded. Also, sediment transport is dominated by horizontal velocities. Therefore, it is difficult to interpret results of mesocosm simulations of sediment transport produced by an up-and-down oscillating plunger (Oviatt *et al.*, 1981). The results of Oviatt *et al.* (1981), suggesting benthos insensitivity to burial, are in marked contrast, for example, to those of Brenchley (1981). Both the taxa and the simulations are different, however, making the reason for the differences obscure.

There is no doubt that animals which experience daily bedload transport show striking behavioral adaptations (Nowell *et al.*, in press). What is unclear is whether the low frequency (less than daily) of sediment transport at mid-shelf depths will result in strong functional responses. The effects of erosion and deposition might be expected to be most severe in the inshore silt-to-sand transition zone. Because zone shifts occur on a seasonal basis, any sessile organism will experience an annual cycle of changing grain size (Nittrouer, 1978). Again because of the frequency issue, it is unclear whether the animal responses will be behavioral or at a population level.

Large, deep-burrowing infauna are not expected to be seriously affected (in terms of transport or mortality) by large storm events. Numerous large *Brisaster* that have weathered major storms (judging from the age-size relation in this species observed in Puget Sound, cf. Nichols, 1975) are seen on the continental shelf off northern California (Cacchione *et al.*, 1983; Nichols, submitted) in an even more severe wave climate (Cacchione *et al.*, 1987) than that prevailing off Washington. It is not clear what fraction of the benthic fauna and microbes on the Washington shelf will be disturbed by either the major (about 3 cm of sediment eroded and redeposited at 100-m water depth, mean recurrence time about 9.4 y) or minor (about 3 mm reworked at 100-m depth, mean recurrence time about 1.8 y) transport events (estimates from Kachel and Smith, 1989, Chapter 8).

Recent work (Palmer, 1986 and references therein) suggests that meiofauna may often be transported passively along with sediments of similar settling velocity. A large wave-current event removes a net volume of 6.2×10^{6} m of sediments from the shelf between the Columbia River and Quinault Canyon (Kachel and Smith, 1989, Chapter 8). If only 10% of the meiofauna were carried along (conservatively assuming 20 cm^{-3} meiofaunal specimens in the surficial sediments from which meiofauna were eroded), such an event would expatriate about 10^{13} meiofaunal individuals. Local (Oregon shelf) evidence that meiofauna is redistributed by winter storms comes from observed homogenization of the small-scale horizontal distributions (Hogue and Miller, 1981; Hogue, 1982). Hogue (1982) further found a marked faunal boundary in nematode species composition at 25-m water depth, which appears from bottom photographic evidence to be a depth below which wave disturbance becomes much less frequent. His conclusion regarding the cause for this faunal change also is supported by morphological adaptations in the shallower-zone species, apparently for gaining or retaining purchase on the grains surrounding interstices.

If, in any of the above contexts, sediment transport events make resources available (Thistle, 1981) either by inducing mortality (reducing demand) or stimulating microbial production and redistributing it (increasing effective supply), then the large spatial extent of the shelf affected during storm events would argue for the efficacy of planktonic larvae in exploiting those resources. An interesting but as yet undocumented strategy would be to have larval dispersal coincide with the late fall/winter transport season. Besides affording the usual benefits of dispersal (Strathmann, 1974), such a strategy could serve as insurance against adult mortality due to erosion and deposition. Conversely, adults that release larvae after the winter storm season must have effective protection from these sediment transport events.

Kachel and Smith's results (1989, Chapter 8) have other implications for larval transport. The recent hypothesis of passively spread larvae, has already passed several successful tests

(Eckman, 1983; Hannan, 1984; Jackson and Strathmann, 1981; Banse, 1986). While the Kachel and Smith model has little sediment transport interest outside the storm season (because sediments do not move extensively then), it may have considerable application to larval dispersal in other seasons. If larval dispersal occurs during winter, one would expect to see the same sorting and divergence of paths among larvae of varying settling velocities as is seen among sediments of differing grain sizes.

9.3 CONCLUSION AND PROSPECTUS

There has never been a shelf study with sufficiently closely coupled measurements of sediment transport and benthic biological response to explore their relationship in any detail. For many reasons, a shelf with allochthonous sediments, particularly with mid-shelf silt deposits of the U.S. Pacific Northwest from Point Reyes north, is a logical choice for initiating such investigations. It is well studied from the sediment transport perspective (Kachel and Smith, 1989, Chapter 8). Topography is simple, without large bedforms on the mid-shelf, minimizing field measurement problems. Sources (rivers) and sinks (slopes and canyons) of sediments are known. Pulses of organic-matter input from upwelling processes (spring to early fall) are well separated in time from major sediment transport events (late fall to early spring), making responses to them easy to identify without confusion. From the applied perspective, the mid-shelf silts are the site of major demersal fisheries. The three-dimensional, helical circulation of upwelling (Barber and Smith, 1981, Fig. 14) centered over the mid-shelf silts provides a large-scale chromotography column, with sinking fecal pellets and the bottom as the (more or less) stationary phase; hydrophobic pollutants will be concentrated in the mid-shelf silts, where deposit feeders abound. A need to know (Carpenter and Peterson, 1989, Chapter 10) and experimental tractability rarely coincide so well.

ACKNOWLEDGEMENTS

Much of the field work and the subsequent evaluation reviewed here was made possible by extended support from the U.S. Department of Energy and its predecessor, contract DE-AT06-76-EV-75026. Further, we gratefully record partial support for the manuscript preparation by grant No. 86-08157 from the National Science Foundation to P.A. Jumars and A.R.M Nowell and grant No. NAGW-1007 from the National Aeronautics and Space Administration to K. Banse. The patient help of S. Fagerberg is also acknowledged. This is a component of contribution No. 1784 from the School of Oceanography, University of Washington.

REFERENCES

Anderson, D.M. and D. Wall. 1978. The potential importance of benthic cysts of *Gonyaulax tamarensis* and *Gonyaulax excavata* in initiating toxic dinoflagellate blooms. *J. Phycol.*, 14: 224-234.

Anderson, D.M., D.G. Aubrey, M.A. Tyler and D.W. Coats. 1982. Vertical and horizontal distributions of dinoflagellate cysts in sediments. *Limnol. Oceanogr.*, 27: 757-765.

Banse, K. 1986. Vertical distribution and horizontal transport of planktonic larvae of echinoderms and benthic polychaetes in an open coastal sea. *Bull. Mar. Sci.,* 39: 162-175.

Barber, R.T. and R.L. Smith. 1981. Coastal upwelling systems. Pages 31-58 in: A. Longhurst (ed.), *Analysis of Marine Ecosystems.* Academic Press, New York, N.Y.

Boesch, D.F. and M.A. Bowen. In press. Bathymetric distribution of assemblages of macrobenthos in the Middle Atlantic Bight, U.S.A. *Mar. Biol.*

Brenchley, G.A. 1981. Disturbance and community structure: an experimental study of bioturbation in marine soft-sediment environments. *J. Mar. Res.*, 39: 767-790.

Buchanan, J.B. and J.J. Moore. 1986. Long-term studies at a benthic station off the coast of Northumberland. *Hydrobiologia*, 142: 12-127.

Butman, C.A. 1987. Larval settlement of soft-sediment invertebrates: the spatial scales of pattern explained by active habitat selection and the emerging role of hydrodynamical processes. *Oceanogr. Mar. Biol. Ann. Rev.*, 25: 113-165.

Cacchione, D.A., D.E. Drake, W.D. Grant, A.J. Williams, III and G.B. Tate. 1983. Variability of sea-floor roughness within the Coastal Ocean Dynamics Experiment (CODE) region. Woods Hole Oceanogr. Inst., Woods Hole, Mass., Tech. Rep. WHOI-83-25, 44 pp.

Cacchione, D.A., W.D. Grant, D.E. Drake and S.M. Glenn. 1987. Storm-dominated bottom boundary layer dynamics on the northern California continental shelf: measurements and predictions. *J. Geophys. Res.*, 92: 1817-1827.

Carey, A.G., Jr. 1972. Ecological observations on the benthic invertebrates from the central Oregon continental shelf. Pages 422-443 in: A.T. Pruter and D.L. Alverson (eds.), *The Columbia River Estuary and Adjacent Ocean Waters: Bioenviormental Studies.* Univ. of Washington Press, Seattle, Wa.

Carpenter, R. and M.L. Peterson. 1989. Chemical cycling in Washington's coastal zone. Pages 366-510 in: M.L. Landry and B.M. Hickey (eds.), *Coastal Oceanography of Washington and Oregon.* Elsevier, Amsterdam, Neth.

Christensen, J.P, A.H. Devol and W.M. Smethie, Jr. 1984. Biological enhancement of solute exchange between sediments and bottom water on the Washington continental shelf. *Cont. Shelf Res.*, 3: 9-23.

Christensen, J.P, W.M. Smethie, Jr. and A.H. Devol. 1987a. Benthic nutrient regeneration and denitrification on the Washington continental shelf. *Deep-Sea Res.*, 34: 1027-1047.

Christensen, J.P, J.W. Murray, A.H. Devol and L.A. Codispoti. 1987b. Denitrification in continental shelf sediments has a major impact on the oceanic nitrogen budget. *Global Biogeochem. Cycles*, 1: 97-116.

Curray, J.R. 1965. Late Quaternary history, continental shelves of the United States. Pages 723-735 in: H.E. Wright, Jr. and D.G. Frey (eds.), *The Quaternary of the United States.* Princeton Univ. Press, Princeton, N.J.

Dauer, D.M., C.A. Maybury and R.M. Ewing. 1981. Feeding behavior and general ecology of several spionid polychaetes from the Chesapeake Bay. *J. Exp. Mar. Biol. Ecol.*, 54: 21-38.

DeFlaun, M.F. and L.M. Mayer. 1983. Relationships between bacteria and grain surfaces in intertidal sediments. *Limnol. Oceanogr.*, 28: 873-881.

Dobbs, F.C. and J.M. Vozarik. 1983. Immediate effects of a storm on coastal infauna. *Mar. Ecol. Progr. Ser.*, 11: 273-279.

Dortch, Q. and J.R. Postel. 1989. Phytoplankton-nitrogen intereactions. Pages 139-174 in: M.L. Landry and B.M. Hickey (eds.), *Coastal Oceanography of Washington and Oregon.* Elsevier, Amsterdam, Neth.

Eagle, R.A. 1975. Natural fluctuations in a soft bottom benthic community. *J. Mar. Biol. Assoc. U.K.*, 55: 865-878.

Eckman, J.E. 1983. Hydrodynamic processes affecting benthic recruitment. *Limnol. Oceanogr.*, 28: 241-257.

Eckman, J.E., A.R.M. Nowell and P.A. Jumars. 1981. Sediment destabilization by animal tubes. *J. Mar. Res.*, 39: 361-374.

Fanning, K.A., C.L. Carder and P.R. Betzer. 1982. Sediment resuspension by coastal waters: a potential mechanism for nutrient re-cycling on the ocean's margins. *Deep-Sea Res.*, 29: 953-965.

Findlay, R.H., P.C. Pollard, D.J.W. Moriarty and D.C. White. 1985. Quantitative determination of microbial activity and community nutritional status in estuarine sediment: evidence for a disturbance artifact. *Can. J. Microbiol.*, 31: 493-498.

Flint, R.W. 1981. Gulf of Mexico outer-continental shelf benthos: macroinfaunal-environmental relationships. *Biol. Oceanogr.*, 1: 135-155.

Frankenberg, D. and A.S. Leiper. 1977. Seasonal cycles in benthic communities of the Georgia continental shelf. Pages 383-397 in: B.C. Coull (ed.), *Ecology of marine benthos*. Univ. of South Carolina Press, Columbia, S.C.

Fuller, C.M., C.A. Butman and N.M. Conway. In press. Periodicity in fecal pellet production by the capitellid polychaete, *Mediomastus ambiseta*, throughout the day. Proc. 2nd Int. Polychaete Conf., Copenhagen, Aug. 1986.

Gallardo, V.F. 1977. Large benthic microbial communities in sulphide biota under Peru-Chile subsurface countercurrent. *Nature*, 268: 331-332.

Grant, J. 1981. Sediment transport and disturbance on an intertidal sandflat: epifaunal distribution and recolonization. *Mar. Ecol. Progr. Ser.*, 6: 249-255.

Grover, J.J. and B.L. Olla. 1987. Effects of an El Niño event on the food habits of larval sablefish, *Anoplopoma fimbria*, off Oregon and Washington. *Fishery Bull., U.S.*, 85: 71-79.

Hannan, C.A. 1984. Planktonic larvae may act like passive particles in turbulent near-bottom flows. *Limnol. Oceanogr.*, 29: 1108-1116.

Hogue, E.W. 1982. Sediment disturbance and the spatial distributions of shallow water meiobenthic nematodes on the open Oregon coast. *J. Mar. Res.*, 40: 551-573.

Hogue, E.W. and A.G. Carey, Jr. 1982. Feeding ecology of 0-age flatfishes at a nursery ground on the Oregon coast. *Fishery Bull.*, 80: 555-565.

Hogue, E.W. and C.B. Miller. 1981. Effects of sediment microtopography on small-scale spatial distributions of meiobenthic nematodes. *J. Exp. Mar. Biol. Ecol.*, 53: 181-191.

Jackson, G.A. and R.R. Strathmann. 1981. Larval mortality from offshore mixing as a link between precompetent and competent periods of development. *Am. Nat.*, 118: 16-26.

Jumars, P.A. and K. Fauchald. 1977. Between-community contrasts in successful polychaete feeding strategies. Pages 1-20 in: B.C. Coull (ed.), *Ecology of Marine Benthos*. Univ. of South Carolina Press, Columbia, S.C.

Jumars, P.A. and A.R.M. Nowell. 1984. Effects of benthos on sediment transport: difficulties with functional grouping. *Cont. Shelf Res.* 3: 115-130.

Jumars, P.A. and R.F.L. Self. 1986. Gut-marker and gut-fullness methods for estimating field and laboratory effects of sediment transport on deposit feeders. *J. Exp. Mar. Biol. Ecol.*, 98: 293-310.

Kachel, N.B. and J.D. Smith. 1989. Sediment transport and Deposition. Pages 287-348 in: M.L. Landry and B.M. Hickey (eds.), *Coastal Oceanography of Washington and Oregon.* Elsevier, Amsterdam, Neth.

Lammens, J.J. 1967. Growth and reproduction in a tidal flat population of *Macoma balthica* (L.). *Neth. J. Sea Res.*, 3: 315-382.

Landry, M.R., J. R. Postel, W.K. Peterson and J. Newman. 1989. Broad scale patterns in the distribution of hydrographic variables. Pages 1-40 in: M.L. Landry and B.M. Hickey (eds.), *Coastal Oceanography of Washington and Oregon.* Elsevier, Amsterdam, Neth.

Lie, U. 1969. Standing crop of benthic infauna in and off the coast of Washington. *J. Fish. Res. Bd. Canada,* 26: 55-62.

Lie, U. and J.C. Kelley. 1970. Benthic infauna communities off the coast of Washington and in Puget Sound: identification and distribution of the communities. *J. Fish. Res. Bd. Canada,* 27: 621-651.

Lie, U. and D.S. Kisker. 1970. Species composition and structure of benthic infauna communities off the coast of Washington. *J. Fish. Res. Bd. Canada,* 27: 2273-2285.

Marcus, N.H. 1984. Recruitment of copepod nauplii into the plankton: importance of diapause eggs and benthic processes. *Mar. Ecol. Progr. Ser.*, 15: 47-54.

Maurer, D. and W. Leathem. 1981. Polychaete feeding guilds from Georges Bank, U.S.A. *Mar. Biol.* 62: 161-171.

McCall, P.L. and M.J.S. Tevesz (eds.). 1982. *Animal-Sediment Relations.* Plenum Press, New York, N.Y., 336 pp.

Miller, D.C. 1985. Interactions of marine sediment transport with deposit feeding and microbial growth. Ph.D. Diss., School of Oceanogr., Univ. of Washington, Seattle, Wa., 156 pp.

Miller, D.C. and P.A. Jumars. 1986. Pellet accumulation, sediment supply and crowding as determinants of surface deposit-feeding rate in *Pseudopolydora kempi japonica* Imajima and Hartman (Polychaeta: Spionidae). *J. Exp. Mar. Biol. Ecol.*, 99: 1-17.

Miller, D.C., P.A. Jumars and A.R.M. Nowell. 1984. Effects of sediment transport on deposit feeding: scaling arguments. *Limnol. Oceanogr.*, 29: 1202-1217.

Miller, D.C. and R.W. Sternberg. In press. The fluid and sediment dynamic environment of an intertidal deposit feeder. *J. Mar. Res.*

Muschenheim, D. 1987. The dynamics of near-bed seston flux and suspension-feeding benthos. *J. Mar. Res.*, 45: 57-82.

Nichols, F.H. 1975. Dynamics and energetics of three deposit-feeding benthic invertebrate populations in Puget Sound, Washington. *Ecol. Monogr.,* 45: 57-82.

Nichols, F.H. 1985. Abundance fluctuations among benthic invertebrates in two Pacific estuaries. *Estuaries*, 8: 136-144.

Nichols, J. and G.T. Rowe. 1977. Infaunal macrobenthos off Cape Blanc, Spanish Sahara. *J. Mar. Res.*, 35: 525-536.

Nittrouer, C.A. 1978. The process of detrital sediment accumulation in a continental shelf environment: an examination of the Washington shelf. Ph.D. Diss., Dept. of Oceanogr., Univ. of Washington, Seattle, Wa., 243 pp.

Nittrouer, C.A. and R.W. Sternberg. 1981. The formation of sedimentary strata in an allochthonous shelf environment: the Washington continental shelf. *Mar. Geol.*, 42: 201-232.

Nowell, A.R.M., P.A. Jumars, R.F.L. Self and J.D. Southard. In press. Bedload transport effects on infauna. *In*: G.R. Lopez and G.L. Taghon (eds.), *Marine Deposit Feeding*. Springer-Verlag, New York, N.Y.

Oviatt, C.A., C.D. Hunt, G.A. Vargo and K.W. Kopchynski. 1981. Simulation of a storm event in marine mesocosms. *J. Mar. Res.*, 39: 605-626.

Paine, R.T. 1986. Benthic community-water column coupling during the 1982-1983 El Niño. Are community changes at high latitudes attributable to cause or coincidence? *Limnol. Oceanogr.*, 31: 351-360.

Palmer, M.A. 1986. Hydrodynamics and structure: interactive effects on meiofaunal dispersal. *J. Exp. Mar. Biol. Ecol.,* 104: 53-68.

Pamatmat, M.M. 1971. Oxygen consumption by the sea bed. VI. Seasonal cycle of chemical oxidation and respiration in Puget Sound. *Int. Rev. ges. Hydrobiol.*, 56: 769-793.

Pamatmat, M.M. 1973. Benthic community metabolism on the continental terrace and in the deep sea in the North Pacific. *Int. Rev. ges. Hydrobiol.*, 58: 345-368.

Pearcy, W.G. and D. Hancock. 1978. Feeding habits of Dover sole, *Microstomus pacificus*; rex sole, *Glyptocephalus zachirus*; slender sole, *Lyopsetta exilis*; and Pacific sanddab, *Citharichthys sordiadus*, in a region of diverse sediments and bathymetry off Oregon. *Fishery Bull., U.S.*, 76: 641-651.

Pearson, T.H., G. Duncan and J. Nuttall. 1986. Long term changes in benthic communities of Loch Linnhe and Loch Eil (Scotland). *Hydrobiologia*, 142: 113-119.

Perry, M.J., J.P. Bolger and D. English. 1989. Primary Production. Pages 117-138 in: M.L. Landry and B.M. Hickey (eds.), *Coastal Oceanography of Washington and Oregon*. Elsevier, Amsterdam, Neth.

Pilson, M.E.Q. 1985. Annual cycles of nutrients and chlorophyll in Narragansett Bay, Rhode Island. *J. Mar. Res.*, 43: 849-873.

Pruter, A.T. and D.L. Alverson (eds.). 1972. *The Columbia River Estuary and Adjacent Ocean Waters*. Univ. of Washington Press, Seattle,Wa., 868 pp.

Purdy, E.G. 1964. Sediments as substrates. Pages 238-271 in: J. Imbrie and N.D. Newell (eds.), *Approaches to Paleoecology*. Wiley, New York, N.Y.

Rees, E.I.S., A. Nicholaidou and P. Laskaridou. 1976. The effects of storms on the dynamics of shallow water benthic associations. Pages 465-474 in: B.F. Keegan, P.O. Ceidigh and P.J.S. Boaden (eds.), *Biology of Benthic Organisms*. Pergamon Press, Oxford.

Rhoads, D.C., P.L. McCall and J.Y. Yingst. 1978. Disturbance and production on the estuarine sea floor. *Am. Sci.*, 66: 577-586.

Rowe, G.T. 1971. Benthic biomass in the Pisco, Peru upwelling. *Invest. Pesq.*, 35: 127-135.

Schaffner, L.C. and D.F. Boesch. 1982. Spatial and temporal resource use by dominant benthic Amphipoda (*Ampeliscidae* and *Corophiidae*) on the Middle Atlantic Bight outer-continental shelf. *Mar. Ecol. Progr. Ser.*, 9: 231-243.

Smethie, W.M., Jr., C.A. Nittrouer and R.F.L. Self. 1981. The use of radon-222 as a tracer of sediment irrigation and mixing on the Washington continental shelf. *Mar. Geol.*, 42: 173-200.

Southwood, T.R.E. 1981. Bionomic strategies and population parameters. Pages 30-52 in: R.M. May (ed.), *Theoretical Ecology.* 2nd ed, Sinauer, Sunderland, Mass.

Strathmann, R.R. 1974. The spread of sibling larvae of sedentary marine invertebrates. *Am. Nat.*, 108: 29-44.

Suess, E. and J. Thiede (eds.). 1983. *Coastal Upwelling: Its Sedimentary Record. Part A: Responses of the Sedimentary Regime to Present Coastal Upwelling.* Plenum Press, New York, N.Y., 604 pp.

Swift, D.J.P. 1970. Quaternary shelves and the return to grade. *Mar. Geol.*, 8: 5-30.

Taghon, G.L., A.R.M. Nowell and P.A. Jumars. 1980. Induction of suspension feeding in spionid polychaetes by high particulate fluxes. *Science,* 210: 562-564.

Thiede, J. and E. Suess (eds.). 1983. *Coastal upwelling: Its Sedimentary Record. Part B: Sedimentary Records of Ancient Coastal Upwelling.* Plenum Press, New York, N.Y., 610 pp.

Thiel, Hj. 1978. Benthos in upwelling regions. Pages 124-138 in: R. Boje and M. Tomczak (eds.), *Upwelling Ecosystems.* Springer-Verlag, Berlin.

Thiel, Hj. 1982 Zoobenthos of the CINECA area and other upwelling regions. *Rapp. P.-v. Cons. int. Explor. Mer.,* 180: 323-334.

Thistle, D. 1981. Natural physical disturbance and communities of marine soft bottoms. *Mar. Ecol. Progr. Ser.*, 6: 223-228.

Veer, H.W. van der, M.J.N. Bergman and J.J. Beukema. 1985. Dredging activities in the Dutch Wadden Sea: Effects on macrobenthic fauna. *Neth. J. Sea Res.*, 19: 183-190.

Weston, D.P. 1988. Macrobenthos-sediment relationships on the continental shelf off Cape Hatteras, North Carolina. *Cont. Shelf. Res.,* 8: 267-286.

Westrich, J.T. and R.A. Berner. 1984. The role of sedimentary organic matter in bacterial sulfate reduction: the G model tested. *Limnol. Oceanogr.*, 29: 236-249.

Yoder, J.A., L.P. Atkinson, S.S. Bishop, E.E. Hofmann and T.N. Lee. 1983. Effect of upwelling on phytoplankton productivity of the outer-southeastern United States continental shelf. *Cont. Shelf Res.*, 1: 385-404.

CHAPTER 10

CHEMICAL CYCLING IN WASHINGTON'S COASTAL ZONE

R. Carpenter and M. L. Peterson
School of Oceanography, University of Washington, Seattle, WA 98195

10.1 INTRODUCTION

The operation of plutonium producing reactors at Hanford, Washington, beginning in 1944, introduced several artificial radionuclides to the Columbia River and subsequently the adjacent continental shelf and slope (Fig. 10.1). An estimated 300,000 Ci y^{-1} of artificial radioactivity was discharged to the river by the mid-1960s (Robertson *et al.*, 1973), but decreasing reactor operations and more stringent discharge controls essentially eliminated radioactive discharges to the river by 1971.

Since the onset of reactor operations, there have been numerous chemical studies to determine the transport paths and fates of both reactor-derived and naturally occurring radionuclides. These studies have evaluated the possible adverse effects of discharged nuclides on the major river, estuarine and adjacent coastal environments. Radioactive decay of short- and moderately-lived radionuclides, sediment export and river flow have combined to reduce artificial nuclide activities to levels which can no longer easily be determined (Beasley and Jennings, 1984).

No adverse impacts upon man or the coastal ecosystem have been noted from these radioactive discharges, but the radionuclides have served as instructive tracers of Columbia River derived material in the adjacent ocean. Seasonal changes in the position of the Columbia River plume were clearly evident in early studies of ^{65}Zn activity (Lewis and Seymour, 1965). ^{65}Zn activity in offshore sediments (Cutshall *et al.*, 1973) showed that the predominant transport of Columbia River derived solids was along a mid-shelf silt deposit trending north northwest from the river mouth toward Quinault submarine canyon. The presence of the short-lived isotopes $^{141}Ce/^{144}Ce$ and $^{95}Zr/^{95}Nb$ in sediments from Astoria Canyon was one of the first clear indications that radionuclides entering the ocean at the surface could be transported to depths of over a kilometer within a few days (Osterberg *et al.*, 1963).

Studies since the early 1970s have focused on chemicals which are of concern due to their persistence and relatively toxic natures, such as the Pu isotopes, ^{210}Po, ^{210}Pb, As, Hg and polynuclear aromatic hydrocarbons. Major improvements in sample collection, preparation and analytical chemistry capabilities have been made for determinations of trace organic compounds, stable metals and their speciation and radionuclides in environmental samples. Most of these improvements have been exploited in studies of chemical cycling off the Washington coast and are described in greater detail throughout this chapter. This chapter will review developments since Pruter and Alverson (1972) in our understanding of the biogeochemical cycling of primarily these toxic and persistent compounds.

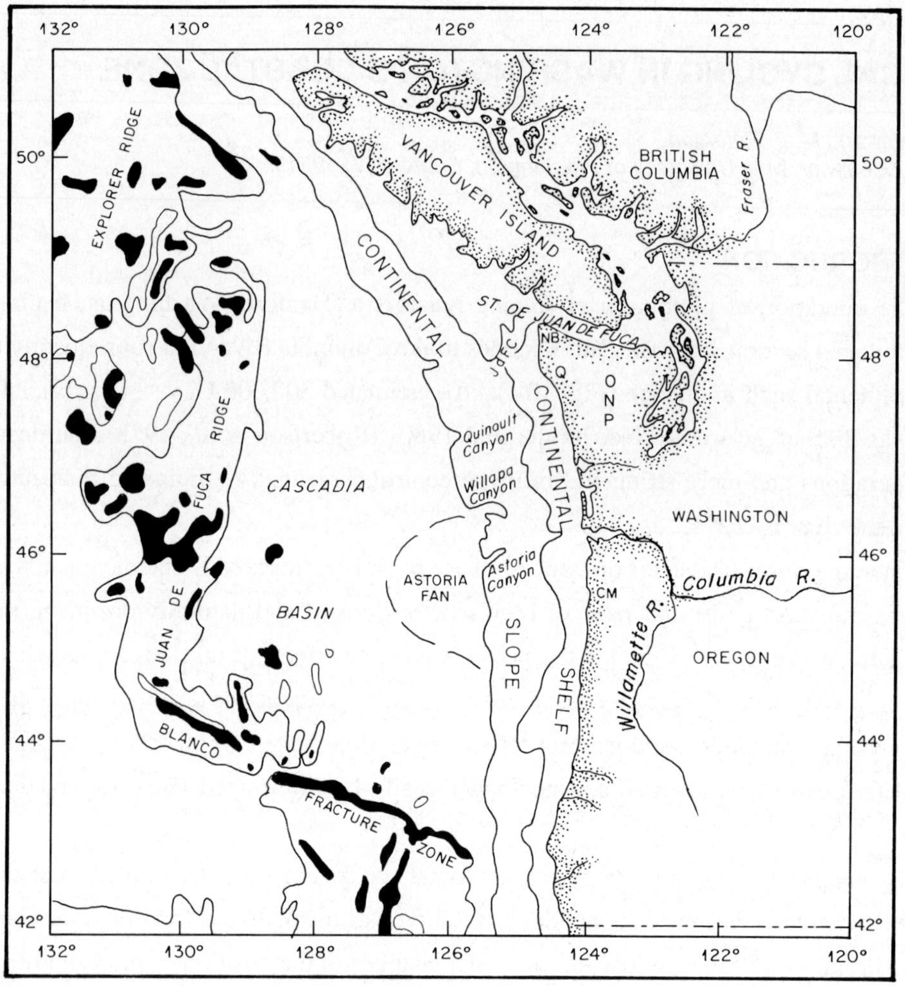

Fig. 10.1 General study area off the Washington and Oregon coasts. NB = Neah Bay/Cape Flattery, Wa., ONP = Olympic National Park, Wa., JdFC = Juan de Fuca submarine canyon, Q = Quillayute, Wa. CM = Cape Meares/Tillamook Head, Ore. Puget Sound is the inland body of water to the east of the Olympic National Park.

Our increased understanding of chemical cycling has come from improvements in both sampling and analytical chemistry techniques as well as our understanding of water circulation and sediment transport off the Washington coast. For example, it is now known that offshore waters upwell onto the shelf from 200-300 m deep (Hickey, 1989, Chapter 2; Small *et al.*, 1989, Chapter 7) and that waters advect into the Strait of Juan de Fuca primarily from the south and southwest rather than from the north. The source of these waters affects the advective supply of chemicals such as radionuclides, whose concentrations and ratios change with water depth.

When advection brings offshore waters with chemicals largely in dissolved form into particle rich areas of the shelf and slope, a fraction of the dissolved chemicals is adsorbed and settles out of the water column. This 'boundary scavenging' of reactive chemicals leads to sediment inventories of ^{210}Pb and Pu which are 4-15 times greater than inventories in sediments of the northwest Atlantic or Gulf of Mexico (Carpenter *et al.*, 1981; Beasley *et al.*, 1982;

Carpenter *et al.*, 1987). The combination of a large supply of reactive, fine-grained particles from the Columbia River with advective supply of dissolved chemicals makes the Washington shelf and slope particularly good areas to study scavenging processes.

The recent discovery of hydrothermal activity at several sites along the Juan de Fuca Ridge system (Fig. 10.1) has excited much interest in ridge crest processes (Canadian-American Seamount Expedition, 1985; and references therein). However, the hydrothermal discharges are believed to have had at most minor effects on the Washington shelf and slope study areas.

Greater numbers of tumors and other diseases in demersal fish from areas of chemically contaminated sediments in parts of Puget Sound have increased public concern about possible adverse effects of chemical discharges to the Sound (Malins *et al.*, 1984). These chemical discharges come from both point sources (sewage treatment plants, smelters, oil refineries and chemical plants) and from numerous diffuse sources. The overall status of Puget Sound was recently reviewed (Puget Sound Water Quality Authority, 1986). In the mid-1970s, concern over the adverse effects of oil spills on marine organisms motivated many of the studies of aliphatic and aromatic hydrocarbon cycling summarized in this chapter. Since the completion of the trans-Alaskan pipeline in 1976 a steady stream of tankers carrying Alaskan oil has been arriving at refineries located at Cherry Point, Ferndale, Anacortes and Tacoma. A large shore-based facility on the Strait of Juan de Fuca capable of handling supertankers was proposed as the western terminus of a pipeline to the midwestern U.S., but was vetoed by Governor John Spellman in 1983.

Major sections of this chapter are organized to reflect processes such as chemical inputs from rivers, advecting seawater, air-sea exchange and hydrothermal activity and outputs to the major sedimentary and biological reservoirs via uptake and exchange of chemicals with the water column. Within each of these sections, we first review cycling of chemicals on the open Washington coast north of the Columbia River mouth; related research in the Strait of Juan de Fuca and in Puget Sound is then compared and contrasted. In each section we consider radionuclides other than ^{14}C (such as Pu, Am, ^{210}Pb, etc.), C isotopes, stable metals (such as As and Hg), aliphatic and aromatic hydrocarbons and other organic compounds. Each section ends with a brief summary of remaining major unanswered questions about chemical cycling.

Much of the research summarized in this chapter has been supported by the Department of Energy. This review also considers all pertinent studies of other research groups and funding agencies of which we are aware.

10.2 RIVERINE CHEMICAL INPUTS

10.2.1 Columbia River Sediment Input to the Washington Coast

Karlin (1980) estimated the amount of Columbia River suspended particulate discharge to the coastal ocean to be 1.4×10^{13} g y^{-1} for the period 1964-1969. Columbia River water discharge during these years was only 70% of the 92-y average flow (U.S. Geological Survey Water Data Reports, 1965-1970), so that Karlin's suspended solid discharge is probably also

Table 10.1. Comparison of concentrations of chemicals in Columbia River suspended and bottom sediments.

Chemical	Concentrations in Bottom Sediments	Concentrations in Suspended Sediments	Reference
Organic C	1.4 ± 0.9%	2.8 ± 1.5%	Prahl (1982)
PAH	see Fig. 10.5	≅ 5 x greater	Prahl (1982)
^{210}Pb	≤ 0.2 dpm g^{-1}	5.1 ± 1.8 dpm g^{-1}	Carpenter *et al.* (1981)
Hg	≅ 11 ppb	≅ 150 ppb	Bothner and Carpenter (1973)
Pu	≤ 1 dpm kg^{-1}	≅ 20 dpm kg^{-1}	Beasley *et al.* (1981)
As*	6.4, 2.7, 2.5, 3.2 and 3.1 ppm	6.6, 29, 40, 17, 7.6 and 15 ppm	Unpublished data of E. Crecelius and R. Carpenter
Sb*	0.81, 0.29, 0.23, 0.22 and 0.24 ppm	0.16, 2.9, 3.0, 3.4, 0.48 and 5.5 ppm	Unpublished data of E. Crecelius and R. Carpenter

*Bottom sediments analyzed for As and Sb were collected between Portland and Astoria, while suspended sediments were from between Tricities and Longview.

about 70% of the 100-y average. Previous estimates of Columbia River suspended solid discharge have ranged from 0.5 to 2.1 x 10^{13} g y^{-1} (Nittrouer, 1978; Karlin, 1980; and references therein). Karlin (1980) concluded that sediments from northern California and Oregon rivers contribute significantly to sediments deposited off southern Oregon. They are insignificant contributors to deposits off Washington, where Columbia River-derived sediments predominate due to year round bottom current transport of Columbia River-derived sediments to the north-northwest (Nittrouer *et al.*, 1979; Kachel and Smith, 1989, Chapter 8). Additional sediment inputs from other Washington rivers are minor (Gross and Nelson, 1966).

The Columbia River transports large amounts of sandy bedload sediments to the adjacent ocean at a rate which is not well established. Karlin (1980) estimated the bedload component to be roughly 10-20% of the total sediment transport. On average, Columbia River estuary bottom sediment is composed of about 1% gravel, 84% sand, 13% silt and 2% clay (Hubbell and Glenn, 1973). The sandy bedload has 5 to 20 times lower concentrations of all chemicals of environmental concern than suspended sediment (Table 10.1). It is also reasonable to assume that the bedload particles will be relatively unreactive in scavenging dissolved chemicals offshore. Thus, bedload particles may be unimportant in budgets and cycling of chemicals, while contributing to higher bulk sediment accumulation rates near the river mouth.

The clay mineral composition of Columbia River sediments changes from predominantly illites (~70%) in upper reservoirs such as Grand Coulee to predominantly montmorillonites

(~55%) in lower reservoirs such as Bonneville and the Dalles (Knebel *et al.*, 1968). Baker (1973; 1976) showed that the montmorillonite/chlorite ratio of suspended sediments was a useful indicator of Columbia River-derived sediments on the adjacent Washington continental shelf and in the upper reaches of submarine canyons indenting the shelf.

Periods of high river water discharge during June (Barnes *et al.*, 1972; Landry *et al.*, 1989, Chapter 1, Fig. 1.1) prevent much accumulation of fine-grained solids in the Columbia River estuary. Several different approaches have suggested that at most one-third of the suspended solids could be trapped in the estuary (Carpenter *et al.*, 1982; and references therein). The most recent study of sediments in the Columbia River estuary also concluded that almost no fine-grained solids were retained within the estuary (Sherwood *et al.*, 1984).

Some Columbia River-derived solids are transported down Astoria submarine canyon located directly west of the river mouth. Major fractions of riverine solids, however, are transported north-northwestward from the river mouth toward Quinault submarine canyon, forming a mid-shelf silt deposit with accumulation rates above 250 mg cm^{-2} y^{-1} (Fig. 10.2). Mt. St. Helens' ash particles supplied by the Columbia River outflow were transported north-northwestward along this deposit at rates of about 80 km y^{-1}, with transport primarily associated with storms (Healy-Ridge and Carson, 1987; Kachel and Smith, 1989, Chapter 8). Table 10.2 shows that rates of shelf sediment transport obtained from dispersal of Mt. St. Helens' ash are in the same range as rates previously estimated by other investigators.

Sediment accumulation rates calculated from excess ^{210}Pb activity profiles in over 100 cores yield a total sediment accumulation rate of 1.4×10^{13} g y^{-1} for the Washington continental shelf (Nittrouer, 1978). Clearly this is most, if not all, of the Columbia River-derived supply of suspended solids and show that only a small fraction is trapped in the estuary or accumulates off Oregon. Sediment accumulation rates have also been reported for the submarine canyons which indent the shelf and for the adjacent open slope areas (Carpenter *et al.*, 1982; Thorbjarnarson *et al.*, 1986). Rates are highest for Astoria and Quinault Canyons, but yearly accumulation in these two canyons amounts to only a few percent of the Columbia River sediment supply.

In summary, sediment accumulation rates and patterns are reasonably well established for the Washington shelf, slope and canyons. They balance Columbia River suspended sediment discharge rates, within an uncertainty estimated at ± 30%. In contrast, few data are available for areas off Oregon.

10.2.2 Riverine Sediment Input to the Puget Sound Region

Freshwater from the large Fraser River (Fig. 10.1) may significantly lower salinities of waters of the Strait of Juan de Fuca and further offshore. Studies of trace chemicals supplied by the Fraser have been few (Fletcher *et al.*, 1983; and references therein). Sedimentation in its estuary has been described by Milliman (1980). A plume of muddy Fraser River water can extend for a considerable distance into the Strait of Georgia and even across it into Saanich Inlet (Fig. 10.3). However, satellite imagery of the Fraser River plume has shown that sediment and

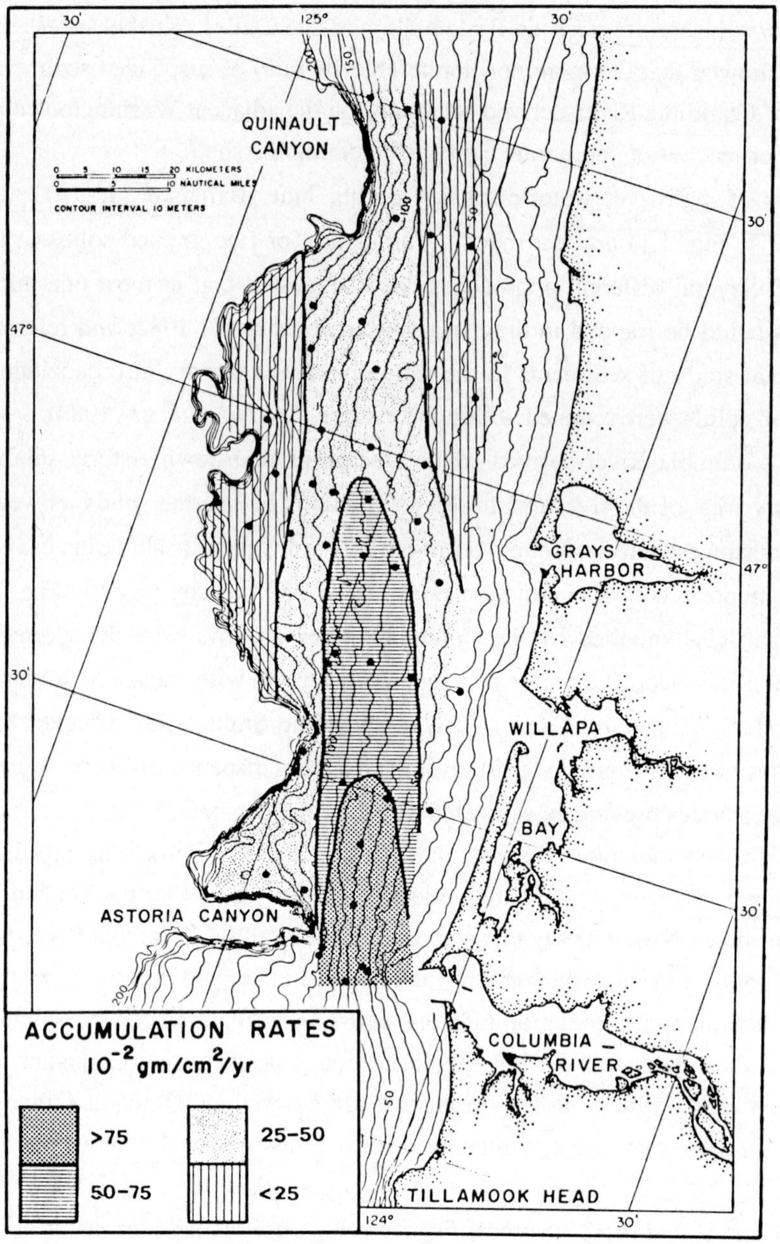

Fig. 10.2 ^{210}Pb-derived recent sediment accumulation rates on the Washington continental shelf. From Nittrouer *et al.* (1979), with permission.

chemicals from the Fraser River are likely to influence only the extreme northern end of the greater Puget Sound region (Feely and Lamb, 1979). Therefore, we have neglected further consideration of sediment and chemical inputs from the Fraser River to the Washington coastal zone.

Freshwater discharges from rivers entering the eastern side of the Puget Sound region (Fig. 10.3) lower the salinity from 32-33 ‰ in the inflowing seawater to 29-30 ‰ in the Sound. Table 10.3 summarizes current best estimates of freshwater and particulates discharged

TABLE 10.2 Comparison between Washington shelf sediment transport rates calculated from Mount St. Helens ash dispersal rates and rates reported from other investigators. Data from Healy (1983) and Healy-Ridge and Carson(1987).

Author	Basis of Rate Calculation	Rate of Sediment Transport	Bearing (Northwestward of isobaths)	Grain Size
Gross and Nelson (1966)	Traced ^{65}Zn, ^{60}Co radionuclides attached to sediment	12 to 31 km y $^{-1}$	12 to 15°	average particle (?)
Smith and Hopkins (1972)	current meter data	40.6 km y $^{-1}$	10°	average particle modes of 125 μm (3 ϕ) and 88 μm (3.5 ϕ)
		80 km y $^{-1}$ (16 km storm^{-1})	10°	suspended load*
Sternberg and McManus (1972)	current meter data	200 km y $^{-1}$ (assuming six storms y $^{-1}$)	19°	suspended load*
		37 km storm $^{-1}$	19°	suspended load*
Kachel (1980)	mathematical model	25 to 80 km (standard storm) $^{-1}$	16°	suspended load*
		60 to 140 m (standard storm) $^{-1}$	23 to 28°	bedload (> 0.63 μm or < 4.0 ϕ)
		330 km (large storm) $^{-1}$	17°	suspended load (< 88 μm or > 3.5 ϕ)
		585 m (large storm) $^{-1}$	26°	bedload (> 88 μm or < 3.5ϕ
Healy (1983)	ash traced to station 3	13 to 45 km y $^{-1}$ (2.7 to 9.0 km storm) $^{-1}$	26°	22 to 7 μm (5.5 to 7.2 ϕ)
	ash traced to station 5	73 to 80 km y $^{-1}$ (14 to 16 km storm) $^{-1}$	12°	12 to 7 μm (6.4 to 7.2 ϕ)
	ash traced to station 7	80 km y $^{-1}$ (16 km storm) $^{-1}$	8°	22 to 7 μm (5.5 to 7.2 ϕ)
	ash traced to station 8	95 to 115 km y $^{-1}$ (18.9 to 23 km storm) $^{-1}$	9°	12 to 7 μm (6.4 to 7.2 ϕ)
	ash traced to station 10	125 to 190 km y $^{-1}$ (24.9 to 38 km storm) $^{-1}$	10°	12 to 7 μm (6.4 to 7.2 ϕ)

*Suspended load represents material < 63 μm (> 4.0 ϕ) unless otherwise indicated.

by the rivers of western Washington entering Puget Sound and compares them to similar parameters for the Columbia and Fraser rivers. Volumes of freshwater discharge and their seasonal trends are fairly well defined. However, particulate loadings have been measured only for the

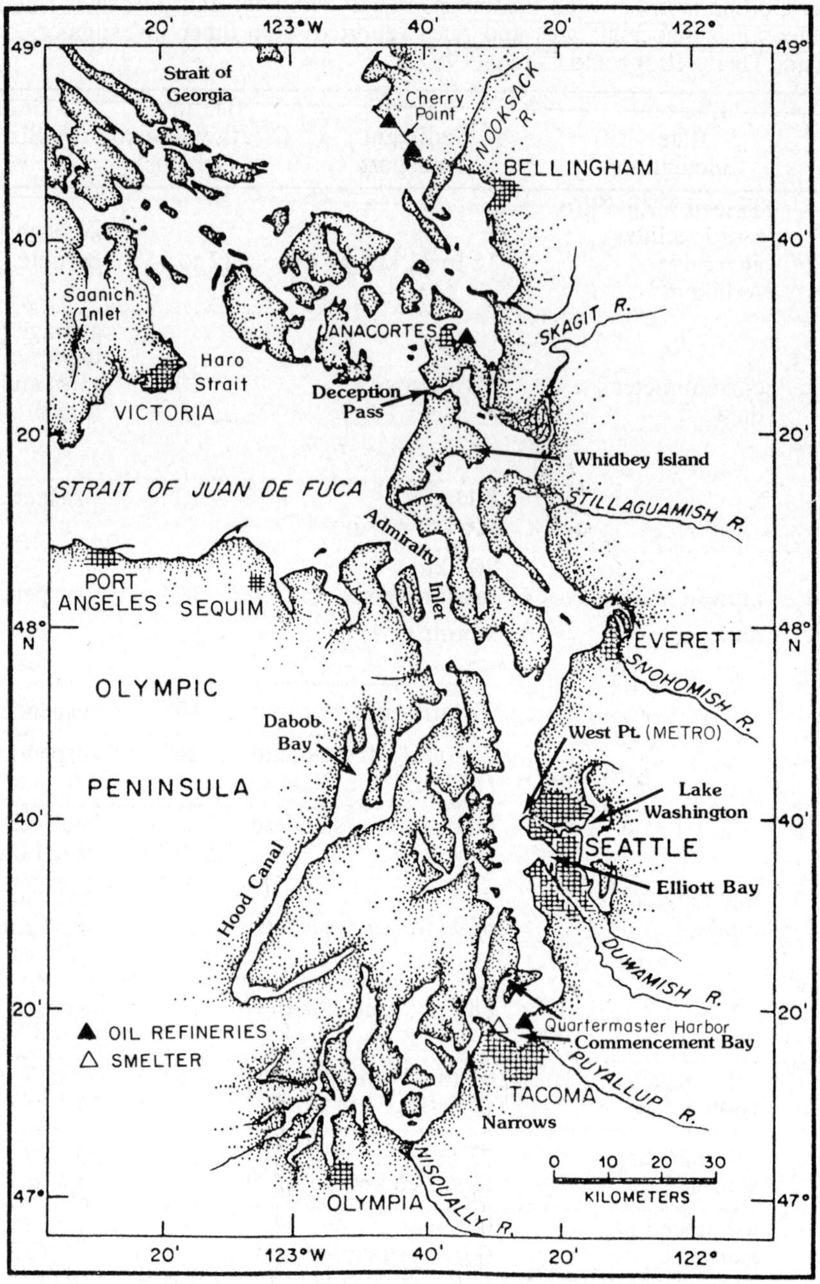

Fig. 10.3 Puget Sound study area. Central Puget Sound is that region bounded by sills at Admiralty Inlet on the north and at the Narrows near Tacoma to the south. Boundary Bay is just off scale to the north of Cherry Point and Pillar Point is off scale to the west along the southern shore of the Strait of Juan de Fuca.

Skagit River during 1967-1977 (Feely and Lamb, 1979) and the Duwamish River during 1964-1965 (Curl, 1982). Estimates of particulates discharged by the Columbia and other Puget Sound rivers are based on semi-empirical relationships of particulate loads to water flow rates. Although riverine particulate loadings are generally proportional to river water flow, there are

TABLE 10.3 Estimates of water and suspended particulate discharges for major rivers entering Puget Sound. Units of discharge are 10^{15} g y $^{-1}$ for water and 10^{11} g y $^{-1}$ for suspended sediments. For comparison, Columbia and Fraser river water and suspended sediment discharges are also given.

River	Water Discharge (U.S.G.S. Water Records)	Suspended Sediment Discharge	
		Dexter *et al.* (1981)	Downing (1983)
Skagit	14	11	12
Stillaguamish	3.5	4.0	0.16
Snohomish	8.1	6.5	4.6
Duwamish	1.3	1.7	1.2
Puyallup	2.8	4.6	5.3
Nisqually	1.5	1.7	1.1
Total	31.2	29.5	24.4
Columbia (Pruter and Alverson,1972)	300	50 to 210	
Fraser (Feely and Lamb, 1979)	83	190	

obviously greater uncertainties in this approach than by directly measuring the loads. Somewhat different and debatable assumptions with respect to the dependence of suspended particulate loads on water discharge rates were made by Dexter *et al.* (1981) and Downing (1983). Differences between the two estimates for the same river in Table 10.3 indicate minimum uncertainties for total riverine particulate discharge. Other major uncertainties are the fractions of discharged particulates which are actually transported to central Puget Sound. Data are also not available to assess the effects of various anthropogenic developments during the past century on riverine sediment discharges. Riverine particulate discharges can support average accumulation rates of muds, sandy muds and muddy sands in central Puget Sound of 610-1,000 mg cm $^{-2}$ y $^{-1}$ even with the extreme assumption that none of the sediment is retained near river mouths in Elliott and Commencement Bays.

Estimates of particulate input from atmospheric deposition (Hardy and Crecelius, 1981), sewage discharge (Barrick, 1982) and primary productivity (Winter *et al.*, 1975) indicate that these sediment sources are negligible. The possible contribution of shoreline erosion to the total sediment input to the main basin of central Puget Sound is unquantifiable at present, but has been roughly estimated by Dexter *et al.* (1981). Coarse materials eroded from shorelines cannot comprise a major fraction of the relatively fine-grained deposits found at most deep locations in central Puget Sound. Transmissometer studies (Baker, 1984) also do not support the idea that significant amounts of fine-grained material eroded from shorelines are continually transported out into deeper waters. We have therefore neglected this possible additional input.

The mean accumulation rate for central Puget Sound determined by Carpenter *et al.* (1985) of 480 ± 110 mg cm $^{-2}$ y $^{-1}$ can be maintained by about two-thirds of the total river particulate discharge, possibly augmented by small inputs from other sources. We consider this an important test of the accuracy of the mean accumulation rate, since no other large sources of fine-grained sediment exist. This conclusion contrasts with recent sediment budgets (Crecelius *et al.*, 1984; Lavelle *et al.*, 1986) which demonstrated sediment accumulation rates much greater than riverine particulate input and invoked a large sediment supply from shoreline erosion to account for the discrepancy.

10.2.3 Columbia River Concentrations and Inputs of Specific Chemicals

Radionuclides

Inventories of Pu isotopes in bottom sediments from Hanford to the Columbia River estuary have been estimated based on activities measured in 50 cores collected in 1977-1978 (Beasley and Jennings, 1984) (see Fig. 10.5 for geographical locations mentioned in the following discussion). ^{240}Pu/^{242}Pu and ^{240}Pu/^{239}Pu isotope ratio data established that the Hanford reactors supply 20-25% of the total Pu in McNary reservoir sediments: fallout from atmospheric testing accounts for the remainder. In contrast, only about 3.5% of the Pu in fine grained sediments in the estuary is from the Hanford reactors (Beasley and Jennings, 1984). The sevenfold decrease in importance of reactor-derived Pu between McNary reservoir and the estuary results from the addition of more fallout-derived Pu to the Columbia River from downstream tributaries, the largest being the Willamette River.

A large quantity of Columbia River sediment from behind McNary Dam was homogenized and prepared as National Bureau of Standards Reference Material 4350B, with activities of Pu and several other radionuclides certified. Analyses of this material test the reliability of radionuclide measurements from different laboratories. The natural radioactivity of U, Th, Ra and K isotopes exceeds that of the artificial radionuclides by nearly an order of magnitude (Beasley and Jennings, 1984).

From Bonneville Dam to the head of the estuary, bottom sediments are coarse-grained sand with uniformly low radionuclide activities. The Trojan Nuclear Power Plant located near river kilometer 140 had been in commercial power production for 2 y at the time of sampling by Beasley and Jennings (1984). Pu, ^{137}Cs and ^{60}Co inventories above and below the plant site were virtually identical, arguing for no consequential additions of these radionuclides to downstream sediments from the Trojan reactor.

Monthly samples of water collected near the mouth of the Columbia River between July 1978 and July 1979 were analyzed for Pu and Am to estimate the quantities of these radionuclides exported to the adjacent coastal zone (Beasley *et al.*, 1981). Most (80-90%) of the Pu and Am was associated with particulates. During this period approximately 70 m Ci of ^{239,240}Pu were carried by the river to the ocean. From the depositional history of Pu recorded in a fine-grained sediment core from Youngs Bay (within the estuary), 4-8 Ci of $^{239,240}Pu$ and 1-2 Ci of ^{241}Am were estimated to have been discharged to the adjacent coastal ocean by the Columbia River since the late 1950s. This Pu and Am discharge is small compared with the total Pu sedimentary inventory on the Washington continental shelf and with estimates of Pu supplied with advecting seawater (Beasley *et al.*, 1982).

Studies of Am export to the northeast Pacific Ocean by Columbia River runoff have been summarized Beasley *et al.* (1981). The mean $^{241}Am/^{239,240}Pu$ activity ratio observed in both total water samples and riverine particulate matter was 0.24 ± 0.06, consistent with ratios measured elsewhere for unfractionated mid-latitude, fallout material (Livingston and Bowen, 1976; Krey *et al.*, 1976). The mean ratio in seven horizons of the core from Youngs Bay was 0.24 ± .04.

Inventories of ^{241}Am, ^{137}Cs and ^{60}Co were established in the same 50 cores from the Columbia River and its estuary that were analyzed for Pu (Beasley and Jennings, 1984). All the ^{241}Am activity is derived from atmospheric fallout. The $^{241}Am/^{239,240}Pu$ activity ratio determined in seven cores averaged 0.29 (after correction for the small amount of ^{239}Pu of Hanford origin). This ratio is similar to cumulative fallout activity ratios for 1977-1978 (Beasley and Jennings, 1984).

Several lines of evidence led Beasley and Jennings (1984) to conclude that traces of ^{60}Co measured in Columbia River sediments must have come primarily from the activation of cobalt-containing materials in plutonium producing reactors at Hanford prior to 1971. ^{137}Cs was attributed largely to global fallout. No estimates were made of ^{60}Co export by the river to the adjacent ocean. Rough estimates of ^{137}Cs export (Beasley *et al.*, 1982) indicate that riverine input could supply the entire sedimentary inventory of ^{137}Cs on the Washington shelf, even allowing for some desorption of Cs from riverine particles.

^{210}Pb and ^{210}Po were determined in 21 samples of suspended particulates collected from the freshwater regime of the Columbia River near its mouth during 1977-1979 (Carpenter *et al.*, 1981). No effort was made to determine dissolved ^{210}Pb since the dissolved fraction was found to be minor near the mouths of the Susquehanna (Lewis, 1977) and Connecticut (Benninger, 1978) rivers. ^{210}Pb activities in the Columbia River suspended particulates averaged 5.1 ± 1.8 dpm g^{-1}, similar to activities in particulates from the Mississippi (Shokes, 1976) and Connecticut (Benninger, 1978) rivers and from rivers discharging into Puget Sound (Carpenter *et al.*, 1984). Analyses of eight grab samples of sandy sediments from the Columbia River bottom near the mouth revealed very low ^{210}Pb activities, with an average of only 0.2 dpm g^{-1} (Twiss *et al.*, 1973).

If the riverine ^{210}Pb flux calculated from the total Columbia River suspended sediment discharge and the average ^{210}Pb activity of suspended particles is entirely deposited on the Washington continental shelf, the magnitude of the depositional flux would be similar to that of atmospheric fallout (Carpenter *et al.*, 1981). Both ^{210}Pb fluxes are much smaller than fluxes observed to shelf sediments or estimated for dissolved ^{210}Pb supplied by seawater advecting across the shelf. ^{210}Pb activities and $^{210}Pb/Al$ and $^{210}Pb/C$ ratios in offshore sediments are greater than in Columbia River suspended particulates, clearly revealing a major, non-riverine source of excess ^{210}Pb to Washington coastal sediments.

Fourteen of the 21 Columbia River particulate samples have $^{210}Po/^{210}Pb$ activity ratios significantly below the secular equilibrium ratio of 1.0 (Carpenter *et al.*, 1981). No convincing explanation for such ^{210}Po deficiencies has been found, but they have also been reported in well and ground waters of Illinois (Holtzmann, 1964). The average ratio of 0.8 ± 0.3 in the freshwater particulates could be produced if 80% of the isotopes came from soils and older snowmelt-derived water with ratios of 1.0, and 20% came from recent precipitation and surface runoff with ratios close to 0.1 (Poet *et al.*, 1972). The remaining seven particulate samples from the Columbia River with ratios above 1.0 are believed to contain plankton, which preferentially scavenge ^{210}Po (Cherry and Shannon, 1974).

Dissolved Nutrients

Concentrations of dissolved inorganic phosphate, nitrate plus nitrite and silicate in the main stream of the Columbia River and its estuary generally are greatest in winter and early spring and lowest in summer. For example, the nitrate plus nitrite concentration often ranges as high as 35 μM in early spring and drops to less than 1 μM in mid-summer. The spring to summer decrease in nutrients is largely due to phytoplankton utilization evident as a concomitant increase in chlorophyll *a* (Frey *et al.*, 1984). Nutrient concentrations are higher in estuary bottom waters in summer at flood tide due to upwelling offshore. In winter, when coastal upwelling usually does not occur, concentrations are higher in overriding river water.

Annual fluxes for dissolved phosphate, nitrate and silica near the mouth of the Columbia River were comparable for 1966, 1967, 1973-1975 and 1980-1981 (Carpenter, 1987). There was no indication of any large increase in nutrient concentrations or fluxes during the 15-y period due to anthropogenic activities in the river drainage basin, which includes some heavily fertilized agricultural areas. Total nitrate and phosphate fluxes via rivers to Pacific Northwest coastal zones increased by 6 and 34% respectively between 1974 and 1981. No data for temporal changes in ammonia and urea fluxes are available, due to problems with sample storage and larger small-scale spatial and temporal variabilities than for other nutrients. The available data contradict the claim of Walsh (1984) that Columbia River concentrations and inputs of inorganic N have increased by a factor of at least three in recent decades.

Fluorine

Dissolved fluorine concentrations in the Columbia River and its estuary were determined in 1972 by S. G. Wakeham using an F^- ion selective electrode with a total ionic strength buffer and a known additions technique (Fig. 10.4A). Dissolved F^- exhibits conservative behavior during mixing of river and seawater in the Columbia River estuary. This is similar to the F^- mixing behavior reported for Georgia estuaries (Windom, 1971) and for Chesapeake Bay (Warner, 1972). A simple end-member mixing hypothesis explains dissolved F^- concentrations rather well in all three estuarine systems. In all three cases dissolved F^- concentrations were about 0.1 ppm in freshwater and 1.3 ppm in the seawater (assuming 35 $^o/_{oo}$ salinity). Dissolved F^- concentrations in Columbia River water are not detectably higher than those in other rivers,

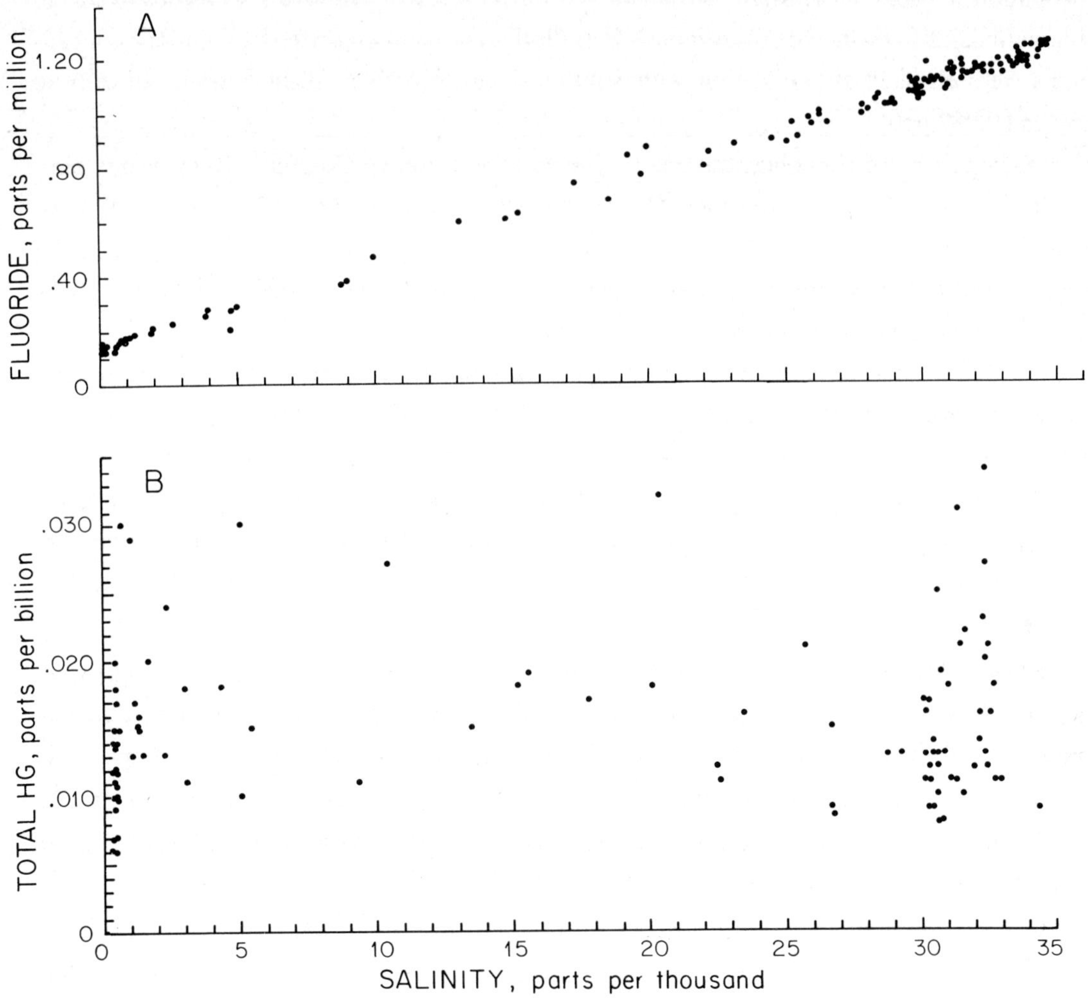

Fig. 10.4 Total concentrations of (A) fluorine and (B) mercury versus salinity for water samples from the Columbia River estuary. Fluorine determinations were made by S.G. Wakeham and mercury determinations by M.H. Bothner.

even though some F^- discharges occur in the effluents of several large Al smelters located along the rive

Mercury

Dissolved Hg concentrations in Columbia River water were determined in 1972 up and downstream of a chloralkali plant at Longview, Washington, to assess the plant's contribution to locally high Hg concentrations in the river water (Bothner and Carpenter, 1973). Mercury concentrations of several ppb, which far exceed levels in most other rivers, had been reported (U.S.G.S., 1970). Careful attention was given to minimizing contamination during sample collection followed by immediate analysis using cold vapor atomic absorption. Total Hg concentrations in unfiltered water from different areas of the Columbia River and its major tributaries were uniformly low. The average value for 100 analyses was 0.01 ppb, with a range of only 0.005 to 0.015 ppb and a coefficient of variation of 20% based on replicate samples taken at the same station in rapid succession. No significant differences in Hg levels were observed between periods of low (January) and high (May) discharge. On two different dates Hg concentrations were elevated at one station within 20 m of the chloralkali plant outfall, but only to 0.050 and 0.087 ppb.

Comparison of Hg concentrations in filtered and unfiltered Columbia River water samples showed that 47-58% of the total Hg was associated with suspended particulate phases. Suspended matter plays a significant role in the transport of Hg in the Columbia River, even though typical Columbia River suspended loads of 10-20 mg l^{-1} (Conomos and Gross, 1972; Karlin, 1980) are much less than those of most other rivers.

Concentrations of Hg in the medium to coarse sandy bottom sediments of the Columbia River downstream from Portland, Oregon, ranged from 7.4 to 17.3 ppb dry weight (Table 10.1, Bothner and Carpenter, 1973). Samples of suspended particulates collected from the lower 100 km of the river during periods of high runoff contained 126 to 176 ppb dry weight. Washington continental shelf sediments, with grain sizes intermediate between riverine suspended and bottom sediments, contained an average of 33 ppb dry weight Hg (Bothner and Carpenter, 1973).

The influence of Columbia River suspended matter on the fate of dissolved Hg in the river and estuary was investigated by spiking Columbia River water with ^{203}Hg labeled mercuric nitrate and methyl mercury chloride under laboratory conditions and following the reactions of the radioactive species (Bothner and Carpenter, 1973). Between 50-75% of both the natural mercury and the added radioactive species was associated with the riverine particulates after about a 10-h exposure.

Desorption experiments showed that at least half of the adsorbed Hg was easily desorbed by filtered river water of lower total Hg concentration. Subsequent desorptions of spiked particulates in seawater removed little additional Hg during a 1-wk contact time. Total Hg concentrations were also determined in waters collected across the salinity gradient in the Columbia River estuary (Fig. 10.4B). Results showed no clear dependence of Hg concentrations upon salinity. Concentrations were all low and probably more dependent upon suspended particle concentrations and types than other factors. A similar conclusion was reach by Rae and Aston (1982), who reported that Hg was not readily desorbed during mixing processes in the Wyre estuary in northwest England.

That little particle-bound Hg desorbs in seawater is surprising given the thermodynamically favorable reaction of Hg^{2+} to form $HgCl_4^{2-}$. This observation also contradicts the hypothesis of DeGroot (1973) that most Hg bound to sediments from the Rhine and Ems estuaries was solubilized upon contact with seawater. In addition, DeGroot's (1973) results for Zn and Mn desorption are inconsistent with the findings of Johnson *et al.* (1967) for these metals in the Columbia River system. Muller and Forstner (1975) have pointed out that the metal solubilization claimed by DeGroot could be the result of mixing between relatively non-contaminated North Sea and contaminated Rhine sediments.

Chromium

Extensive studies of ^{51}Cr (28 d half-life) behavior in the Columbia River and adjacent ocean were made because the isotope is released from the plutonium production reactors at the Hanford reservation. The neutron flux in the reactors irradiates both the Cr originally in the coolant river water and the sodium dichromate added to it as a corrosion inhibitor. Forster (1972a and b) concluded that 90% of the ^{51}Cr added to the river remained in the soluble CrO_4^{2-} form. Dissolved ^{51}Cr was found in ocean waters more than 500 km from the river mouth (Cutshall *et al.*, 1966), where it became undetectable due to dilution and decay. Traces of ^{51}Cr were also found in sediments up to 115 km from the river mouth (Gross *et al.*, 1965). Evans and Cutshall (1973) concluded that CrO_4^{2-} behaved conservatively as river and seawaters mixed within the estuary.

A few measurements have been made recently of stable dissolved Cr species and particulate Cr concentrations in the Columbia River using 'clean' sampling and analytical techniques (Cranston and Murray, 1980). The Columbia River and estuary water contain about 5 nM total Cr; Cr^{6+} is the dominant form and more than 30% of the total is particulate Cr. These data agree with thermodynamic predictions that Cr^{6+} should be the dominant species in oxygenated waters and that variations in Cr^{3+} and particulate Cr are related to biological or flocculation processes. Dilution plots also suggest that CrO_4^{2-} exhibits conservative behavior within the estuary.

Arsenic and Antimony

We are aware of no measurements of dissolved concentrations of these elements in Columbia River water since the work of Silker (1964) on As. However, total concentrations of As and Sb in Columbia River suspended particles and bottom sediments as well as Washington shelf sediments, have been determined by instrumental neutron activation techniques described by Crecelius *et al.* (1975). These previously unpublished results are summarized in Tables 10.1 and 10.6 in Section 10.6. Both As and Sb concentrations are several times greater in finer suspended particles than coarser bottom sediments. Concentrations in offshore Washington shelf sediments are intermediate, reflecting the fact that the offshore deposits are intermediate in grain size between riverine suspended and bottom sediments.

Other Stable Trace Metals

Riedel *et al.* (1984) determined dissolved and suspended concentrations of Cd, Cu, Fe, Mn, Ni, Pb and Zn in the Columbia River estuary following the 18 May 1980 eruption of Mt. St. Helens, which injected large amounts of volcanic ash and mud into the river. Soluble concentrations of these metals were *not* substantially elevated during this period, except for slightly higher concentrations of Mn and Cu. Soluble Cd, Ni, Pb and Zn were in many cases below the limit of detection. Dissolved Mn concentrations in the estuary showed the mid-salinity maximum previously observed in several other estuaries. Fe concentration trends with salinity showed the loss of Fe from solution observed in most other estuaries. Riedel *et al.* (1984) also

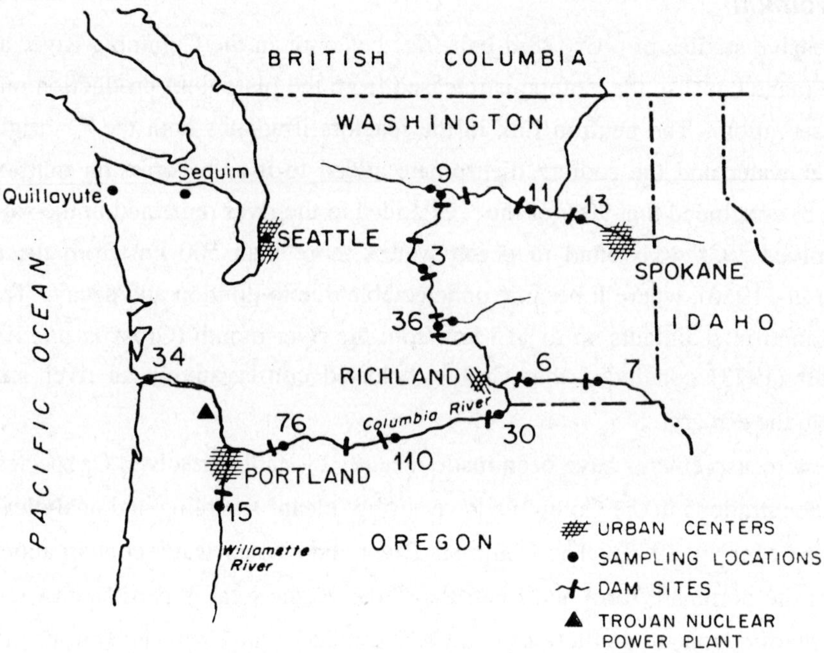

Fig. 10.5 Sum of nine individual combustion-derived polynuclear aromatic hydrocarbon (PAH) concentrations in riverine bottom sediments collected at various sites within the Columbia River drainage basin. Data are from Prahl (1982) and are given in units of μg (g organic C) $^{-1}$. Sediments were collected with a small grab sampler. Bonneville Dam is the first dam upstream from Portland and McNary Dam is the first dam downstream of Richland. The Hanford Nuclear Reservation is just north of Richland on the west side of the Columbia River.

tabulated the available data on concentrations of these metals in the Columbia River and its estuary.

Aliphatic and Polynuclear Aromatic Hydrocarbons

Aliphatic and polynuclear aromatic hydrocarbons (PAH) have been determined along with organic C and C/N ratios in Columbia River suspended and bottom sediments at several locations throughout the river drainage basin (Prahl, 1982) (Fig. 10.5). Plant wax alkanes dominate the sedimentary aliphatic hydrocarbon composition, with only trace amounts of the C_{15} and C_{17} compounds derived from plankton activity in the various reservoirs. Few, if any, diasteranes and triterpanes were present. The unresolved complex mixture characteristic of fossil fuel contamination was also of minor importance in most sediments. Columbia River suspended particulate material also contains a series of long chain saturated *n*-aldehydes likely derived from higher plant debris (Prahl and Pinto, 1987). The total *n*-alkanes are 1.5-1.9 times as abundant as the total *n*-aldehydes.

PAH concentrations were surprisingly high at certain sites where large Al smelters are located on or near the river. The elevated concentrations are attributed to PAH leakage from the large graphite electrodes used in the reduction process which converts molten Al ore into Al metal. Pyrolytic PAH are readily produced from such carbon electrodes (Neff, 1979) and the Columbia River basin is the production site for over one third of the U.S. supply of Al (Canby,

1978). Palmork *et al.* (1973) showed that sediments in a Norwegian fjord receiving effluent from an Al smelter contained up to 1,000 mg total PAH (kg dry sediment) $^{-1}$.

PAH concentrations in samples of suspended particulates collected near the Columbia River mouth showed that the PAH discharge of the river is large relative to estimates of off-shore PAH atmospheric deposition. The riverine input of PAH essentially balances PAH accumulation rates measured in shelf sediments, establishing the Columbia River as the major source of PAH to the adjacent Washington coastal zone (Prahl *et al.*, 1984).

Lignins and Sedimentary Organic Matter

Organic C sources and transport in the Columbia river have been examined by Hedges *et al.* (1984) based on analyses of lignin oxidation products, C/N elemental ratios and $^{13}C/^{12}C$ compositions of sedimentary organic matter. Vascular plant tissues account for an average of one-third of the total sedimentary organic matter in bottom sediments from 16 sites within the Columbia River and its tributaries. Nonwoody angiosperm tissues and gymnosperm woods are the major types of this plant debris.

Hedges *et al.* (1984) found a clear relationship between the types of local vegetation and the sedimentary lignin composition of the Columbia River tributaries. For example, results for the Crab Creek tributary showed the dominance of lignins from grasses and herbaceous plants, the characteristic vegetation of this arid portion of the drainage basin. Crab Creek lignins were quite different from lignins in sediments from the Willamette and Methow river tributaries which drain basins with extensive coniferous forests.

Hedges and Mann (1979b) presented results of similar measurements for seven cores collected on the southern Washington continental shelf which receives detrital constituents almost exclusively from the Columbia River. The same types of vascular plant remains found in river sediments also occurred in offshore sediments. However, the offshore sediments contained about twice as much lignin as results for river bottom sediments would predict. This lignin enrichment in coastal sediments could be due to preferential degradation of lignin-free organic matter. There was no evidence for significant chemical alteration of the lignin components in offshore sediments for periods up to 400 y. Alternatively, organic matter discharged by the river may be more lignin rich than that which accumulates in bottom sediments of the river.

Dahm *et al.* (1981) measured total (TOC), dissolved (DOC) and particulate (POC) organic C levels in the Columbia River at Bonneville Dam (230-km upstream) and at Kalama, Washington (128-km upstream) monthly from May 1973 to December 1974. They found that 89% of the exported TOC was dissolved. The average annual TOC contribution from the Columbia River drainage to the northeastern Pacific was estimated at 4.9 x 10 ^{10} mol C. The average TOC concentration was about 200 mM and concentrations showed minimal diel variations or upstream/downstream gradients. The relative constancy of the DOC during the year suggested that a significant fraction was refractory. The TOC flux of 4.9 x 10 ^{10} mol C y $^{-1}$ was less than 20% of that carried by the Mississippi River (Malcolm and Durum, 1976). The major difference is the 15-fold lower concentration of suspended POC in the Columbia River.

The fate of riverine DOC and POC in the estuary and the Columbia River plume has not yet been determined.

10.2.4 Riverine Inputs of Specific Chemicals to the Puget Sound Region

Radionuclides

^{210}Pb activities of suspended particulates collected near the mouths of seven rivers entering Puget Sound averaged 4.6 ± 2.3 dpm g^{-1} and were not significantly different among rivers (Carpenter *et al.*, 1984). Unlike the Washington coast, riverine input is a significant source of ^{210}Pb to Puget Sound sediments, accounting for about one-half the total supply. No data are available on riverine inputs of other radionuclides such as U, Th, or Pu isotopes.

Stable Metals

Arsenic concentrations in the Skagit, Stillaguamish and Snohomish rivers, which supply over 75% of the freshwater to Puget Sound, averaged 0.62 ppb dissolved As and 0.19 ppb particulate As (Table 10.4) (Carpenter *et al.*, 1978). Combining the average As concentration for each river and its flow rate gives a total river influx of As to Puget Sound of $3 \pm 1 \times 10^{7}$ g As y^{-1}, less than 20% that of inflowing seawater. Particulate matter from rivers with drainage basins downwind of the large smelter near Tacoma, Washington (Fig. 10.3) were contaminated with As (Carpenter *et al.*, 1978; Peterson and Carpenter, 1986; and references therein). The discharge-weighted, average As concentrations of suspended particulate matter from rivers with drainage basins downwind (northeast) of the smelter was 30 ± 9 ppm, while rivers to the south and west averaged 6.5 ± 1.2 ppm (Peterson and Carpenter, 1986; Crecelius *et al.*, 1975). Paulson *et al.* (1984) examined the concentrations and behavior of Fe, Mn, Cu and Cd in the Duwamish River downstream of the discharge of a large sewage treatment plant and in the estuarine mixing zone of this river. In the estuarine mixing zone, dissolved Fe and Cu were removed from solution, while Mn and Cd were desorbed from the riverine particles. The chemical forms of the metals rather than their source appeared to determine the extent of their participation in sorption-desorption reactions in the estuary.

These measurements of riverine stable metal concentrations are almost the only riverine data for trace metals based on collection and analytical techniques adequate to ensure that samples were not contaminated. Data for other stable metals were reported by Curl (1982); however, these data consist of scattered samples from various rivers and are not adequate to allow recognition of any regional trends or confident assessments of yearly-averaged, discharge-weighted, riverine inputs to Puget Sound.

Organic Compounds

Aliphatic hydrocarbon abundances in the Green-Duwamish River system have been partially characterized by Hamilton *et al.* (1984). Concentrations were measured in suspended matter and sediments in an effort to document sources, transport paths and fates of these compounds. Hydrocarbons derived from algae, higher plants, highway runoff, a secondary sewage treatment plant and sedimentary sources were distinguished.

TABLE 10.4 Total As, Mn and Al concentrations and As/Mn and As/Al ratios in solid phase materials. See Figure 10.13 for most station locations.

Material Type Source	As (ppm)	Mn (ppm)	Al (%)	As/Mn (x 10 2)	As/Al (x 10 4)	Ref.
Crustal materials						
granite (low Ca)	1.5	390	7.2	0.38	0.21	0
granite (high Ca)	1.9	540	8.2	0.35	0.23	0
shale	13	850	8.0	1.9	1.6	0
basalt	2.0	1500	7.8	0.13	0.26	0
Sediments						
deep sea						
carbonates	1.0	1000	2.0	0.10	0.50	0
clay	13	6700	8.4	0.19	1.6	0
Lake Washington						
average 0-1 cm	95	6200	5.5	1.5	17	1
station 11 surf.	99	6500	5.6	1.5	18	1
deep	11	1800	4.1	0.60	2.7	1
Puget Sound						
station 98 surf.	24	1200	7.8	2.0	3.1	1
deep	15	720	9.2	2.1	1.6	1
station 51 surf.	14	1200	7.8	1.2	1.8	1
deep	6.6	690	7.7	1.0	0.87	1
station 56 surf.	24	760	8.8	3.2	2.7	1
deep	11	560	--	2.0	--	1
station 39 surf.	37	340	11	7.4	5.0	1
deep	5.2	510	7.7	1.0	0.68	1
Dabob Bay surf.	9.0	1400	7.5	0.64	1.2	2
deep	7.2	620	6.6	1.2	1.1	2
Washington coast						
station 30 surf.	7.2	390	7.0	1.9	1.0	1
deep	9.5	420	7.9	2.3	1.2	1
station 67 surf.	9.0	530	8.3	1.7	1.1	1
deep	6.9	450	8.3	1.5	0.83	1
station 91 surf.	8.7	530	7.6	1.6	1.1	1
deep	6.6	350	6.7	1.9	1.0	1
Saanich Inlet						
station A surf.	9.4	890	10	1.1	0.94	1
4-7cm	13	1400	--	0.93	--	1
deep	4.3	500	--	0.86	--	1
station E surf.	16	1200	5.7	1.4	2.9	1
deep	1.3	370	--	0.35	--	1

(continued)

TABLE 10.4 Continued.

Material Type Source	As (ppm)	Mn (ppm)	Al (%)	As/Mn (x 10^{2})	As/Al (x 10^{4})	Ref.
Riverine suspended particulate material						
Columbia	19					3
Discharge to L. Wash.						
weighted average	18	900	6.0	2.0	3.0	4
Discharge to Puget Sound (north-northeast of smelter)						
Skagit	20		8.5		2.4	5,6
Stillaguamish	32		6.5		4.9	5,6
Snohomish	39		8.1		4.8	5,6
Duwamish	26		5.8		4.5	5,6
weighted average	30 ± 8.7					
Discharge to Puget Sound (south and east of smelter)						
Puyallup	5.8		7.9		0.60	5,6
Nisqually	8.6		8.1		1.1	5,6
Dosewallips	6.5	980	7.9	0.66	0.82	2
weighted average	6.5 ± 1.2					
Riverine bedload material						
Columbia	3.4					3
Water column suspended material						
Lake Washington						
0-2 meters	78	2400	1.7	3.3	46	4
50 meters	160	2300	2.6	7.0	62	4
Puget Sound						
Qtr. Mstr. Hbr.	73					7
Dabob Bay	9.9	900	6.0	0.53	1.6	2
Saanich Inlet (station A)						
within Mn PL	45	23000		0.02		8
below O_2/H_2S interface	45	880		5.1		8

PL = particulate layer.
(0) Turekian and Wedepohl (1961)
(1) Peterson and Carpenter (1986)
(2) Bennett (1980)
(3) This work
(4) Crecelius (1975)
(5) Crecelius *et al.* (1975)
(6) Carpenter *et al.* (1984)
(7) Crecelius (1974)
(8) Peterson and Carpenter (1983)

No data on amounts and types of PAH contributed by rivers entering Puget Sound have been published. The presence of coal fragments in some Puget Sound sediments was indicated by similarities in geochemical marker compounds and in compositional patterns of PAH and azaarenes in the sediments and in 16 western Washington coals (Barrick *et al.*, 1984). Sedimentary coal fragments were most likely transported from terrestrial coal beds to adjacent drainage basins by natural erosional processes, possibly accelerated by mining activities.

Measurements of chlorinated organic compounds in rivers entering Puget Sound have been almost exclusively limited to polychlorinated biphenyl (PCB) determinations in the

Duwamish River which discharges into Elliott Bay (Hafferty *et al.*, 1977). These PCB studies were motivated by the spillage of 1,170 liters (720 kg) of pure Aroclor 1242, a mixture of predominantly 3- and 4-chloro biphenyls, into the river near its mouth in September 1974. In 1976-1977 the PCB concentration of Duwamish River water averaged 22 ± 13 (n = 60) ng l^{-1}, 3-7 times greater than observed in Puget Sound surface waters (Pavlou and Dexter, 1979). PCB concentrations decreased linearly with increasing salinity, reflecting the mixing of the less saline, but more PCB-contaminated, river water with that of the bay (Pavlou and Dexter, 1979).

PCB concentrations in Duwamish River sediments collected in 1976-77 ranged from 0.01 to 7.0 ppm, with a mean of 2.0 ppm (Pavlou *et al.*, 1982). PCB concentrations in surface sediments of Elliott Bay decrease with distance from the river mouth (Pavlou and Dexter, 1979). Within Elliot Bay, PCB concentrations were approximately 10^{4} times greater on suspended particles than in ambient water. Additional PCB sources other than the 1974 spillage probably occur along the highly industrialized Duwamish River and its estuary but have not been evaluated. We are not aware of samplings more recent than 1977. However, the rate at which the Duwamish River estuary is cleansing itself of the 1974 PCB release is unknown.

10.2.5 Remaining Questions and Needed Research Concerning Riverine Inputs

Studies outlined above have provided only first order estimates of riverine inputs for some energy related chemicals of greatest concern. No data are available for chlorinated organic compounds in Columbia River water, suspended particulates, or bottom sediments. However, both the Columbia River itself and some of its major tributaries, such as the Willamette, drain agricultural regions where various DDT containing pesticides were extensively employed, and PCB emissions seem likely from urban areas such as Portland.

Departures from linearity in estuarine mixing curves, generally interpreted as indicating non-conservative chemical behavior (internal sources or sinks of the chemical), may also be generated simply by variations in end member concentrations and conservative mixing (Loder and Reichardt, 1980). This possibility has not yet been examined for any Washington coastal rivers but may be important. For example, the Columbia River's flow (and possibly its chemical properties) varies diurnally, with periodic storm events and seasonally. Chemical properties of the oceanic end member may also vary over different time scales due to seasonal differences in upwelling intensity.

No data were found on U, Th, or Ra isotopes in Columbia River waters and suspended sediments. ^{228}Ra may be a particularly useful tracer of near-bottom sediment water exchange processes on the adjacent shelf (Li *et al.*, 1979). Only the limited stable metal data mentioned above for the Columbia River were obtained with contamination controls equivalent to those now deemed necessary for environmental samples. Consequently, there is a shortage of reliable data for most stable metals in the Columbia River and its estuary.

Data on the extent to which trace chemicals associated with riverine particulate matter reequilibrate with different aquatic environments or are available to biota remain scant. Studies of Hanford-derived radionuclides revealed extensive ^{54}Mn desorption from Columbia River sediments upon contact with seawater, while ^{65}Zn remained largely bound to the sediments.

Partial desorption of Hg and As bound to particulates is dependent on their dissolved concentrations (Bothner and Carpenter, 1973; Carpenter *et al.*, 1978). Major fractions of the PAH in Washington shelf sediments were associated with relatively large, low density particles, such as fly ash, soot and charred wood and likely cannot equilibrate with surrounding waters (Prahl and Carpenter, 1983).

Many of the same gaps in knowledge of Columbia River inputs are also found in our current understanding of riverine chemical inputs to the Puget Sound region. Sediment discharges of rivers entering Puget Sound are poorly known and all efforts to set up budgets of sediments and associated chemicals are limited by lack of knowledge of the fraction of the largest river discharge (the Skagit) which reaches central Puget Sound. We are aware of no data on U, Th, Pu, or Ra isotopes in these river systems and almost no reliable stable trace metal data. There is no published information on aliphatic and aromatic hydrocarbons for the major rivers and data on chlorinated organic compounds are limited to PCBs in the Duwamish River system.

10.3 CHEMICAL INPUTS FROM ADVECTING SEAWATER

10.3.1 Overview

Seawater advection has long been recognized as the major source of nutrients and salt to the open Washington coast and Puget Sound. Nutrients supplied by wind-induced seasonal upwelling support high planktonic productivity in summer (see Landry *et al.*, 1989, Chapter 1 Hickey, 1989, Chapter 2; Perry *et al.*, 1989, Chapter 2; Hermann *et al.*, 1989, Chapter 6) and salt plays an important role in establishing density gradients. Concentration profiles of nutrients and O_2 off the open Washington coast have not changed since at least 1961 (Fig. 10.6). The O_2 minimum lies within the depth range of the continental slope and exerts an influence on processes such as respiration within the underlying sediments. Central Puget Sound data show no discernable change in concentration of dissolved phosphate or O_2 between 1934 and 1973 (Duxbury, 1975).

Temperature, salinity and ^{14}C data demonstrate that the Strait of Juan de Fuca and associated inland waters, such as Puget Sound and Saanich Inlet, are supplied with northeast Pacific water from an average depth of about 250 m (Buddemeier, 1969). Deep water entering the mouth of the Strait of Juan de Fuca originates from the south and along Juan de Fuca Canyon from the southwest (Fig. 10.1) rather than from the north or northwest. This inflow path is supported by the observation that bottom drifters released off the mouth of the Columbia River were retrieved three months later inside the Strait (Barnes *et al.*, 1972). Temperature, salinity and current measurements of Hickey (1979; 1989, Chapter 2) have identified a poleward jet-like subsurface flow at depths of 200-300 m seaward of the Washington and Oregon shelf break. This water provides nutrients, via upwelling, to the Washington coast and probably most of the flow into the Strait of Juan de Fuca (see Hickey, 1989, Chapter 2).

10.3.2 Lead, Plutonium and Uranium Radionuclides

Advective supply of dissolved trace chemicals to the Washington coastal zone was assumed minor relative to Columbia River input and neglected until the mid-1970s. It is now

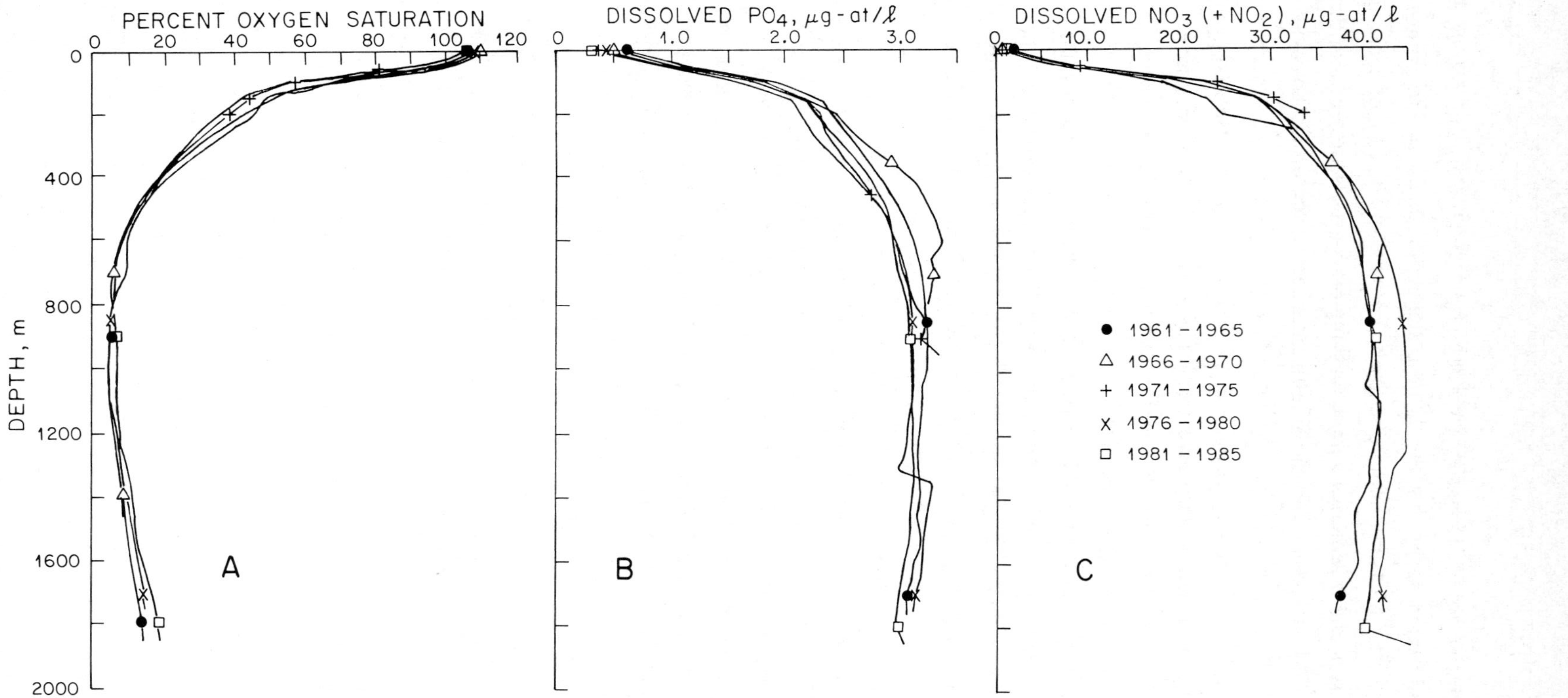

Fig. 10.6 Summary of 5-y mean concentrations of (A) dissolved O_2 expressed as $\%O_2$ saturation, (B) phosphate, and (C) nitrate + nitrite as a function of water depth off the Washington coast. The sum of nitrate + nitrite was used since methods employed in earlier years did not distinguish between the two forms. The same basic methods were used for O_2 and phosphate over the entire time period. Data were grouped for 50-m intervals, and symbols are placed on curves merely for visual reference. The profiles were kindly provided by J. Postel.

clear from ^{210}Pb (Carpenter *et al.*, 1981) and Pu (Beasley *et al.*, 1982; Carpenter *et al.*, 1987 studies, however, that advection is by far the largest source of these two chemicals to Washington coastal sediments. Measured atmospheric and riverine inputs cannot account for observed ^{210}Pb and Pu fluxes and inventories within Washington coastal sediments. Although data on dissolved Pu or ^{210}Pb advecting into the Washington coastal region are not available, conservative estimates based on water volumes advecting through the region and concentrations observed in other North Pacific waters suggest that the advective supply of these isotopes considerably exceeds their observed sedimentary inventories and fluxes. Sedimentary inventories of these chemicals on the Washington coast are therefore limited by boundary-layer scavenging reactions rather than by supply of dissolved chemicals.

In open ocean waters deeper than ~100 m, where particulate concentrations are below 0.1 mg l^{-1}, particle reactive elements like Pb and Pu are primarily in dissolved rather than particulate form. Even when the equilibrium partition coefficient (K_D) of an element between particulate and dissolved phases is as high as 10^7, over 50% of the element should still be in dissolved form when particle concentrations are ≤ 0.1 mg l^{-1} (Beasley *et al.*, 1982). Significant scavenging of particle reactive, dissolved chemicals occurs when open ocean water enters the Washington coastal zone and encounters 10-100 times higher particulate concentrations due to the Columbia River discharge and the biologically-productive near-shore environment. Frequent resuspension of bottom sediments by storms, bottom currents and benthic organisms also increases particle concentrations and enhances scavenging in near-bottom waters. Boundary scavenging, first suggested by Bacon *et al.* (1976), has only been confirmed for ^{210}Pb and Pu off the Washington coast, but should also be an effective mechanism for removal of other particle reactive elements such as Be (Brown *et al.*, 1985).

Because of boundary scavenging, sedimentary inventories and fluxes of excess ^{210}Pb (Carpenter *et al.*, 1981) and Pu (Beasley *et al.*, 1982; Carpenter *et al.*, 1987) off Washington are 4 to 15 times greater than off the U.S. east coast. This enhanced boundary scavenging off the Washington coast is attributed to a combination of the following: 1) concentrations of many dissolved chemicals are greater in mid-depth North Pacific than North Atlantic waters; 2) upwelling and advection of deep water occurs along the Washington coast than off the east coast; 3) particulate concentrations, fluxes and sediment accumulation rates are higher in the Washington coastal region; 4) Columbia River fine-grained particles are relatively rich in montmorillonite (Knebel *et al.*, 1968) and hence are more reactive than east coast continental margin sediments that are dominated by non-expandable clays (Hathaway, 1972); 5) the fraction of biological particles produced in the overlying water column which accumulates in Washington slope sediments may be greater than on the east coast, because O_2 concentrations are much lower at mid-depths off the Washington coast; and 6) lower O_2 concentrations and higher organic C fluxes to Washington coast sediments may lead to greater dissolved Mn fluxes from the sediment to the overlying water, enhancing trace element scavenging by hydrous Mn oxides near the sea floor (Carpenter *et al.*, 1981). Advecting seawater is less important as a source of dissolved particle reactive elements like ^{210}Pb to Puget Sound because of in-transit scavenging along the relatively particle-rich, 140-km long Strait of Juan de Fuca.

Advection is the major source of dissolved U to both the open Washington coast and to Puget Sound, since U is not efficiently scavenged onto particles under oxic conditions (Carpenter *et al.*, 1984; 1987). A fraction of this advected U is removed to coastal sediments by sediment pore-water exchange with overlying water, followed by U reduction in pore waters to the less soluble tetravalent state. This flux of U from the open ocean into margin sediments is believed to be an important sink for U on a global scale.

10.3.2 Carbon Isotopes

No data are currently available for advective carbon isotope supply to the Washington coastal zone. ^{14}C measurements were made in 1966-1968 on waters collected from the Canadian Gulf Islands, Saanich Inlet and Haro and Juan de Fuca straits (Buddemeier, 1969) (Figs. 10.1 and 10.3). Studies of the varied sediments in Saanich Inlet revealed that, prior to 1950, surface waters of the Strait of Georgia, and presumably Puget Sound, had $\Delta^{14}C$ values of about –110 ‰, corresponding to an apparent age of about 1,000 y $^{-1}$ (Buddemeier, 1969; Yang, 1971) (see Table 10.7, Section 10.6, below for unit definitions). Water with similar age characteristics occurs at a depth of several hundred meters in the northeast Pacific (Fairhall and Young, 1985). Therefore, older water appears to flow into Saanich Inlet from offshore via the Straits of Juan de Fuca and Georgia.

Water movement in and out of the Strait of Juan de Fuca is dominated by an inflow of high salinity (33.4-33.95 ‰) seawater at depth and an outflow of lower salinity water on the surface. Dilution of surface seawater arises primarily from the Fraser and other rivers along the eastern boundary of the Strait. Circulation in the Strait of Georgia tends to be counterclockwise, with low salinity water exiting via Haro Strait and the northern side of the Strait of Juan de Fuca (Schumacher *et al.*, 1978; and references therein). Thus, relatively ^{14}C poor water flows into the Strait of Juan de Fuca at depth and is gradually upwelled and mixed with low salinity water before flowing back out as a highly modified surface layer. ^{14}C activities in waters collected in Juan de Fuca Strait in 1975-1977 were 10-12% higher than in waters collected in 1966-1968 (Buddemeier, 1969) due to some addition of ^{14}C from weapons testing to intermediate depth waters offshore.

^{14}C activities in inflowing waters strongly depend on the season. During summer months, wind-induced coastal upwelling offshore feeds cold, dense, deep water low in ^{14}C into the Strait of Juan de Fuca and Puget Sound. During winter months winds predominantly from the southwest produce downwelling and a northward current along the Washington (Hickey, 1989, Chapter 2). The stronger winds also enhance air-sea exchange of CO_2. Both effects tend to enrich water entering the Strait in ^{14}C during winter months.

From May 1975 to December 1977, at about four-month intervals, 128 water samples were collected at stations shown in Figure 10.7. The samples were analyzed for carbon isotopes by A.W. Fairhall and J. Erickson. Results are partially summarized by Swanson (1980). The greatest variability observed was in surface waters from the eastern part of the Juan de Fuca and

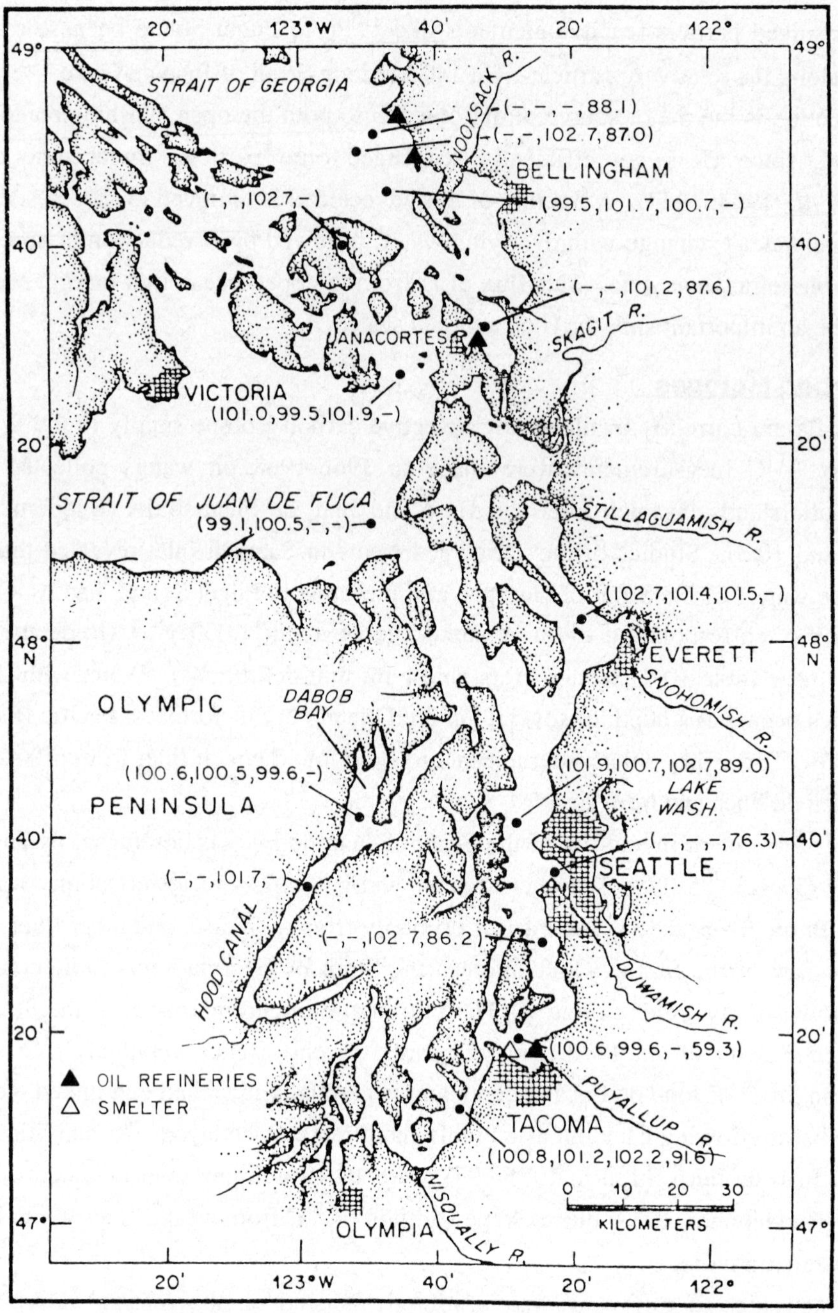

Fig. 10.7 Radiocarbon activities in seawater, plankton and surface sediments of the Puget Sound region collected during 1975-77. Sample activities are given as the % of the activity of the contemporary NBS oxalic acid standard x 0.95. Activities are listed in order for: total dissolved CO_2 in surface waters; total dissolved CO_2 in bottom waters; total C in planktonic organisms; and total C in surface sediments. A dash in the sequence indicates no data for that parameter. This figure combines data in Figures 11, 14 and 18 of Swanson (1980).

Rosario straits, where $\Delta^{14}C$ values as low as –27 ‰ and as high as +31 ‰ were measured ($\Delta^{14}C$ range = 6%). Low values were associated with the tongue of ^{14}C poor water entering the

Strait during the period of strong coastal upwelling in summer. High values were observed when upwelling was absent and the surface layers were stabilized with freshwater inputs to the Puget Sound system. Elsewhere, the extreme values measured were: –2 to +28 ‰ off Seattle, +6 to +25 ‰ near Anacortes, –4 to +34 ‰ off Cherry Point and +22 to +66 ‰ off Everett.

Puget Sound surface and near bottom waters in 1975-1977 had mean ^{14}C activities of 102.0 ± 4.9% and 100.8 ± 1.6%, respectively, of the activity of the NBS oxalic acid radiocarbon standard x 0.95. Input of ^{14}C from nuclear weapons testing during the late 1950s and early 1960s has caused atmospheric ^{14}C activities in excess of 133% during the past 25 y. Mean activities of around 101% for Puget Sound waters clearly reveal that inflowing seawater must come from several hundred meters deep and cannot be offshore surface water. Since uncertainties in mean ^{14}C activities in Puget Sound surface waters are at most ± 3%, ^{14}C data on organic matter extracted from organism growing in these waters should reveal if fossil fuel contamination comprises more than a few percent of the total C. ^{14}C data for Puget Sound sediments and organisms are summarized in Sections 10.6 and 10.7, respectively.

10.3.3 Trace Metals

Jones and Murray (1984) reported total concentrations of Ni, Cd and Cu in unfiltered seawater along a 500-km transect normal to the Washington coast at 47° N latitude. Ni concentrations increased from an average of 5.5 nmol kg^{-1} at the surface to 10 nmol kg^{-1} in deep waters and showed a dual covariance with concentrations of dissolved phosphate and silica. An additional Ni source from the sediments was invoked to explain a strong concentration gradient between coastal and open ocean water. Biological cycling of organic matter controlled the geochemistry of Cd in these waters. Cd concentrations were depleted to 0.1 mol kg^{-1} at the surface, but increased ten-fold to a maximum of around 1.3 nmol kg^{-1} at the depth of the phosphate and nitrate maxima. Cd-phosphate and Cd-nitrate correlations were significantly different in near-shore and offshore waters. Cu data were more scattered and showed no correlation with nutrient concentrations. An apparent surface maximum may result from upwelling and offshore advection of Cu-rich shelf water. Cu is rapidly stripped from subsurface waters and returned to bottom waters via dissolved Cu from sediment pore water.

Jones and Murray (1985) reported dissolved (< 0.4 μm) and total dissolvable Mn in waters along the transect normal to the Washington coast. Vertical profiles exhibited the same general features observed elsewhere in the Pacific, namely,aximum concentrations at the surface, in the O_2 minimum zone and in deep waters near the hydrothermally-active Juan de Fuca ridge system. Mn concentrations throughout the water column increased markedly nearer the continental margin. The surface water maximum in Mn was attributed to a combination of Columbia River input (with about 75 nmol kg^{-1}) and to upwelling of Mn-enriched bottom water on the shelf. The Mn concentration maximum coinciding with the O_2 minimum appeared to primarily result from lateral advective-diffusive transport of Mn diffusing out of mildly reducing slope and shelf sediments. Regeneration of Mn from sinking biogenic particles was judged to be of lesser importance. Pore water and solid phase Mn data support the hypothesis that Mn is being actively remobilized under reducing conditions in slope sediments. Deep water concentration maxima near the Juan de Fuca Ridge (Fig. 10.1) were attributed to the injection of hydrothermal fluids rich in dissolved Mn (Jones *et al.*, 1981).

Cranston (1983) reported amounts and speciation of Cr in the water column of Cascadia Basin off the Washington coast (Fig. 10.1). All particulate Cr concentrations were less than the detection limit of 0.24 nM, as were 49 of the 53 Cr^{3+} concentrations. The Cr^{3+} presence in 4 of the 53 samples was attributed to some leakage or excretion from biological material, coupled with slow oxidation of the thermodynamically unstable Cr^{3+} to Cr^{6+}. Dissolved Cr^{6+} thus accounted for 95 ± 5% of the total Cr in the Cascadia Basin water column.

Total Cr concentrations were in the 1.8-4.0 nM range and tended to increase with depth in the water column. Cr correlations with nutrients were positive, the highest correlation being with dissolved Si. An increase in total Cr approaching the sediment-water interface was seen at all stations, and deep Cr concentrations increased toward the continental margin. This behavior suggests diffusion of Cr into bottom water from sediments, possibly as a result of dissolution of siliceous tests.

10.3.4 Arsenic

Amounts and forms of arsenic have been of special interest in the Washington coastal zone because As is toxic and large amounts have been introduced to this environment by a smelter near Tacoma, mining operations in the upper Columbia River drainage basin and possibly by hydrothermal activity along the Juan de Fuca Ridge spreading center. The two main valence states of concern, As^{3+} and As^{5+}, have different toxicities, particle reactivities and stabilities in oxidizing and reducing environments. While the As atom itself has a positive valence state, As exists in several neutral or anionic species depending on the Eh or pH of the environment. Hence, the terms *arsenate* for oxidized species containing the AsO_4^{3-} anion ($H_2AsO_4^-$, $HAsO_4^{2-}$) and *arsenite* for reduced species containing the AsO_3^{3-} anion (H_3AsO_3, $H_2AsO_3^-$). As^{3+} may also exist in a soluble sulfidic form, AsS_2^-.

Determinations of dissolved As speciation in seawater and pore water were based on a pH selective reduction to arsines coupled with boiling point difference separation (Braman *et al.*, 1977) and photoionization or emission detection. Accuracy of the speciation method was confirmed by comparing the sum of arsenite and arsenate concentrations with total As concentrations independently determined by a neutron activation technique. Shipboard determinations avoid problems with contamination or speciation changes during storage. Results for water columns and pore waters of the Pacific Northwest are given by Peterson and Carpenter (1983, 1986).

Analyses made prior to 1980 employed a DC-arc emission detection system. Subsequently a photoionization detection (PID) system was developed to improve sensitivity and eliminate the need for in-line traps for CO_2 and H_2S, both of which strongly interfere with the arsenic emission signal. The PID system can easily be operated on board ship, in contrast to atomic absorption detectors used by Andreae (1979). The system, which is shown schematically in Figure 10.8, was designed to minimize dead volume (especially after the AsH_3 trap),

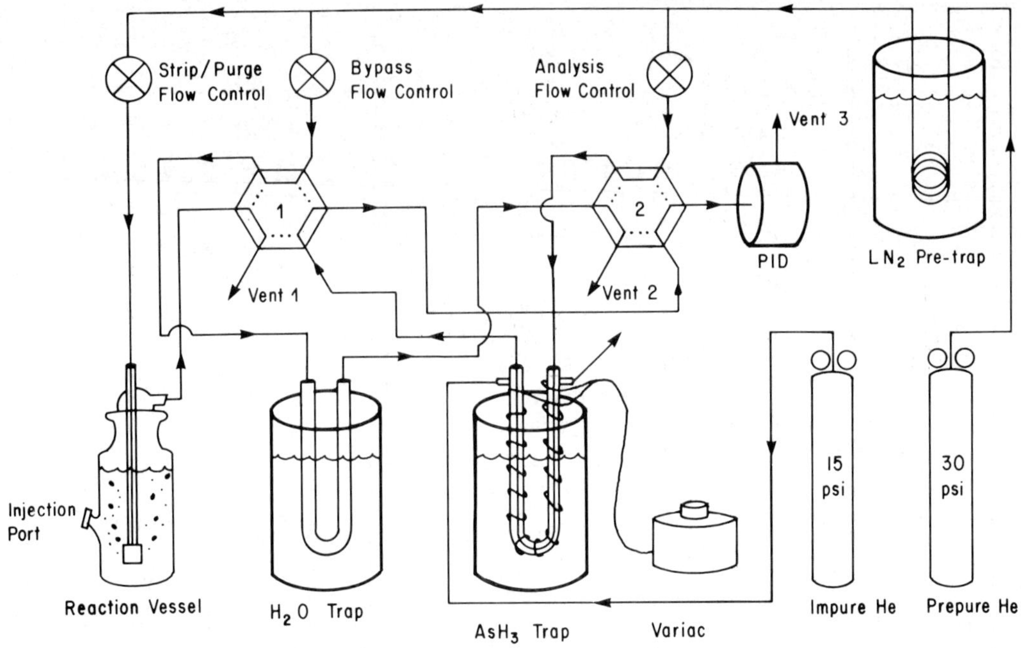

Fig. 10.8 Gas flow diagram for As speciation apparatus shown in the sample-purge/detector-analysis position. All tubing is 3.2-mm o.d., 1.5-mm i.d. Teflon; the six-way valves are Rheodyne and all fittings are Alltex or Swagelok. Each gas flow rate can be measured at vents 1, 2 and 3. An inline water trap, located immediately downstream of the reaction vessel, consists of a 10-mm i.d. glass U-tube immersed in a dry ice-isopropyl alcohol bath (–80° C). The AsH_3 trap was made by inserting the 3.2 mm Teflon tubing through a 70-cm long glass U-tube (6-mm o.d., 4-mm i.d.) wrapped with 24 gauge nichrome wire (5 turns cm^{-1}). The downstream end of the glass U-tube was adapted so that a He gas flow (400 ml min^{-1}) could be maintained around the Teflon tubing inside. This allows for more uniform heating of the column after its removal from the liquid N_2. The downstream half of the Teflon U-tube was packed with glass beads (100 mesh).

reduce the flow of substances other than those of interest (primarily water vapor, HCl and H_2S) through the PID and maintain a constant He flow through the PID at all times. For maximum efficiency the system allows for different He flow rates: a sample purging and stripping flow (150 ml min^{-1}), a bypass flow (250 ml min^{-1}) and an analysis flow through the PID (50 ml min^{-1}).

Initially 50 ml of sample solution (or a few ml of pore water diluted to 50 ml) is placed in the reaction vessel and attached to the system's gas dispersion tube via a ground glass connection. The two six-way valves are set in the sample-purge/detector-bypass configuration for 2 min to remove air and other volatile species from the system. The liquid nitrogen (LN_2) trap is placed around the column. Valve (1) is changed to the sample-strip position and 5 ml of a 0.34 M sodium citrate in 0.1 N HCl buffer solution is injected into the vessel to adjust the pH to between 5-6. Three 1 ml injections of 2% $NaBH_4$ in 0.02 N KOH are made 2 min apart. Three minutes after the last injection, both valves are rotated to the sample-purge/detector analysis configuration (shown in Fig. 10.8). The LN_2 AsH_3 trap is removed and the variac switched on, driving off the AsH_3 gas through the PID. After elution of the AsH_3 peak both valves are again rotated leaving the system in the sample-strip/detector-bypass mode, the variac is switched off

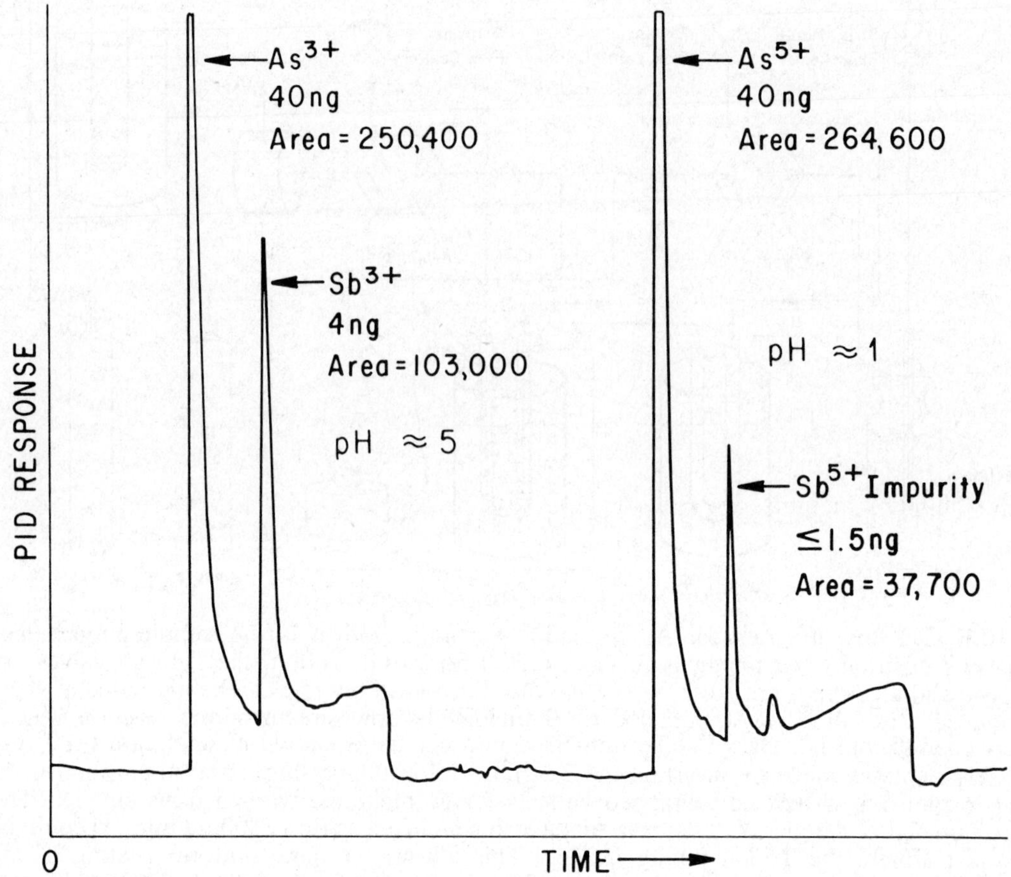

Fig. 10.9 Separation of nanogram amounts of As $^{3+}$, Sb $^{3+}$, As $^{5+}$ and Sb $^{5+}$ achieved with the pH selective reduction procedures described in the text and with the gas flow system shown in Figure 10.8.

and the LN_2 trap replaced around the AsH_3 trap column. Reduction of As $^{5+}$ is then begun by adjusting the pH to 1.0 with 5 ml of 2N HCl followed by three 2 ml injections of the $NaBH_4$ solution at 2-min intervals. From this point on, the analysis for As $^{5+}$ is the same as for As $^{3+}$. After elution of the AsH_3, valve (2) is rotated to the detector-bypass position and the system allowed to dry out in the sample-purge/detector-bypass mode while the next sample is attached to the system and purged.

Arsenic is one of several elements occurring in different valence states which can be isolated from one another using this combination of pH dependent hydride formation and boiling point separation. For example, Sb $^{3+}$ and Sb $^{5+}$ are resolved from each other and from As $^{3+}$ and As $^{5+}$ by this technique (Fig. 10.9). Sb is about 5 times more sensitive to photoionization than As – a fortunate situation since Sb seawater concentrations are usually about 10 times lower than As. This approach has been used to characterize different forms of Se across the redox transition in the Saanich Inlet water column (Cutter, 1982). Significant amounts of organo-Ge compounds have been reported in deep-sea waters (Hambrick *et al.*, 1984; Andreae and

Froelich, 1984). The compounds could be distinguished separately from inorganic Ge forms by minor modifications of the above system.

Total As and As species concentrations in water columns at three Washington coast stations are shown in Figure 10.10 (See Peterson and Carpenter, 1986 for station locations). Total As concentrations increase from about 1.0 ppb (13 nM) in surface waters to between 1.5 and 2.0 ppb at depth, probably due to uptake and release by plankton. Changes in As concentrations with depth are generally smaller than those of phosphate. This is consistent with the observation of Andreae (1979) that phosphate is more depleted in open Pacific Ocean surface water than As, a result of As incorporation into the cellular structures of primary producers being less efficient than P incorporation.

As^{5+} is thermodynamically favored in oxygenated waters and comprises most of the As in waters off Washington. However, small amounts of thermodynamically unstable As^{3+} are commonly present in these, as well as other waters (Waslenchuk, 1978; Andreae, 1979; Sanders, 1980; and references therein). The presence of As^{3+} is attributed to leakage from plankton and microbes, which attempt to detoxify As^{5+} by transforming it to As^{3+}. The good agreement between the two independent total As determinations supports the conclusion that methylated forms of As are not significant in Washington coastal waters, in contrast to the euphotic zones off California (Andreae, 1979) and Georgia (Waslenchuk, 1978; Sanders, 1980; and references therein). In Washington coastal waters, generally higher phosphate concentrations result in less As uptake and methylation by plankton.

Similar As concentrations and speciation were observed in oxygenated waters of Puget Sound (Carpenter *et al.*, 1978). Speciation of As changed from predominantly As^{5+} to predominantly As^{3+} when waters became anoxic, as at depth in Saanich Inlet (Peterson and Carpenter, 1983) and the Baltic Sea (Andreae and Froelich, 1984). This speciation change is in the direction predicted for thermodynamic equilibrium at the different pHs, but equilibrium As^{3+}/As^{5+} concentration ratios were not observed.

10.3.5 Mercury

Bothner (1973) determined total Hg in seawater from a station 2,250-m deep on the Washington slope using samples collected and treated to minimize gain or loss of Hg during storage. All concentrations were less than 20 $^{o}/_{oo}$ (0.1 nM), barely above blank levels with the cold vapor atomic absorption technique employed. Trends in Hg concentrations with water depth were not evident but may have been obscured by the relatively poor signal-to-noise ratio. Hg studies elsewhere have reported differently shaped Hg profiles, different covariations with nutrient concentrations, differing fractions of reactive Hg and even order of magnitude differences in Hg concentrations (Bloom and Crecelius, 1983a; Dalziel and Yeats, 1985; and references therein). To improve sensitivity, future studies of Hg cycling in Washington coastal waters will require a preconcentration step such as the gold amalgamation technique (Bloom and Crecelius, 1983; Dalziel and Yeats, 1985). However, the gold amalgamation procedure probably gives less than true values for total Hg in waters with more than a few mg l^{-1}

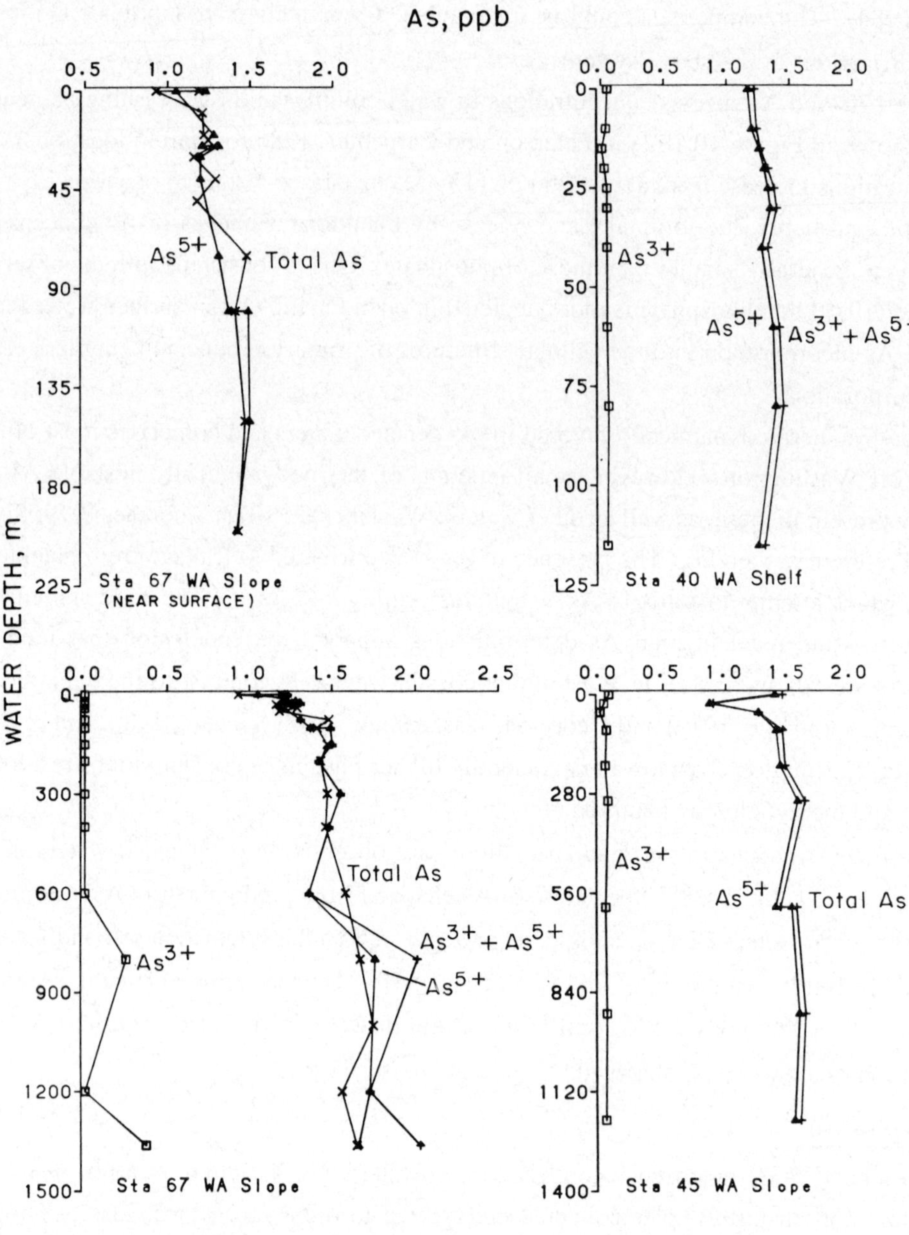

Fig. 10.10 Concentration profiles of total arsenic, As $^{3+}$ and As $^{5+}$ in seawater off the Washington coast. Concentrations of As $^{3+}$ and As $^{5+}$ were determined by pH selective generation of arsines and photoionization detection, while total As was independently determined by instrumental neutron activation analysis (see text).

particulate concentrations, since an unknown fraction of particulate Hg is recovered during sample digestion. Easily reducible Hg was less than 0.5 ng l $^{-1}$ and total Hg determined by three different techniques was less than 1.0 ng l $^{-1}$ in water samples collected in 1980-81 from rural Sequim Bay, the Pillar Point area on the Strait of Juan de Fuca and industrialized Puget Sound (Bloom and Crecelius, 1983; Fig. 10.3).

10.3.6 Other Trace Metals

Paulson and Feely (1985) used clean collection and analytical techniques to determine total dissolved Cu, Cd, Zn, Ni and Pb concentrations in waters at the Admiralty Inlet entrance to Puget Sound and at several sites within the Sound. Concentrations of Cu, Zn and Pb were 4 to 5 times greater in surface waters near Seattle and Tacoma than in inflowing seawaters at Admiralty Inlet. In contrast, concentrations of Ni and Cd appeared unaffected by pollutant discharges from Seattle and Tacoma.

10.3.7 Summary

In summary, for waters advecting into the Washington coastal region, reliable metal concentration data exist only for As, Cr, Cu, Cd, Ni and Mn. Concentrations of these metals are affected to varying degrees by biological removal from surface waters and partial regeneration and release to deeper waters. Near the coast, Ni, Cr, Cu and Mn have increasing concentrations in near-bottom waters, suggesting release from slope sediments. Speciation determinations have been made only for As and Cr. Results indicate that thermodynamic equilibrium between different oxidation states of these metals is approached but not perfectly achieved.

Reliable measurements of total dissolved As concentrations of 1.6 to 1.7 $\mu g\ kg^{-1}$ in seawater at the eastern end of the Strait of Juan de Fuca, which may advect into Puget Sound, have been made by Carpenter *et al.* (1978). Stukas and Wong (1981) report dissolved stable Pb concentrations of about 20 $ng\ kg^{-1}$ in waters at the eastern end of this strait. True total Hg concentrations in Washington coastal waters appear less than the < 10-20 $ng\ l^{-1}$ reported by Bothner (1973). The only other data on concentrations or forms of other metals in waters of the eastern end of the Strait are for total concentrations of Cu, Cd, Ni, Zn and Pb (Paulson and Feely, 1985).

10.3.8 Remaining Questions and Needed Research Concerning Advective Supplies of Trace Chemicals

Concentrations of dissolved radionuclides (^{210}Pb and Pu), other trace metals and trace organic compounds like the PAH all need to be determined in open ocean waters which advect into the Washington coastal zone. Such measurements should be made on waters collected off southern Oregon in close collaboration with physical oceanographic studies to ensure sampling of the seawater end member before it is affected by Columbia River discharge. PAH fluxes from the Columbia River to adjacent Washington shelf sediments suggest that advecting seawater is an unimportant source of PAH to the Washington coast. However, no data on dissolved PAH or other trace organic compounds in seawaters are available to test this hypothesis.

Determinations of dissolved and particulate metal concentrations in Washington coastal waters and associated equilibrium partitioning coefficients would allow better modeling of particle reactive chemicals in this region where particulate concentrations vary at least 100-fold. They would also allow more precise estimation of metal inventories and budgets.

Collection and analytical techniques for stable trace metals dissolved in seawater were greatly improved during the 1970s (Wong *et al.*, 1983; Bruland, 1983). Most dissolved trace metal results for seawater prior to circa 1975 are now believed too high because of contamination during sample collection and workup. Procedures for the analyses of U, Th, ^{210}Pb, Ra and even Pu radioisotopes are also reasonably well established. Determinations of transuranic nuclides in seawater require large volumes of water, painstaking attention to minimize blanks and long counting periods. However, Mann *et al.* (1984) recently reported the use of MnO_2 impregnated fiber filters in a pumping system for *in situ* chemisorption of transuranic nuclides from large volumes of seawater. Determination of chemical speciation is less advanced and requires further development since speciation may control biological effects and inorganic reactivity. Reactivity of dissolved Pu is particularly dependent upon its valence states. A start at quantifying this dependency can be made based on methods of Nelson and Lovett (1978), Wahlgren *et al.* (1978) and Orlandini *et al.* (1986).

A generally accepted protocol does not exist for the collection of large volumes of open ocean water needed for analysis of trace organic compounds. Plastic hydro-bottles developed for trace metal studies may contribute unacceptably high blanks for some trace organic compounds. Organic-rich sea surface slicks must be prevented from affecting deeper water samples. Storage of large water volumes without changing the trace organic molecular composition is difficult, if not impossible, but extractions aboard ship remain highly susceptible to contamination from the prevalent fumes and exhaust. Methods based upon *in situ* pumping of large water volumes through various kinds of resins selective for different classes of compounds appear promising and may solve some of these difficulties. The operational distinction between dissolved and particulate forms is even more arbitrary for organic compound classes than for metals, because many organic colloids and bacteria pass through most filters.

Estimation of the advective supply of dissolved chemicals to Puget Sound is limited by the lack of reliable data on trace chemical concentrations in waters at the eastern end of the Strait of Juan de Fuca. There is also a large uncertainty in the extent to which outflowing surface water is mixed with inflowing deeper water across the sill at Admiralty Inlet (Ebbesmeyer and Barnes, 1980). Recycled surface waters may be as much as one to two thirds of the total inflowing water. Physical oceanographic studies aimed at reducing this uncertainty are hampered because a tracer specific to recycled surface waters has not yet been identified.

10.4 AIR-SEA EXCHANGE OF CHEMICALS

10.4.1 Overview

No studies of chemical exchange across the air-sea interface were reported in Pruter and Alverson (1972). Since then, development of electron capture and flame photometric detectors for gas chromatography has greatly improved detection sensitivities for many trace gases, enabling reliable quantitation of low molecular weight gases in air and seawater samples. A surprising number of low molecular weight gases, including CO, CH_3 SCH_3, CH_3 I and even $CHCl_3$, are now known to be produced in significant quantities within surface seawaters by a combination of biological and photochemical processes. Freons, which may destroy part of the stratospheric ozone layer and some trace S- and N-containing gases (both naturally and anthropogenically derived), which may contribute significantly to the the acidity of atmospheric precipitation, can now be routinely measured. Depth profiles of synthetic, low molecular weight gases with known production histories (i.e., Freons) are used as tracers for rates at which fossil CO_2 may be entering the surface waters of the ocean and mixed to depth.

The net air-to-sea transport of continentally-derived chemicals to the Washington coast north of the Columbia River is expected to be of relatively small importance for two reasons. First, during most of the year, westerly winds blow from offshore onto the land. Transport of chemicals during the relatively infrequent easterly winds is partially blocked by the 2,000-2,500-m high Olympic mountain range. Easterly winds only occasionally blow down the Columbia River gorge and carry terrestrial debris out over the Washington coast. Second, large chemical inputs from the Columbia River reduce the relative importance of air-to-sea fallout. Air-to-sea transport of anthropogenic chemicals is relatively more important within the Strait of Juan de Fuca and Puget Sound regions, which are partially bordered by urban areas.

Only limited studies have been made of sea-to-air transport of chemicals off the Washington coast. However, the Washington coast location would be particularly well suited for such studies because: 1) there is a lack of large anthropogenic emissions on or near the coast; 2) the coastal ocean is an area of high biological productivity, but one with significant seasonal and spatial variations which should enhance our ability to identify major chemical producing processes; 3) prevailing westerly winds off the ocean and high precipitation along the coast enhance deposition of compounds introduced to the atmosphere from the ocean on the adjacent land; and 4) many areas in the Pacific Northwest may be ecologically sensitive to acidic precipitation, because the surface geology consists of thin topsoil and resistant rock with little buffering capacity (Logan *et al.*, 1982).

10.4.2 Air-to-Sea Transport of Specific Chemicals

Lead-210 and Stable Lead

Atmospheric depositional fluxes of excess ^{210}Pb to the Washington coast were estimated from integrated inventories of ^{210}Pb in eight undisturbed terrestrial soil cores collected along the coast at seven locations between the Columbia River mouth and Neah Bay (Fig. 10.1) (Carpenter *et al.*, 1981). This approach has the advantage of integrating both wet and dry deposition of ^{210}Pb over a long time period. The ratio of the ^{210}Pb flux determined from soil cores to the average rainfall for the Washington coast was similar to that for other west coast locations (Fuller and Hammond, 1983), suggesting that rainfall probably dominates ^{210}Pb deposition. The average ^{210}Pb flux for the soil cores (0.73 ± 0.31 dpm cm $^{-2}$ y $^{-1}$) agrees well with the average of 0.65 dpm cm $^{-2}$ y $^{-1}$ based on 12 direct analyses of precipitation in Seattle by Schell (1977) and with the seven-year mean atmospheric flux in Seattle of 0.44 ± 0.24 dpm cm $^{-2}$ y $^{-1}$ (Nevissi, 1985). Excess ^{210}Pb fluxes of this magnitude are small compared with total fluxes to sediments of the Washington continental shelf and with the input from advecting seawater (Carpenter *et al.*, 1981).

Fluxes of excess ^{210}Pb to Puget Sound estimated from analyses of rain and snow, advecting seawaters and riverine discharges indicate that atmospheric input of ^{210}Pb is relatively more important to the Sound than to the Washington shelf and slope (Carpenter *et al.*, 1984). This is primarily due to smaller inputs to Puget Sound by inflowing seawater.

Profiles of stable Pb concentrations in Puget Sound sediments commonly show increases in recently deposited layers compared to those deposited prior to 1850 (METRO TPSS Report 1984; Bloom and Crecelius, 1987 and references therein). The Puget Sound Air Pollution Control Agency reported quarterly averages during 1978-79 for total atmospheric particulate concentrations and their Pb contents for several sites around the Puget Sound main basin. Stable Pb atmospheric depositional fluxes calculated by combining these concentration data with literature values for air-to-sea particulate deposition velocities indicated that a major fraction of the total stable Pb flux to recently deposited sediments of Puget Sound is atmospherically derived. Similar calculations for Cd, Ag, Fe, Cu and Hg revealed that the atmosphere contributes only a minor fraction of these metal fluxes to Puget Sound sediments.

Plutonium and Beryllium-7

Atmospheric deposition of Pu isotopes has been estimated from integrated inventories in terrestrial soil cores collected at sites around the world covering a range of latitudes (Hardy *et al.*, 1973). A single core from Puyallup, Washington (47° N, near Tacoma) gave a total $^{239,240}Pu$ inventory of 1.4 mCi km $^{-2}$. Other sites at 40-50° N latitude had somewhat higher inventories and brought the global mean Pu fallout inventory for this latitude band to 2.2 ± 0.5 mCi km $^{-2}$. Pu inventories of this magnitude are small relative to total depositional fluxes for Washington continental shelf sediments and to fluxes in advecting seawater (Beasley *et al.*, 1982).

Samples of atmospheric particulates collected with high-volume, air-filtration systems at Quillayute (Fig. 10.1) have been analyzed to determine concentrations of ^{7}Be and other metals (Crecelius, 1981). These concentrations were combined with literature estimates of particle deposition velocities to calculate air-to-ground fluxes of these chemicals. Airborne concentrations and dry deposition velocities have also been reported for Al, Ba, Ca, Fe, Mg, Na, Ti, Ag, As, Cd, Cu, Pb and Zn at several stations in the Olympic National Park of western Washington (Fig. 10.1) (Davidson *et al.*, 1985).

Determination of the ^{7}Be (53-d half-life) fallout rate is valuable because both Be and many anthropogenic contaminants are primarily associated with submicron-sized air particulates. In addition, ^{7}Be is easily measured by gamma spectroscopy. Concentration ratios of anthropogenic metals to Be in particulates are then sufficient to estimate metal depositional velocities (Crecelius, 1981).

^{7}Be associated with atmospheric particulates was rapidly solubilized upon contact with seawater (Bloom and Crecelius, 1983b). Within a period of days a near-equilibrium distribution of ^{7}Be between suspended and dissolved phases is achieved. Because of this limited affinity for suspended matter in seawater, ^{7}Be is not a suitable tracer for the fate of more strongly particle adherent chemicals such as Pb, Hg and PAHs. In addition, variations in the suspended particulate load should control dissolved ^{7}Be concentrations in the coastal zone, rendering it of questionable value as a quantitative tracer for sedimentation in coastal areas. Suspended loads can vary by at least a factor of two over time and location in estuarine and coastal waters, so that deposition of ^{7}Be is probably not in steady state. For similar reasons, one must question the accuracy of methods using ^{7}Be to determine sediment focusing and reworking in coastal sediments (Krishnaswami *et al.*, 1987).

Trace Metals

Studies of As concentrations in atmospheric dust and precipitation collected in Seattle by Crecelius (1974) first revealed that the second largest input of As to Puget Sound was probably atmospheric precipitation enriched in As from stack dust discharged by a large smelter located near Tacoma (Fig. 10.3). Subsequent analysis of atmospheric precipitation samples collected at 11 locations around Puget Sound over a two-year period showed that As concentrations decreased with distance north of the Tacoma smelter, consistent with the fact that southerly winds usually accompany rain storms in this region (Carpenter *et al.*, 1978). Elevated As concentrations were still found in precipitation samples collected 50-km downwind of the smelter. However, dissolved As concentrations in Puget Sound seawaters were not higher than background values except within 1-2 km of the smelter because the As input from advecting seawater exceeds atmospheric input by at least a factor of 20 (Carpenter *et al.*, 1978). The smelter ceased operations in spring 1985 and a follow-up study of atmospheric As concentrations would make an interesting comparison with results from 1974-78.

Chemical species in atmospheric particulates that are at least moderately soluble in seawater may be expected to have the greatest effects on coastal ocean chemical composition and biological ecosystems. Crecelius (1980) determined that the leachable fraction of several metals in Quillayute air particulates ranged from 30 to 80% for As, Cu, Pb and Zn. These solubilities in seawater are similar to those of anthropogenic-rich aerosols collected in southern California and to those of continental dust-rich aerosols collected in Baja California (Hodge *et al.*, 1978).

Seawater solubilities of trace metals associated with atmospheric particulates from Quillayute, Seattle and St. Louis were further examined by Hardy and Crecelius (1981) as part of a study of the effects of airborne particulate matter on marine phytoplankton productivity. Their results suggest that, even if the 1 mg l^{-1} of particles typically found in surface seawaters were entirely composed of urban particulates, photosynthesis would be inhibited by at most 10%. Analogous studies of solubilities of potentially toxic organic compounds such as PAH in atmospheric particulates were not performed. Thus it is unclear whether the inhibition of photosynthesis is due to metals or to organic compounds.

Hardy *et al.* (1985a) used laboratory microcosm studies to derive a model to predict the behavior of six metals from atmospheric particulate matter in the 50-µm thick sea surface microlayer for various conditions of biotic enrichment, wind mixing and atmospheric deposition rate. Predicted enrichments of metal concentrations in the microlayer over concentrations in deeper waters agreed well with metal enrichments measured in the field in Puget Sound and Sequim Bay (Fig. 10.3) (Hardy *et al.*, 1985b). More recent data on sea surface contamination in Puget Sound are given by Hardy *et al.* (1988).

Polynuclear Aromatic Hydrocarbons

Concentrations of PAH and stable lead were measured once a month for a year in atmospheric particulates collected at Quillayute, Sequim and Seattle (Prahl *et al.*, 1984). Average individual PAH concentrations ($\mu g\ g^{-1}$) were 2.5 to 6.0 times greater in atmospheric particulates from highly urbanized Seattle than in corresponding compounds in airborne particles from the rural areas of Sequim and Quillayute. In addition, average annual atmospheric particle concentrations ($\mu g\ m^{-3}$) were 2.1 to 2.4 times greater in Seattle. PAH concentrations, therefore, were 5 to 12 times greater in Seattle's air than at the other two sites.

Atmospheric deposition rates for individual PAH were estimated using measured air particulate concentrations and assuming a PAH atmospheric deposition velocity of 0.4 cm s^{-1}. This value is within the range of 0.1-0.5 cm s^{-1} measured in the field for chlorinated pesticides and polychlorinated biphenyls (Eisenreich *et al.*, 1981). Aliphatic and PAH atmospheric deposition rates for the Washington shelf based on Quillayute air samples are small relative to Columbia River input (Prahl *et al.*, 1984). Estimates of atmospheric PAH deposition rates to Puget Sound based on Seattle air samples were comparable to PAH discharges from Seattle's large METRO primary sewage treatment plant (Barrick, 1982).

We know of no reliable estimates of atmospheric fluxes of any other trace organic compound classes, including EPA's priority pollutants, to either the Washington coast or to Puget Sound. Indirect evidence from sediment cores suggests that atmospheric fluxes of primarily anthropogenically-derived azaarenes are significant for the Puget Sound region (Furlong and Carpenter, 1982). Atmospheric fluxes of some primarily naturally-produced organo-sulfur compounds, such as dibenzothiophene, may also be significant (Bates and Carpenter, 1979a).

Freon

Chlorofluoromethanes, CCl_3F (^{11}Freon) and CCl_2F_2 (^{12}Freon), have no known natural sources. Freons in the atmosphere have increased from 1940-1970, similar to the atmospheric increase of fossil CO_2 and on time-scales characteristic of upper ocean processes. Freons are extremely stable in the lower atmosphere and ocean, being destroyed only in the stratosphere where they are suspected of causing a partial destruction of the ozone layer. They are only slightly soluble in seawater. However, because of the lack of other rapid removal processes, perhaps 20% of the annual removal of ^{11}Freon from the atmosphere may be due to dissolution in the ocean (Khalil and Rasmussen, in press). These properties make Freon gases useful tracers of surface ocean mixing and the transfer of CO_2 from surface to deeper waters.

The severe sample contamination problems which slowed early efforts to measure Freons in seawater were first overcome by sample collection and measurement techniques developed by J. Cline of NOAA's Pacific Marine Environmental Laboratory, Seattle. These techniques were then used to make the first reliable vertical concentration profiles of these transient tracers in marine waters at two locations in the northeast Pacific, 46° N, 125° W off the Washington-Oregon coast and 50° N, 140° W in the Gulf of Alaska (Gammon *et al.*, 1982).

Freon concentrations at both sites decreased exponentially with increasing depth below the mixed layer. Surface concentrations were in saturation equilibrium with measured atmospheric concentrations. Freon concentration profiles were interpreted with a one-dimensional model for the vertical diffusion and advection of a conservative tracer whose input into a bottomless ocean had an exponential time dependence. The model allowed simple scaling from one exponentially-driven, transient tracer (Freon), to another (fossil fuel CO_2), and led to a predicted mean depth of penetration of fossil fuel CO_2 of approximately 300 m in the eastern North Pacific. Subsequently, Khalil and Rasmussen (in press) reported that Pacific Ocean surface waters collected in 1983 were slighly undersaturated with respect to measured atmospheric concentrations of Freon.

Calculations of global Freon fluxes have relatively large uncertainties due to near- equilibrium dissolved Freon concentrations and to changes in atmospheric Freon discharge rates. The flux of Freon from air-to-sea may have decreased during the last decade in response to recent reductions in ^{11}Freon release to the atmosphere (Liss, 1983). In calculating Freon fluxes it is important to use recent measurements of Henry's law constants (Wisegarver and Cline, 1985) rather than earlier values which are now believed to be incorrect.

10.4.3 Sea-to-Air Transport of Specific Chemicals

Dimethylsulfide and Carbonyl Sulfide

Dimethyl sulfide (DMS) has been reported recently in surface seawaters from several regions (Barnard *et al.*, 1982; Andreae and Raemdonck, 1983; Cline and Bates, 1983; Andreae and Barnard, 1984). DMS is formed through enzymatic decomposition of β-dimethyl-propiothetin by a variety of marine algae (Vairavamurthy *et al.*, 1985; and references therein). Biologically-controlled production causes supersaturation of surface waters and subsequent loss to the atmosphere. The volatile DMS plays an important role in the global S budget. Its oxidation within the atmosphere to sulfurous and sulfuric acids may be a significant contributor to the acidity of precipitation, especially in coastal areas devoid of large anthropogenic discharges of S and N oxides.

Bates and Cline (1985) determined DMS in seawater samples collected from transects along and perpendicular to the west coast of the United States during 1983-1984, including several stations off the Washington and Oregon coasts. Concentrations of DMS were highest either at the surface or in a subsurface maximum at 20-30-m deep and decreased exponentially to less than 1 ng S l^{-1} below about 150 m. In surface waters, average DMS concentrations were three times greater in summer than winter, presumably due to greater biological productivity. However, despite its biological origin, DMS concentrations did not correlate with chlorophyll concentrations, nutrient concentrations or other indicators of upwelling. The flux of

DMS from the ocean to the continent was calculated to be approximately 13% of the combined total anthropogenic DMS emissions from western California, Oregon and Washington. Additional DMS flux data are given by Bates *et al.* (1987).

Atmospheric carbonyl sulfide (OCS) is believed to affect the stratospheric sulfate layer and thus the earth's energy balance and climate. Rasmussen *et al.* (1982) reported the first measurements of OCS in seawater showing that the oceans are a global source of OCS rather than a sink. Average concentrations in surface seawaters from coastal regions were 2.5 times higher than concentrations from surface waters of the open ocean, suggesting that OCS is related to biological productivity. OCS may also be produced by photochemical processes (Ferek and Andreae, 1984). It is not clear how many of the samples of Rasmussen *et al.* (1982) were from coastal Washington or Oregon. Johnson (1985) found that OCS concentrations off Washington were only a few percent of DMS concentrations (Bates and Cline, 1985), as also reported by Rasmussen *et al.* (1982). The importance of the oceanic OCS source was pointed out by Khalil and Rasmussen (1984), who compared global estimates for other oceanic natural and anthropogenic OCS emissions.

Nitrous Oxide

From measurements made at Cape Meares, Ore. (45° N, Fig. 10.1) and Cape Grim, Tasmania (42° S) over a three-year period, Khalil and Rasmussen (1983) confirmed that N_2O concentrations in the air are increasing slowly in both northern and southern hemispheres. Their estimates of natural and anthropogenic sources of N_2O to the atmosphere suggested that the oceans are a net source of N_2O, although not a major one. Despite the extensive time series of measurements of N_2O atmospheric concentrations on the Oregon coast, we are aware of no studies of N_2O concentrations in coastal ocean waters off either Oregon or Washington.

Chloroform and Methyl Chloride

From observed concentrations of chloroform ($CHCl_3$) in the air and surface seawaters, Khalil *et al.* (1983) showed that the tropical oceans are a relatively large source of $CHCl_3$ to the atmosphere. This was surprising given the many man-made sources of this chemical. The authors reported an average $CHCl_3$ concentration for ground level air at Cape Meares, Ore., based on 56 measurements between 1977 and 1982. Apparently very few measurements have been made of $CHCl_3$ in surface ocean waters off Oregon and Washington. Better definition of the magnitude of the $CHCl_3$ flux from productive coastal regions seems important, since, on a global basis, it may equal Freon discharges as a source of Cl atoms, and the latter affect the stratospheric ozone layer (Liss, 1983).

Methyl chloride (CH_3Cl) is the most abundant halocarbon in the earth's atmosphere. It is primarily natural in origin, being released from the ocean as a result of processes not yet understood (Khalil and Rasmussen, 1981). Fairly extensive time series of measurements of CH_3Cl concentrations in air at Cape Meares, Ore. and six other locations around the world were reported by Khalil and Rasmussen (1981). We are aware of no data on concentrations of CH_3Cl in coastal waters off Washington and Oregon.

Radon

^{222}Rn deficiencies relative to ^{226}Ra in surface open ocean water column profiles have been often used to estimate gas transfer rates across the air-sea interface (Broecker and Peng, 1982 and references therein). Peng *et al.* (1974) summarized results of a time series of ^{222}Rn and ^{226}Ra measurements under stormy winter conditions at weather station PAPA (50° N, 145° W). The gas exchange rate increased with wind speed but in a complex manner. Similar measurements have not been carried out on the Washington shelf, in Puget Sound, or in the Strait of Juan de Fuca.

Smethie (1979) observed large ^{222}Rn excesses in river and stream waters flowing into fjords along the west coast of Canada (Lake Nitinat, Princess Louisa Inlet and Narrows Inlet). Rivers and streams are fed in part by ground water, which has extremely high excess Rn concentrations. Degassing of ^{222}Rn is not rapid enough to deplete the freshwater of Rn before it enters these fjords. In certain fjords this excess ^{222}Rn in surface waters may provide a tracer for vertical mixing in the near-surface water column.

10.4.4 Remaining Questions and Needed Research in Air-Sea Chemical Exchange

The fraction of organic chemicals contained in atmospherically-deposited particles which is soluble enough to enter the chemical and biological cycles of the coastal ocean is unknown. Studies by Crecelius (1980) and Hardy and Crecelius (1981) addressed the question of bioavailability for metals, but analogous studies have not yet been performed for organics such as PAHs and PCBs. The fact that large fractions of PAH in Washington coastal sediments are associated with relatively large and light organic particles resembling soot or fly ash (Prahl and Carpenter, 1984) suggests that they may be transported through the water column with little alteration and subsequently buried in underlying sediments. The temporal and spatial extents of and possible adverse biological effects of concentrations of contaminants in the sea surface microlayer need more investigation.

Literature values for depositional velocities of trace organic compounds like PAH, DDT (and its degradation products) and PCBs vary by a factor of almost ten, even for the same compound class (Eisenreich *et al.*, 1981). Deposition velocities estimated from ^{7}Be may be reasonable approximations for other metals and ^{210}Pb, but seem inappropriate for organic compounds with different volatilities and associations with particles of different sizes, shapes and densities. Large uncertainties in appropriate depositional velocities introduce correspondingly large uncertainties in the calculated air-to-sea transfer rates of these important classes of chemicals.

Variability in atmospheric particulate concentrations and fluxes off the Washington coast may be significantly different from that measured at Quillayute or other coastal sampling stations. Fox and Ludwick (1976) argued that strictly oceanic air is depleted in particulate Pb relative to average Quillayute air by at least a factor of five. The lack of suitable offshore sampling platforms, such as weather ships, lightships, or oil production platforms, will continue to limit our ability to sample oceanic atmospheric deposition. ^{7}Be total depositional velocities have been determined at many locations in the open ocean (Young and Silker, 1974, 1980) allowing inferences about deposition variations in other metals with similar particle affinities.

Atmospheric fluxes of chlorinated organics such as PCBs and DDT (and its degradation products) to both Puget Sound and the Washington coast need to be determined. Better characterizations of azaarenes in air particulates and their fluxes also seem desirable, at least for Puget Sound. More data on concentrations of the volatile compounds like Freon, N_2O, CH_3Cl, $CHCl_3$, DMS and OCS in surface waters off the Washington and Oregon coasts, along with concurrent measurements of atmospheric concentrations, are needed. Seasonal measurements as a function of location (i.e., within or outside of upwelling areas or the Columbia River plume) are required in order to identify and quantify processes dominating the production of these compounds in surface waters. Despite their potential value, additional Freon concentration profiles have not been published for the Washington coast or Puget Sound.

Shipboard methods to confirm constant atmospheric conditions for several days prior to water sampling must be developed. Equilibration between the air and sea before sampling is essential to prevent incorrect conclusions about the direction or magnitude of chemical fluxes across the air-sea interface. Also, difficulties persist in establishing the true relationship between wind velocity and the exchange rate of gases across the air-sea interface (Liss and Slinn, 1983).

Concentrations of chemicals vary widely in atmospheric samples as a function of such factors as length and type of precipitation. The length of time needed to obtain a well-integrated, representative sample is still unknown. The relative importances of dry *vs.* wet deposition and exchange of gases *vs.* particulate phases are not understood. Gas exchange may be especially important for organic compounds with relatively high vapor pressures and low particle association tendencies.

10.5 CHEMICAL FLUXES ASSOCIATED WITH HYDROTHERMAL ACTIVITY

Hydrothermal activity has been observed at several 'hot spots' along the Juan de Fuca Ridge spreading center located approximately 300-km west of the Washington coast (Fig. 10.1) (Jones *et al.*, 1981; Normark *et al.*, 1982; Tivey and Delaney, 1985; Canadian-American Seamount Expedition, 1985; Crane *et al.*, 1985; Baker and Massoth, 1986, *ibid*, 1987; Baker *et al.*, 1987; Feely *et al.*, 1987). Tubeworms, crabs and other biota which resemble those in the biological ecosystems of the Galapagos and 21°N sites on the East Pacific Rise have been investigated. Water from black and grey 'smokers' has been sampled with temperatures up to 385-400° C (MERGE, 1984). One megaplume due to a cataclysmic event has been recognized (Baker *et al.*, 1987). Chemical reactions between hot, acidic hydrothermal solutions and cold, alkaline seawater and basaltic rocks are of major importance in chemical mass balances of several major and trace elements in seawater (McDuff and Morel, 1980). Associated metalliferous deposits have attracted economic interest because of their relatively high concentrations of valuable metals such as silver and gold.

Only a limited number of chemical analyses have been reported for seawaters from the Juan de Fuca Ridge area. Jones *et al.* (1981) first reported elevated Mn concentrations in waters 100-200-m above the bottom at five stations along a 400-km section of the ridge axis. Near-bottom concentrations along the ridge axis were comparable to values previously found near the Galapagos and 21° N vent fields. Mn concentrations were a factor of 10 lower at two off-axis stations 60-80 km from the ridge axis. Excess ^3He in deep waters along the ridge has been measured by Lupton *et al.* (1981). Ge concentrations have not yet been reported in waters from the Juan de Fuca area, but broad hydrothermal plumes with higher than normal Ge concentrations have been traced over the southern East Pacific Rise (Mortlock and Froelich, 1986).

Wheat and Carpenter (in press) used an electron spin resonance (ESR) approach to determine the extent of ion pair formation by Mn^{2+} and several other metals at temperatures up to 200° C. Association constants determined at elevated temperatures should be useful in predicting and modeling chemical reactions and mineral solubilities in hydrothermal vent waters at mid-ocean ridge spreading centers.

Figure 10.11 shows increases in Mn^{2+} complexation by both Cl^- and SO_4^{2-} with increasing ligand concentration and temperature, typical of most metal-ligand reactions. The thermodynamic association constants were calculated from the slope of the Mn-ligand/Mn^{2+} *vs.* free ligand plot and activity coefficients were computed using the extended Debye-Huckel equation. The $MnSO_4^0$ association constant increased 5.5-fold with a temperature increase from 25° C to 100° C. The $MnCl^+$ association constant increased 4.6-fold with a temperature increase from 25° C to 140° C. Greater Mn complex formation at higher temperatures helps explain the relatively high Mn concentrations in vent waters.

Determinations of total Hg concentrations (Carpenter, unpublished data) in water samples from the same stations as Jones *et al.*'s (1981) Mn data showed total Hg concentrations less than 10 ppt (50 pmol l^{-1}), with no enrichment in waters containing higher Mn concentrations. In fact, Hg concentrations appeared somewhat lower in waters influenced by hydrothermal activity. These observations are consistent with reports that Hg concentrations in near-bottom waters are not elevated in the FAMOUS area of the mid-Atlantic Ridge (Dalziel and Yeats, 1985; Carr *et al.*, 1975) and near Iceland (Bruegmann *et al.*, 1982; Olafsson, 1983).

Elevated concentrations of particulate Fe and As have been reported in waters near the hot spots along the Juan de Fuca Ridge (Massoth *et al.*, 1983). Dissolved Fe precipitates out rather quickly as these fluids cool and mix with ambient waters. Hydrous iron oxides are excellent scavengers of disolved As (Carpenter *et al.*, 1978). Total dissolved As concentrations in waters from the same depths as Jones *et al.*'s (1981) Mn data are either normal (1.5-1.8 ppb) or slightly depleted (M.L. Peterson and R. Carpenter, unpublished data). This suggests that the higher particulate As concentrations reported by Massoth *et al.* (1983) are more likely the result of efficient scavenging of dissolved As from ambient deep waters by the Fe-rich particulates than the result of injection of hyrothermal fluids enriched in As.

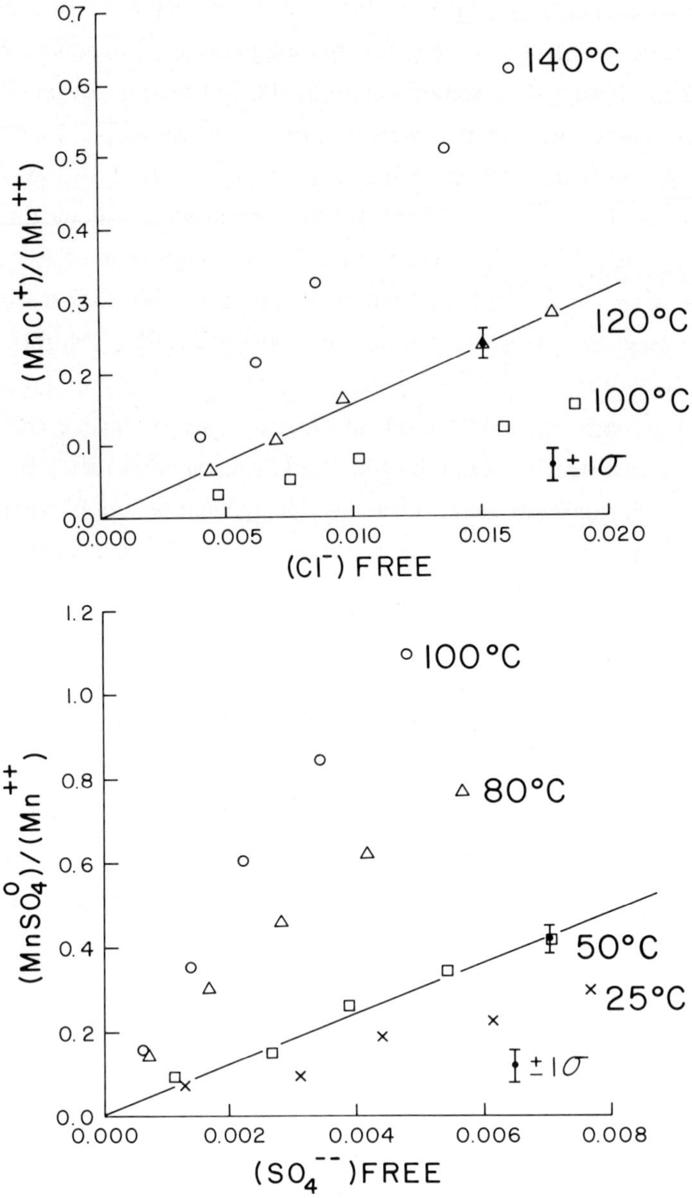

Fig. 10.11 Increases in $MnCl^+$ and $MnSO_4{}^0$ complex formation with increasing temperature and free ligand concentration, determined by the electron spin resonance technique. Based on data in Wheat and Carpenter (1988).

Several hydrothermally-vented substances have been determined in water-column samples collected along a 200-km transect normal to and bisected by the southern Juan de Fuca Ridge (Massoth *et al.*, 1984). The composition and dissolution of black smoker particulates from active vents on the Juan de Fuca Ridge also have been examined (Feely *et al.*, 1987). Profiles of particulate and dissolved Mn, Fe and methane and light scattering were acquired from the hydrothermally-altered water column at 12 cross-axial stations. The water-column hydrothermal plume, characterized by coincident maxima in light scattering and particulate Fe/Al and Mn/Al ratios, extended several hundred meters vertically above the ridge crest. The largest hydrothermal gradients were observed within 20 km of the spreading center, although elevated particulate parameters were still detected some 100 km from the ridge crest (Massoth *et al.*, 1984). Thus, it appears that seawater with hydrothermally-altered chemistry should be barely detectable at the extreme western edge of the Washington continental slope. Therefore such water cannot significantly affect either water or sediment within the Washington shelf and slope regions.

10.6 SEDIMENTARY FLUXES, INVENTORIES AND CHEMICAL PROFILES

10.6.1 Bulk Sediment Characteristics, Accumulation and Mixing Rates

Most of the sediment accumulation and mixing rates for the Washington coast and Puget Sound were determined using the ^{210}Pb dating technique. These ^{210}Pb-derived accumulation rates were based on cores carefully collected to minimize artifacts caused by coring gear and procedures. Sediment mixing is believed to cause inaccuracies of less than 30% in the ^{210}Pb-derived accumulation rates for Washington shelf sediments (Carpenter *et al.*, 1982; Nittrouer *et al.*, 1984). Magnitudes and geographic trends of ^{210}Pb-derived accumulation rates in Puget Sound agree with limited data on sediment inputs and transport paths (Carpenter *et al.*, 1985). Their accuracy has been supported by several independent approaches, including comparison with water-column particle fluxes determined in year-long time series of sediment trap deployments at two sites.

The Washington Shelf and Slope Regions

Major features of shelf topography and sediment texture near the Columbia River mouth were described by McManus (1972). Nittrouer (1978) and Nittrouer *et al.* (1979) have reported more grain-size measurements and seismic profiling results which have refined, but not substantially altered, previous conclusions. There is an inner shelf sand wedge with undetectable accumulation of fine-grained materials, a mid-shelf silt deposit (MSSD) (Fig. 10.2) of Columbia River-derived material and an outer shelf zone of sands and silts which are largely relict and reworked material. The thickness of the MSSD decreases in a direction north-northwest from the river mouth.

Gross *et al.* (1972) summarized the distribution and abundance of organic C in surface sediments from the shelf and slope. Factors controlling organic C concentrations include grain size, rates of sediment accumulation and dissolved O_2 concentrations in near-bottom waters. A pronounced O_2 minimum, with concentrations only 5 to 10% of saturation, impinges upon the slope near 800 m (Fig. 10.6) and is partially responsible for the fact that organic C concentrations are 2 to 3 times greater in slope sediments than in shelf or deeper sediments (Carpenter *et al.*, 1981). Table 10.5 summarizes surface sediment concentrations of organic C, Al and Mn and C/N ratios in different geologic regimes off the Washington coast.

Major improvements in understanding accumulation and mixing rates in Washington shelf and slope sediments have come from having the most extensive data set of excess ^{210}Pb activity profiles for any coastal area in the world (Nittrouer *et al.*, 1979; Carpenter *et al.*, 1981; 1982). Shelf sediment accumulation rates are low south of the Columbia River and north of

TABLE 10.5 Sediment parameter for nine subregions off the Washington coast. From Carpenter *et al.* (1981), with permission.

	Surface ^{210}Pb Activity (dpm g^{-1})	Accumulation Rate (mg cm^{-2} yr^{-1})	^{210}Pb Flux (dpm cm^{-2} yr^{-1})	Total Carbon (%)	C/N (g g^{-1})	Aluminum (%)	Manganese (ppm)
Canyons							
Quinault	110 ± 48 (8)[a]	220 ± 140 (8)	18 ± 2.8 (8)	3.3 ± 0.4 (6)	12.4 ± 1.0 (6)	7.1 ± 0.3 (6)	353 ± 22 (6)
Grays	99 ± 58 (7)	95 ± 82 (7)	12 ± 13 (7)	2.6 ± 0.6 (6)	11.0 ± 0.9 (5)	7.3 ± 0.5 (5)	409 ± 180 (5)
Willapa	120 ± 50 (9)	94 ± 26 (9)	12 ± 5.3 (9)	2.8 ± 0.8 (9)	10.3 ± 0.9 (9)	7.5 ± 0.4 (9)	425 ± 133 (9)
Astoria	80 ± 38 (8)	410 ± 350 (8)	30 ± 19 (8)	3.0 ± 0.3 (6)	10.8 ± 0.8 (6)	7.1 ± 1.0 (6)	349 ± 20 (6)
Canyon mean	100 ± 49 (32)	220 ± 230 (32)	18 ± 13 (32)	2.9 ± 0.5 (27)	11.4 ± 0.2 (27)	7.3 ± 0.6 (26)	388 ± 111 (26)
Slopes							
Copalis Beach	130 ± 21 (8)	41 ± 19 (8)	5.4 ± 2.5 (8)	3.1 ± 0.6 (8)	10.6 ± 0.7 (8)	7.0 ± 0.7 (8)	502 ± 200 (8)
S. Grays Harbor	100 ± 65 (5)	59 ± 27 (5)	6.4 ± 5.7 (5)	3.0 ± 0.5 (4)	10.8 ± 1.2 (4)	7.3 ± 0.7 (4)	395 ± 50 (4)
Willapa Bay	87 ± 50 (10)	59 ± 28 (10)	4.3 ± 1.5 (10)	2.2 ± 0.9 (8)	10.2 ± 0.9 (6)	7.1 ± 0.9 (6)	512 ± 142 (6)
Slope mean	110 ± 48 (23)	53 ± 25 (23)	5.2 ± 3.1 (23)	2.7 ± 0.8 (20)	10.5 ± 0.9 (18)	7.1 ± 0.7 (18)	482 ± 158 (18)
Shelf							
Outside MSSD[b]	23 ± 15 (35)	260 ± 150 (35)	4.7 ± 2.0 (35)	1.1 ± 0.5 (23)	11.4 ± 1.8 (22)	7.1 ± 0.7 (22)	436 ± 84 (22)
Within MSSD	16 ± 9 (50)	380 ± 160 (50)	4.9 ± 1.7 (50)	1.1 ± 0.5 (28)	12.5 ± 1.7 (28)	7.4 ± 0.7 (28)	463 ± 64 (28)
Shelf mean	18 ± 12 (85)	330 ± 170 (85)	4.8 ± 1.8 (85)	1.1 ± 0.5 (51)	12.0 ± 1.8 (50)	7.3 ± 0.7 (50)	451 ± 74 (50)

[a]Number of samples used to compute mean ± one standard deviation.

[b]MSSD = Mid-Shelf Silt Deposit

Quinault Canyon, indicating that most modern accumulation of Columbia River sediment occurs between these boundaries. The predominant feature is the tongue of high accumulation rates extending from the river mouth north-northwestward toward Quinault Canyon (Table 10.5, Fig. 10.2). The axis of this feature, with accumulation rates above 250 mg cm $^{-2}$ y $^{-1}$, corresponds to the axis of the MSSD. The general decrease in accumulation rates away from the river mouth along the axis agrees with the seismically-determined thinning of the MSSD in the same direction (Nittrouer, 1978).

The study of recent shelf sediment accumulation and mixing rates has been extended to the continental slope and four submarine canyons which indent the slope (Carpenter *et al.*, 1982). The fact that mixing may cause ^{210}Pb derived accumulation rates for slope and canyon sediments to be at most 2-3 times higher than true accumulation rates was demonstrated with four different approaches. Total sediment accumulation rates in these canyons are about twice those on the nearby open slope, even though canyons comprise < 40% of the depositional area. This is because ^{210}Pb derived accumulation rates for Astoria and Quinault canyons are 3 to 4 times greater than those for Willapa and Grays canyons and intermediate open slope areas, supporting the hypothesis that Columbia River-derived solids are preferentially transported off the shelf via Astoria and Quinault canyons (see Kachel and Smith, 1989, Chapter 8). More extensive studies of currents, sediment transport and deposition in Quinault canyon were conducted following recognition of its key role in funneling Columbia River-derived solids to the deep sea (Carson *et al.*, 1986 and references therein; Hickey, 1989, Chapter 2)

Large-scale features of Quaternary tectonics and sedimentation on the Washington continental slope and in Cascadia Basin (Fig. 10.1) have been examined by Barnard (1973, 1978) and Carson (1971, 1973), respectively. Carson found that sediments of Cascadia Basin consist of interlayered turbidites and hemipelagic deposits. Sediments were partitioned into Holocene (13,000 y B.P.) and late Pleistocene sections on the basis of planktonic foraminiferan-radiolarian biostratigraphy and radiocarbon age determinations. Planktonic foraminifera were dominant during the late Pleistocene, but declined rapidly in the Holocene, reaching minimal concentrations in sediments by 9,000 y B.P. This decrease in foraminiferan tests may be a result of increased dissolution during the Holocene. The abundance of radiolarian tests increases markedly in Holocene sediments, apparently in response to increased productivity. Sediment properties and relatively low accumulation rates (4 mg cm $^{-2}$ y $^{-1}$) indicate that hemipelagic deposition was the dominant mode of sedimentation in nearly all portions of the study area during the late Holocene (2,800-9,000 y B.P.). In the early Holocene (9,000-13,000 y B.P.), however, turbidity current deposition was common, being largely confined to major channels where sedimentation rates on the order of 20 mg cm $^{-2}$ y $^{-1}$ are found. Turbidite deposition dominated nearly all slope regions during the late Pleistocene resulting in accelerated accumulation rates in both channel (25-45 mg cm $^{-2}$ y $^{-1}$) and interchannel (20-37 mg cm $^{-2}$ y $^{-1}$) areas.

A marked change in color and in certain chemical properties has been found at the 1-3-m depth in sediments from the Washington and Oregon continental slopes and Cascadia Basin (Carson, 1971; Barnard, 1978). The transition is from recent, olive colored deposits with 2-3%

organic C to a distinct grey deposit with only 0.3-0.5% organic C and roughly half the water content of the recent layers. The transition has been radiocarbon-dated as occurring at about 10,000 years B.P., at the end of the last glacial period (Carson, 1971; Table II of Barnard, 1978). Reasons for the change are not yet known.

Barnard determined average postglacial sediment accumulation rates on the continental slope by means of radiocarbon dating and lithologic and paleontologic changes in the sediments. Most of these sediments were probably deposited as turbidity currents during periods of glaciation on the adjacent continent. Seismic reflection profiles provided no evidence for major shifts in routes of sediment dispersal during the Pleistocene.

Slope sediments collected with piston cores consisted of interlayered turbidite and hemipelagic deposits. This sediment was divided into glacial and post-glacial sections based on a change from foraminiferan-rich, grey lutite to a radiolarian-rich, olive lutite. This faunal change, which began about 10,500 y B.P. on the lower slope, is time transgressive in the northeast Pacific and may separate geologic-climatic units. However, it cannot be used to precisely date the Pleistocene-Holocene stratigraphic boundary. Postglacial sediment deposited between 10,500 y B.P. and the present can be divided into early and late sections based on the deepest horizon of Mt. Mazama glass, which was deposited about 6,750 y B.P.

During the last glacial period, the lower continental slope was apparently deluged with turbidity currents producing the deep-sea fans of Cascadia Basin. All canyons indenting the upper slope were probably actively transporting sediments to Cascadia Basin until about 10,500 y B.P. The importance of hemipelagic sedimentation increased with the onset of post-glacial conditions during which transport of sediment off the shelf via canyons was probably less active and turbidity currents were less frequent and were composed of finer-grained material. Over the last 6,750 y Astoria, Grays and Juan de Fuca canyons have been accumulating hemipelagic mud and some turbidites. Willapa Canyon remained an active avenue for erosional turbidity currents until about 5,000 y B.P. All these canyons now appear to be accumulating fine-grained material, derived primarily from the Columbia River. Quinault Canyon apparently acts as a funnel for a large amount of material moving northwestward along the continental shelf. This material accumulates at the canyon head. Much of the material is flushed down the canyon by large, periodic turbidity currents occurring with a frequency of about 500 y. Between these events material falls into the canyon via intermediate nepheloid layer formation during shelf resuspension events (Hickey, 1989, Chapter 2).

Excess ^{210}Pb profiles also revealed greater surface sediment mixing coefficients in the canyons off the Washington coast than in adjacent open slope areas (Carpenter *et al.*, 1982). This phenomenon was initially attributed to greater sediment resuspension by bottom currents. However, Rowe *et al.* (1982) have recently reported greater macrobenthic biomass, activity and mixing in upper Hudson Canyon sediment relative to nearby open slope sediments. Mixing coefficients for Washington coast slope sediments decrease sharply below about 800-m water depth (Fig. 10.12; Carpenter *et al.*, 1982). Low dissolved O_2 concentrations in waters over the

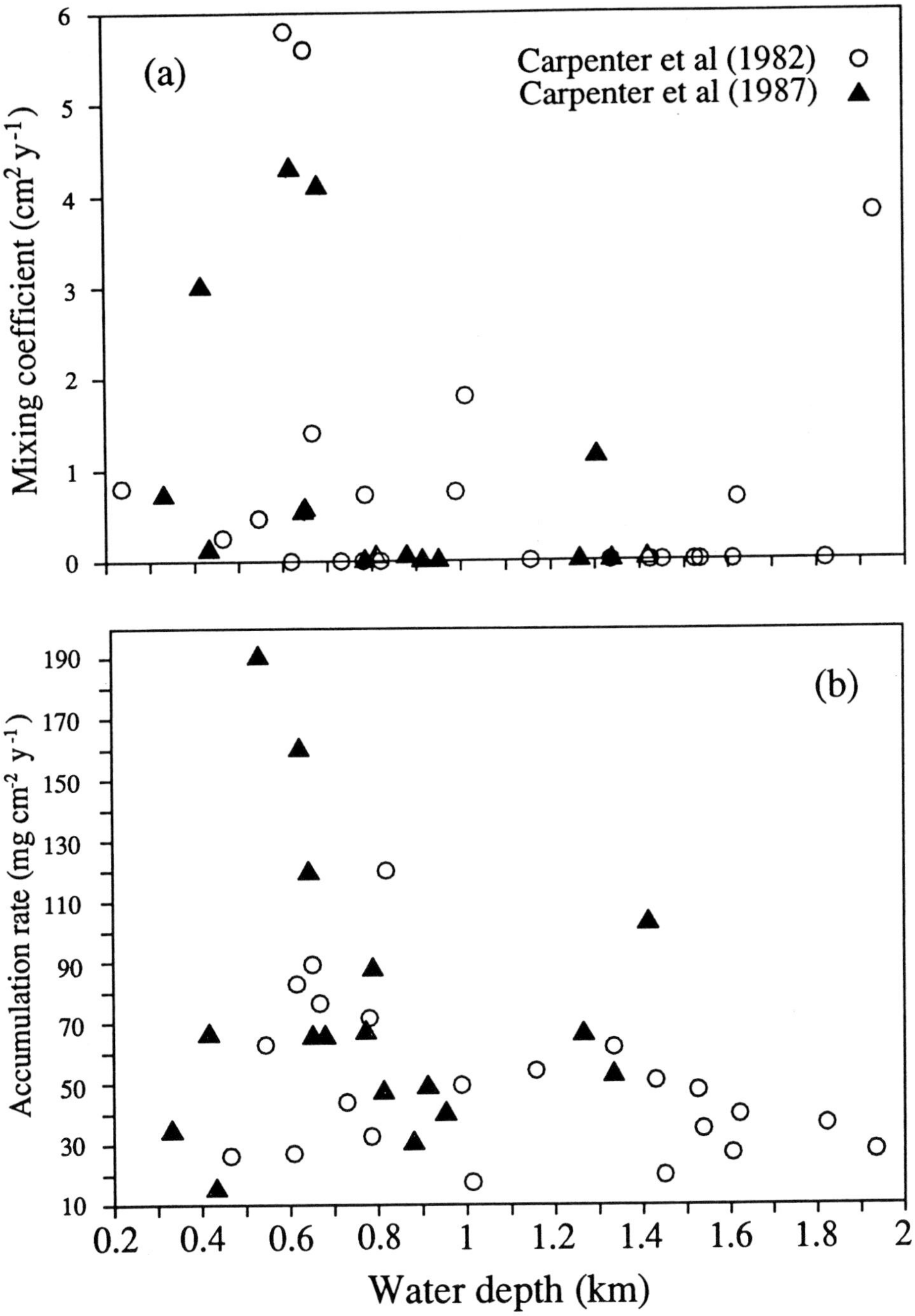

Fig. 10.12 Dependency of Washington slope ^{210}Pb-derived sediment accumulation and mixing rates upon depth of overlying water.

slope (below 10% of saturation) from about 600 to 1,400 m may inhibit macrobenthic activity and hence reduce sediment mixing.

Mixing processes in Washington shelf and slope sediments appear to affect ^{239,240}Pu and excess ^{210}Pb activities in similar but not identical manners (Carpenter *et al.*, 1987). ^{239,240}Pu activities are generally uniform in surface mixed layers of uniform ^{210}Pb activity and then decrease fairly smoothly to detection limits. Minor irregularities in ^{239,240}Pu activities within surface layers of uniform ^{210}Pb activity occasionally result because particle residence times of several years or longer are required to completely homogenize variations in ^{239,240}Pu activities on freshly depositing particles. Occasional subsurface maxima in activities of ^{210}Pb or ^{239,240}Pu occur in Washington slope sediments and have also been reported in sediments from other regions (Smith *et al.*, 1986). Such features have been attributed to subsurface egestion of surficial material by certain large infaunal sipunculid worms (Smith *et al.*, 1986). The abundance of such organisms in Washington slope sediments is unknown, but their feeding behavior cannot be simulated by multi-layered eddy diffusive mixing models.

Strait of Juan de Fuca and the Puget Sound Region

Sediments in the Strait of Juan de Fuca are relatively coarse-grained due to winnowing of finer particles by strong bottom currents. Studies of bottom photographs, dredge hauls, grabs and piston cores (Anderson, 1967) showed that surface layers of much of the Strait consist of reworked, relict gravel and cobbles. Sands, generally the finest sediments, are limited to only a few areas (Roberts, 1979). The Strait of Juan de Fuca may be characterized as a region of possible transport of fine-grained solids and dissolved chemical scavenging, but little accumulation at the sea floor. Finer particulate fractions and their associated contaminants are either carried westward out to sea in surface waters or eastward in the near-bottom drift along the southern side of the Strait. Because of the coarse nature of the bottom sediments, excess ^{210}Pb activities are not sufficient to allow sediment accumulation rate determinations.

Strong sediment grain-size variations and significant focusing of hydrodynamically equivalent particles also occur in certain regions of Puget Sound (Wang, 1955; Roberts, 1979). Sediment types range from silty clays on main basin floors, to fine sand or gravel on slopes and to cobbles and gravel in narrows and across sills where currents are particularly strong. Areas which accumulate fine-grained sediments comprise approximately 2.6 x 10^{12} cm ^{2} or one-third of the main Puget Sound basins (Roberts, 1979) and are assumed to be the most active regions for deposition of particle-associated chemical contaminants. Noxious chemical discharges from the urbanized areas of Seattle and Tacoma enter the central Puget Sound area between sills at Admiralty Inlet and the Narrows near Tacoma and tend to accumulate in the underlying sediments. Mean particle residence times of less than 15 d in the central Puget Sound water column are indicated by both sediment inventories of excess ^{234}Th (24-d half-life), which do not depend upon the accuracy of sediment accumulation rates and by long-term time scale sediment accumulation rates calculated from excess ^{210}Pb profiles (Carpenter *et al.*, 1984; 1985). Sediment accumulation and mixing rates and particulate residence times and fluxes for central Puget Sound have been determined because of their importance in chemical budgets (Carpenter *et al.*, 1984, 1985).

Puget Sound surface sediment concentrations of C, Al and Mn and surface areas are summarized by Carpenter *et al.* (1985). Total Fe in Puget Sound sediments ranged from 2 to 6% (mean of 3.5%), with no obvious geographic trends. Al concentrations varied from 4 to 8%, with a mean of 6.4%, essentially the same mean and range found in suspended sediments of the major rivers entering Puget Sound (Crecelius *et al.*, 1975).

Recent sediment accumulation rates of 46 to 1,200 mg cm $^{-2}$ y $^{-1}$ have been calculated from excess ^{210}Pb activity profiles in 44 sediment cores from 27 sites in the greater Puget Sound region (Carpenter *et al.*, 1985) (Fig. 10.13). Rates calculated for 9 stations in the main channel of central Puget Sound average 480 ± 110 mg cm $^{-2}$ y $^{-1}$ and are greatest between Seattle and Tacoma (Carpenter *et al.*, 1985). Additional ^{210}Pb-derived accumulation rates for Puget Sound are given by Lavelle *et al.* (1986).

Surficial sediments of Puget Sound are generally oxygenated and contain a rich benthic infauna (Lie, 1968, 1969, 1974; Nichols, 1974) which extensively mix the upper sediment layers. Benthic mixing produces surface mixed layers of 2 to 22 cm in excess ^{210}Pb profiles in Puget Sound sediments. Sediment mixing coefficients in these surface layers average at least 43 cm ^{2} y $^{-1}$, four times greater than mixing coefficients similarly calculated for Washington shelf sediments (Carpenter *et al.*, 1985). The difference is attributed to the greater biological biomass and activity in Puget Sound sediments (Lie, 1968, 1969, 1974).

Excess ^{234}Th profiles in Puget Sound sediments detail the depth-dependency, dynamic nature and recent history of solid phase mixing processes better than excess ^{210}Pb profiles (Carpenter *et al.*, 1984). ^{234}Th profiles show that mixing within the ^{210}Pb defined surface mixed layer is, in many cases, depth dependent. ^{234}Th derived mixing rates are often highest several centimeters below the sediment-water interface, indicating greater macro-benthic activity at these depths. Depth-dependent mixing coefficients derived from the best fit of a four-layer, advection-diffusion-decay model to the ^{234}Th data are consistent with ^{210}Pb profiles determined for the same sediments, strongly suggesting that ^{234}Th and ^{210}Pb are mixed equivalently and in a multilayered manner.

Surface sediment horizons with excess ^{234}Th lack excess ^{228}Th which might be used to ascertain sediment accumulation rates over the past decade. Instead, deficiencies of ^{228}Th due to loss of soluble ^{228}Ra persist to 20 to 30 cm, revealing that exchange of soluble chemicals between pore and overlying waters reaches substantial depths in Puget Sound sediments (Carpenter *et al.*, 1984).

Particle residence times in surface mixed layers of Puget Sound sediments vary from 1.5 to 68 y and introduce corresponding uncertainties in ages of horizons beneath the mixed layers. For constituents whose stability and phase associations resemble those of ^{210}Pb, particle residence in the mixed layer will cause a 20 to 30 y time-lag in achieving a proportional response by surficial sediment chemical concentrations to changes of chemical concentrations associated with freshly-depositing particles. Such time-lags limit the ability of sediment chemistry monitoring programs to recognize recent changes in chemical composition and fluxes to sediments. They also limit the accuracy of fluxes based on the product of surface sediment chemical concentration and accumulation rates for components whose input concentrations have changed in the last 20 to 30 y.

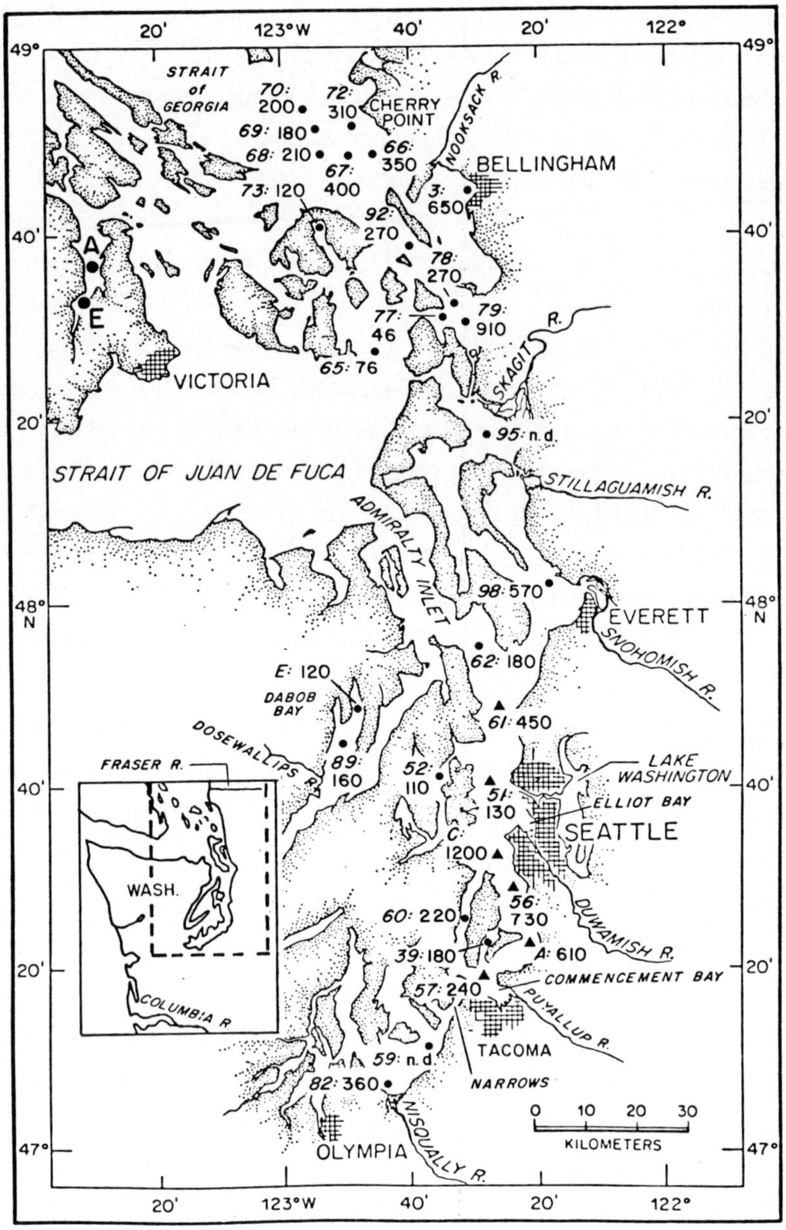

Fig. 10.13 Recent ^{210}Pb derived sediment accumulation rates (mg cm $^{-1}$ y $^{-1}$) for the Puget Sound region. Slanted numbers represent station identifications which are referred to throughout this chapter. From Carpenter *et al.* (1985), with permission.

10.6.2 Chemical Profiles, Inventories and Fluxes

Radionuclides other than Carbon-14

Zinc-65. ^{65}Zn activities in surface sediments of the Washington shelf measured during 1971-1972 were reported by Cutshall *et al.* (1973). Highest activities were found along the north-northwestward trending MSSD. Integrated ^{65}Zn activities were at least two thirds of the total ^{65}Zn discharged by the Columbia River, suggesting that at least this fraction of fine-grained riverine suspended sediments is transported offshore and not trapped in the estuary.

Lead-210. Excess ^{210}Pb activities have been determined in over 100 vertical core profiles from the Washington shelf and slope (Carpenter *et al.*, 1981; 1987). ^{210}Pb surface activities and fluxes (Fig. 10.14) determined from integrated ^{210}Pb inventories are 2 to 7 times greater than those reported for other coastal regions and greatest in submarine canyons indenting the shelf and slope. Sedimentary fluxes are limited by scavenging reactions rather than by supply of dissolved ^{210}Pb, since advection of dissolved ^{210}Pb produced *in situ* from ^{226}Ra provides an input several times larger than sedimentary fluxes. Selective chemical leaching of Washington coastal sediments revealed that both ^{210}Pb (Carpenter *et al.*, 1981) and Pu isotopes (Beasley *et al.*, 1982) are predominately bound to hydrous Mn and Fe oxide phases, rather than clays or organic phases.

Once the effect of Mn cycling on ^{210}Pb and Pu sedimentary fluxes was recognized, a new electron spin resonance (ESR) technique was developed to determine speciation of dissolved Mn in overlying and interstitial waters (Carpenter, 1983). Essentially all dissolved Mn was found to be in the +2 valence state, with an average of 75% free hydrated Mn^{2+}, 12% complexed with chloride and 11% complexed with sulfate. Organic, carbonate, bicarbonate and bisulfide were determined to be unimportant ligands for Mn^{2+}, even in anoxic pore waters. Thus, most of the dissolved Mn^{2+} is available to form Mn-oxyhydroxide phases which could scavenge Pb and Pu.

Plutonium, Americium and Cesium-137. The Pu and Am analytical technique consists of chemical extraction and separation procedures developed by T. M. Beasley followed by alpha spectroscopy. The sample workup involves several precipitation-dissolution and solvent extraction-back extraction steps to ensure complete separation of Pu and Am from the much more abundant Th and Po isotopes. It is especially important to exclude ^{228}Th, since its alpha particle energies overlap those of ^{238}Pu and ^{241}Am. ^{137}Cs was purified by ion exchange techniques following total sediment dissolution of the same samples analyzed for Pu and Am. Cesium chloroplatinate was then precipitated and its activity determined. Procedures for these fallout nuclides yielded results in excellent agreement with those from other laboratories in several intercalibration exercises (Beasley *et al.*, 1982). A few mass spectrometric determinations of Pu isotopic ratios have also been reported (Beasley *et al.*, 1982; Carpenter *et al.*, 1987). Ratios of $^{242}Pu/^{240}Pu$ and $^{240}Pu/^{239}Pu$ are potentially valuable because they vary greatly in high and low yield nuclear weapons tests. Unfortunately, facilities capable of these measurements are extremely limited in number.

Pu activities, vertical profiles and inventories (Fig. 10.15) have been determined in sediments of the Washington shelf and slope by Beasley *et al.* (1982) and Carpenter *et al.* (1987), respectively. The Washington coast is an ideal region for studying Pu cycling, because it is in the latitude band of high atmospheric fallout of artificial nuclides and because nuclide scavenging reactions are pronounced due to the large Columbia River discharge of fine grained particles. Pu inventories in Washington shelf sediments are 4 to 15 times inventories in

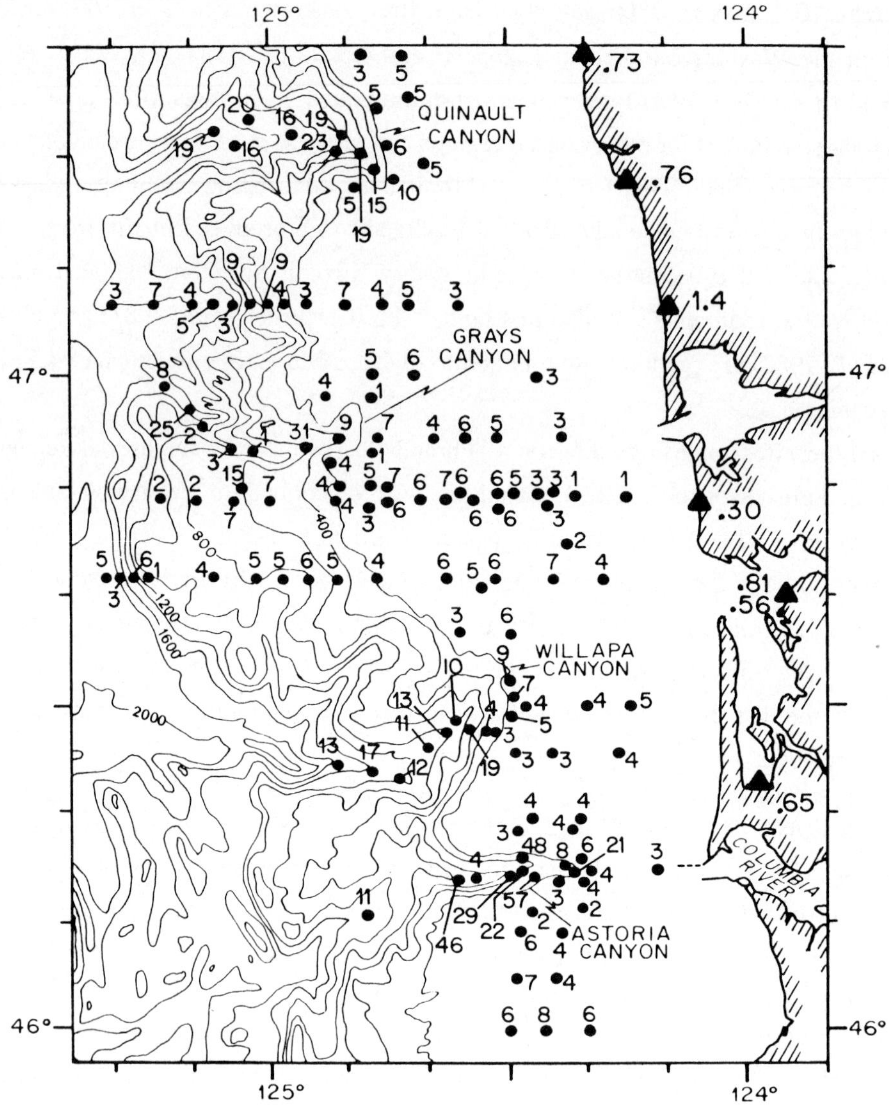

Fig. 10.14 Steady state, excess ^{210}Pb fluxes (dpm cm^{-2} y^{-1}) in sediments of the continental shelf, slope and submarine canyons off the Washington coast. The atmospheric flux determined for terrestrial soil cores is also given at locations indicated by closed triangles on the coast. From Carpenter *et al.* (1981), with permission.

sediments of the northwest Atlantic (Buesseler *et al.*, 1986; and references therein) and in the Gulf of Mexico (Scott *et al.*, 1983). Higher Pu inventories are not primarily due to additional Pu input from reactors at the Hanford Reservation on the Columbia River, but rather to scavenging of dissolved Pu in upwelled waters by fine-grained particles (Beasley *et al.*, 1982).

$^{239,240}Pu$ activities in the 0-1 cm of Washington slope sediments average 380 ± 170 dpm kg^{-1}, twice the mean activity in adjacent shelf sediments. The higher specific activities almost offset fourfold lower average sediment accumulation rates on the slope, so that slope sediment integrated Pu inventories are at most a factor of two less than inventories in shelf sediments. Pu activities and inventories in Washington shelf and slope sediments have strong

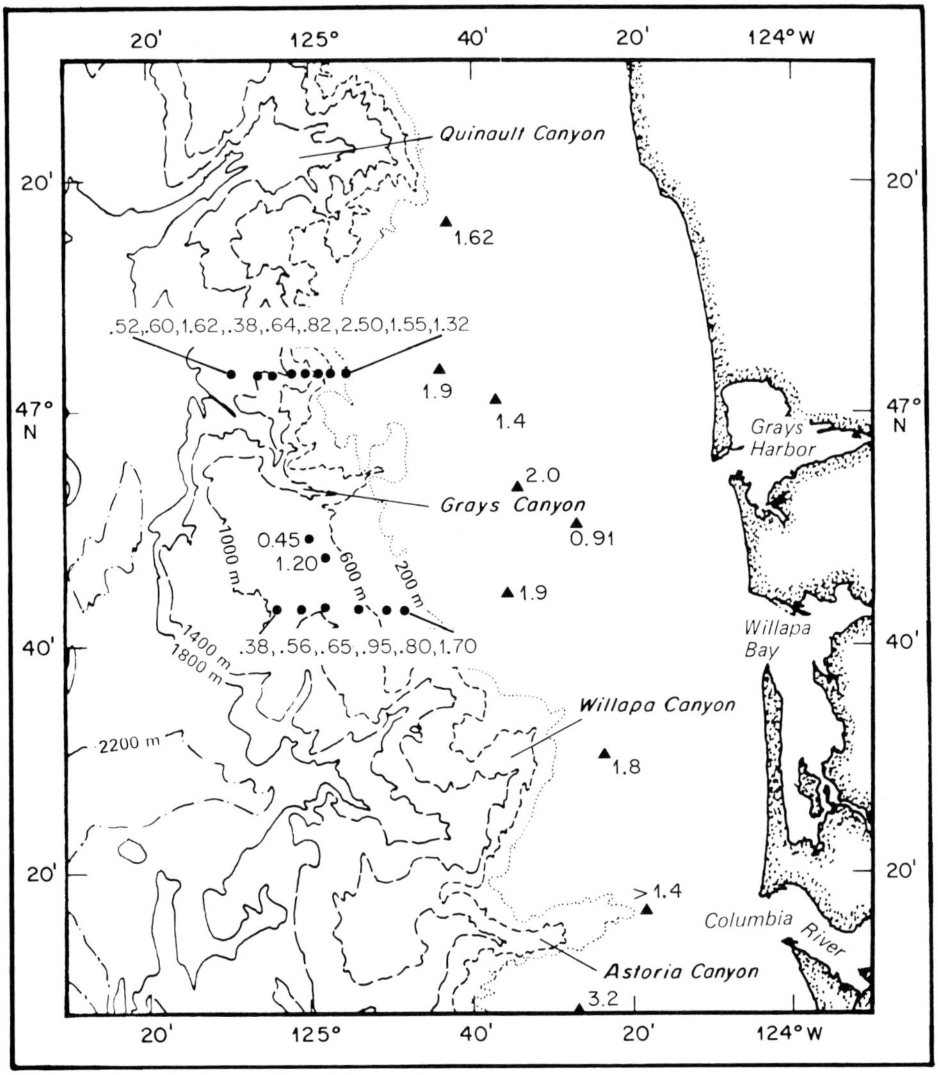

Fig. 10.15 Integrated sedimentary inventories of $^{239,240}Pu$ (dpm cm $^{-2}$) for Washington shelf (triangles, Beasley *et al.*, 1982) and slope (circles, Carpenter *et al.*, 1986) locations.

correlations with excess ^{210}Pb activities and inventories, respectively (Fig. 10.16). $^{210}Pb/Pu$ inventory ratios tend to increase from about 100 in shelf sediments to 300-380 in deeper slope sediments. This increase is attributed to the higher ratio of these dissolved nuclides in waters supplied to the slope. Selective chemical leachings of sediments have shown that Pu and ^{210}Pb are both preferentially associated with hydrous Mn and Fe oxides (Carpenter *et al.*, 1981; Beasley *et al.*, 1982). This similar phase association enables prediction of Pu inventories and accumulation rates for Washington shelf and slope sediments to within a factor of two from relatively easily measured ^{210}Pb data (Carpenter *et al.*, 1987).

$^{240}Pu/^{239}Pu$ atom ratios in surface layers of Washington slope and shelf sediments are typically close to 0.20 (Beasley *et al.*, 1982; Carpenter *et al.*, 1987), slightly greater than the mean ratio of 0.18 in atmospheric fallout, but identical to ratios in stratospheric air masses

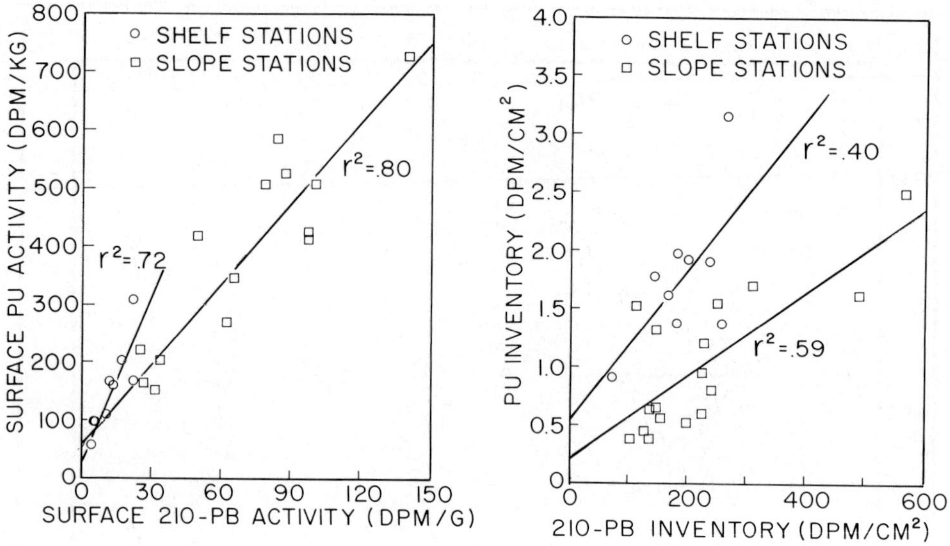

Fig. 10.16 Relationships between excess ^{210}Pb and ^{239,240}Pu: (A) activities in surface sediments and (B) integrated inventories in Washington shelf and slope sediments. Separate lines are drawn for least squares regressions of shelf and slope data. Least squares regressions of shelf and slope sediment data combined have $r^2 = 0.87$ for the surface activity relationship and 0.26 for the integrated inventory relationship.

during 1977. Ratios do not change with depth in shelf sediments, but decrease significantly in three of four slope cores. Decreases in ^{240}Pu/^{239}Pu atom ratios with depth are due to mixing of more recent fallout debris, which has higher ratios, with debris from early, low yield, low neutron flux tests, which produce fewer multiple neutron capture reactions necessary to produce the heavier isotope. These early, low yield tests occurred both at the Nevada test site, where ratios as low as 0.03 were reported, and at the Pacific and Asian test ranges of the United States and the U.S.S.R. Low ratios have been previously reported in Gulf of Mexico sediments (Scott *et al.*, 1983) and Atlantic sediments (Buesseler and Sholkovitz, 1987) and attributed to debris from Nevada. Nevada debris is probably of secondary importance off the Washington coast, where westerly winds predominate. Horizons with ratios below 0.16 contained less than 10% of the total Pu inventory in four slope cores (Carpenter *et al.*, 1987). Thus, close-in fallout from the early, low yield tests has increased total Pu activities in Washington slope sediments over those resulting from atmospheric fallout by at most 10%.

The high power Ivy-Mike tests of 1952 injected debris with ^{240}Pu/^{239}Pu ratios of 0.36 into the stratosphere and caused elevated ratios at depth in polar ice sheets (Koide *et al.*, 1985). Such high ratios are not evident at depth in Washington slope sediments. Each centimeter of slope sediment represents several years of deposition and the additional mixing in most surficial sediments plus the Pu residence time in the water column combine to decrease the resolution of the fallout record preserved in sediments relative to that in polar ice sheets.

^{241}Am activity profiles and inventories have also been determined in both Washington shelf (Beasley *et al.*, 1982) and slope (Carpenter *et al.*, 1987) sediments. ^{241}Am specific activities in surface deposits average four times higher in slope than shelf sediments, due to more efficient scavenging by finer grained particles and likely higher concentrations of the dissolved nuclide in slope waters. Higher ^{241}Am specific activities offset lower bulk sediment accumulation rates on the slope, so that integrated sedimentary inventories are similar for shelf and slope deposits.

^{239,240}Pu and ^{241}Am sedimentary inventories both tend to decrease with water depth, but have some suggestion of maxima around 600-m depth (Fig. 10.17), where pronounced maxima in dissolved Pu and probably Am concentrations occur throughout much of the north Pacific (Bowen *et al.*, 1980). However, sediment accumulation and mixing rates are also greatest in this depth range (Fig. 10.12), so the effects of the water column maxima cannot be isolated from increased scavenging due to greater particle flux and sediment mixing.

Average ^{241}Am/^{239,240}Pu inventory ratios are twice as large in slope sediments relative to shelf deposits (Carpenter *et al.*, 1987). This increase is due to a greater particle affinity for dissolved Am than Pu (and hence enhanced Am deposition) and possible higher dissolved Am/Pu ratios in deeper water. Am/Pu activity ratios change little with depth in either slope or shelf sediments, suggesting insignificant separation of the two nuclides during diagenesis.

^{137}Cs activities in the surface 0-1 cm of Washington slope sediments average 560 ± 250 dpm kg^{-1} (Carpenter *et al.*, 1987), slightly less than activities in adjacent Washington shelf sediments (Beasley *et al.*, 1982). The finer grained, more clay-rich deposits of the slope do not contain more ^{137}Cs than mid-shelf silt deposits. The difference in ^{137}Cs activity between shelf and slope sediments is opposite that for ^{210}Pb, Pu isotopes and ^{241}Am activities, which are all higher in slope than shelf sediments.

^{137}Cs inventories (Fig. 10.17) in both Washington shelf and slope sediments result primarily from residual Cs retention on particles discharged by the Columbia River. ^{137}Cs is readily measurable in slope sediments collected at 1,415-m depth, even though it is barely detectable below about 700 m in typical oceanic water columns. Sedimentary ^{137}Cs must be acquired from surface waters or land and partially retained during transport to slope environments. ^{239,240}Pu/^{137}Cs inventory ratios are 2 to 3 times greater in slope than shelf sediments, due to preferential Pu scavenging and loss of ^{137}Cs through desorption and decay. Interestingly, Pu/Cs activity and inventory ratios are not markedly higher in sediments from around 600 m, where there is a large water-column maximum in dissolved Pu but not Cs.

^{239,240}Pu/^{137}Cs activity ratios decrease with depth in most Washington slope sediments (Carpenter *et al.*, 1987), resembling Atlantic shelf sediments (Livingston and Bowen, 1979). This pattern contrasts with Washington shelf sediments (Beasley *et al.*, 1982) which exhibit constant ratios with depth (Fig. 10.18). The differences are attributed to preferential downward migration of the more soluble ^{137}Cs in higher porosity Washington slope and Atlantic shelf sediments, while ^{239,240}Pu remains essentially immobile after deposition. Some preferential downward diffusion of ^{137}Cs must have occurred even in the Washington shelf deposits, to prevent higher Pu/Cs ratios at depth due to decay of ^{137}Cs (30-y half-life).

Upward diffusion of Pu in near-shore and coastal sediments was originally advanced by Livingston and Bowen (1979) to explain the deeper penetration of ^{137}Cs than Pu isotopes.

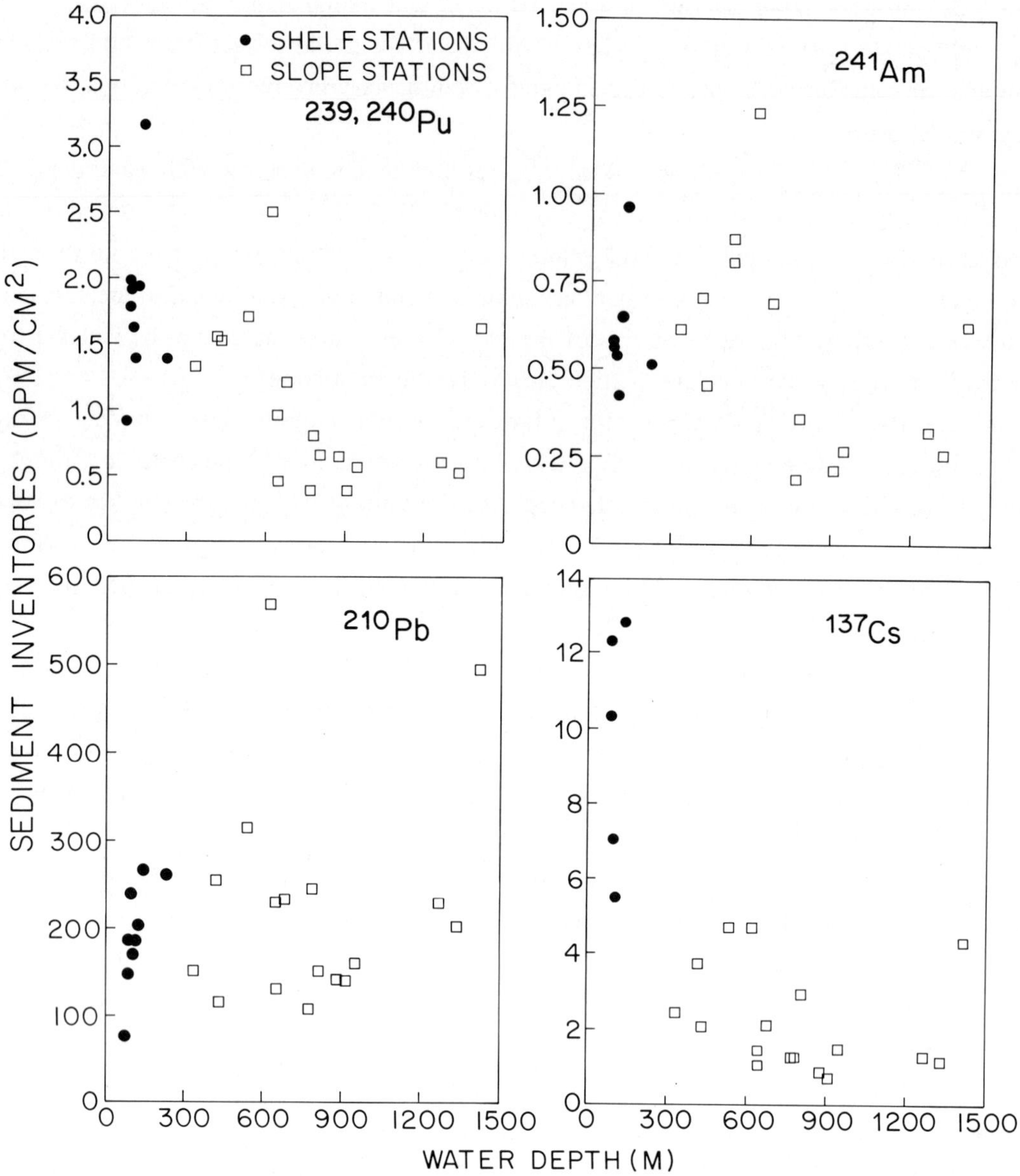

Fig. 10.17 Variations in integrated inventories of $^{239,240}Pu$, ^{241}Am, ^{137}Cs and ^{210}Pb in Washington slope sediments with depth of overlying water. From Carpenter *et al.* (1987), with permission.

However, Edgington (1981) and Carpenter and Beasley (1981) pointed out that preferential downward migration of ^{137}Cs was equally possible and chemically more reasonable. The Pu, Am and Cs data for both Washington shelf and slope sediments strongly argue against any preferential Pu migration, but for porosity-dependent, downward ^{137}Cs migration. Subsequently, the conclusion that Pu was not mobile has been supported by studies in the MERL microcosm and Narragansett Bay (Santschi *et al.*, 1983). Also, investigations based on pore water measurements of $^{239,240}Pu$ and ^{137}Cs (Sholkovitz *et al.*, 1983; Sholkovitz and Mann, 1984) have concluded that Pu remobilization is negligible in near-shore and coastal sediments representing a

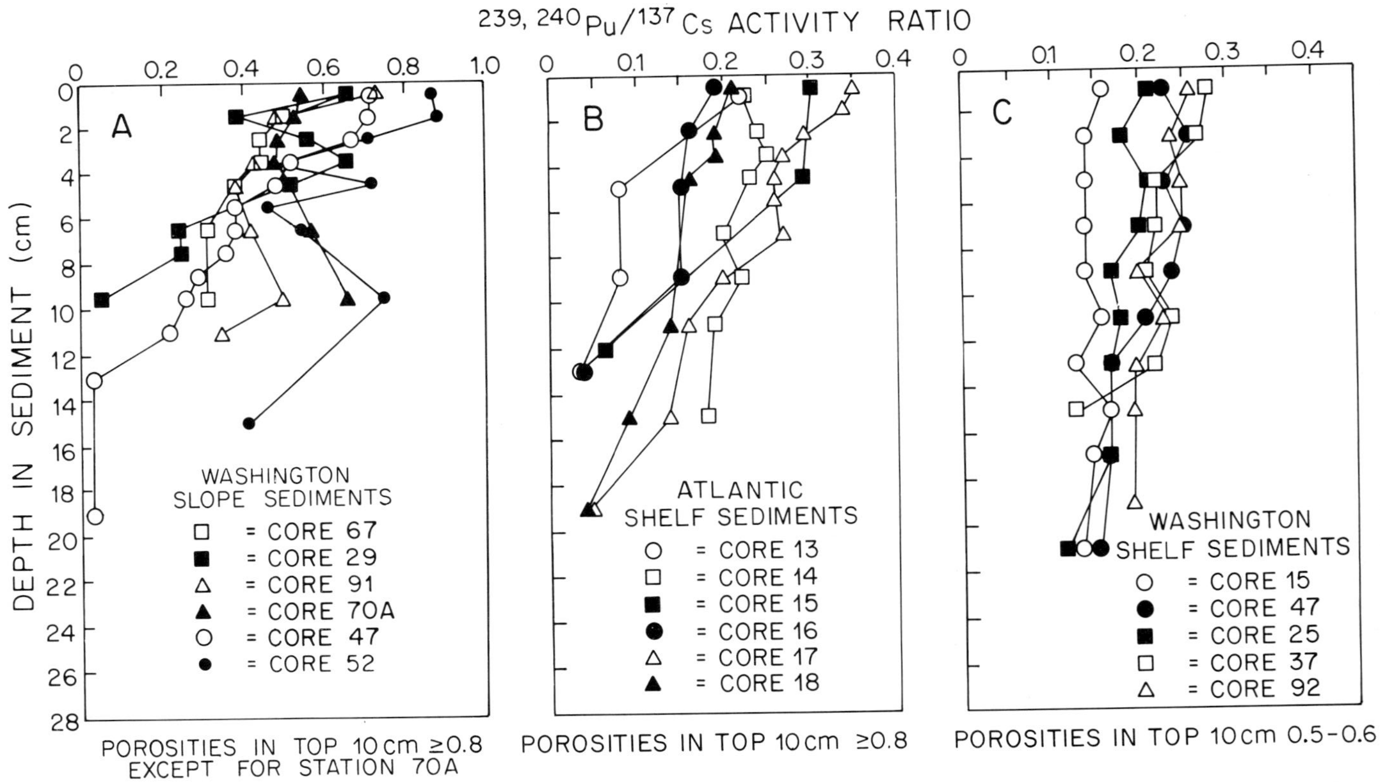

Fig. 10.18 Comparison of $^{239,240}Pu/^{137}Cs$ activity ratio vertical profiles in (A) six representative Washington slope sediment cores (Carpenter *et al.*, 1986) with those in (B) Atlantic (Livingston and Bowen, 1979) and (C) Washington (Beasley *et al.*, 1982) shelf sediment cores. Data for 11 other Washington slope cores are given in Carpenter *et al.* (1987).

variety of redox conditions. The extent of Pu mobility in deep-sea sediments remains controversial and extremely important because disposal of high level radioactive wastes by burial in the deep-sea floor is under serious consideration (Hinga *et al.*, 1982). Migration of Pu from sediments back into overlying waters has been proposed as the cause of elevated near-bottom Pu concentrations reported by Bowen *et al.* (1980) in Pacific GEOSECS water samples.

Beasley *et al.* (1982) noted that small fractions of total Pu inventories were present in horizons of ^{210}Pb-dated Washington shelf sediment cores which pre-date atmospheric nuclear testing (pre-1944). They pointed out, however, that this apparent discrepancy could be due to the fact that small amounts of downtrained artificial radionuclides are more detectable at depth since no supported or background activities exist. Beasley *et al.* (1982) also pointed out that no significant changes in Pu/Cs or Am/Pu activity ratios occurred between pre- and post-1944 horizons. Thus, processes responsible for greater-than-expected penetration depths of fallout isotopes were not unique to Pu. Selective downward mixing of all artificial nuclides relative to ^{210}Pb now seems the most likely explanation. Greater penetration depths of Pu than excess ^{210}Pb have also been reported for abyssal sediments from the Atlantic and have been attributed to differential particle mixing (Stordahl *et al.*, 1985). The important effects that subsurface egestion of surficial sediments can have on profiles of ^{210}Pb and ^{239,240}Pu have recently been demonstrated (Smith *et al.*, 1986). Such feeding and excretion behavior cannot be accurately simulated by models assuming only eddy diffusive mixing.

In the Puget Sound region, ^{239,240}Pu and ^{137}Cs profiles have been determined on only one core from Dabob Bay (Bennett, 1980) (Fig. 10.19). Pu was again found at slightly deeper depths than expected from ^{210}Pb derived accumulation and mixing rates. The integrated Pu inventory was less than in Washington shelf sediments due to the smaller advective input of dissolved Pu to Puget Sound.

Uranium and Thorium Isotopes. ^{234}U, ^{238}U, ^{232}Th, ^{230}Th, ^{228}Th and ^{234}Th activity profiles have been determined in seven cores from the Washington slope (Carpenter *et al.*, 1987). Small amounts of authigenic U were evident and there were no indications of any authigenic ^{230}Th or ^{232}Th. ^{228}Th/^{232}Th activity ratios slightly less than 1.00 indicated some loss of the soluble intermediate ^{228}Ra, but the loss appeared less than in the more extensively bioturbated Puget Sound sediments (Carpenter *et al.*, 1984). Greater deficits in ^{228}Th relative to ^{232}Th and larger increases in U concentrations with depth are expected as the exchange with depth of pore and overlying waters increases. Because these sediments did not have markedly different Pu, Am and Cs profiles, pore water exchange has had little effect on artificial nuclide profiles. However, solid phase mixing depths and rates deduced from excess ^{210}Pb and ^{234}Th profiles greatly influence penetration depths and scavenging of fallout nuclides.

Uranium activities as a function of depth increased by up to factors of two in both Washington slope and Puget Sound sediments (Carpenter *et al.*, 1984, 1987). Relatively low Th/U activity ratios in Washington coastal sediments also suggested modest preferential scavenging of U from sea water relative to Th. The largest increases in U with depth occurred in layers with zero excess ^{234}Th. These data can be explained by benthic pumping of sea water

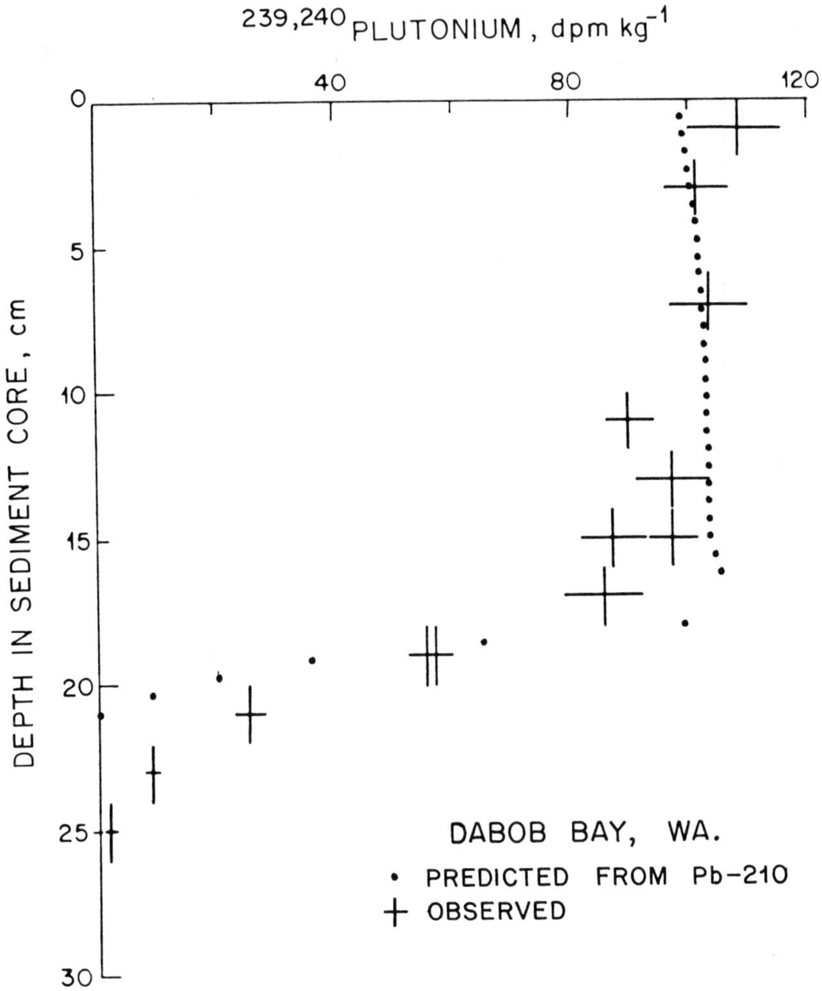

Fig. 10.19 Observed and predicted vertical distributions of $^{239,240}Pu$ activity in bottom sediments of Dabob Bay, Wa. From Bennett (1980). The predicted vertical profile was constructed using ^{210}Pb-derived sediment accumulation and mixing rates, the model of Guinasso and Schink (1975) and the time-dependent atmospheric fallout function reported by Robbins and Edgington (1975). Horizontal error bars for measured values are ± 1 standard deviation based on counting statistics.

with about 2.0 dpm l^{-1} of dissolved U, but little dissolved Th several cm down into sediment horizons which undergo nitrate and sulfate reduction (Christensen *et al.*, 1987a and b). Enhanced U deposition onto particles in reducing environments is well known. Such post-depositional removal of U from sea water in continental margin sediments could be important in global U cycling, because of the common occurrence of mildly reducing conditions at shallow depths in margin sediments.

Measured sedimentary inventories of excess ^{234}Th were 11 to 82% of the ^{234}Th production in overlying water column, except for one station where the measured inventory exceeded the water column production by a factor of two. This situation is similar to that reported by Carpenter *et al.* (1984) for excess ^{234}Th inventories in Puget Sound sediments. Differences are attributed to variations in the extent of horizontal movement and resuspension of particles in the water column and hence, the extent of scavenging. Excess ^{234}Th activities were found at surprisingly deep depths, down to 14 to 16 cm in several cores of slope sediments.

Activity profiles of excess ^{234}Th, excess ^{210}Pb, ^{232}Th, ^{230}Th, ^{234}U and ^{238}U and $^{228/232}Th$ ratios have also been determined at six sites in central Puget Sound (Carpenter *et al.*, 1984). The profiles provide new insights into the dynamic nature of solid phase mixing in surface sediments, the exchange of ^{228}Ra and other soluble species across the sediment-water interface as well as the cycling of U, Th and ^{210}Pb in this coastal zone. Surface Puget Sound sediment horizons with excess ^{234}Th have deficiencies of ^{228}Th relative to its parent ^{232}Th, due to loss of soluble ^{228}Ra from pore water to the overlying water. Solid phase U isotope concentrations in Puget Sound sediments tend to increase by up to a factor of two with depth in sediments because dissolved U is biologically pumped into sediments where it is partially removed under reducing conditions. ^{232}Th and ^{230}Th activities and $^{230}Th/^{232}Th$ ratios are constant with depth in sediments, indicating constant detrital phase compositions and essentially no authigenic ^{230}Th. Steady state ^{210}Pb depositional activities in and fluxes to Puget Sound sediments average only about one half those for sediments of the open Washington coast north of the Columbia River mouth. This is primarily because seawaters advecting into Puget Sound have a much lower supply of dissolved ^{210}Pb.

Trace metals

Arsenic, Antimony and Mercury. Total As, Sb and Hg concentrations in surface sediments from the Washington continental shelf and the North Pacific are given in Table 10.6. All sediments were collected with precautions against trace metal contamination and alteration of surface layers during coring and stored frozen in plastic until subsequent analyses. None of the sediments was from the area of the Juan de Fuca spreading center; two samples were from the bottom of the Aleutian Trench. The deep-sea sediments can be described as red clays containing little, if any, $CaCO_3$. Several contained small manganese nodules.

Arsenic concentrations in the deep-sea sediments were between 3 and 18 ppm, similar to values for East Pacific sediments far removed from active oceanic ridge systems (Bostrom and Valdes, 1969). Most Hg concentrations ranged from 100 to 200 ppb. Bostrom and Fisher (1969) reported Hg concentrations in east Pacific sediments varied from 1 to 400 ppb, with highest values in sediments near the East Pacific Rise. The two Aleutian trench samples had typical As and Sb and high Hg concentrations.

Concentrations of As, Sb and Hg in shelf sediments were between values for Columbia River suspended matter and bed load (see Section 10.2). Concentration variations in shelf sediments are primarily attributed to relative proportions of riverine suspended and bottom sediments. Generally higher concentrations of Sb and Hg and possibly As in deep-sea relative to shelf sediments are believed partially due to grain size and surface area variations. No anthropogenic contributions of these three metals was evident in shelf surface sediments since concentrations in surface layers were comparable to those observed in deep, non-contaminated horizons in Puget Sound cores (Crecelius *et al.*, 1975).

Table 10.6 Total concentrations of As, Sb and Hg in Washington shelf and North Pacific deep-sea sediments. As and Sb contents were determined by nondestructive neutron activation analysis (Crecelius *et al.*, 1975). Hg concentrations were determined by flameless atomic absorption (Bothner *et al.*, 1980).

Latitude	Longitude	Depth (m)	As (ppm)	Sb (ppm)	Hg (ppb)
		Washington Shelf Stations			
48° 17.5'N	124° 00.0'W	205	9.2 ± 1.8	0.95 ± 0.19	59 ± 1
48° 25.0'N	124° 30.0'W	240	9.5 ± 0.4	0.81 ± 0.03	55 ± 1
			6.7 ± 0.7	0.49 ± 0.04	77 ± 2
					51 ± 1
48° 8.0'N	124° 45.0'W	34	9.5 ± 0.8	0.52 ± 0.13	7 ± 0.5
48° 6.5'N	124° 51.0'W	50	5.2 ± 0.3	0.27 ± 0.01	19 ± 1
48° 4.5'N	125° 5.1'W	135	--	--	91 ± 2
48° 2.0'N	125° 15.0'W	150	5.2 ± 0.6	0.63 ± 0.04	46 ± 1
47° 29.0'N	124° 29.0'W	22	4.5 ± 0.4	0.30 ± 0.03	28 ± 1
47° 29.0'N	124° 33.0'W	50	5.1 ± 0.3	0.25 ± 0.06	15 ± 0.5
47° 29.0'N	124° 42.0'W	91	5.2 ± 0.4	0.54 ± 0.02	39 ± 1
47° 9.0'N	124° 39.0'W	95	7.6 ± 0.3	0.97 ± 0.07	46 ± 1
			5.9 ± 0.4	0.68 ± 0.03	
47° 8.0'N	124° 24.1'W	50	--	--	7 ± 0.5
46° 40.5'N	124° 37.0'W	164	7.7 ± 0.4	0.83 ± 0.07	61 ± 1
46° 38.1'N	124° 26.9'W	110	5.2 ± 0.3	0.30 ± 0.06	24 ± 1
			4.2 ± 0.3	0.35 ± 0.03	
46° 37.8'N	124° 9.8'W	28	3.8 ± 0.3	0.43 ± 0.09	11 ± 0.5
Mean			6.3 ± 1.9	0.55 ± 0.25	40 ± 2.5
		Deep Sea Stations			
42° 20.0'N	129° 36.0'W	3394	7.2 ± 0.3	1.7 ± 0.1	191 ± 4
36° 8.0'N	140° 27.0'W	5270	16 ± 1	2.0 ± 0.1	
			18 ± 1	2.4 ± 0.1	
			18 ± 1	2.6 ± 0.1	
41° 56.4'N	150° 43.0'E	5087	8.8 ± 0.8	1.8 ± 0.1	--
50° 56.4'N	149° 42.1'W	4799	6.9 ± 0 6	1.9 ± 0.1	106 ± 2
48° 42.0'N	127° 29.5'W	2554	5.9 ± 0.6	4.1 ± 0.2	158 ± 3
49° 7.0'N	132° 60.0'W	3144	5.7 ± 0.3	2.6 ± 0.1	113 ± 2
47° 36.0'N	165° 48.0'E	5719	7.8 ± 0.5	1.5 ± 0.07	80 ± 2
51° 13.0'N	174° 27.0'E	7005*	3.1 ± 0.3	0.87 ± 0.04	270 ± 6
50° 58.0'N	171° 33.0'W	6976*	4.5 ± 0.2	0.38 ± 0.04	286 ± 6
			5.0 ± 0.3	0.47 ± 0.04	
49° 41.0'N	141° 1.0'W	3668	3.9 ± 0.	0.90 ± 0.05	47 ± 1
48° 48.0'N	130° 3.0'W	2744	5.3 ± 0.6	4.3 ± 0.2	202 ± 4
49° 6.0'N	133° 11.0'W	3204	--	--	139 ± 3
45° 49.0'N	159° 31.0'E	4720	6.8 ± 0.4	0.76 ± 0.07	40 ± 1
			5.6 ± 0.4	0.62 ± 0.07	
44° 39.8'N	127° 13.6'W	2862	--	--	326 ± 7
					320 ± 7
Mean			8.0 ± 4.8	1.8 ± 1.2	175 ± 100

*Aleutian Trench sites

Table 10.6 summarizes total Mn, Al and As concentrations in sediments of the Washington coastal zone. Total Mn concentrations in Washington slope and shelf sediments were relatively low, averaging 400 to 500 ppm and varied by less than a factor of two with depth. These sediments had less pronounced surface maxima in Mn concentrations than Puget Sound sediments. They also had much lower Mn concentrations than deep-sea clays, which average about 6,700 ppm.

Jones and Murray (1985) argued from bottom and interstitial Mn data that there is a source of dissolved Mn in or near the sediments of the Washington continental margin and that this Mn undergoes horizontal advective-diffusive transport away from the coast. They concluded that mildly reducing conditions exist in sediments of the Washington continental margin and that active remobilization of Mn occurs by a combination of diffusion from pore to overlying waters and resuspension of bottom sediments. The sediments have only a thin oxidizing surface layer, which may not be an effective trap for upwardly diffusing Mn or other redox sensitive ions. Mn recycling at the sediment water interface is believed to play an important role in enhancing near-bottom scavenging of dissolved Pb-210 (Carpenter *et al.*, 1981) and Pu isotopes (Beasley *et al.*, 1982).

Arsenic concentrations in non-contaminated Puget Sound surface sediments and at depths in other cores ranged from 3 to 15 ppm (Crecelius *et al.*, 1975; Peterson and Carpenter, 1986). Sb concentrations in these same samples ranged from 0.3-1.0 ppm. The lowest Hg concentrations (between 0.01 and 0.1 ppm) are found in Hood Canal and Skagit Bay, away from industrial activities and the Strait of Juan de Fuca and Admiralty Inlet, areas where sediments are coarse due to strong tidal currents (Crecelius *et al.*, 1975).

Sedimentary As concentrations in parts of Puget Sound have increased since 1890 due to discharges from a large smelter which began operation at that time and only recently ceased operation in early 1985 (Crecelius *et al.*, 1975; Carpenter *et al.*, 1978; Peterson and Carpenter, 1986). Smelter emissions have had no inpact on Al or Mn concentrations in Puget Sound sediments. Table 10.6 lists typical As/Mn ratios for average crustal rock types and marine deposits as well as solids from rivers, coastal sediments and water column particulate matter within the study area. Relationships between solid phase total concentrations of As and Mn for the Washington coast and Puget Sound are summarized in Figure 10.20.

Near-surface sediment at Puget Sound station 39 in Quartermaster Harbor (Figs. 10.3, 10.13) had the highest As/Mn and As/Al ratios of any site studied, reflecting proximity to the smelter and deposition of its atmospheric particulate emissions. At Station 56, in East Passage between Seattle and Tacoma and more distant from the smelter, near surface As/Mn and As/Al ratios were lower, but still above ratios in sediment horizons predating smelter construction. The most plausible explanation for these data is direct incorporation of significant amounts of smelter stack dust, which had As concentrations as high as 55%. Surface sediments at stations 98, 51 and 11 have high As/Al ratios, indicating that these sediments are contaminated with As. However, their low As/Mn ratios suggest that a mechanism other than direct, smelter-derived solid phase As input is responsible for the elevated As levels.

Cores collected at Puget Sound stations show about a 30-y time lag between sedimentary horizons dated at 1890 and those with consistently higher As concentrations

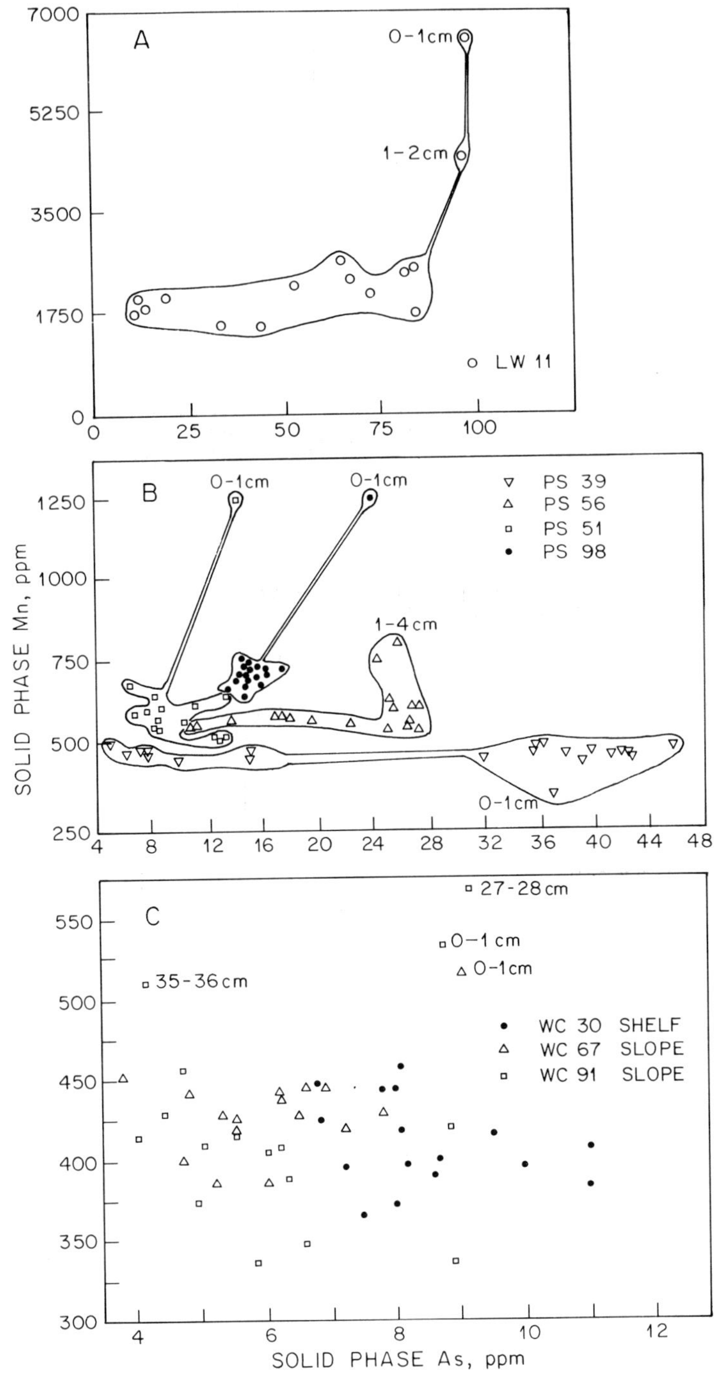

Fig. 10.20 Relationships between total solid phase As and Mn concentrations in sediments of: (A) Lake Washington; (B) Puget Sound; and (C) the Washington coast.

(Peterson and Carpenter, 1986). This is similar to the residence time calculated for particles within ^{210}Pb derived surface mixed layers of these locations (Carpenter *et al.*, 1985), suggesting that near-surface mixing has decreased the resolution of the sedimentary As input record by a time roughly equal to the mixed layer residence time. Sediments younger than 1920 from Puget

Sound stations 51, 98, 56 and 39 averaged 13 ± 1.4, 16 ± 2.4, 25 ± 2.3 and 39 ± 3.6 ppm As, respectively. Arsenic concentrations in intervals older than 1890 from cores collected at stations 51 and 39 averaged 7.7 ± 0.8 and 7.1 ± 0.9 ppm, respectively. Thus, As concentrations in central Puget Sound sediments have increased by factors of 2 to 5 as a result of smelter emissions.

Crecelius (1975) reported average suspended particulate As concentrations in Lake Washington surface and bottom waters of 37 to 159 ppm and 62 to 363 ppm, respectively. He suggested that precipitation of Fe- and Mn-rich solid phases within the water column was the most important mechanism removing As to the sediments. The average concentration of As above 12-cm depth in a core collected at station 11 was 78 ppm, similar to water-column particulate matter concentrations. Both solid phase As and Mn exhibited large surface maxima. Large variations in As concentrations within the surface mixed layer suggest that As is redistributed within this region of the sedimentary column (Peterson and Carpenter, 1986). Manganese oxide precipitation in the oxidizing, near-surface sediments is supported by upward Mn^{2+} diffusion from the reducing pore water at depth. A concomitant solid phase As maxima could occur through adsorption or coprecipitation with hydrous Mn oxides.

Puget Sound station 98 near Everett shows large increases in both Mn and As in surface sediments, indicating the importance of diagenetic recycling processes. Data for Puget Sound station 51 and Washington coast stations 67 and 91 (duplicate occupations of a location at the base of the continental slope shown in Peterson and Carpenter (1983)) also show evidence for some Mn and possibly As, diagenetic recycling (Figs. 10.20b and c). Station 39 has essentially no change in Mn concentrations (Fig. 10.20b), indicating that increased Mn scavenging is not responsible for increased As concentrations in surface sediments. Washington coast and station 39 sediments have similar Mn concentrations: the much higher As levels at station 39 clearly reflect smelter-derived As. Station 56 has increased Mn near the surface, but no authigenic As due to increased Mn scavenging is detectable above the high As values of 25 ppm As. The high values are largely smelter-derived.

Arsenic concentrations in sediments from Lake Washington station 11 show both anthropogenic and diagenetic influences: (1) As increases due to smelter inputs unaccompanied by Mn increase except very near the sediment-water interface (Fig. 10.20a); and (2) Mn and probably As diagenetic recycling near the sediment surface. Diagenetically altered As concentrations are not as dramatic as those at some Puget Sound locations because of high As concentrations below 2 to 3 cm. However, As does increase from about 80 to 100 ppm in surface horizons. Sediments at station 11 have much higher As concentrations than sediments at station 39, despite its greater distance from the smelter. This implies either more efficient As removal to Lake Washington sediments or better physical focusing of As deposition. Runoff of atmospheric precipitation is certainly more focused into the lake than into Puget Sound sediments, and lake sediments have much more Mn and Fe than Puget Sound sediments.

Sediments of two Saanich Inlet stations show completely different behavior in that the Mn/As ratio is essentially constant throughout the cores (Fig. 10.21). This is attributed to As

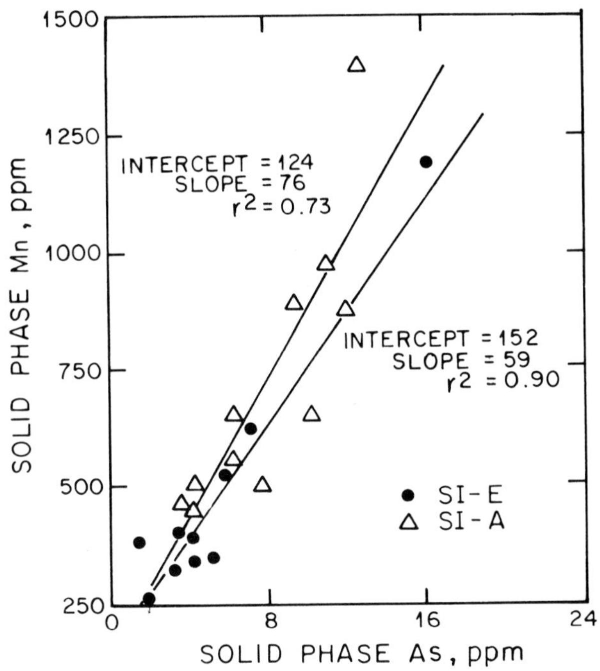

Fig. 10.21 Relationships between solid phase total As and Mn concentrations in anoxic sediments from two Saanich Inlet stations.

scavenging during Mn precipitation in the water column, followed by subsequent return of both Mn and As to the water column as the Mn oxide precipitate dissolves in response to anoxic conditions within the sedimentary column.

Vertical, particulate As and Mn fluxes (1.1 μg cm $^{-2}$ yr $^{-1}$ and 200 μg cm $^{-2}$ yr $^{-1}$, respectively) measured in the Dabob Bay water column in a year-long time series of sediment trap deployments were highly correlated and could only be partially accounted for by riverine particulate sources (50 ± 6 and 40 ± 5% of the total As and Mn fluxes, respectively) (Bennett, 1980). Bennett concluded that the excess Mn flux was most likely due to Mn oxide coatings formed on detrital particles, either at the sediment water interface or within the water column. The excess solid phase As flux was believed to be supplied through the adsorption of dissolved As onto particles such as clays and hydrous Mn and Fe oxides. This is in agreement with the low As/Mn and high As/Al ratios for the trapped material (Table 10.6). Bennett also determined that only 58 and 33% of the water column As and Mn fluxes, respectively, were preserved in the underlying sediment of Dabob Bay (see Table 10.11 in Section 10.7).

Peterson and Carpenter (1986) found that porewater total As concentrations exhibit subsurface maxima with magnitudes 10 to 60 times greater than overlying bottom waters. In marine environments their magnitudes were proportional to solid phase As concentrations at the same depths (Fig. 10.22). These maxima consistently occurred within ± 1 cm of the bottom of the ^{210}Pb-derived surface mixed layer, suggesting solid phase As release to pore water in response to a redox gradient across this mixing boundary. Dissolved As beneath these maxima indicated As removal to solid phases in marine but not lacustrine porewaters. Porewater

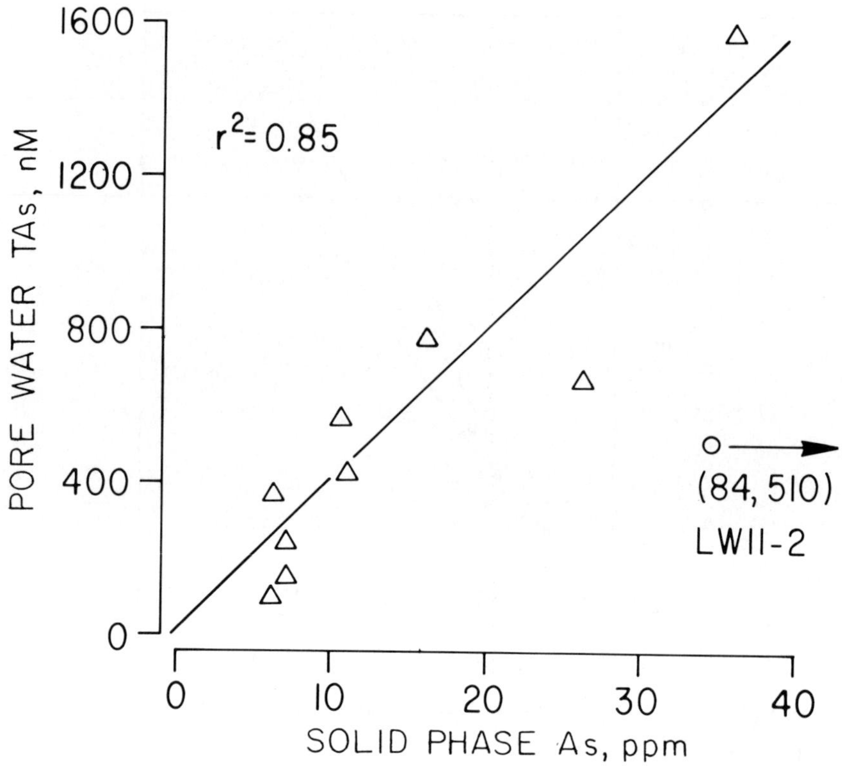

Fig. 10.22 Relationship between solid phase and total dissolved As concentrations at the depth of the subsurface porewater As maximum in sediments from Puget Sound and the Washington coast. The datum point for Lake Washington sediments (LW11-2) falls off scale to the right well below the linear relationship found for marine samples. From Peterson and Carpenter (1986), with permission.

As^{3+} / As^{5+} ratios were 1.0 for Washington coastal sediments and usually between 1.0 and 4.0 for Puget Sound and Lake Washington sediments (Peterson and Carpenter, 1986). These ratios suggest equilibrium pε values similar to those believed to exist about the oxic/anoxic transition in the Saanich Inlet water column (Emerson *et al.*, 1979).

Sedimentary As geochemistry for the Washington coast and Puget Sound regions may be summarized as follows. Typically, strong correlations between solid phase As and Mn concentration maxima in surface sediments are the result of either intense As scavenging in overlying waters by hydrous Mn and Fe oxides or of solid phase As dissolution at depth in sediments followed by upward diffusion and coprecipitation with Mn and Fe oxides near the sediment water interface. Atypically, high sedimentary As concentrations occur without accompanying high Mn levels, due to anthropogenic As input. Sedimentary As fluxes are an order of magnitude greater in Puget Sound and Lake Washington than off the Washington coast or in Saanich Inlet, probably reflecting the large anthropogenic which was present until 1985 in the Puget Sound region.

Highest concentrations of Hg in Puget Sound sediments were found in Bellingham Bay deposits collected in the early 1970's from near the outfall of a mercury-chloralkali plant

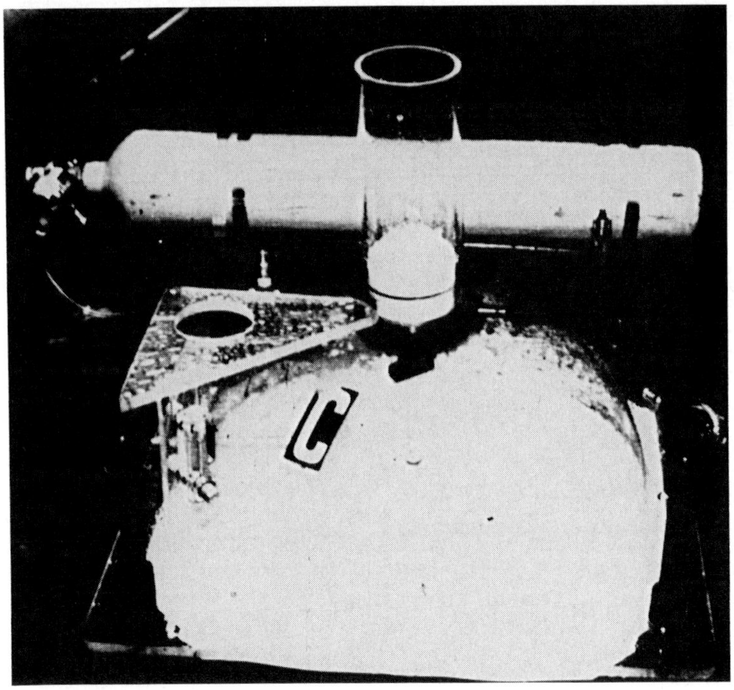

Fig. 10.23a Bell jar used in studies of chemical fluxes out of Washington coastal sediments. Photograph shows unit prior to deployment in the 'blank' configuration, with a teflon base plate in position to prevent any chemical influx from sediments. Air tank and regulators to control stirring air are visible, as are the rubber septum through which water samples are taken with syringes, the collapsible plastic 'bellows' in the neck which compensates for volume changes during sampling and the one-way valve at the top of the bellows through which bubbles exit.

(Bothner *et al.*, 1980). Discharges estimated at 4.5 to 9 kg Hg d^{-1} between 1965-1970 resulted in surface sediments within 7 km of the plant with Hg concentrations 10 to 20 times background values. Hg concentrations in near-surface sediments have decreased surprisingly rapidly since the Hg discharge was essentially eliminated in 1970 (Bothner *et al.*, 1980).

Many particle reactive chemical contaminants tend to accumulate in coastal sediments. It is of major importance to determine whether the contaminants will remain buried in the sediments or if they will be remobilized into pore or overlying waters where they are more likely to interact with the biota. The possibility of As and Hg remobilization was investigated in bell jar studies by Carpenter *et al.* (1978).

Bell jars (Figs. 10.23a and b) were constructed out of the top halves of 12 gallon Pyrex carboys and attached to a plexiglass base plate which limited jar penetration into the bottom to about 5 cm and increased stability. When placed on the sea floor by SCUBA divers, the jars covered an area of 1240 cm^2 and contained about 20 liters of bottom water. Compressed gas was passed through Tygon tubing and fed into the jars through a glass frit which generated small bubbles. A regulating valve on the plexiglass frame allowed divers to adjust the flow of gas to gently mix waters in the jars without disturbing the enclosed sediment. Tests with dyes

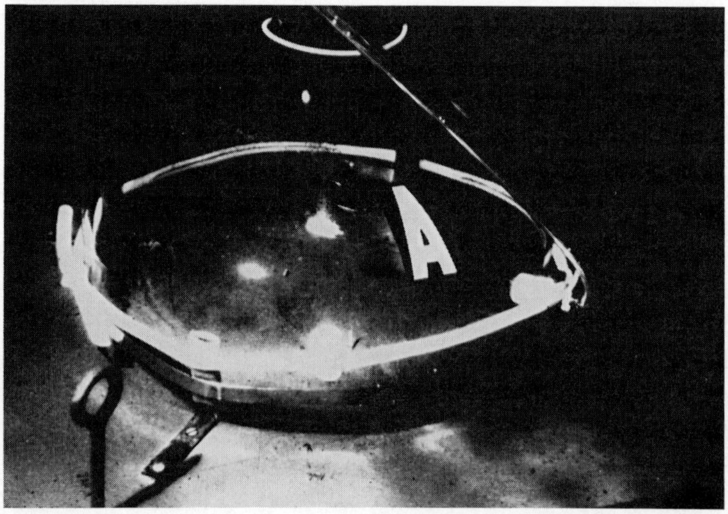

Fig. 10.23b Bell jar used in studies of chemical fluxes out of Washington coastal sediments. Photograph shows a unit of a slightly different design from that shown in Figure 10.23a, sitting on the sea floor with stirring air bubbles gently rising.

confirmed the thoroughness of mixing. Bubbles rose into the neck of the bell jar and escaped through a one-way valve. Breathing quality air was used to maintain oxidizing conditions typical of Puget Sound bottom waters, but other gases such as N_2 could easily be substituted. In one experiment the water inside the jar was deliberately allowed to go anoxic and a Teflon paddle machined into one port of the bell jar allowed divers to gently mix the enclosed water.

Chemical concentrations in the enclosed seawater periodically were monitored by analyzing aliquots of water collected by divers during the experiments. Water from inside the bell jars was collected by divers with 100-ml glass hypodermic syringes equipped with stainless steel needles inserted through a rubber septum fastened to the bell jar wall. To compensate for the volume of water removed, one of two methods was used: 1) water from outside the bell jar was allowed to flow in through a replacement inflow tube (2 mm x 200 cm); or 2) a collapsible bladder was attached to the top of the jar. This prevented any vacuum from developing and sucking pore water into the jars. Blank runs were made with the apparatus at the sea floor enclosing bottom water, but resting on Pyrex or Teflon plates, preventing any chemical influx from the sediment. Devol (1987) has recently described another type of *in situ* benthic chamber which has been used to measure fluxes of nutrients and alkalinity. The system has two important additional features: 1) the automated overlying water sampling system does not require divers and 2) underlying sediments are recovered at the end of the experiment.

Results of bell jar studies are given by Carpenter *et al.* (1978) for As and Bothner *et al.* (1980) for Hg. Fluxes of As out of contaminated Quartermaster Harbor sediments were found to be small and could be explained by molecular diffusion without augmentation by biological stirring or pumping. Arsenic concentrations in surface pore waters appear dependent upon redox conditions, being smaller under oxidizing conditions which precipitate hydrous Mn and Fe oxides. Hg fluxes out of Bellingham Bay sediments were substantially increased when both the water inside the bell jar and sediments were anoxic. Instead of forming highly insoluble cinnabar (HgS), apparently polysulfide or other complexes of Hg were formed. The latter are more soluble and mobile.

Chromium, Cobalt and Other Trace Metals in Puget Sound Sediments.

Crecelius *et al.* (1975) observed no elevated Cr or Co concentrations in Puget Sound that could be attributed to the activities of man. Cr concentrations ranged from 40 to 200 ppm, with a mean of 90 ppm, while Co concentrations ranged from 5 to 30 ppm, with a mean of 15 ppm.

Central Puget Sound surface sediments have 10 to 12 times higher concentrations of Ag, 5 to 6 times higher concentrations of Pb and Hg and 2 times higher concentrations of Cu than sediments deposited over 100 y ago (Bloom and Crecelius, 1987). In some areas of central Puget Sound, sediment accumulation rates and mixing depths are so great that the depth of contaminated sediments exceeds 200 cm. Subsurface concentration maxima are observed for Pb, Hg and possibly Ag, but not Cu and Cd. This likely result from reduction in contaminant discharges to the Sound since about 1960.

Feely *et al.* (1983) and Paulson *et al.* (in press) examined the processes which remove trace metals from Puget Sound waters to underlying sediments. Metal uptake and release by hydrous manganese oxides in both sediments and suspended particles was shown to have a significant affect on distributions of several trace metals. Feely *et al.* (1986) measured seasonal and vertical variations in total concentrations and phase associations of several trace metals in suspended sediments at a single site in the main basin of Puget Sound. Samples were collected bimonthly for a year. Vertical fluxes were measured and combined with dissolved metal profiles to compute scavenging residence times for Cr, Mn, Fe, Ni, Cu, Zn and P.

Concern over possible adverse biological effects motivated several other studies of stable metals in Puget Sound sediments (e.g., Schell and Nevissi, 1977; Curl, 1982; METRO TPSS report, 1984). Unfortunately, data from these studies are difficult to interpret in terms of geochemical processes. In particular, (1) some sediments were collected by devices which lose, disturb, or mix the near surface layers; (2) analytical techniques employed were unreliable for certain metals at low concentrations found in relatively uncontaminated samples; (3) sediment sampling schemes were heavily biased toward atypical, highly-contaminated hotspots, such as Hylebos Waterway and Commencement Bay; (4) inadequate consideration was given to the likelihood that geographic variations in concentrations were due in part to variations in sediment texture or grain size rather than to chemical inputs; (5) inadequate consideration was given to effects of benthic mixing of surface layers on chemical profiles; and (6) insufficient ancillary information was obtained on parameters such as sediment accumulation rate and bulk organic carbon content.

Total and Organic Carbon

Gross *et al.* (1972) provided the first summary of the distributions and concentrations of organic C in surface sediments from the Washington shelf and slope. Table 10.5 summarizes more recent results on surface sediment concentrations of total C, Al, Mn and C/N ratios in different geologic regimes of the Washington coast. Factors controlling sedimentary organic C concentrations include grain size, rates of sediment accumulation and dissolved O_2 concentrations in near-bottom waters. A pronounced O_2 minimum with concentrations only 5 to 10% of saturation impinges on the slope at around 800-m depth (Fig. 10.6). This O_2 minimum results in preservation of 2 to 3 times more organic C concentrations in slope than shelf or deeper sediments (Gross *et al.*, 1972).

Carbon determinations before and after acidification with HCl on other sediments of the region show that almost all the total C is in organic forms rather than $CaCO_3$. Distributions of plant wax *n*-alkanes (Prahl and Carpenter, 1984) and lignins (Hedges and Mann, 1979b) suggest that at least 74 ± 10% of the organic C in Washington slope sediments is of marine origin. Total C concentrations in Washington slope sediments are in the 2 to 3% (g g^{-1}) range, typical of most slope sediments.

Figure 10.24 summarizes vertical total C profiles in Washington continental slope sediments from 19 locations with water depths between 332 to 1415 m. At most, 8 of the 19 slope cores have higher C concentrations in more recently deposited layers. In only three cores is the change greater than a factor of 2 and in two of those three cores the relatively large change is due to recovery at depth of a C-poor, 20,000-y old grey clay layer (Barnard, 1978). Thus, these data do not show any region-wide increase in organic C in recent decades, as postulated by Walsh (1984). The modest decreases in organic C with depth can be explained by organic matter decomposition.

C/N weight ratios in surface layers of these 19 cores average 12.5 ± 1.7 and change by more than 1 unit with depth only in the three cores (stations 29, 45, 70A and 72) which penetrate into the Pleistocene grey clay (Carpenter, 1987). Constant C/N ratios in depth profiles also suggest little change in organic matter residue composition in recent decades due to changes in efficiency of recycling of organic residues as hypothesized by Walsh (1984).

Accumulation rates of organic C in sediments of different areas of the Washington coast can be calculated from the product of bulk sediment accumulation rates and C contents given in Table 10.5, assuming negligible changes in C concentration with core depth and assuming that ^{210}Pb-derived accumulation rates are not affected by mixing. Mean C accumulation rates so calculated are 6.4, 1.4 and 3.6 mg C cm^{-2} y^{-1} for sediments of canyons, open slope and shelf areas, respectively (Carpenter, 1987). Sedimentary C accumulation rates are lower on the slope than the shelf, even though slope sediments have two to three times greater C concentrations, because sediment accumulation rates are lower on the slope.

Carpenter *et al.* (1982) summarized several lines of evidence suggesting that ^{210}Pb derived sediment accumulation rates for the Washington slope are within a factor of 2 or 3 of true, long-term rates. There is no evidence for any dramatic increase in sedimentation rate in recent decades. This conclusion is supported by radiocarbon dating of long-term accumulation rates

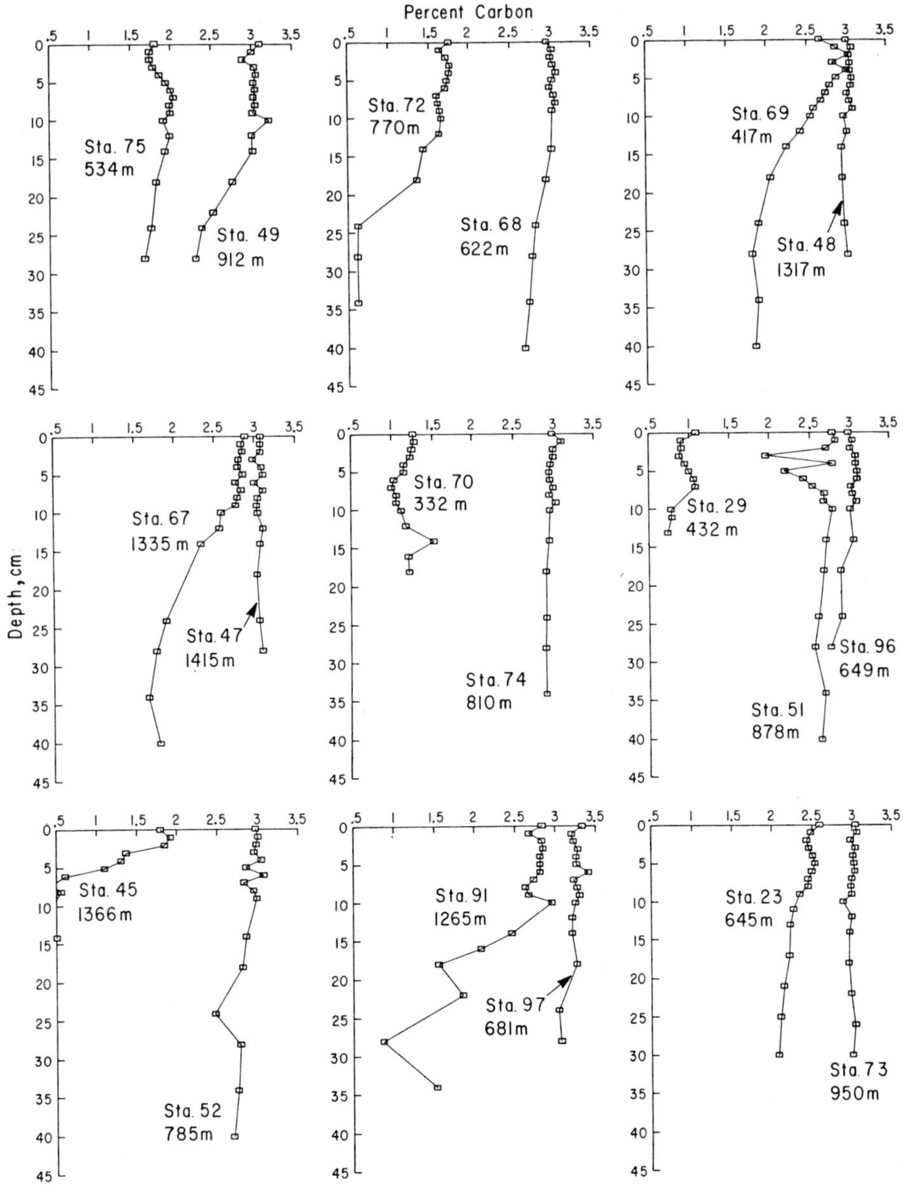

Fig. 10.24 Profiles of total organic carbon concentrations in sediments from 19 locations on the Washington continental slope at various water depths. From Carpenter (1987), with permission.

on the Oregon slope (Kulm and Scheidegger, 1979). The mean sedimentary C flux calculated with the ^{210}Pb-derived sedimentation rates for the Washington slope (1.4 mg C cm $^{-2}$ y $^{-1}$) is in reasonable agreement with that estimated from the productivity of slope waters and the fraction which survives to burial depths (Suess, 1980).

Walsh (1984) originally calculated a global mean carbon accumulation rate of 0.9 x 10 ^{9} tons y $^{-1}$ in slope sediments, using ^{210}Pb-derived sediment accumulation rates for the Bering Sea, the Mid-Atlantic Bight and the Peruvian continental slopes. This rate is large enough that slope sediments would be significant depositories of anthropogenically released fossil CO_2. In fact, up to 15% of the CO_2 emitted to the atmosphere by fossil fuel burning would reside as organic C in surface slope sediments. Sediment accumulation rates used by Walsh *et al.* (1985) for other slope areas are at least a factor of ten higher than rates estimated by other methods for either the whole Pleistocene epoch or since the last Wisconsin glaciation. Walsh *et al.* (1985) recognized that such rates could not represent true long-term sediment accumulation, but argued that they reflected true, recent, short-term storage rates for slope sediments.

^{210}Pb-derived accumulation rates used by Walsh *et al.*(1985) could be inflated because the ^{210}Pb activity profiles were not adequately corrected for mixing effects. In fact Anderson and Biscaye (1985) found that Pu profiles in five cores from the mid-Atlantic slope constrained recent accumulation rates to be less than one fourth the maximum rates indicated by ^{210}Pb profiles. ^{14}C data from some slope cores led Walsh *et al.* (1985) to correct the ^{210}Pb profiles for mixing and to reduce slope accumulation rates by factors of 2 to 3. The correction reduced their global slope C flux to 0.3-0.5 x 10^{9} tons y^{-1} (or about 1.4 mg organic C cm^{-2} y^{-1}). This global mean is equal to the sedimentary C flux calculated above for the Washington slope. However, high bulk sediment accumulation rates due to the proximity of the Columbia River and the pronounced O_2 water column minimum should favor greater accumulation of organic C in Washington slope sediments relative to other slope regions of the world (Gross *et al.*, 1972; Muller and Suess, 1979; Emerson, 1984). Therefore, Walsh *et al.*'s mean global organic C flux for slope sediments appears to be several times greater than the true flux, a result of mixing-caused inaccuracies in ^{210}Pb-derived sediment accumulation rates. Anderson *et al.* (in press) reached this same conclusion for slope sediments off Cape Cod.

Carbon Isotopes

In estimating sediment accumulation rates during the past 20,000 y, Carson (1971; 1973) and Barnard (1973; 1978) obtained ^{14}C ages at various depth horizons in piston cores of Cascadia Basin and Washington slope sediments. Apparent ^{14}C derived ages of the topmost intervals (usually 20-cm thick) ranged from 1,500 to 4,300 y and averaged 2,900 y for 17 cores collected between 1968 and 1972. Sediment accumulation rates calculated from ^{14}C ages, the depth of Mt. Mazama ash, changes in foraminifera/radiolaria ratio (Barnard, 1978) and excess ^{210}Pb profiles (Carpenter *et al.*, 1982) all indicate that the top 20 cm of typical Washington slope sediments have accumulated in the past 200 to 400 y. This sets an upper limit for the mean ^{14}C age of surface sediments.

Three of Barnard's cores gave apparent ages of 8,435 ± 160, 8,365 ± 135 and 8,550 ± 210 y for the lowest sediment horizons with Mt. Mazama glass shards. The Mt. Mazama eruption has been confidently dated at 6,600 y B.P. Therefore, Barnard concluded that all slope sediment layers at the sea-sediment interface had apparent ^{14}C ages of 1,700 y. This was supported by the subsequent determination that the topmost intervals of cores collected in 1972 from Quinault and Willapa canyons (0-20 and 0-3 cm thick, respectively) had large excess ^{210}Pb activities. The excess ^{210}Pb activities confirmed recovery of material deposited within the past 50 y or so and yet these intervals had apparent ^{14}C ages of 1,535 ± 85 and 1,895 ± 85 y, respectively. The similarity of Mazama ash 'excess ages' (1,700 y) and newly deposited sediment ages (1,500-1,900 y) strongly suggests that little extra ^{14}C was input into recent Washington slope sediments (Carpenter, 1986). ^{14}C-derived apparent ages of a few thousand years B.P. have also been reported for recent sediments of the New York Bight (Turekian *et al.*, 1980), the Santa Barbara Basin (Emery and Bray, 1962), the Gulf of California (Calvert, 1966) and sediment south of Martha's Vineyard (Bothner *et al.*, 1981). Unexpectedly old ages are not unique to Washington slope sediments.

Two explanations advanced for such old apparent ages in surface sediments from other areas can be ruled out for the Washington slope. Excess ^{210}Pb results confirm recovery of recently-accumulating material, so the old ages cannot be attributed to lack of recovery of recently-deposited sediments. Aliphatic and aromatic hydrocarbon compositions of several cores from the Washington slope and shelf (Prahl and Carpenter, 1984) show very little indication of fossil hydrocarbons in surface sediments, ruling out the possibility that the old ages are due to fossil coal or petroleum material.

$\Delta^{14}C$ activities of +100 to +200 ‰ have been measured in surface Washington waters collected since 1960. These may be compared to values of around −50 ‰ estimated for prebomb activities (Buddemeir, 1969; Swanson, 1980; Fairhall and Young, 1985). No data on ^{14}C activity of surface waters and plankton are available for the Washington slope. However, in Puget Sound, Saanich Inlet (Buddemeier, 1969) and other areas (Williams *et al.*, 1970; Williams and Linck, 1975), the ^{14}C activity of marine organisms has been shown to be essentially the same as the activity of dissolved CO_2 in ambient waters in which they grew. Thus, plankton growing in Washington slope surface waters in recent decades should have been significantly labeled with bomb-derived ^{14}C.

After ^{14}C uptake in primary productivity, oxidation and recycling of organic material leads to release back into the water column of some of the bulk C and ^{14}C associated with the organisms, fecal pellets and other organic debris. The fraction of fecal material surviving to a given depth depends greatly upon size and shape of the pellets, which is species-dependent. Suess (1980) summarized data from numerous sediment trap deployments in other areas which suggest that organic C fluxes to depths typical of the Washington slope (300 and 1,500 m) most probably would represent only about 10 and 3%, respectively, of surface primary productivity.

Prahl *et al.* (1981) showed that planktonic hydrocarbons are rapidly and preferentially (relative to terrestrial hydrocarbons) remineralized in Dabob Bay, Washington. The mean residence time of planktonic C in the top 5 cm of Long Island Sound sediments is about 2 y relative to metabolic loss (Turekian *et al.*, 1980; and references therein). Extensive remineralization of recent matter derived from plankton is thus one likely reason why ^{14}C labeled organic matter is not obvious in Washington slope surface sediments.

Dilution with either terrigenous organic C or fossil fuel C will tend to increase apparent ages of organic matter in surface slope sediments and buffer any sudden increases in ^{14}C activity due to bomb radiocarbon associated with plankton debris. Lignin (Hedges and Mann, 1979) and hydrocarbon characterizations (Prahl and Carpenter, 1984) have shown that terrestrial organic C may comprise some 50% of total C in Washington shelf sediments, but less than 25%

in slope sediments. Part of this terrestrial organic C may be equilibrated with atmospheric CO_2 rich in bomb ^{14}C, with the remainder being derived from humus or other materials out of equilibrium with atmospheric CO_2. No data are available on ^{14}C activities in terrestrial organic matter from the Columbia or any other Washington river, but data from other areas typically have indicated ages of a few hundred to a few thousand years (Martel and Paul, 1974; Wakeham and Carpenter, 1976; Erlenkhauser *et al.*, 1974; Benoit *et al.*, 1979).

Mixing processes in surface sediments dilute that fraction of bomb ^{14}C labeled planktonic organic matter which is rapidly transported to the sea floor and survives remineralization with older organic matter containing less ^{14}C. Carpenter *et al.* (1982) reported that the mean residence time for particles in ^{210}Pb-derived surface mixed layers of Washington slope sediments averaged 74 ± 64 y. Assuming the ^{14}C particles are mixed similarly to ^{210}Pb labeled particles, time lags equivalent to the surface mixed layer residence times will occur before surface sediment ^{14}C activities respond fully to changes in activities in freshly depositing particles. However, unless ^{14}C associated particles are mixed much differently than ^{210}Pb associated particles, mixing alone should not have caused surface ages as old as 1,700 y B.P.

Prebomb, bulk organic matter in Washington slope surface sediments seems to have had apparent ^{14}C ages of about 1,700 y B.P., corresponding to $\Delta^{14}C$ activities of about -200 ‰, whereas $\Delta^{14}C$ activities in plankton since weapons testing began should have been +100 to +200 ‰ (Carpenter, 1987 and references therein). A simple mass balance calculation using these two end members shows that future (i.e., positive) ^{14}C ages for bulk organic matter in surface layers of Washington slope sediments can be expected only when bomb ^{14}C-labeled planktonic debris exceeds 50% of the total organic matter. The fact that surface layer ages are still so old means that much less than 50% of the organic matter is recent. Although the failure to find future apparent ^{14}C ages in surface slope sediments argues against the slope organic C sink model of Walsh *et al.* (1981, 1985), it is not as severe a criticism as Peng and Broecker (1984) claim.

Table 10.7 summarizes $\Delta^{14}C$ and $\delta^{13}C$ values for tops and bottoms of several Puget Sound sediment cores collected in 1975-1977. Despite the high planktonic productivity of this estuary, ^{14}C activities in surface sediments do not show the positive values expected for organic matter synthesized from surface waters perturbed by the bomb effect. $\delta^{13}C$ values of Puget Sound sediments fall in the range of −24.6 to −26.8 ‰. Terrestrial organic C has an average $\delta^{13}C$ value of about −25 ‰ (Degens, 1969). The close correspondence suggests that most of the sedimentary organic C in Puget Sound is of terrestrial origin.

Aliphatic and aromatic hydrocarbon studies have shown that parts of the Puget Sound region have received increasing amounts of fossil hydrocarbons since about 1900 and that terrestrially-derived hydrocarbons are important contributors to lipids in most Puget Sound sediments (Barrick *et al.*, 1980; Barrick and Hedges, 1981). Extremely old ^{14}C values for station 57 sediments (Fig. 10.13) are due to the presence of coal fragments throughout this core (Furlong and Carpenter, 1982; Barrick *et al.*, 1984). Remineralization of planktonic organic C, dilution with terrestrial C and fossil fuel C contamination give Puget Sound surface sediments

TABLE 10.7 ^{14}C and ^{13}C contents of bulk organic matter from tops and bottoms of Puget Sound sediment cores. Station locations are shown in Figure 10.13. Units are defined in footnote.

Station	depth in core (cm)	$\Delta^{14}C$ (‰)	$\delta^{13}C$ (‰)
51	0-5	−131	−25.0
51	25-30	−217	−24.6
56	0-5	−140	−25.8
56	29-34	−226	−25.2
57	0-5	−404	−26.8
57	31-36	−595	−26.4
72	0-5	−179	−24.7
78	0-5	−131	−25.0
78	25-30	−217	−24.6

The ^{14}C standard activity is taken as 95% of the observed net activity of N.B.S. oxalic acid. Sample activities are given as the percentage of this standard, or 100 x R defined in equation (1) below. When $^{13}C/^{12}C$ ratios were also determined, sample ^{14}C activities are expressed as $\Delta^{14}C$, defined by equation (4) below. The $\Delta^{14}C$ expression represents the isotope fractionation corrected ^{14}C concentration relative to the contemporary standard. Each $\Delta^{14}C$ unit difference corresponds to a time difference of 8 y.

(1) R = Activity of sample/0.95 x Activity of NBS oxalic acid

(2) $\delta^{14}C = (R-1) \times 1000$ (in per mille)

(3) $\delta^{13}C = [(^{13}C/^{12}C)_{sample}/(^{13}C/^{12}C)_{PDB} - 1] \times 1000$ (in per mille)

(4) $\Delta^{14}C = \delta^{14}C - (2\delta^{13}C + 50)(1 + \delta^{14}C/1000)$ (in per mille)

(5) Age (years before 1950) = $(1/\lambda) \times \ln[(\Delta^{14}C/1000) + 1]^{-1}$, where λ = the ^{14}C decay constant, 1.210×10^{-4} y^{-1}, which corresponds to the 5,730-y value for half-life.

^{14}C activities well below those of overlying water and plankton and ages of several hundred y B.P. (Fig. 10.7). ^{14}C activities in surface sediments ranged from 59.3% of the contemporary standard at the coal contaminated station near Tacoma to 91.6% in southern Puget Sound. Activities of the other seven surface sediments averaged 86.5%. Swanson (1980) also reported more detailed depth profiles of ^{14}C activities in sediment cores from station 51 in the main channel of central Puget Sound, station 69 near the Cherry Point oil refineries and station 50 in Elliott Bay near Seattle. Increasing ^{14}C activities in sediments deposited since 1950 at stations 51 and 69 were attributed to some additions of bomb ^{14}C. However, even the surfacemost layers of these two stations had ^{14}C activities of only about 88%, while overlying waters and plankton had activities of 100 to 102%. The sediment profile from Elliott Bay, on the other hand, showed *decreasing* ^{14}C activities since about 1950 to the present due to fossil C overriding bomb ^{14}C inputs.

Swanson (1980) also determined ^{14}C activity of aliphatic and polynuclear aromatic hydrocarbon (PAH) fractions extracted from 15 Puget Sound sediment horizons. PAH extracts had very low ^{14}C activities, confirming that they were essentially all fossil-derived. Aliphatic hydrocarbon fractions gave ^{14}C activities of 10 to 50%. These intermediate activities reflect the presence of a mixture of recently biosynthesized planktonic material, terrigenous material and the fossil hydrocarbons which are known to be present in these sediments (Barrick *et al.*, 1980; Barrick and Hedges, 1981).

Aliphatic and Aromatic Hydrocarbons

Two important features of hydrocarbon studies conducted off the Washington coast and in Puget Sound are that hydrocarbon concentrations are given per gram organic C in sediments, rather than in per gram dry weight of bulk sediment and that hydrocarbon analyses have been made on cores dated by the excess ^{210}Pb method. Normalization compensates for varying ratios or organic to inorganic materials in texturally dissimilar sediments common to this and most coastal regions. Physical sorting of sediment compounds must be accounted for before conclusions can be drawn about concentration changes with geographic location or sediment depth. ^{210}Pb-derived sediment accumulation and mixing rates allow calculations of depositional fluxes, dating of and rates of change between sediment horizons.

Washington Coast Sediments. Sources and distributions of PAH and aliphatic hydrocarbons have been characterized by analyzing sediments from 17 locations off the southwestern coast of Washington (Prahl and Carpenter, 1984). The major hydrocarbons are land-derived, introduced as preformed compounds and displaying long-term stability in sediment cores. Recent hydrocarbon fluxes to Washington shelf sediments can be balanced by measured fluxes from the Columbia River (Prahl *et al.*, 1984); therefore, atmospheric and advective inputs of these hydrocarbons are probably of secondary importance.

The percentage of total organic matter in the MSSD which is of terrestrial origin was estimated to be 69 ± 25% using plant wax *n*-alkane data from coastal and Columbia River sediments (Prahl and Carpenter, 1984). This value compares well with an estimate of over 50% made from lignin data by Hedges and Mann (1979b). A similar calculation showed 26 ± 10% of the bulk organic matter in slope sediments was terrestrially-derived, substantially larger than the 5% estimate of Hedges and Mann(1979b) which was based on one slope sediment from a more southerly location.

The PAH composition in Washington coastal sediments is dominated by PAH of fossil fuel combustion origin and two naturally-derived compounds, retene and perylene. Plant wax *n*-alkanes are the major aliphatic hydrocarbon constituents and their concentrations change little with depth in sediments. Aliphatic hydrocarbons of marine origin, pristane and a series of four acyclic, multibranched C_{25} olefins, are also observed in many sediments. Concentrations of these marine-derived olefins rapidly decrease to negligible levels with increasing sediment depth, suggesting degradation. Washington coast sediments (Fig. 10.25) lack the unresolved complex mixture, diasteranes and triterpanes suggestive of fuel oil contamination. The latter are prominent hydrocarbon components in Puget Sound surface sediments (see Fig. 10.29 below).

One triterpene, hop-22(29)-ene (diploptene), is widespread in marine sediments along the Washington coast (Prahl, 1985). It occurs along with a series of plant wax *n*-alkanes

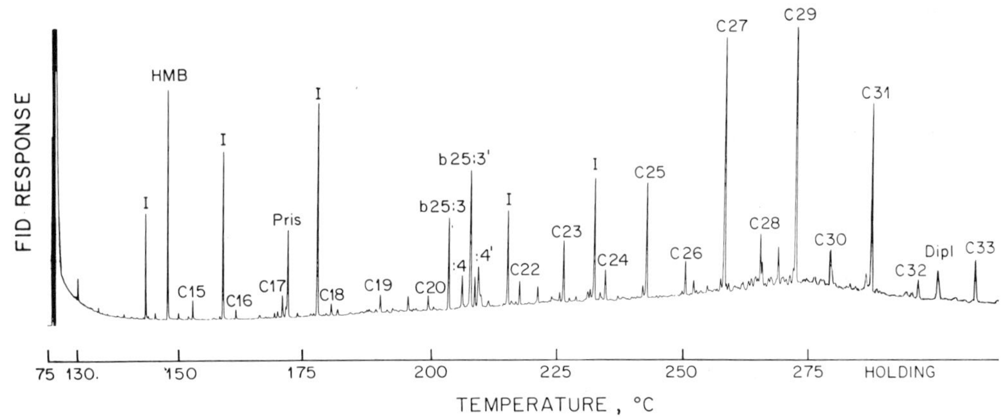

Fig. 10.25 Gas chromatogram of aliphatic hydrocarbons in the 0-2 cm sediment horizon near the head of Quinault canyon. Hydrocarbon components are identified by carbon number for *n*-alkanes, Pris for pristane, Dipl for diploptene and b25: 3, b25: 3', : 4, : 4' for a series of four 25 carbon acyclic tri- and tetraolefins. Peaks labeled HMB and I are hexamethyl benzene, the GC injection standard and a series of five branched alkane recovery standards. From Prahl and Carpenter (1984), with permission.

(C_{23} to C_{33}) at concentrations similar to *n*-C_{33}. Freshwater sediments throughout the Columbia River basin and suspended particulate matter from the mouth of the Columbia River contain the same aliphatic hydrocarbon distributions. This suggests that diploptene is introduced erosionally to the Washington coastal region as a refractory component of land-derived detritus and is not produced by *in situ* biological activity. The major land-derived hydrocarbons are concentrated in the MSSD. A less abundant natural series of phenanthrene homologs, also of terrestrial origin, is preferentially advected further offshore and deposited in continental slope sediments.

The concentration of diploptene relative to plant wax *n*-alkanes in surface sediments increases systematically with distance offshore, but remains constant in surface sediments along the MSSD (Prahl, 1985). Certain lignin and PAH compounds measured in the same samples show analogous distributional trends. These data are evidence that diploptene, lignin characteristic of non-woody vascular plant tissue and a fossil organic material contained within weathered rock debris are distributed in sediments across the Washington continental shelf and slope in a geographic pattern distinct from that of other river-derived materials. These areal distributional differences result from the association of these organic chemicals with different particle types which follow different dispersal pathways following their discharge at the mouth of the Columbia River. These distributions are consistent with recognized particle associations of these compounds and sediment dispersal paths in this coastal environment. Sediment core records suggest that the present pattern of dispersal of naturally derived compounds has persisted for at least the past century and possibly since the late Pleistocene. Combustion-derived PAH concentrations in shelf sediments increase two-to threefold from preanthropogenic to recent sedimentary horizons (Fig. 10.26).

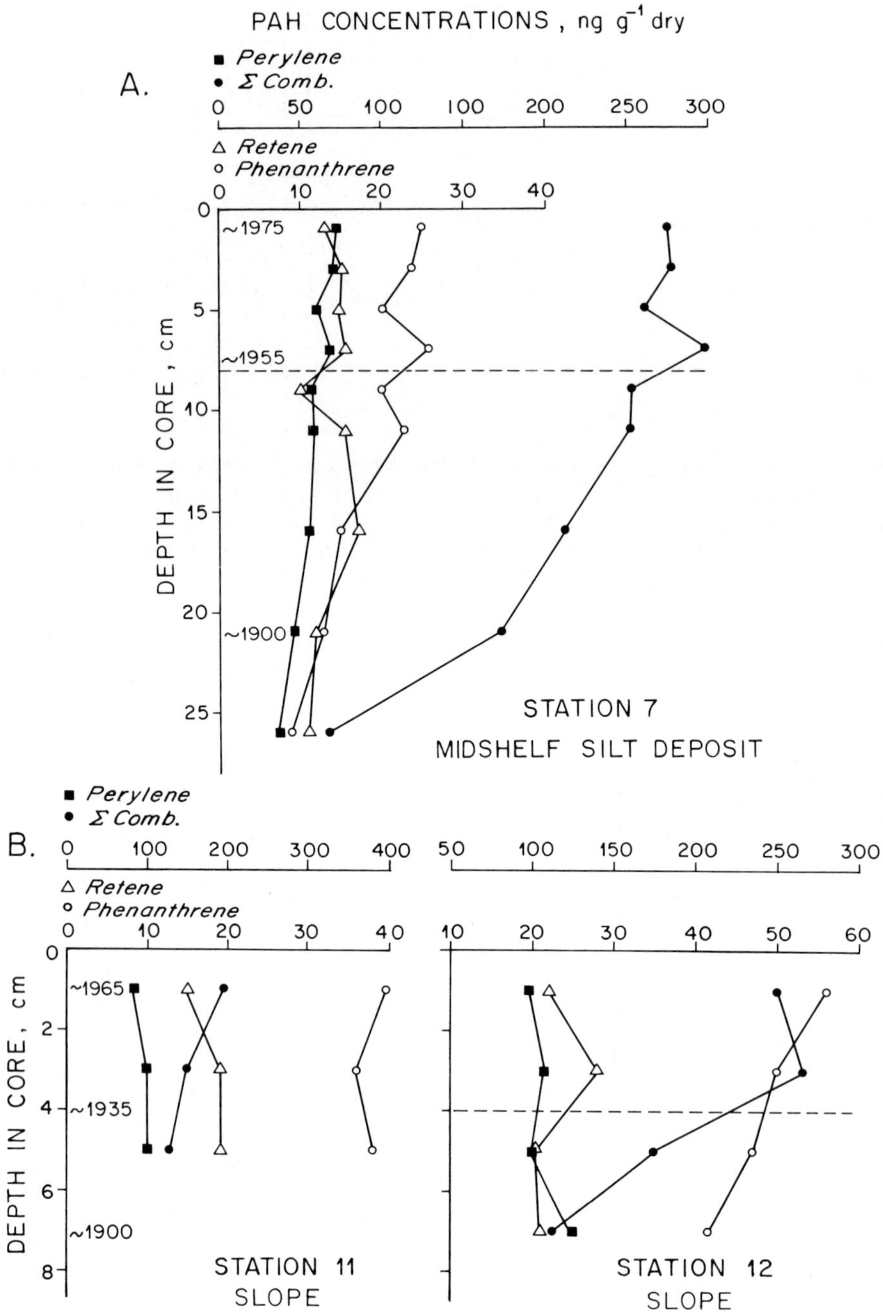

Fig. 10.26 Sedimentary PAH concentration profiles from (A) the mid-shelf silt deposit and (B) continental slope off the Washington coast. Σcomb. is the concentration sum of nine, primarily combustion-derived PAH. Approximate dates for specific sediment horizons are based on ^{210}Pb-derived accumulation rates and are not corrected for mixing within the ^{210}Pb-defined surface mixed layer (dashed line). From Prahl and Carpenter (1984), with permission.

Prahl and Carpenter (1984) made a start at characterizing predominant phase association of organic compounds in Washington coastal sediments. PAH and aliphatic hydrocarbon compositions, organic carbon, nitrogen and lignin contents were determined in whole, unfractionated sediment from the Washington continental shelf and in discrete sediment fractions separated by particle size and density. This first use of both size and density fractionation in the study of PAH phase associations showed that all three PAH types (combustion-derived, retene, perylene and fossil) are highly enriched in different components of the bulk sediment. At least 20 to 25% of perylene and PAH derived from pyrolytic processes and 50% of the retene measured in whole sediment were contained within organic C- and lignin-rich particles of density < 1.9 g cm^{-3}. These particles, which include primarily vascular plant remains and bits of charcoal, comprised less than 1% of the total sediment weight. Retene almost certainly was contained in higher plant debris, reflecting its diagenetic origin from resin acids. These low density particles have not been specifically identified. In contrast, a series of methylated phenanthrene homologs, possibly of fossil origin, were concentrated in the more dense, lithic matrix components of the sediment. None of the three PAH types in these coastal sediments were significantly associated with humic substances which constitute a major proportion of the sedimentary organic C.

Washington coastal sediments contained combustion-derived PAH mixtures similar in composition to those in sediments from most other locations. Combustion-derived PAH are not preferentially enriched in finer-grained bulk sediment particles of greater surface area, as would be expected if PAH geochemical cycling was controlled by equilibrium sorption/desorption partitioning between molecules in solution and solid surfaces. These results point out the need to establish the particle associations of hydrophobic compounds prior to applying equilibrium sorption/desorption models (Steen and Karickoff, 1981) in order to predict the fate and behavior of such compounds in aquatic environments. Two-phase equilibrium models of PAH sorption/desorption behavior implying exchange of dissolved molecules onto small particles of high surface area do not appear applicable in this coastal environment.

Puget Sound Sediments. Since 1975 an extensive data set has been developed on the amounts and distributions of aliphatic, aromatic, S- and N-containing hydrocarbons in ^{210}Pb dated sediment cores from Puget Sound (Prahl and Carpenter, 1979; Bates and Carpenter, 1979a; Barrick *et al.*, 1980; Prahl *et al.*, 1981; Barrick and Hedges, 1981; Barrick, 1982; Furlong and Carpenter, 1982; Bates *et al.*, 1984; Barrick *et al.*, 1984; Barrick and Prahl, 1987). Sedimentary hydrocarbon distributions and processes responsible for them have been examined more thoroughly in sediments from Puget Sound than from most other coastal areas.

PAH compositions were determined in Dabob Bay cores and a year-long time series of sediment trap particles (Prahl and Carpenter, 1979). PAH associated with the predominantly zooplankton fecal material collected in the traps quantitatively accounted for PAH fluxes measured in the underlying sediments. The quantitative agreement between trapped and sedimentary fluxes showed that PAH were neither significantly degraded during sediment deposition nor produced after deposition within the roughly 100-y period represented by sediment column examined. Preservation of PAH with four or more rings during transport through the water column to the sediments was substantiated by Gschwend and Hites (1981).

PAH profiles in Dabob Bay sediments (Fig. 10.27) first revealed a pronounced subsurface concentration maximum in horizons deposited around 1950. Subsequently,

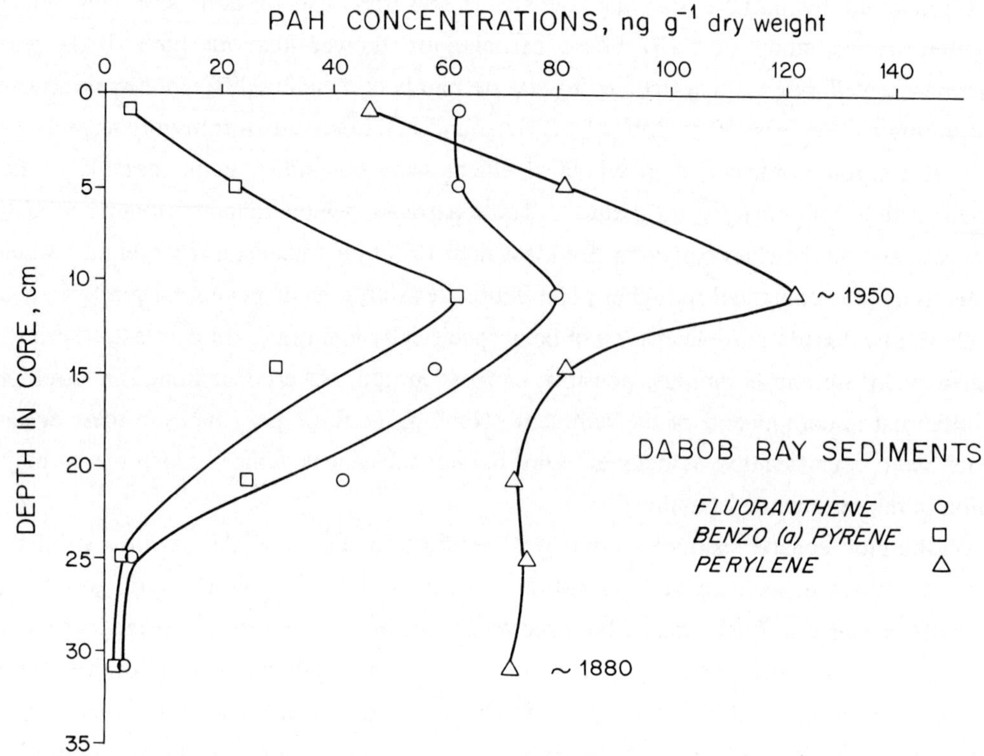

Fig. 10.27 Concentration profiles of three polynuclear aromatic hydrocarbons (PAH) in a sediment core from Dabob Bay, Wa. From Prahl and Carpenter (1979), with permission.

subsurface maxima in PAH concentrations have been found in most, but not all, Puget Sound sediment cores (Bates *et al.*, 1984; Barrick and Prahl, 1987) (Fig. 10.28). Maxima in all cores occur in layers deposited between 1945-1960. Subsurface PAH maxima appear to be more pronounced in Puget Sound than Washington shelf sediments (Prahl and Carpenter, 1984). These PAH maxima often occur within surface mixed layers of uniform excess ^{210}Pb activity, suggesting that PAH and ^{210}Pb-containing particles are mixed in a different manner. Similar subsurface maxima in PAH concentrations have been observed in sediments of Lake Constance, Switzerland (Grimmer and Bohnke, 1977) and remote and urban sites in the northeastern United States (Gschwend and Hites, 1981). The ubiquity of PAH subsurface maxima in both oxic and anoxic sediments led Gschwend and Hites to suggest that they reflect the transition from coal to oil and gas usage in home heating. Residential coal burning is an inefficient combustion process which produces relatively high PAH discharges compared to other fuels or combustion processes. This hypothesis is supported by the fact that the number of dwelling units using coal for home heating in Washington state decreased by a factor of 10 between 1950 and 1970 (U.S. Bureau of Census, 1970).

Depth profiles of two naturally-produced PAH, perylene and retene, reveal different chemical behaviors and sources relative to the other PAH (Prahl and Carpenter, 1979; Bates *et al.*, 1984). In Dabob Bay (and probably most other Puget Sound sediments) perylene and retene

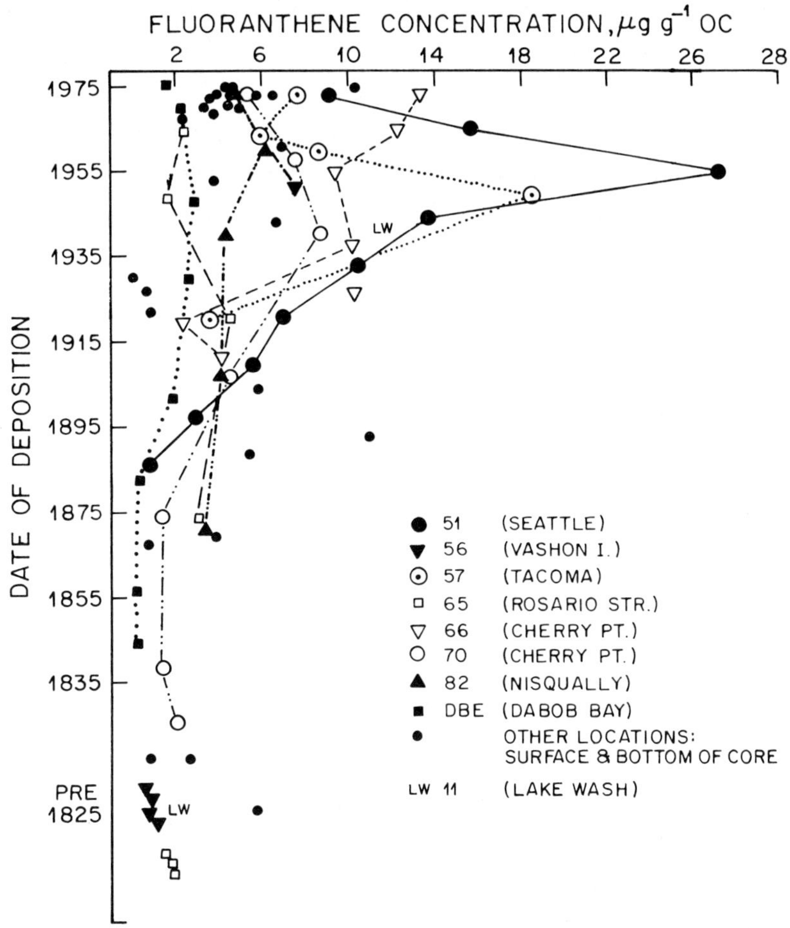

Fig. 10.28 Age profiles of fluoranthene concentrations in eight cores from the Puget Sound region. Data points are also shown for the top and bottom intervals of cores from 16 other Puget Sound stations and one Lake Washington station. Dates of deposition are ^{210}Pb-derived, assuming no surficial mixing of the fluoranthene. From Barrick and Prahl (1987), with permission.

are deposited as preformed PAH (the result of still undefined natural processes), with no postdepositional formation during the roughly 100-y sedimentary record examined.

PAH compositions have been characterized by capillary gas chromatography in 96 sediment samples from sites throughout greater Puget Sound (Barrick and Prahl, 1987). Detailed profiles were determined for eight sites, while only core tops and bottoms were analyzed at 16 sites. The combustion of fossil fuel materials, recognizable by a distinct signature of nonalkylated compounds with 3 to 6 condensed aromatic rings, was found to be a major source of PAH. Total combustion PAH concentrations in Puget Sound deposits ranged from 16 to 2,400 ng (g dry sediment) $^{-1}$ or 8,000 to 73,000 ng (g organic C) $^{-1}$. Concentrations of combustion PAH decreased with depth in all cores, indicating an increasing anthropogenic contribution in more contemporary times. Nonetheless, measurable concentrations in non-contaminated sediments at depth demonstrate that natural burning processes also contribute combustion PAH.

The concentration range for combustion PAH in surface sediments was about one order of magnitude, similar to that observed for aliphatic hydrocarbons (Barrick *et al.*, 1980). Combustion PAH and aliphatic hydrocarbon concentrations displayed a distinct geographical difference. Highest aliphatic concentrations were found throughout sediments from central Puget Sound and were attributed to enhanced urban stormwater and sewage discharge in this region (Barrick, 1982). Combustion PAH, on the other hand, displayed high, localized concentrations in sediments within close proximity to major municipalities of central Puget Sound and near two relatively rural areas of northern Puget Sound. Industrial activities of aluminum smelting and oil refining near these northern Sound locations were believed responsible for the high PAH levels. Combustion processes produce strikingly similar mixtures of PAH regardless of what fuel is burned (Lee *et al.*, 1977). However, statistical analysis of the PAH data set in surface sediments suggested that the mixture of combustion PAH currently being deposited in central Puget Sound sediments can be distinguished from that in northern Puget Sound and in older, non-contaminated sediments. The possibility that the ratio of indenopyrene to fluoranthrene allows this distinction and therefore may be useful in identifying PAH sources, warrants further study.

Perylene was present in all sediments at concentrations of about 330 ng (g dry sediment) $^{-1}$ or 16,600 ng (g organic C) $^{-1}$. Constant concentrations with depth in cores and a spatial correlation with plant wax *n*-alkanes both argue that perylene is introduced erosionally from natural, land-based sources as a preformed hydrocarbon. No suggestion of production within sediments by diagenesis of its natural, still unidentified precursor was evident from the data.

Distributional information for phenanthrene and C_1^-, C_2^-, C_3^- and C_4^- substituted phenanthrenes indicated at least three sources contribute to this suite of compounds. Combustion processes have introduced primarily phenanthrene and relatively minor amounts of the C_1- substituted compound throughout the region. A fossil series of largely C_1^- and C_2^- substituted phenanthrenes became evident as the anthropogenic combustion PAH signal lessened with increasing depth in cores. This series of natural origin is likely associated with lithic debris (Prahl and Carpenter, 1983) which has been chronically introduced throughout Puget Sound by erosional processes. Finally, a complete series of C_1^-, C_2^-, C_3 and C_4^- substituted phenanthrenes was enriched in several sediment horizons in cores from Commencement Bay and near the mouth of the Nisqually River (Fig. 10.3). Barrick *et al.* (1984) showed that coal fragments within these sediments are the carrier of these compounds. Coal fragments also account for the presence of several aliphatic diterpenes and a series of chrysenes and picenes in the same sediments. Although all of these compounds can be related diagenetically to resins found in conifers and other higher plant natural products, no evidence for diagenesis of such precursors has been observed in Puget Sound cores.

Aliphatic hydrocarbon compositions of surface Puget Sound sediments are dominated by pristane, *n*-alkanes and the unresolved complex mixture (UCM) (Fig. 10.29). Various terpenoids, phytadienes, isoprenoid alkanes and multibranched acyclic C_{25} hydrocarbons are

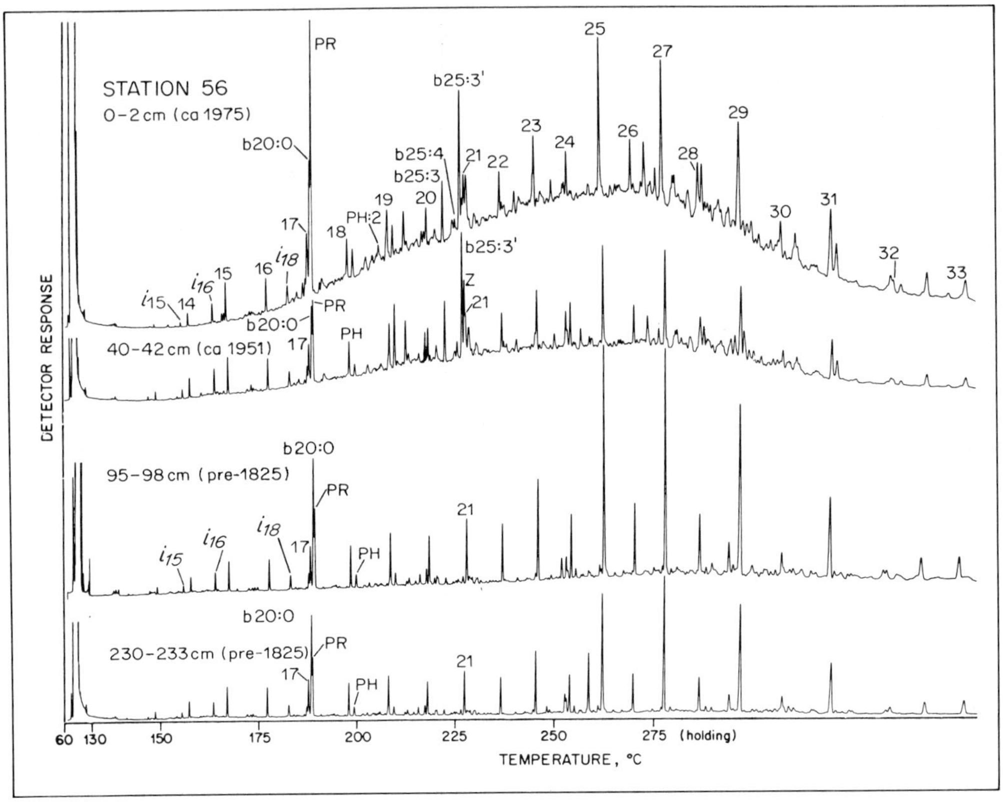

Fig. 10.29 Gas chromatograms of aliphatic hydrocarbons in four sediment horizons from Puget Sound station 56 between Seattle and Tacoma. Numbers on the figure refer to *n*-alkanes; PR = pristane, PH = phytane, i_{15}, i_{16} and i_{18} = regular isoprenoid alkanes with subscripts indicating carbon number. From Barrick *et al.* (1980), with permission.

present at lower concentrations. Sources of these compounds are summarized by Barrick *et al.* (1980), Barrick and Hedges (1981) and in Table 10.8. Total aliphatic hydrocarbons, the UCM and individual hydrocarbons such as certain diasteranes and hopanes which serve as molecular markers of weathered fossil hydrocarbons, all have concentration profiles in Puget Sound sediments which either decrease from the sea-sediment interface with no subsurface maxima, or which have subsurface maxima at more shallow depths than PAH subsurface maxima (Barrick *et al.*, 1980; Barrick and Hedges, 1981; Bates *et al.*, 1984). Near-surface concentrations and increases are greater near the urban areas of Seattle and Tacoma than in other areas. The highest total aliphatic hydrocarbon concentrations measured in Puget Sound surface sediments are about 10% surface concentrations in New York Bight sediments (Farrington and Tripp, 1977). Hydrocarbon increases beginning around 1900 coincide with rapid development of the Seattle-Tacoma urban areas and clearly predate opening of major oil refineries in northern Puget Sound in the 1950s and the first operation of METRO's large primary sewage outfall into Puget Sound in 1966. Hydrocarbon increases are therefore attributed to chronic inputs from a variety of sources rather than single point discharges from more recent sources.

TABLE 10.8 Summary of distributions, sources and probable phase associations of hydrocarbons in sediments of the main basin of Puget Sound.

Polycyclic Aromatic Hydrocarbons	Non-aromatic Hydrocarbons
Major Components	
1. Contaminant PAH [1,9] 2. Natural PAH a. retene/perylene (major) [9] b. PAH from coal fragments (minor)[8,9] origin [2,3,4]	1. Contaminant hydrocarbons (diasteranes, triterpanes and an unresolved complex mixture) [2,4] 2. Natural compounds of terrestrial 3. Natural compounds of marine origin [2,3]
Distributions	
1. Contaminant PAH increase 5-10-fold since 1900; some indications of subsurface maxima around 1950 [1,7,9] 2. Natural PAH fairly uniform both geographically and with age of sediments [1,9]	1. Contaminant hydrocarbons increase ten-fold since 1900 [2,4] 2. Natural terrestrial compounds fairly uniform distributions both geographically and with age [2] 3. Concentrations of natural marine origin decrease rapidly below sediment interface [2,3]
Sources	
1. Combustion products from urban areas transported via rivers and/or air (primarily nonalkylated structures) [1,7,9] Fuel oil dumping, especially phenanthrenes and alkylated homologs which are partially degraded in sediment environment [5] 2. a. Natural sources outside of sediment environment; not *in situ* formation [9] b. Both natural erosion and mining activities (primarily alkylated structures) [8,9]	1. Urban-derived contaminants transported via rivers, atmospheric fallout and sewer discharge [2,4,5,7] 2. Two major sources including: a. Terrestrial vegetation [2,3,4] b. Natural fossil alkanes likely from erosion of rocks and coal [2,8] 3. Two major sources including: a. Plankton (pristane) [2,3] b. Microbial hydrocarbons (possibly antibiotic hydrolysis products) [4]
Probable Phase Associations	
1.Primarily associated with low density ($\leq$ 1.9 g cc $^{-1}$) soot and fly ash, [6] or primarily associated with sediments as surface coatings 2. a.Primarily associated with low density ($\leq$ 1.9 g cc $^{-1}$) material such as plant debris [6] b. Associated with coal fragments [8]	1. Primarily associated with sediments as surface coatings 2. a. Primarily associated with low density plant material [4] b. Associated with denser matrix of particles and coal fragments [1,8] 3. Associated with relatively large rapidly biologically-cycled zooplankton fecal pellets [1]

[1]Prahl and Carpenter (1979). [2]Barrick *et al.* (1980). [3]Prahl *et al.* (1981). [4]Barrick and Hedges (1981). [5]Barrick (1982). [6] Prahl and Carpenter (1983). [7]Bates *et al.* (1984). [8]Barrick *et al.* (1984). [9]Barrick and Prahl (1987).

In addition to the UCM, Barrick *et al.* (1980) identified four compositionally distinct, but often overlapping, acyclic hydrocarbon assemblages: 1) a predominant assemblage of primarily odd C numbered, high molecular weight *n*-alkanes derived from the leaf waxes of vascular land plants. Once deposited, these terrestrially-derived substances are relatively resistant to chemical alteration; 2) a suite of naturally-derived, fossil hydrocarbons which has a minimal UCM and, possibly because of protective physical matrices, has not been subject to extensive microbial degradation; 3) an assemblage of plankton-derived hydrocarbons of unexpectedly low concentrations in view of the high primary productivity in overlying waters, because plankton remains are rapidly and preferentially degraded near the sea-sediment interface (see below); and 4) a suite of seven different multibranched acyclic C_{20} and C_{25} compounds which appear to have a common, natural marine origin, yet unidentified. Decreases in concentrations of the C_{25} compounds with depth in sediments are attributed to natural degradation and instability due to their multiple double bonds, rather than to increased sedimentation rates in recently deposited layers.

Cyclic components of the 'aliphatic' hydrocarbon mixtures extracted from Puget Sound sediment cores have been characterized by Barrick and Hedges (1981). Distinct suites of cyclic compounds include: 1) 10 naturally-derived diterpenoid compounds and three naturally-derived triterpenoids including diploptene, all believed ultimately derived from terrestrial plants; 2) a bicyclic C_{25} diene and a novel group of five C_{30} polyenes, all apparently natural and marine-derived; and 3) 5 diasteranes and 4 hopanes which occur in relatively constant proportions as components of a highly weathered fossil hydrocarbon assemblage. These cycloalkanes, along with the strongly covarying UCM, have been introduced to Puget Sound from adjacent urban areas at increasing levels during the past 100 y.

Barrick *et al.* (1980), Barrick (1982) and Bates *et al.* (1984) have considered possible effects on hydrocarbon distributions in Puget Sound sediments caused by Seattle's large METRO primary sewage treatment facility and its discharge into Puget Sound. Prior to 1973, wastes discharged by METRO, including virtually all the storm water component, flowed untreated into Puget Sound. Since 1973, primary sewage treatment and sludge removal have reduced hydrocarbon discharges to some extent, but the UCM and *n*-alkane inputs from METRO's aqueous discharges between December 1977 and August 1979 were still sufficient to entirely account for fluxes of these compounds to central Puget Sound sediments (Barrick, 1982).

Barrick *et al.* (1980) observed no major changes in aliphatic hydrocarbon profiles from mid-channel sediment horizons in central Puget Sound sediment horizons on a time-scale corresponding to the onset of METRO's Westpoint discharge in 1966. However, some indication of decreases in aliphatic hydrocarbon concentrations in the most recently deposited layers were evident. In a core from the same area, Bates *et al.* (1984) found clear evidence for subsurface maxima in concentrations of UCM and *n*-alkanes. Sediment horizons with maximum UCM and *n*-alkane concentrations were deposited during 1965 to 1970, if a ^{210}Pb-derived accumulation rate measured at a nearby location (~1.3 km) is applied to this core and no mixing of the hydrocarbon containing particles is assumed. Additional hydrocarbon profiles in dated cores are required to firmly establish the existence of aliphatic hydrocarbon subsurface maxima and their relationship to the METRO facility.

Aliphatic hydrocarbon concentrations were determined in sediment profiles and in settling particulate material collected in a one-year time series of sediment trap deployments in Dabob Bay, Washington (Prahl *et al.*, 1981). Results showed that aliphatic hydrocarbons and bulk organic matter in sedimenting particles underwent rapid, quantitatively significant and differential degradation early in their depositional history. The most pronounced decreases were measured for planktonically-derived hydrocarbons, such as pristane and two unsaturated compounds, which were rapidly remineralized at or near the sediment water interface. Consequently, the product of the sediment accumulation rate and compound concentrations in surface sediments is not necessarily a quantitative indicator of their depositional fluxes. The *n*-alkanes characteristic of terrestrial plant waxes became the predominant aliphatics by 4 to 6–cm depth in Dabob Bay sediments, with reasonably constant net accumulation below this interval. C/N ratios and ^{13}C isotopic measurements also supported the conclusion that organic matter from marine plankton was preferentially remineralized, causing the residual organic matter in Dabob Bay bottom sediments to closely resemble terrestrial organic matter. Subsequently, Bates *et al.* (1984) repeated the approach of Prahl *et al.* (1981) at a mid-channel station in the main basin of central Puget Sound and found the flux of aliphatic hydrocarbons determined from a sediment trap time series to be greater than corresponding hydrocarbon accumulation rates in underlying sediments.

Pore Water Hydrocarbon Concentrations. Aliphatic and PAH in the dissolved fraction of interstitial waters of coastal sediments may be more likely to be incorporated into benthic organisms and to cause adverse biological effects than hydrocarbons associated with sediment particles (Roesijadi *et al.*, 1978; Varanasi *et al.*, 1985; Eadie *et al.*, 1982; and references therein). Since hydrocarbon concentrations in pore water are low and difficult to measure without contamination during workup, several scientists have proposed using equilibrium partitioning models to calculate them from more easily measured bulk sediment concentrations (Kenaga and Goring, 1980; Karickhoff, 1981; and references therein). These partitioning models assume that the system is at equilibrium and that the pore water concentrations are related to concentration in sediment by a compound-specific partitioning coefficient. The coefficients are normalized to sedimentary organic carbon content because sediment sorption is dependent on the organic carbon concentration (Means *et al.*, 1980; Kenaga and Goring, 1980; and references therein).

There are several reasons why pore water hydrocarbon concentrations may not be reliably predicted by equilibrium partitioning models. Pore water concentrations *lower* than predicted may occur because pollutants incorporated into particle matrices may never reach equilibrium with pore waters (Kenaga and Goring, 1980; Prahl and Carpenter, 1983a; Karickhoff and Morris, 1985). Dissolved hydrocarbon concentrations *greater* than predicted may occur through hydrocarbon associations with dissolved organic matter (Boehm and Quinn, 1973; Landrum *et al.*, 1984; McCarthy and Jiminez, 1985) and colloids (Wijayaratne and Means, 1984; Brownawell and Farrington, 1986).

Socha and Carpenter (1987) presented data for aliphatic and PAH concentrations in sediments and associated pore waters from two areas of Puget Sound which have been contaminated to varying degrees with hydrocarbons from different sources. One station, in the central Puget Sound main basin, has received hydrocarbon contaminants mainly of combustion origin. The other, Eagle Harbor on Bainbridge Island, is a site with additional contamination from creosote-wood treatment plant. Demersal fish from Eagle Harbor have very high incidences of liver tumors and other diseases (Malins *et al.*, 1985). Field data from these areas were used to test equilibrium partitioning models and examine the importance of other factors controlling hydrocarbon concentrations in Puget Sound pore waters. Pore waters were separated from sediments by high speed centrifugation and filtration. Hydrocarbons were extracted using XAD-2 resin columns, purified using Sephadex columns and quantified by gas chromatography.

Equilibrium partition coefficients calculated from field pore water and solid phase data for creosote-contaminated sediments agreed with laboratory-derived coefficients to within factors of $\pm$ 4. The most heavily contaminated sediments appeared to show saturation concentrations in pore waters for some PAH. PAH were not detected in pore waters of sediments whose PAH were primarily derived from combustion and natural sources, even though pore water concentrations predicted from sediment concentrations and equilibrium partitioning models were above detection limits. Measured pore water aliphatic hydrocarbon concentrations tended to be much higher than concentrations predicted by equilibrium partitioning models, due to enhanced solubility by dissolved organic matter and colloids. These initial field data suggest that pore water concentrations of most hydrocarbons cannot be predicted to better than a factor of four with the two phase equilibrium partitioning approach.

Other Organic Compounds

Polychlorinated Biphenyls. We are aware of no measurements of DDT or its degradation products, polychlorinated biphenyls (PCBs), or any other chlorinated organic compounds in sediments of the Washington shelf or slope. Apparently PCBs are the only class of chlorinated organic compounds whose concentrations in Puget Sound sediments have been reported in refereed literature (Hafferty *et al.*, 1977; Pavlou and Dexter, 1979; Pavlou *et al.*, 1982). These concentrations, obtained using packed data columns in gas chromatographs, are limited exclusively to surface horizons.

In 1974, 720 kg of Arachlor 1242 was accidentally discharged into the highly industrialized Duwamish River estuary. Following this release, surfacial sediments from this estuary contained the highest PCB concentrations observed in the Puget Sound region (Pavlou and Dexter, 1979). Elliott Bay, which receives the Duwamish discharge, also had sediments with elevated PCB concentrations. PCB concentrations in surface sediments decreased with distance from the river mouth. Rural areas within the main basin of Puget Sound, the southern Sound and Hood Canal had lower PCB concentrations in surface sediments. Within certain subregions, such as the inner harbors at Everett and Sinclair Inlet, elevated PCB concentrations were reported, reflecting the fact that PCBs can enter the environment from localized sources. Spatial inhomogeneities in sedimentary PCB concentrations also resulted, due to differences in sediment texture, grain sizes and organic C contents between various bays and inlets of Puget Sound.

Organo-sulfur Compounds. Hydrocarbons with S atoms replacing one or more C atoms in the rings are relatively toxic, environmentally persistent and present in high concentrations in Prudhoe Bay crude oils which are routinely shipped through the Washington coastal zone. Bates and Carpenter (1979b) developed a method to quantitatively extract and characterize many of these organo-sulfur compounds (OSC) in coastal sediments. The saponification step often used following extraction of marine sediments was found to produce numerous S-containing hydrocarbons as artifacts. Individual OSC were quantified using a gas chromatograph equipped with a sulfur specific flame photometric detector.

OSC were characterized in one sediment core from the Astoria deep-sea fan off the Oregon coast (45° 52.4' N, 125° 47.2' W) and in several Puget Sound cores (Bates and Carpenter, 1979a). OSC concentrations and compositions within deep-sea fan sediments were comparable to those observed in many Puget Sound sediments, with dibenzothiophene being the predominant single compound. Neither OSC concentrations nor assemblages changed over the 42-cm length of this core, even though sediments at depth were estimated to be 4,000-y old. Results from this core imply that the bulk of the OSC are derived from natural sources such as forest fires and are extremely stable following deposition.

This conclusion is supported by more detailed analyses of several Puget Sound sediment cores. Dibenzothiophene concentrations were essentially constant with depth in central Puget Sound sediments (Fig. 10.30), even in cores where total aliphatic hydrocarbons, the UCM and diasteranes have increased steadily since around 1900. Concentrations of most PAH approach zero near 37 cm in the core shown in Figure 10.30, but OSC concentrations and dibenzothiophene remain essentially constant down to at least 50 cm. Only in sediments from Elliott Bay was there any indication of higher OSC concentrations due to human activities. Thus, while concentrations of dibenzothiophene or other OSC may be good indicators of fossil fuel contamination in some areas, they are not good indicators in Washington coastal sediments.

Three OSC had subsurface concentration maxima near depths where sediments became anoxic (Bates and Carpenter, 1979a). They are apparently created *in situ* near depths where hydrogen sulfide production begins and are probably associated with microbially-mediated conversions of various sulfur forms during sulfate reduction. Although electron impact and chemical ionization mass spectra were obtained for these compounds (Bates and Carpenter, 1979a), they have not yet been identified due to a shortage of authentic organo-sulfur standards.

Azaarenes. Azaarenes are structurally analogous to PAH but have one or more N atoms substituted in the ring structure. The N atom confers basic chemical characteristics to azaarenes which can be exploited in their isolation and analysis, and imparts mutagenicity and carcinogenicity to some of these compounds. No measurements have been made on azaarenes in Washington shelf or slope sediments, but the first quantitative measurements of azaarenes in coastal marine sediments were made in Puget Sound (Furlong and Carpenter, 1982). The

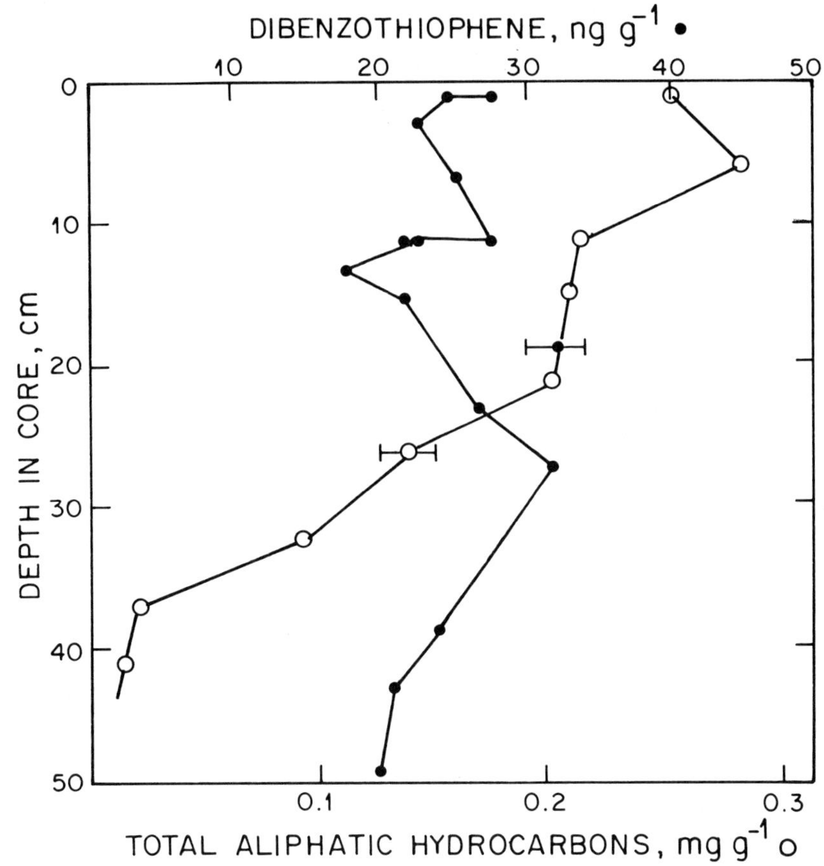

Fig. 10.30 Concentration profiles of dibenzothiophene and total aliphatic hydrocarbons in a sediment core from central Puget Sound station 51, just north of Seattle. Error bars show one standard deviation about measured concentrations. From Bates and Carpenter (1979a), with permission.

method involved Soxhlet extraction of sediments with benzene-methanol, a combination of acid-base partitioning and Sephadx LH-20 gel permeation-size exclusion chromatography and glass capillary gas chromatography with N-selective detection. Identification of many peaks was hampered by the lack of authentic standards. Concentrations of 2- to 4-ring azaarenes as low as 50 ng (g organic C) $^{-1}$ were determined.

Surface sediments off Seattle contained a complex mixture of azaarenes eluting in the range of quinolines and methylquinolines, with much lower amounts of higher molecular weight components (Fig. 10.31). Total azaarene concentrations in upper layers of Puget Sound sediments increased with increasing proximity to urban areas, suggesting important anthropogenic sources. Some low level natural sources were indicated by the trace quantities of azaarenes found in sediment layers deposited prior to 1900.

Azaarene compositions in most Puget Sound sediment horizons resembled those in New York City air particulates; i.e., they have a predominance of lighter, 2- and 3-ring compounds. In contrast, air particulates, street dusts and sediments from Belgium and

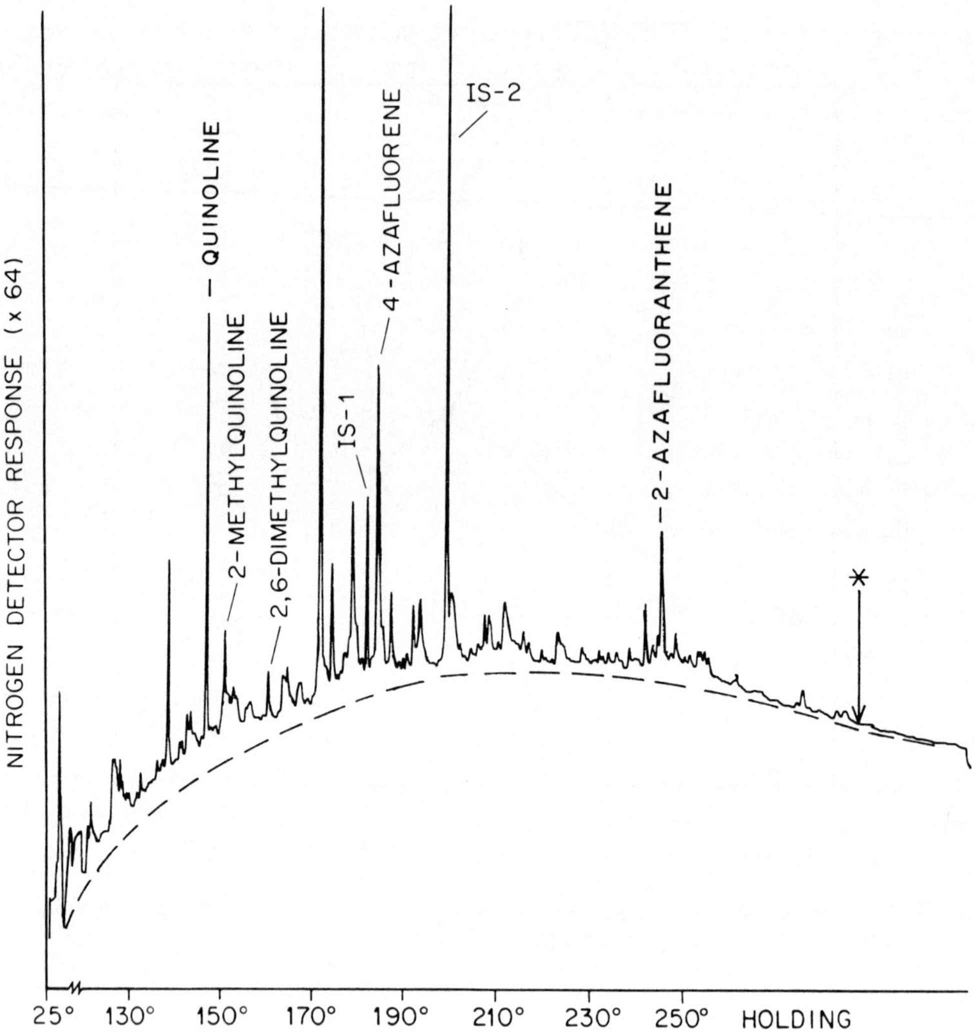

Fig. 10.31 Gas chromatogram of N-containing hydrocarbons contained in the basic lipid fraction. Extracted sediment was from the 0-2 cm horizon of a core collected at station 51 in central Puget Sound. IS-1 and IS-2 are injection standards, while the asterisk gives the approximate elution time of 1-azapyrene. Dashed line indicates the thermal response of the N-selective detector. From Furlong and Carpenter (1982), with permission.

Switzerland contained azaarenes with maximum concentrations centered around 4- and 5-ring compounds. The difference is attributed to combustion of primarily petroleum in North America and coals in Europe. A highly concentrated and complex azaarene mixture was found in sediment horizons deposited around 1920 near the Puyallup River delta (Fig. 10.2). These sediments contained black particles identified microscopically as coal fragments, as well as higher abundances of 3- and 4-ringed azaarenes, relative to 2-ring compounds.

Coal particles may be introduced either by natural erosional processes or by spillage during mining operations or trans-shipment from the port of Tacoma. Several coal-bearing strata have been exposed by the Puyallup River and its tributaries. Commercial coal production began in 1854, peaked around 1918 and has since steadily declined to the point that only one western Washington coal mining company operated in 1982 (Beikman *et al.*, 1961). At least one coal company has introduced fine coal particles into several Puyallup River tributaries in the past and the port of Tacoma was Washington's main coal trans-shipment port (Morgan, 1979). A new, major coal exporting plant is nearing completion near Kalama on the Columbia River. Another has been proposed for Tacoma and expansion of coal mining has been extensively discussed.

The possibility that concentration ratios of 2- to 3-ring (and heavier) azaarenes might be useful in distinguishing coal from petroleum sources and even possibly in identifying particular coal sources, was further examined by Barrick *et al.* (1984). Determinations of hydrocarbon and azaarene concentrations and compositions in 16 western Washington coals, ranging in rank from lignite to anthracite, confirmed that these coals generate an overall lipid signal distinct from other fossil fuel sources including street runoff and municipal sewage discharges. None of the coals had a UCM in the aliphatic extracts, unlike many petroleum products and urban runoff from the Puget Sound region. Complex azaarene compositions are found in all but the lowest and highest ranked coals. Coals could be divided into four groups based on azaarene contents. Also, a decrease in 2-ring relative to 3-ring azaarene dominance occurs with increasing vitrianite reflectance.

Lipid extracts from bulk marine sediments collected near the Puyallup and undeveloped Nisqually River deltas, coal particles isolated from the Puyallup delta sediments and certain coals within the adjacent drainage basins display similar azaarene fingerprints and hydrocarbon suites. Thus, these compounds give molecular evidence of the transport of mined and naturally eroded coal deposits to Washington coastal sediments.

Lignins. Lignin compounds are phenolic polymers abundant in the tissues of vascular land plants, but essentially absent from marine organisms. When chemically oxidized, lignin polymers produce characteristic phenolic compounds which can be used to classify and quantify land-derived organic matter in marine sediments on the basis of angiosperm versus gymnosperm and woody versus non-woody character (Hedges and Mann, 1979; and references therein). Lignin oxidation products and stable carbon isotope distributions have been used to investigate the sources, transport and chemical stability of land-derived organic matter in eight dated cores of modern sediment from the southern Washington continental shelf and slope (Hedges and Mann, 1979b). They found no evidence for significant chemical alteration of lignin compounds in these sediments for time periods of up to 400 y. Subsequently, longer cores in the MSSD indicated little change in bulk lignin composition over a 2,000-y period (Ertel, 1985). However, the base soluble lignin component, i.e., the humic substance fraction which represents only about 10% of the bulk lignin, does appear to reflect minor *in situ* lignin diagenesis over a 2,000-y period. The observation that lignin compounds can be preserved essentially intact in coastal marine sediments for up to 2,000 y suggests that lignin, like pollen, may offer detailed information of past vegetative and climatic variations.

Gymnosperm woods and nonwoody angiosperm tissues account for most of the land-derived organic matter in the Washington coastal deposits. These land plant remains have an average $\delta^{13}C$ of approximately –25.5 ‰ and are concentrated in the MSSD. Marine organic matter having an approximate $\delta^{13}C$ of –21.5 ‰ predominates in most other shelf and slope sediments. Net fluxes of marine organic matter into the surface sediments are highest in environments which favor the preservation of organic matter but still correspond to less than 1% of primary productivity in overlying waters.

Lignin characterizations have also been made by Prahl (1985) in six sediment cores along an east-west transect just north of the track of Hedges and Mann (1979b). Once again lignin was most enriched in the MSSD. Although both lignin and plant wax alkanes originate on land and are introduced at the mouth of Columbia River, Prahl (1985) noted that the linear correlation between lignin parameters and sum of plant wax *n*-alkanes was poor ($r^2 = 0.02$) in Washington coastal sediments. This poor correlation suggests that lignin and plant wax hydrocarbons have different physical relationships to the bulk plant debris component of these sediments or degradation rates.

Hedges and Mann (1979b) showed that gymnosperm wood represents 50 to 67% of the lignin in Washington coastal sediment and that the remaining lignin derived from nonwoody angiosperm tissue. Plant wax *n*-alkanes are more likely associated with nonwoody parts of higher plants such as leaves and stems (Caldicott and Eglinton, 1973). This difference is believed responsible for the poor correlation between plantwax *n*-alkanes and lignin. It is supported by the fact that the hydrocarbon retene, which is physically associated with lignin containing plant debris in these sediments (Prahl and Carpenter, 1984), has a strong correlation with lignin concentrations ($r^2 = 0.79$).

Long Chain Aldehydes. A series of C_{20}- to C_{32}-saturated *n*-aldehydes has been reported in a variety of Washington coastal sediments of recent to glacial ages, in suspended particulate material from the Columbia River mouth and in several important plants indigenous to the Pacific Northwest (Prahl and Pinto, 1987). The *n*-aldehydes are predominately introduced to the Washington coastal zone 'preformed' within eroded terrestrial plant debris amid the detritus discharged by the Columbia River. The *n*-aldehydes seem to be another molecular remnant of higher plant waxes that has survived transport from terrestrial plant sources to these offshore sedimentary sinks. There is no evidence to support formation of *n*-aldehydes within coastal sediments as a result of alteration of some other organic components. Comparison of lipid compositions in surface and deeper intervals of continental shelf and slope cores which differed in age by a few hundred years indicated that plant wax derived *n*-aldehydes may slowly degrade postdepositionally, but preferentially degrade relative to plant wax *n*-alkanes.

Carbohydrates. Individual neutral sugars have been measured in sediments, particulates collected in sediment traps and major potential biological sources in Dabob Bay, Wa. (Cowie and Hedges, 1984). Biochemically-consistent differences in carbohydrate composition were found between marine and terrestrial source materials, between angiosperms and gymnosperms plants and even between woody and nonwoody vascular plant tissues. Plankton and bacteria could not be distinguished based on their carbohydrate compositions alone. Dabob

Bay sediments and sediment trap particulates contained primarily marine-derived carbohydrates, with increased levels of terrigenous polysaccharides evident in the water column during winter. Carbohydrate concentrations and compositions were essentially constant over the top 50 cm of a Dabob Bay sediment core, suggesting surprisingly little *in situ* degradation below the surface mixed layer in periods of a few hundred years.

Pigments. Furlong (1986) measured photosynthetic pigments and their alteration products in oxic (Dabob Bay) and anoxic (Saanich Inlet) marine and lacustrine (Lake Washington) sediments of the Pacific Northwest, to determine preservation, degradation and alteration of these phytoplanktonic biomarkers over time periods up to 10,000 y. Sediment anoxia led to greater preservation of pigments, but in all three depositional environments studied, 95% or more of the pigment flux collected in sediment traps moored below the euphotic zone was degraded and did not accumulate in the top 2 cm of underlying sediments.

Pheopigment concentrations decreased exponentially with depth in oxic sediments of Dabob Bay. ^{210}Pb-derived sediment accumulation rates and mixing coefficients were used to determine decomposition constants from pigment profiles modeled with a one-dimensional, advective-sedimentation diffusive-bioturbation model (Furlong and Carpenter, 1988). Pigments were decomposed more rapidly than hydrocarbons and fatty acids in coastal marine sediments with half-lives of ~40 d, similar to published rates for plankton degradation. Pheopigment decomposition rates are orders of magnitude faster than bulk carbon decomposition rates measured in deep-sea or continental margin sediments. These are the first, directly estimated constants for organic decomposition at the molecular level using the constraints provided by mixing and sediment accumulation rates determined from isotope geochronology.

Total pheopigment concentrations did *not* decrease over the past 150 y in anoxic sediments of Saanich Inlet. However, compositional changes expressed as increased percentages of sedimentary pheopigments as post depositional derivatives were observed. Anoxic sediments deposited over 10,000 y contained measurable total organic C and total pheopigment decreases with depth which could be modeled as a function of sedimentation and decomposition alone (i.e., no bioturbation). Pigment decomposition calculated for ^{14}C dated anoxic sediments had half-lives of about 1,500 y. This value is 3 to 4 orders of magnitude larger than observed for oxic marine sediments and larger than decomposition constants calculated for deep-sea sediment organic C.

Lake Washington sedimentary pheopigment concentrations deposited over the past 150 y reflect historical phytoplankton changes resulting from anthropogenic activities. Pigment concentrations were highly variable in Lake Washington sediments deposited over the past 10,000 y. Concentration maxima at 1,900 and 7,000 y B.P. resulted from either enhanced productivity or enhanced preservation. Molar pheopigment a/pheopigment c ratios suggested development of increasingly diverse algal populations with lake maturation. Post depositional derivatives were an increasing fraction of total sedimentary pheopigments with depth (age). They are likely intermediates in the conversion of chlorophylls to petroporphyrins.

Furlong (1986) also modified a chromic acid oxidation technique to produce low, but reproducible yields of maleimides (dicarbonyl pyrroles generated from the pyrrole components of the chlorophyll ring) from synthetic porphyrins and purified sedimentary pigments. The neutral maleimide oxidation products were then analyzed by gas chromatography using an N-selective detector. Maleimides derived from humic substances and acetone-extracted sediments of Dabob Bay showed that a large part of the particulate chlorophyll flux deposited in the sediment becomes associated with humic substances, a form not accounted for in previous sedimentary pheopigment studies. This previously unquantified pheopigment sink is a small fraction of humic substance N, but makes up 25 to 75% of total preserved sedimentary pheopigments.

The humic associated fraction of sedimentary pigments alone can account for the petroporphyrin plus phorbin yield typically estimated in Quarternary sediments (Louda and Baker, 1981). Some Pleistocene, Miocene and older sediments have significantly higher solvent-extractable petroporphyrin concentrations than can be derived from pheopigment precursors measured in acetone extracts of oxic or anoxic surface sediments. In these sediments, petroporphyrin generation from acetone extractable pheopigments alone would require significantly lower pigment remineralization or higher preservation prior to or after initial sediment deposition. Furlong's measurements suggest that an alternate source of petroporphyrins could be from humic substances, supporting the kerogen cracking hypothesis of Hoering (1969). The data also suggest that petroporphyrin concentration differences may not reflect original C source as much as the degree to which humic associated pheopigments are altered and released.

10.6.3 Remaining Questions and Needed Research Concerning Sedimentary Chemical Deposition

The almost complete lack of concentration profiles for chlorinated and other synthetic organic chemicals, such as phthalate esters and PCBs, in sediments of the Washington coast needs to be remedied. Besides being important in their own right, certain of these compounds may prove useful markers of specific inputs, such as the METRO sewage discharge to Puget Sound. Among the possible organic molecular markers of sewage discharges are coprostanol (Brown and Wade, 1984), linear alkylbenzenes (Eganhouse and Kaplan, 1982), a tocopheryl acetate (Eganhouse and Kaplan, 1985) and organosilicones (Pellenbarg, 1985). For maximum interpretive value, these profiles should be determined on cores for which excess ^{210}Pb and possibly other radionuclide profiles are determined. Based on observations of PAH subsurface maxima within ^{210}Pb well mixed surface layers, it appears some of the organic compounds may be mixed less extensively and hence retain a more precise chronological record than ^{210}Pb.

The extent of particle selective mixing needs more study. The Washington coast is an interesting area to study this topic, since possibly different discrimination mechanisms exist between Puget Sound sediments, where biologically-induced mixing seems to predominate and Washington coast sediments, where physical processes seem more important. Evaluation of possible particle selective mixing will require better definition of the sizes and types of particles which carry major fractions of the chemicals of interest. In Washington coastal sediments, ^{210}Pb, Pu and PAH are definitely associated with different sizes and types of particles and may well be differentially mixed.

The basic question of the extent to which potentially harmful chemicals in sediments are actually available to organisms needs much further study. Determination of predominant phase associations is only a part of this problem. Progress will also require input from both chemists and biologists to design new and realistic exposure methods for key organisms over their entire life cycles.

Suggestions that pore water concentrations of trace organic compounds such as PCBs can be predicted using more readily measured bulk sediment concentrations and a simple equilibrium partitioning model employing laboratory-derived equilibrium constants need to be more thoroughly tested with concordant field data on pore water and solid phase concentrations. This is important since soluble phases may be more available to organisms. Organic colloids in pore waters may allow concentrations above those calculated for equilibrium partitioning with inorganic seawater. Concentrations of combustion-derived PAH may be less than predicted, because their predominant solid phase carriers are relatively large fly ash and soot particles which may not equilibrate with pore waters.

Although methods for separation of several milliliters of pore water from undisturbed sediment cores for trace metal analyses have been greatly improved since 1970, collection of enough pore water from closely spaced sediment horizons for trace organic compound analyses remains a challenge. Reliable data on pore water concentration profiles of chlorinated organics, such as PCBs, are still needed. The approach of Sholkovitz and Mann (1984), which yielded 1-2 liters of pore water from 2-cm thick intervals, may be satisfactory for some compounds, but the method involved the extensive use of plastic, so organic blanks must be closely monitored.

Data for Washington shelf sediments strongly argue against significant soluble Pu remobilization from or within these coastal sediments. The single core analyzed for Pu to date from Puget Sound (from Dabob Bay) also argues against resolubilization of Pu. However, additional Pu profiles in Puget Sound sediments are needed. The Washington shelf data have clearly revealed that Pu remobilization cannot be determined from ^{210}Pb profiles alone.

Bell jar studies have directly evaluated the remobilization of Hg and As out of Puget Sound sediments. Water column studies suggest detectable amounts of Mn, Ni, Cu (Jones and Murray, 1984; 1985) and Cr (Cranston, 1983) may flux out of slope sediments. U and Th isotope data in sediments suggest a major flux of ^{228}Ra (and other Ra isotopes) out of these sediments. On the other hand, data for other elements (Pu and ^{210}Pb) suggest enhanced near-bottom scavenging of dissolved elements and their removal to slope sediments. Further field studies are needed to identify which metals go into and which come out of Washington coastal sediments and under what conditions the flux occurs.

Piston cores from the Washington slope and Cascadia Basin appear to have reasonably well defined sequences of sediments deposited during the past 20,000 y. The transition from Holocene into Pleistocene sediments is often marked by obvious color change from drab olive to grey clay. Careful geochemical analyses of layers through the glacial-interglacial transitions would test models of the effects of glacial waxing and waning on ocean chemistry. The Washington slope seems an area especially well-suited area to test the hypothesis of lower dissolved O_2 concentrations during glacial times, because O_2 saturation levels in slope waters are currently only about 10% and small decreases in O_2 would lead to anoxic conditions which should be reflected in sedimentary profiles of certain chemicals.

10.7 CHEMICAL INTERACTIONS WITH BIOTA

10.7.1 Overview

Since 1970 much has been learned and much dogma invalidated concerning interactions of potentially toxic chemicals with marine organisms. Woodwell *et al.* (1967) showed that concentrations of DDT and its degradation products increase fairly regularly in successively higher trophic levels of coastal marine food webs. This led to the concern and expectation that many other toxic chemicals would behave similarly. However, data subsequently compiled for other chemicals often did *not* show such concentration biomagnification in higher organisms (PCBs, Clayton *et al.*, 1977 and references therein; Hg, Windom *et al.*, 1972; As, see below; PAH, Clark, 1983). Clearly it is not possible to reliably infer food web behavior of other chemicals from the behavior of DDT.

In 1970, Hg concentrations in swordfish steaks and canned tuna often were found to exceed the maximum permissible level of 0.5 ppm set by the U.S. Food and Drug Administration. Much effort was then spent monitoring these and other organisms for potentially toxic chemicals. It quickly became clear, however, that these Hg concentrations, as well as other examples such as high As concentrations in shrimp and lobster, were the result of poorly understood natural processes. It is difficult, if not impossible, to relate chemical concentrations in mobile, field caught organisms to biogeochemical processes, or to concentrations in water, sediments, or food organisms at the collection sites. Difficulties in interpreting data for mobile organisms were part of the rationale for establishing the national Mussel Watch Program to monitor concentrations of bioavailable chemicals in coastal zones. Mussel Watch results for samples from six sites around the Washington coast region are included in this section.

During the 1970s, the importance of microbially-mediated reactions, large vertical chemical fluxes associated with fecal material and effects of benthic organisms on chemical fluxes across the sediment-water interface were recognized. Bioturbation complicates efforts to calculate accurate sediment accumulation rates from profiles of radionuclides such as excess ^{210}Pb. Interest in microbially-mediated reactions grew with concern over effects of oil spills in the coastal zone and recognition that the natural cleansing capacity of the coastal zone is heavily dependent upon microbial degradation of the oil (National Academy of Sciences, 1985).

Since 1970 it has also become increasingly clear that sediments are a, if not the, major reservoir of many potentially toxic chemicals in the coastal zone. However, the extent to which the sedimentary chemicals are actually available for uptake or other interactions with organisms has been determined in only a few cases (Anderson *et al.*, 1977; Seelye *et al.*, 1982; Landrum and Scavia, 1983; Rubinstein *et al.*, 1984; Malins *et al.*, 1985). Confident assignment of cause and effect is also complicated by large, natural variabilities in abundances and behaviors of organisms.

A series of major oil spills worldwide in the mid-1970s led to much concern over effects of oil on marine organisms and to numerous studies of effects of oil (and other chemicals) on various marine organisms. Many of the studies were poorly designed and used such unrealistic exposure conditions that their results have been of limited value in assessing the real hazards and effects likely under reasonable field conditions. Our understanding of petroleum interactions with marine ecosystems has recently been reviewed (National Academy of Sciences, 1985).

10.7.2 Microbial Modifications of Chemical Processes

The discovery that Hg in fish tissues was present largely in the methylated form so toxic to humans led to much interest in determining the processes responsible for methylation and also determining what other metals might also undergo methylation (see Table 10.9). Subsequent studies have shown that natural microbial activity in sediments can both methylate and demethylate Hg (Spangler *et al.*, 1973). However, methyl-Hg forms are only small fractions of the total Hg in sediments and interstitial waters of the Washington coastal zone (Bothner *et al.*, 1980) and other coastal areas (Lindberg and Harriss, 1974). Bell jar studies of Hg fluxes out of contaminated sediments of Bellingham Bay, Washington found that less than 3% of the Hg was in methylated forms (Bothner *et al.*, 1980). Most Hg methylation may therefore be occurring within fish.

Methylated forms of As have been found in waters of the euphotic zone off Georgia and southern California (Waslenchuk, 1978; Andreae, 1979), possibly produced by microbial reactions. However, methylated forms of As were not found in bacteria-rich, interstitial waters collected off California (Andreae, 1979) or in laboratory experiments with marine muds (McBride *et al.*, 1978). Methylated forms of As also represent a very small fraction, if any, of the As in pore waters of Puget Sound, Saanich Inlet, Lake Washington and the Washington coast (Peterson and Carpenter, 1986). Thus, microbial production of methylated As compounds independent of plankton or other organisms, appears minor. The presence of methylated As in the euphotic zone is apparently due to leakage from plankton, with methylate As as a detoxification mechanism for arsenate being taken up inadvertently along with phosphate. In contrast to results for Georgia and California waters, methylated forms of As represent very small fractions of dissolved As in the water column of Puget Sound (Carpenter *et al.*, 1978) and the Washington coast (Fig. 10.10).. This difference is attributed to the higher water-column phosphate concentrations normally found in Washington coastal waters, which reduce As uptake by plankton.

TABLE 10.9 Summary of observations of methylated and other organic forms of several metals in Washington coast and other ocean areas.

Metal	Washington Coast Organic forms	Reference	Other Areas Organic Forms	Reference
Hg	Methylated forms < 3% of total in Bellingham Bay waters and pore waters	Bothner *et al.* (1980)	Methylated forms only small % of total Hg in sediments and pore waters	Lindberg and Harriss (1974)
As	Methylated forms < 5% of total As in both oxic and anoxic pore and overlying waters of Puget Sound, Lake Washington, Saanich Inlet and Washington coast	Perterson and Carpenter (1983, 1986)	Methylated forms common in surface waters off California and Georgia	Andreae (1979) Waslenchuk (1978)
			No methylated forms in pore waters off California or in lab incubations of marine muds	Andreae (1979) McBride *et al.* (1978)
	High molecular weight organo-As forms not yet sought		Complex high molecular weight organic-As forms in marine organisms as detoxification mechanism	Penrose *et al.* (1977) Klump and Peterson (1981) & reference therein
Se	In Saanich Inlet waters present as high molecular weight organic forms, most likely as Se-amino acids in peptides, not methylated	Cutter (1982) Cutter and Bruland (1984)	Organic selenite important in surface waters and sub-oxic waters of Pacific	Cutter and Bruland (1984) & references therein
Ge	Not yet sought		Mono- and di-methylated forms reported in estuarine and open ocean waters	Hambrick *et al.* (1984) Lewis *et al.* (1985)
Sn	Not yet sought		Several di- and tri-methylated forms observed, relative magnitude of input from organic Sn biocides from ship hulls and from natural production is unclear	Hodge *et al.* (1979) Byrd and Andreae (1982) Hallas *et al.* (1982) Matthias *et al.* (1986)
Sb	Not yet sought		Several methylated forms observed in waters	Andreae *et al.* (1981)

Rates of Mn^{2+}, Co^{2+} (Tebo *et al.*, 1984) and As^{3+} (Peterson and Carpenter, 1983) oxidation in seawater collected from just above the redox transition in the Saanich Inlet water column are greatly slowed by the addition of poisons. The poisons suppress microbial activity. The rate of inorganic Mn^{2+} oxidation is several orders of magnitude slower than microbially-catalyzed rates. Slow, inorganic reaction kinetics explain why thermodynamically-unstable concentrations of Mn^{2+} and As^{3+} can persist in oxic seawater.

Information on numbers and types of bacteria found in Washington shelf sediments was available in the early 1970s (Wiebe and Liston, 1972). Only subsequently has the widespread importance of microbial processes in the biogeochemical cycling of numerous chemicals in coastal sediments become clear. Diagenetic reactions ultimately driven by bacterial and macrobenthic organism respiration are responsible for remobilization of Mn^{2+} (Jones and Murray, 1985) and possibly other metals such as Cr, Ni and Cu (Cranston, 1983; Jones and Murray, 1984) from mildly reducing sediments of the Washington continental slope to the overlying water. Subsequent oxidation and overlying water. Subsequent oxidation and precipitation of the Mn^{2+} as hydrous oxides play important roles in scavenging chemicals such as ^{210}Pb (Carpenter *et al.*, 1981) and Pu isotopes (Beasley *et al.*, 1982). Microbially-controlled sulfate reduction in sediments also affects pore- water, trace metal concentrations and fluxes out of Puget Sound sediments (Bothner *et al.*, 1980; Emerson *et al.*, 1984; Peterson and Carpenter, 1986).

Several unidentified organo-sulfur compounds are formed in relatively large amounts just above the oxic-anoxic transition in Puget Sound sediments (Bates and Carpenter, 1979a). Microbially-mediated sulfate reduction reactions are very likely responsible for producing these compounds. Microbial (and other) processes also preferentially degrade planktonically over terrestrially-derived hydrocarbons in Puget Sound (Prahl *et al.*, 1981).

10.7.3 Chemical Concentrations in Plankton

Results for only a few analyses of potentially toxic chemicals in plankton from Washington coastal waters have been published (Bennett and Carpenter, 1979; As data below and Clayton *et al.*, 1977). Analyses of Puget Sound neuston samples are of special interest because of the recent discovery that effluent discharged at depth from Seattle's METRO primary sewage treatment plant can occasionally reach the sea surface and accumulate as slicks (Word *et al.*, 1988). Sea surface slicks from several Puget Sound locations have higher concentrations of several potentially toxic chemicals than underlying waters (Hardy *et al.*, 1985, 1988). The extents of possible adverse effects of the elevated contaminant concentrations upon the eggs and juvenile forms of numerous species which also accumulate in the sea surface microlayer are just beginninning to be determined (Hardy and Antrim, 1988; and references therein).

Clayton *et al.* (1977) showed that PCB concentrations in zooplankton from Puget Sound were highly correlated with the lipid contents of the organisms as well as the PCB concentrations in seawater, suggesting that PCB bioaccumulation is predominately controlled by equilibrium partitioning of these chemicals between the ambient water and the organism's internal lipid pool. Lipid-water amplification factors of about 10^6 were relatively uniform over a wide range of spatial and temporal regimes, species compositions and lipid contents. Results showed that food chain biomagnification is *not* a controlling factor in attaining observed PCB levels in marine organisms.

Clayton *et al.* (1977) pointed out the potential use of environmentally-determined, lipid-water partition coefficients to predict PCB concentrations in aquatic ecosystems. The idea that zooplankton might serve as useful indicators of PCB concentrations in waters is appealing in view of the sampling and analytical difficulties in measuring the low PCB concentrations normally encountered in seawaters. The validity of using partition coefficients between water and various organic solvents (octanol, triolein, etc.) to predict concentrations and bioaccumulation of various organic chemicals remains a topic of considerable interest and debate (Gschwend and Wu, 1985; Chiou, 1985). This approach seems most likely to be valid for compounds which enter the ocean as, or are quickly converted to dissolved forms, such as the more soluble components of fuels and oils. It seems invalid for combustion-derived PAH, which are largely associated with bits of flyash and charcoal-like material (Prahl and Carpenter, 1983) and seem less likely to equilibrate with surrounding waters. Lack of equilibrium could also result from vertical migration of the zooplankton between waters of different PCB contents.

Crecelius (1974) determined As concentrations in plankton collected at seven locations in the Puget Sound region, including a site only 1 km from a smelter near Tacoma. The smelter had discharged large amounts of As into Puget Sound. Concentrations of As in the 63-333 μm size fraction which contained mostly diatoms ranged from 2.8 to 18 ppm dry weight, with a mean of 8.0 ppm. Concentrations of As in the > 333 μm fraction, which contained mainly copepods, ranged from 1.0 to 21 ppm dry weight, with a mean of 6.0 ppm. These As concentrations in Puget Sound plankton are similar to those reported for plankton from open ocean locations farremoved from major pollution sources (Robertson *et al.*, 1972; Windom, 1972; Bohn, 1975; Bohn and McElroy, 1976). Direct effects of even relatively high dissolved As^{5+} concentrations upon zooplankton are small. However, low concentrations of As^{5+} can alter phytoplankton species composition and succession patterns, affect normal prey-predator relationships and indirectly greatly reduce zooplankton grazing, survival and reproductive success (Sanders, 1986).

10.7.4 Carbon Isotopes in Marine Organisms

Organic compounds synthesized since about 1963 have a ^{14}C activity about 20% higher than pre-1950 levels, due to contributions from bomb produced ^{14}C. Petroleum, on the other hand, contains no ^{14}C. Thus, ^{14}C measurements may be useful indicators of whether or not an organic fraction is mainly of recent biological origin or a fossil fuel contaminant (Wakeham and Carpenter, 1976; Pearcy and Stuiver, 1984; and references therein). The measurement is usually made on bulk organic material, but more specific information is gained when enough material can be obtained to analyze individual compound classes, such as aliphatic or aromatic hydrocarbons. $^{13}C/^{12}C$ isotope ratios also provide some source information and their determination requires much less material, although the end members are not as distinct from one another as in the case of ^{14}C.

A. W. Fairhall and J. Erickson completed ^{14}C and $^{13}C/^{12}C$ ratio measurements on total organic C in 33 samples of plankton collected during 1975-1977 at the same Puget Sound stations as the water samples discussed in Section 10.3. Standard precautions and procedures for collection and preservation of samples for pollutant analyses in general and hydrocarbon analyses in particular, were followed. Nets were towed from booms extending from the side of the research vessel and upwind of the ship to minimize contamination.

Organisms living in the top 20-30 cm of the water column were collected with a 335 μm neuston net. A series of vertical tows from depths of 200 m or near bottom to the surface were made with 64 μm and 335 μm nets. Organisms collected in these tows were divided into size classes by wet sieving with seawater through solvent-washed stainless steel screens having mesh openings of 1700, 850, 300 and 64 μm. Organisms on each screen were transferred to solvent-washed glass bottles with a minimum of seawater and frozen on board ship. Prior to sieving, a 1/16 split of the sample was removed and preserved in buffered formalin for taxonomic studies. Subdividing the organisms into size classes proved useful because often one species organism was overwhelmingly predominant in a particular size class. Analysis of isolated size classes better defines how chemicals are distributed between various organisms comprising the overall planktonic community.

^{14}C activities for most of the plankton samples were the same as the ambient seawater (Table 10.10) to within the experimental error and natural variability observed in the waters. This observation is supported by similar measurements elsewhere (Williams *et al.*, 1970; Williams and Linck, 1975). Data in Table 10.10 and Swanson (1980) indicate a mean ^{14}C activity of 102.1 ± 1.5% of the contemporary NBS standard in Puget Sound organisms collected in 1975-1977. Their $\delta^{13}C$ contents averaged -20.3 ± 1.4 ‰. Isotopic distributions were reasonably homogeneous among the organisms and no geographic trends were discerned (Fig. 10.7). These ^{14}C activities closely reflect those of dissolved inorganic C in Puget Sound waters, which average 102.0% for surface and 100.8% for deep waters (see Section 10.3). Puget Sound plankton fractionate against ^{13}C by an average of 18.7 ‰ relative to dissolved, inorganic C.

Occasionally the large size fraction of plankton, particularly those from the neuston tows, had an isotope fractionation-corrected $\Delta^{14}C$ activity which is significantly higher than that of the dissolved, inorganic C of the water. This was particularly noticeable in a few instances when samples contained significant amounts of plant material, possibly of terrestrial origin, which reflects the higher ^{14}C activity of the atmosphere.

There were a few instances where organisms appear to have significantly lower $\Delta^{14}C$ value than the seawater in which they were collected (stations 98, 59 and 83; Fig. 10.13). If such decreases in $\Delta^{14}C$ activity were due to petroleum contamination, $\Delta^{14}C$ measurements on solvent-extracted hydrocarbon fractions from these organisms should have significantly lower $\Delta^{14}C$ activity than the total organic C of the organisms. Enough organisms were obtained at three stations to extract their lipid fractions and determine the $\Delta^{14}C$ and $\delta^{13}C$ contents (Swanson, 1980). Differences between $\Delta^{14}C$ and $\delta^{13}C$ in lipids relative to bulk organic matter were small, but in the direction expected for traces of fossil fuel contamination. However, at station 98 the change was definitely outside propagated uncertainties of the two measurements. Now that development of accelerator techniques for ^{14}C determinations has reduced required sample size by a factor of about 100, more ^{14}C isotope measurements on specific organic compound classes should be feasible.

TABLE 10.10 Comparison of ^{14}C activities for plankton samples and dissolved CO_2 in waters from which they were collected. Standard deviation of individual measurements is about ± 10 ‰ and seasonal variability of CO_2 in waters from same area is at most ± 3% about the mean (see Section 10.3). Most station numbers are located in Figure 10.13.

Date	Station	Net Size (μm) and Type of Tow	Sieve Mesh Size (μm)	$\Delta^{14}C$ (‰) Organism	$\Delta^{14}C$ (‰) Water
8/75	62 Pt. No Point	335 Neuston	300	+28	+23
		"	"	+22	
5/76	94 North of Lumni Is.	335 Vertical	1700+850	+42	–4
			300	–12	
"	69 Cherry Point	64 Vertical	1700+850	+21	
		"	300	–8	+16
		"	64	+16	
"	65 Rosario Strait	335 Vertical	1700+850	+19	+16
		"	300	+21	
"	73 Orcas Is.	335 Vertical	300	+27	+26
"	98 Everett	335 Vertical	1700	+48	
		"	850	+15	+66
		"	300	+26	
"	75 Near Anacortes	64 Vertical	300	+33	
		335 Vertical	1700+850	+9	+6
		"	300	–11	
8/76	59 South of Tacoma	64 Vertical	300	+10	
		"	64	–4	
		335 Vertical	1700+750	+29	+31
		"	300	+44	
		335 Neuston	1700+850+300	+79	
"	83 Near Olympia	64 Vertical	1700+850+300	0	
		"	64	–6	
		335 Vertical	300	+47	+22
		335 Neuston	1700+850	+110	
		"	300	+22	
"	56 East Passage between Seattle & Tacoma	64 Vertical	300	+22	
		"	64	+15	
		335 Vertical	1700	+55	
		"	850	+21	–9
		"	300	+36	
		335 Neuston	1700	+34	
		"	300	+38	

10.7.5 Role of Fecal Pellets in Chemical Cycling

During the 1970s it became clear that conventional seawater sampling bottles did not adequately sample the relatively large and more rapidly sinking particles, such as zooplankton fecal pellets and that these large particles constitute a major flux of chemicals from ocean surface layers to deeper waters and underlying sediments. Accordingly, several different sizes and shapes of sediment traps were developed by various oceanographic research groups to better sample this large particle flux (Honjo, 1978; Knauer *et al.*, 1979; Bruland *et al.*, 1981; Dymond *et al.*, 1981; Lorenzen *et al.*, 1981; Moore *et al.*, 1981; Baker and Milborn, 1983; and references therein). Sinking rates of 100 m d^{-1} and more are now commonly accepted for such particles (Deuser *et al.*, 1981; 1983; Bacon *et al.*, 1985; and references therein). Such particles are undoubtedly important in removing particle-associated chemicals from surface waters, but decomposition and disaggregation in deeper waters prevent many of them from reaching the deep-sea sediment-water interface.

Chemical and geochemical studies of the role of fecal pellets in Washington coastal waters have been carried out primarily with cylindrical sediment traps described by Lorenzen *et al.* (1981). These traps included the following features: (1) the presence of a grating in the mouth to reduce turbulence, help prevent washout and exclude large animals; (2) deployment on a taut-wire mooring with a subsurface float so traps could be suspended throughout the water column; (3) remote deployment as a self-contained unit with trap closures at predetermined times; (4) prevention of surface water and particles entering the traps during deployment; and (5) addition of a $HgCl_2$ brine preservative to minimize microbial activity, prevent ingestion of collected material by trapped animals and reduce washout.

The fact that Lorenzen's traps worked well on moorings in Dabob Bay deep waters does not mean such traps will work equally well in waters with stronger advective flows over the traps. In an effort to get around the advection problem, free floating assemblies of smaller traps of the same basic design were developed and have been applied to study pigment cycling in and just below the euphotic zone (see Landry and Lorenzen, 1989, Chapter 5). The small mass of material collected during the short deployments of these smaller traps limits the chemical analyses which can be made.

Baker and Milburn (1983) subsequently developed a sequentially sampling sediment trap which collects ten separate samples during a single deployment. It has been deployed on moorings in central Puget Sound along with standard Aanderaa current meters and beam transmissometers to simultaneously monitor horizontal particle fluxes. Samples collected with this sediment trap instrument system have been analyzed for aliphatic hydrocarbons and PAH (Bates *et al.*, 1984).

Debate continues over the most appropriate poison to add to such traps to prevent microbial decay or alteration of chemical composition. Larger 'swimmers' are periodically found in traps and chemical contributions from their carcasses introduce uncertainties in determinations of true vertical fluxes due to fecal material. In areas of bottom sediment resuspension, such as central Puget Sound, there is still no good way to separate freshly sinking from resuspended particles in traps placed near the sea floor.

Fecal pellets may contain significant concentrations of trace metals and radionuclides, including Pu, Pb, Zn, Ce, Cd and Th (Beasley *et al.*, 1978; Bacon *et al.*, 1985 and references therein). Thus sinking of fecal material is strongly implicated in the removal of such elements from surface waters. Elder and Fowler (1977) reported high concentrations of PCBs in laboratory-collected fecal pellets from natural populations of euphausiids and suggested these particles provide an important mechanism transporting PCBs into the deep-sea. From the visual presence of oil droplets in guts and fecal pellets of filter-feeding zooplankton after an oil spill, Conover (1971) calculated that copepod fecal pellets can provide a significant removal pathway for oil from surface waters to bottom sediments.

Oxidation and recycling of organic fecal material releases some of the associated trace elements. Hoffman *et al.* (1981) concluded that less than 1% of fecal pellets survived to a depth of 35 m on the Atlantic shelf. The fraction surviving to a given depth will depend greatly on the size and shape of the fecal pellets, which is highly species dependent. Bacon *et al.* (1976) observed that approximately 50% of the particulate ^{210}Po was regenerated in the water column as particulate material sank through the thermocline, while only about 5% of the particulate ^{210}Pb was recycled. Correlations of dissolved Cd and Zn with phosphate and silicate, respectively, in the water column indicated that Cd and Zn are incorporated into organic matter and released following oxidation (Bruland *et al.*, 1978a,b).

Low ^{14}C activities in most deep-sea animals collected off Oregon indicated that the majority of their carbon was not derived directly from sea-surface food chains labelled with bomb ^{14}C (Pearcy and Stuiver, 1984). This suggested that rapidly sinking particles from surface waters, such as fecal pellets, are not the major source of organic C for deep-sea animals and large benthic invertebrates. This conclusion is surprising given the abundant evidence from sediment trap studies for the importance of rapidly sinking fecal pellets and large particles in transporting material from surface to deep water. It has been challenged by Williams *et al.* (1987).

Studies of the role of fecal pellets in the cycling of potentially harmful trace chemicals in Washington coastal waters were begun by Bennett (1980) in Dabob Bay, Puget Sound. This area, which is removed from direct inputs of major rivers, is characterized by minimal horizontal advection below the thermocline and macrozooplankton-dominated sedimentation processes (Shuman, 1978; Welschmeyer, 1982; and references therein). Traps described by Lorenzen *et al.* (1981) were deployed 50-m above the bottom at water depth of 110 m. The nepheloid layer extends at most 10-m above the bottom at this site; hence collection of resuspended sediment was insignificant. Particles collected in a year long series of sediment trap deployments were predominately fecal pellets and other fecal matter.

Comparison of expected and observed excess ^{210}Pb fluxes gave the first *in situ* calibration of the collection efficiency (118 ± 13%) of sediment traps (Lorenzen *et al.*, 1981). Flux variations of factors of 2 to 5 during the year were observed for most chemicals, including pigments (Shuman, 1978), organic C and aliphatic hydrocarbons (Prahl *et al.*, 1981), PAH (Prahl and Carpenter, 1979) and several trace metals (Bennett, 1980). Maximum trace chemical fluxes in Dabob Bay were associated with a fall bloom in September and October. Greater pristane concentrations and fluxes indicated populations and grazing of calanoid copepods were greater during the fall than spring bloom.

The importance of vertical transport for different chemicals by fecal pellets was quantitatively established by comparing time-integrated fluxes from the traps with the annual fluxes in bottom sediments determined from surface-sediment concentrations (uncorrected for surficial mixing) and ^{210}Pb derived recent sediment accumulation rates. The PAH flux associated with fecal pellets essentially accounted for 100% of the PAH accumulating in the underlying sediments (Prahl and Carpenter, 1979). The observation that PAH were neither produced nor degraded during relatively rapid transport through the water column has been subsequently confirmed elsewhere (Gschwend and Hites, 1981; Bates *et al.*, 1984). Comparison of hydrocarbon fluxes collected in traps with fluxes in underlying sediments clearly revealed rapid and preferential remineralization of planktonic hydrocarbons relative to those of terrestrial origin (Prahl *et al.*, 1981). An apparent excess flux of organic C in the traps was balanced by benthic oxygen consumption rates in Dabob Bay. Only about 1/3 of the organic C flux arriving at the sea floor of Dabob Bay is buried in sediments, while some 2/3 is utilized by benthic organisms (Bennett, 1980).

Table 10.11 compares mass and elemental fluxes from time-integrated trap collections with fluxes accumulating in bottom sediments. Results for relatively unreactive metals (Al, Cr, Co, Fe and V) show that indeed the traps collected essentially all of the flux of these species to the sea floor. Trapped fluxes of several other metals, including Mn and As, exceeded fluxes determined from concentrations accumulating in the sediments. In several other coastal regions, such as Narragansett Bay, Mn has been shown to diffuse back out of sediments to overlying waters (Graham *et al.*, 1976). This recycling may explain the higher Mn fluxes observed in the traps than in the sediments.

The Dabob Bay sediment trap data certainly support the conclusion that fecal pellets are the predominant vertical transport agent at this location. However, because of extensive benthic reworking and utilization of sedimenting organic matter, chemicals carried to the sea floor with the organic components need not remain associated with them. In fact, in surface sediments of Dabob Bay, ^{210}Pb is preferentially released upon selective dissolution of hydrous Mn and Fe oxides and not organic matter (Carpenter *et al.*, 1984). Thus, phase associations in bottom sediments need not reflect the main vertical transporting agents.

The fact that fecal pellets dominate the vertical transport in Dabob Bay does not mean that these particles will be as important in all other Washington coastal waters. Dabob Bay was chosen for study because it offered the best chance to observe and document vertical transport of chemicals by fecal pellets. The area is in fact somewhat unique in the extent to which utilization of primary production is dominated by macrozooplankton. As detailed Landry and Lorenzen (1989, Chapter 5), microzooplankton play a more important role in the utilization of primary production on the outer coast of Washington. Because microzooplankton do not produce rapidly sinking fecal material, one might expect chemical remineralization and cycling of chemicals within the euphotic zone of coastal Washington to be relatively more important.

Table 10.11 Comparison between water column vertical fluxes in Dabob Bay, Wa. obtained by (1) integrating collections in a year long series of sediment trap deployments and (2) multiplying bottom sediment accumulation rates and surface sediment concentrations. Uncertainties are propagated standard deviations. Data are from Bennett (1980).

Constituent	Vertical Flux in Water Column (dry material cm^{-2} yr^{-1})		Accumulation Rate in Bottom Sediment (dry material cm^{-2} yr^{-1})	Labile Portion of Vertical Flux (dry material cm^{-2} yr^{-1})	% Non-Terrestrial Material that is Labile
	Total	Terrestrial Portion			
	1	2	3	1–3	$\left(\frac{1-3}{1-2}\right) \times 100$
Particulate Matter (mg)	109 ± 5	82.9 ± 5.5	93.8 ± 5.5	15.2 ± 1.8	58 ± 23
$CaCO_3$ (mg)	3.90 ± 0.35	0.0	1.16 ± 0.15	2.74 ± 0.38	70 ± 12
N (mg)	1.03 ± 0.38	0.17 ± 0.04	0.24 ± 0.02	0.79 ± 0.39	92 ± 61
Organic C (mg)	6.92 ± 0.59	2.22 ± 0.23	2.50 ± 0.21	4.42 ± 0.63	94 ± 18
Al (mg)	6.55 ± 0.51	6.55 ± 0.51	6.55 ± 0.51	0.00 ± 0.72	0 ±
As (μg)	1.08 ± 0.09	0.53 ± 0.05	0.63 ± 0.08	0.45 ± 0.12	81 ± 26
Cr (μg)	9.97 ± 0.80	9.49 ± 0.97	8.89 ± 1.11	1.08 ± 1.37	227 ± 635
Co (μg)	1.65 ± 0.10	2.02 ± 0.17	1.61 ± 0.18	0.04 ± 0.21	−17 ± 57
Fe (mg)	4.13 ± 0.11	4.36 ± 0.49	3.94 ± 0.33	0.19 ± 0.35	−82 ± 235
Mn (μg)	203 ± 19.0	81 ± 8	66 ± 9	137 ± 21	12 ± 26
V (μg)	12.8 ± 2.9	15.6 ± 1.8	12.9 ± 2.8	−0.10 ± 3.29	4 ± 118

10.7.6 Intertidal Organisms

Seymour and Nelson (1973) summarized several years of study of ^{65}Zn activity in mussels collected along the coast of Washington from the Columbia River mouth north to Neah Bay (Fig. 10.1). ^{65}Zn activities in mussels decreased with distance northward from the river mouth. In 1971, the last of the 'single pass' Hanford reactors, which induced radioactivity into the coolant water, was deactivated. ^{65}Zn activities in mussels declined by a factor of six by the end of the following year.

Samples of the mussels *Mytilus edulis* or *Mytilus californianus* were collected once a year during 1976, 1977 and 1978 at six relatively remote sites along the Washington coast chosen to avoid obvious localized contaminant inputs (Palmieri *et al.*, 1984 and Farrington *et al.*, 1983). These suspension feeding bivalves pump large amounts of overlying water and particulates through their bodies and concentrations in their tissues should reflect amounts of biologically-available chemicals in the water and particulates. Chemical concentrations in these bivalves originate from a combination of natural and anthropogenic inputs and upwelling of intermediate-depth water of the North Pacific enriched in fallout Pu and Am. Concentrations of PCBs, DDT and its degradation products, stable metals and fallout radionuclides Pu, Am and ^{137}Cs, have been reported for these organisms (Goldberg *et al.*, 1983; Farrington *et al.*, 1982; 1983; Palmieri *et al.*, 1984).

Mussels from the remote sites from Washington (and Oregon) had significantly <u>lower</u> concentrations of chlorinated organics (PCBs and DDEs), the unresolved complex mixture and PAH than did mussels from sites from the heavily urbanized northeastern region United States (Farrington *et al.*, 1982; 1983). Concentrations of PCBs and DDEs in Washington mussels ranged between 7-42 and 2-10 ppb dry weight, respectively. No significant changes in PCBs or DDE concentrations were noted between mussels collected during 1976 and 1978 from the Columbia River mouth and at Boundary Bay, Washington near the Canadian border. Seasonal changes in concentrations due to spawning or other factors were not evaluated for Washington coastal samples, but monthly variations of factors between 2 and 10 were noted in mussels from Narragansett Bay.

Concentrations of aliphatic hydrocarbons and PAH also varied randomly by at least a factor of two in nine major species of intertidal organisms collected quarterly for four years from a remote, rocky intertidal community on the Strait of Juan de Fuca (Clark, 1983). A variety of plants and animals were analyzed representing different feeding strategies and trophic levels. Although seasonal or yearly trends were hard to distinguish from natural variability, PAH concentrations dropped significantly after the first year and then remained more or less constant. Saturated hydrocarbon patterns in the organisms seemed to reflect food sources combined with products of biosynthetic and metabolic processes within the organisms and their digestive tract microflora. PAH concentrations in some organisms had greater similarity to PAH patterns in seawater than foods, suggesting that water may be a more important PAH source. PAH concentrations may be good indicators of short term petroleum uptake in species without enzymatic detoxification mechanisms, but were poor indicators in carnivores which can metabolize, PAH.

Selected organisms from the rocky intertidal community were also subjected to a simplified laboratory food chain study using ^{3}H- labeled naphthalene (Clark, 1983). The ^{3}H-naphthalene was added at levels below its theoretical solubility in filtered seawater to a culture of *Thalassiosira weissflogii* diatoms, one of three principal components of phytoplankton blooms in the area. The diatoms took up sufficient naphthalene for use in further feeding studies, but did not metabolize measurable quantities of this compound.

The diatoms were gently harvested, rinsed and resuspended in clean filtered seawater to serve as food for filter feeding mussels (*Mytilus* sp.). Over 80% of the available diatom radioactivity was removed from seawater by the filter feeding mussels; 93% of the radioactivity within the mussels was unmetabolized ^{3}H-naphthalene. The majority (54%) of the radioactivity was associated with the digestive system, with the remainder distributed among the gills (26%), gonads (13%), mantle (6%) and abductor muscle (1%). Insignificant amounts of radioactivity were detected in shell or pseudofeces during these short term (< 18 hr) exposures.

Oregon triton snails, when fed the ^{3}H-naphthalene-labeled mussels, assimilated sufficient radioactivity to allow measurement of its distribution in various organs. However, in contrast to mussels and diatoms, the snails apparently metabolized the compound, since higher concentrations of polar compounds were formed and the bioaccumulation factor was low (0.07 for exposures of less than 24 hr and 0.009 for exposures longer than 48 hr). The bulk (89%) of the radioactivity in snails was found in digestive glands, with less in other organs (6%), foot (4%) and ctenidia (1%). The overall accumulation of ^{3}H-naphthalene and its radioactive metabolites at the carnivore (snail) level was 0.14% of the initially available aromatic hydrocarbon. Thus, PAH concentrations did *not* steadily increase in higher trophic levels of the food web.

When exposed to membrane-filtered medium containing dissolved ^{3}H-naphthalene, Oregon triton snails, leafy hornmouth snails and wrinkled purple snails displayed greater total radioactivity (bioaccumulation factors of 24 to 94) than the same species fed labelled mussels. Mussels were also grown in waters in which ^{3}H-naphthalene was available only in dissolved form. Bioaccumulation of ^{3}H-naphthalene from the water was 60 times greater than bioaccumulation from labeled diatoms. The radioactivity obtained from dissolved naphthalene was also distributed differently than that obtained from the labeled diatoms. Greater uptake by mantle and gonads and reduced levels in gills and other organs was observed. These data confirm that the equilibration of dissolved PAH (and possibly other petroleum hydrocarbons) across exposed organism membranes may be one of the most important uptake mechanisms for most aquatic animals, since they process large volumes of water during food collection and respiration.

In contrast to organic compounds, Pu concentration in mussels collected along the Washington coast were up to ten times greater than in mussels from the U.S. east coast (Goldberg *et al.*, 1983; Farrington *et al.*, 1983; Palmieri *et al.*, 1984). Elevated Pu concentrations in west coast mussels are attributed to greater Pu exposure from upwelled, intermediate-depth North Pacific waters enriched in fallout Pu. Pu concentrations were even higher in mussels from areas of more intense upwelling off the central California coast. Pu concentrations in mussels from Cape Flattery (Fig. 10.1) and from other west coast sites did not steadily drop from 1976 to 1978 despite the lack of Pu input from atmospheric testing (Goldberg *et al.*, 1983). This confirms the importance of continuious Pu input with the advecting seawater (Beasley *et al.*, 1982).

^{241}Am concentrations and ^{241}Am/Pu ratios were also about a factor of ten higher in west coast mussels than east coast mussels. Elevated ^{241}Am values were attributed to association of Am with open ocean, fine-grained particles, combined with greater exposure of west coast mussels to these particles than east coast mussels, which were collected from more sheltered locations (Palmieri *et al.*, 1984). ^{241}Am concentrations were more variable and only partially reflected the Pu trends. This result is a paradox since ^{241}Am also has a concentration maximum at mid-depths in the North Pacific water column. Seasonal variations in Am concentrations of factors of 2 to 5 were observed in mussels collected monthly from Bodega Head and Narragansett Bay, but were not determined for Washington coastal samples. Variations were due to influences of both spawning and seasonal changes in upwelling intensity. Nielsen (1982) examined concentrations of ^{239,240}Pu and ^{241}Am in *Mytilus californianus* from the region of Coos Bay, Oregon and concluded that the high Am/Pu ratio in this species is a species-specific, physiological response.

In contrast to ^{241}Am and Pu isotopes, stable Pb and ^{137}Cs concentrations in Washington coast mussels were relatively low, because upwelling does not supply higher dissolved concentrations of these chemicals. Stable Pb concentrations were consistently less than 1.6 ppm dry weight of edible tissue. Ag concentrations were substantially lower and concentrations of Ni, Cu, Cd and Zn were about the same or lower in Washington coast mussels than mussels from other west coast sites (Goldberg *et al.*, 1983).

The broad, low gradient, sandy beaches from Point Grenville, Washington, south to Tillamook Head, Oregon (Fig. 10.1), have abundant surf diatom populations comprised almost exclusively of *Chaetocerus armatum* and *Asterionella socialis* (see Lewin *et al.*, 1989, Chapter 12). These diatoms constitute the major food of razor clams, *Siliqua patula*, which inhabit these beaches and are an important recreational and commercial fishery resource of Washington state.

Temporal variations in concentrations of ^{210}Pb, total ^{210}Po, unsupported ^{210}Po and Al were determined in surf zone diatoms, sediments and benthic infauna from Copalis Beach, Washington, over a 21 month period (Bennett and Carpenter, 1979). ^{210}Po was of special interest because of its preferential accumulation by marine organisms and because this alpha-emitter often contributes the major dose of internal radiation to marine organisms (Cherry and Shannon, 1974; Cherry and Heyraud, 1982).

A high correlation between Al and ^{210}Pb in the surf diatoms suggested that the ^{210}Pb and supported ^{210}Po were also associated with the clay size particles which naturally coat the *Chaetoceros* cells. From fall through spring, unsupported ^{210}Po concentrations in surf diatoms were correlated with input of ^{210}Po by precipitation. A summer maximum in diatom unsupported ^{210}Po concentrations was attributed to higher dissolved ^{210}Po in surf waters. The higher ^{210}Po was caused by greater upwelling of offshore waters than in late spring and fall. Assuming surf diatoms to be the major food of the benthos, food chain magnification of ^{210}Pb and ^{210}Po by benthic organisms was not observed. However, razor clams exhibited higher $^{210}Po/^{210}Pb$ activity ratios than the diatoms, indicating ^{210}Po was accumulated preferentially over ^{210}Pb.

A drastic decrease in razor clam abundance on these beaches occurred in 1982. This event forced complete closure of the large recreational fishery based on these shellfish, with adverse economic consequences for coastal communities. Chemical pollution of some sort was initially suspected. However, the cause has since been traced to the proliferation of a parasite caused by warmer coastal waters associated with the El Niño event of 1980-1981 (K. Chew, personal communication, 1985). With the return of normal, cooler waters, parasite abundances decreased and by fall 1985 the razor clam population had recovered enough to allow a limited recreational harvest. A precipitous decline in Dungeness crab abundances off the Washington coast was noted in late 1984 and also may be due to poorly understood natural fluctuations.

In Puget Sound, common blue mussels (*Mytilus edulis*) and kelp (*Nereocystis leutkeana*) had similar As concentrations regardless of distance from a large copper smelter near Tacoma which discharged large amounts of As into the Puget Sound region from 1890-1985 (Crecelius, 1974). Dry weight As concentrations in mussels ranged from 6 to 14 ppm, with a mean of 10 ppm, similar to values reported by Lunde (1970) and Bertine and Goldberg (1972). Dry weight As concentrations in kelp ranged from 33 to 65 ppm with a mean of 48 ppm.

10.7.7 Benthic organisms

Chemical analyses of smaller benthic organisms such as polychaete worms and heart urchins are rare. Several studies, however, have documented that benthic organisms can have important effects on chemical fluxes and distributions near the sediment-water interface, even without incorporating chemicals into their tissues (Aller, 1982; Carpenter *et al.*, 1984; Christensen *et al.*, 1984; Emerson *et al.*, 1984; Karickhoff and Morris, 1985; and references therein). For example, Renfro (1973) reported that losses of ^{65}Zn were 3 to 7 times higher from sediments with polychaete worms (*Nereis diversicolor*) burrowing through them than sediments without worms. Karickhoff and Morris (1985) reported tubificid oligochaetes can have an important impact on organic pollutant transport in bottom sediments. Solid phase mixing process for Washington shelf and slope and Puget Sound are described in detail in Section 10.6.

Resuspension and mixing of solids and pore waters due to both physical and biological processes lead to increased scavenging of excess ^{210}Pb from deeper waters to sediments of canyons indenting the Washington slope and shelf (Carpenter *et al.*, 1981). Similar processes appear active in increasing Pu scavenging and inventories in Washington slope sediments (Carpenter *et al.*, 1987) and probably affect other particle reactive chemicals as well.

Mixing by benthic organisms in Washington coastal sediments helps keep surface layers oxygenated, causing dissolved Mn and Fe to precipitate as oxyhydroxides. The diffusive flux of As out of Washington coast and Puget Sound sediments is limited by scavenging of dissolved As by Mn and Fe oxyhydroxides (Carpenter *et al.*, 1978; Peterson and Carpenter, 1986). Pore water As concentrations increase sharply at the base of the ^{210}Pb defined surface mixed layer, due to release from Mn and Fe oxyhydroxides, which begin to dissolve as oxygen is depleted. In contrast with As, fluxes of other trace metals which form insoluble sulfides (such as Cu, Cd and Ni) may be greater from well-mixed and oxic layers than from anoxic layers (Emerson *et al.*, 1984). However, soluble metal-polysulfide complexes may form and cause the solubility and fluxes of certain metals such as Hg out of anoxic sediments to be much greater than would be predicted on the basis of simple metal-sulfide solubilities alone (Bothner *et al.*, 1980).

A number of benthic organisms continually pump pore water out of surficial sediments and replace it with overlying seawater. It has been suggested that such 'irrigation' activities of macrobenthic organisms in Puget Sound sediments are responsible for: (1) peaks in pore water concentrations of nitrate at depths of 20 to 30 cm (Grundmanis and Murray, 1977); (2) increases in solid phase U isotope concentrations at depths (Carpenter *et al.*, 1984); (3) greater fluxes of ^{222}Rn out of sediments than could be accounted for by molecular diffusion (Smethie, 1979); (4) minima in pore water Mn concentrations at depth (Carpenter, 1983); (5) deficiencies of ^{228}Th relative to its parent ^{232}Th due to loss of the soluble intermediate ^{228}Ra and presumably other Ra isotopes as well (Carpenter *et al.*, 1984); and (6) enhanced fluxes of tritiated water and dissolved silica across the sediment-water interface in Quartermaster Harbor (Emerson *et al.*, 1984). The important point is that exchange of soluble chemicals between overlying and interstitial waters can extend to depths of 20 to 30 cm in the extensively bioturbated Puget Sound sediments, even though the ^{210}Pb defined well-mixed surface layer usually does not exceed 15 cm in depth (Carpenter *et al.*, 1985). Pore water profiles of ^{222}Rn (Smethie *et al.*, 1981) and sulfate (Christensen *et al.*, 1984), as well as U and Th isotope profiles in sediments of the Washington slope (Carpenter *et al.*, 1987) indicate such irrigation effects also occurs in offshore sediments. However, the extent and effects of irrigation appear less significant in coastal than in Puget Sound sediments because of the lower coastal benthic biomass.

10.7.8 Chemical Concentrations in Nekton

Determinations of chemical concentrations in mobile organisms collected off the Washington coast have not received high priority since 1973, because of declining levels of Columbia River-derived radionuclides and difficulties in interpreting results in terms of specific processes.

Zinc-65

Several studies have documented the decline in ^{65}Zn activities in nekton following the gradual shutdown of the single pass nuclear reactors at Hanford between 1965 and 1971. Pearcy and Vanderpleog (1973) showed that activities of that ^{65}Zn declined markedly in several species of benthic fish between 1965 and 1971. Specific activities also decreased noticeably with increasing water depth of collection sites, both for individual species inhabiting broad depth ranges and for different species inhabiting specific depths. ^{65}Zn activities in benthic invertebrates also decreased with distance from the river mouth, increasing water depth and time since 1965 (Carey and Cutshall, 1973).

Pearcy *et al.* (1977) determined ^{65}Zn specific activities (^{65}Zn : Zn ratio) in pelagic and benthic animals collected off Oregon as a function of capture depth, to estimate ^{65}Zn vertical transport rates. Significant decreases in radioactivity with increasing depth of capture were found despite uncertainties over whether or not a fish caught at a given depth was representative of that depth. Vertical transport of ^{65}Zn appeared slower in waters above 500 m, suggesting recycling of Zn within biological communities. Results suggested about 2 y are required for transport of a typical Zn atom from near surface to abysobenthic animals. Differences in average ^{65}Zn specific activities among benthic animals collected on the continental shelf and abyssal plain off central Oregon were equivalent to a ^{65}Zn surface to abyssal transit time of about 3 y for fishes (Pearcy and Vanderploeg, 1973) and 2.4 y for invertebrates (Carey, 1972). Low ^{65}Zn activities in deep-sea deposit feeders (Osterberg *et al.*, 1963; Carey, 1972) also argue that little ^{65}Zn rapidly reached the deep-sea floor, in contrast to the behavior of $^{95}Zr/^{95}Nb$ and $^{141}Ce/^{144}Ce$ ratios measured in sediments. Thus, differences in vertical transport rates of various radiochemicals are evident. The relatively long vertical transport time for ^{65}Zn also contrasts with the shorter times estimated for transport of other elements by zooplankton fecal pellets (Bacon *et al.*, 1985 and references therein). This difference raises questions about the importance of fecal pellets as a mechanism for rapidly transporting biologically required materials to the deep-sea.

Renfro *et al.* (1975) found that shrimp and crabs accumulated more ^{65}Zn from seawater than from food organisms and that equilibration of ^{65}Zn between adult organism pools and the surrounding food or water was very slow. Both Renfro *et al.* (1975) and Willis and Sunda (1984) concluded that food was the main source for ^{65}Zn taken up by several marine fish.

Arsenic

Crecelius (1974) determined total As concentrations in benthic fish and invertebrates living in southern Puget Sound where a copper smelter (Fig. 10.3) was a large source of anthropogenic As from 1890 to 1985. Total As concentrations in these organisms are given in Table 10.12 and resemble those of other metals in varying widely between different species as well as within a single species. Small crustaceans living on the slag pile had higher As concentrations than similar organisms from the control station, possibly due to ingestion of slag particles. Arsenic concentrations in Dungness crabs, ratfish, shrimp and sole from Puget Sound are similar to those in corresponding organisms from uncontaminated areas on the Pacific coast of Canada (LeBlanc and Jackson, 1973) and off the Washington and Oregon coasts (Robertson *et al.*, 1972). Puget Sound organisms resembled those from other areas in having higher As concentrations in marine crustaceans such as crabs and shrimp than other organisms from the same location (LeBlanc and Jackson, 1973). There was no simple, steady increase in total As concentrations for higher trophic level organisms (Robertson *et al.*, 1972; Windom *et al.*, 1973). The fact that higher-than-background As concentrations were found only in organisms living very near the smelter agrees with observations that elevated As concentrations in Puget Sound waters are limited to locations near this anthropogenic source (Carpenter *et al.*, 1978; Crecelius *et al.*, 1975).

TABLE 10.12 Arsenic concentrations in organisms from Puget Sound, based on data of Crecelius (1974).

		As concentration (ppm dry weight)			
		Location*			
Type of Organism	Body Part Analyzed	1	2	3	4
Dungeness Crab	Muscle	--	--	47	58
Hermit Crab	Whole body	181	14	--	--
Rat Fish	Muscle	29	41	--	--
Sea Cucumber	Muscle	--	16	2.8	5.0
Sea Pen	Top of Fan	--	5.4	5.5	--
Sea Slug	Whole Body	--	12	2.3	--
Shrimp	Whole body	58	21	--	77
Shrimp	Muscle	--	--	10	22
Sole	Muscle	31	23	10	17
Starfish	Arm	12	7.7	1.0	3.8

*Location 1 was just off the Tacoma smelter slag pile at 30-m depth, where tidal currents are moderate and the bottom is covered primarily with sand- and gravel-sized material, much of which is smelter slag. Location 2 was near Fox Island, 12 km southwest of the slag pile in 30 m of water. Here the currents are also moderate, the bottom sandy and the general oceanographic conditions similar to location 1. Location 3 was in Quartermaster Harbor, 8 km northeast of the smelter in 15 m of water. This location is downwind of the smelter during most of the year and received smelter dust with relatively high As concentrations. Soils surrounding Quartermaster Harbor have abnormally high As concentrations (Crecelius *et al.*, 1974). Surface sediments have As concentrations 5 times greater than concentrations in horizons deposited prior to 1890 when the smelter began operations (Carpenter *et al.*, 1978). Location 4 is in Carr Inlet, 14 km west of the smelter in 15 m of water. This inlet represents a location similar to Quartermaster Harbor, but without smelter dust contamination.

Improved techniques during the 1970s for maintaining salmon and shrimp for long periods of time in seawater aquaculture systems offered an opportunity to overcome some of the difficulties in interpreting data for field-caught specimens by sampling well defined populations feeding on controlled diets. J. Engblom and R. Carpenter determined As concentrations in salmon and shrimp at different stages of development obtained from either the National Marine Fisheries Puget Sound Aquaculture Project at Manchester, Washington (courtesy of C. Mahnken and E. Prentiss), or from the University of Washington, School of Fisheries (courtesy of W. Hershberger). Aquaculturists were glad to cooperate because of their concern that high concentrations of toxic metals may affect either the survival or growth rate of the young, or restrict the commercial value of the older organisms.

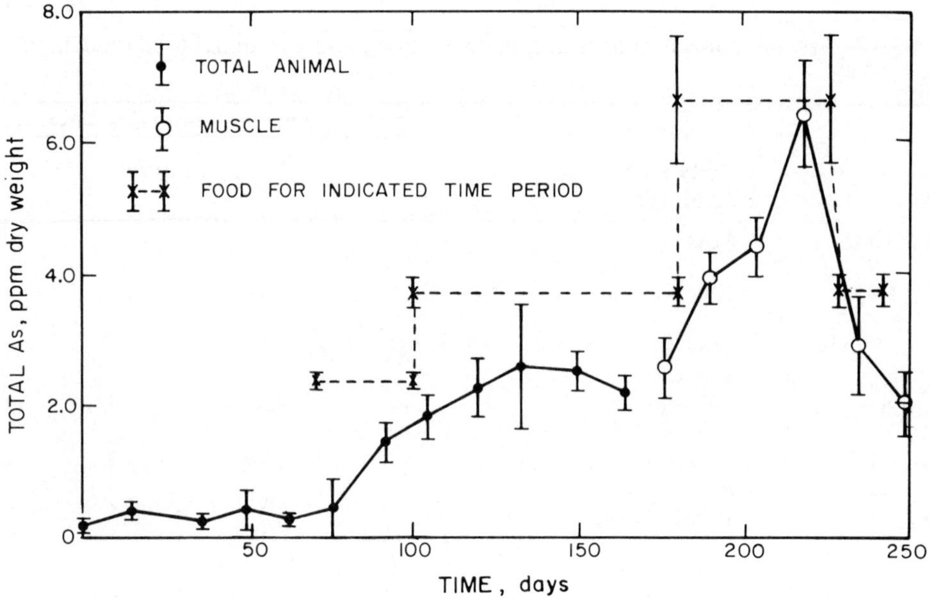

Fig. 10.32 Variations of total As concentrations in hatchery-reared coho salmon with time from date eggs were obtained. Values presented are the mean ±1 standard deviation for concentrations in 3-5 organisms. Arsenic concentrations in the diet were controlled for the indicated time intervals at levels shown by the dashed line. For larger fish, As concentrations in liver and muscle tissues were essentially equivalent.

Figure 10.32 shows the results of the salmon studies. Arsenic concentrations in salmon eggs were low and remained low (about 0.03 ppm As dry weight) as long as the fry fed on the yolk sac. Once the young began feeding on the prepared food (about day 70), their As concentrations began to reflect the As concentration in their food. Arsenic concentrations in the salmon increased rapidly when concentrations in food were raised on days 70 and 180 and decreased when As concentrations in food were reduced on day 210. It is not clear why As concentrations in the fish decreased below concentrations in their food after day 240. What is clear is that a surprisingly rapid time of only about 15 to 20 d is needed for As concentrations in the young salmon to approach As concentrations in their food. No major changes in As concentrations were observed when the young fish were transferred from fresh to salt water, which had about ten times greater dissolved As concentrations.

In adult Chinook salmon (both male and female) concentrations of As were slightly higher (≤ 2x) in liver than muscle tissue and eggs had about 10% the As concentrations of liver tissue. Spawning Chinook salmon had significantly (~5x) higher levels of As in their eggs, muscle and liver tissues than Coho salmon. In growing Coho salmon, As concentrations were consistently higher in muscle than in liver tissue. This was also the case for Arctic cod (Bohn and McElroy, 1976). Reasons for As distributional differences in tissues of different fish species are not known, but dietary differences may be partially responsible.

Arsenic concentrations were also monitored in a wild-caught, mixed population of 1 to 3 y old spot prawns (*Pandalus platyceros*) collected from Hood Canal, Washington

(Fig. 10.3), an area with no obvious anthropogenic As inputs. Fifteen wild-caught prawns of varying sizes had As concentrations of 10 to 65 ppm dry weight in muscle tissue, with an average of 38 ppm total As. Seven laboratory hatched and reared prawns had only 11 to 13 ppm total As. The high As concentrations in wild-caught prawns are apparently the result of natural processes, but it is not known what food source could account for such enrichment. Arsenic concentrations were consistently higher (~3x) in edible muscle tissue than in exoskeletal material of both wild caught and laboratory reared prawns.

After capture the spot prawns were maintained in two groups. One group was fed dead Coho salmon (with less than 1 ppm dry weight As) and the other group geoduck clam flesh (with much higher As levels of 26 ± 17 ppm dry weight). Over the subsequent 200 d period, both groups of prawns showed a slow decline in As concentrations (Fig. 10.33). More than 200 d were required for a factor of 2 decline from the initial As concentrations of 45 ppm.

It is clear that young salmon respond much more rapidly to changes in dietary As concentrations than the more mature spot prawns. Reasons for the difference are not known. It may simply require more time for As in body tissues of the mature, slower-growing prawns to 'equilibrate' with As concentrations in a new food supply than it does for the rapidly growing young salmon. The prawns could also acquire a much higher fraction of their As from the water.

In these initial studies the mass of food eaten per day and the chemical form of As in various foods, were not determined because of logistical limitations. Nevertheless, the results indicate this approach has the potential for obtaining biological turnover times for a variety of chemicals in commercially-important organisms at different life stages and to relate observed chemical concentrations in the organisms to known concentrations in food sources. Determinations of biological turnover times are needed to predict how rapidly organism chemical concentrations will respond to changes in water and food concentrations.

At the concentrations observed in salmon and shrimp, most of the As must be present in forms less toxic than inorganic As $^{5+}$ and As $^{3+}$. Methylated and other organic forms are produced by marine algae (Andreae and Klumpp, 1979; and references therein). Several other still unidentified lipid and water soluble organo-As compounds are produced by various higher marine organisms (Klumpp and Peterson, 1981; and references therein). Penrose *et al.* (1977) confirmed that a complex, relatively stable and high molecular weight organic form of As accounts for all the As found in muscle tissue of shrimp and that readily excreted organic As forms are also produced by sea urchins.

Thus, As appears to be incorporated into marine organisms in a variety of inorganic and organic forms in varying proportions. The behavior of As in marine food webs is complex and hard to unravel because of this complex chemistry (Phillips and Depledge, 1985). Arsenic does not simply accumulate in higher organisms, because some organisms metabolize it or excrete certain forms. In addition, the relative importance of As uptake from food *vs.* water is usually unknown. Langston (1984) showed that particulate As is the predominant form available to the

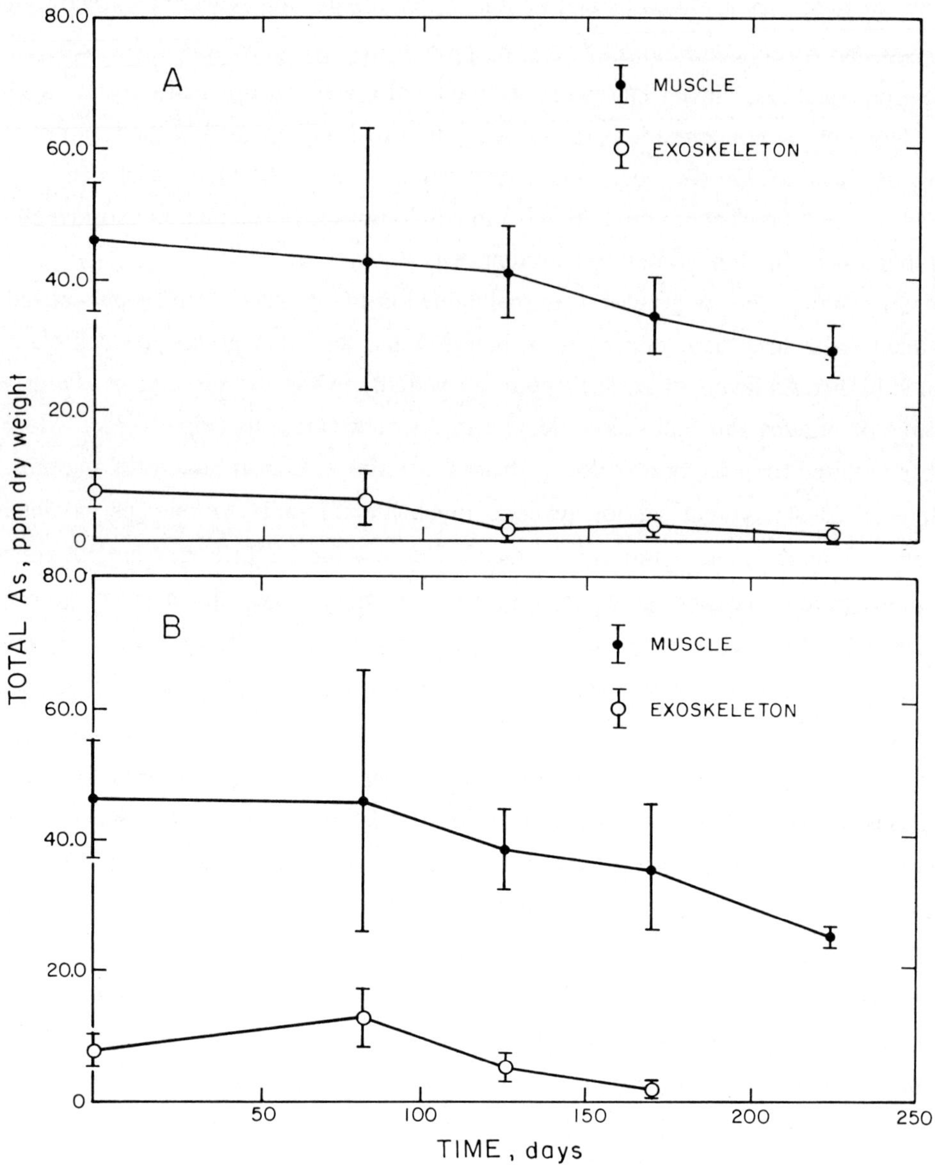

Fig. 10.33 Variations in total As concentrations with time in wild-caught, hatchery-reared spot prawns (*Pandalus platycerous*). (A) Prawns fed geoduck clams which contained 27 ± 16 ppm As in edible tissue. (B) Prawns fed whole dead coho salmon containing less than 1 ppm As. Values are the mean ± 1 standard deviation for analyses of 6-8 organisms.

deposit feeding bivalve *Scrobiculatia plana*, whereas As concentrations in the seaweed *Fucus vesiculosus* were related to dissolved As concentrations.

Polonium-210

^{210}Po was determined in the prawns because it is preferentially accumulated by marine organisms and often contributes the major dose of an organism's internal radiation (Cherry and Heyraud, 1982; and references therein). ^{210}Po activities in exoskeletal material of 21 prawns averaged 0.17 ± 0.03 dpm (g dry weight)$^{-1}$. Activities in muscle tissue were always at least a factor of 10 greater than in exoskeletal material. Not enough samples were analyzed to test for relationships between ^{210}Po in tissues and diets, or size or dry weights of the organism.

10.7.9 Remaining Questions and Needed Research Concerning Chemical Interactions with Biota

Without concentration data for potentially harmful chemicals in plankton of Washington coastal waters there is no way to judge whether anthropogenic chemical inputs are adversely affecting organisms at the bases of food webs. An example of the importance of this can be seen in the Mussel Watch data which showed that, despite elimination of most atmospheric testing of nuclear weapons, Pu levels in Washington coast mussels have not markedly declined, due to continuing Pu input with advecting seawaters. The lack of data on planktonic chemical concentrations is particularly serious for Puget Sound, given recent publicity and concern over the degree to which it is polluted.

The importance of uptake and transport of dissolved chemicals by zooplankton fecal material relative to inorganic phases needs further study. ^{210}Pb and Pu studies off the Washington coast strongly argue that inorganic processes enhanced near the sea floor are of greatest importance, whereas in Dabob, Bay zooplankton fecal pellet transport is clearly the dominant process. ^{65}Zn and ^{14}C activity trends in organisms caught at different water depths off Oregon suggest recycling of biologically important elements in the upper water column delays their eventual transport to the deep sea. The relative importance of these processes probably varies considerably from place to place, depending upon water depth, predominant zooplankton, bottom currents and topography. One cannot assume zooplankton fecal pellets dominate vertical transport of chemicals to the sea floor everywhere in Washington coastal waters.

Information concerning the relative importance of food *vs.* water in supplying chemicals found in organisms and biological turnover times for these chemicals is sparse at best. This is because it is difficult, if not impossible, to culture most marine organisms throughout their entire life cycles using realistic exposure conditions which approach those found in the field. No widely applicable method now exists for the determination of whether organisms acquire most of their body burden of potentially toxic chemicals through their food or from surrounding waters. Limited studies to date have considered only a few organisms over parts of their life cycles exposed to only a few chemicals. A start has been made with determinations of As in salmon and prawns at various life stages under partially defined conditions, but many more such studies need to be done. For example, it is still unclear why larger halibut have Hg concentrations which exceed the U. S. maximum permissible limit of 0.5 ppm for commercial food products more often than smaller halibut. More data on biological turnover times are also necessary to establish the response of organisms to increases and decreases in pollutants.

The sparse data available for bioaccumulation of various chemicals in different organisms reveal that this process is much more complex than initially implied by early DDT studies. A related question concerns the validity of using equilibrium partitioning coefficients such as octanol:water ratios to predict pollutant uptake by organisms. Better analytical methods are also needed to determine polar metabolites of nonpolar organics such as PCBs and PAHs, which are not volatile and so are not readily amenable to the capillary GC and GC-MS systems so sensitive for their parent compounds.

Accelerator mass spectrometric determinations of ^{14}C may reduce required sample sizes sufficiently to allow routine measurements on various compound classes (proteins, sugars, fatty acids, hydrocarbons, etc.). The few ^{14}C determinations in lipid extracts of organisms to date suggest such results will be useful in identifying the main source of C in specific organic compound classes.

Few studies of the bioavailability of sedimentary chemicals to benthic organisms have been carried out in any coastal zone. Studies of phase associations in Washington coastal sediments have been carried out for Pb, Pu and the PAH, but they need to be combined with biological studies to determine the bioavailability of these chemicals. A major gap in our understanding of organic pollutant bioavailability is the lack of pore water data for concentrations of potentially harmful organic compounds such as PCBs and PAHs. This information is important since the dissolved forms probably are more available for biological incorporation. Data on the bioavailability of specific chemicals will enable regulatory agencies to set up reliable criteria for: (1) defining sedimentary chemical concentrations at which adverse biological effects may occur; (2) estimating relative toxicities of sediments from various sites; and (3) ranking chemicals or chemical combinations according to their toxicity. Knowledge of chemical interactions with biota will become increasingly critical during the next 10 to 15 y. Acquiring this kind of information will require interdisciplinary research by biologists, chemists and biochemists.

ACKNOWLEDGEMENTS

Much of the research summarized in this chapter has been supported by DOE contracts DE-AT06-76-700024 to R. Carpenter at the University of Washington and –700030 to T.M. Beasley at Oregon State University. Collaboration with Dr. Beasley has been particularly enjoyable and stimulating. We also appreciate the diligence, enthusiasm and creativity of numerous students and research associates, including M.H. Bothner, E.A. Crecelius, S.G. Wakeham, J.T. Bennett, F.G. Prahl, T.S. Bates, E.T. Furlong, R.C. Clark, R. Barrick and M Strom. This is a component of Contribution No. 1784 from the School of Oceanography, University of Washington.

REFERENCES

Aller, R.C. 1982. The effects of macrobenthos on chemical properties of marine sediment and overlying water. Pages 53-102 in: P.L. McCall and M.J.S. Tevesz (eds.), *Animal-sediment Relations*. Plenum Press, New York, N.Y.

Anderson, F.E. 1967. Stratigraphy of late Pleistocene and Holocene sediments from the Strait of Juan de Fuca. Ph.D. Diss., Dept. of Oceanogr., Univ. of Washington, Seattle, Wa. 168 pp.

Anderson, J.W., L.J. Moore, J.W. Blaylock, D.L. Woodruff and S.L. Kiesser. 1977. Bioavailability of sediment-sorbed naphthalenes to the sipunculid worm *Phascolosoma agassizii*. Pages 276-285 in: D.A. Wolfe (ed.), *Fate and Effects of Petroleum Hydrocarbons in Marine Organisms and Ecosystems*. Pergamon Press, New York, N.Y., pp: 276-285.

Anderson, R.F. and P. Biscaye. 1985. Mixing rates of continental slope sediments. *EOS, Trans. Am. Geophys. Union*, 66: 284.

Anderson, R.F., R.F. Bopp, K.O. Buesseler and P.E. Biscaye. 1988. Mixing of particles and organic constituents in sediments from the continental shelf and slope off Cape Cod: Seep-I results. *Cont. Shelf Res.*, 8: 925-946.

Andreae, M.O. 1979. Arsenic speciation in seawater and interstitial waters: the influence of biological-chemical interactions on the chemistry of a trace element. *Limnol. Oceanogr.*, 24: 440-452.

Andreae, M.O., J.F. Asmode, P. Foster and L. Van'tdack. 1981. Determination of antimony(III), antimony(V) and methylantimony species in natural waters by atomic absorption spectrometry with hydride generation. *Anal. Chem.*, 53: 1766-1771.

Andreae, M.O. and W.R. Barnard. 1984. The marine chemistry of dimethylsulfide. *Mar. Chem.*, 14: 267-279.

Andreae, M.O. and P.N. Froelich. 1984. Arsenic, antimony and germanium biogeochemistry in the Baltic Sea. *Tellus*, 36B: 101-117.

Andreae, M.O. and D. Klumpp. 1979. Biosynthesis and release of organoarsenic compounds by marine algae. *Environ. Sci. Technol.*, 13: 738-741.

Andreae, M.O. and H. Raemdonck. 1983. Dimethylsulfide in the surface ocean and the marine atmosphere: a global view. *Science*, 221: 744-747.

Bacon, M.P., C.A. Huh, A.P. Fleer and W.G. Deuser. 1985. Seasonality in the flux of natural radionuclides and plutonium in the deep Sargasso Sea. *Deep-Sea Res.*, 32: 273-286.

Bacon, M.P., D.W. Spencer and P.G. Brewer. 1976. $^{210}Pb/^{226}Ra$ and $^{210}Po/^{210}Pb$ disequilibrium in seawater and suspended particulate matter. *Earth Plan. Sci. Lett.*, 32: 277-296.

Baker, E.T. 1973. Distribution and composition of suspended sediment in the bottom waters of the Washington continental shelf and slope. *J. Sed. Petrol.*, 43: 812-821.

Baker, E.T. 1976. Distribution, composition and transport of suspended particulate matter in the vicinity of Willapa submarine canyon, Wa. *Geol. Soc. Am. Bull.*, 87: 625-632.

Baker, E.T. 1984. Patterns of suspended particle distribution and transport in a large fjord like estuary. *J. Geophys. Res.*, 89: 6553-6566.

Baker, E.T. and G.J. Massoth. 1986. Hydrothermal plume measurements: a regional perspective. *Science*, 234: 980-982.

Baker, E.T. and G.J. Massoth. 1987. Characteristics of hydrothermal flumes from two vent fields on the Juan de Fuca Ridge, northeast Pacific Ocean. *Earth Plan. Sci. Lett.,* 85: 59-73.

Baker, E.T., G.J. Massoth and R.A. Feely. 1987. Cataclysmic hydrothermal venting on the Juan de Fuca Ridge. *Nature*, 329: 149-151.

Baker, E.T. and H.B. Milborn. 1983. An instrument system for the investigation of particle fluxes. *Cont. Shelf Res.*, 1: 425-435.

Barnard, W.D. 1973. Late Cenozoic sedimentation on the Washington continental slope. Ph.D. Diss., Dept. of Oceanogr., Univ. of Washington, Seattle, Wa., 255 pp.

Barnard, W.D. 1978. The Washington continental slope: Quaternary tectonics and sedimentation. *Mar. Geol.*, 27: 79-114.

Barnard, W.S., M.O. Andreae, W.E. Watkins, H. Bingemer and H.W. Georgii. 1982. The flux of dimethylsulfide from the oceans to the atmosphere. *J. Geophys. Res.*, 87: 8787-8794.

Barnes, C.A., A.C. Duxbury and B.A. Morse. 1972. Circulation and selected properties of the Columbia River effluent at sea. Pages 41-80 in: A.T. Pruter and D.L. Alverson (eds.), *The Columbia River Estuary and Adjacent Ocean Waters: Bioenvironmental Studies.* Univ. of Washington Press, Seattle, Wa.

Barrick, R.C. 1982. Flux of aliphatic and polycyclic aromatic hydrocarbons to central Puget Sound from Seattle (Westpoint) Primary Sewage Effluent. *Environ. Sci. Technol.*, 16: 682-692.

Barrick, R.C., E.T. Furlong and R. Carpenter. 1984. Hydrocarbon and azaarene markers of coal transport to aquatic sediments. *Environ. Sci. Technol.*, 18: 846-854.

Barrick, R.C. and J.I. Hedges. 1981. Hydrocarbon geochemistry of the Puget Sound region-II. Sedimentary diterpenoid, steroid and triterpenoid hydrocarbons. *Geochim. Cosmochim. Acta,* 45: 381-392.

Barrick, R.C, J.I. Hedges and M.L. Peterson. 1980. Hydrocarbon geochemistry of the Puget Sound region-I. Sedimentary acyclic hydrocarbons. *Geochim. Cosmochim. Acta*, 44: 1349-1362.

Barrick, R.C. and F.G. Prahl. 1987. Hydrocarbon geochemistry of the Puget Sound region-III. Polycyclic aromatic hydrocarbons in sediments. *Est. Coast. Shelf Sci.*, 25: 175-192.

Bates, T.S. and R. Carpenter. 1979a. Organo-sulfur compounds in sediments of the Puget Sound basin. *Geochim. Cosmochim. Acta*, 43: 1209-1221.

Bates, T.S. and R. Carpenter. 1979b. Determination of organosulfur compounds extracted from marine sediments. *Anal. Chem.*, 51: 551-554.

Bates, T.S. and J.D. Cline. 1985. The role of the ocean in a regional sulfur cycle. *J. Geophys. Res.*, 90: 9168-9172.

Bates, T.S., S.E. Hamilton and J.D. Cline. 1984. Vertical transport and sedimentation of hydrocarbons in the central main basin of Puget Sound, Wa. *Environ. Sci. Technol.*, 18: 299-305.

Bates, T.S., J.D. Cline, R. Bammon and S. Kelly-Hanson. 1987. Regional and seasonal variations in the flux of oceanic dimethyl sulfide to the atmosphere. *J. Geophys. Res., 92:* 2930-2938.

Beasley, T.M., L.A. Ball, J.E. Andrews and J.E. Halverson. 1981. Hanford derived plutonium in Columbia River sediments. *Science*, 214: 913-915.

Beasley, T.M., R. Carpenter and C.D. Jennings. 1982. Plutonium, ^{241}Am and ^{137}Cs ratios, inventories and profiles in Washington and Oregon continental shelf sediments. *Geochem. Cosmochim. Acta,* 46: 1931-1946.

Beasley, T.M., M. Heyraud, J.J.W. Higgo, R.D. Cherry and S.W. Fowler. 1978. ^{210}Po and Pb-210 in zooplankton fecal pellets. *Mar. Biol.,* 44: 325-328.

Beasley, T.M. and C.D. Jennings. 1984. Inventories of ^{239,240}Pu, ^{241}Am, ^{137}Cs and ^{60}Co in Columbia River sediments from Hanford to the Columbia River estuary. *Environ. Sci. Technol.,* 18: 207-212.

Beikman, H.M., H.D. Gower and T.A.M. Dana. 1961. Coal Reserves of Washington. Washington State Division of Mines and Geology, Olympia. Bull. No. 47., 115 pp.

Bennett, J.T. 1980. The biogeochemical significance of zooplankton fecal material in a biologically productive, temperate fjord. Ph.D. Diss., Dept. of Oceanogr., Univ. of Washington, Seattle, Wa., 258 pp.

Bennett, J.T. and R. Carpenter. 1979. Concentrations and temporal variation of ^{210}Po, ^{210}Pb and Al in the surf zone ecosystem of Copalis Beach, Wa. *Est. Coast. Mar. Sci.* 8: 127-140.

Benninger, L.K. 1978. ^{210}Pb balance in Long Island Sound. *Geochim. Cosmochim. Acta,* 42: 1165-1174.

Benoit, G.J., K.K. Turekian and L.K. Benninger. 1979. Radiocarbon dating of a core from Long Island Sound. *Est. Coast. Mar. Sci.,* 9: 171-180.

Bertine, K.K. and E.D. Goldberg. 1972. Trace elements in clams, mussels and shrimp. *Limnol. Oceanogr.,* 17: 877-884.

Bloom, N.S. and E.A. Crecelius. 1983a. Determination of mercury in seawater at sub-nanogram per liter levels. *Mar. Chem.,* 14: 49-59.

Bloom, N. and E.A. Crecelius. 1983b. Solubility behavior of atmospheric ^{7}Be in the marine environment. *Mar. Chem.,* 12: 323-331.

Bloom, N.S. and E.A. Crecelius. 1987. Distribution of silver, mercury, lead, copper and cadmium in central Puget Sound sediments. *Mar. Chem.,* 21: 377-390.

Boehm, P.D. and J.G. Quinn. 1973. Solubilization of hydrocarbons by the dissolved organic matter in seawater. *Geochim. Cosmochim. Acta,* 37: 2459-2477.

Bohn, A. 1975. Arsenic in marine organisms from West Greenland. *Mar. Poll. Bull.,* 6: 87-89.

Bohn, A. and R.O. McElroy. 1976. Trace metals (As, Cd, Cu, Fe and Zn) in Arctic cod, *Boreogadus saida* and selected zooplankton from Strathcona Sound, Northern Baffin Island. *J. Fish. Res. Board Can.,* 33: 2836-2840.

Bostrom, K. and D.E. Fisher. 1969. Distribution of mercury in east Pacific sediments. *Geochim. Cosmochim. Acta,* 33: 743-745.

Bostrom, K. and S. Valdes. 1969. Arsenic in ocean floors. *Lithos,* 2: 351-360.

Bothner, M.H. 1973. Mercury, some aspects of its marine geochemistry in Puget Sound. Ph.D. Diss., Dept. of Oceanogr., Univ. of Washington, Seattle, Wa., 126 pp.

Bothner, M.H. and R. Carpenter. 1973. Sorption-desorption reactions of mercury with suspended matter in the Columbia River. Pages 73-88 in: *Radioactive Contamination of the Marine Environment, Proceedings of a Symposium,* Seattle, 1972. I.A.E.A., Vienna.

Bothner, M.H., R.A. Jahnke, M.C. Peterson and R. Carpenter. 1980. Rate of mercury loss from contaminated estuarine sediments. *Geochim. Cosmochim. Acta*, 44: 273-285.

Bothner, M.H., E.C. Spiker, P.P. Johnson, R.R. Rendigs and P.J. Aruscavage. 1981. Geochemical evidence for modern sediment accumulation on the continental shelf off southern New England. *J. Sed. Petrol.*, 51: 281-292.

Bowen, V.T., V.E. Noshkin, H.D. Livingston and H.C. Volchok. 1980. Fallout radionuclides in the Pacific Ocean: vertical and horizontal distribution largely from GEOSECS stations. *Earth Plan. Sci. Lett.*, 49: 411-434.

Braman, R.S., D.L. Johnson, C.C. Foreback, J. Ammons and J. Bicker. 1977. Separation and detection of sub-nanogram quantities of inorganic arsenic and methylarsenic compounds. *Anal. Chem.*, 44: 621-625.

Broecker, W.S. and T.-H. Peng. 1982. *Tracers in the Sea*. Lamont-Doherty Geological Observatory, Palisades, N.Y., 690 pp.

Brown, L., J. Klein and R. Middleton. 1985. Anomalous isotopic concentrations in the sea off California. *Geochim. Cosmochim. Acta*, 49: 153-157.

Brown, R.C. and T.L. Wade. 1984. Sedimentary coprostanol and hydrocarbon distribution adjacent to a sewage outfall. *Water Res.*, 18: 621-632.

Brownawell, B.J. and J.W. Farrington. 1986. Biogeochemistry of PCBs in the interstitial waters of a coastal marine sediment. *Geochim. Cosmochim. Acta*, 50: 157-169.

Bruegmann, L., L. Danielsson, B. Magnusson and S. Westerlund. 1982. Data from an expedition with R/V "Alexander von Humboldt" on the Atlantic Ocean, June, 1981. Reports on the chemistry of seawater, Dept. of Analytical and Marine Chemistry, Chalmers Univ. of Technology and Univ. of Goteborg, Sweden.

Bruland, K.W. 1983. Trace elements in sea-water. Pages 157-221 in: J.P. Riley and R. Chester (eds.), *Chemical Oceanography*. Vol. 8, Academic Press, New York, N.Y.

Bruland, K.W., R.P. Franks, W.M. Landing and A. Soutar. 1981. Southern California inner-basin sediment trap calibration. *Earth Plan. Sci. Lett.*, 53: 400-408.

Bruland, K.W., G.A. Knauer and J.H. Martin. 1978a. Cadmium in northeast Pacific waters. *Limnol. Oceanogr.*, 23: 618-625.

Bruland, K.W., G.A. Knauer and J.H. Martin. 1978b. Zinc in north-east Pacific water. *Nature*, 271: 741-743.

Buddemeier, R.W. 1969. A radiocarbon study of the varved marine sediments of Saanich Inlet, British Columbia. Ph.D. Diss., Dept. of Chem., Univ. of Washington, Seattle, Wa., 125 pp.

Buesseler, K.O., H.D. Livingston and E.R. Sholkovitz. 1986. $^{239,240}Pu$ and excess ^{210}Pb inventories along the shelf and slope of the northeast U.S.A. *Earth Plan. Sci. Lett.*, 76: 10-22.

Buesseler, K.O. and E.R. Sholkovitz. 1987. The geochemistry of fallout plutonium in the North Atlantic: $^{240}Pu/^{239}Pu$ ratios and their significance. *Geochim. Cosmochim. Acta*, 51: 2623-2638.

Byrd, J.T. and M.O. Andreae. 1982. Tin and methyltin species in seawater: concentration and fluxes. *Science*, 218: 565-569.

Caldicott, A.B. and G. Eglinton. 1973. Surface waxes. Pages 162-194 in: L.P. Miller (ed.), *Phytochemistry*. Vol. III, Van Nostrand Reinholdt Co., New York, N.Y.

Calvert, S.E. 1966. Accumulation of diatomaceous silica in the sediments of the Gulf of California. *Geol. Soc. Am. Bull.*, 77: 569-596.

Canadian-American Seamount Expedition. 1985. Hydrothermal vents on an axis seamount of the Juan de Fuca Ridge. *Nature*, 313: 212-214.

Canby, T.Y. 1978. Aluminum, the magic metal. *Natl. Geogr.*, 154: 186-211.

Carey, A.G. 1972. ^{65}Zn in benthic invertebrates off the Oregon coast. Pages 833-842 in: A.T. Pruter and D.L. Alverson (eds.), *Columbia River Estuary and Adjacent Ocean Waters: Bioenvironmental Studies*. Univ. of Washington Press, Seattle, Wa.

Carey, A.G. and N.H. Cutshall. 1973. ^{65}Zn specific activities from Oregon and Washington continental shelf sediments and benthic invertebrate fauna. Pages 287-305 in: *Radioactive Contamination of the Marine Environment*, Proceedings of a symposium, Seattle, 1972. I.A.E.A., Vienna.

Carpenter, R. 1983. Quantitative electron spin resonance (ESR) determinations of forms and total amounts of Mn in aqueous environmental samples. *Geochim. Cosmochim. Acta*, 47: 875-885.

Carpenter, R. 1987. Has man altered the cycling of nutrients and organic C on the Washington continental shelf and slope? *Deep-Sea Res.*, 34: 881-896.

Carpenter, R. and T.M. Beasley. 1981. Plutonium and americium in anoxic marine sediments: evidence against remobilization. *Geochim. Cosmochim. Acta*, 45: 1917-1930.

Carpenter, R., T.M. Beasley, D. Zahnle and B.L.K. Somayajulu. 1987. Cycling of fallout (Pu, ^{241}Am, ^{137}Cs) and natural (U, Th, ^{210}Pb) radionuclides in Washington continental slope sediments. *Geochim. Cosmochim. Acta*, 51: 1897-1921.

Carpenter, R., J.T. Bennett and M.L. Peterson. 1981. ^{210}Pb activities in and fluxes to sediments of the Washington continental slope and shelf. *Geochim. Cosmochim. Acta*, 45: 1155-1172.

Carpenter, R., M.L. Peterson and J.T. Bennett. 1982. ^{210}Pb derived sediment accumulation and mixing rates for the Washington continental slope. *Mar. Geol.*, 48: 135-164.

Carpenter, R., M.L. Peterson and J.T. Bennett. 1985. ^{210}Pb derived sediment accumulation and mixing rate for the greater Puget Sound region. *Mar. Geol.*, 64: 291-312.

Carpenter, R., M.L. Peterson, J.T. Bennett and B.L.K. Somayajulu. 1984. Mixing and cycling of uranium, thorium and ^{210}Pb in Puget Sound sediments. *Geochim. Cosmochim. Acta*, 48: 1949-1963.

Carpenter, R., M.L. Peterson and R.A. Jahnke. 1978. Sources, sinks and cycling of arsenic in the Puget Sound region. Pages 459-480 in: M.L. Wiley (ed.), *Estuarine Interactions*. Academic Press, New York, N.Y.

Carr, R.A., M.M. Jones, T.B. Warner, C.H. Cheeh and E.R. Rurs. 1975. Variation in time of mercury anomalies at the Mid-Atlantic Ridge. *Nature*, 258: 588-589.

Carson, B. 1971. Stratigraphy and depositional history of Quaternary sediments in northern Cascadia Basin and Juan de Fuca abyssal plain, Northeast Pacific Ocean. Ph.D. Diss., Dept. of Oceanogr., Univ. of Washington, Seattle, Wa., 249 pp.

Carson, B. 1973. Acoustic stratigraphy, structure and history of Quaternary deposition in Cascadia Basin. *Deep-Sea Res.*, 20: 387-396.

Carson, B., E.T. Baker, B.M. Hickey, C.A. Nittrouer, D.J. DeMaster, K.W. Thorbjarnarson and G.W. Snyder. 1986. Modern sediment dispersal and accumulation in Quinault submarine canyon – a summary. *Mar. Geol.*, 71: 1-14.

Cherry, R.D. and M. Heyraud. 1982. Evidence of high natural radiation doses in certain mid-water oceanic organisms. *Science*, 218: 54-56.

Cherry, R.D. and L.V. Shannon. 1974. The alpha radioactivity of marine organisms. *At. Energy Rev.*, 12: 3-46.

Chiou, C.T. 1985. Partition coefficients of organic compounds in lipid-water systems and correlations with fish bioconcentration factors. *Environ. Sci. Technol.*, 19: 57-63.

Christensen, J.P., A.H. Devol. and W.M. Smethie. 1984. Biological enhancement of solute exchange between sediments and bottom water on the Washington continental shelf. *Cont. Shelf Res.*, 3: 9-23.

Christensen, J.P., W.M. Smethie and A.H. Devol. 1987a. Benthic nutrient regeneration and denitrification on the Washington continental Shelf. *Deep-Sea Res.*, 34: 881-896.

Christensen, J.P., J.W. Murray, A.H. Devol and L.A. Codispoti. 1987b. Denitrification in continental shelf sediments has a major impact on the oceanic nitrogen budget. *Global Biochem. Cyc.,* 1: 97-116.

Clark, R.C. 1983. The biogeochemistry of aromatic and saturated hydrocarbons in a rocky intertidal marine community in the Strait of Juan de Fuca. Ph.D. Diss., Dept. of Oceanogr., Univ. of Washington, Seattle, Wa., 268 pp.

Clayton, J.R., S.P. Pavlou and N.F. Breitner. 1977. Polychlorinated biphenyls in coastal marine zooplankton: bioaccumulation by equilibrium partitioning. *Environ. Sci. Technol.*, 11: 676-681.

Cline, J.D. and T.S. Bates. 1983. Dimethylsulfide in the equatorial Pacific. *Geophys. Res. Lett.*, 10: 949-952.

Conomos, T.J. and M.G. Gross. 1972. River-ocean suspended particulate matter relations in summer. Pages 176-202 in: A.T. Pruter and D.L. Alverson (eds.), *The Columbia River Estuary and Adjacent Ocean Waters: Bioenvironmental Studies*. Univ. of Washington Press, Seattle, Wa.

Conover, R.J. 1971. Some relations between zooplankton and Bunker C oil in Chedabucto Bay following the wreck of the tanker "Arrow". *J. Fish. Res. Board Can.*, 28: 1327-1330.

Cowie, G.L. and J.I. Hedges. 1984. Carbohydrate sources in a coastal marine environment. *Geochim. Cosmochim. Acta*, 48: 2075-2087.

Crane, K., F. Aikman, R. Embley, S. Hammond, A. Malahoff and J. Lupton. 1985. The distribution of geothermal fields on the Juan de Fuca Ridge. *J. Geophys. Res.*, 90: 727-744.

Cranston, R.E. 1983. Chromium in Cascadia Basin, northeast Pacific Ocean. *Mar. Chem.*, 13: 109-125.

Cranston, R.E. and J.W. Murray. 1980. Chromium species in the Columbia River and estuary. *Limnol. Oceanogr.*, 25: 1104-1112.

Crecelius, E.A. 1974. The geochemistry of arsenic and antimony in Puget Sound and Lake Washington. Ph.D. Diss., Dept. of Oceanogr., Univ. of Washington, Seattle,Wa.. 133 pp.

Crecelius, E.A. 1975. The geochemical cycle of arsenic in Lake Washington and its relation to other elements. *Limnol. Oceanogr.*, 20: 441-450.

Crecelius, E.A. 1980. The solubility of coal fly ash and marine aerosols in seawater. *Mar. Chem.*, 8: 245-250.

Crecelius, E.A. 1981. Prediction of marine atmospheric deposition rates using total Be deposition velocities. *Atmos. Environ.*, 15: 579-532.

Crecelius, E.A., N.S. Bloom and G. Massoth. 1984. Sediment mass balance for the main basin of Puget Sound. *EOS, Trans. Am. Geophys. Union*, 64: 1104.

Crecelius, E.A., M.H. Bothner and R. Carpenter. 1975. Geochemistries of arsenic, antimony, mercury and related elements in sediments of Puget Sound. *Environ. Sci. Technol.*, 9: 325-333.

Curl, H. 1982. FY82 Annual Report – Estuarine and coastal pollutant transport and transformation: the role of particulates. Pacific Marine Environmental Laboratory, NOAA, Seattle, Wa., 228 pp.

Cutshall, N., V. Johnson and C. Osterberg. 1966. ^{151}Cr in seawater: chemistry. *Science*, 152: 202-203.

Cutshall, N., W.C. Renfro, D.W. Evans and V.G. Johnson. 1973. ^{65}Zn in Oregon-Washington continental shelf sediments. Pages 294-702 in: P.J. Nelson (ed.), *Radionuclides in Ecosystems: Proceedings of Third National Symposium on Radioecology.* CONF 710501-P2, Oak Ridge, Tenn.

Cutter, G.A. 1982. Selenium in reducing waters. *Science*, 217: 829-831.

Cutter, G.A. and K.W. Bruland. 1984. The marine biogeochemistry of selenium: a re-evaluation. *Limnol. Oceanogr.*, 29: 1179-1192.

Dahm, C.N., S.V. Gregory and P.K. Park. 1981. Organic carbon transport in the Columbia River. *Est. Coast. Shelf Sci.*, 13: 645-658.

Dalziel, J.A. and P.A. Yeats. 1985. Reactive mercury in the central North Atlantic Ocean. *Mar. Chem.*, 15: 357-361.

Davidson, C.I., W.D. Goold, T.P. Mathison, G.B. Wiersma, K.W. Brown and M.T. Reilly. 1985. Airborne trace elements in Great Smoky Mountains, Olympic and Glacier national parks. *Environ. Sci. Technol.*, 19: 27-35.

Degens, E.T. 1969. Biogeochemistry of stable carbon isotopes. *Organic Geochemistry.* Springer-Verlag, New York, N.Y., pp. 304-329.

De Groot, A.J. 1973. Occurrence and behavior of heavy metals in river deltas. Pages 308-326 in: E.D. Goldberg (ed.), *North Sea Science.* M.I.T. Press, Cambridge, Mass.

Deuser, W.G., E.H. Ross and R.F. Anderson. 1981. Seasonality in the supply of sediment to the deep Sargasso Sea and implications for the rapid transfer of matter to the deep ocean. *Deep-Sea Res.*, 28: 495-505.

Deuser, W.G., P.G. Brewer, T.D. Jickells and R.F. Commean. 1983. Biological control of the removal of abiogenic particles from the surface ocean. *Science*, 219: 388-391.

Devol, A.H. 1987. Verification of flux measurements made with *in situ* benthic chambers. *Deep-Sea Res.*, 34: 1007-1026.

Dexter, R.N., D.E. Anderson, E.A. Quinlan, L.S. Goldstein, R.M. Strickland, S.P. Pavlou, J.R. Clayton, R.M. Kocan and M. Landolt. 1981. A summary of knowledge of Puget Sound related to chemical contaminants. NOAA Tech. Mem. OMPA-13., 435 pp.

Downing, J. 1983. The Coast of Puget Sound: Its processes and development. Univ. of Washington Press, Seattle, Wa., 126 pp.

Duxbury, A.C. 1975 Orthophosphate and dissolved oxygen in Puget Sound. *Limnol. Oceanogr.*, 20:270-274.

Dymond, J.R., W. Gardner and W. Berger. 1981. A sediment trap intercomparison study in the Santa Barbara Basin. *Earth Plan. Sci. Lett.*, 53: 409-418.

Eadie, B.J., P.F. Landrum and W. Faust. 1982. Polycyclic aromatic hydrocarbons in sediments, pore water and the amphipod *Pontoporeia hoyi* from Lake Michigan. *Chemosphere*, 11: 847-858.

Ebbesmeyer, C.C. and C.A. Barnes. 1980. Control of a fjord basin's dynamics by tidal mixing in embracing sill zones. *Est. Coast. Mar. Sci.*, 11: 311-330.

Edgington, D.N. 1981. A review of the persistence of long-lived radionuclides in the marine environment: sediment-water interactions. Pages 67-91 in: *Impacts of Radionuclide Releases into the Marine Environment, Proceedings of a Symposium*, 1980. I.A.E.A., Vienna.

Eganhouse, R.P. and I.R. Kaplan. 1982. Extractable organic matter in municipal wastewaters-2. Hydrocarbons: molecular characterization. *Environ. Sci. Technol.* 16: 541-551.

Eganhouse, R.P. and I.R. Kaplan. 1985. a-tocopheryl acetate as an indicator of municipal waste contamination in the environment. *Environ. Sci. Technol.*, 19: 282-285.

Eisenreich, S.J., B.B. Looney and J.D. Thornton. 1981. Airborne organic contaminants in the Great Lakes ecosystem. *Environ. Sci. Technol.*, 15: 30-38.

Elder, D.L. and S.W. Fowler. 1977. Polychlorinated biphenyls: penetration into the deep ocean by zooplankton fecal pellet transport. *Science,* 197: 459-460.

Emerson, S. 1984. Organic carbon preservation in marine sediments. Pages 78-87 in: E.T. Sundquist and W.S. Broecker (eds.), *The Carbon Cycle and Atmospheric* CO_2*: Natural Variations Archean to Present.* Geophysical Monograph 32, American Geophysical Union, Washington, D.C., pp: 78-87.

Emerson, S., R.E. Cranston and P.S. Liss. 1979. Redox species in a reducing fjord: equilibrium and kinetic considerations. *Deep-Sea Res., 26:* 859-878.

Emerson, S., R.A. Jahnke and D. Heggie. 1984. Sediment-water exchange in shallow water estuarine sediments. *J. Mar. Res.*, 42: 709-730.

Emery, K.O. and E.E. Bray. 1962. Radiocarbon dating of California Basin sediments. *Bull. Am. Assoc. Pet. Geol.*, 46: 1839-1856.

Erlenkhauser, H., E. Suess and H. Willkonm. 1974. Industrialization affects heavy metal and carbon isotope concentrations in recent Baltic Sea sediments. *Geochim. Cosmochim. Acta,* 38: 823-842.

Ertel, J.R. 1985. The lignin geochemistry of sedimentary and aquatic humic substances. Ph.D. Diss., Dept. of Oceanogr., Univ. of Washington, Seattle, Wa., 139 pp.

Evans, D.W. and N.H. Cutshall. 1973. Effects of ocean water on the soluble-suspended distribution of Columbia River radionuclides. Pages 125-140 in: *Radioactive Contamination of the Marine Environment, Proceedings of Symposium*, Seattle, July 1972. I.A.E.A., Vienna.

Fairhall, A.W. and A.W. Young. 1985. Historical ^{14}C measurements from the Atlantic, Pacific and Indian oceans. *Radiocarbon*, 27: 473-507.

Farrington, J.W., E.D. Goldberg, R.W. Risebrough, J.H. Martin and V.T. Bowen. 1983. U.S. Mussel Watch 1976-1978: an overview of the trace metal, DDE, PCB, hydrocarbon and artificial radionuclide data. *Environ. Sci. Technol.*, 17: 490-496.

Farrington, J.W., R.W. Risebrough, P.L. Parker, A.C. Davis, B. deLappe, J.K. Winters, D. Boatwright and N.M. Frew. 1982. Hydrocarbons in polychlorinated biphenyls and DDE in mussels and oysters from the U.S. coast 1976-1978 – The Mussel Watch. WHOI Tech. Rep. 82-42, 106 pp.

Farrington, J.W. and B.W. Tripp. 1977. Hydrocarbons in western North Atlantic surface sediments. *Geochim. Cosmochim. Acta*, 41: 1627-1641.

Feely, R.A. and M.F. Lamb. 1979. A study of the dispersal of suspended sediment from the Fraser and Skagit rivers into northern Puget Sound using Landsat imagery. DOC/EPA Interagency Energy/Environment R&D Program Rep. EPA-600/7-79-165, 46 pp.

Feely, R.A., G.J. Massoth, A.J. Paulson and J.F. Gendron. 1983. Possible evidence for enrichment of trace metals in the hydrous Mn oxide phases of suspended matter from an urbanized embayment. *Est. Coast. Shelf Sci., 17:* 693-708.

Feely, R.A., G.J. Massoth, E.T. Baker, J.F. Gendron, A.J. Paulson and E.A. Crecelius. 1986. Seasonal and vertical variations in the elemental composition of suspended and settling particulate matter in Puget Sound, Wa. *Est. Coast. Shelf Sci.*, 22: 215-239.

Feely, R.A., M. Lewison, G. Massoth, G. Robert-Baldo, J.W. Lavelle, R.H. Byrne, K.L. von Damm and H.C. Curl. 1987. Composition and dissolution of black smoker particulates from active vents on the Juan de Fuca Ridge. *J. Geophys. Res., 92:* 11347-11363.

Ferek, R.J. and M.O. Andreae. 1984. The photochemical production of carbonyl sulfide in marine surface water. *Nature,* 307: 148-150.

Fletcher, W.K., G.S. Holmes and A.G. Lewis. 1983. Geochemistry and biological availability of iron and trace elements in the upper Fraser River estuary. *Mar. Chem.*, 12: 195-217.

Forster, W.O. 1972a. Radioactive and stable nuclides in the Columbia River and adjacent northeast Pacific Ocean. Pages 663-700 in: A.T. Pruter and D.L. Alverson (eds.), *The Columbia River Estuary and Adjacent Ocean Waters: Bioenvironmental Studies.* Univ. of Washington Press, Seattle, Wa.

Forster, W.O. 1972b. Radionuclide distribution in Columbia River and adjacent Pacific shelf sediments. Pages 701-735 in: A.T. Pruter and D.L. Alverson (eds.), *The Columbia River Estuary and Adjacent Ocean Waters: Bioenvironmental Studies.* Univ. of Washington Press, Seattle, Wa.

Fox, T.D. and J.D. Ludwick. 1976. Lead concentrations associated with 1,000 mb geostrophic back trajectories at Quillayute, Wa. *Atmos. Environ.*, 10: 799-803.

Frey, B.E., R. Lara-Lara and L.C. Small. 1984. Water column primary production in the Columbia River Estuary. Final Report to Columbia River Estuary Data Development program, CREST, Astoria, Ore., 133 pp.

Fuller, C. and D.E. Hammond. 1983. The fallout rate of ^{210}Pb on the western coast of the United States. *Geophys. Res. Lett.*, 10: 1164-1167.

Furlong, E.T. 1986. Sediment geochemistry of photosynthetic pigments in oxic and anoxic marine and lacustrine sediments: Dabob Bay, Saanich Inlet and Lake Washington. Ph.D. Diss., Dept. of Oceanogr., Univ. of Washington, Seattle, Wa., 214 pp.

Furlong, E.T. and R. Carpenter. 1982. Azaarenes in Puget Sound sediments. *Geochim. Cosmochim. Acta*, 46: 1385-1396.

Furlong, E.T. and R. Carpenter. 1988. Pigment preservation and remineralization in oxic coastal marine sediments. *Geochim. Cosmochim. Acta*, 52: 87-99.

Gammon, R., J. Cline and D. Wisegarver. 1982. Chlorofluoromethanes in the Northeast Pacific Ocean: measured vertical distributions and applications as transient tracers of upper ocean mixing. *J. Geophys. Res.*, 87: 9441-9454.

Goldberg, E.D., M. Koide, V. Hodge, A.R. Flegal and J. Martin. 1983. U.S. Mussel Watch: 1977-1978 results on trace metals and radionuclides. *Est. Coast. Shelf Sci.*, 16: 69-93.

Graham, W.F., M.L. Bender and G.P. Klinkhammer. 1976. Manganese in Narragansett Bay. *Limnol. Oceanogr.*, 21: 665-673.

Grimmer, G. and H. Bohnke. 1977. Investigations of drilling cores of sediments of Lake Constance-I. Profiles of the polycyclic aromatic hydrocarbons. *A. Naturforsch.*, 32C: 703-711.

Gross, M.G., C.A. Barnes and G.K. Riel. 1965. Radioactivity of the Columbia River effluent. *Science*, 149: 1088-1090.

Gross, M.G., A.G. Carey, G.A. Fowler and L.D. Kulm. 1972. Distribution of organic carbon in surface sediment, Northeast Pacific Ocean. Pages 254-264 in: A.T. Pruter and D.L. Alverson (eds.), *The Columbia River Estuary and Adjacent Ocean Waters: Bioenvironmental Studies*. Univ. of Washington Press, Seattle, Wa.

Gross, M.G. and J.L. Nelson. 1966. Sediment movement on the continental shelf near Washington and Oregon. *Science,* 154: 879-885.

Grubel, K.A. and C.S. Martens. 1984. ^{222}Rn tracing of sediment-water chemical transport in an estuarine sediment. *Limnol. Oceanogr.*, 29: 587-597.

Grundmanis, V. and J.W. Murray. 1977. Nitrification and denitrification in marine sediments from Puget Sound. *Limnol. Oceanogr.*, 22: 804-813.

Gschwend, P.M. and R.A. Hites. 1981. Fluxes of polycyclic aromatic hydrocarbons to marine and lacustrine sediments in the northeastern United States. *Geochim. Cosmochim. Acta*, 45: 2359-2367.

Gschwend, P.M. and S. Wu. 1985. On the constancy of sediment-water partition coefficients of hydrophobic organic pollutants. *Environ. Sci. Technol.*, 19: 90-96.

Guinasso, N.L. and D.R. Schink. 1975. Quantitative estimates of biological mixing rates in abyssal sediments. *J. Geophys. Res.*, 80: 3032-3043.

Hafferty, A.J., S.P. Pavlou and W. Horn. 1977. Release of polychlorinated biphenyls (PCBs) in a salt-wedge estuary as induced by dredging of contaminated sediments. *Sci. Total Environ.*, 8: 229-239.

Hallas, L.E., J.C. Means and J.J. Cooney. 1982. Methylation of tin by estuarine microorganisms. *Science*, 215: 1505-1507.

Hambrick, G.A., P.N. Froelich, M.O. Andreae and B.L. Lewis. 1984. Determination of methylgermanium species in natural waters by graphite furnace atomic absorption spectrometry with hydride generation. *Anal. Chem.*, 56: 421-424.

Hamilton, S.E., T.S. Bates and J.D. Cline. 1984. Sources and transport of hydrocarbons in the Green-Duwamish River, Wa. *Environ. Sci. Technol.*, 18: 72-79.

Hammond, D.E. and C. Fuller. 1979. The use of ^{222}Rn to estimate benthic exchange and atmospheric exchange rates in San Francisco Bay. Pages 213-229 in: T.J. Conomos (ed.), *San Francisco Bay, the Urbanized Estuary*. Symp. Pac. Div. Am. Assoc. Adv. Sci

Hammond, D.E., H.J. Simpson and G. Mathieu. 1977. ^{222}Rn distribution and transport across the sediment-water interface in the Hudson River. *J. Geophys. Res.*, 82: 3913-3920.

Hardy, E.P., P.W. Krey and H.L Volchuck. 1973. Global inventory and distribution of fallout plutonium. *Nature*, 241: 444-445.

Hardy, J.T. and E.A. Crecelius. 1981. Is atmospheric particulate matter inhibiting primary productivity? *Environ. Sci. Technol.*, 15: 1103-1105.

Hardy, J.T., C.W. Apts, E.A. Crecelius and G.W. Fellingham. 1985a. The sea-surface microlayer: fate and residence time of atmospheric metals. *Limnol. Oceanogr.*, 30: 93-101.

Hardy, J.T., E.A. Crecelius and N.S. Bloom. 1985b. Sea-surface microlayer metals enrichments in an urban and rural bay. *Est. Coast. Shelf Sci.*, 20: 299-312.

Hardy, J.T. and L. Antrim. 1988. Distribution and biological effects of sea surface contamination in Puget Sound. Pages 400-408 in: *Proceedings, First Annual Meeting on Puget Sound Research.* Puget Sound Water Quality Authority, 217 Pine St., Seattle, Wa.

Hardy, J.T., E.A. Crecelius, C.W. Apts and J.M. Gurtisen. 1988. Sea surface contamination in Puget Sound. II. Concentrations and distributions of contaminants. *Marine Environmental Res., 23:* 251-272.

Hathaway, J.C. 1972. Regional clay mineral facies in estuaries and continental margin of the U.S. east coast. *Geol. Soc. Am. Mem.*, 133: 293-316.

Healy, M.J. 1983. Dispersal of Mount St. Helens ash across the Washington Continental shelf. M.S. Thesis, Dept. of Geological Sciences , Lehigh Univ., Bethelehem, Penn., 183 pp.

Healy-Ridge, M.J. and B. Carson. 1987. Sediment transport on the Washington continental shelf: estimate of dispersal rate from Mount St. Helen ash. *Cont. Shelf Sci.*, 7: 759-772.

Hedges, J.I. and D.C. Mann. 1979a. The characterization of plant tissues by their lignin oxidation products. *Geochim. Cosmochim. Acta*, 43: 1803-1807.

Hedges, J.I. and D.C. Mann. 1979b. The lignin geochemistry of marine sediments from the southern Washington coast. *Geochim. Cosmochim. Acta*, 43: 1809-1818.

Hedges, J.I., H.J. Turin and J.R. Ertel. 1984. Sources and distributions of sedimentary organic matter in the Columbia River drainage basin, Washington and Oregon. *Limnol. Oceanogr.*, 29: 35-46.

Hermann, A. J., B.M. Hickey, M.L. Landry and D.F. Winter. 1989. Coastal upwelling dynamics. Pages 211-254 in: M.L. Landry and B.M. Hickey (eds.), *Coastal Oceanography of Washington and Oregon.* Elsevier, Amsterdam, Neth.

Hickey, B.M. 1979. The California Current system – hypotheses and facts. *Progr. Oceanogr.* 8: 191-279.

Hickey, B. M. 1989. Patterns and process of circulation over the shelf and Slope. Pages 41-116 in: M.L. Landry and B.M. Hickey (eds.), *Coastal Oceanography of Washington and Oregon.* Elsevier, Amsterdam, Neth.

Hinga, K.R., G.R. Heath, D.R. Anderson and C.D. Hollister. 1982. Disposal of high-level radioactive wastes by burial in the sea floor. *Environ. Sci. Technol.*, 16: 28A-37A.

Hodge, V., S.R. Johnson and E.D. Goldberg. 1978. Influence of atmospherically transported aerosols on surface ocean water composition. *Geochem. J.*, 12: 7-20.

Hodge, V.F., S.L. Seidel and E.D. Goldberg. 1979. Determination of tin (IV) and organotin compounds in natural waters, coastal sediments and macroalgae by atomic absorption spectrometry. *Anal. Chem.*, 51: 1256-1259.

Hoffmann, E.E., J.M. Klinck and G.A. Paffenhofer. 1981. Concentrations and vertical fluxes of zooplankton fecal pellets on a continental shelf. *Mar. Biol.*, 61: 327-335.

Holtzmann, R.B. 1964. ^{210}Pb and ^{210}Po in Potable Waters in Illinois Pages 227-237 in: J.A.S. Adams and W.M. Lowder (eds.), *The Natural Radiation Environment.* Univ. of Chicago Press.

Honjo, S. 1978. Sedimentation of materials in the Sargasso Sea at a 5,367-m deep station. *J. Mar. Res.*, 36: 469-492.

Hubbell, D.W. and J.L. Glenn. 1973. Distribution of radionuclides in bottom sediments of the Columbia River Estuary. U.S. Geological Survey Prof. Paper 433-L, pp: 1-35.

Johnson, J. 1985. The role of the oceans in the atmospheric cycle of carbonyl sulfide. Ph.D. Diss., Dept. of Atmospheric Sciences, Univ. of Washington, Seattle, Wa., 137 pp.

Johnson, V.G., N.H. Cutshall and C.L. Osterberg. 1967. Retention of ^{65}Zn by Columbia River sediments. *Water Resources Res.*, 3: 99-102.

Jones, C.J., H.P. Johnson and J.R. Delaney. 1981. Distribution of hydrothermal manganese over the Juan de Fuca Ridge. *Geophys. Res. Lett.*, 8: 873-876.

Jones, C.J. and J.W. Murray. 1984. Nickel, cadmium and copper in the northeast Pacific off the coast of Washington. *Limnol. Oceanogr.*, 29: 711-720.

Jones, C.J. and J.W. Murray. 1985. The geochemistry of manganese in the northeast Pacific Ocean off Washington. *Limnol. Oceanogr.*, 30: 81-92.

Kachel, N.B. 1980. A time dependent model of sediment transport and strata formation on a continental shelf. Ph. D. Diss., Dept of Oceanogr., Univ. of Washington, Seattle, Wa., 123 pp.

Kachel, N.B. and J.D. Smith. 1989. Sediment transport and Deposition Pages 287-365 in: M.L. Landry and B.M. Hickey (eds.), *Coastal Oceanography of Washington and Oregon.* Elsevier, Amsterdam, Neth.

Karickhoff, S.W. 1981. Semi-empirical estimation of sorption of hydrophobic pollutants on natural sediments and soils. *Chemosphere*, 10: 833-846.

Karickhoff, S.W. and K.R. Morris. 1985. Sorption dynamics of hydrophobic pollutants in sediment suspension. *Environ. Tox. Chem.*, 4: 469-479.

Karlin, R. 1980. Sediment source and clay mineral distributions off the Oregon coast. *J. Sed. Petrol.*, 50: 543-560.

Kenaga, E.A. and C.A.I. Goring. 1980. Relationship between water solubility, soil sorption, octanol-water partitioning and concentration of chemicals in biota. Pages 78-115 in: J.G. Eaton, P.R. Parrish and A.C. Hendricks (eds.), *Aquatic Toxicology.* Third Symposium, ASTM-STP707.

Khalil, M.A.K. and R.A. Rasmussen. 1981. Atmospheric methylchloride (CH_3Cl). *Chemosphere*, 10: 1019-1023.

Khalil, M.A.K. and R.A. Rasmussen. 1983. Increase and seasonal cycle of nitrous oxide in the earth's atmosphere. *Tellus*, 35B: 161-169.

Khalil, M.A.K. and R.A. Rasmussen. 1984. Global sources, lifetimes and mass balances of carbonyl sulfide (OCS) and carbon disulfide (CS_2) in the earth's atmosphere. *Atmos. Environ.*, 18: 1805-1813.

Khalil, M.A.K. and R.A. Rasmussen. In press. Latitudinal distributions, sources and sinks of halocarbons: observations from an oceanic cruise. in: *Proceedings of Specialty Meeting of the APCA Conference on Environmental Impact of Natural Emissions.*

Khalil, M.A.K., R.A. Rasmussen and S.D. Hoyt. 1983. Atmospheric chloroform ($CHCl_3$): ocean-air exchange and global mass balance. *Tellus*, 35B: 266-274.

Klumpp, D. and P.J. Peterson. 1981. Chemical characteristics of arsenic in a marine food chain. *Mar. Biol.*, 62: 297-305.

Knauer, G.A., J.H. Martin and K.W. Bruland. 1979. Fluxes of particulate carbon, nitrogen and phosphorus in the upper water column of the northeast Pacific. *Deep-Sea Res.*, 26: 97-108.

Knebel, H.J., J.C. Kelley and J.T. Whetten. 1968. Clay minerals of the Columbia River: a qualitative, quantitative and statistical evaluation. *J. Sed. Petrol.*, 38: 600-611.

Koide, M., K.K. Bertine, T.J. Chow and E.D. Goldberg. 1985. The $^{240}Pu/^{239}Pb$ ratio, a potential geochronometer. *Earth Plan. Sci. Lett.*, 72: 1-8.

Krey, P.W., E.P. Hardy, C. Pachucki, F. Rowike, J. Coluzza and W.K. Benson. 1976. Mass isotopic composition of global fallout plutonium in soil. Pages 671-678 in: *Transuranium Radionuclides in the Environment.* STI/PUB/410. I.A.E.A., Vienna,.

Krishnaswami, S., L.K. Benninger, R.C. Aller and K.L. von Damm. 1980. Atmospherically derived radionuclides as tracers of sediment mixing and accumulation in nearshore marine and lake sediments: evidence from ^{7}Be, ^{210}Pb and $^{239,240}Pu$. *Earth Plan. Sci. Lett.*, 47: 307-318.

Kulm, L.D. and K.F. Scheidegger. 1979. Quaternary sedimentation on the tectonically active Oregon continental slope. Society of Economic Paleontologists and Mineralogists Special Publication No. 27, pp: 247-263.

Landrum, P.F. and D. Scavia. 1983. Influence of sediment on anthracene uptake, depuration and biotransformation by the amphipod *Hyabella azteca. Can. J. Fish. Aquat. Sci.*, 40: 298-305.

Landrum, P.F., S.R. Nihart, B.J. Eadie and W.S. Gardner. 1984. Reverse-phase separation method for determining pollutant binding to Aldrich humic acid and dissolved organic carbon of natural waters. *Environ. Sci. Technol.*, 18: 187-192.

Landry, M.R and C.J. Lorenzen. 1989. Utilization and transformation of primary production by zooplankton. Pages 175-210 in: M.L. Landry and B.M. Hickey (eds.), *Coastal Oceanography of Washington and Oregon.* Elsevier, Amsterdam, Neth.

Landry, M.R., J. R. Postel, W.K. Peterson and J. Newman. 1989. Broad scale patterns in the distribution of hydrographic variables. Pages 1-40 in: M.L. Landry and B.M. Hickey (eds.), *Coastal Oceanography of Washington and Oregon.* Elsevier, Amsterdam, Neth.

Langston, W.J. 1984. Availability of arsenic to estuarine and marine organisms: field and laboratory evaluation. *Mar. Biol.*, 80: 143-154.

Lavelle, J.W., G.J. Massoth and E.A. Crecelius. 1986. Accumulation rates of recent sediments in Puget Sound, Washington. *Mar. Geol.*, 72: 59-70.

LeBlanc, P.J. and A.L. Jackson. 1973. Arsenic in marine fish and invertebrates. *Mar. Poll. Bull.*, 4: 88-90.

Lee, M.L., G.P. Prado, J.B. Howard and R.A. Hites. 1977. Source identification of urban airborne polycyclic aromatic hydrocarbons by GC/MS and HRMS. *Biomed. Mass Spectrum.*, 4: 182-186.

Lewin, J., C.T. Schaefer and D.F. Winter. 1989. Surf -zone ecology and dynamics. Pages 565-592 in: M.L. Landry and B.M. Hickey (eds.), *Coastal Oceanography of Washington and Oregon.* Elsevier, Amsterdam. Neth.

Lewis, B.L., P.N. Froelich and M.O. Andreae. 1985. Methylgermanium in natural waters. *Nature*, 313: 303-305.

Lewis, D.M. 1977. The use of ^{210}Pb as a heavy metal tracer in the Susquehanna River system. *Geochim. Cosmochim. Acta*, 41: 1557-1564.

Lewis, G.B. and A.H. Seymour. 1965. Distribution of ^{65}Zn in plankton from offshore waters of Washington and Oregon, 1961-1962. Pages 956-967 in: *Transactions of the Joint Conference*, Marine Technology Society and American Society of Limnology and Oceanography, Washington, D.C., June 1965,

Li, Y.-H., H.W. Feely and P.H. Santschi. 1979. ^{228}Th/^{228}Ra radioactive disequilibrium in the New York Bight and its implications for coastal pollution. *Earth Plan. Sci. Lett.*, 42: 13-26.

Lie, U. 1968. A quantitative study of benthic infauna in Puget Sound, Washington, U.S.A., in 1963-1964. *Fiskeridir. Skr. Ser. Havanders*, 14: 229-556.

Lie, U. 1969. Standing crop of benthic infauna in Puget Sound and off the coast of Washington. *J. Fish. Res. Board Can.*, 26: 55-62.

Lie, U. 1974. Distribution and structure of benthic assemblages in Puget Sound, Washington, U.S.A. *Mar. Biol.*, 26: 203-223.

Lindberg, S.E. and R.S. Harriss. 1974. Mercury-organic matter associations in estuarine sediments and interstitial water. *Environ. Sci. Technol.*, 8: 459-462.

Liss, P.S. 1983. Gas transfer: experiments and geochemical implications. Pages 241-298 in: P.S. Liss and W.G. Slinn (eds.), *Air-Sea Exchange of Gases and Particles.* D. Reidel Publishing Co., Boston, Mass.

Liss, P.S. and W.G.N. Slinn. 1983. *Air-sea exchange of gases and particles.* D. Reidel Publishing Co., Boston, Mass., 553 pp.

Livingston, H.D. and V.T. Bowen. 1976. Americium in the marine environment: relationships to plutonium. Pages 107-130 in: M.W. Miller and J.N. Stannard (eds.), *Toxicity of Aquatic Radionuclides: Models and Mechanisms.* Ann Arbor Science Publishers.

Livingston, H.D. and V.T. Bowen. 1979. Pu and ^{137}Cs in coastal sediments. *Earth Plan. Sci. Lett.*, 43: 29-45.

Loder, T.C. and R.P. Reichardt. 1980. The dynamics of conservative mixing in estuaries. *Estuaries* , 4: 64-69.

Logan, R.M., J.C. Derby and L.C. Duncan. 1982. Acid precipitation and lake susceptibility in the central Washington Cascades. *Environ. Sci. Technol.*, 16: 771-775.

Lorenzen, C.J., F.R. Shuman and J.T. Bennett. 1981. *In situ* calibration of a sediment trap. *Limnol. Oceanogr.*, 26: 580-585.

Louda, J.W. and E.W. Baker. 1981. Geochemistry of tetrapyrrole, carotenoid and perylene pigments in sediments from the San Miguel Gap (site 467) and Baja California Borderland (site 471), Deep Sea Drilling Project Leg 63, Init. Repts. DSDP, 63: 785-818.

Lunde, G. 1970. Analysis of arsenic and selenium in marine raw materials. *J. Sci. Fd. Agric.*, 21: 242-247.

Lupton, J.E., H.P. Johnson and J.R. Delaney. 1981. Hydrothermal ^{3}He on the Juan de Fuca Ridge. *EOS, Trans. Am. Geophys. Union*, 62: 913.

Malcolm, R.L. and W.H. Durum. 1976. Organic carbon and nitrogen concentrations and annual organic load of six selected rivers of the United States. U.S. Geological Survey Water Supply Paper 1817-F, 21 pp.

Malins, D.C., M.M. Krahn, M.S. Myers, L.D. Rhodes, D.W. Brown, C.A. Krone, B.B. McCain and S. Chan. 1985. Toxic chemicals in sediments and biota from a creosote-polluted harbor: relationships with hepatic neoplasms and other hepatic lesions in English sole (*Parophrys vetulus*). *Carcinogenesis*, 6: 1463-1470.

Malins, D.C., B. McCain, D. Brown, S. Chan, M. Myers, J. Landall, P. Prohaska, A. Friedman, L. Rhodes, D. Burrows, W. Gronlund and H. Hodgins. 1984. Chemical pollutants in sediments and diseases of bottom dwelling fish in Puget Sound, Wa. *Environ. Sci. Technol.*, 18: 705-713.

Mann, D.R., L.D. Surprenant and S.A. Casso. 1984. *In situ* chemisorption of transuranics from seawater. *Nucl. Instr. Meth. Phys. Res.*, 223: 235-238.

Martel, Y.A. and E.A. Paul. 1974. The use of radiocarbon dating of organic matter in the study of soil genesis. *Soil Sci. Soc. Am. Proc.*, 38: 501-506.

Massoth, G.J., E.T. Baker, R.A. Feely and H.C. Curl. 1984. Hydrothermal signals away from the southern Juan de Fuca Ridge. *EOS, Trans. Am. Geophys. Union*, 65: 1112.

Massoth, G.J., R.A. Feely, H.C. Curl and E.T. Baker. 1983. The geochemistry of particulate hydrothermal plumes over the Juan de Fuca Ridge. *EOS, Trans. Am. Geophys. Union*, 64: 722-723.

Matthias, C.L., J.M. Bellama, G.J. Olson and F.E. Brinckman. 1986. Comprehensive method for determination of aquatic butyltin and butylmethyltin species at ultra trace levels using simultaneous hybridization-extraction with gas chromatography-flamephotometric detection. *Environ. Sci. Technol.*, 20: 609-615.

McBride, B.C., H. Merilee, W.R. Cullen and W. Pickett. 1978. Anaerobic and aerobic alkylation of arsenic. *ACS Symp. Series*, 82: 94-115.

McCarthy, J.F. and B.D. Jiminez. 1985. Interactions between polycyclic aromatic hydrocarbons and dissolved humic material: binding and dissociation. *Environ. Sci. Technol.*, 19: 1072-1076.

McDuff, R.E. and F.M.M. Morel. 1980. The geochemical control of seawater (Sillen revisited). *Environ. Sci. Technol.*, 14: 1182-1186.

McManus, D.A. 1972. Bottom topography and sediment texture near the Columbia River mouth. Pages 251-253 in: A.T. Pruter and D.L. Alverson (eds.), *The Columbia River Estuary and Adjacent Ocean Waters: Bioenvironmental Studies*. Univ. of Washington Press, Seattle, Wa.

Means, J.C., S.G. Wood, J.J. Hassett and W.L. Banwart. 1980. Sorption of PAH by sediments and soils. *Environ. Sci. Technol.*, 14: 1524-1528.

MERGE. 1984. Regional setting and local character of a hydrothermal field/sulfide deposit on the Endeavour segment of the Juan de Fuca Ridge. *EOS, Trans. Am. Geophys. Union*, 65: 1111.

METRO TPPS. 1984. Presence, distribution and fate of toxicants in Puget Sound and Lake Washington. in: G.P. Romberg *et al.* (eds.), *METRO Toxicant Program.* Rep. No. 6A, Municipality of Metropolitan Seattle, 231 pp.

Milliman, J.D. 1980. Sedimentation in the Fraser River and its estuary, southwestern British Columbia (Canada). *Est. Coast. Mar. Sci.*, 10: 609-633.

Moore, W.S., K.W. Bruland and J. Michel. 1981. Fluxes of uranium and thorium series isotopes in the Santa Barbara Basin. *Earth Plan. Sci. Lett.*, 53: 391-399.

Morgan, M.D. 1979. Puget's Sound: A Narrative of Tacoma and the Southern Sound. University of Washington Press, Seattle, Wa., 370 pp.

Mortlock, R.A. and P.N. Froelich. 1986. Hydrothermal germanium over the southern East Pacific Rise. *Science*, 231: 43-45.

Muller, G. and U. Forstner. 1975. Heavy metals in sediments of the Rhine and Elbe estuaries: mobilization or mixing effect. *Environ. Geol.*, 1: 33-39.

Muller, P.J. and E. Suess. 1979. Productivity, sedimentation rate and sedimentary organic matter in the ocean-I. Organic carbon preservation. *Deep-Sea Res.*, 26: 1347-1367.

National Academy of Sciences. 1985. Oil in the sea: Inputs, fates and effects. National Academy Press, Washington, D.C., 601 pp.

Neff, J. 1979. Polycyclic aromatic hydrocarbons in the aquatic environment: Sources, fates and biological effects. Applied Science Publ., London, 262 pp.

Nelson, D.M. and M.B. Lovett. 1978. Oxidation state of plutonium in the Irish Sea. *Nature,* 276: 599-601.

Nevissi, A.E. 1985. Measurement of ^{210}Pb atmospheric flux in the Pacific Northwest. *Health Phys.*, 48: 169-174.

Nichols, F.H. 1974. Sediment turnover by a deposit feeding polychaete. *Limnol. Oceanogr.*, 19: 945-950.

Nielsen, R.D. 1982. Americium and plutonium in water, biota and sediments from the central Oregon coast. M.S. Thesis, Dept. of Oceanography, Oregon St. Univ., Corvalis, Ore., 79 pp.

Nittrouer, C.A. 1978. The process of detrital sediment accumulation in a continental shelf environment: an examination of the Washington shelf. Ph.D. Diss., Dept. of Oceanogr., Univ. of Washington, Seattle, Wa., 243 pp.

Nittrouer, C.A., D.J. DeMaster, B.A. McKee, N.H. Cutshall and I.L. Larsen. 1984. The effect of sediment mixing on ^{210}Pb accumulation rates for the Washington continental shelf. *Mar. Geol.*, 54: 201-221.

Nittrouer, C.A., R.W. Sternberg, R. Carpenter and J.T. Bennett. 1979. The use of ^{210}Pb geochronology as a sedimentological tool: application to the Washington continental shelf. *Mar. Geol.*, 31: 297-316.

Normark, W.R., J. Lupton, J. Murray, R. Koski, D. Clague, J. Morton, J. Delaney and H.P. Johnson. 1982. Polymetallic sulfide deposits and water column tracers of active hydrothermal vents on the southern Juan de Fuca Ridge. *Mar. Tech. Soc. J.*, 16: 46-53.

Olafsson, J. 1983. Mercury concentrations in the North Atlantic in relation to cadmium, aluminum and oceanographic parameters. Pages 475-485 in: C.S. Wong *et al.* (eds.), *Trace Metals in Sea Water.* NATO Conference Series, Plenum Press, New York, N.Y.

Orlandini, K.A., W.R. Penrose and D.M. Nelson. 1986. Pu(V) as the stable form of oxidized plutonium in natural waters. *Mar. Chem.*, 18: 49-57.

Osterberg, C.A., A.G. Carey and H. Curl. 1963. Acceleration of sinking rates of radionuclides in the ocean. *Nature*, 200: 1276-1277.

Pak, H., J.R.V. Zaneveld and J. Kitchen. 1980. Intermediate nepheloid layers observed off Oregon and Washington. *J. Geophys. Res.*, 85: 6697-6708.

Palmieri, J., H. Livingston and J.W. Farrington. 1984. U.S. Mussel Watch Program: transuranic element data from Woods Hole Oceanographic Institution 1976-1983. WHOI Tech. Rep. 84-28, 74 pp.

Palmork, K.H., S. Wilhelmson and T. Neppelberg. 1973. The contribution from various industries of polynuclear aromatic hydrocarbons to the marine environment. International Council for Exploration of the Sea. Fisheries Improvement Commission Rep. No. 1973/E33, 20 pp.

Paulson, A.J. and R.A. Feely. 1985. Dissolved trace elements in the surface waters of Puget Sound. *Mar. Poll. Bull.*, 16: 285-291.

Paulson, A.J., R.A. Feely, H.C. Curl and J.F. Gendron. 1984. Behavior of Fe, Mn, Cu and Cd in the Duwamish River estuary downstream of a sewage treatment plant. *Water Res., 18:* 633-641.

Paulson, A.J., R.A. Feely, H.C. Curl, E.A. Crecelius and T. Geiselman. 1988. The impact of scavenging on trace metal budgets in Puget Sound. *Geochim. Cosmochim. Acta,* 52: 1765-1780.

Pavlou, S.P. and R.N. Dexter. 1979. Distribution of polychlorinated biphenyls (PCBs) in estuarine ecosystems: testing the concept of equilibrium partitioning in the marine environment. *Environ. Sci. Technol.*, 13: 65-71.

Pavlou, S.P., W. Hom, R.N. Dexter, D.E. Anderson and E.A. Quinlan. 1982. Release, distribution and impacts of polychlorinated biphenyls (PCBs) induced by dredged material disposal activities at a deep-water disposal site. *Environ. Intl.*, 7: 99-117.

Pearcy, W.G. and H.A. Vanderploeg. 1973. Radioecology of benthic fishes off Oregon. Pages 245-261 in: *Radioactive Contamination of the Marine Environment, Proceedings of a Symposium*, Seattle, 1972. I.A.E.A., Vienna.

Pearcy, W.G., E.E. Krygier and N.H. Cutshall. 1977. Biological transport of ^{65}Zn into the deep sea. *Limnol. Oceanogr.*, 22: 846-855.

Pearcy, W.G. and M. Stuiver. 1983. Vertical transport of ^{14}C into deep-sea food webs. *Deep-Sea Res.*, 30: 427-440.

Pellenbarg, R. 1985. Silicones in estuarine and coastal sediments. Pages 121-126 in: A.C. Sigleo and A. Hattori (eds.), *Marine and Estuarine Geochemistry.* Lewis Publishers, Chelsea, Maine.

Peng, T.H. and W.S. Broecker. 1984. Ocean life cycles and the atmospheric CO_2 content. *J. Geophys. Res.*, 89: 8170-8180.

Peng, T.H., T. Takahashi and W.S. Broecker. 1974. Surface radon measurements in the North Pacific Ocean Station Papa. *J. Geophys. Res.*, 79: 1772-1780.

Penrose, W.R., H.B.S. Conacher, R. Black, J.C. Meranger, W. Miles, H.M. Cunningham and W.R. Squires. 1977. Implications of inorganic-organic interconversion on fluxes of arsenic in marine food webs. *Environ. Health Perspec.*, 19: 53-60.

Peterson, M.L. and R. Carpenter. 1983. Biogeochemical processes affecting total arsenic and arsenic species distribution in an intermittently anoxic fjord. *Mar. Chem.*, 12: 295-321.

Peterson, M.L. and R. Carpenter. 1986. Arsenic distributions in pore water and sediments of Puget Sound, Lake Washington, the Washington coast and Saanich Inlet, B.C. *Geochem. Cosmochim. Acta*, 50: 353-370.

Phillips, D.J.H. and M.H. Depledge. 1985. Metabolic pathways involving arsenic in marine organisms: a unifying hypothesis. *Mar. Environ. Res.*, 17: 1-12.

Poet, S.E., H.E. Moore and E.A. Martell. 1972. ^{210}Pb, ^{210}Bi and ^{210}Po in the atmosphere: accurate ratio measurement and application to aerosol residence time determination. *J. Geophys. Res.*, 77: 6515-6527.

Prahl, F.G. 1982. The geochemistry of polycyclic aromatic hydrocarbons in Columbia River and Washington coastal sediments. Ph.D. Diss., Dept. of Oceanogr., Univ. of Washington, Seattle, Wa., 209 pp.

Prahl, F.G. 1985. Chemical evidence of differential particle dispersal in the southern Washington coastal environment. *Geochim. Cosmochim. Acta*, 49: 2533-2539.

Prahl, F.G., J.T. Bennett and R. Carpenter. 1981. The early diagenesis of aliphatic hydrocarbons and organic matter in sedimentary particulates from Dabob Bay, Wa. *Geochim. Cosmochim. Acta*, 44: 1967-1976.

Prahl, F.G. and R. Carpenter. 1979. The role of zooplankton fecal pellets in the sedimentation of polycyclic aromatic hydrocarbons in Dabob Bay, Wa. *Geochim. Cosmochim. Acta*, 43: 1959-1972.

Prahl, F.G. and R. Carpenter. 1983. Polycyclic aromatic hydrocarbon (PAH)-phase associations in Washington coastal sediment. *Geochim. Cosmochim. Acta*, 47: 1013-1023.

Prahl, F.G. and, R. Carpenter. 1984. Hydrocarbons in Washington coastal sediments. *Est. Coast. Shelf Sci.*, 18: 703-720.

Prahl, F.G., E.A. Crecelius and R. Carpenter. 1984. Polycyclic aromatic hydrocarbons in Washington coastal sediments: an evaluation of atmospheric and riverine routes of introduction. *Environ. Sci. Technol.*, 18: 687-693.

Prahl, F.G. and L.A. Pinto. 1987. A geochemical study of long chain *n*-aldehydes in Washington coastal sediments. *Geochim. Cosmochim. Acta*, 51: 1573-1582.

Pruter, A.T. and D.L. Alverson. 1972. *The Columbia River estuary and adjacent ocean waters: Bioenvironmental studies.* Univ. of Washington Press, Seattle, Wa., 868 pp.

Puget Sound Water Quality Authority. 1986. State of the Sound: 1986 Rep. Puget Sound Water Quality Authority, 217 Pine Street, Suite 1100, Seattle, Wa., 84 pp.

Rae, J.E. and S.R. Aston. 1982. The role of suspended solids in the estuarine geochemistry of mercury. *Water Res.*, 16: 649-654.

Rasmussen, R.A., M.A.K. Khalil and S.D. Hoyt. 1982. The oceanic source of carbonyl sulfide (OCS). *Atmos. Environ.*, 16: 1591-1594.

Renfro, W.C. 1973. Transfer of ^{65}Zn from sediments by marine polychaete worms. *Mar. Biol.*, 21: 305-316.

Renfro, W.C., S.W. Fowler, M. Heyraud and J. LaRosa. 1975. Relative importance of food and water in long-term Zn-65 accumulation by marine biota. *J. Fish. Res. Board Can.*, 32: 1339-1345.

Riedel, G.F., S.L. Wilson and R.L. Holton. 1984. Trace metals in the Columbia River estuary following the 18 May 1980 eruption of Mt. St. Helens. *Pacific Sci.*, 38: 340-349.

Robbins, J.A. and D.N. Edginton. 1975. Determination of recent sedimentation rates in Lake Michigan using ^{210}Pb and ^{137}Cs. *Geochim. Cosmochim. Acta*, 39: 285-304.

Roberts, R.W. 1979. Sediment distribution maps for the Puget Sound region. Unpublished. Available at Dept. of Oceanography, Univ. of Washington, Seattle.

Robertson, D.E., L.A. Rancitelli, J.C. Langford and R.W. Perkins. 1972. Battelle-Northwest contributions to the IDOE Baseline Study. Pages 231-273 in: *BaselineStudies of Pollutants in the Marine Environment.* Init. Decade of Oceanogr., Brookhaven National Laboratory, pp. 231-273.

Robertson, D.E., W.B. Silker, J.C. Langford, M.R. Peterson and R.W. Perkins. 1973. Transport and depletion of radionuclides in the Columbia River. Pages 141-158 in: *Radioactive Contamination of the Marine Environment, Proceedings of a Symposium*, Seattle, 1972. I.A.E.A., Vienna.

Roesijadi, G., J.W. Anderson and J.W. Blaylock. 1978. Uptake of hydrocarbons from marine sediments contaminated with Prudhoe Bay crude oil: influence of feeding, type of test specie and availability of PAH. *J. Fish. Res. Board Can.*, 35: 608-614.

Rowe, G.T., P.T. Polloni and R.L. Haedrich. 1982. The deep-sea macrobenthos on the continental margin of the northwest Atlantic Ocean. *Deep-Sea Res.*, 29: 257-278.

Rubenstein, N.I., W.T. Gillian and N.R. Gregory. 1984. Dietary accumulation of PCBs from a contaminated sediment source by a demersal fish (*Leiostomus xanthurus*). *Aquat. Toxicol.*, 5: 331-342.

Sanders, J.G. 1980. Arsenic cycling in marine systems. *Mar. Environ. Res.*, 3: 257-266.

Sanders, J.G. 1986. Direct and indirect effects of arsenic on the survival and fecundity of estuarine zooplankton. *Can. J. Fish. Aquat. Sci.*, 43: 694-699.

Santschi, P.H., Y.H. Li, D.M. Adler, M. Amourer, J. Bell and U. Nyffeler. 1983. The relative mobility of natural (Th, Pb and Po) and fallout (Pu, Am, Cs) radionuclides in the coastal marine environment: results from model ecosystems (MERL) and Narragansett Bay. *Geochim. Cosmochim. Acta*, 47: 201-210.

Schell, W.R. 1977. Concentrations, physico-chemical states and mean residence times of ^{210}Pb and ^{210}Po in marine and estuarine waters. *Geochim. Cosmochim. Acta*, 41: 1019-1031.

Schell, W.R. and A. Nevissi. 1977. Heavy metals from waste disposal in central Puget Sound. *Environ. Sci. Technol.*, 11: 887-893.

Schumacher, J.D., C.A. Pearson, R.L. Charnell and N.P. Laird. 1978. Regional response to forcing in the southern Strait of Georgia. *Est. Coast. Mar. Sci.*, 7: 79-91.

Scott, M.R., P.F. Salter and J.E. Halverson. 1983. Transport and deposition of plutonium in the ocean: evidence from Gulf of Mexico sediments. *Earth Plan. Sci. Lett.*, 63: 202-222.

Seelye, J.G., R.J. Hesselberg and M.J. Mac. 1982. Accumulation by fish of contaminants released from dredged sediments. *Environ. Sci. Technol.*, 16: 459-464.

Seymour, A.H. and V.A. Nelson. 1973. Decline of ^{65}Zn in marine mussels following the shut-down of Hanford reactors. Pages 277-286 in: *Radioactive Contamination of the MarineEnvironment, Proceedings of a Symposium*, Seattle, 1972. I.A.E.A., Vienna.

Sherwood, C.R., J.S. Creager, E.H. Roy, G. Gelfenbaum and T. Dempey. 1984. Sedimentary processes and environments in the Columbia River Estuary. Final Report to Columbia River Estuary Data Development Program, CREST, Astoria, Ore., 318 pp.

Shokes, R.F. 1976. Rate dependent distributions of ^{210}Pb and interstitial sulfate in sediments of the Mississippi River delta. Ph.D. Diss., Dept. of Oceanography, Texas A&M Univ., 122 pp.

Sholkovitz, E.R., J.K. Cochran and A.E. Carey. 1983. Laboratory studies of the diagenesis and mobility of $^{239,240}Pu$ and ^{137}Cs in nearshore sediments. *Geochim. Cosmochim. Acta*, 47: 1369-1379.

Sholkovitz, E.R. and D.R. Mann. 1984. The pore water geochemistry of $^{239,240}Pb$ and ^{137}Cs in sediments of Buzzards Bay, Massachusetts. *Geochim. Cosmochim. Acta*, 49: 1107-1114.

Shuman, F.R. 1978. The fate of phytoplankton chlorophyll in the euphotic zon – Washington coastal waters. Ph.D. Diss., Dept. of Oceanogr., Univ. of Washington, Seattle,Wa., 243 pp.

Silker, W.B. 1964. Variations in elemental concentrations in the Columbia River. *Limnol. Oceanogr.*, 9: 540-545.

Smethie, W.M. 1979. An investigation of vertical mixing rates in fjords using naturally occurring ^{222}Rn and salinity as tracers. Ph.D. Diss., Dept. of Oceanogr., Univ. of Washington, Seattle, Wa., 247 pp.

Smethie, W.M., C.A. Nittrouer and R.F.L. Self. 1981. The use of ^{222}Rn as a tracer of sediment irrigation and mixing on the Washington continental shelf. *Mar. Geol.*, 42: 173-200.

Smith, J.N., B.P. Boudreau and V. Noshkin. 1986. Plutonium and ^{210}Pb distributions in northeast Atlantic sediments: subsurface anomalies caused by non-local mixing. *Earth Plan. Sci. Lett.*, 81: 15-28.

Smith, J.D. and T.S. Hopkins. 1972. Sediment transport on the continental shelf off Washington and Oregon in light of recent current measurements. Pages 143-180 in: J. D. Swift, D.B. Duane and O. H. Pilkey (eds.), *Shelf Sediment Transport*. Dowden, Hutchison and Ross, Sroudsburg, Penn.

Smith, R.A., R.B. Alexander and M.G. Wolman. 1987. Water quality trends in the nation's rivers. *Scienc,e* 235: 1607-1615.

Socha, S.B. and R. Carpenter. 1987. Factors affecting pore water hydrocarbon concentrations in Puget Sound sediments. *Geochim. Cosmochim. Acta*, 51: 1273-1284.

Spangler, W.J., J.L. Spigarelli, J.M. Rose and H.M. Miller. 1973. Methylmercury: bacterial degradation in lake sediments. *Science*, 180: 192-193.

Steen, W.C. and S.W. Karickhoff. 1981. Biosorption of hydrophobic organic pollutants by mixed microbial populations. *Chemosphere*, 10: 27-32.

Sternberg, R.W. and D.A. McManus. 1972. Implications of sediment dispersal from long-term bottom-current measurements on the continental shelf of Washington. Pages 143-180 in: J. D. Swift, D.B. Duane and O. H. Pilkey (eds.), *Shelf Sediment Transport*. Dowden, Hutchison and Ross, Sroudsburg, Penn.

Stordal, M.C., J.W. Johnson, N.L. Guinasso and D.R. Schink. 1985. Quantitative evaluation of bioturbation rates in deep ocean sediments. II. Comparison of rates determined by ^{210}Pb and ^{239,240}Pb. *Mar. Chem.*, 17: 99-114.

Stukas, V.J. and C.S. Wong. 1981. Stable lead isotopes as a tracer in coastal waters. *Science,* 211: 1424-1427.

Suess, E. 1980. Particulate organic carbon flux in the oceans – surface productivity and oxygen utilization. *Nature,* 288: 260-263.

Swanson, J.R. 1980. Carbon isotope analysis of carbonaceous compounds in Puget Sound and Lake Washington. Ph.D. Diss., Dept. of Chem., Univ. of Washington, Seattle,Wa., 217 pp.

Tebo, B.M., K.H. Nealson, S. Emerson and L. Jacobs. 1984. Microbial mediation and Mn(II) and Co(II) precipitation at the O_2/H_2S interfaces in two anoxic fjords. *Limnol. Oceanogr.*, 29: 1247-1258.

Thorbjarnarson, K.W., C.A. Nittrouer and D.J. DeMaster. 1986. Accumulations of modern sediment in Quinault submarine canyon. *Mar. Geol.*, 71: 107-124.

Tivey, M.K. and J.R. Delaney. 1985. Sulfide deposits from the Endeavour segment of the Juan de Fuca Ridge. *Mar. Mining*, 5: 165-179.

Turekian, K.K. and H. Wedepohl. 1961. Distribution of the elements in some major units of the earths crust. *Geol. Soc .Amer. Bul.*, 72: 175-192.

Turekian, K.K., G.J. Benoit and L.K. Benninger. 1980. The mean residence time of plankton-derived carbon in a Long Island Sound sediment core: a correction. *Est. Coast. Mar. Sci.*, 11: 583.

Twiss, E.C., R. Carpenter and W.R. Schell. 1973. ^{210}Po in sediments off the Washington coast. *EOS, Trans. Am. Geophys. Union*, 54: 341.

U.S. Geological Survey. 1970. Mercury in the environment. U.S. Geological Survey Prof. Paper 713, 67 pp.

Vairavamurthy, A., M.O. Andreae and R.C. Iverson. 1985. Biosynthesis of dimethylsulfide and dimethylpropiothetin by *Hymenomonas carterae* in relation to sulfur source and salinity variations. *Limnol. Oceanogr.*, 30: 59-70.

Varanasi, U., W.L. Reichert, J.E. Stein, D.W. Brown and H.R. Sanborn. 1985. Bioavailability and biotransformation of aromatic hydrocarbons in benthic organisms exposed to sediment from an urban estuary. *Environ. Sci. Technol.*, 19: 836-841.

Wahlgren, M.A., K.A. Orlandini and D.M. Nelson. 1978. Determination of plutonium oxidation state in large volumes of natural water. Pages 74-76 in: *Radiological and Environmental Research Division Annual Report.* Argonne National Laboratory ANL-78-65, Part III.

Wakeham, S.G. and R. Carpenter. 1976. Aliphatic hydrocarbons in sediments of Lake Washington. *Limnol. Oceanogr.*, 21: 717-723.

Walsh, J.J. 1984. The role of ocean biota in accelerated ecological cycles: a temporal view. *Bioscience,* 34: 499-507.

Walsh, J.J., E.T. Premuzic, J.S. Gaffney, G.T. Rowe, G. Harbottle, W.L. Balsam, P.R. Petzer and S.A. Macko. 1985. Organic storage of CO_2 on the continental slope off the Mid-Atlantic Bight, the southeastern Bering Sea and the Peru coast. *Deep-Sea Res.*, 32: 853-883.

Walsh, J.J., G.T. Rowe, R.L. Iverson and C.P. McRoy. 1981. Biological export of shelf carbon is a sink of the global CO_2 cycle. *Nature*, 291: 196-201.

Wang, F. 1955. Recent sediments in Puget Sound and portions of Washington Sound and Lake Washington. Ph.D. Diss., Dept. of Oceanogr., Univ. of Washington, Seattle, Wa., 160 pp.

Warner, T.B. 1972. Mixing model prediction of fluoride distribution in Chesapeake Bay. *J. Geophys. Res.*, 77: 2728-2732.

Waslenchuk, D.G. 1978. The budget and geochemistry of arsenic in a continental shelf environment. *Mar. Chem.*, 7: 39-52.

Welschmeyer, N.A. 1982. The dynamics of phytoplankton pigments: implications for zooplankton grazing and phytoplankton growth. Ph.D. Diss., Dept. of Oceanogr., Univ. of Washington, Seattle, Wa., 176 pp.

Wheat, C.G. and R. Carpenter. 1988. Determinations of $Mn\text{-}Cl^{++}$ and $Mn\text{-}SO_4$ association constants to 170° C. *J. Sol. Chem.*, 17: 467-480.

Wiebe, W.J. and J. Liston. 1972. Studies of the aerobic, nonexacting, heterotrophic bacteria of the benthos. Pages 281-312 in: A.T. Pruter and D.L. Alverson (eds.),*The Columbia River Estuary and Adjacent Ocean Waters: Bioenvironmental Studies*. Univ. of Washington Press, Seattle, Wa.

Wijayaratne, R.D. and J.C. Means. 1984. Sorption of polycyclic aromatic hydrocarbons by natural estuarine colloids. *Mar. Environ. Res.*, 11: 77-89.

Williams, P.A. and T.W. Linck. 1975. Cycling of organic carbon in the ocean: use of naturally occurring radiocarbon as a short term tracer. Pages 153-166 in: *Isotope Ratios as Pollutant Source and Behavior Indicators*. International Atomic Energy Agency, Vienna, pp: 153-166.

Williams, P.M., J.A. McGowan and M. Stuiver. 1970. Bomb ^{14}C in deep-sea organisms. *Nature*, 227: 375-376.

Williams, P.M., E.R.M. Druffel and K.L. Smith. 1987. Dietary carbon sources for deep-sea organisms as inferred from their organic radiocarbon activities. *Deep-Sea Res.*, 34: 253-266.

Willis, J.N. and W.G. Sunda. 1984. Relative contributions of food and water in the accumulation of zinc by two species of marine fish. *Mar. Biol.*, 80: 273-279.

Windom, H.L. 1971. Fluoride concentrations in coastal and estuarine waters of Georgia. *Limnol. Oceanogr.*, 16: 806-810.

Windom, H.L. 1972. Arsenic, cadmium, copper, lead, mercury and zinc in marine biota--North Atlantic Ocean. Pages 121-147 in: *Baseline Studies of Pollutants in the Marine Environment*. Intl. Decade of Oceanogr., Brookhaven National Laboratory,.

Windom, H.L., R. Stickney, R. Smith, D. White and F. Taylor. 1973. Arsenic, cadmium, copper, mercury and zinc in some species of North Atlantic finfish. *J. Fish. Res. Board Can.*, 30: 275-279.

Winter, D.F., K. Banse and G.C. Anderson. 1975. The dynamics of phytoplankton blooms in Puget Sound, a fjord in the northwestern United States. *Mar. Biol.*, 29: 139-176.

Wisegarver, D.P. and J.D. Cline. 1985. Solubility of trichlorofluoromethane (F-11) and dichlorodifluoromethane (F-12) in seawater and its relationship to surface concentrations in the North Pacific. *Deep-Sea Res.*, 32: 97-106.

Wong, C.S., E. Boyle, K.W. Bruland, J.D. Burton and E.D. Goldberg. 1983. Trace metals in sea water. Plenum Press, New York, N.Y., 920 pp.

Woodwell, G.M., C.F. Wurster and P.A. Isaacson. 1967. DDT residences in an East Coast estuary: a case of biological concentration of a persistent insecticide. *Science*, 156: 821-824.

Word, J.Q., C.C. Ebbesmeyer, C.D. Boatman, R.E. Finger, S. Fischnaller and Q.J. Stober. In press. The vertical transport of effluent material to the surface of marine waters. in: I. Duedall and D. Baumgarner (eds.), *Proceedings of Fifth International Ocean Disposal Symposium*, Corvallis, Ore.

Yang, A.I.C. 1971. Variations of natural radiocarbon during the last 11 millennia and geophysical mechanisms for producing them. Ph.D. Diss., Dept. of Chem., Univ. of Washington, Seattle, Wa., 102 pp.

Young, J.A. and W.B. Silker. 1974. The determination of air-sea exchange and ocean mixing rate using ^{7}Be during the Bomex experiment. *J. Geophys. Res.*, 79: 4481-4489.

Young, J.A. and W.B. Silker. 1980. Aerosol deposition velocities on the Pacific and Atlantic Ocean calculated from ^{7}Be measurements. *Earth Plan. Sci. Lett.*, 50: 92-104.

CHAPTER 11

OCEANOGRAPHIC INFLUENCES ON THE DYNAMICS OF COMMERCIALLY FISHED POPULATIONS

L. W. Botsford[1]*, D. A. Armstrong*[2] *and J. M. Shenker*[3]
[1] *Department of Wildlife and Fisheries Biology, University of California, Davis, CA 95616*
[2] *School of Fisheries WH-10, University of Washington, Seattle, WA 98195*
[3] *University of California, Bodega Marine Laboratory, P.O. Box 247, Bodega Bay, CA 94923*

11.1 INTRODUCTION

In the U.S. Pacific Northwest, the coastal ocean supports a number of valuable fisheries that are influenced by oceanographic conditions. Total fishery landings along Washington, Oregon and California in 1986 were 312 metric tons valued at 312 million dollars. Abundance in these fisheries varies from year to year by as much as a factor of ten. Over the past 40 y there has been an increasing realization that fisheries management might be improved by better understanding of oceanographic influences on fluctuations in fish abundance (e.g., Anon., 1978; Parsons *et al.*, 1978; Wooster, 1983; Cushing, 1982; May. 1984). Consequently, there has been an increasing focus on oceanography in the fisheries research conducted by national and international agencies (Anon., 1985; Bakun *et al.*, 1982; Sherman *et al.*, 1983). However, we are not to the point at which information on oceanographic influences on fish abundance is used directly in the management process.

Previous chapters in this book have described increases in synoptic data and advances in understanding of oceanographic processes in the coastal zone off Washington, Oregon and Northern California. This chapter is intended to give oceanographers some idea of how such information can be used to better understand the dynamic behavior of exploited populations. We do this through three examples from the Pacific Northwest: an anadromous fish, coho salmon (*Oncorhynchus kisutch*), a benthic flatfish, English sole (*Parophrys vetulus*), that spawns in nearshore areas and has pelagic larvae and a benthic crustacean, Dungeness crab (*Cancer magister*) that also has pelagic larvae and utilizes near-shore and estuarine areas. Because the dependence of these species on oceanographic conditions is only partially understood, we are limited to presenting an incomplete picture (i.e., lines of evidence for competing hypotheses and statistical relationships with conflicting explanations). As an aid to assessing the implications of these various pieces of evidence, we preface the examples with a brief introduction to the approaches researchers have taken to understanding environmental influences on exploited marine populations.

11.2 PROBLEMS, APPROACHES AND MECHANISMS

Evaluating the effect of oceanographic processes on abundance of fish populations typically follows one of two basic approaches: statistical analyses of long term interannual variability, or mechanistic analyses of mathematical models and field data. Effective integration of these approaches considers the appropriate temporal and spatial scales and ecological levels of organization. We briefly outline here the philosophy behind this process. As further background for the three examples, we also review mechanisms by which oceanographic conditions have been shown to affect fish populations.

11.2.1 Approaches

The statistical approach to understanding the effects of environment on exploited populations seeks statistical relationships between annual population data and environmental variables. It typically involves using correlation coefficients or linear regression to determine the degree of covariation between time series of annual measures of physical and population variables. The population variables are often indirect, noisy measures of the true variable of interest. For example, a catch record is distorted by fluctuations in fishing effort and is only a weighted average of the variable of usual interest, recruitment. Similarly, the physical oceanographic variables used are usually indirect indicators of a variety of oceanographic processes. As examples, changes in sea surface temperature reflect several physical phenomena such as upwelling and longshore flow and a variable such as upwelling index is inferred from atmospheric pressure measurements.

Interpretation of results obtained by the statistical approach requires close attention to the difficulties involved in demonstrating that the detected relationships did not arise from chance alone. The first problem, that intra-series correlation decreases the effective number of degrees of freedom, is probably more familiar to physical oceanographers (Davis, 1976; Chelton, 1983) than to biologists, but has recently become more widely appreciated (Chelton, 1984; Botsford, 1987; Botsford and Wainwright, unpublished). A second statistical problem is that examination of all of the many time series of environmental variables that could influence the population can easily lead to spurious correlations. In some cases one is unable to discriminate between two or more causes that provide equally good 'fits' to the available data. A third problem is lack of stationarity. For example, the time series could include a point at which a substantial, long term oceanographic change occurred (e.g., as has been proposed for central California near 1958, Namais and Huang 1972) or an unusually strong year class was recruited (e.g., the 1960-61 y class of English sole and several other species, Section 11.4.2).

The mechanistic approach attempts to identify the actual causal mechanism through which the environment influences a fished population. Rather than inductive, statistical comparisons using arbitrary linear models, realistic models are used to test hypothetical physical and biological mechanisms on a hypothetico-deductive basis (cf. Platt, 1964). Information on proposed mechanisms, such as bottom temperatures decreasing reproduction or surface currents transporting larvae to unsuitable settlement areas, is obtained through direct sampling and incorporated in a population model. Those mechanisms that do not produce the observed catch series are rejected. One major problem with this hypothetico-deductive approach is that results can be "over-interpreted" when a model mechanism that reproduces observed behavior is accepted as the mechanism rather than one of many possible mechanisms. Conversely, the true mechanism can be falsely rejected because a model which is incorrect in other aspects does not produce the observed behavior (e.g., Botsford, 1981b).

11.2.2 Temporal and Spatial Scales

Evidence from the two approaches can be integrated by focusing on the appropriate scales of variability at each of the important levels of integration. Analyses and sampling must both be constrained to appropriate temporal and spatial scales in population studies in general (e.g., Levin, 1987) and, in particular, the study of marine fish populations (Smith, 1978; Walsh, 1978; Steele, 1984) and their food (Steele, 1978; Horne and Platt, 1984). The lowest scale of interest, the maximum rate of variability, determines the sampling rate and the largest scale of interest, the extent of the phenomenon, determines the extent of the sampling. A program of sampling and analysis that covered all scales of possible importance to fish population dynamics would require an almost-impossible amount of effort.

Different temporal and spatial scales are relevant at different levels of ecological organization (the organism, population, community and ecosystem) (cf. Steele, 1984; Rothschild, 1986). The level of direct interest in fisheries is the population level, which involves variables such as biomass, numbers and age structure. However, because behavior at one level depends heavily on interactions at the next lower level, the individual level is also important (and for some fish, the school level must be considered, Smith, 1978). The individual level involves growth, reproduction and mortality. Although these processes are important throughout life, we focus here on survival through the younger stages, the larval and post-metamorphosis stages (or smolt stages in salmon), because these are the stages at which year class abundance is typically determined and oceanographic conditions have greatest effect. We refer to these early processes as recruitment.

At the population level, the fundamental lower limit of temporal variability at temperate latitudes is typically one year. Recruitment of the young is an annual event for most populations at these latitudes. The upper end of the time scale depends on the population's age structure or demographics (e.g., Horwood and Shepherd, 1981; Reed, 1983; Botsford, 1986b). For example, because catch is the sum of several year classes, high frequency variability in recruitment is usually "averaged out" and hence does not appear in the catch record. The degree to which this occurs depends on individual life history, with longer lived species showing less variability in catch than shorter lived species. While this averaging effect is widely appreciated among fishery scientists, the fact that resonance phenomena can also arise (e.g., from density-dependent recruitment, Botsford, 1986b; 1987) is not. As an example, the time scales of variability in recruitment survival that are important to Dungeness crab populations are on the order of 5 to 10 y depending on exploitation rate, width of a year-class' size distribution and degree of density-dependent recruitment (Fig. 11.1). The time scales over which these processes take place are on the order of the inverse of the mortality rate for averaging and twice the life span of the fish for resonance, a range that includes time scales of important interannual variability in oceanographic conditions in the California Current (e.g., ENSO events, coastal winds). In addition to these effects, populations also respond to long-term, environmentally-induced changes in recruitment survival.

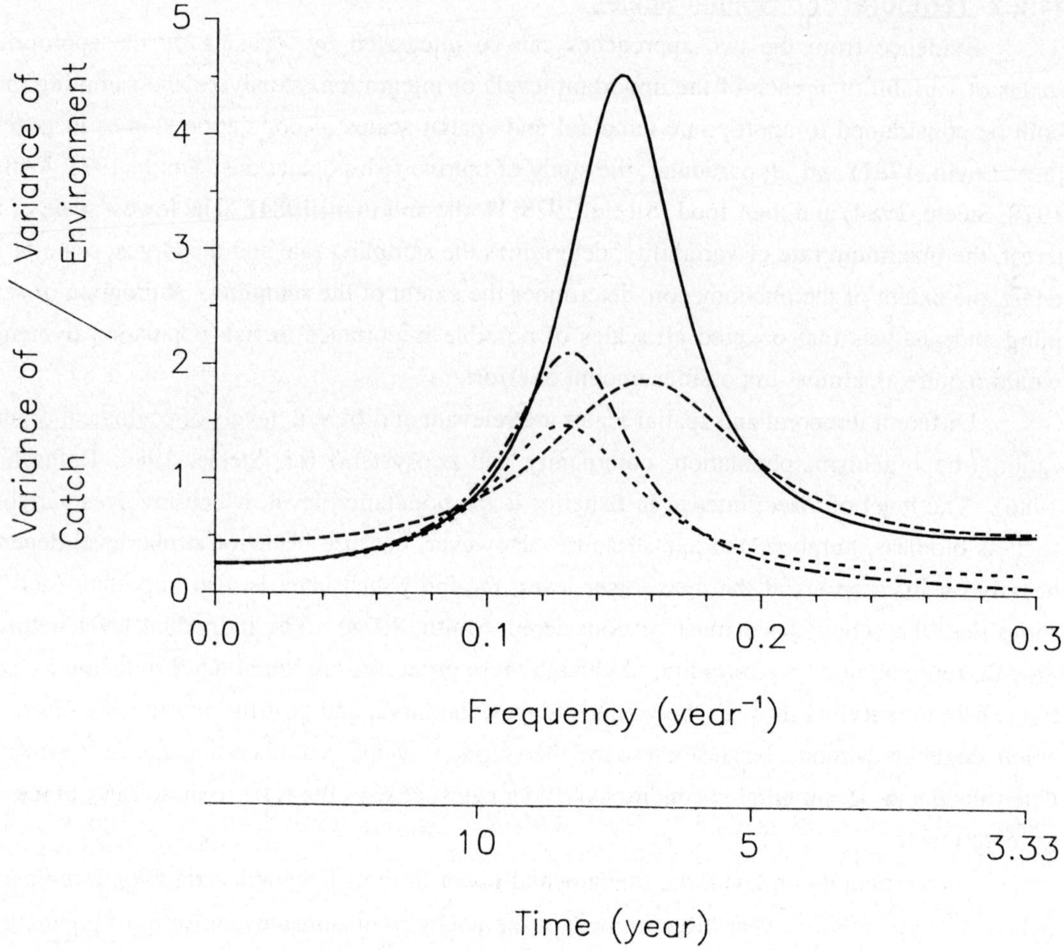

Fig. 11.1 An example of the relative response of recruitment in a population with age structure and density-dependent recruitment to environmentally-induced variability in recruitment survival for (———) a baseline model with harvest rate = 0.8, year classes completely entering the fisheries in one year, and a moderate change in recruitment survival with slope = −2.0, (— · — ·) year classes growing into the fishery over 4 years, (- - - - -) a harvest rate of 0.5 and (— — —) more positive slope of recruitment survival. For this population, Dungeness crab, oceanographic variability on time scales of 5 to 10 years are most important. Redrawn from Botsford (1986b).

The fundamental lower limit of spatial scales at the population level is usually on the order of kilometers. Populations (though not necessarily individuals), can usually integrate over variations in physical environment and food availability at shorter scales because they move over those distances easily. However, adult fish sometimes aggregate to feed at sharp boundaries along oceanic fronts (e.g., albacore; Laurs and Lynn, 1977). The upper limit of spatial variability at the population level is determined by the extent of individual breeding stocks of fish, the fundamental unit for management.

The actual mechanisms by which the oceanographic environment affects populations involve interactions at the individual level. The annual variability in recruitment referred to above is the biological response of planktonic fish and crustacean larvae or young juveniles, their food

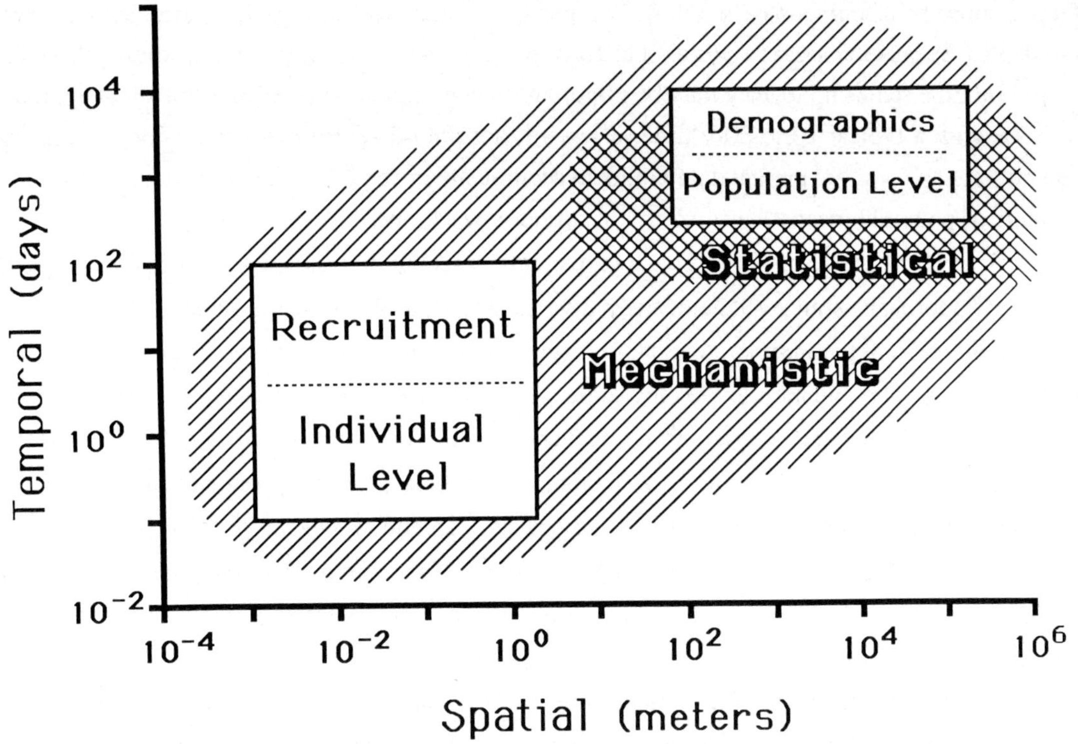

Fig. 11.2 A schematic representation of scales of importance to recruitment processes (individual level) and population level processes for exploited marine populations. The statistical approach is effective only at longer scales, but does not completely cover the scales of interest at the population level. The mechanistic approach integrates information at scales appropriate to both levels of organization.

species and their predators to oceanographic variability. These occur at much shorter temporal and spatial scales. Because the issue of scale in plankton research is treated in other chapters and in recent publications (e.g., Steele, 1978; Denman and Powell, 1984), we do not discuss it in detail here.

The longer time and space scales important at the population level and the shorter scales important at the individual level, the level that governs recruitment, are summarized in Figure 11.2. For most populations in this region, variability in an annual event, recruitment, drives population-level phenomena at spatial scales on the order of kilometers. However, the mechanisms that actually determine recruitment each year occur at much shorter temporal and spatial scales, those of individual larvae.

11.2.3 Integration of Statistical and Mechanistic Approaches

Results reviewed in this chapter from both the statistical and mechanistic approaches can be integrated by considering the time scales and levels of organization appropriate to each. In the statistical approach, problems such as intra-series correlation and multiple environmental series rapidly deplete the available number of degrees of freedom so that one can confidently detect relationships on time scales that are only a small fraction of the length of the available time series. In annual time series of fisheries and oceanographic data in the California Current the effective number of degrees of freedom is typically a small fraction of the series length, about one-fifth. Because analyses such as that for the Dungeness crab (Fig. 11.1) indicate that variability on time scales up to 10 y may be important to commercial species, the statistical approach to cases with a typical 20 to 30-y time series will not provide definitive conclusions. Thus, we can either wait for more annual data points to accumulate or regard statistical results only as a indicators of potential mechanisms worthy of further investigation.

Use of the mechanistic approach to evaluate potential mechanisms identified through the statistical approach is an expedient alternative to waiting. Because the mechanistic approach links phenomena at the individual level, on faster (intra-annual) time scales, to the slower (inter-annually varying) population level, intra-annual sampling over much smaller time periods can be used to confidently reject inappropriate mechanisms. A few years of direct field or laboratory observations may allow rejection of mechanisms that would take many years to discard through the statistical approach. For example, if temperature-dependent egg production is proposed as the mechanism underlying a correlation between ocean temperature and abundance, direct sampling or laboratory studies may be used to show whether the required temperature response of egg production exists.

These considerations of important time scales of variability at the appropriate level of organization (Fig. 11.2) provide a consistent scheme for evaluating the evidence regarding oceanographic influences on recruitment of Dungeness crab, English sole and coho salmon in later sections. The statistical approach should be regarded as exploratory data analysis and results only as suggestions of possible mechanisms. Proposed mechanisms can be confidently evaluated through direct sampling of physical and biological interactions at the individual level and translation of that information to the population level through realistic, mechanistic models (See Ricker, 1975; Bakun and Parrish, 1980; Bakun *et al.*, 1982; and Sissenwine, 1984 for other views of these approaches and Tukey, 1977 and Velleman and Hoaglin, 1981 for discussion of exploratory data analysis.).

11.2.4 Mechanisms Linking Physical and Biological Processes

Most studies of oceanographic influences on fish populations focus on recruitment, although lack of attention to possible effects on other stages (e.g. on adult mortality) is seldom justified. Ideally, before embarking on a study of environmental effects on a population, the age at which the greatest amount of variability in annual mortality occurs should be determined through a statistical comparison of abundance at adjacent stages in a cohort. There is typically a high correlation between all pairs of stages except one, the "critical stage" at which cohort strength is formed. That stage can then be the focus of oceanographic studies.

Research on the determinants of cohort abundance in fish began with the work of Hjort (1914) who proposed that successful recruitment was determined by two conditions: (1) the availability of sufficient food for larvae after the contents of their yolk sac had been consumed and (2) transport of larvae to areas of favorable physical and biological conditions.

Oceanographic conditions affect primary productivity and can therefore influence the food available to larval and early juvenile stages of commercial species, affecting their survival directly (starvation) and indirectly (lower growth rate, hence longer exposure to high predation rate; Harris, 1975). Among fishes, starvation has been observed in nature in northern anchovy (O'Connell, 1980), jack mackerel (Theilacker, 1978; 1981) and sablefish (Grover and Olla, 1986). Larval starvation has also been proposed as the cause of fluctuations in recruitment of Tanner crab (*Chionoecetes opilio* and *C. bairdi;* Incze *et al.*, in review), red king crab (*Paralithodes camchatica*; Paul and Paul, 1980), Dungeness crab (Lough, 1976) and other crabs (Dawirs, 1984; Paul *et al.*, 1979).

Because adequate food is required at the time when larval fish are ready to begin feeding, variations in cohort abundance of temperate latitude fishes could be caused by variability in the relative timing of spawning and production of larval food (Cushing, 1975). According to Cushing's "match/mismatch" hypothesis, relatively small variations in the timing of seasonal peaks in the productivity of food species can have a dramatic effect on larval survival (but see Methot, 1983 for a counter example).

In addition, food concentration must be high enough for efficient consumption and net energy gain. Laboratory studies of larval fish feeding (e.g., Houde, 1975; 1977; 1978) indicate that early larvae usually require denser prey concentrations than have been measured in field studies. Lasker (1975, 1978) and Lasker and Zweifel (1978) proposed that production of prey densities adequate for successful first feeding of larval northern anchovy required relatively calm ocean conditions where dense chlorophyll layers could develop. Poor larval survival was attributed to wind-driven mixing and breakdown of these layers. Subsequent interannual comparisons of larval survival and physical conditions (Lasker, 1981; Peterman and Bradford, 1987a) and mechanistic laboratory and field studies have contributed to the elaboration of this hypothesis (feeding studies: Hunter, 1977; feeding models: Vlymen, 1977; Wroblewski, 1984; verification of starvation: Theilacker, 1978; characterization of intra-annual survival: Methot, 1983; and growth and biochemical composition: Methot and Kramer, 1979).

Sampling on extremely fine spatial scales may be necessary to understand the relationship between oceanographic conditions, prey patchiness and larval feeding and survival (Mullin *et al.*, 1985; Owen, 1981). In addition to turbulence and stratification, other oceanographic phenomena leading to the development of dense prey concentrations include convergence zones along oceanic fronts, Langmuir cells, internal waves and topographically-generated eddies (Alldredge, 1982; Zeldis and Jillett, 1982; Legendre and Demers, 1984; Denman and Powell, 1984).

Because planktonic fish larvae are relatively immotile, transport can profoundly affect year class strength by determining whether larvae are transported to favorable juvenile rearing areas. Larval transport is potentially very important in the Pacific Northwest, but difficult to elucidate because of the highly variable, short-scale nature of nearshore currents. For example, the existence of cross-shelf jets and offshore eddies (Hickey, 1989, Chapter 2; Huyer *et al.*, 1984;

Mooers and Robinson, 1984; Davis, 1985) imply much more complex transport mechanisms than those based on simple two-dimensional upwelling.

There have been numerous studies of the impact of horizontal transport on recruitment (Norcross and Shaw, 1984). A seminal example is Walford's (1938) study of the influence of circulation in the Gulf of Maine on haddock larvae. In the northeast Pacific, both statistical comparisons of upwelling and year class success and measured larval distributions indicate that recruitment of Pacific whiting, *Merluccius productus*, depends on retention of larvae in near-shore areas (Bailey, 1981). Recent studies of capelin and herring larvae in Canada demonstrate that the complex interactions between larval transport and feeding processes can often only be elucidated through direct sampling (Frank and Leggett, 1983a,b; Fortier and Leggett, 1983). Among crustaceans, transport is thought to be important in blue crab, *Callinectes sapidus*, off Chesapeake Bay (Johnson *et al.*, 1984), penaeid shrimp in the Gulf of Carpentaria (Rothlisberg, 1982, Rothlisberg *et al.*, 1983), the edible crab, *Cancer pagurus*, off the east coast of England (Nichols *et al.*, 1982), pandalid shrimp in the Pacific Northwest (Rothlisberg and Miller, 1983) and the rock lobster (*Panulirus cygnus*) off western Australia (Phillips, 1986).

The fact that larvae occur at different depths and can vary their depth through vertical migration has long been known (Russell, 1925; Ahlstrom, 1959). That larvae could "use" this ability to control the horizontal advection resulting from ocean currents was proposed by Hardy (1935). In the northeast Pacific this effect is important in relation to upwelling. That larvae may use vertical migration to move between the Ekman layer and the return (onshore) flow for the purposes of cross-shelf movement underscores the importance of new information regarding the vertical distribution of this return flow (Hickey, 1989, Chapter 2). It has been implicated as part of a mechanism by which plankton avoid offshore Ekman transport at the surface and maintain a distribution close to shore in the spring (Johnson and Gonor, 1982; Peterson *et al.*, 1979; Peterson, 1980; Wroblewski, 1980).

Complex interactions between vertical migration and horizontal advection are often associated with estuaries, which serve as nursery grounds for a variety of marine species (Sulkin, 1984). Vertical migration is used by larvae to select either the seaward outflow at the surface or the inflow near the bottom for the purposes of entering, leaving, or remaining in estuaries (Creutzberg, 1961; Sandifer, 1975; Sulkin and Van Heukelem, 1982; Epifanio *et al.*, 1984; Staples and Vance, 1985; Cronin and Forward, 1986; Sulkin and Epifanio, 1986). This is important in the U.S. Pacific Northwest because of the large estuaries (e.g,. Gray's Harbor and Willapa Bay) and substantial freshwater flows (e.g. Columbia River). Both Dungeness crab and English sole rear in estuaries (see Sections 11.3 and 11.4).

Other specific physical oceanographic phenomena have been implicated in larval transport. The potential impacts of convergence, divergence and upwelling on transport of coastal zone species are described in Bakun and Parrish (1980) and Parrish *et al.* (1981). Similar to the match/mismatch hypothesis for food, timing is also critical for transport processes. This has been demonstrated indirectly by comparisons which show that fish in major upwelling regions tend to spawn in areas and at times with low offshore transport and low turbulence (Bakun and

Parrish, 1982; Parrish *et al.*, 1983). Larvae have also been observed concentrated in windrows which may result from Langmuir circulation. Another intriguing larval transport mechanism is baroclinic currents caused as internal tidal waves progress shoreward across the continental shelf in southern California (Shanks, 1983; 1985).

In addition to food and transport, predation has a major impact on recruitment (Hunter, 1981; Sissenwine, 1984). However, it is difficult to demonstrate that it is the actual cause of a specific variation in recruitment. Cannibalism has been evaluated as a cause of recruitment fluctuations for several species (Dungeness crab – Botsford, 1986a; northern anchovy – Hunter and Kimbrell, 1980) and inter-species predation has been considered for others (e.g., Pacific cod on herring – Walters *et al.* ,1986; chinook and coho salmon on Dungeness crab larvae – Botsford *et al.*, 1982; Thomas, 1985). However, analysis of a predation hypothesis typically depends on demonstrating the appropriate covariation between abundances of the prey and predator. As has been noted by Frank and Leggett (1985), it is difficult to draw firm conclusions from this type of data.

Although measured oceanographic variables such as temperature and salinity generally come into play as proxies indicating mass transport or mixing, they can also have a direct effect on larval growth and metabolism. For example, temperature has been proposed to affect growth and spawning of English sole (see Section 11.4) and spawning of Dungeness crab (see Section 11.3).

Lastly, although we have focused on the effects of the physical environment on recruitment (i.e., the early life stages), environmental factors also affect juvenile and adult fish. The simplest effect is a shift in distribution, such as the northward shift of several species during the 1957 ENSO event (Radovich, 1961). There may also be an effect on migration routes, such as the Fraser River sockeye salmon (*Oncorhynchus nerka*) migrating around the north end of Vancouver Island during ENSO years rather than around the south end (Wickett, 1960). Moreover, oceanographic influences on food distribution can influence distribution and growth of fish. Prey of albacore tuna (*Thunnus alalunga*), for example, congregate near oceanic thermal fronts, enabling fishermen to use satellite and other sea surface temperature data to focus fishing effort (Laurs and Lynn, 1977). Lower weight and survival of chinook (*Oncorhynchus tsawytscha*) and coho salmon (*O. kisutch*) during the 1983-83 ENSO event (Pearcy *et al.*, 1985; Johnson, 1988) may have been caused by shifts in prey distribution. Food availability has also been shown to affect the weight of pink salmon (*O. gorbuscha*) off British Columbia (Peterman, 1987).

11.3 DUNGENESS CRAB

Dungeness crab (*Cancer magister*) are found in fishable abundance from central California through the eastern Aleutians (Meltef, 1985) and they occur as far north as the Pribilof Islands, southeastern Bering Sea (Jensen and Armstrong, 1987). The bulk of the fishery occurs near-shore in less than 50-m depth. Only males greater than about 160 mm carapace width (variable by state) are legally fished, except in British Columbia where females may be

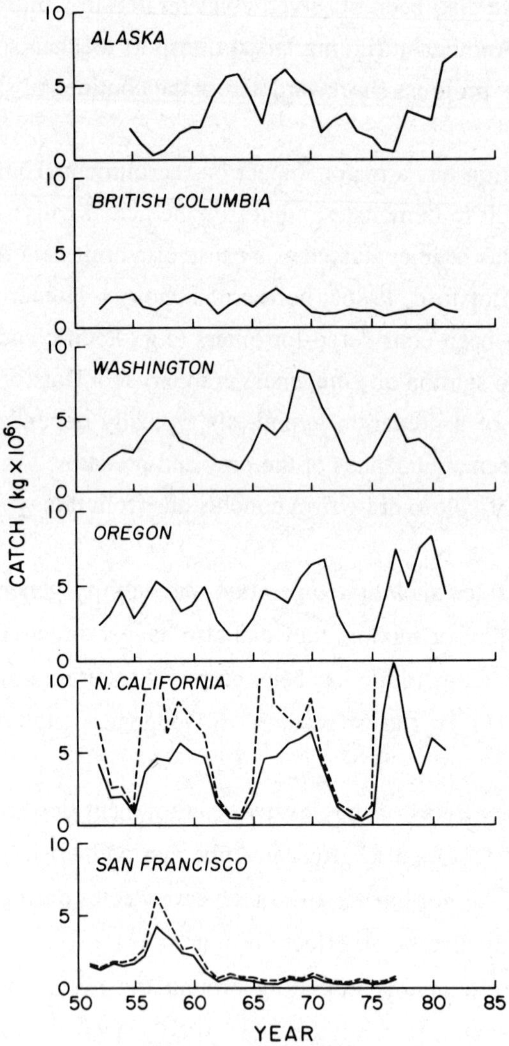

Fig. 11.3 Dungeness crab catch records (solid lines) with estimates of preseason abundance (dashed lines) from Methot and Botsford (1982). From Botsford (1986a).

taken (Methot, 1986; Jamieson, 1986). The striking synchronous cycles in catch of period 9 to 10 y from northern California through Washington (Fig. 11.3) make Dungeness crab an appealing model for study of the effects of oceanographic processes on recruitment (see reviews by Armstrong, 1983; Hankin, 1985; Botsford, 1986a). These cycles may exist in the populations further north, yet may not be apparent in the catch record because of less intense fishing. The cycles were also apparent before the decline in the central California fishery, but diminished in both period and amplitude following the decline (Botsford *et al.*, 1982).

11.3.1 Life History of Dungeness Crab

Dungeness crabs spawn along the open coast; there is no evidence of reproductive activities in coastal estuaries including San Francisco and Humboldt Bay, California and Grays Harbor, Washington (Tasto, 1983; Stevens and Armstrong, 1984; 1985; Armstrong and Gunderson, 1985). Females molt to maturity and breed nearshore in the spring along Washington, Oregon and northern California. They carry sperm for about six months before extruding an egg mass which they carry on their abdomen for approximately three months of embryonic development. Larvae hatch in the winter primarily between December and February and progress through five larval (zoeal) stages between December and March (Fig. 11.4). There is some evidence that larvae occur progressively farther offshore during successive zoeal stages (Lough, 1976; Reilly, 1983) and they have been found as far as 200 km offshore. After five zoeal stages, larvae molt to the final pelagic stage (megalopa). Megalopae have been found concentrated far offshore (Jamieson and Phillips, in press; Jamieson *et al.*, in review) and later megalopal stages are found closer to shore (Hatfield, 1983; Jamieson and Phillips, in press). Megalopae settle nearshore in shallow coastal regions and estuaries (Reilly, 1983; Stevens and Armstrong, 1984; 1985).

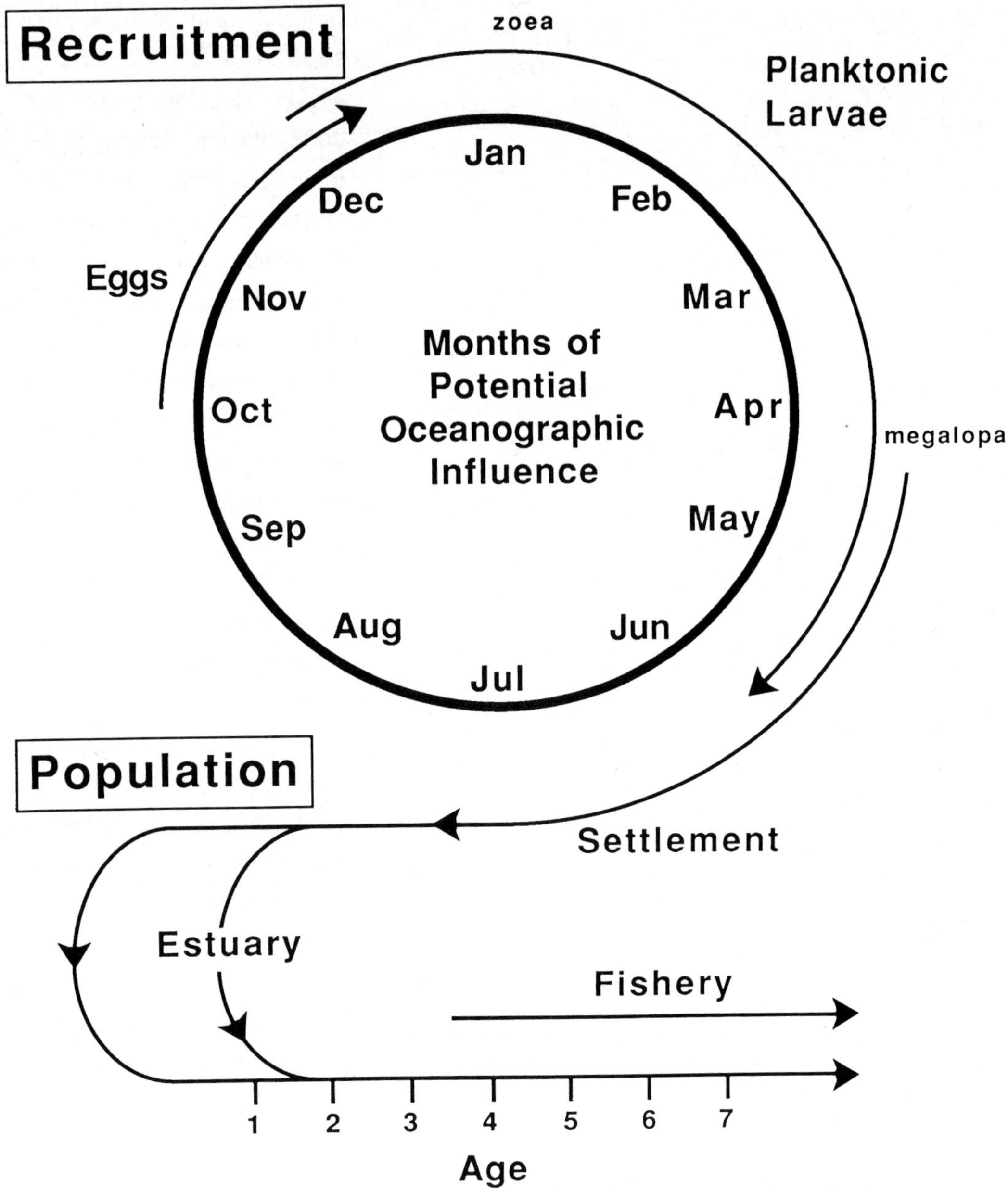

Fig. 11.4 Schematic of Dungeness crab life history showing the time of potential oceanographic influence on eggs and larvae. After the larval stage crabs may enter estuaries, but almost all crabs greater than age two are found nearshore on the coast.

Being nearshore at the time of settlement is apparently critical to year class success. Transport of the larvae depends on their vertical distribution which is poorly known. The megalopal stage has typically been associated with the surface more than the zoeal stages (Reilly, 1983; Booth *et al.*, 1985; and preliminary data from joint NMFS/USSR ichthyoplankton surveys over four years, unpublished). There is some evidence for diel vertical migration of megalopae with most in the neuston at night but as deep as 60 m on bright days (Booth *et al.*, 1985;

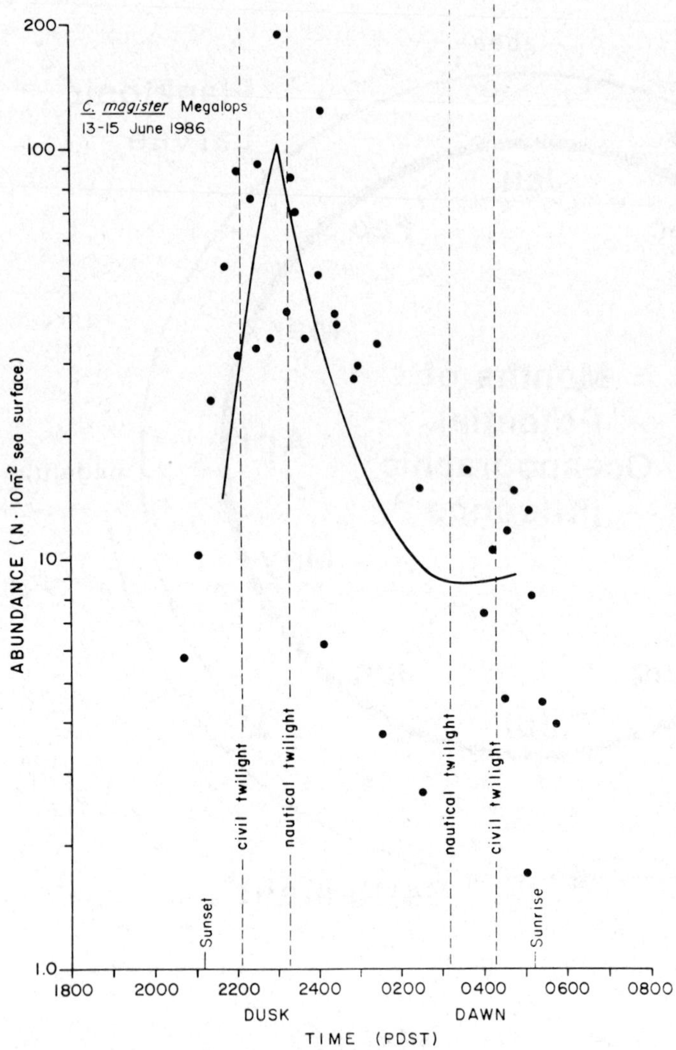

Fig. 11.5 Typical diel pattern of megalopal abundance in the neuston. This pattern is typical of several years of sampling off the coast of Vancouver Island. From Jamieson and Phillips (1988).

unpublished ichthyoplankton data; Jamieson and Phillips, in review; Jamieson *et al.*, in review; see Fig. 11.5). Megalopae have been seen during the day clinging to the pleustonic hydroid, *Velella velella* (Wickham, 1979b; Reilly, 1983; Shenker, in review) and other flotsam and jetsam and aggregating in windrows at the surface (Shenker, in review) on cloudy days. In laboratory studies, Jacoby (1982) found that the larval stages would move toward moderate light (less than the intensity at the surface on bright, sunny days at 37° N) and that they would move up when pressure was increased while they were in the dark.

Although megalopae enter estuaries in high abundance (Stevens and Armstrong, 1984; Armstrong and Gunderson, 1985, Pauley *et al.*, 1986), the fraction that settle in estuaries as opposed to the open coast has only recently been investigated (Armstrong and Gunderson, 1985). Growth of juvenile crab in estuaries is substantially faster than nearshore (Butler, 1961; Poole, 1967; Tasto, 1983; Stevens and Armstrong, 1984; Armstrong and Gunderson, 1985) (Fig. 11.6). This difference is important to oceanographic research because it affects the age at which individuals enter the fishery, hence the expected lags in correlations and temporal scales in population dynamics.

Because the catch record is the only population data series available, it is usually taken as a proxy for recruitment. This is based on the assumptions that: (1) fishing effort is high and constant and (2) only a single year class grows into the fishable size range each year. Fishing effort, in fact, varies from very low (less than 40% of the legal crabs taken) to very high values (84%) (Methot and Botsford, 1982). Estimated preseason abundance of male Dungeness crab show that the time series of recruitment is not as smoothly cyclical as is the catch record (Fig. 11.3).

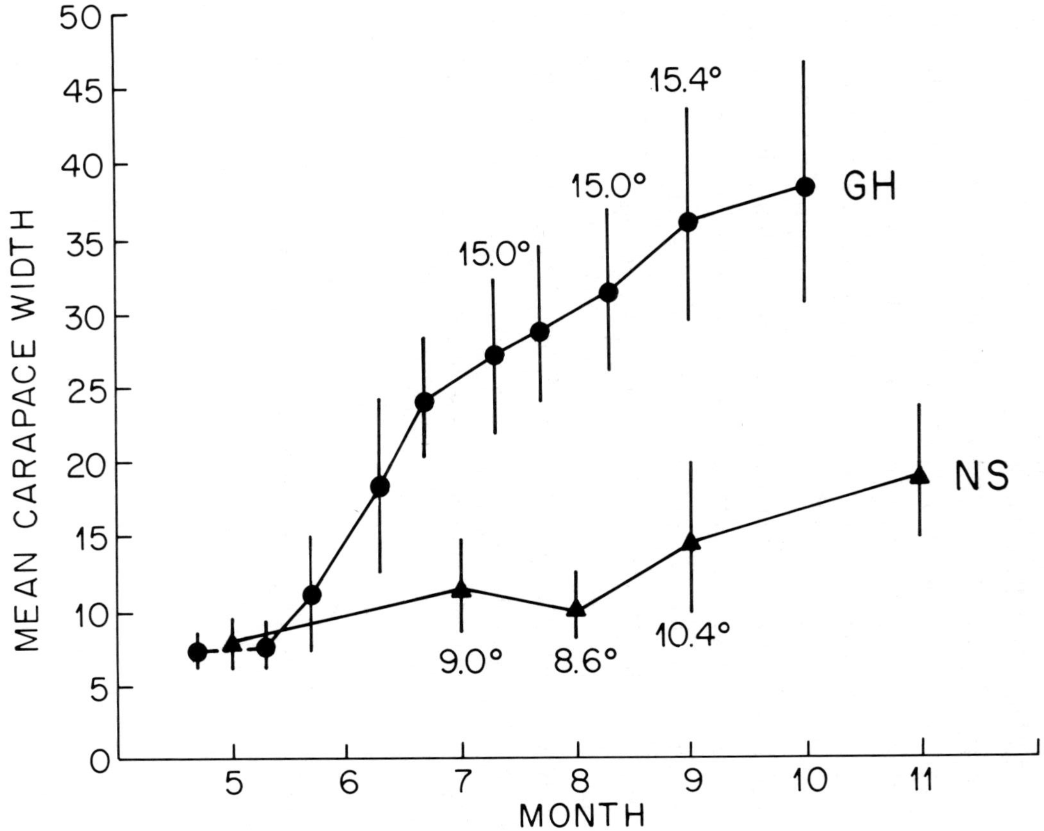

Fig. 11.6 Comparison of juvenile Dungeness crab growth during the first year inside an estuary (Grays Harbor, GH) and nearshore-outside the estuary (NS). Differences in growth may be due to temperature differences shown. (From Armstrong and Gunderson, 1985).

The occasional rapid increases in legal abundance indicate occasional exceedingly strong year classes. With regard to the second assumption, the size distribution of a cohort of crabs when the mode reaches the legal size limit is wide enough that the cohort enters the legal range over 3 or 4 y, with a few entering at age 3 and possibly age 6, but most at ages 4 and 5 along Washington, Oregon and California (Botsford, 1984). Because of the variation in effort and gradual growth of a cohort into the legal size range, the catch record is an imperfect reflection of actual recruitment.

11.3.2 Recruitment of Dungeness Crab

Several investigators have examined statistical relationships between environmental variables and catch records at various lags. From the growth data discussed above, a lag of 4 or 5 y would indicate an influence during the year of egg development, larval hatch and settlement to the juvenile stage (i.e., recruitment). Most studies have shown significant relationships at this lag.

Peterson (1973) found a significant positive correlation between crab landings in a given year and Bakun's (1973) upwelling indices averaged over the spring and summer months 1.5 y earlier in California and Oregon and 0.5 y earlier in Washington (Fig. 11.7). He proposed nutritionally - enhanced survivorship or growth during strong upwelling years. This proposal is difficult to accept because it requires that upwelling benefits older juvenile and subadult crab only 0.5 to 1.5 y before they enter the fishery, rather than during the year of hatch when upwelling could effect either food production or transport of larvae. Botsford and Wickham (1975) showed through autocorrelations that although crab landings were definitely cyclic, upwelling was not. Thus, upwelling was not simply driving crab recruitment.

Mean annual sunspot number was found to be correlated with west coast crab landings by Love and Westphal (1981) (Fig. 11.7). However, their statistical techniques are somewhat suspect and sunspot cycles are of a slightly greater period (11 y) than crab catch (see other comments by Hankin, 1985; Botsford, 1986a).

Wild (1980; 1983; Wild *et al.*, 1983) pointed out that troughs in the crab catch record in northern California corresponded to peaks in local winter sea surface temperatures 4 y earlier (Fig. 11.7). An increase in ocean temperature, possibly accompanied by changes in other oceanographic conditions, was also proposed as an explanation of the decline in the central Californian fishery in the late 1950s (Wild, 1980;1983). Laboratory studies have suggested that higher water temperatures (i.e. greater than 14° C) are associated with lower egg survival during incubation (Wild, 1983). Thus, warm water conditions, such as ENSO events, which occasionally elevate bottom water temperatures above 14° C could be deleterious to reproduction.

Johnson *et al.* (1986) found significant correlations between springtime, southward wind stress (from Caton *et al.*, 1978) and crab landings 4 to 5 y later (Fig. 11.7). Correlations with Bakun winds were not as strong. The mechanism underlying such a relationship is unknown. Megalopae at the surface could be transported southward by northwesterly winds. Onshore

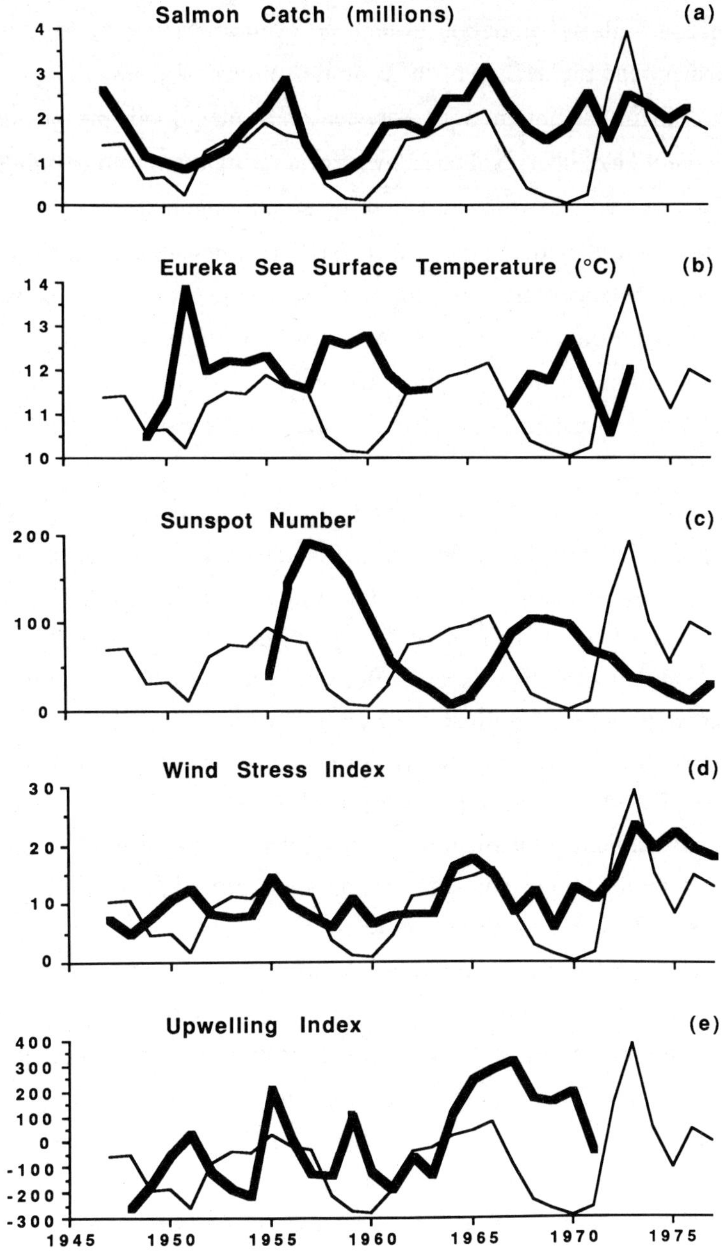

Fig. 11.7 Environmental and biotic variables that have been associated with Dungeness crab catch in northern California (heavy line). For purposes of comparison, crab catch shifted 4 years (the nominal lag from larval stage to the fishery) is shown on each plot (light line). Redrawn from: (a) Botsford *et al.* (1983); (b) Wild (1980); (c) Love and Westphal (1981); (d) Johnson *et al.* (1986); and (e) Peterson (1973); Botsford and Wickham (1975).

transport during the upwelling period could involve light-influenced migration into the layer flowing onshore.

Botsford *et al.* (1982) computed correlations between crab catch and catch records of two known predators on larvae, chinook and coho salmon. The phase relationships in these correlations were not consistent with predation of crabs by these species. Nonetheless, Thomas (1985) has proposed that coho salmon predation influenced cyclic behavior in the northern California Dungeness crab fishery and the decline of the central California fishery.

Mechanisms involving interactions between crab, their food, the physical environment and a crab egg-predator have been evaluated by incorporating them into population models, then testing the models to see whether they satisfied the conditions necessary to cause the observed cycles in abundance. Mathematical analysis of age-structured models with density-dependent recruitment indicate populations are more likely to cycle and are more sensitive to recruitment variability when: (1) recruitment survival changes rapidly with density and (2) older adults have greater effect on recruitment survival than younger adults (Botsford and Wickham, 1978; Botsford, 1986a). These critical aspects of these models have been verified by direct sampling as far as possible.

Cannibalism is a form of density-dependent recruitment that could cause cycles in the fishery (Botsford and Wickham, 1978; Botsford, 1986a). By this mechanism, progeny of an exceptionally strong adult year class would be so heavily cannibalized by large crabs that the resultant year class would lead about four or five years later to a weak fishery. The converse would then occur leading to a strong fishery, after which the 8 to 10 y cycle would repeat itself. Stability conditions require that cannibalism be primarily by 4 or 5 y old crabs on newly settled crabs for cycles to occur. Crabs sampled offshore show the most cannibalism by crabs of this age (Botsford, unpublished). However, Stevens *et al.* (1982) showed that in the Grays Harbor estuary there was substantial cannibalism by crabs of age 1+ newly settled crab. Thus, the conditions for cycles are probably not satisfied for populations reared primarily in estuaries.

Predation on crab eggs by a nemertean worm (*Carcinonemertes errans*) was suggested as a potential cause of the cycles (Botsford and Wickham, 1978; Wickham, 1979a). Because mortality due to predation depends on the number of crab larvae produced, hence female crab density, this mechanism is also a form of density-dependent recruitment. Although this mechanism influences population dynamics, the rate of change of egg survival with changes in worm density is not large enough for the worm to be causing the cycles (Botsford, 1986a; Hobbs and Botsford, in press).

McKelvey *et al.* (1980) concluded that because a model with density-dependent survival of eggs or larvae provided the best fit to crab landings, the cannibalism hypothesis could be rejected. Rejecting cannibalism on this basis involved some very specific assumptions regarding how cannibalism operated. Hence this may be a case of rejecting a mechanism when the model is at fault (see Section 11.2.1; Botsford, 1981b; McKelvey and Hankin, 1981; Hankin, 1985; and Botsford, 1986a). The biological basis of density-dependent egg/larval survival was not satisfactorily addressed by McKelvey *et al.* (1980) and is difficult to demonstrate from field samples.

Adding physical forcing to the models with density-dependent recruitment, Botsford (1986a) showed that windstress could cause the observed catch record only if its effect were non-linear (i.e., stronger winds had a larger than proportional effect than weaker winds) (Fig. 11.8). In that study and in Botsford (1986b) the cyclic character of an environmental influence

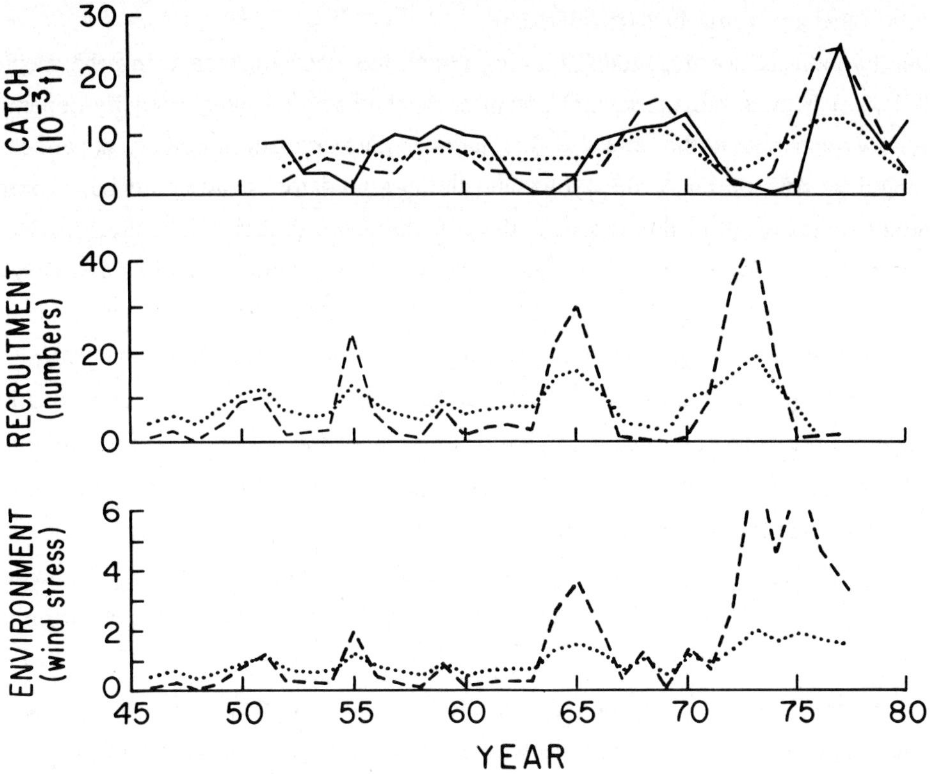

Fig. 11.8 Dungeness crab recruitment (density-dependent) and catch that would result from environmental forcing due to wind stress. In this model, recruitment survival depends on density and the time series termed environment, which, in this example, is wind stress and wind stress cubed. A linear effect of wind stress (dotted line) does not appear to mimic the catch record (solid line), whereas a non-linear effect of wind stress (wind stress cubed, dashed line) does. From Botsford (1986a).

and the relative stability of a population were shown to be "additive" in terms of their propensity to cause cycles. Environmental time series with a relatively minor cyclic component can cause cycles in a population that is stable, through a resonance phenomenon (e.g., Fig. 11.1). This information modifies the criteria used in testing whether a density-dependent recruitment mechanism could cause the cycles (Botsford, 1986b).

11.3.3 Possible Oceanic Influences on Dungeness Crab Recruitment

Research to date has identified several statistical relations that may reflect oceanographic influences, as well as several density-dependent recruitment mechanisms that may operate in conjunction with them. Observed statistical relationships between crab catch and environmental variables are not necessarily independent. For instance, when considered in the light of general relationships between ENSO events, sea temperatures and north-south transport, the correlation between crab catch and spring wind stress is consistent with Wild's (1983) observation that poor recruitment occurred during warmer years (Fig. 11.7). Warmer waters are associated with weaker southward transport which may also be associated with weaker spring winds. This underscores the difficulties involved in extracting a causal mechanism from a number of correlations with variables that are themselves correlated.

One hypothesis, worm predation acting alone, has been rejected using the mechanistic approach, however there remain several potential mechanisms by which density-dependent recruitment may contribute to the observed fluctuations. Cannibalism or density-dependent larval survival could be sole causes of the cycles and the egg-predator worm could be a contributor. An important recent result in this regard is the fact that even though a mechanism may not be strong enough to be the sole cause of cycles, it can contribute to them (Hobbs and Botsford, in press). The impacts on stability are additive; two density-dependent recruitment mechanisms that would not cause cycles when acting alone can cause cycles when both are present.

Identification of the cause of the cycles is hampered by the fact that determination of a critical stage has not been possible for this species. Most of the studies described attempt to relate crab catch to oceanographic, meteorological, or biotic events that occurred some four years earlier, assuming that the greatest determinant of year class strength is a combination of events that affect egg production, larval stages, or survival of first summer juveniles. However, it has never been shown that observed fluctuations in catch are due to events that occur during the first year of life.

Comparisons of crab abundance during the first and second summers also show no consistent trends, indicating that survival following settlement may be as important as egg production or survival through the larval stages. In a coastal area adjacent to the Grays Harbor and Willapa Bay estuaries, population estimates made through the summers of 1983 through 1986 (Armstrong and Gunderson, 1985) show first year crabs (Fig. 11.9a) increasing in abundance from 1983 through 1985, then declining to the 1983 level in 1986, while second year crabs one year later do not show the same pattern either on the coast or inside Grays Harbor (Fig. 11.9b).

There is a high potential for longshore transport of crab larvae by ocean currents. Larvae hatch into the water column nearshore about December, hence encounter a net northward flow in the Davidson Current. Flowing northward at a speed ranging between 8 to 20 km d^{-1} (Hickey, 1989, Chapter 2; Reid and Schwartzlose, 1962; Wyatt *et al.*, 1972) it could carry crab zoeae hundreds of kilometers in the 80 to 100 d required for development (Reed, 1969; Lough, 1976; Reilly, 1983). The prevailing currents and winds after the spring transition, could move the last pelagic stage (megalopae) in the direction of wind (southward and slightly onshore). Alongshore motion, roughly in the same direction as the wind, would be more likely if the larvae are surface-dwelling and the wind is highly variable on hourly time scales (Johnson *et al.*, 1986). Classical Ekman transport (Ekman, 1905) would have them moving 45° to the right of downwind. However, when one considers that (1) transient wind effects are probably more important that steady-state effects (because of the high frequency content of wind variability), (2) a more realistic, variable vertical eddy viscosity leads to a lower angle (Madsen, 1977) and (3) *V. velella* are transported 45° to 60° to left of downwind on this coast (Ricketts *et al*, 1968), the implied direction of transport is closer to the actual wind direction. While sampling the megalopal stage off Washington, Jamieson *et al.* (in review) found that surface drifters moved to the south except when the northwesterly winds relaxed, when they moved onshore.

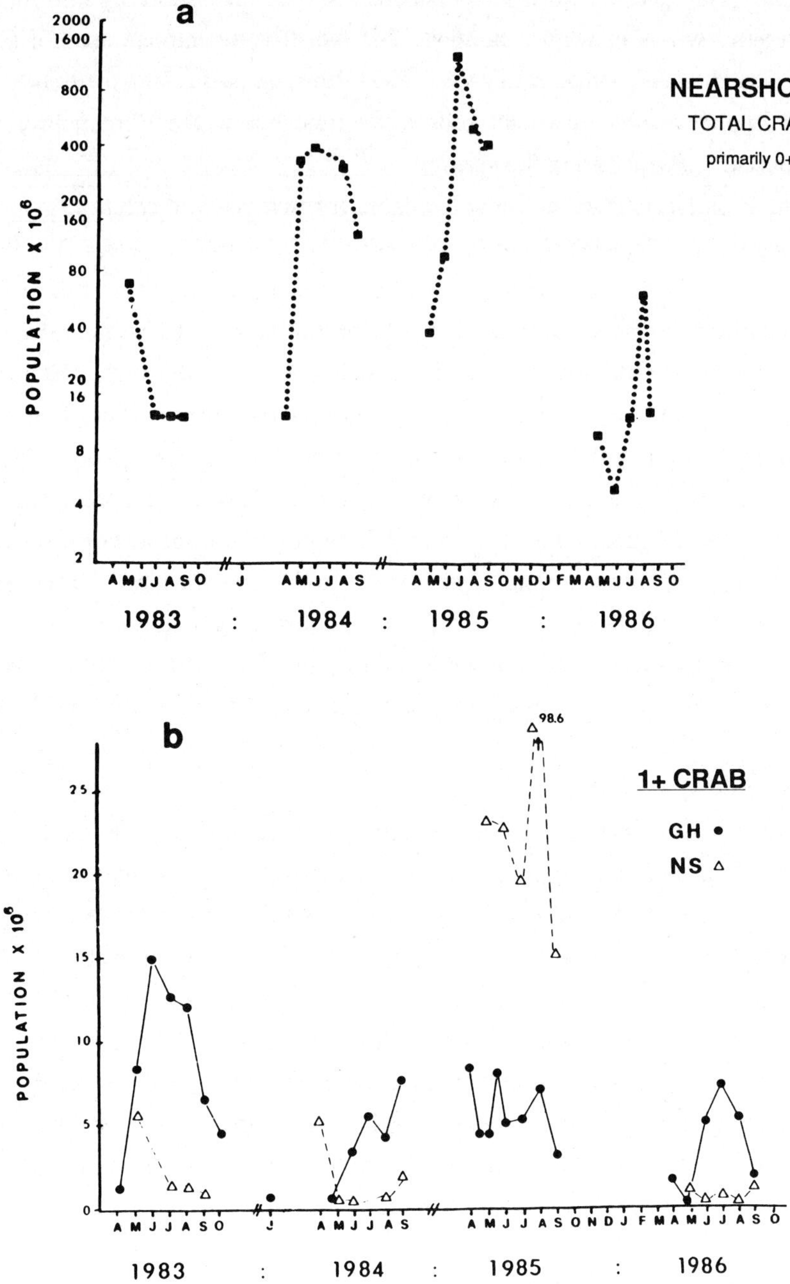

Fig. 11.9 Comparison of abundance of juvenile crabs nearshore during their first summer (a) with abundance of the same cohort during their second summer (b), both inside the estuary (Grays Harbor, GH) and on the nearshore (NS) coast.

The requirement for near-shore settlement presents the question of how larvae are transported or maintained nearshore. Larvae observed as far as 200-km offshore (Reilly, 1983; Lough, 1976) require transport nearshore, unless larvae far from shore are merely lost to the population. Lough (1976) proposed that the bulk of the larval population was retained nearshore within 15-17 km (8-9 nautical miles) of the coast. If zoeae, which are photopositive to moderate light (Jacoby, 1982), are near the surface, they could be held near-shore by currents due to prevailing southwesterly winds in winter months. The fact that megalopae are not at the surface during the day (Booth *et al.*, 1985; Jamieson and Phillips, in press; Botsford and Shenker, unpublished data) implies they may spend substantial time below the Ekman layer where they would be transported onshore during the spring.

Information on larval distribution is available for two years of the four year study of juvenile crabs along the southern Washington coast (Fig. 11.9a; Armstrong and Gunderson, 1985). Canadian zooplankton studies off the west coast of Vancouver Island in both 1985 and 1986 (Jamieson and Phillips, in press; Jamieson *et al.*, in review) found comparable densities of Dungeness crab megalopae offshore (> 50 km) in both years (actually higher in 1986), but low densities nearshore in 1985 and virtually no megalopae nearshore in 1986. This may indicate that the exceedingly low population of 0+ crab measured inshore in 1986 (Fig. 11.9a) was caused by failure of megalopae to reach (or be retained in) shallow coastal areas and estuaries.

ENSO events could affect crab recruitment through larval transport or production of crab prey. In the four years of sampling nearshore benthic juveniles off Grays Harbor, Wa. (Armstrong and Gunderson, 1985), one of the lowest summer population abundances of small instar crab (7 to 15 mm) occurred during the El Niño year of 1983 (Fig. 11.9a). Wild's (1980; 1983) inverse temperature relationship is consistent with poor settlement during ENSO years.

11.4 ENGLISH SOLE

English sole (*Parophrys vetulus*) is a commercially important flatfish along the northwest coast of the United States and Canada with substantial interannual variability in catch (Ketchen, 1956; Hayman *et al.*, 1980). This species is captured in trawl nets along with related bottom fish. The fishery is regulated by setting mesh sizes and total allowable catches for each of several coastal regions. English sole have a pelagic larval stage of 50-70 d (Fig. 11.10; Laroche *et al.*, 1982; Rosenberg and Laroche, 1982) and like the Dungeness crab, the larvae must return to near-coastal and estuarine nursery grounds for successful settlement to the benthic habitat (Rosenberg, 1982; Rogers, 1985; Krygier and Pearcy, 1986; Boehlert and Mundy, 1987). The extended planktonic larval stage and the requirement of a coastal benthic nursery habitat subject recruitment of this species to variations in oceanographic conditions.

11.4.1 Life History of English Sole

Adult English sole typically spawn in late fall and winter, with a 4-12-d egg stage (Alderdice and Forrester, 1968). There is significant interannual variability in the timing of spawning activity. Data from adult gonadal indices (Harry, 1959; Hewitt, 1980), ichthyoplankton surveys (Richardson, 1977; Richardson and Pearcy, 1977; Laroche and Richardson, 1979; Mundy, 1983) and benthic juvenile surveys (Myers, 1980; Rosenberg, 1982; Rogers, 1985; Krygier and Pearcy, 1986) indicate that one or two spawning periods can occur during late fall and winter (Fig. 11.10; Kruse and Tyler, 1983). Ichthyoplankton surveys have found that English sole larvae are generally confined to within 30 km and are often most abundant within

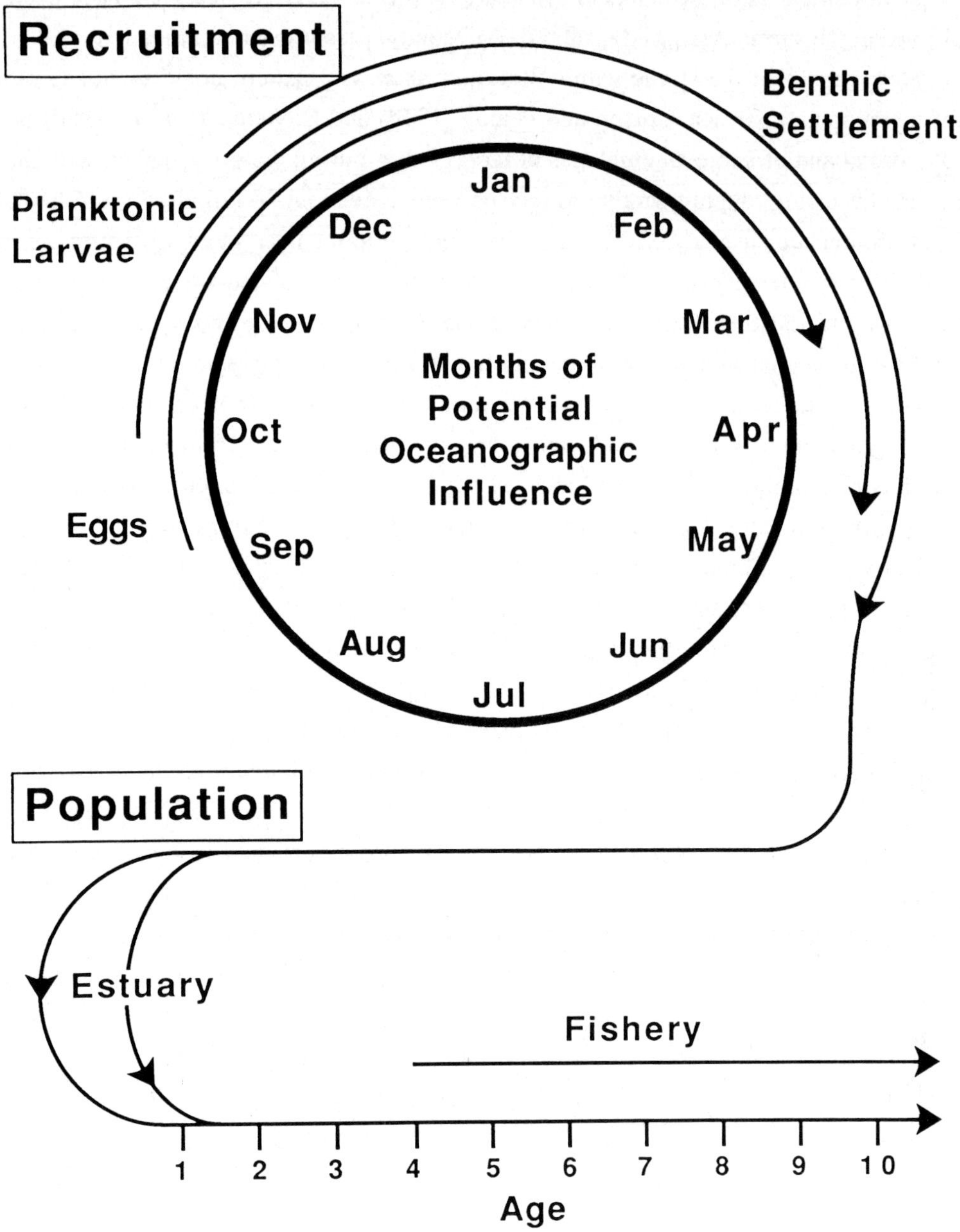

Fig. 11.10 Schematic of English sole life history showing the time of potential oceanographic influence on eggs and larvae. There is substantial overlap of times during which the various stages can be found. All life history stages occur in the ocean except for the indicated estuarine period.

10 km, of the coasts of Washington, Oregon and northern California (Richardson and Pearcy, 1977; Laroche and Richardson, 1979; Richardson *et al.*, 1980; Mundy, 1983; Shenker, in review). The relatively small amount of vertical sampling effort has not identified a distinct diel stratification or migration of larvae, although Boehlert and Mundy (1987) suggest that vertical movements assist larval immigration into estuarine nurseries. Large larvae near metamorphosis

are frequent inhabitants of the neustonic zone at night (Shenker, in review) and have been visually observed at the surface on cloudy, calm days (Mundy, personal communication).

Maintenance of the larvae within 30 km of shore is a pattern noted for many other species of ichthyoplankton. Richardson and Pearcy (1977) and Richardson *et al.* (1980) described distinct coastal and offshore assemblages of larval fishes during the early spring, with the transitional zone occurring roughly parallel to the shelf break at about 30 km offshore. They attributed the interannual consistency of the coastal assemblage to the near-shore spawning grounds utilized by these species and the onshore component of motion of the surface layer prior to the spring transition. Distinct fronts that form offshore, parallel to the coast during the upwelling period, were also considered to inhibit the offshore movement of larvae.

Dramatic interannual variability in the numbers of larval sole is apparent. However, because ichthyoplankton surveys conducted along the Oregon coast from 1969-1984 utilized a wide variety of sampling schemes and focused on different spatial and temporal scales using different types of collecting gear in different habitats, it is difficult to draw direct comparisons between the data sets. The most intensively studied larval year-classes, from 1969-1975, provide an indication of this variation. Mundy (1983) determined that the larvae of the 1969-70 and 1970-71 cohorts were more abundant than the 1971-72 year class. Two spawning peaks were observed in November and February, 1969-70. Large (> 15mm) larvae were abundant later in the spring, indicating relatively high larval survival. A single large spawning peak was detected during February 1971, but subsequent survival was poor. Much lower larval densities were observed during the 1971-2 spawning year. Similar patterns in abundance for the 1970-71 and 1971-72 year classes were observed in the broader onshore/offshore study of Richardson and Pearcy (1977).

The timing of spawning and abundance of larvae during early winter were not assessed during the following years, but March-April sampling from 1972-1976 quantified late winter/spring larval densities along coastal Oregon. Sampling was generally too late to collect newly hatched fish, but large larvae were abundant only during 1973 and 1975 (Laroche and Richardson, 1979).

Settlement of metamorphosing English sole occurs along the coast, generally at depths less than 30 m and in estuaries. Some metamorphosing planktonic larvae settle directly into estuaries while a significant proportion of the coastal benthic juveniles appear to migrate into nearby estuaries (Rogers, 1985; Krygier and Pearcy, 1986; Boehlert and Mundy, 1987). Boehlert and Mundy's (1987) data indicated that larval sole entered the Yaquina Bay estuary from February through June 1982, with three distinct peaks in March, April and June.

The highly specific diet of larval English sole makes them very vulnerable to mismatches with the temporal or spatial distribution of their prey. Gadomski and Boehlert (1984) demonstrated that 97% of the diet of all sizes of larval sole during the late winter and spring of 1973, a year identified as having abundant larvae, was appendicularians (primarily *Oikopleura* spp). Conversely, only 66% of the diet of larvae from January to April, 1971, a year with low larval survival, was comprised of appendicularians and gut fullness levels were lower in 1971 than in

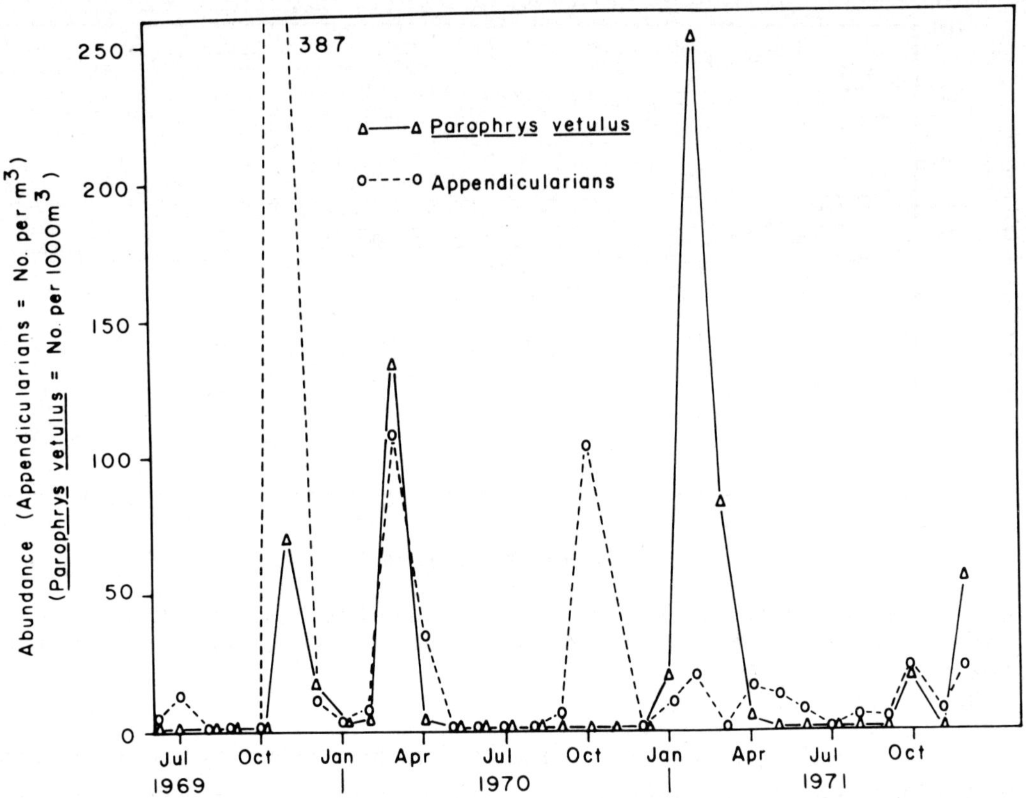

Fig. 11.11 English sole larvae (information from Mundy, 1983) and appendicularian mean abundances (information from Peterson and Miller, 1976) from June 1969 through December 1971. From Gadomski and Boehlert (1984).

1973. Data from Mundy (1983) and Peterson and Miller (1976) indicate that peak appendicularian abundance in the winter of 1970-71 occurred prior to peak larval sole abundance. A good match between predator and prey abundances occurred during 1969 to 1970 (Fig. 11.11, Gadomski and Boehlert, 1984).

The reliance of larval English sole on appendicularians as their dominant food source strongly suggests that prey abundance is a primary determinant of larval survival. Peterson and Miller (1976) found appendicularians (primarily *Oikopleura* spp.) to be most abundant in the fall and winter off central Oregon, with occasional short-lived peaks in abundance (e.g., Fig. 11.11) which were correlated with low surface salinity (< 33.0 ppt) indicating that they seldom occurred during active upwelling. Peterson and Miller (1976) suggested that the appendicularians were characteristic of the offshore low salinity water of the Columbia River plume. However, because their samples with high densities of *Oikopleura* spp. were collected from 1.6-16 km of the coastline, the occurrence of *Oikopleura* was more likely associated with low salinity coastal water influenced by freshwater runoff from the land during the rainy winter season.

Actual abundance of *Oikopleura* spp. and the factors that influence their abundances remain to be determined. Appendicularian populations have very rapid growth rates and can occur in dense aggregations. Alldredge (1982) documented the aggregation of spawning *O. longicau-*

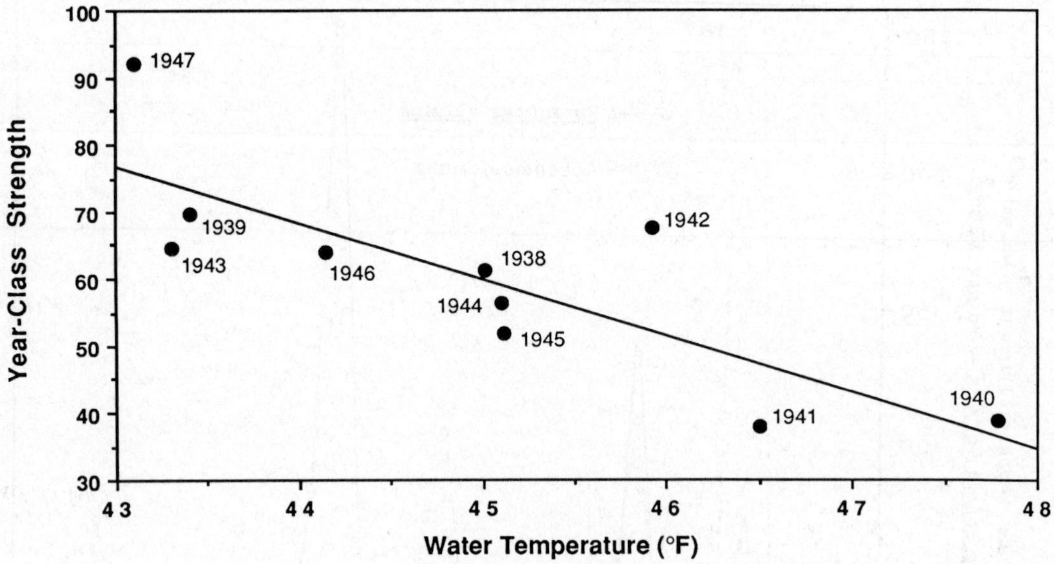

Fig. 11.12 Relationship between cohort strength of English sole and water temperature at the approximate time of the pelagic stage (February-April, inclusive). Redrawn from Ketchen (1956).

da in densities reaching 35,600 l $^{-1}$ in surface windrows created by Langmuir vortices. Generation times for this species can be as short as even days (Fenaux and Palazzoli, 1979), hence populations can respond to environmental conditions on short time scales. Oceanic conditions favoring phytoplankton productivity and stability of this water column could thus trigger blooms of the primary food source of larval sole.

11.4.2 Recruitment of English Sole

In an early statistical analysis of the influence of oceanographic conditions on sole stocks in Hecate Strait, B.C., Ketchen (1956) showed that female year class strengths from 1938-1947 were inversely correlated with mean sea surface temperature, but not with salinity, from February through April of the spawning year (Fig. 11.12). Ketchen attributed the greater recruitment during colder years to a temperature-induced delay in egg and larval development rates which would result in a longer period of drift in the northerly currents, allowing a larger proportion of the cohort to reach the nursery area in northern Hecate Strait before metamorphosis.

Hayman *et al.* (1980) utilized both female age distribution and catch-per-unit-effort (CPUE) data to produce estimates of cohort strength for the 1955 to 1966 year-classes (Fig. 11.13) in the Pacific Marine Fisheries Commission (PFMC) Area 3A (northern Oregon -central Washington). The 1960-61 year-class was exceptionally large. Hayman and Tyler (1980) looked for statistical relationships between cohort strength and environmental variables such as sea surface temperature, upwelling indices, wind and Columbia River discharge parameters. Early fall upwelling, barometric pressure and sea surface temperature accounted for 84% of the variability in cohort strength for 1956-66, with the large 1960-61 year class omitted. They

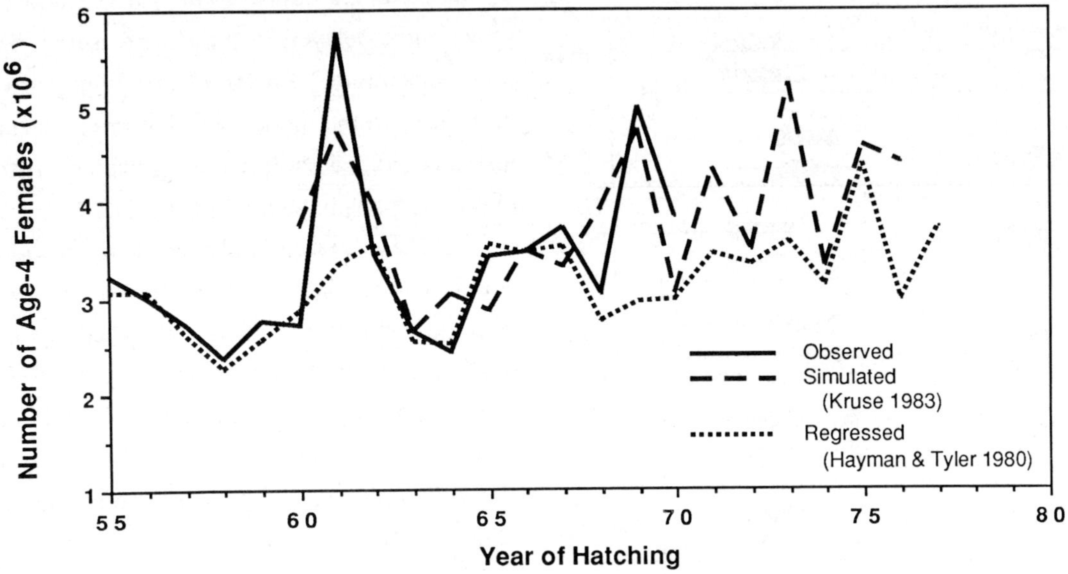

Fig. 11.13 Comparison of observed cohort strength of English sole from cohort analysis (———) with regression (-------) and simulation (— — —) results. Redrawn from Kruse (1983).

proposed that these environmental factors were linked to bottom temperature and that extension of upwelling into the fall delayed spawning and led to stronger cohorts.

High recruitment in 1960-61 does not fit the Hayman-Tyler relationship. Unlike the other years in their statistical analysis, the fall and winter of 1960-1961 was characterized by frequent, weak storms. These conditions may have provided an ideal environment for enhancing primary productivity through frequent nutrient enrichment of the surface layer and sufficient calm periods for formation of aggregations of biomass. The conditions resulted in exceptionally high recruitment for other species as well as English sole [e.g., Pacific whiting *(Merluccius productus)*; Pacific halibut *(Hippoglossus stenolepis)* and Pacific ocean perch *(Sebastes alutus)*; Dark (1975); Gunderson *et al.* (1977)].

A statistical relationship between oceanographic conditions and growth of English sole has also been noted. Kreuz *et al.* (1982) found a significant correlation between bottom temperature (Kruse and Huyer, 1983) and age one growth. Peterman and Bradford (1987b) showed that this growth rate also depended on cohort abundance.

Kruse and Tyler (1983) developed a simulation model which they used to test hypothetical recruitment mechanisms using 13 y of data on spawning activity and four environmental variables: barometric pressure, upwelling index, photoperiod and bottom temperature. In contrast to the statistical approaches (e.g. correlation, regression and principal component analyses), this approach allowed investigation of complex, non-linear relationships between the variables and included the effects of variations in timing of events. Three hypotheses were tested together as a biologically meaningful explanation of the observed pattern in spawning: '(1) the rate of gonadal development is inversely related to summer bottom temperatures; (2) spawning is inhibited

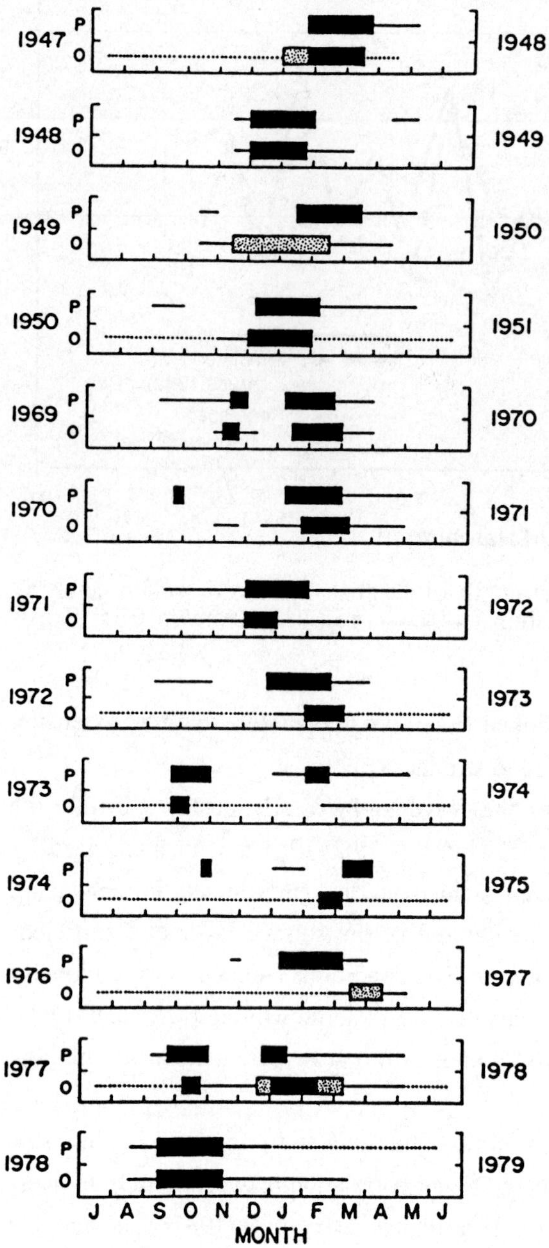

Fig. 11.14 Comparison of observed (O) spawning records and those predicted (P) when all three temperature hypotheses are incorporated into a simulation model driven by bottom temperature estimates. A distinction is made between spawning (solid line), peak spawning (solid shaded areas) and probable peak spawning (stippled areas). A dotted line indicates no data. From Kruse and Tyler (1982).

by temperatures below about 7.8° C; and (3) spawning is delayed by rapid increases in bottom temperature.' Predicted spawning periods derived from the model with the three mechanisms combined show close agreement to the observed spawning activity (Fig. 11.14).

These results indicate that spawning could be strongly influenced by bottom temperature, which in turn is controlled by upwelling circulation (Kruse and Huyer, 1983). Cold bottom temperatures during the summer upwelling could enhance gonadal maturation allowing peak somatic growth of adults (Kreuz *et al.*, 1982). The breakdown of the upwelling circulation in the fall leads to a gradual warming of the bottom waters, which may induce gonadal maturation. However, the temperature may increase too rapidly for acclimation (> approximately 1° C mo $^{-1}$), resulting in a delay in spawning.

A more elaborate model which included potential effects of the oceanic environment on early life stages of English sole in PMFC Area 3A was subsequently developed by Kruse (1983) to test hypotheses involving oceanographic influences on stock sizes (Fig. 11.15). The interactions tested included: (1) influence of temperature and adult stock size on maturity and fecundity, (2) influence of temperature on hatching success, (3) transport resulting from horizontal and vertical distribution of larvae, wind-driven advective processes and the presence/absence of an offshore convergence/divergence zone and (4) dependence of larval feeding and survival on incident radiation, primary productivity, storm strength and storm frequency. Data on the abundance of the preferred appendicularian prey of the larvae were not available, so Kruse indirectly modeled prey abundance from incident radiation and primary production. His best model in-

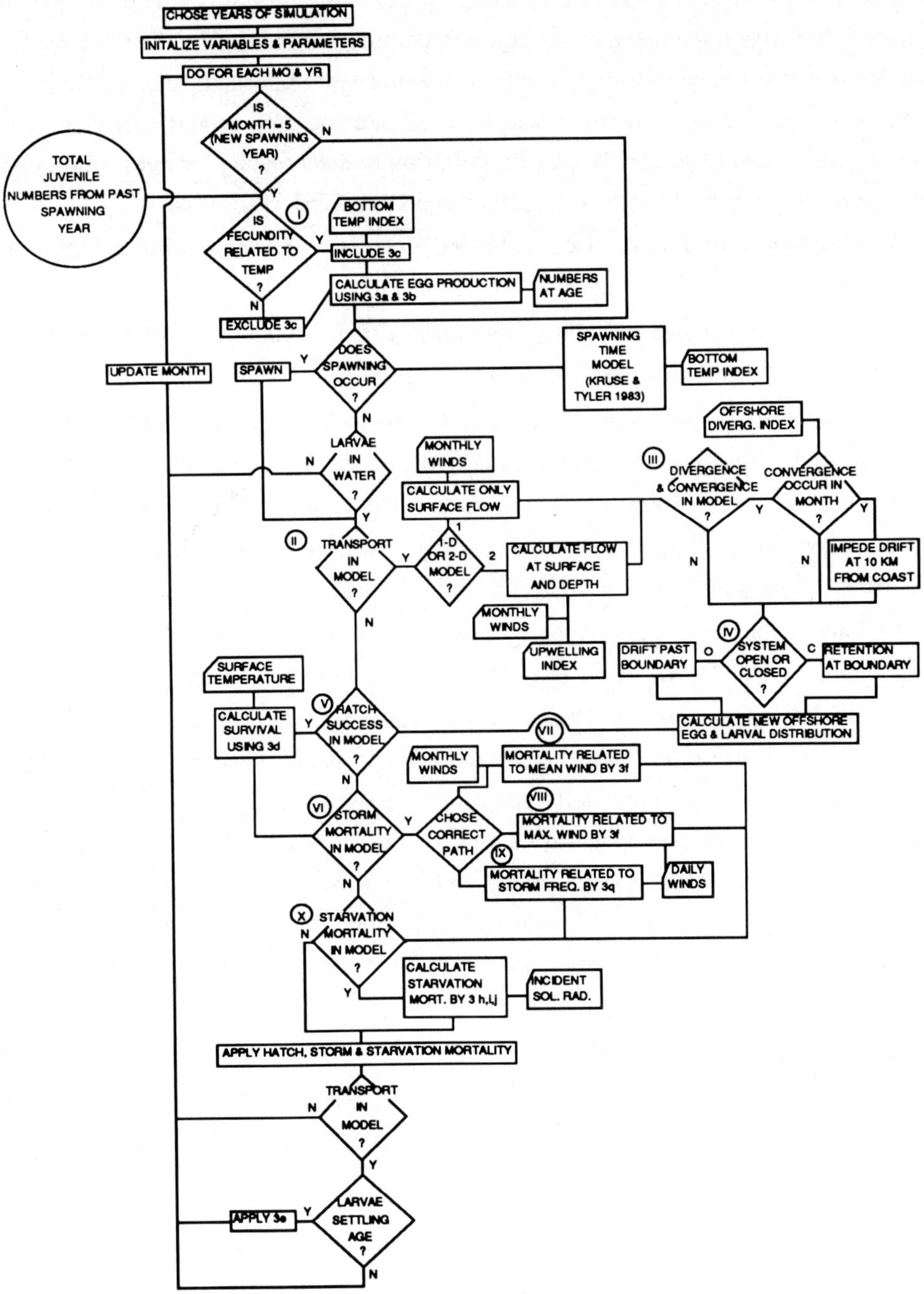

Fig. 11.15 An example of the level of detail that can be incorporated into population models that attempt to explain the influence of oceanographic variability, Kruse's model of English sole. (From Kruse, 1983).

cluded realistic values for population egg production and the relationships between fecundity and bottom temperature, hatching success and surface temperature and the effects of wind-driven, two-dimensional circulation on distance offshore at settlement (Fig. 11.15).

Kruse's (1983) model indicated that egg production, timing of spawning and larval advection are of primary importance in determining recruitment success, while the effects of water-column stability and productivity on larval survival are less important. The optimal model provided estimates of year-class strength which accounted for 53% (r = 0.73) of the interannual variation in abundance (Fig. 11.13). Again, the primary discrepancy between simulated and observed cohort sizes occurred for the unusually large 1960-1961 year class.

Peterman *et al.* (in press) further extended Kruse's (1983) model by including the potential for age-dependent mortality, density-dependent growth and mortality of larvae and age-dependent mortality of juvenile and adult life stages. These hypotheses were tested against the population abundance of four-year old sole during 1968-1974 (hatching years 1964-1970) in PMFC. Inclusion of density-dependent juvenile growth and age-dependent mortality components improved the 'fit' to these data, but density-dependent larval mortality did not appear to be important over the observed range of abundances. The results of the optimal model explained 61% of the interannual abundance (r = 0.78), a marginal improvement over the previous model (Kruse, 1983). Peterman *et al.* (in press) did not attempt to explain the abundant 1960-61 year class. This anomalous year reduced the correlation between Kruse's model and the abundance data and would presumably affect the fit of the model of Peterman, *et al.*(in press).

In summary, Hayman and Tyler's (1980) statistical approach and the models of Kruse and Tyler (1983), Kruse (1983) and Peterman *et al.* (in press) have demonstrated that one or more processes such as environmentally-controlled timing of benthic spawning, an effect of surface temperatures on embryonic and larval development, onshore transport of larvae for settlement and age- and density-dependent growth and mortality of juveniles and early adults could control year-class strength of English sole. The depletion of statistical degrees of freedom in-these analyses of many series of limited length could be partially alleviated by further testing the hypotheses with more recent cohort data now available.

Direct sampling has indicated that a suitable larval feeding environment is probably important, but sufficient information about factors controlling prey abundance is not available. The development of an extremely large year class in 1960-61 has not been explained or accounted for by either statistical or mechanistic analyses. It may have resulted from the fortuitous co-occurrence of optimal conditions for each of many variables.

The least understood component of the models, the dynamics of larval predator-prey relationships, is a significant question in fisheries research. The temporal and spatial variability of planktonic prey is relatively difficult to evaluate for many species of larvae that feed on a relatively wide diversity of prey. The strong preference of sole larvae for appendicularians make this an attractive system for research that can not only provide information on sole recruitment dynamics, but could also be be used as a more complex model of larval fish feeding dynamics.

11.5 COHO SALMON

The coho salmon (*Oncorhynchus kisutch*) is another example of a commercially important fish with wide fluctuations in abundance potentially linked to oceanographic conditions.

However, unlike Dungeness crab and English sole, salmon enter the ocean as competent juveniles, hence do not have a precarious meroplanktonic larval stage. Coho salmon have been commercially fished along the west coast of the United States since the mid-1800s and catch records go back to before the turn of the century (Gunsolus, 1978). They are currently managed by specifying season closures and a minimum size limit.

In 1960, catch between coastal Washington and California had declined to about one-fifth of its level during the 1920s. In response to this decline and the deteriorating coastal habitat, state agencies (primarily in Oregon and Washington) began releasing artificially reared coho salmon. Hatchery releases in this area increased from less than a million per year in the 1950s to more than 30 million per year by 1968 (Nickelson, 1986). The advent of private hatcheries in the late 1970s further increased releases and by 1981 releases peaked at 60 million. The fact that catch at first increased as releases increased, but then appeared to fluctuate about a constant level (Fig. 11.16) and eventually declined, drew considerable research attention to the possibility of a limit on salmon production in the coastal ocean. Catch levels between 1977 and 1985 have been below the former high catch levels of the 1960s and early 1970s.

11.5.1 Life History of Coho Salmon

Coho salmon, like most Pacific salmon (i.e. the genus *Oncorhynchus*), are anadromous. Naturally produced young hatch from eggs in coastal streams during the winter and spring and most juvenile fish spend the next year and several months in fresh water before going to sea at the beginning of their second summer (Fig. 11.17). As a prelude to marine life they undergo physiological changes that enable them to tolerate ocean salinities and enter the ocean as smolts. Because of physiological stress during smoltification they are sensitive to environmental changes. Most hatchery-reared fish are released in fresh water as smolts, at the same age or time of year that natural fish enter the ocean.

Most adults return to their native streams to spawn after their second ocean summer (age 3), but some spawn precociously after the first summer (age 2). The latter are mostly males and are termed jacks. All Pacific salmon die after spawning.

For coho salmon, the critical stage is the smolt stage. Variability in catch (mostly 3 y olds) is much greater than variability in the number of smolts and there is no positive covariation between them. However, there is a significant positive correlation between jacks and catch in the following year. This correlation is very good for the years since 1977 with the exception of the recent ENSO year 1983, during which the predicted catch was low by approximately 50% (Pearcy *et al.*, 1985). On this basis the most likely age of oceanographic influence, with the exception of extreme ENSO events, is the first summer of ocean life.

11.5.2 Recruitment of Coho Salmon

Several potential explanations of the variability in coho salmon abundance have been evaluated and discounted. The impact of fishing is unlikely because smolts are too small to be caught by fishermen (Gunsolus, 1978). From the unusual number of declines in catch immedi-

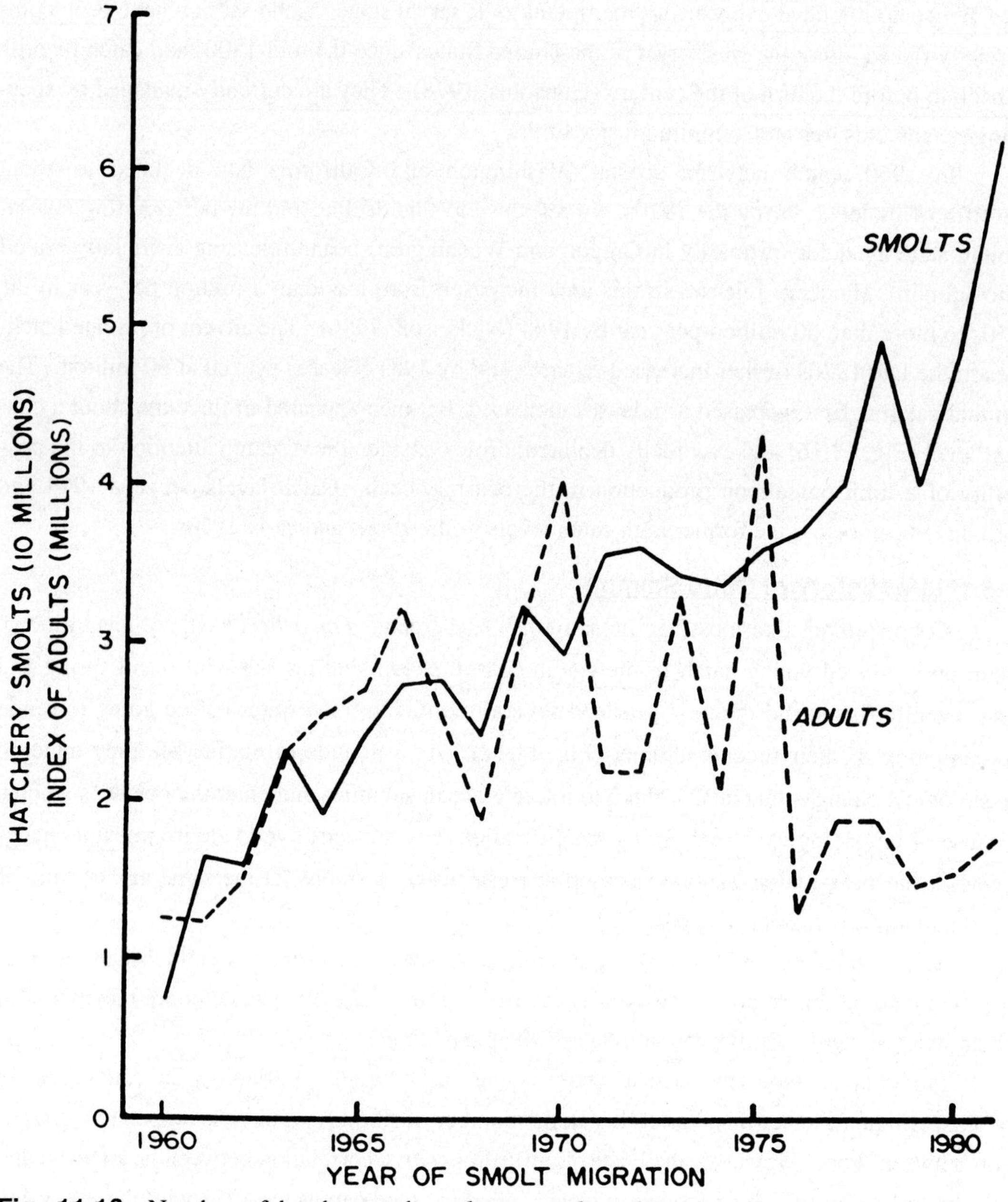

Fig. 11.16 Number of hatchery coho salmon smolts released and estimated abundance of hatchery and wild adults produced the following year in the Oregon Production Area. From Nickelson (1986).

ately following years of high catch, Gunsolus (1978) postulated smolt cannibalism by adults could be important, but this mechanism would be difficult to demonstrate from the existing time series. Streamflow in Puget Sound streams was shown to covary with catch two years later in coastal Washington for the years before 1955 (Smoker, 1955) and later (Mathews and Olson, 1980) and a similar relationship held in Oregon for the years 1942 to 1962 (Scarnecchia, 1981), presumably because of its affect on rearing conditions for natural fish. Correlation between

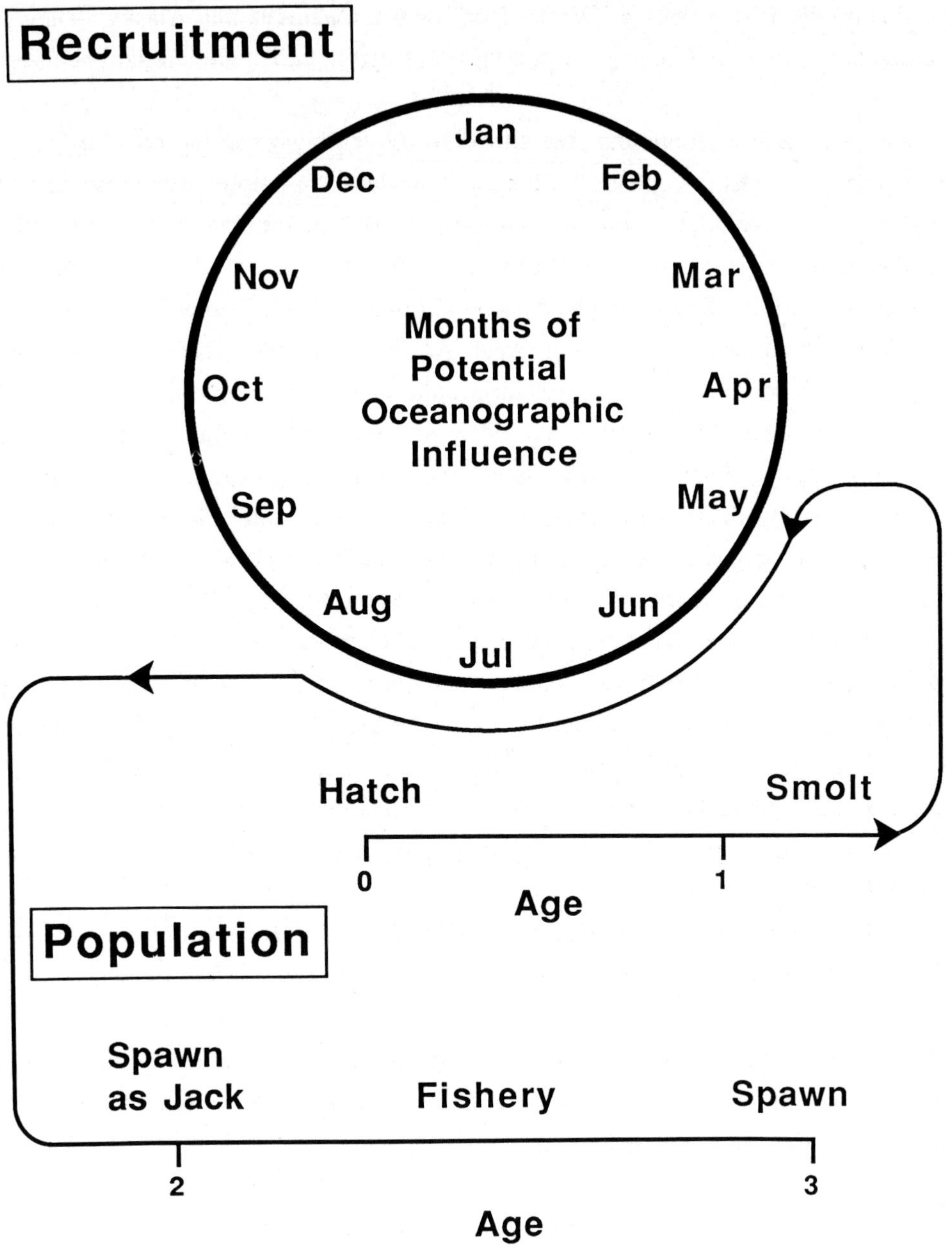

Fig.11.17 Schematic of coho salmon life history showing the time of potential oceanographic influence on smolts after they enter the ocean. Individuals not spawning as jacks may also be affected by extreme ENSO events, although this is not shown. Life between smoltification and spawning occurs in the ocean.

streamflow and coho catch has been poor since then, possibly because of the increased reliance on hatchery produced fish. Although coho salmon prey on larval Dungeness crab, salmon catch does not appear to be related to crab abundance at the proper time lag (Gunsolus, 1978; Botsford *et al.*, 1983).

In an analysis of spatial and temporal covariation, Nickelson and Lichatowich (1984) compared catch records in Oregon, Washington, British Columbia and Alaska. They found no correlation except between Oregon and coastal Washington, which are similar oceanographically.

Studies of coho salmon since the late 1970s have drawn varying conclusions regarding the relative roles of density-dependence (i.e., a dependence of smolt survival on smolt or adult numbers) and oceanographic conditions. Gunsolus (1978) noted that catch was leveling off in spite of the continued increase in smolt releases. He further observed a close relationship between catch and an arbitrarily weighted sum of Bakun's (1973) upwelling index in various months during the summer of the previous year. Scarnecchia (1981) found a significant correlation between catch from 1947 to 1962 (which would include primarily wild fish) and upwelling index during April through June of the previous year. In response to Gunsolus (1978), Peterman (1981) pointed out that if the standard deviation of catch was proportional to catch (i.e., the variable was lognormally distributed), then there was no demonstrable leveling off in the catch record. This same issue was addressed by McCarl and Rettig (1983) who concluded that the standard deviation was proportional to the square of smolt releases. Their analysis showed significant density-dependence and a dependence of smolt survival on upwelling.

Clark and McCarl (1983) used a unique approach to this problem based on Leibig's Law of the Minimum. The model fit includes two regressions of smolt survival, one on upwelling and one on smolt releases and the predicted survival each year is the minimum of results from these regressions. Although this model explained more of the variance than others tried and both regressions were significant, they concluded that upwelling was significant, but that density-dependence was not.

Because the number of natural smolts is difficult to estimate, most of the previous studies concentrated on hatchery fish. More recent studies disaggregate both biological (i.e., wild *vs.* hatchery fish) and oceanographic (i.e., high *vs.* low upwelling years) variables. McGie (1984) described the changing composition of the stock by noting that natural escapement had been decreasing at an annual rate of 9.3% in coastal tributaries and 20.5% in Columbia River tributaries. He stressed the importance of including wild fish in measures of density. By separating upwelling into low and high years, he observed that density-dependence was evident only in years of low upwelling, which is biologically reasonable if density-dependence is food-related.

Nickelson (1986) disaggregated both oceanographic and biological data. He found that survival of hatchery fish was low following low upwelling and high following high upwelling and that survival was negatively related to near-shore ocean temperature during high upwelling (Fig. 11.18). Nickelson (personal communication) has recently shown that for weak upwelling years since 1971 there is a negative correlation between spring ocean temperature and adult abundance the following year. In Nickleson (1986) wild fish did not show a dependence on upwelling, but since this contradicted Scarnecchia's (1981) result from pre-hatchery data, this result was qualified. Wild fish did respond negatively to ocean temperature during high upwelling years. By disaggregating the biological data, Nickelson (1986) determined that wild fish, private

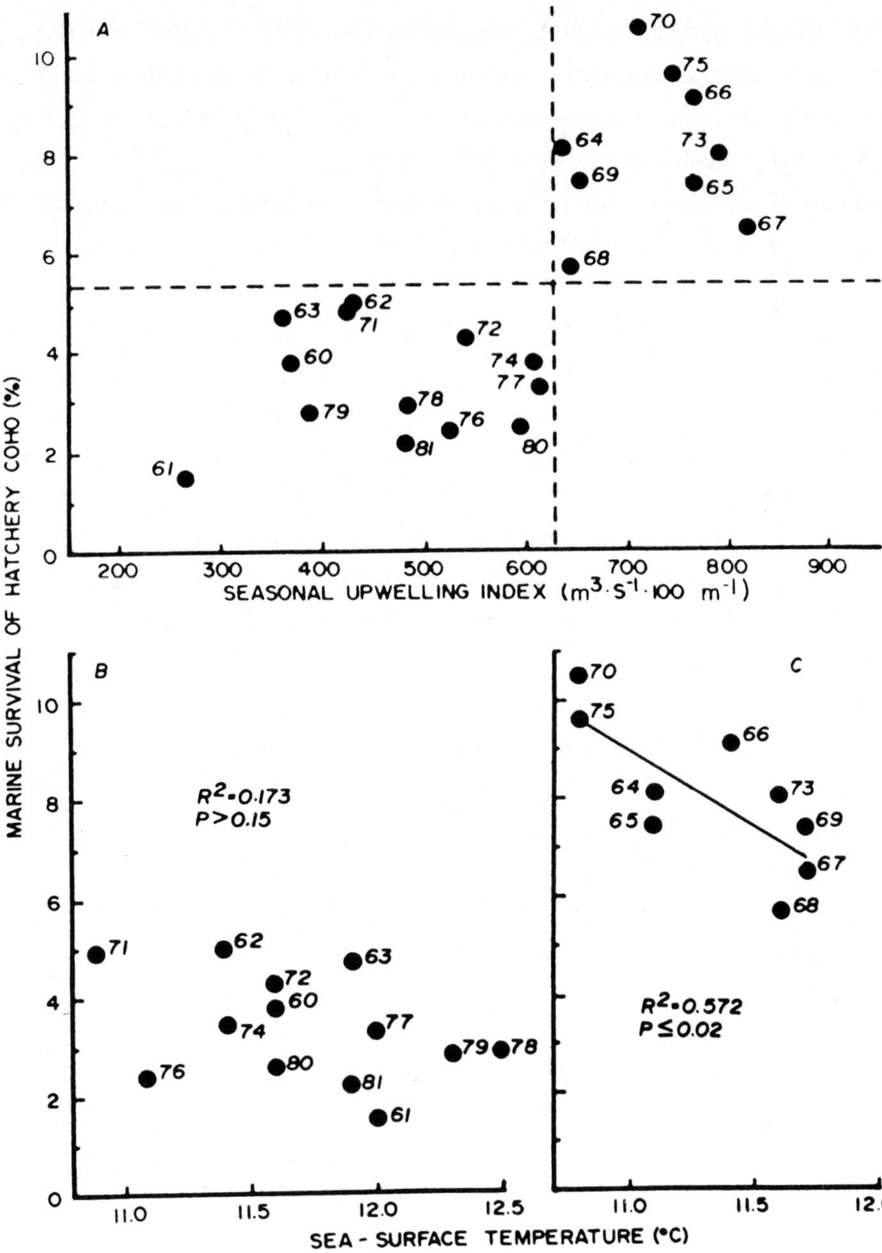

Fig. 11.18 Relationships between the estimated survival of hatchery coho smolts and (a) seasonal upwelling index during their first summer in the ocean, (b) sea surface temperature in weak upwelling years and (c) sea surface temperature in strong upwelling years. Numbers indicate year. From Nickelson (1986).

hatchery fish and public hatchery fish, when considered separately do not exhibit density-dependence. His explanation for the apparent decline in survival of the total population was therefore a gradual replacement of wild fish with high survival by hatchery fish with lower survival rate combined with a shift from a period of strong upwelling to a period of weak upwelling.

Most of the direct information on coho salmon has been obtained from purse seine samples taken off Oregon during the spring and summer of 1981 through 1985 (Pearcy, 1984; Pearcy *et al.*, 1985). Because this period includes low to moderate upwelling as well as a major ENSO event, it provides a good opportunity to directly examine the effects of varying oceanographic conditions on juvenile salmon survival.

A primary consideration in a sampling program to determine oceanographic effects on coho salmon is their geographic location at various life history stages. The dogma from early studies in the 1960s in which substantial numbers of juvenile coho tagged in the Gulf of Alaska were recovered off California, Oregon and Washington was that coho salmon from these states migrate northward into the Gulf of Alaska during their first summer in the ocean (Hartt and Dell, 1986; Loeffel and Forster, 1970). However, more recent studies indicate that some juveniles may be a relatively non-migratory type. The results from recent seining of juveniles off the Oregon coast imply that many coho salmon never make distant migrations (Fisher and Pearcy, 1987; 1988). Also, the majority of marked juvenile coho salmon that were reported in ocean fisheries from 1972 to 1983 were caught in the same area in which they were released (Table 11.1). These apparent differences in migratory patterns may be a fairly recent change attributable to the shift from wild to hatchery fish (Fisher and Pearcy, 1988).

The 1981-85 purse seine samples also allowed refinement of the critical stage for coho salmon. Based on the high correlation between jack survival and abundances of juvenile coho salmon caught in purse seines in June (Fig. 11.9a), Fisher and Pearcy (1988) concluded that cohort abundance is established by June, roughly a month after migrating to sea.

Analyses of field data to determine the potential effects of upwelling on coho salmon in early marine life has focused on two hypotheses: (1) an effect of upwelling on food production, hence growth and survival and (2) an effect of upwelling on transport carrying recent smolts away from near-shore predators. With regard to the food production hypothesis, Peracy and Fisher (in press) found no evidence for a relationship between food supply and early ocean survival rates of Oregon coho salmon. They found little variability in growth, condition and gut fullness of juvenile coho collected during 1981 through 1985, years in which smolt survival as determined from smolt releases and subsequent jack returns (Fig. 11.19a) varied considerably and covaried with upwelling between March and June (Fig. 11.19b). Growth rates of Oregon coastal jacks and sizes of jacks from coastal streams, however, did vary with upwelling. This may have been due to a difference in location between jacks and others in their cohort who would mature in the next year.

Alongshore transport could also influence coho distribution. Purse seine catches of marked fish indicate that new smolts first travel south briefly, then travel predominantly north (Pearcy and Fisher, in press). They may be attempting to swim north throughout this period but may simply not be able to overcome the prevailing southward currents until they grow larger. Their distribution varies with the strength of alongshore currents (e.g., it was more northward during 1983). Juvenile coho salmon are found in the top 20 m of the water column, hence could be affected by offshore Ekman transport. However, there is no apparent correlation between

TABLE 11.1. Recoveries of coded wire tagged juvenile coho in the ocean fisheries 1977-1983 (Data from Pacific Marine Fisheries Commission, Pearcy and Fisher, 1988).

	Landing Area						
Release Area	S.E Alaska	British Columbia*	Puget Sound	Washington Coast (Areas 2-4)	Columbia River (Area 1 + Astoria)	Oregon Coast (Garibaldi & south)	California
S.E. Alaska	20 (39)	0	0	0	0	0	0
British Columbia	2 (2)	1086 (1735)	24 (90)	2 (8)	0	0	0
Puget Sound	0	201 (729)	1352 (5262)	9 (40)	1 (2)	2 (9)	0
Washington Coast	1 (5)	71 (316)	7 (42)	125 (451)	42 (151)	2 (77)	0
Columbia River	2	24 (133)	4 (14)	39 (162)	67 (310)	31 (164)	0
Oregon Coast	0	18 (107)	3 (13)	21 (83)	62 (213)	308 (1137)	10 (45)
California	0	0	0	0	1 (4)	138 (552)	19 (200)

*Sports catches are not expanded. This number = expanded commercial catch + actual number CWT's recovered in the sports fisheries. Preliminary data.

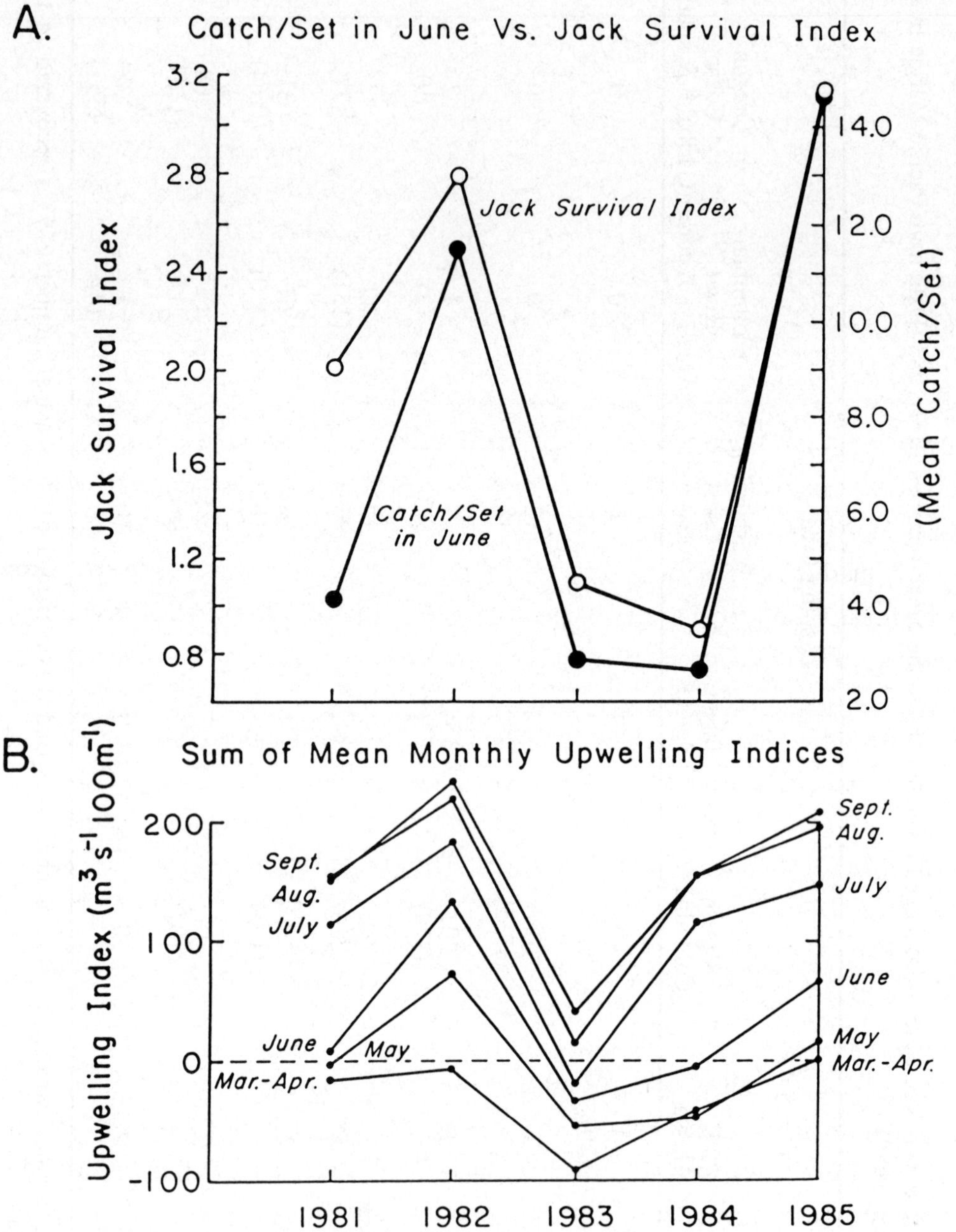

Fig 11.9 A comparison of (a) observed index of smolt abundance (catch/set) and jack survival index, (b) sum of mean monthly upwelling index at 45° N from March through September. From Fisher and Pearcy (1988).

cross-shelf distribution and upwelling in the 1981-85 data set (W. Pearcy, personal communication). With regard to the predation hypothesis, seabirds are known to prey on juvenile salmon and converge on release areas during artificial releases of private salmon ranchers (Bayer, 1986). The results of two out of five attempts to reduce private hatchery mortality due to predation by towing net pens of salmon offshore before release suggest that there is less predation offshore in some years (W. Pearcy, personal communication).

The effects of the 1982-83 ENSO event varied with salmon age. The effects on adults were lower weight per unit length (i.e., a lower condition factor) and a lower 1983 catch than predicted on the basis of jacks in the previous year (Pearcy *et al.*, 1985; Johnson, 1988). Conversely, juvenile growth did not appear to be adversely affected. Gut fullness and growth rates of juveniles were no lower in 1983 than in other years (Pearcy *et al.*, 1985; Fisher and Pearcy, 1988). However, survival of hatchery smolts released in 1983 was one of the lowest on record and this effect may have lingered into 1984 (Nickelson, personal communication). Also, a northward shift in juvenile distribution was implied by the lower than normal catch per sample off Oregon and higher catch per sample near the northern end of the sampling grid (Pearcy *et al.*, 1985).

In summary, studies to date have indicated that oceanographic conditions affect the survival of coho salmon but do not provide a concrete picture of the underlying mechanisms. From statistical analyses, oceanographic conditions related to Bakun's (1973) upwelling index influence early marine survival. An effect of density on marine survival is less likely. Growth and survival in later life appear to be influenced only by major ENSO events. Direct sampling has provided a test of one hypothetical mechanism by which upwelling could influence early marine survival, via variable food production, with a negative result.

The nature of the existing data sets places an inherent limit on the results possible through the statistical approach. For the coho salmon analysis, low and high upwelling values are somewhat contiguous in time with a low period from 1960 to 1963, a high period from 1964 to 1970, then a low period from 1976 to 1981. The fact that these periods correspond to an initial period of low catch, an increase in catch, then a decline in catch lead to a high correlation, but its statistical significance cannot be judged using the sample size as the basis for the number of degrees of freedom.

11.6 DISCUSSION

These case studies of three different fisheries on the northern California/Oregon/Washington coast provide a comparative evaluation of oceanographic processes that may be important to different fisheries as well as the kinds of evidence used to demonstrate an oceanographic influence on recruitment. The first important difference between the three species is that identification of the critical recruitment stage was possible only for coho salmon. Because salmon abundance-at-age can be followed to show the age at which oceanographic conditions affect survival, we can have much more confidence that oceanographic conditions affect coho salmon recruitment than we can recruitment of the other two species. For the Dungeness crab and

English sole, variation in survival after the larval stage may still be the cause of variability in catch. Large settlement of Dungeness crab can result in a low number of crabs of age 1+ in the following year (Fig. 11.9).

In general, there is no guarantee that identified oceanographic influences on early stages will be evident in relative cohort abundance measured at older ages (i.e., there may be more than one 'critical period'). For example, despite the demonstrated importance of wind influences on successful first feeding of 19-d old northern anchovy, *Engraulis mordax*, abundance at age 19 d is not correlated with abundance at age one year (Peterman *et al.*, in press). Results such as these contribute to a growing disenchantment with Hjort's (1914) concept of a critical period after the yolk sac has been consumed (see discussion in Peterman *et al.*, in press).

The dynamics of the juvenile stage can also obscure oceanographically induced variability in the larval stage. Limited carrying capacity of the post-larval environment can dampen variability in recruitment due to fluctuations in larvae. As examples, the effects of oceanographic variability on year class size of the Australian rock lobster and the American lobster are diminished because of a saturation relationship between larval settlement and subsequent recruitment to the fishery (Chittleborough and Phillips, 1979; Morgan *et al.*, 1982; Fogarty and Idoine, 1986; also see discussion of 'fine tuning of a year class' in Bakun *et al.*, 1982).

Given its limitations, it is not surprising that the statistical approach identified one or more potential environmental or biotic influences on recruitment for each species. The relationship between coho salmon survival and the upwelling index (and variables) has held except during strong ENSO years, but the relationship with density -ependence is not as clear-cut. An extensive search among many time series led to early identification of a relationship between the upwelling index and English sole recruitment (except for 1961). This analysis is being extended to more recent recruitment data (E. Pikitch, pers. comm.). Several relationships were identified for Dungeness crab, some of which could be discounted on the basis of lag structure and others of which may be linked through covarying environmental parameters.

Some of the relationships identified statistically have been further evaluated through the mechanistic models and direct sampling of critical parameters. Because of its simple life history and known critical stage, population modeling has not been necessary for coho salmon. Direct sampling has shown convincingly that food production does not underlie the statistical relationship with upwelling. Work on the English sole has involved extensive, elaborate modeling with less direct sampling. A variety of mechanisms, egg production, timing of spawing and advection of larvae were used to explain the observed variability in recruitment (except for 1961). Oceanographic sampling indicates that food may be more important than was implied by the modeling work. For the Dungeness crab, models have indicated the critical parameters for various hypotheses and some of these have been sampled (e.g., the age dependence of cannibalism and the change in egg mortality with density of the egg predator worm). The cyclic behavior provides an advantage in that one can formulate testable stability conditions. Little direct testing has been done of hypotheses involving environmental influences on crab recruitment.

In these examples, the distinction between mechanistic and statistical approaches has become blurred. In the recent modeling work on English sole, for example, models are selected on the basis of statistical fit, rather than being tests of independently derived hypotheses, hence the problems associated with limited numbers of degrees of freedom are incurred. This may not be an insurmountable drawback, depending on how the results are used. To simply conclude that the model with the closest fit contains the true mechanism and manage on that basis would be shortsighted. Better management will be attained more efficiently if the fitting process is used to reject those mechanisms which fit poorly and suggest further sampling or experiments to further differentiate among the remaining alternatives.

Although we have only a partial understanding of the processes affecting recruitment in these three species, comparison of progress provides some indication of promising approaches for further advances. Results for these species and the failure of other established statistical relationships (Sissenwine, 1984) emphasize the advantage of treating the statistical approach as merely exploratory, rather than definitive. Furthermore, direct sampling of parameters critical to a hypothesis (possibly determined through modeling) appears necessary to establish mechanisms. Critical parameters would be in the form of necessary conditions in a hypothetical-deductive scheme. As examples, an effect of food on growth or condition is necessary if upwelling affects coho salmon through food production, cannibalism must be by older individuals if it is to cause cycles in the Dungeness crab, larvae must exhibit specific patterns of vertical distribution for proposed transport schemes. By appropriate use of models, critical parameters can be sampled on shorter time and space scales rather than waiting for annual data to accumulate.

There may be some value in extending the focus of future studies on oceanographic influences on fisheries recruitment to a wider range of populations, species, or ecosystems than the one of immediate interest. For example, statistical analyses of several abundance series of the same species at different locations can alleviate problems of limited degrees of freedom (cf. Anderson and Wilen, 1985; Walters and Staley; in press). They can also unveil spatio-temporal mechanisms such as the propagation of tuna distributions and sea surface temperature in the southern Atlantic (Mendelssohn and Roy, 1987). A second reason for wider focus is its potential for comparison of species or ecosystems (cf., Bakun, 1986). For instance, comparison of population and community dynamics in the four major eastern boundary current regions has led to some generalities regarding the timing of spawning relative to times of offshore transport and wind mixing (Bakun and Parrish, 1982; Parrish *et al.*, 1983).

Even the preliminary results for these three species can help identify the kinds of oceanographic research that will best contribute to understanding oceanographic influences on commercial stocks on the northern California-Oregon-Washington coast. The requirement for nearshore settlement of Dungeness crab and English sole and the relationship between upwelling index and coho salmon imply that transport is an important process. Investigation of physical processes nearshore, on short time and space scales, therefore has great potential. In the Pacific Northwest, very few physical measurements have been made on the inner shelf, the region perhaps most critical to recruitment (Hickey, 1989, Chapter 2). Correlations with upwelling and winds suggest that local processes are important, but the effects of strong ENSO events need also be considered. A good understanding of physical effects on recruitment requires at least some concurrent physical and biological sampling. For example, defining effects on larval transport will require sampling the larvae over time as well as measuring the current field moving them.

Some of the biological-physical interactions identified thus far may provide fertile starting points for further research on transport. For example, Peterson *et al.* (1979) identified several zones (in the two -dimensiona, cross-shelf plane) having specific species of zooplankton during the upwelling period which they explained with two-celled upwelling circulation. However, Wroblewski (1980) showed with a simulation model that two-celled circulation was not necessary. The distributions could be explained by onshore surface flow during relaxation periods between upwelling events. Moreover, Hickey (1989, Chapter 2) and Huyer (1984) find little evidence for the two-celled circulation off either Washington or Oregon. On the other hand, data from the CODE experiment off central California indicate little, if any, onshore flow during upwelling 'relaxation' (Send *et al.*, 1987).

Recent advances in physical oceanography, such as finer characterization of the three dimensional upwelling response to wind forcing (Hickey, 1989, Chapter 2; Winant *et al.*, 1987) and incorporation of the effects of eddies and jets (Mooers and Robinson;1984; Davis, 1985; Huyer *et al.*, 1984; Hickey, 1989, Chapter 2), will contribute to better understanding of the effects of transport on commercial stocks. Better characterization of the nature and timing of important events such as the spring transition (Lentz, 1987; Strub *et al.*, 1987) will allow comparison with biological events to investigate transport, match-mismatch hypotheses. Studies of the impact of temperature on English sole recruitment imply that, although it is much less dramatic, the fall transition may also be important.

Improvements in the characterization of physical conditions in the near-shore ocean will not only improve our ability to gain an understanding of larval processes, but will also lead to a shift in the methods used. For example, laboratory results on larval feeding, swimming behavior and responses to light, pressure and other physical variables can be put to greater use in describing responses of the larval phase to oceanographic conditions through a reductionist approach, once physical conditions are better described (c.f., Sulkin, 1984). The success of this approach to anchovy and the apparent benefit of more laboratory information on Dungeness crab and English sole (as suggested by Kruse and Tyler, 1983) argues strongly for this approach. Another approach that will be more effective as oceanographic descriptions improve is numerical modeling of transport processes and physical conditions. These could range from simple plots of hypothetical results of wind-forced drift (e.g., Rothlisberg and Miller, 1983) to more complex descriptions including vertical migration and the three-dimensional current fields (cf., Rothlisberg *et al.*, 1983; Power 1986; Sundby, 1983; Jackson and Strathmann, 1981; Jackson, 1986).

The primitive state of our understanding of processes controlling recruitment and the lack of use of existing oceanographic information in management, lead to the questions of whether better understanding of oceanographic influences will ever be of use in management and how it will be used. There are three general ways in which this information could aid harvest management: (1) annual prediction of fishable abundance, (2) better understanding of population dynamics and (3) explanation of long -rm changes. The first is the most commonly recognized, although its value is not universally acknowledged. On the positive side, Mathews (1971) computed the positive value of prediction in salmon fisheries. Prediction of catch in another crustacean fishery, the western rock lobster fishery in western Australia, has been extremely valuable in projecting requirements for fishing resources (Phillips, 1986; Caputi and Brown, 1986) and was recommended as a goal for other fisheries (Caddy and Campbell, 1986). In a computer simulation study, Swartzman *et al.* (1983) showed that a management scheme for Pacific whiting using sea temperature to specify management quotas outperformed other methods. Buckingham and Walters (1975) found preseason abundance predictions to be of little value because of the remaining uncertainties (e.g., catchability, response of fishermen to regulations, etc.). On the negative side, preseason predictions of northern California Dungeness crab catch based on preseason sampling were discontinued because of their unpopular influence on price.

The second aid to management involves determining the relative effect of environment and density on recruitment. The latter is important to management because the economic decision to allow a fish to be caught hinges on its current market value relative to its future value if left in the population, which in turn includes its contribution to density (Botsford, 1981a). The third aid to management is less tangible. It involves knowing enough about how the population works not to be completely surprised by rare events such as strong ENSO events or episodic changes in oceanographic regime. In addition to the direct benefit of this knowledge to catch, there is also the benefit of not losing the confidence of those being managed (i.e., the fishermen).

The way in which the three species described herein are currently managed does not involve oceanographic information on recruitment.and only one of them incorporates prediction. Management of coho salmon each year is based on the jack index from the previous year. Because the index is a good positive predictor and is inexpensively available, it is doubtful that oceanographic information will be used much in prediction, with the exception of strong ENSO years. Dungeness crab are not actively managed (Methot, 1986). Size, sex and season limits are set to protect reproductive capacity, but effort is not varied in response to fluctuating recruitment and catchable abundance is not predicted. Management of English sole also does not respond to recruitment predictions. Year-class strength is computed from catch-at-age information (i.e., after the year class is in the fishery) and is used to set an "acceptable biological catch." However, there are no specified changes in fishing effort each year. Such a policy would be difficult because English sole is part of a multispecies fishery. It appears that more intensive, prediction-based management of Dungeness crab and possibly English sole would be useful. However, whether preseason biological sampling is less expensive in the long run than determining how knowledge of oceanographic information could be used is unclear.

Turning this question around, management may actually be useful in determining oceanographic influences on recruitment rather than vice versa. The response of a population to varying management can reflect critical population characteristics. Adaptive management is a relatively recent development (Walters and Hilborn, 1976; Walters, in press). An example appropriate to the issue addressed here would be adjusting management to maintain the population at a fixed density in order to determine how much of the recruitment variability is not density-dependent. For coho salmon, management could provide experimental analysis through control of smolt releases (Peterman and Routledge, 1983).

In addition to effects on recruitment, other oceanographic effects can be used in fisheries. Knowledge of oceanographic influences on distribution is used in targeting harvest and multi-species management. Examples are use of oceanographic information by fishermen to locate albacore tuna, quota adjustments for sockeye salmon based on whether the Fraser River stock is returning north or south of Vancouver Island and use of temperature information to avoid salmon bycatch in the flying squid (*Ommastrephes bartromi*) gill net fishery in the North Pacific. Another potential use of oceanographic information would be to determine optimal conditions for release of stocked fish. For example, this would be useful in coho salmon if offshore transport due to upwelling increases survival.

In summary, identification of the influence of oceanographic conditions on fish stocks is at a relatively primitive stage in the Pacific Northwest as elsewhere. Oceanographic issues of apparent importance to fish stocks are some of those of current independent interest to physical oceanographers (short time and space scales, nearshore currents). Although better understanding may not always be of direct management use in prediction of abundance, it will contribute more generally to better fisheries management in less direct ways.

ACKNOWLEDGEMENTS

We dedicate this contribution to the the memory of Dr. Reuben Lasker in honor of his many contributions to the understanding of fish recruitment and in appreciation for his active interest in the work of others in that field.

We would like to thank G.S. Jamieson, R.D. Methot, G.H. Kruse, T.E. Nickelson, W.G. Pearcy, R.M. Peterman, T.M. Powell, P.E. Smith and A.V. Tyler for their reviews of this work. We also thank the editors of this volume for their contribution to the readability of our chapter. This research was supported in part by the Army Corps of Engineers and by the Washington Sea Grant Program, grant NA84AA900011. This is a component of Contribution No. 1784 from the School of Oceanography, University of Washington.

REFERENCES

Ahlstrom, E.H. 1959. Vertical distribution of pelagic fish eggs and larvae off California and Baja California. *Fish. Bull.*, U.S., 60: 106-146.

Alderdice, D.F. and C.R. Forrester. 1968. Some effects of salinity and temperature on early development and survival of the English sole (*Parophrys vetulus*). *J. Fish. Res. Bd. Can.*, 25: 495-521.

Alldredge, A.L. 1982. Aggregations of spawning appendicularians in surface windrows. *Bull. Mar. Sci.*, 32: 250-254.

Anderson, J.L. and J.E. Wilen. 1985. Estimating the population dynamics of coho salmon (*Oncorhynchus kisutch*) using pooled time-series and cross-sectional data. *Can. J. Fish. Aquat. Sci.*, 42: 459-467.

Anon. 1978. Climate and fisheries. Proceedings from a workshop on the influence of environmental factors on fisheries production. Center for Ocean Management Studies, Univ. of Rhode Island, 135 pp.

Anon. 1985. IOC-FAO guiding group of experts on the programme of ocean science in relation to living resources. Intergovernmental Oceanographic Commission Rep. SC-85/WS/18.

Armstrong, D.A. 1983. Cyclic crab abundance and relationships to environmental causes. Pages 102-110 in: W. Wooster (Ed.), From year to year: Interannual variability of the environment and fisheries of the Gulf of Alaska and eastern Bering Sea. Washington Sea Grant Publ. WSG-WO 83-3.

Armstrong, D.A. and D.R. Gunderson. 1985. The role of estuaries in Dungeness crab early life history: A case study in Grays Harbor, Washington. Proc. of Symp. on Dungeness crab biology and management. Univ. of Alaska, Anchorage, Sea Grant Rep. No. 85, 3: 145-170.

Bailey, K.M. 1981. Larval transport and recruitment of Pacific hake. *Merluccius productus*. *Mar. Ecol. Prog., Ser.* 6: 1-10.

Bakun, A. 1973. Coastal upwelling indices, west coast of North America, 1946-71. U.S. Dept. Commer., NOAA Tech. Rep. NMFS SSRF-671, 103 pp.

Bakun, A. 1986. Definition of environmental variability affecting biological processes in large marine ecosystems. Pages 89-108 in: K. Sherman and L. M. Alexander (eds.), Variability and Management of Large Marine Ecosystems. Am. Assoc. Advance. of Science Sel. Symp. 99.

Bakun, A., J. Beyer, D. Pauly, J.G. Pope and G.D. Sharp. 1982. Ocean sciences in relation to living resources. *Can. J. Fish. Aquat. Sci.*, 39: 1059-1070.

Bakun,A. and R.H. Parrish. 1980. Environmental inputs to fishery population models for eastern boundary current regions. Pages 67-104 in: G.D. Sharp (ed.), *Workshop on the Effects of Environmental Variation on the Survival of Larval Pelagic Fishes.* IOC Workshop Rep. 28.

Bakun, A. and R. Parrish. 1982. Turbulence, transport and pelagic fish in the California and Peru Current systems. CalCOFI Rep. 23, pp: 99-112.

Bayer, R.D. 1986. Seabirds near an Oregon estuarine salmon hatchery in 1982 and during the 1983 El Niño. *Fish. Bull., U.S.*, 84: 279-286.

Boehlert, G.W. and B.C. Mundy. 1987. Recruitment dynamics of metamorphosing English sole, *Parophrys vetulus*, to Yaquina Bay, Ore. *Est. Coast. Shelf Sci.*, 25: 261-281.

Booth, J., A. Phillips and G.S. Jamieson. 1985. Fine scale spatial distribution of *Cancer magister* megalopae and its relevance to sampling methodology. Pages 273-286 in: Melteff, B. (ed.), *Proc. Symp. Dungeness Crab Biology and Management.* Univ. Alaska Sea Grant Rep. No. 85-3.

Botsford, L.W. 1981a. Optimal fishery policy for size-specific, density-dependent population models. *J. Math. Biol.*, 12: 265-293.

Botsford, L.W. 1981b. Comment on cycles in the northern California Dungeness crab population. *Can. J. Fish. Aquat. Sci.,* 38: 1295-1296.

Botsford, L.W. 1984. Effects of individual growth rates on expected behavior of the northern California Dungeness crab (*Cancer magister*) fishery. *Can. J. Fish. Aquat. Sci.,* 42: 99-107.

Botsford, L.W. 1986a. Population dynamics of the Dungeness crab (*Cancer magister*). Pages 140-153 in: G.S. Jamieson and N. Bourne (eds.), *North Pacific Workshop on Stock Assessment and Management of Invertebrates. Can. Spec. Publ. Fish. Aquat. Sci.* 92.

Botsford, L.W. 1986b. Effects of environmental forcing on age-structured populations: northern California Dungeness crab (*Cancer magister*) as an example. *Can. J. Fish. Aquat. Sci.*, 43: 2435-2352.

Botsford, L.W. 1987. Analysis of environmental influences on population dynamics. Pages 54-64 in: T.L. Vincent, Y. Cohen, W.J. Grantham, G.P. Kirkwood and J.M. Skowronski (eds.), *Modeling and Management of Resources Under Uncertainty*. Springer-Verlag, New York, N.Y.

Botsford, L.W., R.D. Methot Jr. and W.E. Johnston. 1983. Effort dynamics of the northern California Dungeness crab (*Cancer magister*) fishery. *Can. J. Fish. Aquat. Sci.*, 40: 337-346.

Botsford, L.W., R.D. Methot and J.E. Wilen. 1982. Cyclic covariation in the California king salmon, *Oncorhynchus tshawytscha*, silver salmon, *O. kisutch* and Dungeness crab, *Cancer magister*, fisheries. *Fish. Bull. U.S.*, 80: 791-801.

Botsford, L.W. and D.E. Wickham. 1975. Correlation of upwelling index and Dungeness crab catch. *U.S. Fish. Wildl. Serv. Fish. Bull.*, 73: 901-907.

Botsford, L.W. and D.E. Wickham. 1978. Behavior of age-specific, density-dependent models and the northern California Dungeness crab (*Cancer magister*) fishery. *J. Fish. Res. Board Can.*, 35: 833-943.

Buckingham, S.L. and C.J. Walters. 1975. A control system for intraseason salmon management. Proc. IIASA Workshop on Salmon Management, Laxemburg. Austria, CP-75-2, pp. 105-137.

Butler, T.H. 1961. Growth and age determination of the Pacific edible crab, *Cancer magister* Dana. *J. Fish. Res. Board Can.*, 18: 873-891.

Caddy, J.F. and A. Campbell. 1986. Summary of session 9: summary of research recommendations. *Can J. Fish. Aquat. Sci.*, 43: 2394-2396.

Caputi, N. and R.S. Brown. 1986. Relationships between indices of juvenile abundance and recruitment in the western rock lobster (*Panulirus cygnus*) fishery. *Can. J. Fish. Aquat. Sci.*, 43: 2131-2139.

Caton, F.G., M.J. Cummings and B.R. Mendenhall. 1978. *A Northern Hemisphere History of Marine Wind-Based Parameters*. Meteorology Intern. Inc., Monterey, Calif., 115 pp.

Chelton, D.B. 1983. Effects of sampling errors in statistical estimation. *Deep-Sea Res.*, 30: 1083-1103.

Chelton, D.B. 1984. Commentary: short-term climatic variability in the northeast Pacific ocean. Pages 87-99 in: W.G. Pearcy (ed.), *Influence of Ocean Conditions on the Production of Salmonids in the North Pacific*. Oregon St. Univ, Sea Grant College Prog, (ORESU-W-83-001).

Chittleborough, R.G. and B.F. Phillips. 1979. Recruitment to the catchable stock of the western rock lobster to the catchable stock of the western rock lobster *Panulirus longipes cygnus* George. *Rapp. P.-v Reun. Cons. Int. Explor. Mer.*, 175: 139-142.

Clark, J. and B. McCarl. 1983. An investigation of the relationship between Oregon coho salmon (*Oncorhynchus kisutch*) hatchery releases and adult production utilizing law of the minimum regression. *Can. J. Fish. Aquat. Sci.*, 40: 516-523.

Creutzberg, F. 1961. On the orientation of migrating elvers (*Anguilla vulgaris* Turt.) in a tidal area. *Neth. J. Sea Res.*, 1: 257-338.

Cronin, T.W. and R.B. Forward, Jr. 1986. Vertical migration cycles of crab larvae and their role in larval dispersal. *Bull. Mar. Sci.*, 39: 192-201.

Cushing, D.H. 1975. *Marine Ecology and Fisheries*. Cambridge Univ. Press, Cambridge, Mass., 278 pp.

Cushing, D.H. 1982. *Climate and Fisheries*. Academic Press, London, 373 pp.

Dark, T.A. 1975. Age and growth of Pacific hake, *Merluccius productus*. *Fish. Bull.U.S.*, 73: 336-355.

Davis, R.E. 1976. Predictability of sea surface temperature and sea level pressure anomalies over the North Pacific Ocean. *J. Phys. Oceanogr.*, 6: 249-266.

Davis, R.E. 1985. Drifter observations of coastal surface currents during CODE: The method and descriptive view. *J. Geophys. Res.*, 90: 4741-4755.

Dawirs, R.R. 1984. Influence of starvation on larval development of *Carcinus maenas L. (Decapoda: Portunidae)*. *J. Exp. Mar. Biol. Ecol.*, 80: 47-66.

Denman, K.L. and T.M. Powell. 1984. Effects of physical processes on planktonic ecosystems in the coastal ocean. *Oceanogr. Mar. Biol. Ann. Rev.*, 22: 125-168.

Ekman, V.W. 1905. On the influence of the earth's rotation on ocean currents. *Arch. Math. Astron. Phys.* 2, No. 11.

Epifanio, C.E., C.C. Valenti and A.E. Pembroke. 1984. Dispersal and recruitment of blue crab larvae in Delaware Bay, U.S.A. *Estuar. Coast. Shelf Sci.*, 18: 1-12.

Fenaux, R. and I. Palazzoli. 1979. Estimation *in situ* d'une population d'*Oikopleura longicauda* (Appendicularia) a l'aide de deux filets de maille differente. *Mar. Biol.*, 55: 197-200.

Fisher, J.P. and W.G. Pearcy. 1987. Movements of coho, *Oncorhynchus kisutch* and chinook, *Ontshawytscha*, salmon tagged at sea off Oregon, Washington and Vancouver Island during the summers, 1982-85. *Fish. Bull., U.S.*, 85: 819-826.

Fisher, J.P. and W.G. Pearcy. 1988. Growth of juvenile coho salmon (*Oncorhynchus kisutch*) in the ocean off Oregon and Washington, U.S.A, in years of differing coastal upwelling. *Can. J. Fish. Aquat. Sci.*, 45: 1036-1044.

Fogarty, M.J. and J.S. Idoine. 1986. Recruitment dynamics in an American lobster (*Homarus americanus*) population. *Can. J. Fish. Aquat. Sci.*, 43: 2368-2376.

Fortier, L. and W.C. Leggett. 1983. Vertical migrations and transport of larval fish in a partially mixed estuary. *Can. J. Fish. Aquat. Sci.*, 40: 1543-1555.

Frank, K.T. and W.C. Leggett. 1983a. Multispecies larval fish associations: accident or adaptation? *Can. J. Fish. Aquat. Sci.*, 40: 754-762.

Frank, K.T. and W.C. Leggett. 1983b. Survival value of an opportunistic life-stage in capelin (*Mallotus villosus*). *Can. J. Fish. Aquat. Sci.*, 40: 1442-1448.

Frank, K.T. and W.C. Leggett. 1985. Reciprocal oscillations in densities of larval fish and potential predators: a reflection of present or past predation? *Can. J. Fish. Aquat. Sci.*, 42: 1841-1849.

Gadomski, D.M. and G.W. Boehlert. 1984. Feeding ecology of pelagic larvae of English sole *Parophrys vetulus* and butter sole *Isopsetta isolepis* off the Oregon coast. *Mar. Ecol. Prog. Ser.*, 20: 1-12.

Gunsolus, R.T. 1978. The status of Oregon coho and recommendations for managing the production, harvest and escapement of wild and hatchery-reared stocks. Oregon Dept. of Fish. Wildl. Rep., 59 pp.

Grover, J.J. and B.L. Olla. 1986. Morphological evidence for starvation and prey size selection of sea-caught larval sablefish, *Anoplopoma fimbria*. *Fish. Bull. U.S.*, 84: 484-489.

Gunderson, D.R., S.J. Westrheim, R.L Demory and M.E. Fraidenburg. 1977. The status of Pacific ocean perch (*Sebastes alutus*) stocks off British Columbia, Washington and Oregon in 1974. *Can. Fish. Mar. Serv. Tech. Rep.* 690.

Hankin, D.G. 1985. Proposed explanations for fluctuations in abundance of Dungeness crabs: a review and critique. Pages 305-326 in: *Proceedings of the Symposium on Dungeness Crab Biology and Management*. Univ. of Alaska Sea Grant Rep. 85-3.

Hartt, A.C. and M.B. Dell. 1986. Early oceanic migrations and growth of juvenile Pacific salmon and steelhead trout. *Int. North. Pac. Fish. Invest.*, 46, 106 pp.

Hardy, A.C. 1935. The plankton community, the whale fisheries and the hypothesis of animal exclusion. *Discovery Rep.*, 11: 273-360.

Harris, J.G.K. 1975. The effect of density-dependent mortality on the shape of the stock and recruitment curve. *J. Cons. Int. Explor. Mer.*, 36: 144-149.

Harry, G.Y., Jr. 1959. Time of spawning, length of maturity, and fecundity of the English, petrale and Dover soles (*Parophrys vetulus*, *Eopsetta jordani* and *Microstomus pacificus*, respectively). *Oregon Fish. Comm. Res. Briefs,* 7: 5-13.

Hatfield, S.E. 1983. Distribution of zooplankton in association with Dungeness crab, *Cancer magister*, larvae in California. Pages 97-123 in: P. Wild and R. Tasto (eds.), *Life History, Environment and Mariculture Studies of the Dungeness Crab,* Cancer Magister, *With Emphasis on the Central California Fishery Resource. CDFG Fish Bull.* 172.

Hayman, R.A. and A.V. Tyler. 1980. Environment and cohort strength of Dover sole and English sole. *Trans. Am. Fish. Soc.,* 109: 54-70.

Hayman, R.A., A.V. Tyler and R.L. Demory. 1980. A comparison of cohort analysis and catch per unit effort for Dover sole and English sole. *Trans. Am. Fish. Soc.*, 109: 35-53.

Hewitt, G.R. 1980. Seasonal changes in English sole distribution: an analysis of the inshore trawl fishery off Oregon. M.S. Thesis, Oregon St. Univ., Corvallis, Ore., 59 pp.

Hickey, B. M. 1989. Patterns and process of circulation over the shelf and Slope. Pages 41-116 in: M.L. Landry and B.M. Hickey (eds.), *Coastal Oceanography of Washington and Oregon.* Elsevier, Amsterdam, Neth.

Hjort, J. 1914. Fluctuations in the great fisheries of northern Europe viewed in the light of biological research. *Rapp. P.-V. Reun. Comm. Int. Explor. Mer*, 20: 1-228.

Horne, E.P.W. and T. Platt. 1984. The dominant space and time scales of variability in the physical and biological fields on continental shelfs. *Rapp. P.-v. R'eun. Cons. int. Explor. Mer.,* 183: 8-19.

Hobbs. R.L. and L.W. Botsford. In press. Dynamics of Dungeness crab and nemertean worm interaction: an age-structured prey with density--and predation-dependent recruitment. *Theor. Pop. Biol.*

Horwood, J.W. and J.G. Shepherd. 1981. The sensitivity of age-structured populations to environmental variability. *Math. Biosci.,* 57: 59-82.

Houde, E.D. 1975. Effects of stocking density and food density on survival, growth and yield of laboratory-reared larvae of sea bream *Archosargus rhomboidalis* (L.) (Sparidae). *J. Fish. Biol.,* 7: 115- 127.

Houde, E.D. 1977. Food concentrations and stocking density effects on survival and growth of laboratory-reared larvae of bay anchovy *Anchoa mitchilli* and lined sole *Achirus lineatus*. *Mar. Biol.*, 43: 333-341.

Houde, E.D. 1978. Critical food concentrations for larvae of three species of subtropical marine fishes. *Bull. Mar. Sci.*, 28: 395-411.

Hunter, J.R. 1977. Behavior and survival of northern anchovy *Engraulis mordax* larvae. *Calif. Coop. Oceanic Fish. Invest. Rep.*, 19: 138-146.

Hunter, J.R. 1981. Feeding ecology and predation of marine fish larvae. Pages 33-79 in: R. Lasker (ed.), *Marine Fish Larvae*. Univ. of Washington Press, Seattle,Wa.

Hunter, J.R. and C.A. Kimbrell. 1980. Egg cannibalism in the northern anchovy, *Engraulis mordax*. *Fish. Bull., U.S.*, 78: 811-816.

Huyer, A. 1984. Hydrographic observations along the CODE central line off northern California, 1981. *J. Phys. Oceanogr.*, 14: 1647-1658.

Huyer, A., R.L. Smith and B.M. Hickey. 1984. Observations of a warm-core eddy off Oregon, January to March 1978. *Deep-Sea Res.*, 31(2): 97-117.

Incze, L.S., D.A. Armstrong and S.L. Smith. In review. Abundance of larval tanner crabs (*Chionoecetes spp.*) in relation to adult female stocks and regional oceanography of the southeastern Bering Sea. *Can. J. Fish. Aquat. Sci.*

Jackson, G.A. 1986. Interaction of physical and biological processes in the settlement of planktonic larvae. *Bull. Mar. Sci.*,39: 202-212.

Jackson, G.A. and R.R. Strathmann. 1981. Larval mortality from offshore mixing as a link between precompetent and competent periods of development. *Am. Nat.*, 118: 16-26.

Jacoby, C. 1982. Behavior responses of the larvae of *Cancer magister* Dana (1852) to light, pressure and gravity. *Mar. Behav. Physiol.*, 8: 267-283.

Jamieson, G.S. 1986. Implications of fluctuations in recruitment in selected crab populations. *Can. J. Fish. Aquat. Sci.*, 43: 2085-2098.

Jamieson, G.S. and A. C. Phillips. In press. Occurrence of *Cancer* crab (*C. magister* and *C. oregonensis*) megalopae off the west coast of Vancouver Island, British Columbia. *Fish. Bull., U.S.*

Jamieson, G.S., A.C. Phillips and W.S. Hugget. In review. Effects of ocean variability on the abundance of Dungeness crab larvae.

Jensen, G. and D. Armstrong. 1987. Range extension of some northeastern Pacific Decapoda. *Crustaceana,* 52: 215-217.

Johnson, D.F., L.W. Botsford, R.D. Methot, Jr. and T.C. Wainwright. 1986. Wind stress and cycles in Dungeness crab (*Cancer magister*) catch off California, Oregon and Washington. *Can. J. Fish. Aquat. Sci.*, 43: 838-845.

Johnson, D.R., B.S. Hester, J.R. McConangha. 1984. Studies of a wind mechanism influencing the recruitment of blue crabs in the middle Atlantic Bight. *Cont. Shelf Res.*, 3: 425-437.

Johnson, G.E. and J.J. Gonor. 1982. The tidal exchange of *Callianassa californiensis* (Crustacea, Decapoda) larvae between the ocean and the Salmon River Estuary, Oregon. *Estuar., Coast. and Shelf Sci.,* 14: 501-516.

Johnson, S.L. 1988. The effect of the 1983 El Niño on Oregon's coho (*Oncorhynchus kisutch*) and chinook (*O. tshawytscha*) salmon. *Fish. Res.,* 6: 105-123.

Ketchen, K.S. 1956. Factors influencing the survival of the lemon sole (*Parophrys vetulus*) in Hecate Strait, British Columbia. *J. Fish. Res. Bd. Can.*, 13: 647-694.

Kreuz, K.L, A.V. Tyler, G.H. Kruse and R.L. Demory. 1982. Variation in growth of Dover sole and English sole as related to upwelling. *Trans. Am. Fish. Soc.*, 111: 180-192.

Kruse, G.H. 1983. A simulation model of English sole (*Parophrys vetulus*) recruitment mechanisms. Ph.D. Diss., Oregon St. Univ., Corvallis, Ore., 103 pp.

Kruse, G.H. and A. Huyer. 1983. Relationships among shelf temperatures, coastal sea level and the coastal upwelling index off Newport, Ore. *Can. J. Fish. Aquat. Sci.*, 40: 238-242.

Kruse, G.H. and A.V. Tyler. 1983. Simulation of temperature and upwelling effects on the English sole (*Parophrys vetulus*) spawning season. *Can. J. Fish. Aquat. Sci.*,40: 230-237.

Krygier, E.E. and W.G. Pearcy. 1986. The role of estuarine and offshore nursery areas for young English sole, *Parophrys vetulus* Girard, of Oregon. *Fish. Bull., U.S.*, 84: 119-132.

Laroche, J.L. and S.L. Richardson. 1979. Winter-spring abundance of larval English sole, *Parophrys vetulus*, between the Columbia River and Cape Blanco, Oregon, during 1972-1975, with notes on occurrences of three other pleuronectids. *Estuar. Coast. Mar. Sci.*, 8: 455-476.

Laroche, J.L., S.L. Richardson and A.A. Rosenberg. 1982. Age and growth of a pleuronectid, *Parophrys vetulus*, during the pelagic larval period in Oregon coastal waters. *Fish. Bull., U.S.*, 80: 93-104.

Lasker, R. 1975. Field criteria for survival of anchovy larvae: the relation between inshore chlorophyll maximum layers and successful first feeding. *Fish. Bull., U.S.*, 73: 453-462.

Lasker, R. 1978. The relation between oceanographic conditions and larval anchovy food in the California Current: identification of factors contributing to recruitment failure. *Rapp. P.-v. Reun. Cons. Int. Explor. Mer.*, 173: 375-388.

Lasker, R.L. 1981. Factors contributing to variable recruitment of the northern anchovy (*Engraulis mordax*) in the California current: contrasting years, 1975 through 1978. *Rapp.P.-v. Reun. Cons. Int. Explor. Mer.,* 173: 212-230.

Lasker, R. and J. R. Zweifel. 1978. Growth and survival of first-feeding northern anchovy larvae (*Engraulis mordax*) in patches containing different proportions of large small prey. Pages 329-354 in: J. H. Steele (ed.), *Spatial Pattern in Plankton Communities*. Plenum Press, New York, N.Y.

Laurs, R.M. and R.J. Lynn. 1977. Seasonal migration of albacore, *Thunnus alalunga*, into North American coastal waters, their distribution, relative abundance and association with transition zone waters. *Fish. Bull., U.S.*, 75: 795-822.

Legendre, L. and S. Demers. 1984. Towards dynamic biological oceanography and limnology. *Can. J. Fish. Aquat. Sci.*, 41: 2-19.

Lentz, S.J. 1987. A description of the 1981 and 1982 spring transitions over the Northern California shelf. *J. Geophys. Res.*, 92: 1545-1567.

Levin, S. 1987. Scale and predictability in ecological modeling. Pages 2-8 in: T.L. Vincent, Y. Cohen, W.J. Grantham, G.P. Kirkwood and J.M. Skowronski (eds.), *Modeling and Management of Resources Under Uncertainty*. Springer-Verlag, New York, N.Y.

Loeffel, R.E. and W.O. Forster. 1970. Determination of movement and identity of stocks of coho salmon in the ocean using the radionucleide zinc-65. *Res. Rep. Fish. Comm. Oreg.*, 2: 1-12.

Lough, R.G. 1976. Larval dynamics of the Dungeness crab, *Cancer magister*, off the central Oregon coast, 1970-1971. *Natl. Mar. Fish. Serv., Fish Bull., U.S.*, 74: 353-376.

Love, M.S. and M.V. Westphal. 1981. A correlation between annual catches of Dungeness crab, *Cancer magister*, along the west coast of North America and mean annual sunspot number. *Fish. Bull., U.S.*, 79: 794-796.

Madsen, O.S. 1977. A realistic model of the wind induced Ekman boundary layer. *J. Phys. Oceanogr.*, 7: 248-255.

Mathews, S..B. 1971. Economic evaluation of forecasts of sockeye salmon (*Oncorhynchus nerka*) runs to Bristol Bay, Alaska. FAO Fisheries Tech. Paper No. 103, 17 pp.

Mathews, S.B. and F.W. Olson. 1980. Factors affecting Puget Sound coho salmon (*Oncorhynchus kisutch*). *Can. J. Fish. Aquat. Sci.*, 37: 1373-1378.

May, R.M. (ed.). 1984. *Exploitation of Marine Communities.* Springer Verlag, New York, N.Y., 366 pp.

McCarl, B.A. and R.B. Rettig. 1983. Influence of hatchery smolt releases on adult salmon production and its variability. *Can. J. Fish. Aquat. Sci.,* 40: 1880-1886.

McGie, A.M. 1984. Commentary: Evidence for density dependence among coho salmon stocks in the Oregon Production Index area. Pages 37-429 in: W.G. Pearcy (ed.), *The Influence of Ocean Conditions on the Production of Salmonids in the North Pacific: a Workshop.* Oregon St. Univ. Press, Corvallis, Ore.

McKelvey, R. and D. Hankin. 1981. Reply to comment on cycles in the northern California Dungeness crab population. *Can. J. Fish. Aquat. Sci.,* 1296-1297.

McKelvey, R., D. Hankin, K. Yanosko and C. Snygg. 1980. Stable cycles in multistage recruitment models: an application to the northern California Dungeness crab (Cancer magister) fishery. *Can. J. Fish. Aquat. Sci.,* 37: 2323-2345.

Meltef, B. 1985. Proc. symp. Dungeness crab biology and management. Univ. of Alaska Sea Grant Rep. 85-3, 424 pp.

Mendelssohn, R. and C. I. Roy. 1987. Environmental influences on the French, Ivory Coast, Senegalese and Moroccan tuna catches in the Gulf of Guinea. Pages 170-188 in: P.E.K. Symons, P.M. Miyake and G.T. Sakagawa (eds.) *Proceedings of the ICCAT Conference on the International Skipjack Year Program.*

Methot. R.D. Jr. 1983. Seasonal variation in survival of larval *Engraulis mordax* estimated from the age distributions of juveniles. *Fish. Bull., U.S.*, 81 741-750.

Methot, R.D. Jr. 1986. Management of Dungeness crab fisheries. Pages 326-334 in: G.S. Jamieson and N. Bourne (ed.), *North Pacific Workshop on Stock Assessment and Management of Invertebrates. Can. Spec. Publ. Fish. Aquat. Sci.*, 92.

Methot, R.D. Jr. and L.W. Botsford. 1982. Estimated preseason abundance in the California Dungeness crab (*Cancer magister*) fisheries. *Can. J. Fish. Aquat. Sci.*, 39: 1077-1083.

Methot, R.D. Jr. and D. Kramer. 1979. Growth of northern anchovy, *Engraulis mordax*, larvae in the sea. *Fishery Bull., U.S.,* 77: 413-429.

Mooers, C.N.K. and A.R. Robinson. 1984. Turbulent jets and eddies in the California Current and inferred cross-shore transports. *Science,* 223: 51-53.

Morgan, G.R., B.R. Phillips and L.M. Joll. 1982. Stock and recruitment relationships in *Panulirus cygnus*, the commercial rock (spiny) lobster of western Australia. *Fish. Bull., U.S.,* 80: 475-486.

Mullin, M.M., E.R. Brooks, F.M.H. Reid, J. Napp and E.F. Stewart. 1985. Vertical structure of nearshore plankton off southern California: a storm and a larval fish food web. *Fish. Bull., U.S.*, 83: 151-170.

Mundy, B.C. 1983. Yearly variation in the abundance and distribution of fish larvae in the coastal zone off Yaquina Head, Ore., from June 1969 to August 1972. M.S. Thesis, Oregon St. Univ., Corvallis, Ore., 158 pp.

Myers, K.W. 1980. An investigation of the utilization of four study areas in Yaquina Bay, Ore., by hatchery and wild juvenile salmonids. M.S. Thesis, Oregon St. Univ., Corvallis, Ore., 234 pp.

Namais, J. and J.C.K. Huang. 1972. Sea level at Souther California: a decadal fluctuation. *Science*, 177: 351-353.

Nelson, R.N., M.C. Ingham and W.E. Schaaf. 1977. Larval transport and year-class strength of Atlantic menhaden, *Brevoortia tyrannus. Fish. Bull., U. S.*, 75: 23-41.

Nichols, J.H., B.M. Thompson and M. Cryer. 1982. Production, drift and mortality of the planktonic larvae of the edible crab (*Cancer pagurus*) off the northeast coast of England. *Neth. J. Sea Res.*, 16: 173-184.

Nickelson, T.E. 1986. Influences of upwelling, ocean temperature and smolt abundance on marine survival of coho salmon (*Oncorhynchus kisutch*) in Oregon production area. *Can J. Fish. Aquat. Sci.*, 43: 527-535.

Nickelson, T.E. and J.A. Lichatowich. 1984. The influence of the marine environment on the interannual variation in coho salmon abundance: an overview. Pages 24-36 in: W.G. Pearcy (ed.), *The Influence of Ocean Conditions on the Production of Salmonids in the North Pacific: a Workshop.* Oregon St. Univ. Press, Corvallis, Ore.

Norcross, B.L. and R.F. Shaw. 1984. Oceanic and estuarine transport of fish eggs and larvae: a review. *Trans. Amer. Fish. Soc.* 113: 153-165.

Owen, R.W. 1981. Microscale plankton patchiness in the larval anchovy environment. *Rapp. P.-v. Reun. Cons. Int. Explor. Mer,* 178: 364-368.

Parsons, T.R., B.O. Jansson, A.R.Longhurst and G. Saetersdal (eds.). 1978. Marine ecosystems and fisheries oceanography. *Rapp. P.-v. Reun. Cons. Int. Explor. Mer*, 173: 240 pp.

Parrish, R.H., A. Bakun, D.M. Husby and C.S. Nelson. 1983. Comparative climatology of selected environmental processes in relation to eastern boundary current pelagic fish reproduction. Pages 731-778 in: G.D. sharp and J. Csirke (ed.), *Proceedings of the Expert Consultation to Examine Changes in Abundance and Species of Neritic Fish Resources.* FAO Fish. Rep., 291.

Parrish, R.H., C.S. Nelson and A. Bakun. 1981. Transport mechanisms and reproductive success of fishes in the California Current. *Bio. Oceanogr.*, 1: 175-203.

Paul, A.J. and J.M. Paul. 1980. The effect of early starvation on later feeding success of king crab zoeae. *J. Exp. Mar. Biol. Ecol.*, 44: 247-251.

Paul, A.J., J.M. Paul, P.A. Shoemaker and H.M. Feder. 1979. Prey concentrations and feeding response in laboratory-reared stage on zoeae of king crab, snow crab and pink shrimp. *Trans. Amer. Fish. Soc.*, 108: 440-443.

Pauley, G.B., D.A. Armstrong and T.W. Heun. 1986. Species profiles: life histories and environmental requirements of coastal fishes and invertebrates (Pacific Northwest). *U.S. Fish. Wildl. Serv. Biol. Rep.* 82, 19 pp.

Pearcy, W.G. 1984. Where do all the coho go? The biology of juvenile coho salmon off the coasts of Oregon and Washington. Pages 50-60 in: W.G. Pearcy (ed.), *Influence of Ocean Conditions On the Production of salmonids in the North Pacific.* Oregon St. Univ. Press, Corvalis, Ore.

Pearcy, W.G. and J.P. Fisher. In press. Migrations of coho salmon (*Oncorhynchus kisutch* during their first summer in the ocean. *Fish Bull.,U.S.*

Pearcy, W.G., J. Fisher, R.Brodeur and S. Johnson. 1985. Effects of the 1983 El Niño on coastal nekton off Oregon and Washington. Pages 188-204 in: W.S.Wooster and D.L Fluharty (eds.), *El Niño North*. Washington Sea Grant Program, Seattle, Wa.

Peterman, R.M. 1981. Form of random variation in salmon smolt-to-adult relations and its influence on production estimates. *Can. J. Fish. Aquat. Sci.*, 38: 1113-1119.

Peterman, R.M. 1987. Review of the components of recruitment of Pacific salmon. *American Fish. Soc. Symp.,* 1: 417-429.

Peterman, R.M. and M.J. Bradford. 1987a. Wind speed index correlates with mortality rate of a marine fish, the northern anchovy. *Science,* 235: 354-356.

Peterman, R.M. and M.J. Bradford. 1987b. Density-dependent growth of age 1 English sole (*Parophrys vetulus*) in Oregon and Washington coastal waters. *Can. J. Fish. Aquat. Sci.*, 44: 48-53.

Peterman, R.M. and R.D. Routledge. 1983. Experimental management of Oregon coho salmon (*Oncorhynchus kisutch*): designing for yield of information. *Can J. Fish. Aquat. Sci.,* 40: 1212-1223.

Peterman, R.M., M.J. Bradford and G.H. Kruse. In press. Simulation model of English sole (*Parophrys vetulus*) population dynamics in Washington and Oregon coastal waters. *Can. J. Fish. Aquat. Sci.*

Peterson, W.T. 1973. Upwelling indices and annual catches of Dungeness crab, *Cancer magister*, along the west coast of the United States. *Fish. Bull., U.S.*, 71: 902-910.

Peterson, W.T. 1980. Life history and ecology of *Calanus marshallae* frost in the Oregon upwelling zone. Ph. D. Diss., Oregon St. Univ., Corvalis, Ore., 200 pp.

Peterson, W.T. and C.B. Miller. 1976. Zooplankton along the continental shelf off Newport, Oregon, 1969-1972: distribution, abundance, seasonal cycle and year-to-year variations. Oregon St. Univ., Sea Grant Publ. ORESU-t-76-002, 111 pp.

Peterson, W.T., C.B. Miller and A. Hutchinson. 1979. Zonation and maintenance of copepods in the Oregon upwelling zone. *Deep-Sea Res.,* 26: 467-494.

Phillips, B.F. 1986. Prediction of commercial catches of the western rock lobster *Panulirus cygnus*. *Can J. Fish. Aquat. Sci.*, 43: 2126-2130.

Platt, J.R. 1964. Strong inference. *Science*, 146: 347-353.

Poole, R.L. 1967. A description of laboratory-reared zoeae of *Cancer magister* Dana and megalopae taken under natural conditions (Decapoda, Brachyura). *Crustaceana,* 11: 83-97.

Power, J.H. 1986. A model of the drift of northern anchovy *Engraulis mordax*, larvae in the California Current. *Fish. Bull.,U.S.,* 84: 585-603.

Radovich, J. 1961. Relationships of some marine organisms of the northeast Pacific to water temperature, particularly during 1957 through 1959. *Fish. Bull.,Calif. Dept. Fish Game,* 112 pp.

Reed, P.H. 1969. Culture methods and effects of temperature and salinity on survival and growth of the Dungeness crab (*Cancer magister*) larvae in the laboratory. *J. Fish Res. Board Can.*, 26: 389-397.

Reed, W.J. 1983. Recruitment variability and age structure in harvested animal populations. *Math. Biosci.*, 65: 239-268.

Reid, J.L. Jr. and R.A. Schwartzlose. 1962. Direct measurement of the Davidson Current off Central California. *J. Geophys. Res.*, 67: 2491-2497.

Reilly, P.N. 1983. Dynamics of Dungeness crab, *Cancer magister*, larvae off central and northern California. Pages 57-84 in: P.W. Wild and R.N. Tasto (eds.), Life history, Environment and Mariculture Studies of the Dungeness Crab, Cancer magister, With Emphasis on the central California Fishery Service. *Fish. Bull.,Calif. Dept. Fish Game,* 172.

Richardson, S.L. 1977. Larval fishes in ocean waters off Yaquina Bay, Oregon: abundance, distribution and seasonality. January 1971 to August 1972. Sea Grant Publ. ORESU-T-77-003, 73 pp.

Richardson, S.L, J.L. Laroche and M.D. Richardson. 1980. Larval fish assemblages and associations in the north-east Pacific Ocean along the Oregon coast, winter-spring 1972-1975. *Est. Coast. Mar. Sci.*, 11: 671-699.

Richardson, S.L. and W.G. Pearcy. 1977. Coastal and oceanic fish larvae in an area of upwelling off Yaquina Bay, Oregon. *Fish. Bull., U.S.*, 75: 125-145.

Ricker, W.E. 1975. Computation and interpretation of biological statistics of fish populations. *Bull. Fish., Res. Board Can.* 191, 382pp.

Ricketts. E.R., J. Calvin and J. W. Hedgepeth. 1968. Between Pacific Tides. 4th ed. Stanford Univ. Press, Stanford, Calif., 614 pp.

Rogers, C. 1985. Population dynamics of juvenile flatfish in the Grays Harbor estuary and adjacent nearshore area. M.S. Thesis, Univ. of Washington, Seattle, Wa., 195 pp.

Rosenberg, A.A. 1982. Growth of juvenile English sole, *Parophrys vetulus*, in estuarine and open coastal nursery grounds. *Fish. Bull., U.S.*, 80: 245-252.

Rosenberg, A.A. and J.L. Laroche. 1982. Growth during metamorphosis of English sole, *Parophrys vetulus. Fish. Bull., U.S.*, 80: 150-153.

Rothschild, B.J. 1986. Dynamics of marine fish populations. Harvard Univ. Press, Cambridge, Mass., 277 pp.

Rothlisberg, P.C. 1982. Vertical migration and its effect on dispersal of penaeid shrimp larvae in the Gulf of Carpentaria, Australia. *Fish. Bull., U.S.*, 80: 541-554.

Rothlisberg, P.C., J.A. Church and A.M.G. Forbes. 1983. Modeling the advection of vertically migrating shrimp larvae. *J. Mar. Res.*, 41: 511-538.

Rothlisberg, P.C. and C.B. Miller. 1983. Factors affecting the distribution, abundance and survival of *Pandalus jordani* (decapoda, pandalidae) larvae off the Oregon coast. *Fish. Bull., U.S.*, 81: 455-472.

Russell, F.S. 1925. The vertical distribution of marine macroplankton. An observation on diurnal changes. *J. Mar. Biol. Assoc. U.K.*, 13: 769-809.

Sandifer, P.A. 1975. The role of pelagic larvae in recruitment to populations of adult decapitate crustaceans in the York River Estuary and adjacent lower Chesapeake Bay, Virginia. *Estuar. Coast. Mar. Sci.*, 3: 269-279.

Scarnecchia, D.L. 1981. Effects of streamflow and upwelling on yield of wild coho salmon (*Oncorhynchus kisutch*) in Oregon. *Can. J. Fish. Aquat. Sci.*, 38: 471-475.

Send, U., R.C. Beardsley and C.D. Winant. 1987. Relaxation from upwelling in the Coastal Ocean Dynamics Experiment. *J. Geophys. Res.*, 92: 1683-1698.

Shanks, A.L. 1983. Surface slicks associated with tidally forced internal waves may transport pelagic larvae of benthic invertebrates and fishes shoreward. *Mar. Ecol. Prog. Ser.*, 13: 311-315.

Shanks, A.L. 1985. Behavioral basis of internal-wave-induced shoreward transport of megalopae of the crab *Pachygrapsus crassipes. Mar. Ecol. Prog. Ser.*, 24: 289-295.

Shenker, J.M. In review. Oceanographic associations of neustonic larval and juvenile fishes and Dungeness crab megalopae off Oregon. *Fish. Bull., U.S.*

Sherman, K. R. Lasker, W. Richards and A. Kendall, Jr. 1983. Ichthyoplankton and fish recruitment studies in large marine ecosystems. *Mar. Fish. Rev.*, 45: 1-25.

Sissenwine, M.P. 1984. Why do fish populations vary? Pages 59-94 in: R.M. May (ed.), *Exploitation of Marine Communities*. Springer Verlag, New York, N.Y.

Smith, P.E. 1978. Biological effects of ocean variability: time and space scales of biological response. Pages 117-127 in: Marine Ecosystems and Fisheries Oceanography. *Rapp. P.-v. Reun. Cons. Intern. Explor. Mer*, 173.

Smoker, W.A. 1955. Effects of stream flow on silver salmon production in western Washington. Ph. D. Diss., Univ. of Washington, Seattle, Wa., 175 pp.

Staples, D.J. and D.J. Vance. 1985. Short-term and long-term influences on the immigration of postlarval banana prawns, *Penaeus merguiensis*, into a mangrove estuary of the Gulf of Carpentaria, Australia. *Mar. Ecol. Prog. Ser.*, 23: 15-29.

Steele, J.H. 1978. Some comments on plankton patches. Pages 1-20 in: H. Steele (ed.), *Spatial Pattern in Plankton Communities*. Plenum Press, New York, N.Y.

Steele, J.H. 1984. Kinds of variability and uncertainty affecting fisheries. Pages245-262 in: R.M. May (ed.) *Exploitation of Marine Communities*. Springer Verlag, New York, N.Y.

Stevens, B. and D.A. Armstrong. 1984. Distribution, abundance and growth of juvenile Dungeness crabs, (*Cancer magister*), in Grays Harbor estuary, Washington U.S.A. *Fish. Bull., U.S.*, 82: 469-483.

Stevens, B.G. and D.A. Armstrong. 1985. Ecology, growth and population dynamics of juvenile Dungeness crab, *Cancer magister* Dana, in Grays Harbor, Washington, 1980-1981. Pages 119-134 in: B. Metlef (ed.), *Proceedings of the Symposium on Dungeness Crab Biology and Management, Oct. 1984*. Univ. of Alaska Sea Grant Rep. No. 85-3.

Stevens, B.G., D.A. Armstrong and R. Cusimano. 1982. Feeding habits of the Dungeness crab *Cancer magister* as determined by the index of relative importance. *Mar. Biol.*, 72: 135-145.

Strub, P.T., J.S. Allen, A. Huyer and R.L.Smith. 1987. Large scale structure of the spring transition in the coastal ocean off western North America. *J. Geophys. Res.*, 92: 1527-1544.

Sulkin, S.D. 1984. Behavioral basis of depth regulation in the larvae of brachyuran crabs. *Mar. Ecol. Prog. Ser.*, 15: 181-205.

Sulkin, S.D. and C.E. Epifanio. 1986. A conceptual model for recruitment of the blue crab, *Callinectes sapidus* Rathbun, to estuaries of the Middle Atlantic Bight. Pages 59-94 in: G.S. Jamieson and N. Bourne (eds.), North Pacific Workshop on Stock Assessment and Management of Invertebrates. *Can. Spec. Publ. Fish. Aquat. Sci.*, 92: 117-123.

Sulkin, S.D. and W.F. VanHeukelem. 1982. Larval recruitment in the crab *Callinectes sapidus* Rathbun: an amendment to the concept of larval retention in estuaries. Pages 459-475 in: V. Kennedy (ed.), *Estuarine Comparisons*. Academic Press, New York, N.Y.

Sundby, S. 1983. A one-dimensional model for the vertical distribution of pelagic fish eggs in the mixed layer. *Deep-Sea Res.*, 30: 645-661.

Swartzman, G.L., W.M. Getz, R.C. Francis, R.T. Haar and K. Rose. 1983. A management analysis of the Pacific whiting (*Merluccius productus*) fishery using an age-structured stochastic recruitment model. *Can. J. Fish. Aquat. Sci.*, 40: 524-539.

Tasto, R.N. 1983. Juvenile Dungeness crab, *Cancer magister*, studies in the San Francisco area. Pages 135-154 in: P.W. Wild and R.N. Tasto (eds.). Life History, Environment and Mariculture Studies of the Dungeness Crab, *Cancer magister*, with Emphasis on the Central California Fishery Resource, *Fish Bull.,Calif. Dept. Fish Game,* 172.

Theilacker, G.H, 1978. Effect of starvation on the histological and morphological characteristics of jack mackerel, *Trachurus symmetricus*, larvae. *Fish. Bull., U.S.*, 76: 403-414.

Theilacker, G.H. 1981. Effect of feeding history and egg size on the morphology of jack mackerel, *Trachurus symmetricus*, larvae. *Rapp. P.-v Reun. Cons. Int. Explor. Mer,* 178: 432-440.

Thomas, D.H. 1985. A possible link between coho (silver) salmon enhancement and a decline in central California Dungeness crab abundance. *Fish. Bull., U.S.*, 83: 682-691.

Tukey, J.W. 1977. *Exploratory Data Analysis.* Addison-Wesley, Reading, Mass., 688 pp.

Velleman, P.F. and D.C. Hoaglin. 1981. *Applications, Basics and Computing of Exploratory Data Analysis.* Duxbury Press, Boston, Mass., 354 pp.

Vlymen, W.J. 1977. A mathematical model of the relationship between larval anchovy (*Engraulis mordax*) growth, prey microdistribution and larval behavior. *Environ. Biol. Fish.*, 2: 211-233.

Walford, L. A. 1938. Effect of currents on distribution and survival of the eggs and larvae of the haddock (*Melanogrammus aeglefinus*) on Georges Bank. *Fish. Bull.,U. S.*, 49: 1-73.

Walsh, J.J. 1978. The biological consequences of interaction of the climatic, El Niño and event scales of variability in the eastern tropical Pacific. Pages 182-191 in: Marine Ecosystems and Fisheries Oceanography. *Rapp. P.-v Reun. cons. int. Explor. Mer*, 173.

Walters. C.J. In press. Perspectives on adaptive policy design in fisheries management. *In:* S.K. Jain and L.W. Botsford (eds.), *Applied Population Biology.* W. Junk Publ., Hingham, Mass.

Walters, C.J. and R. Hilborn. 1976. Adaptive control of fishing systems. *J. Fish. Res. B. Can.*, 33: 145-159.

Walters, C.J. and M.J. Staley. In press. Evidence against the existence of cyclic dominance in Fraser River sockeye salmon (Oncorhynchus nerka). *Can. J. Fish. Aquat. Sci.*

Walters, C.J., M. Stocker, A.V. Tyler and S.J. Westrheim. 1986. Interaction between Pacific Cod (*Gadus macrocephalus*) and herring (*Clupea harengus pallasi*) in the Hecate Strait, British Columbia. *Can. J. Fish. Aquat. Sci.*, 43: 830-837.

Wickett, P. 1960. Relationship of coastal water convergence and Fraser River discharge to migration of Fraser River sockeye salmon through Johnstone Strait. Pac. Biol. Station, Nanaimo, B.C. File Rep. 32-7.

Wickham, D.E. 1979a. Predation by the nemertean, *Carcinonemertes errans* on eggs of the Dungeness crab *Cancer magister. Mar. Biol.*, 55: 45-53.

Wickham, D.E. 1979b. The relationship between megalopae of the Dungeness crab, *Cancer magister* and the hydroid, *Velella velella* and its influence on abundance estimates of *C. magister* megalopae. *Fish. Bull., Calif. Dept. Fish Game*, 65: 184-186.

Wild, P.W. 1980. Effects of seawater temperature on spawning, egg development, hatching success and population fluctuations of the Dungeness crab, *Cancer magister. Calif. Coop. Oceanic Fish Invest. Rep.*, 21: 115-120.

Wild, P.W. 1983. The influence of seawater temperature in spawning, egg development, ad hatching success of the Dungeness crab, *Cancer magister*. Pages 197-213 in: P.W. Wild and R.N. Tasto (ed.), Life History, Environment and Mariculture Studies of the Dungeness Crab, *Cancer magister*, with Emphasis on the Central California Fishery Resource. *Fish. Bull., Calif. Dept. Fish Game*, 172.

Wild, P.W., P.M.W. Law and D.R. McLain. 1983. Variations in ocean climate and the Dungeness crab fishery in California. Pages 175-180 in: P.W. Wild and R.N. Tasto (ed.), Life History, Environment and Mariculture Studies of the Dungeness Crab, *Cancer magister*, with emphasis on the central California fishery resource. *Fish. Bull., Calif. Dept. Fish Game,* 172.

Winant, C.D., R.C. Beardsley and R.E. Davis. 1987. Moored wind, temperature and current observations made during Coastal Ocean Dynamics Experiments 1 and 2 over the northern California continental shelf and upper slope. *J. Geophys. Res.*, 92: 1569-1604.

Wooster, W.S. (ed.) 1983 . From year to year. Interannual Variability of the Environment and Fisheries of the Gulf of Alaska and the Eastern Bering Sea. Univ. of Washington. Sea Grant Publ. WSG-WO-83- 3, 208 pp.

Wroblewski, J.W. 1980. A simulation of the distribution of *Acatia clausi* during Oregon upwelling, August, 1973. *J. Plankton Res.*, 2: 43-68.

Wroblewski, J.S. 1984. Formulation of growth and mortality of larval northern anchovy in a turbulent feeding environment. *Mar. Ecol. Prog. Ser.*, 20: 13-22.

Wyatt, B., W.B. Burt and J.G. Pattullo. 1972. Surface currents off Oregon as determined from drift bottle returns. *J. Phys. Oceanogr.*, 2: 286-293.

Zeldis, J.R. and J.B. Jillett. 1982. Aggregation of pelagic *Munidia gregaria* (Fabricius) (Decapoda, Anemura) by coastal fronts and internal waves. *J. Plankt. Res.*, 4: 857.

CHAPTER 12

SURF-ZONE ECOLOGY AND DYNAMICS

J. Lewin[1], C.T. Schaefer[1] and D.F. Winter[2]
[1] School of Oceanography, University of Washington, Seattle, WA 98195
[2] Department of Engineering and Computer Science, University of Redlands, Redlands, CA 92373

12.1 INTRODUCTION

The Pacific Coast of Washington and Oregon is topographically varied, with features ranging from steep rocky shores to extremely broad, gently sloping sandy beaches. The exposed rocky areas provide suitable habitats for attached macroalgae and the seaweed flora of this region is unusually rich and diverse. The focus of our studies, however, has been the extensive sandy beaches and their adjacent surf zones. In general, surf-swept sandy beaches, having a shifting substratum unsuitable for benthic plants, are locales of low primary production. Beaches of the U.S. Pacific Northwest coast, together with those in several distant regions, are a spectacular exception to this generalization (see Lewin and Schaefer, 1983). Persistent phytoplankton populations in the surf along these beaches are so dense as to cause a conspicuous brown coloration that is especially evident in the foam of breaking waves (Fig. 12.1a). Without exception, the algae that form such surf blooms are diatoms.

Although the most intensive investigations of surf-diatom blooms have been conducted at a study site on Copalis Beach, Washington, the phenomenon occurs over a distance of at least 500 km along the U.S. Northwest coast. Blooms have been observed from Point Grenville, Washington (47.3° N) to Cape Blanco, Oregon (42.8° N) (Fig. 12.2). Beyond these limits, the surf diatoms seem to disappear quite abruptly, particularly at the southern end of their range (Garver and Lewin, 1981; Jijina and Lewin, 1983).

Worldwide, only four diatom genera – *Chaetoceros*, *Asterionella*, *Aulacodiscus* and *Anaulus* – occur as the principal constituents of surf blooms. Usually one or two species are strongly dominant in any one region (Lewin and Schaefer, 1983). The major surf species on the Washington and Oregon coasts are *Chaetoceros armatum*. West and *Asterionella socialis* Lewin and Norris (Fig. 12.3). Years of observations at Copalis Beach have shown that *C. armatum* dominates numerically most of the time (Fig. 12.4) and less extensive data from 12 beaches in Oregon suggest that the same generalization applies reasonably well throughout the region (Garver and Lewin, 1981). Because cells of *C. armatum* are much larger than those of *A. socialis* (Fig. 12.3), the former species nearly always accounts for the majority of phytoplankton biomass (Lewin and Hruby, 1973; Lewin and Rao, 1975; Schaefer and Lewin, 1984). Two additional surf diatom species, *Aulacodiscus kittonii* Arnott and *Asterionella glacialis* Castracane (Becking *et al.*, 1927; Thayer, 1935a,b; Lewin and Norris, 1970; Lewin, 1974; Lewin and Schaefer, 1983) also occur on the Northwest coast, but neither currently attains concentrations approaching those of the two dominant species (Jijina and Lewin, 1983).

Fig. 12.1 Surf diatom blooms as seen along the Washington and Oregon coasts. A. Brown water as seen from the beach. B. Aerial view of a surf diatom patch made with infra-red photography.

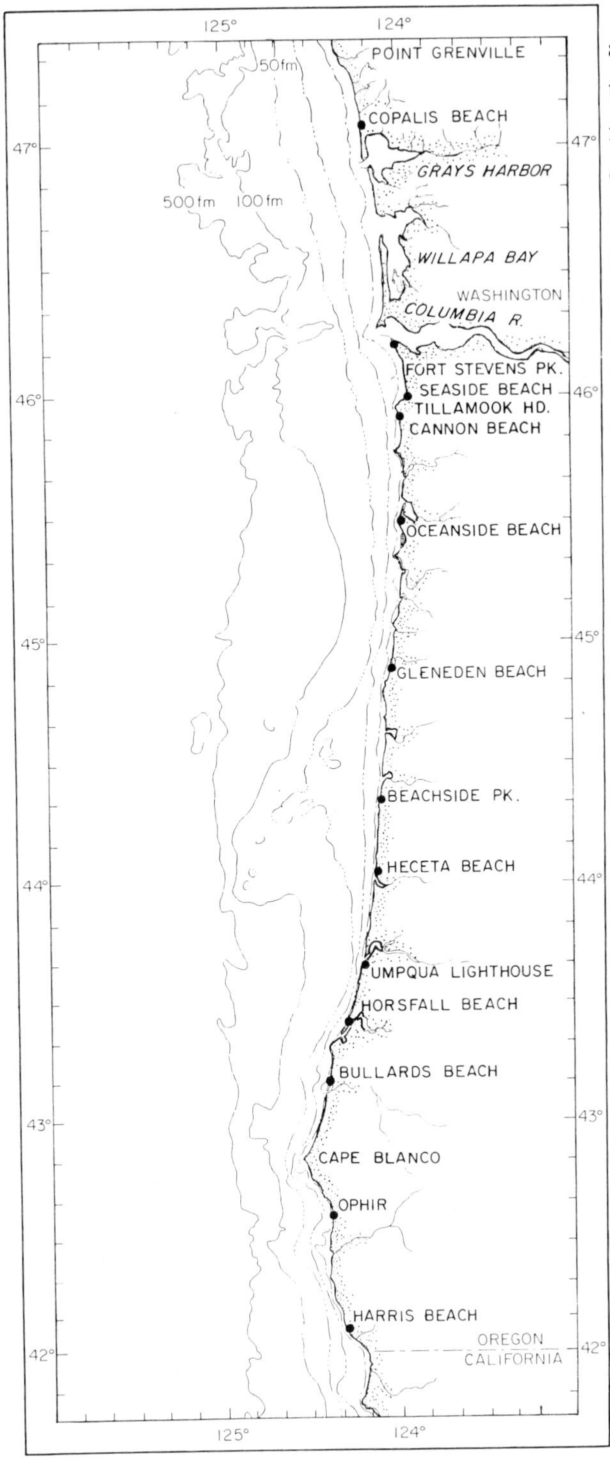

Fig. 12.2 Surf-zone study sites on the coasts of Washington and Oregon.

Surf-diatom abundance, though high at all times of the year, varies in a consistent seasonal pattern. Cell numbers of both major species are lower, typically by an order of magnitude or more, in summer than in other seasons (Fig. 12.4). In fact, *A. socialis* has been completely absent from surf samples for parts of some summers (Lewin *et al.*, 1975; Lewin, 1977, 1978a). The observation of maximal surf-diatom concentrations in winter is quite surprising since this is the season when coastal phytoplanton populations beyond the surf zone virtually disappear due to severe light limitation. The extraordinary abundance of surf diatoms and their anomalous seasonal variations stimulated our interest in elucidating the processes influencing distributions of the populations over space and time.

12.2 HISTORICAL BACKGROUND

Our research program, which operated continuously from 1971 through 1982, was the first attempt by any investigators to achieve a comprehensive understanding of the phenomenon of surf-diatom blooms. It was preceded, however, by several less intensive studies of the subject both here and abroad (see Lewin and Schaefer, 1983). Possibly the earliest published account of an event that we can confidently identify as a surf-diatom bloom is an anecdote by Van Heurck (1896, p. 488), who reported that his friend "found the sea coasts at Banana covered with a greenish bed of *Aulacodiscus africanus* Cott. . . . absolutely pure." (Banana is near the mouth of the Zaire River in Zaire). Blooms on the Washington coast apparently first attracted the attention of scientists sometime around 1920. At

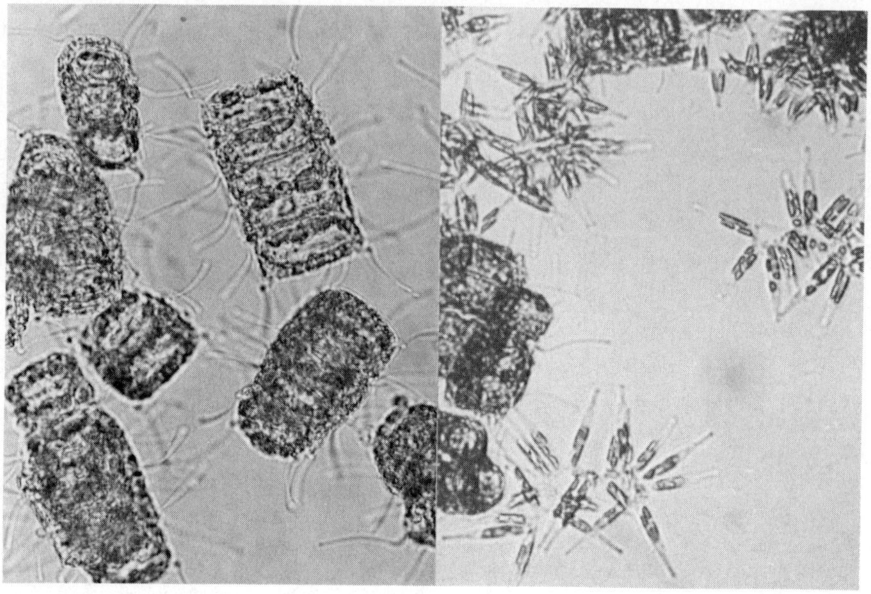

Fig. 12.3 Microphotograph showing cell chains of *Chaetocerus armatum* and colonies of *Asterionella socialis* (Mag. x375).

that time, the interest in the surf diatoms pertained chiefly to their role as food for the Pacific razor clam, *Siliqua patula* Dixon (McMillin, 1924). Shortly thereafter, a group of scientists exploring the hypothesis that coastal petroleum deposits in California were derived chiefly from diatoms mounted a more thorough study of the blooms at Copalis Beach (Becking *et al.*, 1927; Thayer, 1935a,b). Sometime in the late 1930s or the 1940s, Kincaid (1968) collected and examined surf diatoms on a continual basis as an adjunct to a study of conditions affecting growth of cultured oysters in Willapa Bay (Fig. 12.2). These observations extended at least into the mid-1950s, by which time H.C. Tegelberg (personal communication) had begun looking at the diatoms in the course of razor clam research which continued through the 1960s.

Knowledge from these earlier studies has made possible an approximate reconstruction of the history of the Washington surf blooms over the last 60 y (see also Lewin, 1974), revealing some surprising changes. The surf-diatom community during the 1920s was dominated chiefly by *Aulacodiscus kittonii*, mixed in varying proportions with a secondary dominant species identified first as *Synedra nitzschioides* Grun (Becking *et al.*, 1927) and later as *Licmophora socialis* Hanna (Thayer, 1935a). Notably, a thorough taxonomic study by G.D. Hanna, included within Thayer's (1935a) dissertation, contained no mention of *Chaetoceros armatum* nor of any species for which it could have been mistaken. Sometime after 1932, the last year from which

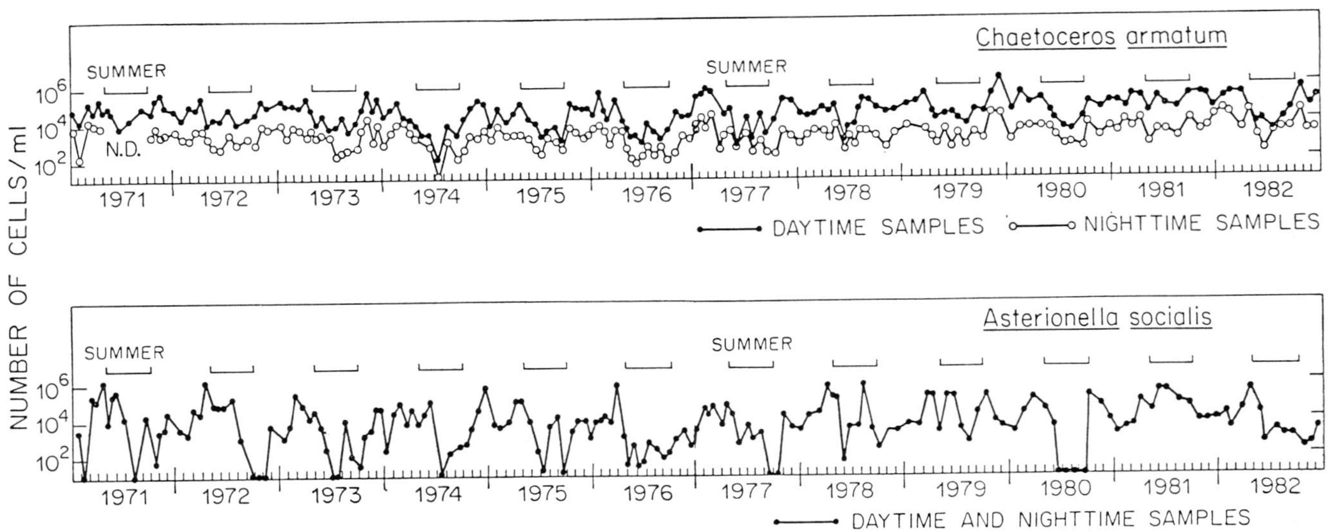

Fig. 12.4 Time series of surf-zone diatom cell densities at Copalis Beach, Washington, from 1971 through 1982.

Thayer reported any observations, *A. kittonii* declined in importance on the Washington coast. In fact, this species possibly disappeared altogether by the time Kincaid began his sampling, since it is not even mentioned in his paper (Kincaid, 1968). During the 1940s, the blooms were dominated by a single species identified by Kincaid (1968) as *Asterionella japonica* Cleve ex Gran, previously identified as *Synedra nitzschioides* and *Licmophora socialis* and now *Asterionella socialis* (Lewin and Norris, 1970; Lewin, 1974). The presence of *Chaetoceros armatum*, never before recorded on the west coast of North America, was first noticed about 1950 (Kincaid, 1968). This species quickly surpassed *A. socialis* in abundance and it has remained dominant in the surf up to the present. Its predecessor, *Aulacodiscus kittonii*, was apparently absent from Copalis Beach from 1971 through 1982 since it was not seen in any of more than 2,000 samples examined.

We cannot be sure of the causes of these dramatic changes in the surf-diatom community, but we can provide some reasonable conjectures. Considering the large influence of Columbia River discharge on conditions along the southern Washington coast (Lewin *et al.*, 1975; Lewin, 1978a, 1978b; Garver and Lewin, 1981; Jijina and Lewin, 1984), the close correspondence in time between the 1937 completion of the first major dam on the Columbia and the post-1932 decline of *A. kittonii* strongly suggests that the two events were related (Jijina and Lewin, 1983). Dams have undoubtedly modified the volume and timing of river discharge as well as the amount and character of its sediment load, any of which might be critically important for the surf diatoms. The apparently sudden arrival of *C. armatum* around 1950, on the other hand, did not coincide with any abrupt changes in environmental conditions of which we are aware. Consequently, the most plausible explanation seems to be that this species is an exotic accidently imported to our coast where it found a favorable environment. One possible source is New Zealand since *C. armatum* has occurred in dense surf blooms there at least since 1938 (Rapson, 1954; Cassie and Cassie, 1960) under conditions strikingly similar to those on the Washington and Oregon coast (Lewin and Norris, 1970; Lewin and Schaefer, 1983).

Inspection of surf samples collected more recently (November 1983, January and March 1984) at Copalis Beach has revealed the reappearance of *A. kittonii* on the Washington coast. Apparently the range of this species has now extended northward from Oregon, where it has been a minor constituent of surf bloom in recent years (Jijina and Lewin, 1983, 1984). Why *A. kittonii* was not reintroduced earlier to Washington beaches is not well understood. The timing of its return sometime between December 1982 and November 1983 suggests that enhancement of northward longshore flows during the 1982-83 El Niño might have been an important causal factor (Schaefer and Lewin, unpublished). To date, however, the abundance of this species at Copalis Beach has remained low.

12.3 PECULIAR CHARACTERISTICS OF SURF-DIATOM SPECIES

Many of the diatom species that form dense blooms in the surf, both here and abroad, appear to be endemic to the surf zone, as evidenced by their absence from phytoplankton samples collected in coastal waters seaward of the breakers. *Chaetoceros armatum*, *Asterionella socialis* and *Aulacodiscus kittonii* are among these endemics (Thayer, 1935a,b; Lewin, 1978a). Apparently all surf-diatom species worldwide share at least one peculiar characteristic that is thought to be an essential factor in their extraordinary proliferation within their particular habitat. They are capable of attaching themselves to air bubbles produced by breaking waves and somehow stabilizing those bubbles, enabling the diatoms to rise to the water surface and remain there for extended periods (Becking *et al.*, 1927; Thayer, 1935b; Rapson, 1954; Lewin and Mackas, 1972; Lewin and Hruby, 1973; Lewin and Rao, 1975; Lewin and Schaefer, 1983). The appearance imparted to floating surf bubbles by certain diatom species is sufficiently distinctive that an experienced observer can sometimes identify the major species present in a bloom simply by noting the color, size and degree of aggregation of the bubbles (McMillin 1924; Becking *et al.*, 1927; Thayer, 1935b; Lewin and Schaefer, 1983). Although the exact mechanism for this flotation remains unknown, it clearly involves a degree of control by the surf diatoms, since some species exhibit a diel periodicity in their flotation that is not synchronized with any known cycle in physical conditions. *C. armatum*, the most-studied example, rises to the water surface before sunrise and sinks out before sunset (Lewin and Hruby, 1973; Lewin and Rao, 1975; Lewin and Schaefer, 1983), a pattern that accounts for the large day-night differences in cell numbers of this species in our surf samples (Fig. 12.4). By contrast, *A. socialis* has been observed floating both day and night (Lewin and Schaefer, 1983); consequently, our samples show no systematic diel variations in cell numbers for this species. Another important observation concerning the flotation mechanism is that it does not involve innate buoyancy of the diatoms themselves, which sink rapidly when removed from the surf and kept in quiescent water (Lewin and Rao, 1975; Lewin and Schaefer, 1983).

Diatom flotation contributes in at least two ways to the maintenance of high diatom concentrations in the surf zone. First, any material floating in the surf tends to be driven landward by breaking waves, resulting in an accumulation near the shoreline. The effects of this physical concentrating mechanism are most strikingly apparent during daytime ebb tides, when large masses of diatoms are left temporarily stranded on the beach (Fig. 12.5), often forming a deposit with a thickness of several centimeters and a longshore extent of several kilometers (Lewin and Hruby, 1973; Lewin and Schaefer, 1983). Provided the diatoms remain healthy during this stranding, as some experimental results appear to indicate (see Lewin and Schaefer, 1983), deposition on the beach further enhances the accumulation effect by ensuring retention of that fraction of the population within the surf and beach ecosystem. Secondly, flotation permits the surf diatoms to take maximal advantage of available sunlight for photosynthesis during periods when light intensities are low. This factor undoubtedly explains why surf-diatom standing stock remains high through the winter despite the diminished population of coastal phytoplankton outside the surf zone (Lewin and Schaefer, 1983; Schaefer and Lewin, 1984).

A second interesting characteristic, evidently unique to surf-diatom species although not universal among them, is the formation of a conspicuous extracellular coat containing very fine inorganic particles. This phenomenon was first recognized in *Chaetoceros armatum* from Washington and New Zealand (Lewin and Norris, 1970). The coat of *C. armatum* from the

Fig. 12.5 Diatom masses left stranded on the beach on a receding tide at Copalis Beach, Washington.

Washington coast is composed chiefly of particles of the clay minerals illite and montmorillonite securely cemented to the outer surface of the frustule by an unidentified substance (Lewin *et al.*, 1979a, 1980). What function such coats might serve remains unknown. In nature, *C. armatum* apparently always has an extracellular coat of some sort (West, 1860; Hendey, 1964; Lewin and Norris, 1970; Lewin and Mackas, 1972; Lewin *et al.*, 1979a; Lewin *et al.*, 1980) but recent evidence indicates that the illite-montmorillonite composition might not be a universal feature (M.J. Kindley, pers. comm.). In laboratory culture, this species can grow satisfactorily without producing any visible coat (Lewin and Mackas, 1972) although it begins to accumulate a clay coat imediately if clay particles are added to the culture medium (Lewin *et al.*, 1980). *A. glacialis*, a species with a very broad geographical distribution in a variety of coastal environments, has been observed with a coat only in samples from the surf along sandy beaches in Brazil (J. Lewin, unpublished observations of samples provided by N.M. Gianuca).

12.4 ENVIRONMENTAL CONDITIONS ASSOCIATED WITH SURF BLOOMS

Throughout our study, one of the paramount objectives was to identify the crucial environmental factors that promote the development of surf-diatom blooms. The logical approach to this problem was to look for the similarities among situations in which dense surf blooms are observed and differences between these and relatively "non-bloom" situations. For the ensuing discussion, it is convenient to classify environmental parameters into two fairly distinct groups. We deal first with conditions that are essentially constant on time scales relevant to phytoplankton blooms. Subsequently, we deal with conditions that are subject to comparatively frequent changes, especially those that vary seasonally.

12.4.1 Physical Characteristics of Beaches

The beaches along the southern part of the Washington coast, from Point Grenville to the Columbia River (Fig. 12.2), are fairly uniform in their physical characteristics and all have prolific surf-diatom blooms. On the Oregon coast, by contrast, the beaches are highly varied. A survey of 12 Oregon beaches, representing a broad range of physical characteristics, has revealed some conspicuous patterns in the relationships between phsyical factors and surf-diatom abundance (Garver, 1979; Garver and Lewin, 1981). These beaches, with one exception, seem to fall naturally into four groups on the basis of diatom populations. The two northernmost beaches, Fort Stevens and Seaside, resemble Copalis Beach, Washington, in that densities of the two major diatom species often exceed 10,000 cells ml $^{-1}$. Five beaches along the central Oregon coast, from Beachside to Bullards, have somewhat lower densities of both species that usually exceed 1,000 cells ml $^{-1}$. Still lower densities, typically around 100 cells ml $^{-1}$, occur at Oceanside and Gleneden, which lie between the preceding two areas. Finally, both species are apparently absent from the two southernmost beaches, Harris and Ophir. Cannon Beach, located between Seaside and Oceanside, does not fit neatly into any of the groups but tends toward the low diatom abundance characteristic of Oceanside and Gleneden (Garver and Lewin, 1981).

Using discriminant analysis, we identified factors other than cell counts that can discriminate among these four groups of beaches. The variables entered into the analysis were physical characteristics of the beaches and coast: bottom slope, percent chlorite, percent montmorillonite, beach length, mean sand grain size and beach face slope. This set of six variables completely discriminates among the groups of beaches, classifies all known cases correctly and places Cannon Beach into the low cell-number group with Gleneden and Oceanside. Moreover, a subset consisting of the first four listed variables produces the same classification and remains statistically significant (Garver and Lewin, 1981).

The beaches between Point Grenville in Washington and Tillamook Head in Oregon are long, wide (200 m from dunes to mid-tide), gently sloping (1°-3°) and composed of fine-grained sand. These beaches can be considered 'high-energy dissipative beaches', typical of regions where abundant fine sand is exposed to high breakers (Short and Wright, 1983). The topographic features of a low beach gradient combined with the very gradual slope of the bottom offshore (0.2°-0.5°, 0.8 km from shore) results in an immense area of shallow water and consequently a very broad surf zone, often extending out for 1000 m during winter storms along the Washington coast. Beaches and surf of this kind appear to be well suited for the development of sustained blooms of the two surf diatoms *C. armatum* and *A. socialis*.

To the south of Tillamook Head, the terrain becomes progressively more rocky and shorter beaches with steeper gradients are prevalent as far south as Newport, Oregon. Although the surf species survive in this region, conditions are not as favorable for the existence of large blooms. Between Newport and Cape Blanco are beaches of varying lengths with fairly low gradients. Large surf blooms occur here but not as persistently as near the Columbia River mouth.

South of Cape Blanco, the physical properties of the beaches that we studied are very different from other regions to the north. Both the beach face slope and the offshore bottom slope are steep, permitting an intense onshore-offshore water exchange. Strong upwelling occurs during much of the year and consequently nitrate concentrations are generally high. Only a few cells of *C. armatum* and *A. socialis* were found in this region and only on a single occasion. On the other hand, *Skeletonema costatum*, a rapidly growing diatom, was often present here during spring and summer (Garver, 1979; Jijina and Lewin, 1983, 1984).

The role that clay minerals may play in controlling diatom distribution and abundance is not completely clear at present. Of the three clay minerals (illite, chlorite and montmorillonite) distributed along the Washington-Oregon coast, montmorillonite appears to be the most suitable for incorporation into the extracellular coat of *C. armatum*, since particles of this clay tend to be more finely dispersed and to sink less readily than do particles of illite and chlorite (Karlin, 1978, 1980). Off the coast of Washington and northern Oregon, montmorillonite, originating from the Columbia River, constituted over 50% of the $< -2\ \mu m$ size fraction of clay particles; off southern Oregon, it represented less than 20% (Karlin, 1978, 1980; Garver and Lewin, 1981).

12.4.2 Meteorologic and Oceanographic Conditions

Beyond the evident requirement of particular beach characteristics for the development of surf-diatom blooms, a variety of meteorologic and oceanographic conditions undoubtedly affect the richness, persistence and species composition of the blooms. The factors that could conceivably be influential are numerous and include air and water temperature, salinity, rainfall, river discharge, wind direction and speed, wave height, upwelling and nutrient concentrations. To assign relative degrees of importance to these environmental variables is a formidable undertaking, made even more complex by the interdependence of most of the factors. We have used two approaches to gain some insight into this problem. One approach was to examine the relationships between diatom populations and environmental variables along 13 beaches over a relatively short time period of 14 mo (June 1977 through August 1978); the other was to examine these relationships at a single beach (Copalis) over an extended time period (1970 through 1982).

A nonparametric correlation analysis of data from field studies showed that cell numbers were positively correlated with rainfall and river discharge and negatively correlated with air temperature, water temperature, salinity, upwelling index, daylength and nitrogen (nitrate and ammonium) concentrations (Jijina and Lewin, 1984). The conditions under which the abundance of surf diatoms was greatest – cool air and water temperatures, short daylength, high rainfall and river discharge, low salinity and absence of upwelling – are those that prevail during late autumn, winter and early spring (generally November through April). Undoubtedly, some of the observed correlations are merely coincidental and reveal nothing about the underlying causes. However, within this set of environmental variables are at least three that we consider critical for the development and maintenance of surf-diatom blooms: winds (of which the upwelling index is an indirect measure), nutrient supply and rainfall.

Observers of surf-diatom blooms in many parts of the world have reported that the densest blooms accompany onshore winds (Lewin and Schaefer, 1983). In coastal regions of Washington and Oregon, the prevailing wind and the resultant transport of surface ocean water vary in a regular seasonal pattern (Duxbury *et al.*, 1966; Barnes *et al.*, 1972; Lewin *et al.*, 1975). Strong southwesterly winds that predominate from October through April produce a net shoreward transport of surface water, thereby augmenting the aforementioned effect of the breakers in concentrating floating diatoms near the shoreline. Weaker northerly and northwesterly winds that prevail from May through September drive surface water seaward, presumably removing part of the surf-diatom populations from the surf zone. The influence of winds on surf-diatom distribution is substantiated by the recent observation of a sharp drop in abundance of diatoms in the surf accompanying an abrupt shift to strong offshore winds (Schaefer and Lewin, 1984). For reasons to be discussed below, we now believe that winds are the most important factor controlling the seasonal changes in the surf blooms on the Washington coast.

The maintenance of high concentrations of growing diatoms obviously requires a large and sustained supply of nutrients. Nutrients are supplied to the surf zone from ocean waters offshore, from river discharge, from rainfall and through recycling of excretion products released by interstitial fauna and by beach and surf macrofauna. Of the principal macronutrients (nitrate, ammonium, phosphate and silicate), only the nitrogenous nutrients have ever been observed to be scarce enough potentially to limit surf-diatom growth. Nitrate is often depleted to undetectable levels for long periods during summer, particularly in years when summer upwelling of nutrient-rich deep ocean water is weak (Lewin *et al.*, 1975; Lewin, 1977, 1978a; Jijina and Lewin, 1984). Ammonium, by contrast, is almost always present in measurable concentrations even when nitrate has disappeared. Evidence indicates that the major source of the ammonium is excretion by beach and surf fauna, especially by the large populations of razor clams inhabiting these beaches (Lewin *et al.*, 1979b). Without recycled ammonium, the surf diatoms would undoubtedly suffer severe nitrogen deficiency during summer. The magnitude of the nutrient demand of the dense surf-diatom populations is evidenced by a regularly observed diel cycle of nutrient concentrations in the surf. Concentrations of nitrate, ammonium and phosphate are typically lower during daylight hours when the diatoms are highly concentrated at the water surface and photosynthetically active. Nutrients are higher at night when *C. armatum* has sunk and photosynthesis has ceased. Thus, nutrient demand evidently accounts for the observed negative statistical correlations between diatom abundance and nutrient concentrations.

Surf-diatom blooms worldwide occur predominantly in areas where rainfall is high (Lewin and Schaefer, 1983). The Pacific Northwest coast is a region of extremely high rainfall, some locations receiving as much as 250 cm y $^{-1}$ (based on 10 y averages). Rainfall likely plays a significant role in the drainage of nutrient-rich interstitial water out of the sand and back into the surf, thereby promoting diatom growth (McLachlan and Lewin, 1981; Lewin and Schaefer, 1983).

12.5 INTERACTIONS BETWEEN SURF DIATOMS AND RAZOR CLAMS

On the Pacific Northwest beaches where dense surf blooms occur, the diatoms constitute practically the entire food supply of the Pacific razor clam, *Siliqua patula* Dixon (Lewin, 1978b; Lewin *et al.*, 1979a), a fact that has been recognized since early in this century (McMillin, 1924). The abundance of appropriate food undoubtedly explains why these beaches are the most productive razor clam beaches on the entire west coast of North America (Lewin *et al.*, 1979a). Even the lower surf-diatom concentrations observed in summer are likely adequate to satisfy the food requirements of the clam population. Studies of razor clam feeding have indicated that adult clams generally maintain water clearance rates exceeding 1 liter h $^{-1}$ over the natural ranges of diatom concentrations (6 x 10 ^{3} to 300 x 10 ^{3} cells ml $^{-1}$ of *C. armatum*) and temperatures (8-17° C) (Lewin, unpublished).

One possible explanation for the annual summer decline in surf-diatom abundance is that grazing on the diatoms intensifies at this time of year, as a result either of increased grazer populations or of higher individual feeding rates in response to higher water temperatures. However, the usual lack of conspicuous zooplankton in the surf has led us to believe that razor clams are the predominant grazers and our experiments revealed no obvious effect of temperature on the clams' filtration rates. Thus, existing evidence does not indicate that grazing can account for the cycle of diatom standing stock.

While the importance of the surf diatoms to the razor clams is obvious, the role of the clams in supplying recycled nutrients that promote diatom growth is more easily overlooked. Amonium is the predominant nutrient excreted by the clams, far exceeding the other nitrogenous nutrients (urea, nitrite and nitrate) as well as phosphate and silicate (Lewin *et al.*, 1979b). This fact is important for the surf diatoms in two respects. First, concentrations of nitrogenous nutrients are potentially growth-limiting during summer. Secondly, marine diatoms utilize ammonium preferentially over other forms of nitrogen. Ammonium excretion by other organisms in the sand community is apparently insignificant in comparison to that by the razor clams (Lewin *et al.*, 1979b), further emphasizing the overwhelming importance of *S. patula* in this ecosystem.

12.6 SURF-DIATOM PRODUCTIVITY IN RELATION TO STANDING STOCK

Our discussion thus far has emphasized physical mechanisms influencing distributions of surf diatoms over space and time. Indeed, these physical factors, in combination with the flotation capability of the surf species, provide reasonable explanations for many aspects of the observed distributions. However, any examination of the dynamics of populations would be far from complete without some consideration of growth rates. The data presented in Figure 12.4

reveal three major characteristics of the standing stock of surf diatoms on the Pacific Northwest coast, each of which raises a question concerning photosynthesis and growth. First, concentrations of diatoms in the surf are extraordinarily high at all times of year. Do the surf diatoms possess exceptional photosynthetic characteristics resulting in unusually high growth rates that are maintained through winter? Second, surf-diatom standing stock is lowest in late spring and summer, the time of year when nitrate is often scarce. Do the diatoms grow more slowly in summer, possibly because of nitrogen limitation? Last, *Chaetoceros armatum* is usually more abundant numerically than *Asterionella socialis*. What are the relative contributions of the two species to total primary production in the surf zone? These questions motivated a 12 mo study of photosynthetic rate measurements by the carbon-14 method at Copalis Beach (Schaefer and Lewin, 1984).

The measured photosynthetic characteristics of the surf diatoms are not extraordinary in relation to those of other marine diatoms. Light-saturated photosynthetic rates (P_{max}), normalized to chlorophyll *a*, ranged roughly 3-8 g C g Chl a^{-1} h $^{-1}$. The same rates normalized to particulate organic carbon (used as an estimate of phytoplankton carbon) fell mostly within the range of 0.09-0.13 g C g C $^{-1}$ h $^{-1}$. In measurements of photosynthesis *versus* light intensity (*P versus I*), the intensity at the onset of saturation (I_K) was approximately 200 μE m $^{-2}$ s $^{-1}$. All of these values are in the realm of typical values reported for other marine diatom species (see Schaefer and Lewin, 1984).

Observed seasonal variations in photosynthetic rates (Fig. 12.6) were inconsistent with the hypothesis that surf diatoms grow more slowly in summer than in winter. In fact, the variation in P_{max} (Fig. 12.6a) showed the opposite pattern with summer values twice as high as winter values. However, the specific growth rate μ would not necessarily follow the same trend as P_{max} since any change in the carbon-to-chlorophyll ratio (C : Chl) would alter the proportionality between μ and P_{max}. A better approximation to μ is the specific carbon incorporation rate μ' (after Li and Goldman, 1981) obtained by normalizing measured photosynthetic rates to estimated phytoplankton carbon. The variation in light-saturated μ' (μ'_{max} Fig. 12.6b) was slight and did not show obvious seasonality. Since μ'_{max} is a light-saturated hourly rate, however, it is likely that daily values of μ are higher in summer than in winter because of the seasonal differences in daylength and light intensity. Therefore, the evidence appears to indicate that the surf diatoms actually grow most rapidly during the season when standing stock is lowest (Schaefer and Lewin, 1984). This conclusion is at odds with earlier results where μ was estimated by an indirect cytological method (Lewin and Rao, 1975), but we now have some doubt regarding the reliability of that method when applied to the surf diatoms (see Schaefer and Lewin, 1984).

Values of μ'_{max} measured during two months when cell numbers of *Asterionella socialis* exceeded those of *Chaetoceros armatum* were not markedly different from values obtained during months when *C. armatum* was numerically dominant. This observation suggests that μ'_{max} (and, by inference, μ) does not differ much between these two species. If this interpretation is correct, then *C. armatum* must strongly dominate primary production in the surf zone since this

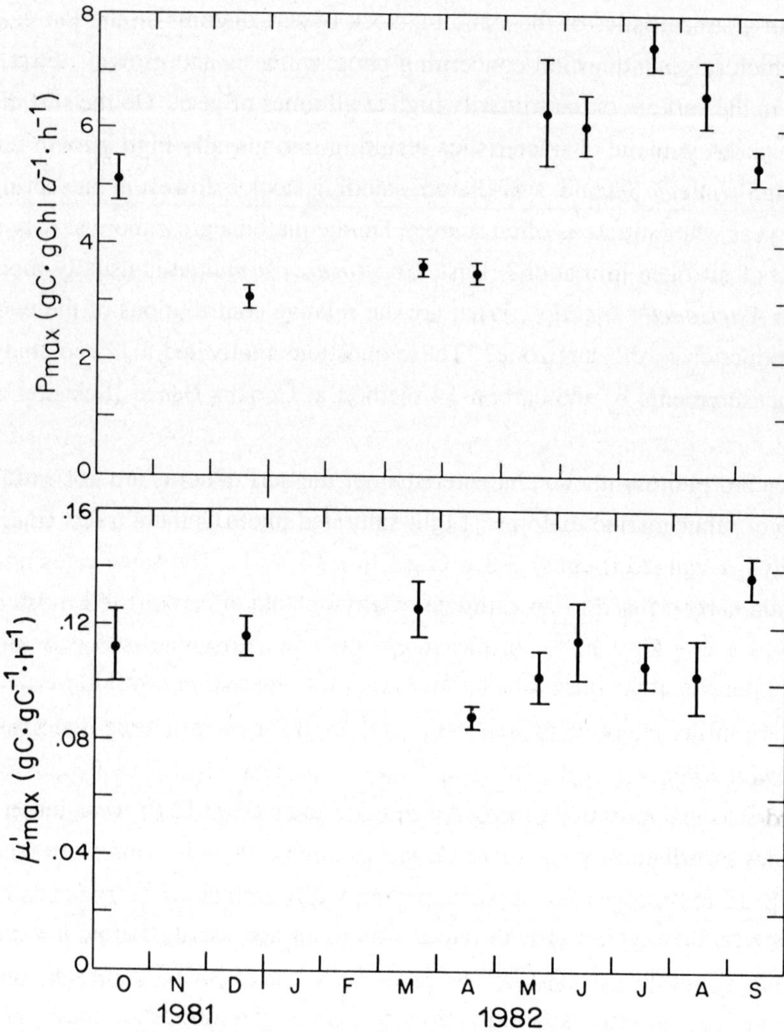

Fig. 12.6 Carbon fixation rates measured for surf diatom populations at Copalis Beach, Washington. Top: mid-day (between 1,000 h and 1,400 h) values for light-saturated chlorophyll-specific photosynthesis (P_{max}) measured throughout the year from October 1981 through September 1982. Bars represent mean ≠2 s.d. Bottom: mid-day values for light-saturated specific carbon incorporation rate (μ'_{max}).

species is usually numerically dominant and has a much higher cellular carbon content than the smaller *A. socialis* (Schaefer and Lewin, 1984).

Evidently surf-diatom growth rates do not hold explanations for the observed patterns of abundance. The persistence of dense blooms throughout the year is probably due chiefly to the flotation ability of the surf diatoms. The seasonal variations in standing stock are best explained by the annual pattern of wind-driven water transport.

12.7 A MATHEMATICAL MODEL OF SURF-ZONE DYNAMICS AND ECOLOGY

Having attained a reasonable qualitative understanding of surf-diatom blooms, we concluded that deeper insights to the dynamics of surf-zone ecosystems might best be gained through mathematical modeling. Before entering the discussion of model development, it is useful to summarize some of the principal characteristics of surf-diatom blooms and the environment in which they occur. Persistent diatom blooms are observed along beaches having exceptionally gentle slopes and, consequently, broad surf zones. Confined almost entirely to the breaker zone, these blooms consist of one or two diatom species in extraordinarily high concentrations. The horizontal distribution of diatoms in the surf is characterized by patches (Fig. 12.1b), the size of which is often many tens of meters (Lewin, *et al.*, 1975). Apparently all diatom species that form such surf blooms are capable of flotation, thereby concentrating a large fraction of the population at or near the water surface and gaining a decisive competitive advantage over other algal species in the turbid water of the surf. Probably because this flotation is dependent on bubbles produced by breaking waves, the few cells of the surf species found beyond the breaker line are never observed to be floating and in fact are most concentrated near the sea bed (see Lewin, 1978a).

The characteristics described above play an important role in the formation of patches within the surf. Another equally important process may be wave-induced circulation throughout the surf zone and the shoaling zone beyond. We propose that patchiness is a consequence of the combined effect of wave-induced advection throughout the nearshore zone, high photosynthetic activity in the shoreward part of the surf zone and net negative growth near the breaker line and beyond. In order to examine this possibility in quantitative terms, a few remarks are in order concerning surf-zone circulation.

It has been shown by several workers (e.g., Bowen, 1969; Mei and Liu, 1977) that circulation in the form of counter-rotating gyres is one of the principal modes of time-averaged flow in the nearshore environment. The flow pattern may be summarized qualitatively as follows. Seaward of the breaker line, variations in depth (and, possibly, wave-wave interactions) tend to produce longshore variations in incoming wave amplitude. Field observations have shown that as waves approach the shore, they steepen to the breaking point when the ratio of the wave height to water depth exceeds a certain critical value. Near the breaker line, in deeper water, higher breaking waves will produce higher mean water set-up, which drives currents and incoming waves into the main surf zone. On the shoreward side of the breaker line, wave energy is dissipated by turbulence and bottom friction. Radiation stress associated with the incoming waves induces a current flow throughout the nearshore zone characterized by a net shoreward flow in the regions of high breaking waves followed by longshore flow toward regions of low breakers. Throughout the surf zone, longshore currents converge toward the latter regions and then flow seaward as rip currents. In summary, the time-averaged water motion in the surf zone is characteristcally a gyre flow pattern as indicated schematically in Figure 12.7.

We can now examine the implications of the circulation on the growth and distribution of diatom populations. Since mean surf-zone flow is characterized by gyres, water in their offshore seaward reaches will move shoreward toward the breaker line and be recirculated into the surf zone. Some fraction of the offshore diatom population is undoubtedly viable (analogous to algal

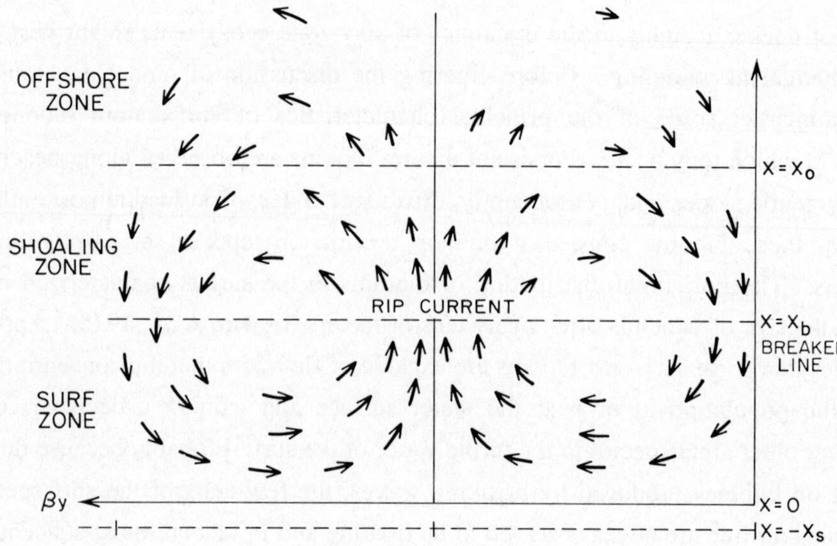

Fig. 12.7 Coordinate system used in the analysis of surf-zone distributions. Mean circulation is indicated by the arrows.

cells in deep water overlying the shelf (Small *et al.*, 1989, Chapter 7) although the net average growth rate may be negative in that region. Water moving shoreward across the breaker line into the surf zone will bring "seed stock" cells into an environment favorable for vigorous photosynthesis. As they are advected shoreward and then in the longshore direction, the seed stock cells begin to grow. Throughout the main part of the surf zone where the cells are near the surface and foam is extant, the specific growth rate is high and positive. The average shoreward and longshore current will be relatively slow when the beach is broad and gently sloping. Hence, the residence time of a water parcel within the surf zone can be long enough to allow development of high cell concentrations in seaward flowing regions of the gyres; specifically, the feeder currents of the rip channels. Even in the vicinity of the rip channels, production will remain high because foam is in ample supply as the region of breakers is approached. However, as the algal population is advected beyond the breaker line, the near-surface environment is no longer characterized by the abundant surface foam which promotes vigorous growth. The cells will enter a respiration phase and the viable population fraction will decline.

In order to test the foregoing qualitative description, we have developed a quantitative model incorporating the hydrodynamic and biological processes. The mathematical details are presented in Winter (1983). A summary of the relevant mathematical aspects of the physical oceanography of the surf zone is presented below, together with the highlights of the biological model.

12.7.1 Near-shore Circulation

We employ a Cartesian coordinate system throughout this analysis, with the *x*-axis positive in the offshore direction and y denoting distance in the longshore direction. The *y*-axis is coincident with the stillwater line (see Fig. 12.7). A non-dimensional shifted offshore coordinate

$\xi = \beta(x + x_s)$ is also introduced, where x_s is the mean beach line and $2\pi/\beta$ is the rip current spacing. Hence, a measure of the width of the circulation gyre is $\xi_b = \beta(x_b + x_s)$, where x_b denotes the position of the breaker line.

Near-shore circulation is assumed to be driven by the excess momentum flux of waves approaching a straight coastline at nearly normal incidence. In the work of Mei and Liu (1977), the bottom topography is assumed to be planar, except for small perturbations which are periodic in the longshore direction. The stillwater depth is represented in the form

$$h = h_0 + h_1 \tag{12.1a}$$

where

$$h_0 = sx \tag{12.1b}$$

and

$$h_1 = \delta h_b f(\xi) \cos\beta y \tag{12.1c}$$

where s represents the mean slope and h_b denotes the water depth at the mean breaker line. In Equation (12.1c), δ is a perturbation parameter whose magnitude is much less than one. The bottom topography modulation function $f(\xi)$ is of the order of one at $\xi = \xi_b$. Note that in this representation, the variation of parameters in the longshore direction is assumed to be sinusoidal.

An incoming wavetrain of amplitude a and circular frequency ω approaching a coastline with bottom topography as represented by Equation (12.1) develops an amplitude variation in the longshore direction due to the variable bottom depth. Two key observations have been made in field studies which are useful in the mathematical developments: (1) the ratio of the wave amplitude a to the total water depth remains approximately constant; i.e.,

$$a = \gamma(h + \eta)$$

where η is the free surface height above stillwater level and $\gamma = 0.4$ and (2) the ratio γ_b of the breaker line wave amplitude to the water depth h is approximately 0.7. The stillwater depth, h_b, at the breaker line can be estimated from this last relation and linearized shoaling theory by

$$h_b = \left[\frac{g a_0^4}{4\gamma_b^4 \omega^2}\right]^{1/5} \tag{12.2}$$

From h_b we can determine the mean position of the breaker line:

$$x_b = h_b / s \tag{12.3}$$

It can also be shown that the location of the mean beach line is approximately

$$x_s = \frac{1}{2}\left[3\gamma^2\left(1 - \frac{1}{4}\gamma_b^2\right) - \frac{1}{2}\gamma_b^2\right]x_b \tag{12.4}$$

The equations which govern the mean flow gyres are the time-averaged vertically-integrated equation of continuity and the two horizontal momentum equations. The horizontal momentum equations express a balance between the hydrostatic pressure gradient, radiation stress and bottom friction. The latter is assumed to be linearly proportional to the velocity $\vec{q}$; i.e., $\vec{\tau} = \rho c_f u_0 \vec{q}$, where ρ represents the mean density, c_f is a friction coefficient and u_0 is a measure of the orbital velocity in the surf zone. A vorticity equation is derived from the two horizontal momentum equations. Outside the breaker line, vorticity is conserved; flow is driven primarily inside the breaker line. Inside the breaker line, circulation is driven by excess momentum flux expressed in terms of radiation stress.

A perturbation analysis of these equations has been carried out by Mei and Liu (1977) who derive linear forms of the relevant equations to the order $0(\delta\gamma^2)$. The continuity equation has the form

$$\vec{\nabla}\cdot\left(H^{\pm}\vec{q}\right) = 0 \tag{12.5}$$

where $H^- = h_0$ beyond the breaker line and $H^+ = \sigma(x + x_s)$ inside the surf zone, with

$$\sigma = s\left(1 + \frac{3}{2}\gamma^2\right)^{-1} \tag{12.6}$$

Transport streamfunctions, $\psi^{\pm}$, are introduced in each region:

$$H^{\pm}\vec{q} = -\,curl\left(\psi^{\pm}\vec{k}\right) \tag{12.7}$$

where $\vec{k}$ is a unit vector perpendicular to the x-y plane and positive upward.

The vorticity equation in the surf zone is a second-order partial differential equation for $\psi(\xi, y)$ with an inhomogeneous term which expresses the effects of the mean hydrostatic gradient, the normal radiation stresses S_{xx} and S_{yy} and the effect of ray deflection θ through the stress component S_{xy}. For nearly normal wave incidence, it can be shown that the deflection angle is given approximately by

$$\theta(\xi, y) = \frac{\delta h_b \beta}{2\sigma} g(\xi)\sin(\beta y) \tag{12.8}$$

where

$$g(\xi) = \xi^{1/2}\int_{\xi_0}^{\xi} \xi^{-3/2} f(\xi)\, d\xi$$

The forcing term in the vorticity equation has a factor of the form

$$\Omega = -\frac{5}{4}\xi f' - \frac{5}{8}f - \frac{1}{4}\left[\frac{5}{4} + \xi^2\right]g$$

In this latter equation, the prime symbol indicates differentiation with respect to ξ. Beyond the breaker line, the vorticity equation has the same form, except that the right hand side is equal to zero since the forcing vorticity is negligible and the variable ξ is replaced by $\xi - \xi_s = \beta x$.

The circulation is determined by the following procedure: first, the bottom topography modulation function f is specified. The vorticity equations are then reduced to ordinary differential equations by separating variables:

$$\psi^{\pm}(\xi,y) = \phi^{\pm}(\xi)\sin\beta y$$

Beyond the breaker line, we have simply

$$\phi^{-}(\xi) = c_2\left(1 + \xi - \xi_s\right)\exp\left(-\xi\right) \tag{12.9a}$$

since ϕ^- vanishes as ξ approaches infinity. The constants c_1 and c_2 are determined by the continuity of $\psi^{\pm}$ and $d\psi^{\pm}/d\xi$ at $\xi = \xi_b$. The inhomogeneous ordinary differential equation for $\phi^+(\xi)$ inside the surf zone can be reduced to quadratures. Linearly independent solutions of the homogeneous form of the equation for ϕ^+ are

$$\phi_1(\xi) = (1 + \xi)\exp(-\xi)$$

and

$$\phi_2(\xi) = (1 - \xi)\exp(\xi)$$

with the Wronskian, $W = -2\xi^2$. Since $\phi^+ = 0$ at $\xi = 0$, the solution for ϕ^+ can be written as

$$\phi^{+}(\xi) = c_1\left(\phi_1 - \phi_2\right) + \frac{1}{2}\psi_0\left[\phi_1\int_0^{\xi}\xi^{-1}\Omega\phi_2\,d\xi - \phi_2\int_0^{\xi}\xi^{-1}\Omega\phi_1\,d\xi\right] \tag{12.9b}$$

where the constant ψ_0 is given by

$$\psi_0 = \frac{\delta\gamma^2 g h_b \sigma^2}{c_f u_0 \beta^2}$$

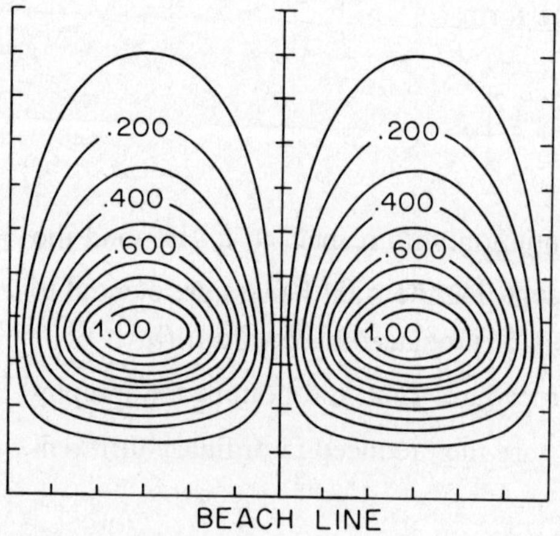

Fig. 12.8 Streamline configuration in adjoining gyres for a beach slope of s = 0.006 and $\xi_b = 2\pi/3$. Other relevant parameters are given in the text.

Although various aspects of the surf-zone biology have been studied, almost no direct observation of the physical circulation at Copalis Beach has been made. From observations at the site and data from nautical charts, the surf zone near Copalis Beach, Washington, is wide, with an exceptionally gentle slope of s = 0.006, the first breaker line is at least 250 m from the beach line and the rip current separation appears to exceed the surf-zone width by a factor of two or three. With an offshore wave amplitude a_0 of 1 m and a wave period of 8 s, one finds $h_b = 1.8$ m from Equation (12.2) and $x_b = 290$ m from Equation (12.3). From Equation 12.4 the location of the mean beach line is approximately $x = -x_s = -30$ m. We have assumed below that $\xi_b = \beta(x_b + x_s)$ is approximately equal to 2 or, more conveniently mathematically, $\xi_b = 2\pi/3$.

Sandy beaches with gentle slopes are usually characterized by a bar near the breaker line. To represent the effect of such a feature at Copalis Beach, we take the modulation function to be

$$
\begin{aligned}
f(\xi) &= \xi/\xi_\beta && 0 < \xi < \xi_\beta \\
&= 2 - \xi/\xi_\beta && \xi_\beta < \xi < 2\xi_\beta \\
&= 0 && \xi > 2\xi_\beta
\end{aligned}
$$

The solution for ϕ^+ inside the breaker line can be expressed in quadrature form convenient for computation (Eq. 12.9b), but the integrated expression is too cumbersome to be reproduced here. The solution for ψ^- is given by Equation (12.9a).

The most difficult parameter to estimate is the bottom friction coefficient c_f. It is an empirically determined coefficient and is assigned a value of 0.02 for beaches of moderately steep slope by Inman *et al.* (1971). However, since horizontal turbulence effects are not explicitly included in the model equations, the stress terms in the momentum equation must account for all dissipative forces. In the case of wide, shallow surf zones, the value of the effective friction coefficent c_f must be larger than 0.02 and in our calculations, we have assigned c_f a value of 0.08. With $c_f = 0.08$, $u_0 = \frac{1}{2}\gamma\sqrt{gh_b}$ and $\delta = 0.1$, the value of ψ_0 is about 2.2. The resulting streamline configuration is exhibited in Figure 12.8, where two gyres are shown.

12.7.2 Diatom Growth and Distribution

A quantitative description having been developed of the circulation environment in which surf-zone diatoms grow, we can proceed with the formulation of an idealized mathematical model for the spatial distribution of cells. Neglecting the effects of nutrient limitation, lateral turbulence and grazing by secondary producers, we can express conservation of the time-averaged vertically-integrated diatom density $P(x,y)$ as a linear first order partial differential equation

$$\vec{\nabla}\cdot\left(H^{\pm}\vec{q}\,P\right)=r(x,y)\,P \tag{12.10}$$

where $r(x,y)$ denotes the net specific growth rate. This equation is to be solved in the domain $-x_s < x < \infty,\ 0 < y < \beta\pi$. We introduce the transport streamfunctions defined in Equation (12.7). Equation (12.10) can then be written in the form

$$\frac{dP}{r(x,y)\,P}=\frac{dx}{-\partial\psi/\partial y}=\frac{dy}{\partial\psi/\partial x}$$

The second of the equalities simply states that the streamline passing through the point $(x_i,\pi/2)$ has the equation

$$\psi(\xi,y)=\psi\left(\xi_i,\frac{\pi}{2}\right)=\psi_i(constant) \tag{12.11}$$

where ψ_i 'identifies' the streamline in question. It is shown in Winter (1983) that from the first equality, a solution of the differential equation for the time-averaged vertically-integrated diatom density $P(x,y)$ can be expressed by an integral in terms of known quantities along a streamline; in terms of the shifted coordinate ξ, we have

$$P(\xi,y)=P_i\exp\left[\pm\frac{1}{\beta^2\Gamma}\int\frac{r(\xi,y)}{\phi^2-\phi_i^2}\,d\xi\right] \tag{12.12}$$

where the lower sign is taken when y lies in the interval $(0,\pi/2)$, Γ is the streamline identified by the constant ψ_i and P_i is a specified value of P at the point $(\xi_i,\pi/2)$ on Γ. In the integral along Γ, ξ is chosen as the integration variable. Since the density P is time-averaged and, by assumption, not influenced by diffusion, the specific growth rate must satisfy the integral constraint

$$\int_{\xi_{min}}^{\xi_{max}}\frac{r(\xi,y)}{\left(\phi^2-\phi_i^2\right)^{1/2}}\,d\xi=0 \tag{12.13}$$

Here, ξ_{max} and ξ_{min} are maximum and minimum values of ξ on Γ.

To carry out the integration in Equation (12.12), some assumptions need to be made regarding the spatial variations of the specific growth rate, r. First, the net specific growth rate is assumed to be independent of the longshore direction. In the offshore direction, when the nutrient supply is adequate, photosynthesis proceeds vigorously throughout the main part of the surf zone. Near the beach line, however, the average photosynthetic rate may be relatively low since algal material is deposited on the beach by the action of waves and wind. Near the breaker line, vigorous vertical mixing probably has the effect of impeding photosynthesis by reducing net exposure to light. Seaward of the breaker line, the cell population enters a respiration phase since the environment no longer provides the conditions to which it is adapted for rapid growth. As a working hypothesis, we assumed the net specific growth rate to vary spatially in the general manner just described. This assumption is speculative, however, since measurements of the offshore spatial variation of primary productivity have not been conducted.

It is consistent with the circulation description to expect that $\phi^+(\xi)$ has no more than one critical point in $(0,\xi_b)$ and in view of the behavior of ϕ^+ as a function of ξ, it is reasonable to represent the specific growth as

$$r(\xi) = p(\phi)\frac{d\phi}{d\xi} \tag{12.14}$$

Here, $p(\phi)$ is a function which can be chosen for convenience. It is important to note that with this choice for $r(\xi)$, the integral constraint is satisfied. For various choices of $p(\phi)$, the distribution of $P(\xi,y)$ can be determined analytically. For example, Winter (1983);

$$p(\phi) = p_{\max}(a\phi + b)^{-1} \tag{12.15}$$

where a and b are arbitrary constants which adjust the shape of the growth rate variation once ϕ has been determined and $p_{\max}$ is the maximum value of $(a\phi + b)^{-1}\, d\phi/d\xi$. The value of $P(\xi,y)$ is now determined in the following way: point (ξ,y) lies on a streamline identified by ψ_i, which determines a value of λ. Now define the quantities

$$\alpha_< = \frac{p_{\max}}{b\beta^2}\left(1-\lambda^2\right)^{-1/2}, \quad 0<\lambda<1 \text{ and} \tag{12.16a}$$

$$\alpha_> = \frac{p_{max}}{b\beta^2}\left(\lambda^2-1\right)^{-1/2}, \quad \lambda>1 \tag{12.16b}$$

where $\lambda = a\psi_i/b$. If λ lies in the interval $0<\lambda<1$, then from Equations (12.12), (12.14) and (12.15), $P(\xi,y)$ is given by

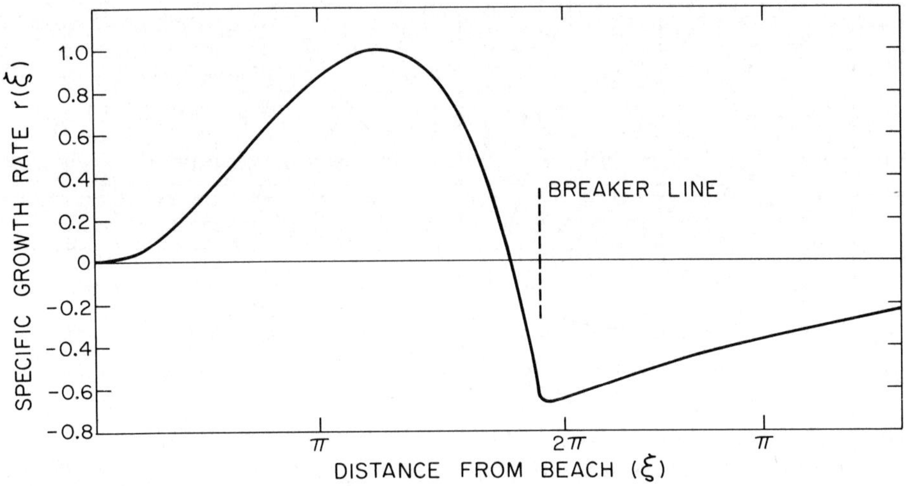

Fig. 12.9 Distribution of normalized, net specific growth rate of surf-zone diatoms as a function of offshore distance ξ.

$$P = P_i \left[\frac{1 + \lambda + \sin\beta y + \cos\beta y \sqrt{(1-\lambda^2)}}{\lambda + \sin\beta y} \right]^{\pm\alpha_<} \tag{12.17a}$$

For $\lambda > 1$, the expression for P is

$$P = P_i \left[\pm\, \alpha_> \operatorname{arcos} \frac{1 + \lambda \sin\beta y}{\lambda + \sin\beta y} \right] \tag{12.17b}$$

Note that the dependence of ξ is embodied in ψ_i, through Equation (12.11). In the foregoing expression the upper sign preceding α is to be taken when βy lies in the interval $(0,\pi/2)$.

As a test calculation, we selected a maximum specific growth rate of 4.86 x 10^{-5} s^{-1}. This choice, made prior to the collection of photosynthetic rate measurements at Copalis Beach, was based on a postulated P_{max} of 3.5 g C g Chl a^{-1} h^{-1} and a nominal value of 20 for the carbon-to-chlorophyll ratio. Both of these values subsequently proved to be within the ranges of values measured at the beach. It is noteworthy that the highest measured values of Schaefer and Lewin's (1984) μ'_{max} were around 0.13 g C g C^{-1} h^{-1} or, equivalently, 3.6 x 10^{-5} s^{-1}. However, given the possibility that μ'_{max} sometimes underestimates the specific growth rate (Schaefer and Lewin, 1984), the assigned value of 4.86 x 10^{-5} s^{-1} is still considered reasonable. In Equation (12.15) for $p(\psi)$, we set $a = 0$ and $b = 1$; the offshore variation of net specific growth rate for this choice is shown in Figure 12.9. The graph of $r(\xi)$ in the figure has been normalized so the maximum value is equal to one; the actual growth rate used in the calculation is obtained by multiplying the ordinate by 4.86 x 10^{-5} s^{-1}.

The solution for the diatom concentration P, which is scaled by the constant P_i, is given by Equation (12.17a) with λ set equal to 0. All that remains is the specification of P_i. Since

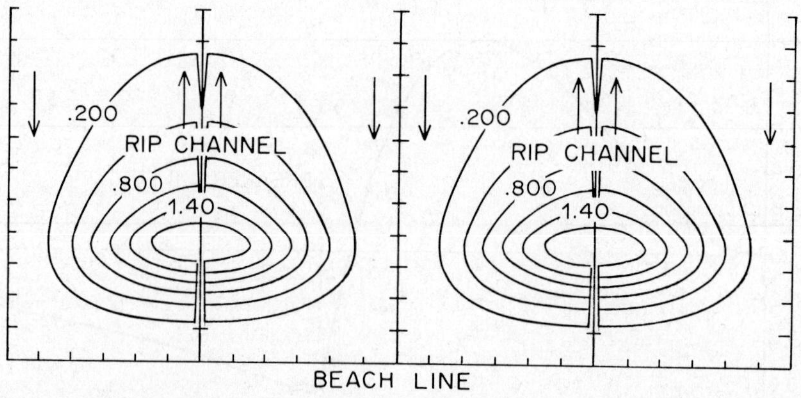

Fig. 12.10 Isopleths of vertically-integrated surf diatom densities in four adjoining gyres for the standard case of high growth rate and gently sloping beach (r^*_{max} = 4.86 x 10^{-6} s^{-1} and slope = 0.006).

measurements of relative diatom concentration along a line from the beach to the center of the circulation gyre have not been made, we assumed that the total number of cells in the water column increases monotonically from the beach line to a point near the breaker line. Since the function $\phi^+(\xi)$ has the appropriate behavior throughout most of the surf zone, for convenience, it was used to specify P_i:

$$P_i = \phi^+(\xi) / \phi^+_{max}$$

Isopleths of relative diatom density were calculated from the formulation described above and are shown in Figure 12.10. Inspection of Figure 12.10 shows that locally high concentrations of cells appear in the vicinity of the rip channel. In fact, the densities there are higher by one or two orders of magnitude than in the regions of shoreward flow. Thus, the mathematical model supports the speculation that the mechanisms of advection and photosynthesis can together produce the patchiness observed in the surf zone along the Washington coastline.

Figure 12.10 shows that all lines of equal concentration are closed within the gyre. Even though high cell density regions are located in the rip currents, a narrow band of low density isopleths appears on the center line of the rip channel. We believe this feature is an artifact associated with the neglect of lateral turbulence. Horizontal mixing would shift the centers of adjacent gyres toward the rip current and produce diffusion of cells across the center line of the rip channel.

12.7.3 Model Discussion and Conclusions

Several numerical experiments were implemented to elucidate the dependence of $P(\xi,y)$ on various parameters. For example, decreasing the growth rate by approximately 40% without altering other variables, including the specification of P_i, produced a somewhat lower density altering other variables, including the specification of P_i, produced a somewhat lower density and the regions of higher concentration moved apart. Nevertheless, diatom patches were still extant in the vicinity of the feeder currents of the rip channels. Isopleth diagrams of various numerical experiments are shown in Winter (1983). Since the phenomenon of proliferating surf-zone species seems to be limited to wide, gently sloping beaches, other numerical experiments were performed with larger values of beach slope s, all other specifications and parameters being the same. In this case, the regions of locally high concentrations of cells disappeared from the rip channels and the population density was generally lower overall. Although diatom density isopleths still followed the streamline configuration, we would not expect surf-zone diatoms to be competitive in such an environment because a narrow surf is characterized by rapid advection and a lesser supply of foam. In another numerical experiment, the constant a in the equation for $p(\phi^+)$ was increased from zero to a positive number greater than b. This had the effect of moving the maximum of the growth rate variation toward the beach. Moreover, the region of high cell concentration in the rip channel became elongated in the seaward direction. The effect of altering the variation of $p(\phi^+)$ to an unrealistic distribution (e.g., zero at the beach line and at the breaker line) produced locally high concentrations, but the distributions are very different from any that have been observed.

In all the numerical experiments conducted in this study, the maxima of the calculated distributions were nearer to the breaker line than has been observed. One possible reason for this discrepancy is that the theory overestimates the photosynthetic rate near the breaker line where the diatoms are less buoyant and the flotation mechanism is less effective in the region of increasingly intense vertical mixing. However, another reason for the discrepancy may be related to the circulation theory which predicts that the center of the circulation gyre is fairly close to the breaker line. In consequence, the greatest accumulation of diatoms occurs closer to the breakers than to the beach line.

In any event, our quantitative developments appear to support the hypothesis that advection in nearshore gyres, coupled with high production within the surf zone and respiration near and beyond the breaker line, is largely responsible for the vigorous growth of diatoms and the formation of patches in the surf at Copalis Beach, Washington. Furthermore, the numerical experiments support the speculation that a broad, well-developed surf with gentle bottom slope is prerequisite to the proliferation of the diatom populations. When advection speeds are low, as will be the case for wide beaches with gentle slopes, the residence time of cells in the surf is relatively long. Moreover, a wide, shallow surf has the necessary supply of surficial foam for cell flotation. Photosynthesis can proceed over a time interval sufficient to produce a substantial increase in the diatom population by the time cell parcels are advected to the region of rip currents.

ACKNOWLEDGMENTS

This research was supported by the Department of Energy under contract DE-AT06-EV-75026. This is a component of Contribution No. 1784 from the School of Oceanography, University of Washington.

REFERENCES

Barnes, C.A., A.C. Duxbury and B.-A. Morse. 1972. The circulation and selected properties of the Columbia River effluent at sea. Pages 41-80 in: D.L. Alverson and A.T. Pruter (eds.), *Bioenvironmental studies of the Columbia River estuary and adjacent ocean regions.* Univ. of Washington Press, Seattle, Wa.

Becking, L.B., C.F. Tolman, H.C. McMillin, J. Field and T. Hashimoto. 1927. Preliminary statement regarding the diatom "epidemics" at Copalis Beach, Washington and an analysis of diatom oil. *Econ. Geol.*, 22: 356-368.

Bowen, A.J. 1969. Rip currents. I. Theoretical investigations. *J. Geophys. Res.*, 74: 5467-5478.

Cassie, R.M. and V. Cassie. 1960. Primary production in a New Zealand West Coast phytoplankton bloom. *N.Z. J. Sci.*, 3: 178-199.

Duxbury, A.C., B.-A. Morse and N. McGary. 1966. The Columbia River effluent and its distribution at sea, 1961-63. Univ. of Washington, Dept. of Oceanogr. Tech. Rep. 156, 105 pp.

Garver, J.L. 1979. A survey of surf diatom blooms along the Oregon coast. M.S. Thesis, Dept. of Oceanogr., Univ. of Washington, Seattle, Wa., 167 pp.

Garver, J.L. and J. Lewin. 1981. Persistent blooms of surf diatoms along the Pacific coast, U.S.A. I. Physical characteristics of the coastal region in relation to the distribution and abundance of the species. *Estuar. Coastal Shelf Sci.*, 12: 217-229.

Hendey, N.I. 1964. An introductory account of the smaller algae of British coastal waters. V. Bacillariophyceae. Her Majesty's Stationery Office, Fish. Invest. Ser. 4, London, 317 pp.

Inman, D.L., J.R. Tait and D.E. Nordstrom. 1971. Mixing in the surf zone. *J. Geophys. Res.*, 76: 3493-3514.

Jijina, J.G. and J. Lewin. 1983. Persistent blooms of surf diatoms (Bacillariophyceae) along the Pacific coast, U.S.A. II. Patterns of distribution of diatom species along Oregon and Washington beaches (1977 and 1978). *Phycologia,* 22: 117-126.

Jijina, J.G. and J. Lewin. 1984. Persistent blooms of surf diatoms (Bacillariophyceae) along the Pacific coast, U.S.A. III. Relationships between diatom populations and environmental variables along Oregon and Washington beaches (1977 and 1978). *Phycologia,* 23: 471-483.

Karlin, R. 1978. Sediment sources and clay mineral distributions off the Oregon coast: evidence for a poleward slope undercurrent. M.S. Thesis, Oregon St. Univ., Corvalis, Ore., 88 pp.

Karlin, R. 1980. Sediment sources of clay mineral distributions off the Oregon coast. *J. Sed. Petrol.*, 50: 543-560.

Kincaid, T. 1968. The ecology of Willapa Bay, Washington, in relation to the oyster industry. Seattle. (Manuscript available in the Univ. of Washington Library.)

Lewin, J. 1974. Blooms of surf-zone diatoms along the coast of the Olympic Peninsula, Washington. III. Changes in the species composition of the blooms since 1925. *Beih. Nova Hedwigia,* 45: 251-256.

Lewin, J. 1977. Persistent blooms of surf diatoms along the Northwest coast. Pages 81-92 in: R.W. Krauss (ed.), *The marine plant biomass of the Pacific Northwest coast.* Oregon St. Univ. Press, Corvalis, Ore.

Lewin, J. 1978a. Blooms of surf-zone diatoms along the coast of the Olympic Peninsula, Washington. IX. Factors controlling the seasonal cycle of nitrate in the surf at Copalis Beach (1971 through 1975). *Estuar. Coastal Mar. Sci.*, 7: 173-183.

Lewin, J. 1978b. The world of the razor-clam beach. *Pac. Search,* 12: 12-13.

Lewin, J., C. Chen and T. Hruby. 1979a. Blooms of surf-zone diatoms along the coast of the Olympic Peninsula, Washington. X. Chemical composition of the surf diatom *Chaetoceros armatum* and its major herbivore, the Pacific razor clam, *Siliqua patula., Mar. Biol.*, 51: 259-265.

Lewin, J., J.R. Colvin and K.L. McDonald. 1980. Blooms of surf-zone diatoms along the coast of the Olympic Peninsula, Washington. XII. The clay coat of *Chaetoceros armatum* T. West. *Bot. Mar.*, 23: 333-341.

Lewin, J., J.E. Eckman and G.N. Ware. 1979b. Blooms of surf-zone diatoms along the coast of the Olympic Peninsula, Washington. XI. Regeneration of ammonium in the surf environment by the Pacific razor clam *Siliqua patula. Mar. Biol.*, 52: 1-9.

Lewin, J. and T. Hruby. 1973. Blooms of surf-zone diatoms along the coast of the Olympic Peninsula, Washington. II. A diel periodicity in buoyancy shown by the surf-diatom species *Chaetoceros armatum* T. West. *Estuar. Coastal Mar. Sci.*, 1: 101-105.

Lewin, J., T. Hruby and D. Mackas. 1975. Blooms of surf-zone diatoms along the coast of the Olympic Peninsula, Washington. V. Environmental conditions associated with the blooms (1971 and 1972). *Estuar. Coastal Mar. Sci.*, 3: 229-41.

Lewin, J. and D. Mackas. 1972. Blooms of surf-zone diatoms along the coast of the Olympic Peninsula, Washington. I. Physiological investigations of *Chaetoceros armatum* and *Asterionella socialis* in laboratory cultures. *Mar. Biol.*, 16: 171-181.

Lewin, J. and R.E. Norris. 1970. Surf-zone diatoms of the coasts of Washington and New Zealand (*Chaetoceros armatum* T. West and *Asterionella* spp.). *Phycologia,* 9: 143-149.

Lewin, J. and V.N.R. Rao. 1975. Blooms of surf-zone diatoms along the coast of the Olympic Peninsula, Washington. VI. Daily periodicity phenomena associated with *Chaetoceros armatum* in its natural habitat. *J. Phycol.*, 11: 330-338.

Lewin, J. and C.T. Schaefer. 1983. The role of phytoplankton in surf ecosystems. Pages 381-389 in: A. McLachlan and T. Erasmus (eds.), *Sandy Beaches as Ecosystems.* W. Junk Publ., Hingham, Mass.

Li, W.K.W. and J.C. Goldman. 1981. Problems in estimating growth rates of marine phytoplankton from short-term ^{14}C assays. *Microb. Ecol.*, 7: 113-121.

McLachlan, A. and J. Lewin. 1981. Observations of surf phytoplankton blooms along the coasts of South Africa. *Botanica Marina,* 24: 553-557.

McMillin, H.C. 1924. The life-history and growth of the razor clam. St. of Wash., Dept. of Fish., Olympia, Wa., 52 pp

Mei, C.C. and P.L-F. Liu. 1977. Effects of topography on the circulation in and near the surf zone – linear theory. *Estuar. Coastal Mar. Sci.*, 5: 25-37.

Rapson, A.M. 1954. Feeding and control of toheroa (*Amphidesma ventricosum* Gray) (Eulamellibranchiata) populations in New Zealand. *Aust. J. Mar. Fresh. Res.,* 5: 486-512.

Schaefer C.T. and J. Lewin. 1984. Persistent blooms of surf diatoms along the Pacific coast, U.S.A. IV. Diatom productivity and its relation to standing stock. *Mar. Biol.,* 83: 205-217.

Short, A.D. and L.D. Wright. 1983. Physical variability of sandy beaches. Pages 133-144 in: A. McLachlan and T. Erasmus (eds.), *Sandy Beaches as Ecosystems.* W. Junk Publ. , Hingham, Mass.

Thayer, L.A. 1935a. Some experiments on the biogenetic origin of petroleum. Ph.D. Diss., Stanford Univ., Stanford, Calif., 357 pp.

Thayer, L.A. 1935b. Diatom water-blooms on the coast of Washington. *Proc. La. Acad. Sci.,* 2: 68-72.

Van Heurck, H. 1896. A treatise on the *Diatomaceae.* Trans. by W.E. Baxter, William Wesley and Sons, London.

West, T. 1860. Remarks on some diatomacea, new or imperfectly described and a new desmid. *Trans. Micr. Soc. London,* 8: 147-153.

Winter, D.F. 1983. A theoretical model of surf zone circulation and diatom growth. Pages 157-167 in: A. McLachlan and T. Erasmus (eds.), *Sandy Beaches as Ecosystems.* W. Junk Publ., Hingham, Mass.

CHAPTER 13

COASTAL OCEANOGRAPHY OF WASHINGTON AND OREGON: A SUMMARY AND A PROSPECTUS FOR FUTURE STUDIES

B.M. Hickey[1] *and M. R. Landry*[2]
[1]*School of Oceanography, University of Washington, Seattle, WA 98195*
[2]*Department of Oceanography and Hawaii Institute of Geophysics, University of Hawaii, Honolulu, HI 96822*

Physical processes with time scales of variability ranging from hours to decades and more drive a complex and dynamic coastal ecosystem off Washington and Oregon. Wind forcing, both remote and local, accounts for seasonal patterns as well as much of the day-to-day variability in shelf circulation, nutrient inputs and the abundances and distributions of organisms and suspended matter (Chapters 1,2,3,5,6,7). Global-scale processes contribute to interannual variability both as dramatic El Niño events and as gradual warming or cooling trends (Chapter 1). Major shelf-sediment removal events are linked to the coincidence of severe winter storms and large ocean swells at decadal frequency (Chapter 8).

Because the coastal wind stress has a strong seasonal modulation, upwelling and hence the nutrient supply as well as the coastal currents and most water properties also have a strong seasonal cycle. Downwelling dominates during winter months when strong coastal winds and currents are both directed poleward on average over most of the shelf and slope (Chapters 1 and 2). Winter storms deepen the mixed layer by a factor of two or three, enriching surface waters with dissolved nutrients. Phytoplankton in the water column are typically light-limited during winter and therefore grow and utilize nutrients at low rates (Chapter 3). In contrast, unique adaptations for floating allow surf-zone diatoms to thrive during this time of year (Chapter 12).

As the winter progresses, periods of clear weather (equatorward winds and currents) become increasingly common and occur over longer intervals. Episodes of upwelling may take place on the inner shelf, although the density field over most of the shelf and slope remains in a downwelling pattern. Phytoplankton blooms occur with the onset of increased near-surface stratification and increased light levels in early spring (Chapter 1). This depletes surface nutrient (nitrate) concentrations which remain low until the winds shift equatorward and the summer upwelling season begins in May. The transition to a seasonal upwelling pattern is dramatic and occurs nearly simultaneously across the whole region (Chapter 2). Shelf currents flow predominantly equatorward during this time, a result of both an alongshelf sea level slope and the equatorward local winds. As the upwelling season progresses, a poleward-flowing current, the California Undercurrent, develops over much of the shelf and slope, with its core over the upper slope. By late summer the undercurrent extends to the inner shelf off Washington. The undercurrent is less developed off Oregon, however, probably because the opposing equatorward wind stress is a factor of two greater off Oregon during summer than off Washington. The California Undercurrent provides the return flow for water exiting the Strait of Juan de Fuca and Puget

Sound. It is also the source of nutrient-rich water advected into the near-shore, euphotic zone during upwelling.

The episodic infusion of dissolved nutrients into the shallow shelf environment during the summer upwelling period stimulates high plankton production and accounts for much of the rich commercial fisheries of the U.S. Pacific Northwest. The upwelling, which is driven by the alongshelf component of the wind stress over the shelf, occurs primarily on the inner half of the shelf, within ~10 km of the coast. Although there are some promontories along the Washington-Oregon coastline that may result in local enhancement of upwelling, the coastline is relatively straight in comparison to the California coastline. Because the structures of the wind field and the sub-surface layer currents driven by the wind field are so large-scale, the upwelling response in this area is much closer to two-dimensional than in other West Coast regions. Nevertheless, satellite sea surface temperature and color images reveal that at least the very near-surface ocean layers contain a rich mesoscale eddy field.

Shelf circulation patterns related to upwelling may have important implications for the distribution of pelagic and benthic organisms and nutrient and chemical cycling. Upwelling of deep water from the outer shelf during early summer probably entrains resting eggs, spores and cysts from the near-bottom, nepheloid layer: these serve as 'seed' stock for plankton communities on the inner shelf (Chapter 7). During late summer, the source of upwelled water is relatively shallow, just below the euphotic zone. Phytoplankton advected in this shallow return flow are physiologically healthy and able to respond rapidly to the favorable light and nutrient conditions in newly-upwelled water. Moreover, the shallow flow provides a mechanism for 'nutrient trapping', i.e., more efficient cycling and utilization of ammonium regenerated below the euphotic zone. The two-layered flow regime appears to be important in maintaining zonation patterns of migrating zooplankton species and in transporting eggs and young to the productive inner shelf. Benthic populations on the mid-shelf silt deposit may also use the shelf current structure for dispersal (in a single-layer alongshelf flow) or to maintain larvae in a favorable region for settlement (in a two-layer flow, e.g., a period of feeding in the surface environment and a period of return flow in the undercurrent). Lastly, the undercurrent has been postulated as a route for poleward transport of migratory fishes.

Large standing stocks of zooplankton, dominated by the copepod *Pseudocalanus* spp., develop in response to high shelf primary production during summer. Zooplankton biomass is at a maximum over the mid-shelf region where grazing demand is largely sufficient to meet, if not exceed, local phytoplankton production. Excess phytoplankton from nearshore is advected to the mid-shelf by upwelling currents where it appears to be completely utilized by zooplankton, sinking to the mid-shelf silt deposit as fecal debris rather than as a senescent algal mass.

The autumn transition, like the spring transition, is abrupt and has a dramatic signature in water properties and the physical structure of the water column. The transition occurs during the first major storm of the year and isopycnals over the entire shelf change from an upward slope toward the coast to a downward slope. Currents become primarily poleward and turbulent mixing increases the depth of the surface mixed layer so that phytoplankton become physically more light deprived at the same time as available solar radiation decreases. Phytoplankton growth rates decrease accordingly.

The retention time of particles in the shelf region is a complex function of particle type and integrity as well as location and depth of the water column. Neglecting the effects of mesoscale eddies and fronts on particle movement, buoyant particles in the near-surface waters at mid-shelf could be retained on the shelf for up to 30 d during summer months. This is sufficient time to be transported from northern Washington to California (about 20 d). Slower moving particles on the inner shelf would still be off Oregon after the same time interval. In contrast to surface particle trajectories, particulates in suspension near the bottom at mid-shelf would tend to move across the shelf and over the slope in only 15 d during summer. During this time, the particles would have been transported about one-third the length of the Washington-Oregon coastline. Retention of organic flux from surface production on the shelf during summer therefore depends on the relatively rapid sinking of fecal pellets from zooplankton. During winter, particles at almost all locations move poleward and onshore in the surface mixed layer or offshore in the bottom boundary layer. Major winter storms are sufficiently strong to resuspend bottom sediments into the advective regime.

The spatial and temporal structures of currents, nutrients and sediments are modified by the presence of the Columbia River, which divides the Washington and Oregon coastal environments into distinct oceanographic regimes. The Columbia River is the source of over 90% of the regional freshwater runoff and sedimentary input. During winter, the Columbia River plume, which is ~20-m deep, lies over the Washington shelf and slope. During summer, the plume lies along the outer edge of the Oregon shelf, compressing the upwelling zone into a sharp front. Circulation on the Washington inner shelf below the surface layer is less poleward during winter than on the Oregon shelf as a result of the presence of the plume. Sediments from the Columbia are deposited on the shelf during the high runoff periods that occur primarily in late spring and are moved poleward under the combined action of waves and currents during the winter to form the mid-shelf silt deposit off Washington. The sediment accumulation rate exceeds 0.75 g cm^{-2} y^{-1} near the mouth of the Columbia River, decreasing to ~0.25 g cm^{-2} y^{-1} near Quinault Canyon. The sediment is moved under the combined action of both waves and currents; typically, material is resuspended primarily via wave-induced boundary shear stress but is mixed upward into the water column and transported along the shelf for distances on the order of tens of kilometers by storm-induced currents. Resuspended material is also transported over canyons and slopes, where it forms nepheloid layers at and below shelf-break depths, and ultimately sinks out of the water column (Chapter 2). Lacking significant sedimentary input, the coast of Oregon does not have a comparable mid-shelf silt deposit (Chapter 8). Consequently, intermediate-depth nepheloid layers are not as well developed off Oregon as off Washington (Chapter 7).

Interannual variations in water properties, standing stocks of phytoplankton and fish stocks are characteristic of most marine ecosystems. The U.S. Pacific Northwest is no exception: interannual variations have been observed in salinity, temperature, chlorophyll *a* and surface nitrate (Chapter 1). In contrast to the California coast, where similar variations were ascribed to large-scale advection in the offshore California Current, interannual variations in the Pacific Northwest are more directly related to interannual variations in local wind-forced coastal upwelling (Chapter 1). However, we note that this apparent difference may represent only spatial differences in sampling: the California data are largely taken from locations seaward of the (narrow) shelf, whereas the Pacific Northwest data are mostly from locations on the shelf, where local wind forcing is more important.

Although interannual variation in coastal circulation off Washington and Oregon affects annual recruitment to harvested fish and crustacean stocks, the underlying mechanisms are not well understood (Chapter 11). While the presence of upwelling is ultimately responsible for the high productivity of these stocks, evidence thus far indicates that interannual variability in recruitment may depend more on the variability in productivity. Further elucidation of these influences will require description of currents on finer temporal and spatial scales, with concurrent biological measurements.

There are many significant gaps in the understanding of physical, chemical and biological processes in the coastal ecosystems of Washington and Oregon. At the most basic level, the available recent data on primary production are so dominated by results from cruises in August 1981 and June 1982, both strong upwelling events, that reliable estimates of annual primary production cannot be made (Chapter 3). Nitrogen uptake studies imply substantially lower production rates (130-450 g C m^{-2} y^{-1} *vs.* the 650 g C m^{-2} y^{-1} from recent ^{14}C studies), but these too are open to question for both methodological and distributional considerations (Chapter 4). The role and magnitude of microzooplankton, specifically protozoan, grazing is another strictly biological process requiring more careful study (Chapter 5). Similarly, there has been no work on the potential importance of bacterial and microbial food webs in organic matter utilization in Washington and Oregon coastal waters. Dissipation of energy by these alternate trophic pathways ultimately constrains the amount of production available to higher trophic levels.

Despite the importance of the Columbia River in regional circulation and sedimentary processes, the dynamics of the plume have not been studied in almost two decades. Modern instrumentation could contribute substantially to the understanding of plume formation and its subsequent modification by mixing and advection. Similarly, although estuaries are important nursery grounds for important fisheries, no comprehensive studies have addressed the interactions between estuarine and shelf processes. Moreover, the importance of near-surface mesoscale features to carbon and particulate export from the shelf need to be examined.

From a purely interdisciplinary point of view, the Washington mid-shelf stands out as an area at the intersection of the most pressing and interesting questions concerning physical, chemical, biological and geological processes and their interactions. The mid-shelf region is an area of high biological production and standing stocks in the upper water column and maximum grazing activity; hence high particulate organic flux through the water-column are expected. From a practical standpoint, the mid-shelf silt deposit contains a rich benthic fauna and productive demersal fisheries, yet it serves as the major shelf repository for surface chemicals advected offshore by upwelling and sedimenting as fecal debris, chemicals associated with Columbia River sediments and inputs of particle-reactive chemicals from deep water advected onto the particle-rich shelf by upwelling (Chapter 10). The interactions of physical circulation, surface production and grazing relationships, particle flux, chemical scavenging and sedimentary inventories, sediment transport mechanisms and animal-sediment relationships are tractable problems in this region. Past studies have provided many "snapshots" of this dynamic ecosystem. Future studies must now emphasize the use of modern instruments, including satellite imagery, which provide synoptic and time-dependent measurements of important parameters.

Subject Index